VD 107/08

THE DESIGN OF CMOS RADIO-FREQUENCY INTEGRATED CIRCUITS

SECOND EDITION

THOMAS H. LEE
Stanford University

CAMBRIDGE
UNIVERSITY PRESS

CAMBRIDGE UNIVERSITY PRESS
Cambridge, New York, Melbourne, Madrid, Cape Town, Singapore, São Paulo

Cambridge University Press
32 Avenue of the Americas, New York, NY 10013-2473, USA

www.cambridge.org
Information on this title: www.cambridge.org/9780521835398

First published 1988
Second edition first publish 2004
Reprinted 2005, 2006

Printed in the United States of America

A catalog record for this publication is available from the British Library.

Library of Congress Cataloging in Publication data

Lee, Thomas H., 1959–
The design of CMOS radio-frequency integrated circuits / Thomas H. Lee – 2nd ed.
p. cm.
ISBN 0-521-83539-9 hardback
1. Metal oxide semiconductors, Complementary – Design and construction. 2. Radio frequency
integrated circuits – Design and construction. 3. Radio – Transmitter-receivers. I. Title.
TK7871.99.M44L44 2004
621.39′732 – dc22   2003061322

ISBN-13 978-0-521-83539-8 hardback
ISBN-10 0-521-83539-9 hardback

To my parents, who had no idea what they were starting when they bought me a pair of walkie-talkies for my sixth birthday

CONTENTS

Preface to the Second Edition   *page* xiii

Preface to the First Edition   xv

1 **A NONLINEAR HISTORY OF RADIO**   1
   1. Introduction   1
   2. Maxwell and Hertz   1
   3. Pre–Vacuum Tube Electronics   2
   4. Birth of the Vacuum Tube   9
   5. Armstrong and the Regenerative Amplifer/Detector/Oscillator   13
   6. Other Radio Circuits   15
   7. Armstrong and the Superregenerator   18
   8. Oleg Losev and the First Solid-State Amplifier   20
   9. Epilog   21
   10. Appendix A: A Vacuum Tube Primer   22
   11. Appendix B: Who Really Invented Radio?   32

2 **OVERVIEW OF WIRELESS PRINCIPLES**   40
   1. A Brief, Incomplete History of Wireless Systems   40
   2. Noncellular Wireless Applications   52
   3. Shannon, Modulations, and Alphabet Soup   56
   4. Propagation   77
   5. Some Closing Comments   83
   6. Appendix: Characteristics of Other Wireless Systems   84

3 **PASSIVE *RLC* NETWORKS**   87
   1. Introduction   87
   2. Parallel *RLC* Tank   87
   3. Series *RLC* Networks   92
   4. Other Resonant *RLC* Networks   92
   5. *RLC* Networks as Impedance Transformers   93

6. Examples   104
Problem Set   107

4 **CHARACTERISTICS OF PASSIVE IC COMPONENTS**   114
1. Introduction   114
2. Interconnect at Radio Frequencies: Skin Effect   114
3. Resistors   120
4. Capacitors   122
5. Inductors   136
6. Transformers   148
7. Interconnect Options at High Frequencies   158
8. Summary   162
9. Appendix: Summary of Capacitance Equations   162
Problem Set   163

5 **A REVIEW OF MOS DEVICE PHYSICS**   167
1. Introduction   167
2. A Little History   167
3. FETs: The Short Story   168
4. MOSFET Physics: The Long-Channel Approximation   169
5. Operation in Weak Inversion (Subthreshold)   180
6. MOS Device Physics in the Short-Channel Regime   183
7. Other Effects   188
8. Summary   190
9. Appendix A: 0.5-μm Level-3 SPICE Models   190
10. Appendix B: The Level-3 Spice Model   191
11. Appendix C: Level-1 MOS Models   195
12. Appendix D: Some Exceptionally Crude Scaling Laws   197
Problem Set   198

6 **DISTRIBUTED SYSTEMS**   202
1. Introduction   202
2. Link between Lumped and Distributed Regimes   204
3. Driving-Point Impedance of Iterated Structures   205
4. Transmission Lines in More Detail   207
5. Behavior of Finite-Length Transmission Lines   212
6. Summary of Transmission-Line Equations   214
7. Artificial Lines   214
8. Summary   218
Problem Set   218

7 **THE SMITH CHART AND S-PARAMETERS**   221
1. Introduction   221
2. The Smith Chart   221
3. S-Parameters   225

    4. Appendix A: A Short Note on Units   227
    5. Appendix B: Why 50 (or 75) Ω?   229
    Problem Set   231

8 BANDWIDTH ESTIMATION TECHNIQUES   233
    1. Introduction   233
    2. The Method of Open-Circuit Time Constants   234
    3. The Method of Short-Circuit Time Constants   254
    4. Further Reading   259
    5. Risetime, Delay, and Bandwidth   259
    6. Summary   265
    Problem Set   266

9 HIGH-FREQUENCY AMPLIFIER DESIGN   270
    1. Introduction   270
    2. Zeros as Bandwidth Enhancers   271
    3. The Shunt–Series Amplifier   282
    4. Bandwidth Enhancement with f_T Doublers   288
    5. Tuned Amplifiers   290
    6. Neutralization and Unilateralization   294
    7. Cascaded Amplifiers   297
    8. AM–PM Conversion   306
    9. Summary   307
    Problem Set   308

10 VOLTAGE REFERENCES AND BIASING   314
    1. Introduction   314
    2. Review of Diode Behavior   314
    3. Diodes and Bipolar Transistors in CMOS Technology   316
    4. Supply-Independent Bias Circuits   317
    5. Bandgap Voltage Reference   318
    6. Constant-g_m Bias   325
    7. Summary   328
    Problem Set   328

11 NOISE   334
    1. Introduction   334
    2. Thermal Noise   334
    3. Shot Noise   342
    4. Flicker Noise   344
    5. Popcorn Noise   347
    6. Classical Two-Port Noise Theory   348
    7. Examples of Noise Calculations   352
    8. A Handy Rule of Thumb   355

9. Typical Noise Performance   356
10. Appendix: Noise Models   357
Problem Set   358

12 LNA DESIGN   364

1. Introduction   364
2. Derivation of Intrinsic MOSFET Two-Port Noise Parameters   365
3. LNA Topologies: Power Match versus Noise Match   373
4. Power-Constrained Noise Optimization   380
5. Design Examples   384
6. Linearity and Large-Signal Performance   390
7. Spurious-Free Dynamic Range   397
8. Summary   399
Problem Set   400

13 MIXERS   404

1. Introduction   404
2. Mixer Fundamentals   405
3. Nonlinear Systems as Linear Mixers   411
4. Multiplier-Based Mixers   416
5. Subsampling Mixers   433
6. Appendix: Diode-Ring Mixers   434
Problem Set   437

14 FEEDBACK SYSTEMS   441

1. Introduction   441
2. A Brief History of Modern Feedback   441
3. A Puzzle   446
4. Desensitivity of Negative Feedback Systems   446
5. Stability of Feedback Systems   450
6. Gain and Phase Margin as Stability Measures   451
7. Root-Locus Techniques   453
8. Summary of Stability Criteria   459
9. Modeling Feedback Systems   459
10. Errors in Feedback Systems   462
11. Frequency- and Time-Domain Characteristics of First- and
    Second-Order Systems   466
12. Useful Rules of Thumb   469
13. Root-Locus Examples and Compensation   470
14. Summary of Root-Locus Techniques   477
15. Compensation   477
16. Compensation through Gain Reduction   478
17. Lag Compensation   481
18. Lead Compensation   484
19. Slow Rolloff Compensation   486

20. Summary of Compensation   487
Problem Set   488

15 **RF POWER AMPLIFIERS**   493
1. Introduction   493
2. General Considerations   493
3. Class A, AB, B, and C Power Amplifiers   494
4. Class D Amplifiers   503
5. Class E Amplifiers   505
6. Class F Amplifiers   507
7. Modulation of Power Amplifiers   512
8. Summary of PA Characteristics   540
9. RF PA Design Examples   541
10. Additional Design Considerations   547
11. Design Summary   555
Problem Set   555

16 **PHASE-LOCKED LOOPS**   560
1. Introduction   560
2. A Short History of PLLs   560
3. Linearized PLL Models   566
4. Some Noise Properties of PLLs   571
5. Phase Detectors   574
6. Sequential Phase Detectors   579
7. Loop Filters and Charge Pumps   588
8. PLL Design Examples   596
9. Summary   604
Problem Set   604

17 **OSCILLATORS AND SYNTHESIZERS**   610
1. Introduction   610
2. The Problem with Purely Linear Oscillators   610
3. Describing Functions   611
4. Resonators   631
5. A Catalog of Tuned Oscillators   635
6. Negative Resistance Oscillators   641
7. Frequency Synthesis   645
8. Summary   654
Problem Set   655

18 **PHASE NOISE**   659
1. Introduction   659
2. General Considerations   661
3. Detailed Considerations: Phase Noise   664
4. The Roles of Linearity and Time Variation in Phase Noise   667

5. Circuit Examples 678
6. Amplitude Response 687
7. Summary 689
8. Appendix: Notes on Simulation 689
Problem Set 690

19 **ARCHITECTURES** 694
1. Introduction 694
2. Dynamic Range 695
3. Subsampling 713
4. Transmitter Architectures 714
5. Oscillator Stability 715
6. Chip Design Examples 716
7. Summary 762
Problem Set 762

20 **RF CIRCUITS THROUGH THE AGES** 764
1. Introduction 764
2. Armstrong 764
3. The "All-American" 5-Tube Superhet 768
4. The Regency TR-1 Transistor Radio 771
5. Three-Transistor Toy CB Walkie-Talkie 773

Index 777

PREFACE TO THE
SECOND EDITION

Since publication of the first edition of this book in 1998, RF CMOS has made a rapid transition to commercialization. Back then, the only notable examples of RF CMOS circuits were academic and industrial prototypes. No companies were then shipping RF products using this technology, and conference panel sessions openly questioned the suitability of CMOS for such applications – often concluding in the negative. Few universities offered an RF integrated circuit design class of any kind, and only one taught a course solely dedicated to CMOS RF circuit design. Hampering development was the lack of device models that properly accounted for noise and impedance at gigahertz frequencies. Measurements and models so conflicted with one another that controversies raged about whether deep submicron CMOS suffered from fundamental scaling problems that would forever prevent the attainment of good noise figures.

Today, the situation is quite different, with many companies now manufacturing RF circuits using CMOS technology and with universities around the world teaching at least something about CMOS as an RF technology. Noise figures below 1 dB at gigahertz frequencies have been demonstrated in practical circuits, and excellent RF device models are now available. That pace of growth has created a demand for an updated edition of this textbook.

In response to many suggestions, this second edition now includes a chapter on the fundamentals of wireless systems. After a brief overview of wireless propagation, we introduce a necessarily truncated discussion of Shannon's famous channel capacity theorem in order to establish a context for discussion of modern modulation methods. To avoid getting bogged down in the arcana of information theory, we cover only those concepts essential to understanding why wireless systems appear as they do. A few illustrative systems – such as IEEE 802.11 wireless LANs, second- and third-generation cellular technology, and emerging technologies such as ultrawideband (UWB) – are briefly examined in this context.

The chapter on passive RLC components now directly precedes a much expanded chapter on passive IC components, rather than following it. The chapter on MOS device physics has likewise been updated to reflect recent scaling trends and augmented to provide readers with an improved understanding of models that are suitable for hand calculation; it also offers some reasonable speculations on how scaling trends are likely to affect technology in the next several years. The related chapter on LNA design includes a detailed discussion of our much-improved understanding of MOS noise mechanisms at radio frequencies. It is with great pleasure that I acknowledge here the kind and timely assistance of Dr. Andries Scholten and his colleagues at Philips for providing a wealth of data just days before the manuscript submission deadline. The clarity and rigor of their work informs much of the new material presented in both the device physics and LNA chapters.

In another rearrangement, the chapter on feedback now precedes that on power amplifiers in order to establish principles necessary for understanding several linearization methods. Readers familiar with the first edition will also note that the power amplifier chapter has been greatly expanded to include much more on the subject of techniques for linearization and efficiency enhancement.

The chapter on transceiver architectures now includes much more detailed coverage of the direct conversion architecture. Persistent, dedicated work by a host of determined engineers has overcome many of the daunting challenges that underlies the pessimism expressed in the first edition.

That chapter now includes illustrative examples to provide a comprehensive overview of how the knowledge of the preceding chapters may be integrated into a coherent whole. It also is a natural place to discuss briefly several topics that – while essential to the realization of practical transceivers – lie somewhat outside of the scope of a book on RF circuits. Additional commentary on practical details such as simulation, floorplanning, packaging, and the like are to be found here as well.

Significant refinements, clarifications, and corrections have been applied to nearly all of the chapters, thanks to a wealth of ongoing research and also to the valuable suggestions of many dedicated students and engineers who have been kind enough to convey their comments over the past five years. The suggestions and corrections provided by Professor Yannis Tsividis and his students at Columbia University have proven especially valuable.

As mentioned in the preface to the first edition, students often take particular delight in pointing out the professor's errors. The rich list of suggestions compiled by graduate students has stimulated substantial improvements in clarity that future students will appreciate. Thanks to all those who identified mistakes and contributed suggestions, this edition has considerably fewer errors than would otherwise have been the case. Alas, experience shows that perfection is unattainable, and I know that I'll hear from readers about some typo, error, or unclear explanation (but please be gentle). That's what *third* editions are for.

Finally, I am most deeply grateful for the love, encouragement and understanding of my brilliant, beautiful and otherwise amazing wife, Angelina, who no doubt hopes that I'm kidding about a third edition, but is much too sweet to say so.

PREFACE TO THE FIRST EDITION

The field of radio frequency (RF) circuit design is currently enjoying a renaissance, driven in particular by the recent, and largely unanticipated, explosive growth in wireless telecommunications. Because this resurgence of interest in RF caught industry and academia by surprise, there has been a mad scramble to educate a new generation of RF engineers. However, in trying to synthesize the two traditions of "conventional" RF and lower-frequency IC design, one encounters a problem: "Traditional" RF engineers and analog IC designers often find communication with each other difficult because of their diverse backgrounds and the differences in the media in which they realize their circuits. Radio-frequency IC design, particularly in CMOS, is a different activity altogether from discrete RF design. This book is intended as both a link to the past and a pointer to the future.

The contents of this book derive from a set of notes used to teach a one-term advanced graduate course on RF IC design at Stanford University. The course was a follow-up to a low-frequency analog IC design class, and this book therefore assumes that the reader is intimately familiar with that subject, described in standard texts such as *Analysis and Design of Analog Integrated Circuits* by P. R. Gray and R. G. Meyer (Wiley, 1993). Some review material is provided, so that the practicing engineer with a few neurons surviving from undergraduate education will be able to dive in without too much disorientation.

The amount of material here is significantly beyond what students can comfortably assimilate in one quarter or semester, and instructors are invited to pick and choose topics to suit their tastes, the length of the academic term, and the background level of the students. In the chapter descriptions that follow are included some hints about what chapters may be comfortably omitted or deferred.

Chapter 1 presents an erratic history of radio. This material is presented largely for cultural reasons. The author recognizes that not everyone finds history interesting, so the impatient reader is invited to skip ahead to the more technical chapters.

Chapter 2 [Chapter 4 in the second edition – ED.] surveys the passive components normally available in standard CMOS processes. There is a focus on inductors because of their prominent role in RF circuits, and also because material on this subject is scattered in the current literature (although, happily, this situation is rapidly changing).

Chapter 3 [2e Ch. 5] provides a quick review of MOS device physics and modeling. Since deep submicron technology is now commonplace, there is a focus on approximate analytical models that account for short-channel effects. This chapter is necessarily brief, and is intended only as a supplement to more detailed treatments available elsewhere.

Chapter 4 [2e Ch. 3] examines the properties of lumped, passive RLC networks. For advanced students, this chapter may be a review and may be skipped if desired. In the author's experience, most undergraduate curricula essentially abandoned the teaching of inductors long ago, so this chapter spends a fair amount of time examining the issues of resonance, Q, and impedance matching.

Chapter 5 [2e Ch. 6] extends into the distributed realm many of the concepts introduced in the context of lumped networks. Transmission lines are introduced in a somewhat unusual way, with the treatment avoiding altogether the derivation of the telegrapher's equation with its attendant wave solutions. The characteristic impedance and propagation constant of a uniform line are derived entirely from simple extensions of lumped ideas. Although distributed networks play but a minor role in the current generation of silicon IC technology, that state of affairs will be temporary, given that device speeds are doubling about every three years.

Chapter 6 [2e Ch. 7] provides an important bridge between the traditional "microwave plumber's" mind-set and the IC designer's world view by presenting a simple derivation of the Smith chart, explaining what S-parameters are and why they are useful. Even though the typical IC engineer will almost certainly not design circuits using these tools, much instrumentation presents data in Smith-chart and S-parameter form, so modern engineers still need to be conversant with them.

Chapter 7 [2e Ch. 8] presents numerous simple methods for estimating the bandwidth of high-order systems from a series of first-order calculations or from simple measurements. The former set of techniques, called the method of open-circuit (or zero-value) time constants, allows one to identify bandwidth-limiting parts of a circuit while providing a typically conservative bandwidth estimate. Relationships among bandwidth, delay, and risetime allow us to identify important degrees of freedom in trading off various parameters. In particular, gain and bandwidth are shown not to trade off with one another in any fundamental way, contrary to the beliefs of many (if not most) engineers. Rather, gain and *delay* are shown to be more tightly coupled, opening significant loopholes that point the way to amplifier architectures which effect that tradeoff and leave bandwidth largely untouched.

Chapter 8 [2e Ch. 9] takes a detailed look at the problem of designing extremely high-frequency amplifiers, both broad- and narrowband, with many "tricks" evolving from a purposeful violation of the assumptions underlying the method of open-circuit time constants.

Chapter 9 [2e Ch. 10] surveys a number of biasing methods. Although intended mainly as a review, the problems of implementing good references in standard CMOS are large enough to risk some repetition. In particular, the design of CMOS-compatible bandgap voltage references and constant-transconductance bias circuits are emphasized here, perhaps a little more so than in most standard analog texts.

Chapter 10 [2e Ch. 11] studies the all-important issue of noise. Simply obtaining sufficient gain over some acceptable bandwidth is frequently insufficient. In many wireless applications, the received signal amplitude is in the microvolt range. The need to amplify such minute signals as noiselessly as possible is self-evident, and this chapter provides the necessary foundation for identifying conditions for achieving the best possible noise performance from a given technology.

Chapter 11 [2e Ch. 12] follows up on the previous two or three chapters to identify low-noise amplifier (LNA) architectures and the specific conditions that lead to the best possible noise performance, given an explicit constraint on power consumption. This power-constrained approach differs considerably from standard discrete-oriented methods, and exploits the freedom enjoyed by IC designers to tailor device sizes to achieve a particular optimum. The important issue of dynamic range is also examined, and a simple analytical method for estimating a large-signal linearity limit is presented.

Chapter 12 [2e Ch. 13] introduces the first intentionally nonlinear element, and the heart of all modern transceivers: the mixer. After identifying key mixer performance parameters, numerous mixer topologies are examined. As with the LNA, the issue of dynamic range is kept in focus the entire time.

Chapter 13 [2e Ch. 15] presents numerous topologies for building RF power amplifiers. The serious and often unsatisfactory tradeoffs among gain, efficiency, linearity, and output power lead to a family of topologies, each with its particular domain of application. The chapter closes with an examination of load-pull experimental characterizations of real power amplifiers.

Chapter 14 provides a review of classical feedback concepts, mainly in preparation for the following chapter on phase-locked loops. Readers with a solid background in feedback may wish to skim it, or even skip it entirely.

Chapter 15 [2e Ch. 16] surveys a number of phase-locked loop circuits after presenting basic operating theory of both first- and second-order loops. Loop stability is examined in detail, and a simple criterion for assessing a PLL's sensitivity to power supply and substrate noise is offered.

Chapter 16 [2e Ch. 17] examines in detail the issue of oscillators and frequency synthesizers. Both relaxation and tuned oscillators are considered, with the latter category further subdivided into LC and crystal-controlled oscillators. Both fixed and controllable oscillators are presented. Prediction of oscillation amplitude, criteria for start-up, and device sizing are all studied.

Chapter 17 [2e Ch. 18] extends to oscillators the earlier work on noise. After elucidating some general criteria for optimizing the noise performance of oscillators, a powerful theory of phase noise based on a linear, time-varying model is presented. The model makes some surprisingly optimistic (and experimentally verified) predictions about what one may do to reduce the phase noise of oscillators built with such infamously noisy devices as MOSFETs.

Chapter 18 [2e Ch. 19] ties all the previous chapters together and surveys architectures of receivers and transmitters. Rules are derived for computing the intercept and noise figure of a cascade of subsystems. Traditional superheterodyne architectures are examined, along with low-IF image-reject and direct-conversion receivers. The relative merits and disadvantages of each of these is studied in detail.

Finally, Chapter 19 [2e Ch. 20] closes the book the way it began: with some history. A nonuniform sampling of classical (and distinctly non-CMOS) RF circuits takes a look at Armstrong's earliest inventions, the "All-American Five" vacuum tube table radio, the first transistor radio, and the first toy walkie-talkie. As with the first chapter, this one is presented purely for enjoyment, so those who do not find history lessons enjoyable or worthwhile are invited to close the book and revel in having made it through the whole thing.

A book of this length could not have been completed in the given time were it not for the generous and competent help of colleagues and students. My wonderful administrative assistant, Ann Guerra, magically created time by handling everything with her remarkable good cheer and efficiency. Also, the following Ph.D. students went far beyond the call of duty in proofreading the manuscript and suggesting or generating examples and many of the problem-set questions: Tamara Ahrens, Rafael Betancourt-Zamora, David Colleran, Ramin Farjad-Rad, Mar Hershenson, Joe Ingino, Adrian Ong, Hamid Rategh, Hirad Samavati, Brian Setterberg, Arvin Shahani, and Kevin Yu. Ali Hajimiri, Sunderarajan S. Mohan, and Derek Shaeffer merit special mention for their conspicuous contributions. Without their help, given in the eleventh hour, this book would still be awaiting completion.

The author is also extremely grateful to the text's reviewers, both known and anonymous, who all had excellent, thoughtful suggestions. Of the former group, Mr. Howard Swain (formerly of Hewlett-Packard), Dr. Gitty Nasserbakht of Texas Instruments, and Professors James Roberge of the Massachusetts Institute of Technology and Kartikeya Mayaram of Washington State University deserve special thanks for spotting typographical and graphical errors, and also for their valuable editorial suggestions. Matt and Vickie Darnell of Four-Hand Book Packaging did a fantastic job of copyediting and typesetting. Their valiant efforts to convert my "sow's ear" of a manuscript into the proverbial silk purse were nothing short of superhuman. And Dr. Philip Meyler of Cambridge University Press started this whole thing by urging me to write this book in the first place, so he's the one to blame.

Despite the delight taken by students in finding mistakes in the professor's notes, some errors have managed to slip through the sieve, even after three years of filtering. Sadly, this suggests that more await discovery by you. I suppose that is what second editions are for.

CHAPTER ONE

A NONLINEAR HISTORY OF RADIO

1.1 INTRODUCTION

Integrated circuit engineers have the luxury of taking for granted that the incremental cost of a transistor is essentially zero, and this has led to the high–device-count circuits that are common today. Of course, this situation is a relatively recent development; during most of the history of electronics, the economics of circuit design were the inverse of what they are today. It really wasn't all that long ago when an engineer was forced by the relatively high cost of active devices to try to get blood (or at least rectification) from a stone. And it is indeed remarkable just how much performance radio pioneers were able to squeeze out of just a handful of components. For example, we'll see how American radio genius Edwin Armstrong devised circuits in the early 1920s that trade *log* of gain for bandwidth, contrary to the conventional wisdom that gain and bandwidth should trade off more or less directly. And we'll see that at the same time Armstrong was developing those circuits, self-taught Soviet radio engineer Oleg Losev was experimenting with blue LEDs and constructing completely solid-state radios that functioned up to 5 MHz, a quarter century before the transistor was invented.

These fascinating stories are rarely told because they tend to fall into the cracks between history and engineering curricula. *Somebody* ought to tell these stories, though, since in so doing, many commonly asked questions ("why don't they do it this way?") are answered ("they used to, but it caused key body parts to fall off"). This highly nonlinear history of radio touches briefly on just some of the main stories and provides pointers to the literature for those who want to probe further.

1.2 MAXWELL AND HERTZ

Every electrical engineer knows at least a bit about James Clerk (pronounced "clark") Maxwell; he wrote those equations that made life extra busy back in sophomore year or thereabouts. Not only did he write the electrodynamic equations[1] that bear his name, he also published the first mathematical treatment of stability in feedback systems ("On Governors," which explained why speed controllers for steam engines could sometimes be unstable[2]).

Maxwell collected all that was then known about electromagnetic phenomena and, in a mysterious[3] and brilliant stroke, invented the displacement (capacitive) current term that allowed him to derive an equation that led to the prediction of electromagnetic wave propagation.

Then came Heinrich Hertz, who was the first to verify experimentally Maxwell's prediction that electromagnetic waves exist and propagate with a finite velocity. His "transmitters" worked on this simple idea: discharge a coil across a spark gap and hook up some kind of an antenna to launch a wave (unintentionally) rich in harmonics.

His setup naturally provided only the most rudimentary filtering of this dirty signal, so it took extraordinary care and persistence to verify the existence of (and to quantify) the interference nulls and peaks that are the earmarks of wave phenomena. He also managed to demonstrate such quintessential wave behavior as refraction and polarization. And you may be surprised that the fundamental frequencies he worked with were between 50 and 500 MHz. He was actually *forced* to these frequencies because his laboratory was simply too small to enclose several wavelengths of anything lower in frequency.

Because Hertz's sensor was another spark gap (integral with a loop resonator), the received signal had to be large enough to induce a visible spark. Although adequate for verifying the validity of Maxwell's equations, you can appreciate the difficulties of trying to use this apparatus for wireless communication. After all, if the received signal has to be strong enough to generate a visible spark, scaling up to global proportions has rather unpleasant implications for those of us with metal dental work.

And then Hertz died – young. Enter Marconi.

1.3 PRE–VACUUM TUBE ELECTRONICS

For his radio experiments Marconi simply copied Hertz's transmitter and tinkered like crazy with the sole intent to use the system for wireless communication (and not incidentally to make a lot of money in the process). Recognizing the inherent limitations of Hertz's spark-gap detector, he instead used a peculiar device that had been developed by Edouard Branly in 1890. As seen in Figure 1.1, the device – dubbed the "coherer" by Sir Oliver Lodge – consisted of a glass enclosure filled with a loosely

---

[1] Actually, Oliver Heaviside was the one who first used the notational conventions of vector calculus to cast Maxwell's equations in the form familiar to most engineers today.

[2] *Proc. Roy. Soc.,* v. XVI, no. 100, pp. 270–83, Taylor & Francis, London, 1868.

[3] Many electricity and magnetism (E&M) texts offer the logical, but historically wrong, explanation that Maxwell invented the displacement current term after realizing that there was an inconsistency between the known laws of E&M and the continuity equation for current. The truth is that Maxwell was a genius, and the inspirations of a genius often have elusive origins. This is one of those cases.

FIGURE 1.1. Branly's coherer.

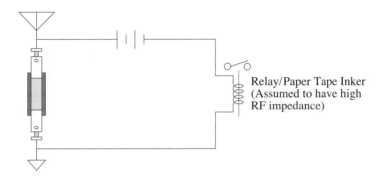

Relay/Paper Tape Inker
(Assumed to have high
RF impedance)

FIGURE 1.2. Typical receiver with coherer.

packed, perhaps slightly oxidized, metallic powder whose resistance turned out to
have interesting hysteretic behavior. Now, it must be emphasized that the detailed
principles underlying the operation of coherers have never been satisfactorily elucidated.[4] Nevertheless, we can certainly describe its behavior, even if we don't fully
understand all the details of how it worked.

A coherer's resistance generally had a large value (say, megohms) in its quiescent
state and then dropped orders of magnitude (to kilohms or less) after an electromagnetic (EM) wave impinged on it. This large resistance change was usually used to
trigger a solenoid to produce an audible click, as well as to ink a paper tape for a permanent record of the received signal. To prepare the coherer for the next EM pulse,
it had to be shaken or whacked to restore the "incoherent" high resistance state. Figure 1.2 shows how a coherer was actually used in a receiver.

As can be seen, the coherer activated a relay (for audible clicks) or paper tape
inker (for a permanent record) when a received signal triggered the transition to a
low-resistance state. It is evident that the coherer was basically a digital device and
therefore unsuitable for uses other than radiotelegraphy.

Marconi spent a great deal of time improving what was inherently a terrible detector and finally settled on the configuration shown in Figure 1.3. He greatly reduced
the spacing between the end plugs (to a minimum of 2 mm), filled the intervening

---

[4] Under large-signal excitation, the filings could be seen to stick together (hence the name "coherer"),
and it's not hard to understand the drop in resistance in that case. However, apparently unknown
to most authors, the coherer also worked with input energies so small that no such "coherence" is
observed, so I assert that the detailed principles of operation remain unknown.

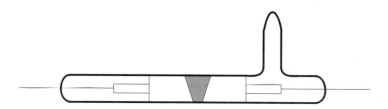

FIGURE 1.3. Marconi's coherer.

space with a particular mixture of nickel and silver filings (in 19 : 1 ratio) of carefully selected size, and sealed the entire assembly in a partially evacuated tube. As an additional refinement in the receiver, a solenoid provided an audible indication in the process of automatically whacking the detector back into its initial state after each received pulse.[5]

As you can imagine, many EM events other than the desired signal could trigger a coherer, resulting in some difficult-to-read messages. Even so, Marconi was able to refine his apparatus to the point of achieving transatlantic wireless communications by 1901, with much of his success attributable to more powerful transmitters and large, elevated antennas that used the earth as one terminal (as did his transmitter), as well as to his improved coherer.

It shouldn't surprise you, though, that the coherer, even at its best, performed quite poorly. Frustration with the coherer's erratic nature impelled an aggressive search for better detectors. Without a suitable theoretical framework as a guide, however, this search sometimes took macabre turns. In one case, even a human brain from a fresh cadaver was used as a coherer, with the experimenter claiming remarkable sensitivity for his apparatus.[6] Let us all be thankful that this particular type of coherer never quite caught on.

Most research was guided by the vague intuitive notion that the coherer's operation depended on some mysterious property of imperfect contacts, and a variety of experimenters stumbled, virtually simultaneously, on the point-contact crystal detector (Figure 1.4). The first patent for such a device was awarded in 1904 (filed in 1901) to J. C. Bose for a detector that used galena (lead sulfide).[7] This appears to be the first

---

[5] The coherer was most recently used in a radio-controlled toy truck in the late 1950s.

[6] A. F. Collins, *Electrical World and Engineer,* v. 39, 1902; he started out with brains of other species and worked his way up to humans.

[7] J. C. Bose, U.S. Patent #755,840, granted 19 March 1904, which describes a detector whose operation depends on a semiconductor's high temperature coefficient of resistance rather than on rectification. It was Ferdinand Braun who first reported asymmetrical conduction in galena and copper pyrites (among others) back in 1874, in "Ueber die Stromleitung durch Schwefelmetalle" [On Current Flow through Metallic Sulfides], *Poggendorff's Annalen der Physik und Chemie,* v. 153, pp. 556–63. The large-area contact was made through partial immersion in mercury, and the other with copper, platinum, and silver wires. None of the samples showed more than a 2 : 1 forward : reverse current ratio. Braun later shared the 1909 Nobel Prize in Physics with Marconi for contributions to the radio art.

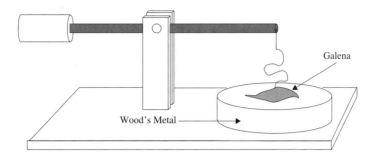

FIGURE 1.4. Typical crystal detector.

patent awarded for a semiconductor detector, although it was not recognized as such (indeed, the word "semiconductor" had not yet been coined). Work along these lines continued, and General Henry Harrison Chase Dunwoody filed for a patent in late 1906 for a detector using carborundum (silicon carbide), followed soon by a patent filing by Greenleaf Whittier Pickard (an MIT graduate whose great-uncle was the poet John Greenleaf Whittier) for a silicon (!) detector. As shown in the figure, one connection to this type of detector consisted of a small wire (whimsically known as a catwhisker) that made a point contact to the crystal surface. The other connection was a large area contact typically formed by a low–melting-point alloy (usually a mixture of lead, tin, bismuth and cadmium, known as Wood's metal, with a melting temperature of under 80°C) that surrounded the crystal. One might call a device made in this way a point-contact Schottky diode, although measurements are not always easily reconciled with such a description. In any event, we can see how the modern symbol for the diode evolved from a depiction of this physical arrangement, with the arrow representing the catwhisker point contact, as seen in the figure.

Figure 1.5 shows a simple crystal[8] radio made with these devices.[9] An LC circuit tunes the desired signal, which the crystal then rectifies, leaving the demodulated audio to drive the headphones. A bias source is not needed with some detectors (such as galena), so it is possible to make a "free-energy" radio![10]

Pickard worked harder than anyone else to develop crystal detectors, eventually trying over 30,000 combinations of wires and crystals. Among these were iron pyrites

---

[8] In modern electronics, "crystal" usually refers to quartz resonators used, for example, as frequency-determining elements in oscillators; these bear absolutely no relationship to the crystals used in crystal radios.

[9] A 1N34A germanium diode works fine and is more readily available, but it lacks the charm of galena, Wood's metal, and a catwhisker to fiddle with. An ordinary penny (dated no earlier than 1983), baked in a kitchen oven for 15 minutes at about 250°C to form CuO, exhibits many of the relevant characteristics of galena. Copper-based currencies of other nations may also work (the author has verified that the Korean 10-won coin works particularly well). The reader is encouraged to experiment with coins around the world and inform the author of the results.

[10] Perhaps we should give a little credit to the human auditory system: the threshold of hearing corresponds to an eardrum displacement of about the diameter of a hydrogen atom!

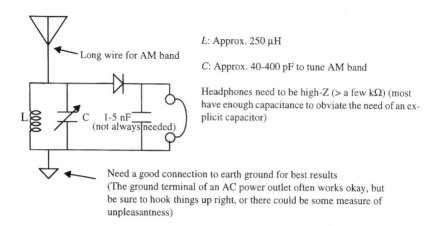

FIGURE 1.5. Simple crystal radio.

(fool's gold) and rusty scissors, in addition to silicon. Galena detectors became quite popular because they were inexpensive and needed no bias. Unfortunately, proper adjustment of the catwhisker wire contact was difficult to maintain because anything other than the lightest pressure on galena destroyed the rectification. Plus, you had to hunt around the crystal surface for a sensitive spot in the first place. On the other hand, although carborundum detectors needed a bias of a couple of volts, they were more mechanically stable (a relatively high contact pressure was all right), and found wide use on ships as a consequence.[11]

At about the same time that these crude semiconductors were first coming into use, radio engineers began to struggle with a problem that was assuming greater and greater prominence: interference.

The broad spectrum of a spark signal made it impractical to attempt much other than Morse-code types of transmissions (although some intrepid engineers did attempt AM transmissions with spark gap equipment, with little success). This broad-band nature fit well with coherer technology, since the varying impedance of the latter made it difficult to realize tuned circuits anyhow. However, the inability to provide any useful degree of selectivity became increasingly vexing as the number of transmitters multiplied.

Marconi had made headlines in 1899 by contracting with the *New York Herald* and the *Evening Telegram* to provide up-to-the-minute coverage of the America's Cup yacht race, and he was so successful that two additional groups were encouraged to try the same thing in 1901. One of these was led by Lee de Forest, whom we'll meet later, and the other by an unexpected interloper (who turned out to be none other than Pickard) from American Wireless Telephone and Telegraph. Unfortunately, with *three* groups simultaneously sparking away that year, *no one* was

---

[11] Carborundum detectors were typically packaged in cartridges and often adjusted through the delicate procedure of slamming them against a hard surface.

able to receive intelligible signals, and race results had to be reported the old way, by semaphore. A thoroughly disgusted de Forest threw his transmitter overboard, and news-starved relay stations on shore resorted to making up much of what they reported.

This failure was all the more discouraging because Marconi, Lodge, and that erratic genius Nikola Tesla had actually already patented circuits for tuning, and Marconi's apparatus had employed bandpass filters to reduce the possibility of interference.[12]

The problem was that, even though adding tuned circuits to spark transmitters and receivers certainly helped to filter the signal, no practical amount of filtering could ever really convert a spark train into a sine wave. Recognizing this fundamental truth, a number of engineers sought ways of generating continuous sine waves at radio frequencies. One group, which included Danish engineer Valdemar Poulsen[13] (who had also invented a crude magnetic recording device called the telegraphone) and Australian–American engineer (and Stanford graduate) Cyril Elwell, used the negative resistance associated with a glowing DC arc to keep an *LC* circuit in constant oscillation[14] to provide a sine-wave radio-frequency (RF) carrier. Engineers quickly discovered that this approach could be scaled up to impressive power levels: an arc transmitter of over 1 *mega*watt was in use shortly after WWI!

Pursuing a somewhat different approach, Ernst F. W. Alexanderson of General Electric (GE) acted on Reginald Fessenden's request to produce RF sine waves at large power levels with huge alternators (*really* big, high-speed versions of the thing that charges your car battery as you drive). This dead-end technology culminated in the construction of an alternator that put out 200 kW at 100 kHz! It was completed just as WWI ended, and was already obsolete by the time it became operational.[15]

The superiority of the continuous wave over spark signals was immediately evident, and it spurred the development of better receiving equipment. Thankfully, the coherer was gradually supplanted by a number of improved devices, including the semiconductor devices described earlier, and was well on its way to extinction by 1910 (although as late as the 1950s there was at least that one radio-controlled toy that used a coherer).

---

[12] Marconi was the only one backed by strong financial interests (essentially the British government), and his British patent (#7777, the famous "four sevens" patent, granted 26 April 1900) was the dominant tuning patent of the early radio days. It was also involved in some of the lengthiest and most intense litigation in the history of technology. The U.S. Supreme Court finally ruled in 1943 that Marconi had been anticipated by Lodge, Tesla, and others.

[13] British Patent #15,599, 14 July 1903. Some sources persistently render his name incorrectly as "Vladimir," a highly un-Danish name!

[14] Arc technology for industrial illumination was a well-developed art by this time. In his Stanford Ph.D. thesis, Leonard Fuller provided the theoretical advances that allowed arc power to break through a 30-kW "brick wall" that had stymied others. Thanks to Fuller, 1000-kW arc transmitters were possible by 1919. In 1931, as the first chair of U.C. Berkeley's electrical engineering department, Fuller arranged the donation of coil-winding machines and surplus 80-ton magnets from Federal for the construction of Ernest O. Lawrence's first large cyclotron.

[15] Such advanced rotating machinery severely stretched the metallurgical state of art.

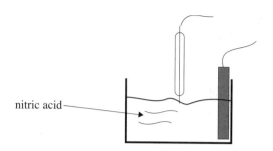

nitric acid

FIGURE 1.6. Fessenden's liquid barretter.

One such improvement, invented by Fessenden, was the "liquid barretter" shown in Figure 1.6. This detector consisted of a thin, silver-coated platinum wire (a "Wollaston wire") encased in a glass rod. A tiny bit of the wire protruded from the rod and made contact with a small pool of nitric acid. This arrangement had a quasiquadratic $V\text{–}I$ characteristic near the origin and therefore could actually demodulate RF signals. The barretter was widely used in a number of incarnations since it was a "self-restoring" device (unlike typical coherers) and required no adjustments (unlike crystal detectors). Except for the hazards associated with the acid, the barretter was apparently a satisfactory detector, judging from the many infringements (including an infamous one by de Forest) of Fessenden's patent.

Enough rectifying detectors were in use by late 1906 to allow shipboard operators on the east coast of the United States to hear, much to their amazement (despite a forewarning by radiotelegraph three days before), the first AM broadcast by Fessenden himself on Christmas Eve.[16] Delighted listeners were treated to a recording of Handel's *Largo* (from *Xerxes*), a fine rendition of *O Holy Night* by Fessenden on the violin (with the inventor accompanying himself while singing the last verse), and his hearty Christmas greetings to all.[17] He used a water-cooled carbon microphone *in series with the antenna* to modulate a 500-W (approximate), 50-kHz (also approximate) carrier generated by a prototype Alexanderson alternator located at Brant Rock, Massachusetts. Those unfortunate enough to use coherers missed out on the historic event, since coherers as typically used are completely unsuited to AM demodulation. Fessenden repeated his feat a week later, on New Year's Eve, to give more people a chance to get in on the fun.

The next year, 1907, was a significant one for electronics. Aside from following on the heels of the first AM broadcast (which marked the transition from radiotelegraphy to radiotelephony), it saw the emergence of important semiconductors. In addition to the patenting of the silicon detector, the LED was also discovered that year! In a brief article in *Wireless World* titled "A Note on Carborundum," Henry J. Round of

---

[16] Aitken (see Section 1.9) erroneously gives the date as Christmas Day.

[17] "An Unsung Hero: Reginald Fessenden, the Canadian Inventor of Radio Telephony," http://www. ewh.ieee.org/reg/7/millennium/radio/radio_unsung.html.

Great Britain reported the puzzling emission of a cold, blue[18] light from carborundum detectors under certain conditions (usually when the catwhisker potential was very negative relative to that of the crystal). The effect was largely ignored and ultimately forgotten, as there were just so many more pressing problems in radio at the time. Today, however, carborundum is actually used in blue LEDs[19] and has been investigated by some to make transistors that can operate at elevated temperatures. And as for silicon, well, we all know how that turned out.

1.4 BIRTH OF THE VACUUM TUBE

The year 1907 also saw the patenting, by Lee de Forest, of the first electronic device capable of amplification: the triode vacuum tube. Unfortunately, de Forest didn't understand how his invention actually worked, having stumbled upon it by way of a circuitous (and occasionally unethical) route.

The vacuum tube actually traces its ancestry to the lowly incandescent light bulb of Thomas Edison. Edison's bulbs had a problem with progressive darkening caused by the accumulation of soot (given off by the carbon filaments) on the inner surface of the bulb. In an attempt to cure the problem, he inserted a metal electrode, hoping somehow to attract the soot to this plate rather than to the glass. Ever the experimentalist, he applied both positive and negative voltages (relative to one of the filament connections) to this plate, and noted in 1883 that a current mysteriously flowed when the plate was positive, but none flowed when the plate was negative. Furthermore, the current that flowed depended on how hot he made the filament. He had no theory to explain these observations (remember, the word "electron" wasn't even coined until 1891, and the particle itself wasn't unambiguously identified until J. J. Thomson's experiments of 1897), but Edison went ahead and patented in 1884 the first electronic (as opposed to electrical) device, one that exploited the dependence of plate current on filament temperature to measure line voltage indirectly. This Rube Goldberg instrument never made it into production since it was inferior to a standard voltmeter; Edison just wanted another patent, that's all (that's one way he ended up with over a thousand of them).

The funny thing about this episode is that Edison arguably had never invented anything in the fundamental sense of the term, and here he had stumbled across an electronic rectifier but nevertheless failed to recognize the implications of what he had found. Part of this blindness was no doubt related to his emotional (and financial) fixation on the DC transmission of power, where a rectifier had no role.

At about this time, a consultant to the British Edison Company named John Ambrose Fleming happened to attend a conference in Canada. He dropped down to the

---

[18] He saw orange and yellow, too. He may have been drinking.

[19] It should be mentioned that LEDs based on GaN offer much higher efficiency, but it was only in 1992–1993 that Shuji Nakamura (then at Nichia Chemical) figured out practical methods for actually manufacturing them.

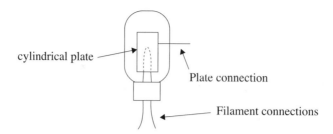

FIGURE 1.7. Fleming valve.

United States to visit his brother in New Jersey and also stopped by Edison's lab. He was greatly intrigued by the "Edison effect" (much more so than Edison, who found it difficult to understand Fleming's excitement over something that had no obvious promise of practical application), and eventually published papers on the Edison effect from 1890 to 1896. Although his experiments created an initial stir, Röntgen's announcement in January 1896 of the discovery of X-rays – as well as the discovery of natural radioactivity later that same year – soon dominated the interest of the physics community, and the Edison effect quickly lapsed into obscurity.

Several years later, though, Fleming became a consultant to British Marconi and joined in the search for improved detectors. Recalling the Edison effect, he tested some bulbs, found out that they worked all right as RF rectifiers, and patented the Fleming valve (vacuum tubes are thus still known as valves in the United Kingdom) in 1905 (Figure 1.7). The nearly deaf Fleming used a mirror galvanometer to provide a visual indication of the received signal, and included this feature as part of his patent.

Although not particularly sensitive, the Fleming valve was at least continually responsive and required no mechanical adjustments. Various Marconi installations used them (largely out of contractual obligations), but the Fleming valve was never popular (contrary to the assertions of some poorly researched histories) – it needed too much power, filament life was poor, the thing was expensive, and it was a remarkably insensitive detector compared to, say, Fessenden's barretter or well-made crystal detectors.

De Forest, meanwhile, was busy in America setting up shady wireless companies whose sole purpose was to earn money via the sale of stock. "Soon, we believe, the suckers will begin to bite," he wrote in his journal in early 1902. As soon as the stock in one wireless installation was sold, he and his cronies picked up stakes (whether or not the station was actually completed) and moved on to the next town. In another demonstration of his sterling character, he outright stole Fessenden's barretter (simply reforming the Wollaston wire into the shape of a spade) after visiting Fessenden's laboratory, and even had the audacity to claim a prize for its invention. In this case, however, justice did prevail, and Fessenden won an infringement suit against de Forest.

Fortunately for de Forest, Dunwoody invented the carborundum detector just in time to save him from bankruptcy. Not content to develop this legitimate invention,[20] though, de Forest proceeded to steal Fleming's vacuum tube diode and actually received a patent for it in 1906. He simply replaced the mirror galvanometer with a headphone and added a huge forward bias (thus reducing the sensitivity of an already insensitive detector). De Forest repeatedly and unconvincingly denied throughout his life that he was aware of Fleming's prior work (even though Fleming published in professional journals that de Forest habitually and assiduously scanned); to bolster his claims, de Forest pointed to his use of bias where Fleming had used none.[21] Conclusive evidence that de Forest had lied outright finally came to light when historian Gerald Tyne obtained the business records of H. W. McCandless, the man who made all of de Forest's first vacuum tubes (de Forest called them "audions").[22] The records clearly show that de Forest had asked McCandless to duplicate some Fleming valves months before he filed his patent. There is thus no room for a charitable interpretation that de Forest independently invented the vacuum tube diode.

His crowning achievement came soon after, however. De Forest added a zigzag wire electrode, which he called the grid, between the filament and wing electrode (later known as the plate), and thus the triode was born (see Figure 1.8). This three-element audion was capable of amplification, but de Forest did not realize this fact until years later. In fact, his patent application described the triode audion only as a detector, not as an amplifier.[23] Motivation for the addition of the grid is thus still curiously unclear. He certainly did not add the grid as the consequence of careful reasoning, as some histories claim. The fact is that he added electrodes all over the place. He even tried "control electrodes" outside of the plate! We must therefore regard his addition of the grid as merely the result of haphazard but persistent tinkering in his search for a detector to call his own. It would not be inaccurate to say that he stumbled onto the triode, and it is certainly true that others had to explain its operation to him.[24]

From the available evidence, neither de Forest nor anyone else thought much of the audion for a number of years (1906–1909 saw essentially no activity on the audion). In fact, when de Forest barely escaped conviction and a jail sentence for stock

---

[20] Dunwoody had performed this work as a consultant to de Forest. He was unsuccessful in his efforts to get de Forest to pay him for it.

[21] In his efforts to establish that he had worked independently of Fleming, de Forest repeatedly and stridently stated that it was his researches into the conductivity properties of flames that informed his work in vacuum tubes, arguing that ionic conduction was the key to their operation. As a consequence, he boxed himself into a corner that he found difficult to escape later, after others developed the superior high-vacuum tubes that were essentially free of ions.

[22] Gerald F. J. Tyne, *Saga of the Vacuum Tube,* Howard W. Sams & Co., Indianapolis, 1977.

[23] Curiously enough, though, his patent for the two-element audion *does* mention amplification.

[24] Aitken (see Section 1.9) argues that de Forest has been unfairly accused of not understanding his own invention. However, the bulk of the evidence contradicts Aitken's generous view.

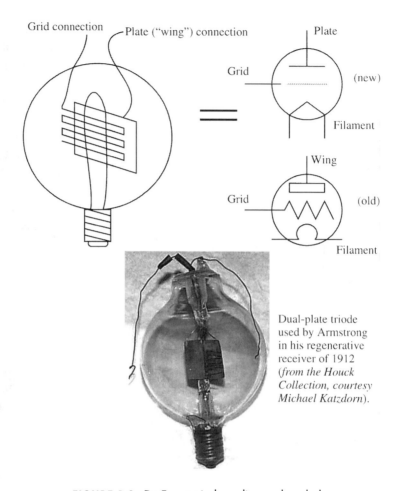

FIGURE 1.8. De Forest triode audion and symbols.

fraud after the collapse of one of his companies, he had to relinquish interest in all of his inventions as a condition of the subsequent reorganization of his companies, with just one exception: the lawyers let him keep the patent for the audion, thinking it worthless.[25]

He intermittently puttered around with the audion and eventually discovered its amplifying potential, as did others almost simultaneously (including rocket pioneer Robert Goddard).[26] He managed to sell the device to AT&T in 1912 as a telephone repeater amplifier, but initially had a tough time because of the erratic behavior of the audion. Reproducibility of device characteristics was rather poor and the tube

---

[25] The recently unemployed de Forest then went to work for Elwell at Federal Telephone and Telegraph in Palo Alto.

[26] Goddard's U.S. Patent #1,159,209, filed 1 August 1912 and granted 2 November 1915, describes a distant cousin of an audion oscillator and actually predates even Armstrong's documented work.

had a limited dynamic range. It functioned well for small signals but behaved badly upon overload (the residual gas in the tube would ionize, resulting in a blue glow and a frying noise in the output signal). To top things off, the audion filaments (made of tantalum) had a life of only about 100–200 hours. It would be a while before the vacuum tube could take over the world.

1.5 ARMSTRONG AND THE REGENERATIVE AMPLIFIER/DETECTOR/OSCILLATOR

Fortunately, some gifted people finally became interested in the audion. Irving Langmuir at GE Labs in Schenectady worked to achieve high vacua, thus eliminating the erratic behavior caused by the presence of (easily ionized) residual gases. De Forest never thought to do this (and in fact warned against it, believing that it would reduce the sensitivity) because he never really believed in thermionic emission of electrons (indeed, it isn't clear he even believed in electrons at the time), asserting instead that the audion depended fundamentally on ionized gas for its operation.

After Langmuir's achievement, the way was paved for a bright engineer to devise useful circuits to exploit the audion's potential. That engineer was Edwin Howard Armstrong, who invented the regenerative amplifier/detector[27] in 1912 at the tender age of 21. This circuit (a modern version of which is shown in Figure 1.9) employed positive feedback (via a "tickler coil" that coupled some of the output energy back to the input with the right phase) to boost the gain and Q of the system simultaneously. Thus high gain (for good sensitivity) and narrow bandwidth (for good selectivity) could be obtained rather simply from one tube. Additionally, the nonlinearity of the tube demodulated the signal. Furthermore, overcoupling the output to the input turned the thing into a wonderfully compact RF oscillator.

In a 1914 paper entitled "Operating Features of the Audion,"[28] Armstrong published the first correct explanation for how the triode worked and provided experimental evidence to support his claims. He followed this paper with another ("Some Recent Developments in the Audion Receiver")[29] in which he additionally explained the operation of the regenerative amplifier/detector and showed how to make an oscillator out of it. The paper is a model of clarity and is quite readable even to modern audiences. De Forest, however, was quite upset at Armstrong's presumption. In a published discussion section following the paper, de Forest repeatedly attacked Armstrong. It is clear from the published exchange that, in sharp contrast with Armstrong, de Forest had difficulty with certain basic concepts (e.g., that the average value of a sine wave is zero) and didn't even understand how the triode, his own invention

---

[27] His notarized notebook entry is actually dated 31 January 1913.
[28] *Electrical World,* 12 December 1914.
[29] *Proc. IRE,* v. 3, 1915, pp. 215–47.

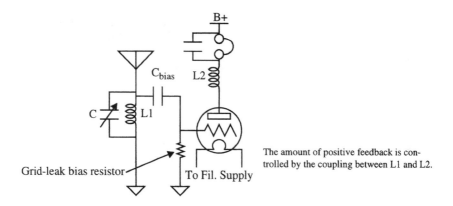

The amount of positive feedback is controlled by the coupling between L1 and L2.

FIGURE 1.9. Armstrong regenerative receiver.

(more of a discovery, really), actually worked. The bitter enmity that arose between these two men never waned.

Armstrong went on to develop circuits that continue to dominate communications systems to this day. While a member of the U.S. Army Signal Corps during World War I, Armstrong became involved with the problem of detecting enemy planes from a distance, and pursued the idea of trying to home in on the signals naturally generated by their ignition systems (spark transmitters again). Unfortunately, little useful radiation was found below about 1 MHz, and it was exceedingly difficult with the tubes available at that time to get much amplification above that frequency. In fact, it was only with extraordinary care that H. J. Round (of blue LED fame) achieved useful gain at 2 MHz in 1917, so Armstrong had his work cut out for him.

He solved the problem by employing a principle originally used by Poulsen and later elucidated by Fessenden. When demodulating a continuous wave (CW) signal, the resultant DC pulse train could be hard to make out. Valdemar Poulsen offered an improvement by inserting a rapidly driven interrupter in series with the headphones. This way, a steady DC level is chopped into an audible waveform. The "Poulsen Tikker" made CW signals easier to copy as a consequence.

Fessenden, whose fondness for rotating machines was well known, used much the same idea but derived his signal from a high-speed alternator that could heterodyne signals to any desired audible frequency, allowing the user to select a tone that cut through the interference.

Armstrong decided to employ Fessenden's heterodyne principle in a different way. Rather than using it to demodulate CW directly, he used the heterodyne method to convert an incoming high-frequency RF signal into one at a lower frequency, where high gain and selectivity could be obtained with relative ease. This signal, known as the intermediate frequency (IF), was then demodulated after much filtering and amplification at the IF had been achieved. The receiver could easily possess enough sensitivity so that the limiting factor was actually atmospheric noise (which is quite

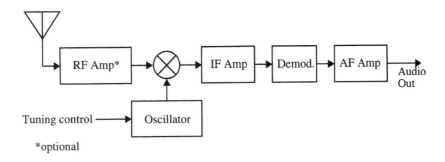

FIGURE 1.10. Superheterodyne receiver block diagram.

large in the AM broadcast band). Furthermore, a single tuning control was made possible, since the IF amplifier works at a fixed frequency.

He called this system the "superheterodyne" and patented it in 1917 (see Figure 1.10). Although the war ended before Armstrong could use the superhet to detect German planes, he continued to develop it with the aid of several talented engineers, finally reducing the number of tubes to five from an original complement of ten (good thing, too: the prototype had a total filament current requirement of ten amps). David Sarnoff of RCA eventually negotiated the purchase of the superhet rights, and RCA thereby came to dominate the radio market by 1930.

The great sensitivity enabled by the invention of the vacuum tube allowed transmitter power reductions of orders of magnitude while simultaneously increasing useful communications distances. Today, 50 kW is considered a large amount of power, yet ten times this amount was the norm right after WWI.

The 1920s saw greatly accelerated development of radio electronics. The war had spurred the refinement of vacuum tubes to an astonishing degree, with the appearance of improved filaments (longer life, higher emissivity, lower power requirements), lower interelectrode capacitances, higher transconductance, and greater power-handling capability. These developments set the stage for the invention of many clever circuits, some designed to challenge the dominance of Armstrong's regenerative receiver.

1.6 OTHER RADIO CIRCUITS

1.6.1 THE TRF AND THE NEUTRODYNE

One wildly popular type of radio in the early days was the tuned radio-frequency (TRF) receiver. The basic TRF circuit typically had three RF bandpass stages, each tuned separately, and then a stage or two of audio after demodulation (the latter sometimes accomplished with a crystal diode). The user thus had to adjust three or more knobs to tune in each station. While this array of controls may have appealed to the tinkering-disposed technophile, it was rather unsuited to the average consumer.

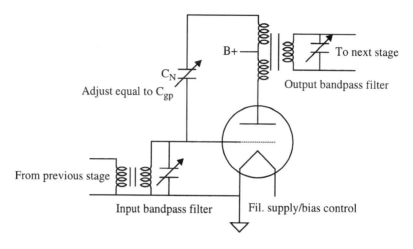

FIGURE 1.11. Basic neutrodyne amplifier.

Oscillation of the TRF stages was also a big problem, caused by the parasitic feedback path provided by the grid–plate capacitance C_{gp}.[30] Although limiting the gain per stage was one way to reduce the tendency to oscillate, the attendant degradation in sensitivity was usually unacceptable.

The problem caused by C_{gp} was largely eliminated by Harold Wheeler's invention[31] of the neutrodyne circuit (see Figure 1.11).[32] Recognizing the cause of the problem, he inserted a compensating capacitance (C_N), termed the neutralizing capacitor (actually, "condenser" was the term back then). When properly adjusted, the condenser fed back a current exactly equal in magnitude but opposite in phase with that of the plate-to-grid capacitance, so that no input current was required to charge the capacitances. The net result was the suppression of C_{gp}'s effects, permitting a large increase in gain per stage without oscillation.[33] After the war, Westinghouse acquired the rights to Armstrong's regeneration patent, negotiated licensing agreements with a limited number of radio manufacturers, and then aggressively prosecuted those who infringed (which was just about everybody). To protect themselves, those "on the outside" organized into the Independent Radio Manufacturers Association and bought the rights to Hazeltine's circuit. Tens of thousands of neutrodyne kits and assembled consoles were sold in the 1920s by members of IRMA, all in an attempt to compete with Armstrong's regenerative circuit.

---

[30] It is left as "an exercise for the reader" to show that the real part of the input impedance of an inductively loaded common-cathode amplifier can be less than zero because of the feedback through C_{gp}, and that this negative resistance can therefore cause instability.

[31] He did this work for Louis Hazeltine, who is frequently given credit for the circuit.

[32] Of course, it should be noted that Armstrong's superheterodyne neatly solves the problem by obtaining gain at a number of different frequencies: RF, IF, and AF. This approach also reduces greatly the danger of oscillation from parasitic input–output coupling.

[33] In some sets, only the middle TRF stage is neutralized.

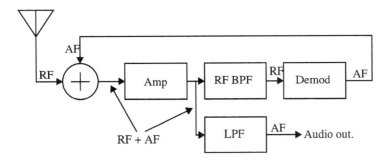

FIGURE 1.12. Reflex receiver block diagram.

Meanwhile, de Forest was up to his old tricks. He bought a company that had a license to make Armstrong's regenerative circuits. Although he knew that the license was nontransferable, he nonetheless started to sell regenerative radios until he was caught and threatened with lawsuits. He eventually skirted the law by selling a radio that *could* be hooked up as a regenerator by the customer simply by reconnecting a few wires between binding posts that had been conveniently provided for this purpose.[34]

1.6.2 THE REFLEX CIRCUIT

The reflex circuit (Figure 1.12) enjoyed popularity with hobbyists in the early 1920s, followed by a commercial resurgence in the mid-1930s. The idea behind the reflex is wonderful and subtle, and perhaps even the inventor of the circuit himself (believed to be French engineer Marius Latour[35]) did not fully appreciate just how marvelous it was. The basic idea was this: pass the RF through some number (say, one) of amplifier stages, demodulate, and then pass the audio back through those *same* amplifiers. A given tube thus simultaneously amplifies both RF and AF signals.

The reason that this arrangement makes sense becomes convincingly clear only when you consider how this connection allows the overall system to possess a gain–bandwidth product that exceeds that of the active device itself. Suppose that the vacuum tube in question has a certain constant gain–bandwidth product limit. Further assume that the incoming RF signal is amplified by a factor G_{RF} over a brickwall passband of bandwidth B and that the audio signal is also amplified by a factor G_{AF} over the same brickwall bandwidth B. The overall gain–bandwidth product is therefore $(G_{RF} G_{AF})B$, while the gain–bandwidth product of the combined RF/AF signal

---

[34] One anecdotal report has it that de Forest sold receivers with a wire that protruded from the back panel, marked with a label that said something like "Do not cut this wire; it converts this receiver into a regenerative one." I have not found a primary source for this information, but it is entirely consistent with all we know about de Forest's character.

[35] It should be noted that Armstrong's second paper on the superheterodyne (published in 1924) contains examples of reflex circuits.

processed by the amplifier is just $(G_{RF} + G_{AF})B$. For the reflex circuit to have an advantage we need only the product of the gains to exceed the sum of the gains, a criterion that is easily satisfied.

The reflex circuit demonstrates that there is nothing fundamental about gain–bandwidth, and that we are effectively fooled into believing that gain and bandwidth must trade off linearly just because they commonly do. The reflex circuit shows us the error in our thinking. For this reason alone, the reflex circuit deserves more detailed treatment than it commonly receives.

1.7 ARMSTRONG AND THE SUPERREGENERATOR

Armstrong wasn't content to rest, although after having invented both the regenerative and superheterodyne receivers he would seem to have had the right.

While experimenting with the regenerator, he noticed that under certain conditions he could, for a fleeting moment, get much greater amplification than normal. He investigated further and developed by 1922 a circuit he called the superregenerator, a circuit that provides so much gain in a *single tube* that it can amplify thermal and shot noise to audible levels!

Perhaps you found the reflex principle a bit abstruse; you ain't seen nothin' yet. In a superregenerator the system is *purposely made unstable,* but is periodically shut down (quenched) to prevent getting stuck in some limit cycle.

How can such a bizarre arrangement provide gain (lots of gain)? Take a look at Figure 1.13, which strips the superregenerator to its basic elements.[36] Now, during the time that it is active (i.e., the negative resistor is connected to the circuit), this second-order bandpass system has a response that grows exponentially with time. Response to what? Why, the initial conditions, of course! A tiny initial voltage will, given sufficient time, grow to detectable levels in such a system. The initial voltage could conceivably even come from thermal or shot-noise processes.

The problem with all real systems is that saturation eventually occurs, and no further amplification is possible in such a state. The superregenerator evades this problem by periodically shutting the system down. This periodic "quenching" can be made inaudible if a sufficiently high quench frequency is chosen.

Because of the exponential growth of the signal with time, the superregenerator trades off *log* of gain for bandwidth. As a bonus, the unavoidable nonlinearity of the vacuum tube can be exploited to provide demodulation of the amplified signal! As you might suspect, the superregenerator's action is so subtle and complex that it has never been understood by more than a handful of people at a given time. It's a

---

[36] The classic vacuum tube superregenerator looks a lot like a normal regenerative amplifier, except that the grid-leak bias network time constant is made very large and the feedback (via the tickler coil) is large enough to guarantee instability. As the amplitude grows, the grid-leak bias also grows until it cuts off the tube. The tube remains cut off until the bias decays to a value that returns the tube to the active region. Thus, no separate quench oscillator is necessary.

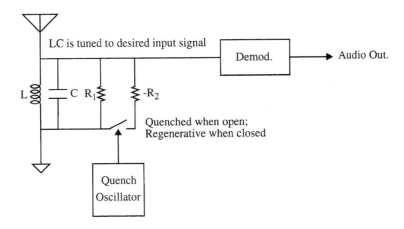

FIGURE 1.13. Superregenerative receiver basics.

quasiperiodically time-varying, nonlinear system that is allowed to go intermittently unstable, and Armstrong invented it in 1922.

Armstrong sold the patent rights to RCA (who shared Armstrong's view that the superregenerator was the circuit to end all circuits), and he became its largest shareholder as a result.[37] Alas, the superregenerator never assumed the dominant position that he and RCA's David Sarnoff had envisioned. The reason is simple for us to see now: every superregenerative amplifier is fundamentally also an oscillator. Therefore, every superregenerative receiver is also a transmitter that is capable of causing interference to nearby receivers. In addition, the superregenerator produces an annoyingly loud hiss (the amplified thermal and shot noise) in the absence of a signal, rather than the relative quiet of other types of receivers. For these reasons, the superregenerator never took the radio world by storm.

The circuit has found wide application in toys, however. When you've got to get the most sensitivity with absolutely the minimum number of active devices, you cannot do better than the superregenerative receiver. Radio-controlled cars, automatic garage-door openers, and toy walkie-talkies almost invariably use a circuit that consists of just one transistor operating as a superregenerative amplifier/detector, and perhaps two or three more as amplifiers of the demodulated audio signal (as in a walkie-talkie). The overall sensitivity is often of the same order as that provided by a typical superhet. On top of those attributes, it can also demodulate FM through a process known as slope demodulation: if one tunes the receiver a bit off frequency so that the receiver gain versus frequency is not flat (i.e., has some slope, hence the name), then an incoming FM signal produces a signal in the receiver whose amplitude varies as the frequency varies; the signal is converted into an AM signal that is demodulated as usual ("it's both a floor wax *and*

---

[37] In a bit of fortuitous timing, Armstrong sold his stock just before the great stock-market crash of 1929.

a dessert topping"). So, if most of the system cost is associated with the number of active devices, the superregenerative receiver provides a remarkably economical solution.

1.8   OLEG LOSEV AND THE FIRST
SOLID-STATE AMPLIFIER

Surely one of the most amazing (and little-known) stories from this era is that of self-taught Soviet engineer Oleg Losev and his solid-state receivers of 1922.[38] Vacuum tubes were expensive then, particularly in the Soviet Union so soon after the revolution, so there was naturally a great desire to make radios on the cheap.

Losev's approach was to investigate the mysteries of crystals, which by this time were all but forgotten in the West. He independently rediscovered Round's carborundum LEDs and actually published about a half dozen papers on the phenomenon. He correctly deduced that it was a quantum effect, describing it as the inverse of Einstein's photoelectric effect, and correlated the short wavelength cutoff energy with the applied voltage. He even noted that the light was emitted from a particular crystalline boundary (which we would call a junction), and he cast doubt on a prevailing theory of a thermal origin by showing that the emission could be electronically modulated up to at least 78 kHz (the limit of his rotating-mirror instrumentation).

Even more startling than his insights into the behavior of LEDs was his discovery of the negative resistance that can be obtained from biased point-contact zincite (ZnO) crystal diodes. With zincite, he actually constructed fully solid-state RF amplifiers, detectors, and oscillators at frequencies up to 5 MHz a full quarter century before the invention of the transistor! Later, he even went on to construct a superheterodyne receiver with these crystals. True, one had to adjust several bias voltages and catwhiskers, but it nevertheless worked (see Figure 1.14). He eventually abandoned the "crystadyne" after about a decade of work, though, because of difficulties with obtaining zincite (it's found in commercially significant quantity in only two mines, and they're both in New Jersey), the fundamental variability of natural crystals in general, as well as the problem of interstage interaction inherent in using two-terminal devices to get gain.

The reason almost no one in the United States has ever heard of Losev is simple. First, almost no one has even heard of Armstrong – it seems there isn't much interest in preserving the names of these pioneers. Plus, most of Losev's papers are in German and Russian, limiting readership. Add the generally poor relations between the United States and the Soviet Union over most of this century, and it's actually a wonder that *anyone* knows who Losev was. Losev himself isn't around because he was one of many who starved to death during the terrible siege of Leningrad, breathing

---

[38] The spelling offered here is the closest English transliteration. A commonly encountered German-based spelling is *Lossev,* where the doubled consonant forces a purely sibilant *s* in that language.

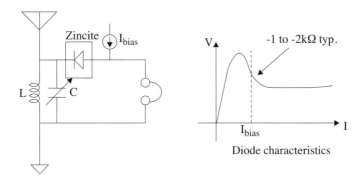

FIGURE 1.14. Losev's crystadyne receiver (single-stage).

his last in January of 1942. His colleagues had advised him to leave, but he was just too interested in finishing up what he termed "promising experiments with silicon." Sadly, all records of those experiments have apparently been lost.

1.9 EPILOG

By the early 1930s, the superhet had been refined to the point that a single tuning control was all that was required. The superior performance and ease of use of the superhet guaranteed its dominance (as well as that of RCA), and virtually every modern receiver, ranging from portable radios to radar sets, employs the superheterodyne principle; it seems unlikely that this situation will change in the near future. It is a tribute to Armstrong's genius that a system he conceived during World War I still dominates at the dawn of the 21st century.

Armstrong, annoyed by the static that plagues AM radio, went on to develop (wideband) frequency modulation, in defiance of theoreticians who declared FM useless.[39] Unfortunately, Armstrong's life did not end happily. In a sad example of how our legal system is often ill-equipped to deal intelligently with technical matters, de Forest challenged Armstrong's regeneration patent and ultimately prevailed in some of the longest patent litigation in history (it lasted twenty years). Not long after the courts handed down the final adverse decision in this case, Armstrong began locking horns with his former friend Sarnoff and RCA in a bitter battle over FM that raged for well over another decade. His energy and money all but gone, Armstrong committed suicide in 1954 at the age of 63. It was the fortieth anniversary of his demonstration of regeneration to Sarnoff. Armstrong's widow, Marian, picked up the fight and eventually went on to win every legal battle; it took fifteen years.

[39] Bell Laboratories mathematician John R. Carson (no known relation to the entertainer) had correctly shown that FM always requires more bandwidth than AM, disproving a prevailing belief to the contrary. But he went too far in declaring FM worthless.

De Forest eventually went legit. He moved to Hollywood and worked on developing sound and color for motion pictures. A few years before he died at the ripe old age of 87, he penned a characteristically self-aggrandizing autobiography titled *The Father of Radio* that sold fewer than a thousand copies. He also tried to get his wife to write a book called *I Married a Genius,* but she somehow never got around to it.

FURTHER READING

The stories of de Forest, Armstrong, and Sarnoff are wonderfully recounted by Tom Lewis in *The Empire of the Air,* a book that was turned into a film by Ken Burns for PBS. Although it occasionally gets into trouble when it ventures a technical explanation, the human focus and rich biographical material that Lewis has unearthed much more than compensates. (Prof. Lewis says that many corrections will be incorporated in a later paperback edition of his book.)

For those interested in more technical details, there are two excellent books by Hugh Aitken. *Syntony and Spark* recounts the earliest days of radiotelegraphy, beginning with pre-Hertzian experiments and ending with Marconi. *The Continuous Wave* takes the story up to the 1930s, covering arc and alternator technology in addition to vacuum tubes. Curiously, though, Armstrong is but a minor figure in Aitken's portrayals.

The story of early crystal detectors is well told by A. Douglas in "The Crystal Detector" (*IEEE Spectrum,* April 1981, pp. 64–7) and by D. Thackeray in "When Tubes Beat Crystals: Early Radio Detectors" (*IEEE Spectrum,* March 1983, pp. 64–9). Material on other early detectors is found in a delightful volume by V. Phillips, *Early Radio Wave Detectors* (Peregrinus, Stevenage, UK, 1980). Finally, the story of Losev is recounted by E. Loebner in "Subhistories of the Light-Emitting Diode" (*IEEE Trans. Electron Devices,* July 1976, pp. 675–99).

1.10 APPENDIX A: A VACUUM TUBE PRIMER

1.10.1 INTRODUCTION

Sadly, few engineering students are ever exposed to the vacuum tube. Indeed, most engineering faculty regard the vacuum tube a quaint relic. Well, maybe they're right, but there are still certain engineering provinces (such as high-power RF) where the vacuum tube reigns supreme. This appendix is intended to provide the necessary background so that an engineer educated in solid-state circuit design can develop at least a superficial familiarity with this historically important device.

The operation of virtually all vacuum tubes can be understood rather easily once you study the physics of the vacuum diode. To simplify the development, we'll follow a historical path and consider a parallel plate structure rather than the more common coaxial structures. The results are easier to derive but still hold generally.

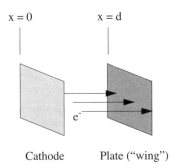

FIGURE 1.15. Idealized diode structure.

1.10.2 CATHODES

Consider the diode structure shown in Figure 1.15. The left-most electrode is the cathode, whose job is to emit electrons. The plate's job is to collect them.

All the early tubes (Edison's and Fleming's diode, and de Forest's triode audion) used directly heated cathodes, meaning that the light-bulb filament did the work of emitting electrons. Physically all that happens is that, at high enough temperatures, the electrons in the filament material are given enough kinetic energy that they can leave the surface; they literally boil off.

Clearly, materials that emit well at temperatures below the melting point make the best cathodes. De Forest's first filaments were made of the same carbon variety used in Edison's light bulbs, although tantalum, which has a high melting point (about 3100 K), quickly replaced carbon. Useful emission from tantalum occurs only if the material is heated to bright incandescence, though, so the early audions were pretty power-hungry. Additionally, tantalum tends to crystallize at high temperature, and filament life is unsatisfactory as a consequence of the attendant increasing brittleness. A typical audion filament had a lifetime as short as 100–200 hours. Some audions were made with a spare filament that could be switched in when the first filament burned out.

Research by W. D. Coolidge (same guy who developed the high-power X-ray tube) at GE allowed the use of tungsten (melting point: 3600 K) as a filament material. He found a way to make filaments out of the unwieldy stuff (tungsten is not ductile, and hence it ordinarily cannot be drawn into wires) and opened the path to great improvements in vacuum tube (and light bulb) longevity because of the high melting point of that material.[40]

Unfortunately, lots of heating power is required to maintain the operating temperature of about 2400 K, and portable (or even luggable) equipment just could not evolve

---

[40] Tungsten is still used in light bulbs today.

until these heating requirements were reduced. One path to improvement (discovered accidentally) is to add a little thorium to the tungsten. If the temperature is held within rather narrow limits (around 1900 K), the thorium diffuses from the bulk onto the surface, where it serves to lower the work function (the binding energy of electrons) and thereby increases emissivity. These thoriated tungsten filaments still find wide use in high-power transmitting tubes, but their filament temperature must be controlled rather tightly. If the temperature is too high, the thorium boils off quickly (leaving a pure tungsten filament behind), and if it is too low, the thorium does not diffuse to the surface fast enough to do any good.

While thoriated tungsten is a more efficient emitter than pure tungsten, it is deactivated by the bombardment of positive ions such as might be associated with any residual gas, or gas that might evolve from the tube's elements during high-temperature operation. Pure tungsten is therefore used in high-voltage tubes (such as X-ray tubes, which may have anode potentials of 350 kV), where any positive ions would be accelerated to energies that would damage a thoriated tungsten filament.

To reduce heater temperatures still further, it is necessary to find ways to reduce the work function even more. This was accomplished with the discovery of a family of barium and strontium oxide mixtures that allow copious emission at a red glow, rather than at full incandescence. The lower temperatures (typically around 1000 K) greatly increase filament life while greatly reducing power requirements. In fact, in most tubes using oxide-coated cathodes, decreased emissivity rather than filament burnout determines the lifetime.

The great economy in power afforded by the oxide-coated cathodes makes practical the use of indirectly heated cathodes. In such tubes, the filament does not do the emitting of electrons. Rather, its function is simply to heat a cylindrical cathode that is coated with the oxide mixture. Such an indirectly heated cathode has a number of advantages. The entire cathode is at one uniform potential, so there is no spatial preference to the emission as there is in a directly heated cathode. Additionally, AC can be used to provide filament power in a tube with a unipotential, indirectly heated cathode, without worrying (much) about the injection of hum that would occur if AC were used in tubes with directly heated cathodes.[41]

The drawback to oxide-coated cathodes is that they are extraordinarily sensitive to bombardment by positive ions. And to make things worse, the cathodes themselves tend to give off gas over time, especially if overheated. Thus, rather elaborate procedures must be used to maintain a hard vacuum in tubes using such cathodes. Aside from pumping out the tube at temperatures high enough to cause all the elements to incandesce, a magnesium or phosphorus "getter" is fired (via RF induction) after assembly to react with any stray molecules of gas that evade the extensive evacuation procedure, or that may evolve over the life of the tube. The getter is easily seen as a

---

[41] Edward S. ("Ted") Rogers obtained the Canadian rights to an early indirectly heated cathode designed by Frederick McCullough (the "mac" in Eimac) in 1924. After considerable refinement, Rogers commercialized the technology, starting in 1925.

mirrorlike metallic deposit on the inner surface of the tube. The sensitivity of oxide cathodes to degradation by positive ion bombardment relegates their use to relatively low-power–low-voltage applications. Tubes that use pure tungsten filaments do not have getters, since they are not nearly as sensitive to trace amounts of gas.

1.10.3 V–I CHARACTERISTICS OF VACUUM TUBES

Now that we've taken care of the characteristics of cathodes, we turn to a derivation of the V–I characteristics of the diode. To simplify the development, assume that the cathode emits electrons with zero initial velocity, and neglect contact potential differences between the plate and cathode. These assumptions lead to errors that are noticeable mainly at low plate–cathode voltages. We will additionally assume that the cathode is capable of emitting an unlimited number of electrons per unit time. This assumption becomes increasingly invalid at lower cathode temperatures and at higher currents.

Further assume that the current flow in the device is space charge–limited. That is, the electrostatic repulsion by the cloud of electrons surrounding the cathode limits the current flow, rather than an insufficiency of electron emission by the cathode.

The anode or plate (originally called the "wing" by de Forest), is located a distance d away from the cathode, and is at a positive voltage V relative to the unipotential cathode. Given our assumption of zero initial velocity, the kinetic energy of an electron at some point x between cathode and plate is simply that due to acceleration by the electric field (SI units are assumed throughout):

$$\tfrac{1}{2}m_e v^2 = q\psi(x), \tag{1}$$

where $\psi(x)$ is the potential at point x. Solving for the velocity as a function of x yields

$$v(x) = \sqrt{\frac{2q\psi(x)}{m_e}}. \tag{2}$$

Now, the current density J (in A/m^2) is just the product of the volume charge density ρ and velocity, and must be independent of x. Hence we have

$$J = \rho(x)v(x) = \rho(x)\sqrt{\frac{2q\psi(x)}{m_e}}, \tag{3}$$

so that

$$\rho(x) = J\sqrt{\frac{m_e}{2q\psi(x)}}. \tag{4}$$

This last equation gives us one relationship between the charge density and the potential for a given current density. To solve for the potential (or charge density) we turn to Poisson's equation, which in one-dimensional form is just

$$\frac{d^2\psi(x)}{dx^2} = -\frac{\rho(x)}{\varepsilon_0}. \tag{5}$$

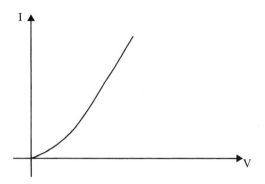

FIGURE 1.16. *V–I* characteristics of diode (space charge–limited).

Combining Eqn. 4 and Eqn. 5 yields a simple differential equation for the potential:

$$\frac{d^2\psi(x)}{dx^2} = -\frac{J}{\varepsilon_0}\sqrt{\frac{m_e}{2q\psi(x)}}, \tag{6}$$

with the boundary conditions

$$\psi(d) = V \tag{7}$$

and

$$E(0) = -\left.\frac{d\psi}{dx}\right|_{x=0} = 0. \tag{8}$$

This last boundary condition is the result of assuming space charge–limited current.

The solution is of the form $\psi(x) = Cx^n$ (trust me). Plugging and chugging yields

$$\psi(x) = V(x/d)^{4/3}. \tag{9}$$

Now, if this last expression is substituted back into the differential equation, we obtain, at long last, the desired *V–I* (or *V–J*) relationship:

$$I = JA = KV^{3/2}, \tag{10}$$

where the (geometry-dependent) constant K is known as the *perveance* and is here given by

$$K = \frac{\varepsilon_0}{d^2}\sqrt{\frac{32q}{81m_e}}. \tag{11}$$

The 3/2-power relationship between voltage V and current I (see Figure 1.16) is basic to vacuum tube operation (even for the more common coaxial structure) and recurs frequently, as we shall soon see.

As stated previously, the *V–I* characteristic just derived assumes that the current flow is space charge–limited.[42] That is, we assume that the cathode's ability to supply electrons is not a limiting factor. In reality, the rate at which a cathode can

---

[42] And, as stated earlier, it also assumes zero initial velocity of electrons emitted from the cathode and neglects contact potential differences between plate and cathode. This correction usually amounts to less than a volt and therefore is important only for low plate-to-cathode voltages.

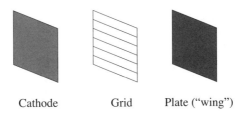

Cathode   Grid   Plate ("wing")

FIGURE 1.17. Idealized planar triode structure.

supply electrons is not infinite and depends on the cathode temperature. In all real diodes, there exists a certain plate voltage above which the current ceases to follow the 3/2-power law because of the unavailability of a sufficient supply of electrons. This regime, known as the emission-limited region of operation, is usually associated with power dissipation sufficient to cause destruction of the device. We will generally ignore operation in the emission-limited regime, although it may be of interest in the analysis of vacuum tubes near the end of their useful life, or in tubes operated at lower-than-normal cathode temperature.

The diode structure we have just analyzed is normally incapable of amplification. However, if we insert a porous control electrode (known as the *grid*) between cathode and plate, we can modulate the flow of current. If certain elementary conditions are met, power gain may be readily obtained. Let's see how this works.

Figure 1.17 shows a triode that is quite similar to the structures in de Forest's first triode audions, and its operation can be understood as a relatively straightforward extension of the diode. The field that controls the current flow will now depend on both the plate-to-cathode voltage and the grid-to-cathode voltage. Let us assume that we may replace the voltage in the diode law with a simple weighted sum of these two voltages. We then write, using notational conventions of the era:

$$I_{\text{plate}} = K \left(E_C + \frac{E_B}{\mu} \right)^{3/2}, \tag{12}$$

where K is the triode perveance, E_C is the grid-to-cathode voltage, E_B is the plate-to-cathode voltage, and μ is a roughly constant (though geometry-dependent) parameter known as the amplification factor. Figure 1.18 shows a family of triode characteristics conforming to this ideal relationship.

Physically what goes on is this: electrons leaving the cathode feel the influence of an electric field that is a function of two voltages. Volt for volt, the more proximate grid exerts a larger influence than the relatively distant plate. If the grid potential is negative then few electrons will be attracted to it, so the vast majority will flow onto the plate. Hence, little grid current flows, and there can be a very large power gain as a consequence.

The negative grid-to-cathode voltage and tiny grid current that characterize normal vacuum tube operation is similar to the negative gate-to-source voltage and tiny

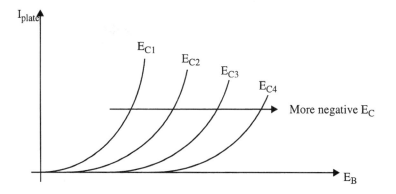

FIGURE 1.18. Triode characteristics.

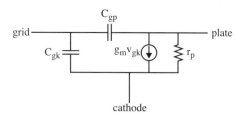

FIGURE 1.19. Incremental model for triode vacuum tube.

gate current of depletion-mode n-channel FETs (field-effect transistors), although this comparison seems a bit heretical to old-timers.

The analogy between FETs and vacuum tubes is close enough that even their incremental models are essentially the same (see Figure 1.19). Approximate equations for the transconductance g_m (sometimes called the "mutual conductance") and incremental plate resistance r_p are readily obtained from the V–I relationship already derived:

$$g_m \equiv \frac{\partial I}{\partial E_C} = \frac{3}{2}K^{2/3}I^{1/3} \tag{13}$$

and

$$r_p \equiv \frac{\partial E_B}{\partial I} = \frac{2}{3}\mu K^{-2/3}I^{-1/3}. \tag{14}$$

Note that the product of g_m and r_p is simply μ, so that μ represents the open-circuit amplification factor. Additionally, note that the transconductance and plate resistance are only weak functions (cube roots) of operating point. For this reason, vacuum tubes generate less harmonic distortion than other devices working over a comparable fractional range about a given operating point. Recall that the exponential V–I relationship of bipolar transistors leads to a linear dependence of g_m on I, and that the square-law dependence of drain current on gate voltage leads to a square-root dependence of g_m on I in FETs. The relatively weak dependence on plate current in vacuum tubes is apparently at the core of arguments that vacuum tube amplifiers are "cleaner" than those made with other types of active devices. It is certainly true that if amplifiers are driven beyond their linear range then a transistor version is likely to produce more (perhaps much more) distortion than its vacuum tube counterpart. However, there is considerably less merit to the argument that audible differences still exist when linear operation is maintained.

The triode ushered in the electronic age, making possible transcontinental telephone and radiotelephone communications. As the radio art advanced, it soon became clear that the triode has severe high-frequency limitations. The main problem is the plate-to-grid feedback capacitance, since it gets amplified as in the Miller effect. In transistors, we can get around the problem using cascoding, a technique that isolates the output node from the input node so that the input doesn't have to charge a magnified capacitance. Although this technique could also be used in vacuum tubes, there is a simpler way: add another grid (called the screen grid) between the old grid (called the control grid) and the plate. If the screen grid is held at a fixed potential, it acts as a Faraday shield between output and input, shunting the capacitive feedback to an incremental ground. In effect, the cascoding device is integral with the rest of the vacuum tube.

The screen grid is traditionally held at a high DC potential to prevent inhibition of current flow. Besides getting rid of the Miller effect problem, the addition of the screen grid makes the current flow even less dependent on the plate voltage than before, since the control grid "sees" what's happening at the plate to a greatly attenuated degree. An equivalent statement is that the amplification factor μ has increased.

All these effects are desirable, yet the tetrode tube has a subtle but important flaw. Electrons can crash into the plate with sufficient violence to dislodge other electrons. In triodes, these secondary electrons always eventually find their way back to the only electrode with a positive potential: the plate.[43] In the tetrode, however, secondary electrons can be attracted to the screen grid whenever the plate voltage is below the potential at the screen. Under these conditions, there is actually a negative plate resistance, since an increase in plate potential increases the generation of secondary electrons, whose current is lost as screen current. The plate current thus behaves roughly as shown in Figure 1.20. The negative resistance region is normally undesirable (unless you're trying to make an oscillator), so voltage swings at the plate must be restricted to avoid it. This limits the available signal power output, making the tetrode a bit of a loser when it comes to making power output devices.

Well, one grid is good, and two are better, so guess what? One way to solve the problem of secondary emission is to add a third grid (called the suppressor grid), and place it nearest the plate. The suppressor is normally held at cathode potential and works as follows. Electrons leaving the region past the screen grid have a high enough velocity that they aren't going to be turned around by the suppressor grid's low potential. So they happily make their way to the plate, and some of them generate secondary electrons, as before. But now, with the suppressor grid in

---

[43] Actually, negative resistance behavior can occur in a triode if the grid is at a higher potential than the plate.

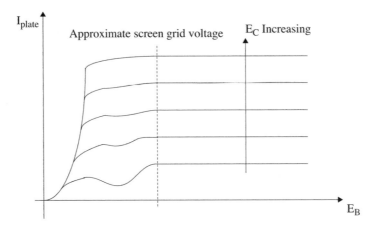

FIGURE 1.20. Tetrode characteristics.

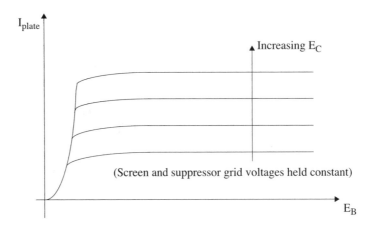

FIGURE 1.21. Pentode characteristics.

place, these secondary electrons are attracted back to the more positive plate, and the negative resistance region of operation is avoided. With the additional shielding provided by the suppressor grid, the output current depends less on the plate-to-cathode voltage. Hence, the output resistance increases and pentodes thus provide large amplification factors (thousands, compared with a typical triode's value of about ten or twenty) and low feedback capacitance (like 0.01 pF, excluding external wiring capacitance). Large voltage swings at the plate are therefore allowed, since there is no longer a concern about negative resistance (see Figure 1.21). For these reasons, pentodes are more efficient as power output devices than tetrodes.

Later, some very clever people at RCA figured out a way to get the equivalent of pentode action without adding an explicit suppressor grid.[44] Since the idea is just to

---

[44] See Otto H. Schade, "Beam Power Tubes," *Proc. IRE*, v. 26, February 1938, pp. 320–64. The first production beam power tube was the famous 6L6, which in moderately updated versions is still in production over 60 years later.

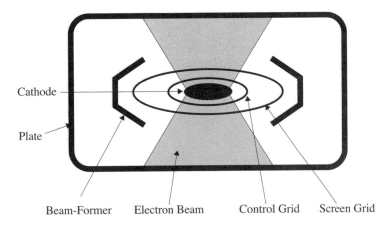

FIGURE 1.22. Beam-power structure (top view).

devise conditions that repel secondary electrons back to the plate, you might be able to exploit the natural repulsion between electrons to do the same job. Suppose, for example, we consider a stream of electrons flowing between two locations. At some intermediate point, there can be a region of zero (or even negative) field if the distance is sufficiently great.

The effect of mutual repulsion can be enhanced if we bunch the electrons together. *Beam-forming electrodes* (see Figure 1.22), working in concert with control and screen grids wound with equal pitch and aligned so that the grid wires overlap, force the electrons to flow in sheets (and away from the screen grid). The concentrated electron beam then generates a negative field region (a virtual suppressor grid) without requiring large electrode spacings. And, as an unexpected bonus, it turns out that the characteristics at low voltages are actually superior in some respects (the plate current and output resistance are higher) to those of true pentodes and are thus actually more desirable than "real" pentodes for power applications.

Well, this grid mania didn't stop at the pentode, or even the hexode. Vacuum tubes with up to seven grids have been made. In fact, for decades the basic superhet AM radio (the "All-American Five-Tuber") had a heptode, whose five grids allowed one tube (usually a 12BE6) to function as both the local oscillator and mixer, thus reducing tube count. For trivia's sake, the All-American Five also used a 35W4 rectifier for the power supply, a 12BA6 IF amplifier, a 12AV6 triode/duo-diode as a demodulator and audio amplifier, and a 50C5 beam-power audio output tube.

Here's some other vacuum tube trivia: for tubes made after the early 1930s, the first numerals in a U.S. receiving vacuum tube's type number indicate the nominal filament voltage, truncated to an integer (with one exception: the "loktals"[45] have numbers beginning with 7, but they are actually 6.3-volt tubes most of the time, as are others whose type number begins with the numeral 6). In the typical superhet mentioned previously, the tube filament voltages sum to about 120 volts, so that no

---

[45] Loktals had a special base that locked the tubes mechanically into the socket to prevent their working loose in mobile operations.

filament transformer was required. The last numbers are supposed to give the total number of elements, but there was widespread disagreement on what constituted an element (e.g., whether one should count the filament), so it is only a rough guide at best. The letters in between simply tell us something about when that tube type was registered with RETMA (which later became the EIA). Not all registered tube types were manufactured, so there are many gaps in the sequence.

In CRTs, the first numbers indicate the size of the screen's diagonal (in inches in U.S. CRTs and in millimeters elsewhere). The last segment has the letter P followed by numbers. The P stands for "phosphor" and the numbers following it tell you what the phosphor characteristics are. For example, P4 is the standard phosphor type for black-and-white TV CRTs, while P22 is the common type for color TV tubes.

The apex of vacuum tube evolution was reached with the development of the tiny Nuvistor by RCA. The nuvistor used advanced metal-and-ceramic construction, and it occupied a volume about double that of a TO-5 transistor. A number of RCA color televisions used them as VHF RF amplifiers in the early 1970s before transistors finally took over completely. RCA's last vacuum tube, a Nuvistor, rolled off the assembly line at Harrison, New Jersey, in 1976, marking the end of about sixty years of vacuum tube manufacturing and, indeed, the end of an era.

1.11   APPENDIX B: WHO REALLY INVENTED RADIO?

The question is intentionally provocative, and is really more of a Rorschach test than purely a matter of history (if there is such a thing). Frankly, it's an excuse simply to consider the contributions of some radio pioneers, rather than an earnest effort to offer a definite (and definitive) answer to the question.

First, there is the matter of what we mean by the words *radio* or *wireless.* If we apply the most literal and liberal meaning to the latter term, then we would have to include technologies such as signaling with smoke and by semaphore, inventions that considerably predate signaling with wires. You might argue that to broaden the definition that much is "obviously" foolish. But if we then restrict the definition to communication by the radiation of Hertzian waves, then we would have to exclude technologies that treat the atmosphere as simply a conductor. This collection of technologies includes contributions by many inventors who have ardent proponents. We could make even finer distinctions based on criteria such as commercialization, practicality, forms of modulation, and so forth. The lack of agreement as to what the words *radio* and *invention* mean is at the core of the controversy, and we will not presume to settle that matter.

One pioneer who has perhaps more than his fair share of enthusiastic supporters is dentist Mahlon Loomis, who patented in 1872 a method for wireless electrical communication.[46] In a configuration reminiscent of Benjamin Franklin's electrical

[46] U.S. Patent #129,971, granted 30 July 1872. This one-page patent has no drawings of any kind. It may be said that one with "ordinary skill in the art" would not be able to practice the invention based on the information in the patent.

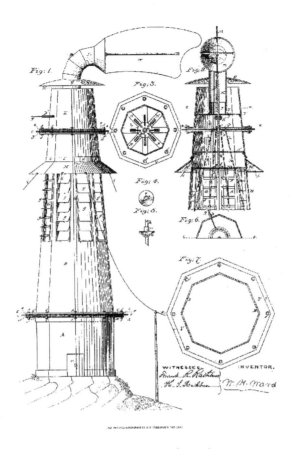

FIGURE 1.23. First page of Ward's patent.

experiments, Loomis proposed a system of kites to hold wires aloft. A sufficiently high voltage applied to these wires would allow electrical signals to be conducted through the atmosphere, where a receiver would detect induced currents using a galvanometer. Allegedly, experiments conducted by Loomis in his home state of West Virginia were successful, but there is no accepted primary evidence to support this claim, and calculations based on modern knowledge cast tremendous doubt in any case.[47] Supporters of Loomis have a more serious difficulty, for William Henry Ward had patented much the same idea (but one using more sophisticated apparatus) precisely three months earlier;[48] see Figure 1.23. Needless to say, reliably conducting

---

[47] Seemingly authoritative reports of successful tests abound (in typical accounts, senators from several states are present for the tests, in which communications between two mountaintops 22 km apart is allegedly demonstrated, then independently verified). However, I have never been able to locate information about these tests other than what Loomis himself provided. Others quoting these same test results apparently have had no better success locating a primary source but continue to repeat them without qualification. See, for example, J. S. Belrose, "Reginald Aubrey Fessenden and the Birth of Wireless Telephony," *IEEE Antennas and Propagation Magazine,* v. 44, no. 2, April 2002, pp. 38–47.

[48] U.S. Patent #126,536, granted 30 April 1872.

enough DC current through the atmosphere to produce a measurable and unambiguous response in a galvanometer is basically hopeless, and neither Loomis nor Ward was able to describe a workable system for wireless telegraphy.

Then there's David Edward Hughes, who noticed that – as a result of a loose contact – his homemade telephone would respond to electrical disturbances generated by other apparatus some distance away. After some experimentation and refinement, he presented his findings on 20 February 1880 to a small committee headed by Mr. Spottiswoode, the president of the Royal Society. The demonstration included a portable wireless receiver, the first in history. One Professor Stokes declared that, although interesting, the phenomenon was nothing more than ordinary magnetic induction in action and was not a verification of Maxwell's predictions. So strong a judgment from his esteemed colleagues caused Hughes to abandon his pursuit of wireless.[49]

That same year, Alexander Graham Bell invented the *photophone,*[50] a device for optical wireless communication that exploited the recent discovery of selenium's photosensitivity.[51] Limited to daylight and line-of-sight operation, the photophone never saw commercial service, and it remains largely a footnote in the history of wireless. Yet Bell himself thought it important enough that four of his eighteen patents are related to the photophone.

Wireless telegraphy based on atmospheric conduction continued to attract attention, though. Tufts University professor Amos Dolbear patented another one of these systems in 1886;[52] see Figure 1.24. This invention is notable chiefly for its explicit acknowledgment that the atmosphere is a shared medium. To guarantee fair access to this resource by multiple users, Dolbear proposed assigning specific time slots to each user. Thus, Dolbear's patent is the first to describe time-division multiple access (TDMA) for wireless communications.

We must also not forget Heinrich Hertz. The apparatus he constructed for his research of 1887–1888 is hardly distinguishable from that used by later wireless pioneers. His focus on the fundamental physics, coupled with his untimely end in 1894 at the age of 36, is the reason that others are credited with the invention of wireless communication.

Sir Oliver Lodge lost to Hertz in the race to publish experimental verifications of Maxwell's equations first, having interrupted his writing to take a vacation (during which he came across Hertz's paper). Like Hertz, Lodge did not initially focus on applications of wireless technology for communications. For example, his

[49] Hughes was sufficiently discouraged that he did not even publish his findings. The account given here is from Ellison Hawks, *Pioneers of Wireless* (Methuen, London, 1927), who in turn cites a published account given by Hughes in 1899.

[50] A. G. Bell and S. Tainter, U.S. Patent #235,496, granted 14 December 1880.

[51] This property of selenium also stimulated numerous patents for television around this time. Such was the enthusiasm for selenium's potential that *The Wireless & Electrical Cyclopedia* (Catalog no. 20 of the Electro Importing Company, New York, 1918) gushed, "Selenium will solve many problems during this century. It is one of the most wonderful substances ever discovered." I suppose that's true, as long as you overlook its toxicity....

[52] U.S. Patent #350,299, granted 5 October 1886.

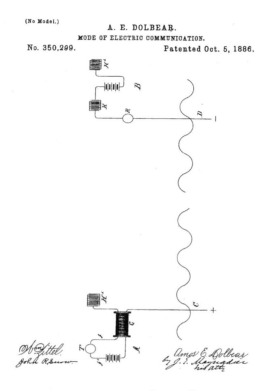

FIGURE 1.24. Front page from Dolbear's patent.

demonstration in 1894 at a meeting of the Royal Institution in London (titled "The Work of Hertz" and which marked the public debut of the coherer in wireless technology) did not involve the transmission or reception of intentional messages.[53] Lodge himself later acknowledged that his initial lack of interest in wireless communications stemmed from two biases. One was that wired communications seemed to be doing just fine. The other was perhaps the result of knowing too much and too little at the same time. Having proven the identity of Hertzian waves and light, Lodge erroneously concluded that wireless would be constrained to line-of-sight communications, limiting the commercial potential of the technology. Lodge was hardly alone in these biases; most "experts" shared his views. Nonetheless, he continued to develop the technology, and he patented the use of tuned antennas and circuits for wireless communication (see Figure 1.25) years before the development of technology for generating continuous waves. He coined the term "syntony" to describe synchronously tuned circuits. As the reader may have noted, the term didn't catch on.

Lodge published extensively, and his papers inspired Alexander Popov to undertake similar research in Russia.[54] Popov demonstrated his apparatus to his colleagues of the Russian Physical and Chemical Society on 7 May 1895, a date which is still

[53] Hugh G. J. Aitken, *Syntony and Spark,* Princeton University Press, Princeton, NJ, 1985.
[54] Also rendered as *Aleksandr Popoff* (and similar variants) elsewhere.

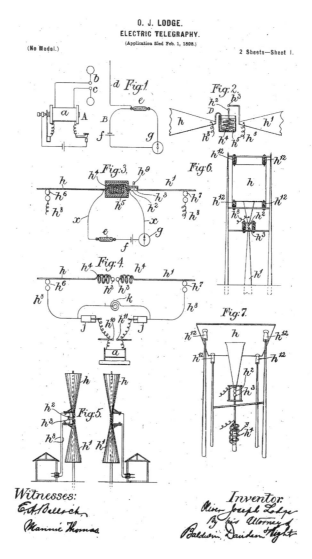

FIGURE 1.25. First page of Lodge, U.S. Patent #609,154
(filed 1 February 1898, granted 10 August 1898).

celebrated in Russia as Radio Day even though his demonstration (like Lodge's a year earlier) did not involve actual communication.

According to anecdotal accounts written down thirty years after the fact, Popov then demonstrated wireless telegraphy on 24 March 1896, with the transmission and reception of the message "Heinrich Hertz" achieved over a distance of approximately 250 meters. He followed this up with the first ship-to-shore communication one year later. Continuing refinements in his apparatus enabled the first wireless-assisted naval rescue in 1899–1900.[55]

---

[55] The range of dates reflects one of the problems with establishing the facts surrounding Popov's contributions. Different sources with apparently equal credibility cite dates ranging from 1899 to

Unlike Hertz, Dolbear, Hughes, Lodge, and Popov, who were all members of an academic elite, young Guglielmo Marconi was a member of a social elite. His father was a wealthy businessman, and his mother was an heiress to the Jameson Irish whiskey fortune. Marconi became aware of Maxwell's equations through his occasional auditing of Professor Augusto Righi's courses at the University of Bologna, but his real inspiration to work on wireless came from reading Righi's obituary of Hertz in 1894. He began to work in earnest that December and had acquired enough knowledge and equipment by early 1895 to begin experiments in and around his family's villa (the Griffone). Ever mindful of commercial prospects for his technology, he applied for patents early on, receiving his first (British #12,039) on 2 June 1896.

The documented evidence indicates that Marconi demonstrated true wireless communications before Popov, although initially to small groups of people without academic or professional affiliations. Neither Marconi nor Popov used apparatus that represented any particular advance beyond what Lodge had produced a year earlier. The chief difference was the important shift from simply demonstrating that a wireless effect could be transmitted to the conscious choice of using that wireless effect to communicate.

So, does the question of invention reduce to a choice between Marconi and Popov? Or between Marconi and Lodge? Lodge and Popov? What about Tesla?

Tesla?

Nikola Tesla's invention of the synchronous motor made AC power a practicality, and the electrification of the world with it. Tesla wasn't content to stop there, however, and soon became obsessed with the idea of transmitting industrially significant amounts of power wirelessly. Based on his experience with gases at low pressure, he knew they were readily ionized and thus rendered highly conductive (this behavior is the basis for neon and fluorescent lights). Just as Loomis and Ward had before him, Tesla decided to use the atmosphere as a conductor. Deducing that the upper atmosphere (being necessarily of low pressure) must also be highly conductive, Tesla worked to develop sources of the exceptionally high voltage necessary to produce a conductive path between ground level and the conductive upper atmosphere. Tesla estimated that he would need tens of megavolts or more to achieve his goals.[56] Ordinary step-up transformers for AC could not practically produce these high voltages. The famous Tesla coil (a staple of high-school science fairs for a century, now) resulted from his efforts to build practical megavolt sources. Based on his

---

1901 for the rescue of the battleship *General-Admiral Apraskin* in the Gulf of Finland. And it is unfortunate that so significant an achievement as allegedly occurred on 24 March 1896 (still other sources give different dates, ranging over a two-week window) would have gone undocumented for three decades. See Charles Susskind, "Popov and the Beginnings of Radiotelegraphy," *Proc. IRE*, v. 50, October 1962.

[56] Later, he would begin construction of a huge tower on Long Island, New York, for transmitting power wirelessly. Designed by renowned Gilded Age architect Stanford White (whose murder was chronicled in *Ragtime*), the Wardenclyffe tower was to feature an impessive array of ultraviolet lamps, apparently to help create a more conductive path by UV ionization. Owing to lack of funds, it was never completed. Parts were eventually sold for scrap, and the rest of the structure was demolished.

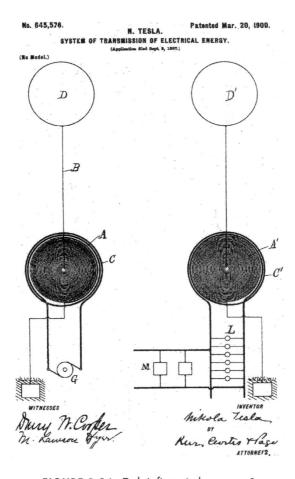

FIGURE 1.26. Tesla's first wireless patent?

deep understanding of resonant phenomena, the Tesla coil uses the significant voltage boosts that tuned circuits can provide.

Tesla's first patent in this series (see Figure 1.26) is U.S. #645,576, filed 9 September 1897 and granted 20 March 1900. It specifically talks about the conduction of electrical energy through the atmosphere – but not about the transmission of intelligence.[57]

This patent is among several cited in a famous 1943 U.S. Supreme Court decision (*320 US 1,* argued April 9–12 and decided on June 21), frequently offered as establishing that Tesla was the inventor of radio. The background for this case is that the Marconi Wireless Telegraph Corporation of America had asserted some of its wireless patents against the U.S. government shortly after the First World War, seeking damages for infringement. The decision says very clearly that *Marconi's patent for*

---

[57] His later patents do discuss transmission of intelligence, but his claims *specifically exclude* the use of Hertzian waves. He was completely obsessed with using the earth as one conductor – and the atmosphere as the other – for the transmission of power.

the four-resonator system is invalid because of prior art. Of three other patents also asserted against the United States, one was held not to be infringed, another to be invalid, and a third to be both valid and infringed, resulting in a judgment against the U.S. government in the trivial sum of approximately $43,000. The 1943 decision put that narrow matter to rest by citing prior inventions by Lodge, Tesla, and one John Stone Stone in invalidating the four-circuit patent (which had begun life as British Patent #7,777). The decision thus certainly declares that Marconi is not the inventor of this *circuit,* but it does not quite say that Marconi didn't invent *radio.* It does note that the four-resonator system enabled the first practical spark-based wireless communications (the four-resonator system is largely irrelevant for continuous-wave systems), but the Court does not then make the leap that either Lodge, Tesla, or Stone was therefore the inventor of radio.[58] The oft-cited decision thus actually makes no affirmative statements about inventorship, only negative ones.

What we can say for certain is that, of these early pioneers, Marconi was the first to believe in wireless communications as more than an intellectual exercise. He certainly did not innovate much in the circuit domain (and more than occasionally, shall we say, *adapted* the inventions of others), but his vision and determination to make wireless communications a significant business were rewarded, for he quickly made the critically important discovery that wireless is not necessarily limited to line-of-sight communication, proving the experts wrong. Marconi almost single-handedly made wireless an important technology by making it an important business.[59]

So, who invented radio? As we said at the beginning of this little essay, it depends on how you choose to define *inventor* and *radio.* If you mean the first to conceive of using some electrical thing to communicate wirelessly, then Ward would be a contender. If you mean the first to build the basic technical apparatus of wireless using waves, then Hertz is as deserving as anyone else (and since light is an electromagnetic wave, we'd have to include Bell and his photophone). If you mean the first to use Hertzian waves to send a message intentionally, either Popov or Marconi is a credible choice (then again, there's Bell, who used the photophone explicitly for communication). If you mean the first to appreciate the value of tuning for wireless, then Lodge and perhaps Tesla are inventors, with Lodge arguably the stronger candidate.

Given the array of deserving choices, it's not surprising that advocacy of one person or another often has nationalistic or other emotional underpinnings, rather than purely technical bases. Situations like this one led President John F. Kennedy to observe that "success has many fathers." Wireless certainly has been a huge success, so it's not surprising that so many claim to be fathers.

---

[58] The reader is invited to verify the author's assertion independently. The entire case is available online at http://www.uscaselaw.com/us/320/1.html.

[59] We must not underestimate the value of his lineage. After the Italian government showed insufficient interest, Marconi's mother used her family's considerable network of connections to gain an audience with British government authorities. Recognizing the value of what Marconi was proposing, William Preece of the British Post Office made sure that Marconi received all necessary support.

CHAPTER TWO

OVERVIEW OF
WIRELESS PRINCIPLES

> The wireless telegraph is not difficult to understand. The ordinary telegraph is like a very long cat. You pull the tail in New York, and it meows in Los Angeles. The wireless is the same, only without the cat.
>
> —Albert Einstein, 1938[1]

2.1 A BRIEF, INCOMPLETE HISTORY OF WIRELESS SYSTEMS

Einstein's claim notwithstanding, modern wireless *is* difficult to understand, with or without a cat. For proof, one need only consider how the cell phone of Figure 2.1 differs from a crystal radio.

Modern wireless systems are the result of advances in information and control theory, signal processing, electromagnetic field theory and developments in circuit design – just to name a few of the relevant disciplines. Each of these topics deserves treatment in a separate textbook or three, and we will necessarily have to commit many errors of omission. We can aspire only to avoid serious errors of commission.

As always, we look to history to impose a semblance of order on these ideas.

2.1.1 THE CENOZOIC ERA

The transition from spark telegraphy to carrier radiotelephony took place over a period of about a decade. By the end of the First World War, spark's days were essentially over, and the few remaining spark stations would be decommissioned (and, in fact, their use outlawed) by the early 1920s. The superiority of carrier-based wireless ensured the dominance that continues to this day. Textbook titles of the time mirror the rapid pace of development. Elmer E. Bucher's *Practical Wireless Telegraphy* of 1917 gives way to Stuart Ballantine's *Radiotelephony for Amateurs* of 1922, for example, also reflecting the rise of *radio* as the more fashionable term (in the U.S., at least). After decades of *radio,* we're back to *wireless.* Plus ça change ....

---

[1] As quoted in the September 2002 issue of *Scientific American* ("Antigravity" by Steve Mirsky).

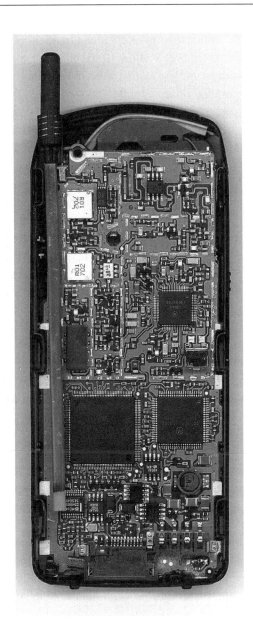

FIGURE 2.1. Typical second-generation cellphone (courtesy C. P. Yue).

Marconi's original vision for wireless was as a largely symmetrical point-to-point communications system, mimicking the cable-based telegraphy it was designed to supplement. Once the technology for radiotelephony was in place, however, pioneering efforts by people like Fessenden and "Doc" Herrold highlighted the commercial potential for wireless as a point-to-*multipoint* entertainment medium.[2] The lack of

---

[2] Charles "Doc" Herrold was unique among radio pioneers in his persistent development of radio for entertainment. In 1909 he began regularly scheduled broadcasts of music and news from a

any historical precedent for this revolutionary idea forced the appropriation of a word from agriculture to describe it: *broadcasting* (the spreading of seeds).

Modulation types in those early systems were chosen on the basis of what could be generated easily (or at all). Spark's limitation to Morse code–type on–off keying (OOK) ensured that the early days of wireless were indeed dominated by radiotelegraphy.[3] Some early continuous-wave (CW) transmitters used OOK as well, and to this day Morse code transmissions are called CW signals, despite their obviously discontinuous nature. Achieving OOK with arc transmitters presented a challenge, however, because of the difficulty of re-establishing an arc quickly once extinguished. Engineers responded by inventing *frequency-shift keying* (FSK), in which the arc is kept continuously alive while two different frequencies (separated by a few percent, and produced by selectively shorting across some turns of an inductor) represent Morse code's dots and dashes.[4] Fessenden's dramatic Christmas Eve demonstration added AM to the mix, so that by 1910 or thereabouts, three forms of modulation had come into use: OOK, AM, and FSK.

Wireless enjoyed rapid growth in the next two decades (due in no small measure to the demands of the First World War), bringing with it increased competition for scarce spectrum, as frequencies commonly in use ranged up to around 1 MHz. A three-way conflict involving radio amateurs ("hams"), government interests, and commercial services was partly resolved by relegating hams to frequencies above 1.5 MHz, a portion of spectrum then deemed relatively unpromising. Persistent development by dedicated hams would later prove the enormous value of this spectrum, which corresponds to wavelengths of 200 meters and below.[5] At the same time, capacity demands in conventional wireline telephony were also increasing. These two pressures stimulated the development of important theoretical concepts concerning the relationships among bandwidth, transmission rates, and noise.

In particular, John R. Carson of AT&T undertook a detailed mathematical study of modulation and, by 1915, had fully comprehended amplitude modulation's inefficiency. He recognized that AM contains an information-free, power-consumptive carrier as well as redundant sidebands. A trio of patents resulting from that work describes circuits for suppressing the carrier and additionally for removing one of the sidebands. He is thus the inventor of both double-sideband, suppressed carrier

succession of transmitters located in San Jose, California, continuing until the 1920s when the station was sold and moved to San Francisco (where it became KCBS). See *Broadcasting's Forgotten Father: The Charles Herrold Story,* KTEH Productions, 1994. The transcript of the program may be found at http://www.kteh.org/productions/docs/doctranscript.txt.

[3] *Keying* remains a part of the modern lexicon, applying even to modulations for which no telegraph key was ever used.

[4] Some arc transmitters nonetheless managed to produce OOK by switching the oscillator output between the antenna and a dummy load. In this way, the frequency could be kept constant, allowing for optimum operation of the arc.

[5] See Clinton B. DeSoto, *Two Hundred Meters and Down,* The American Radio Relay League, 1936. The hams were rewarded for their efforts by having increasing amounts of spectrum taken away from them.

(DSB-SC, often just called SC-AM) and of single-sideband (SSB) forms of AM.[6] He followed up with an important paper that establishes formally the spectral efficiency of SSB. Addressing the general idea of reducing the transmitted bandwidth still further, he concludes that "all such schemes are believed to involve a fundamental fallacy."[7] The paper spends a great deal of time examining frequency modulation in detail, about which he concludes that "this method of modulation inherently distorts without any compensating advantages whatsoever." Such a strongly worded negative pronouncement from so respected an authority essentially killed off further work on frequency modulation for over a decade.

Carson's invention of SSB was quickly put into service, beginning with a wired telephone connection between Baltimore and Pittsburgh in 1918. Transatlantic SSB wireless debuted experimentally in 1923 (thanks to the efforts of Western Electric's Raymond A. Heising, whom we'll meet again in the chapter on power amplifiers), with a commercial radio service inaugurated in 1927 (between New York and London). Elimination of the power-hungry carrier allowed full use of transmitter power to send information, and elimination of the redundant sideband fit well with the narrow absolute bandwidths of the day.[8]

Amidst all that activity, Carson's Bell Laboratories colleagues Harry Nyquist and Ralph V. L. Hartley published papers that formalized the important concepts of time–frequency (or time–bandwidth) duality, placing on a more rigorous mathematical basis what many engineers understood intuitively, if vaguely.[9] Thus having established a relationship between information rate and bandwidth, the stage was set for the explicit accommodation of noise in a comprehensive theory of information. However, this last step took twenty years, finally finding formal expression in the publication of a remarkable paper by Claude Shannon in which the now-famous *channel capacity theorem* makes its debut.[10] We will have more to say about this theorem later, but for now it suffices to observe that Shannon's paper is the first to acknowledge and consider the fundamental role of noise, thereby deriving an upper bound

[6] His U.S. Patents #1,343,306 (filed 5 September 1916, granted 15 June 1920) and #1,343,307 (filed 26 March 1916, granted 15 June 1920) described DSB-SC, and #1,449,382 (filed 1 December 1915, granted 27 March 1923) covers SSB. We see that the invention of SSB actually precedes that of SC-AM.

[7] J. R. Carson, "Notes on the Theory of Modulation," *Proc. IRE,* v. 10, 1922, p. 57.

[8] The low carrier frequencies conspired with the bandpass nature of antennas to produce relatively narrowband (on an absolute basis) systems. For an excellent discussion of SSB and superb expositions of the history and mathematics underlying wireless in general, see Paul J. Nahin, *The Science of Radio,* 2nd ed., AIP Press, New York, 2001.

[9] H. Nyquist, "Certain Factors Affecting Telegraph Speed," *Bell System Tech. J.,* v. 3, April 1924, pp. 324–52, and "Certain Topics in Telegraph Transmission Theory," *AIEE Trans.,* v. 47, April 1928, pp. 617–44. See also R. V. L. Hartley, "Transmission of Information," *Bell System Tech. J.,* v. 7, April 1928, pp. 535–63.

[10] C. E. Shannon, "A Mathematical Theory of Communication," *Bell System Tech. J.,* v. 27, July and October 1948, pp. 370–423 and 623–56. See also http://cm.bell-labs.com/cm/ms/what/shannonday/paper.html for a downloadable version (with corrections).

on the error-free information rate supportable by a channel of a given bandwidth and signal-to-noise ratio. This paper is additionally notable for introducing the word *bit* to most readers.[11]

Unique among radio engineers, Armstrong enthusiastically continued developing frequency modulation and ultimately demonstrated the noise-reducing qualities that *wideband* FM possesses.[12] Every test of his FM system left listeners in awe at the quality of the sound, in a dramatic refutation of Carson's conclusions. Achieving such results through the purposeful use of a bandwidth greatly in excess of the modulation bandwidth not only turned conventional wisdom on its head, it also marked the success of a primitive type of *spread spectrum* modulation, a topic to which we shall return shortly.

In the years during which information theory was taking shape, broadcast radio firmly established its dominance. The promise of wireless seemed limitless. Hundreds of radio manufacturers flooded the market with receivers in the 1920s, and by decade's end the superheterodyne architecture had become important. Stock in RCA shot up from about $11 per share in 1924 to a split-adjusted high of $114 shortly before the big crash of 1929 (then dropped to $3 a share in 1932 as the wireless boom fizzled out).

Although broadcast radio was dominant economically, symmetrical point-to-point services continued to thrive, particularly in communications with vessels at sea. Asymmetrical (broadcast) mobile services began limited deployment in 1936 with police radio receivers made by Motorola.[13] Two-way police radio started operation in 1940, and in that same year Motorola delivered the handheld Handie-Talkie AM transceiver to the U.S. Army Signal Corps (which dubbed it the SCR-536).[14] The demands of the Second World War forced a rapid development in all aspects of the wireless art, mobile and fixed (and we must not forget radar). By 1941, commercial two-way mobile FM communications systems had appeared, with its battlefield counterpart following in 1943 (the 15-kg SCR-300 backpack transceiver, the first to be called a walkie-talkie).[15]

---

[11] He credits the mathematician John W. Tukey (of FFT fame) for suggesting this term, which had been used among a small group of researchers prior to publication of this paper. Tukey was also the first to use the term *software* in print. It must also be noted that Shannon's MIT Master's thesis, "A Symbolic Analysis of Relay and Switching Circuits" (written in 1936, submitted in 1937), had laid the foundation for digital electronics by applying the previously obscure tools of Boolean algebra to the problem of analyzing binary circuits. This work has quite properly been called "perhaps the most important Master's thesis of the 20th century."

[12] E. H. Armstrong, "A Method of Reducing Disturbances in Radio Signaling by a System of Frequency Modulation," *Proc. IRE*, v. 24, no. 5, May 1936, pp. 689–740.

[13] Technically speaking, the company's name at that point was still the Galvin Manufacturing Corporation. Motorola was their trademark and officially became the company's name in 1947.

[14] The designation *SCR* stands for Signal Corps Radio. The Handie-Talkie used a complement of five tubes and operated on a single crystal-selectable frequency between 3.5 MHz and 6 MHz. It would soon become an icon, recognizable in countless movies about the Second World War.

[15] The ever-patriotic Armstrong, who had served in the Army Signal Corps during the First World War, offered his FM patents license-free to the U.S. government for the duration of the war.

2.1.2 THE DEBUT OF MOBILE TELEPHONE SERVICE

Shortly after the war, St. Louis, Missouri, became the first city to enjoy a commercial mobile radiotelephone service, which was dubbed (appropriately enough) the Mobile Telephone Service (MTS).[16] Operating in the 150-MHz band with six channels spaced 60 kHz apart, the transceivers used FDD, *frequency-division duplexing* (i.e., one frequency each for uplink and downlink) and frequency modulation.[17] A single central tower with a 250-W transmitter sent signals to the mobile units. Because the latter were limited to 20 W, five receivers (including one in the main tower) were distributed throughout the city; see Figure 2.2. The outputs of the receivers were monitored, and responsibility for communicating with the mobile unit dynamically passed from one receiver to another as necessary to maintain connectivity as the user roamed from one neighborhood to another. Thus we see the debut of the important network concept of *handoff*. And because different frequencies allowed multiple users to communicate simultaneously, this system also represents an early use of frequency-division multiple access (FDMA) in a mobile wireless network.[18]

The small number of channels (eventually cut to three because of interference problems) – coupled with a single, high-power transmitter – meant that access to the wireless network was limited to a small number of users. Even though this limitation made it quite expensive, long waiting lists nonetheless developed in just about every city where the service was offered. This success stimulated thinking about how to continue the evolution of mobile radiotelephony (and convert the wait-listed into paying customers). The cellular concept emerged from that exercise, starting a bit before 1950. To accommodate a greater number of users, one could subdivide a geographic region into a number of cells. Instead of a single, high-power transmitter, a lower-power base station would serve each cell. The more limited reach of these base stations would allow the spectrum to be re-used in a cell located some distance away. The combination of frequency re-use and handoff constitutes the bulk of the cellular idea. It should be clear from this description that the overall capacity of a wireless network could, in principle, be increased without bound – and without requiring an increase in allocated spectrum – simply by continuing this subdivision of space into ever-smaller regions. Fair comparisons among competing systems and standards can be tricky, because this additional degree of freedom gives the term *capacity* a rather fluid meaning.

---

[16] June 17, 1946, to be precise. See "Telephone Service for St. Louis Vehicles," *Bell Laboratories Record,* July 1946.

[17] Nevertheless, the service offered only half-duplex operation. The user had to push a button to talk and then release it to listen. In addition, all calls were mediated by operators; there was no provision for direct dialing.

[18] To underscore that there really are no new ideas, Bell himself had invented a primitive form of FDMA for his "harmonic telegraph" in which a common telegraph line could be shared by many users, differentiated by frequency. Individually tuned tuning forks assured that only the intended recipient's telegraph would respond.

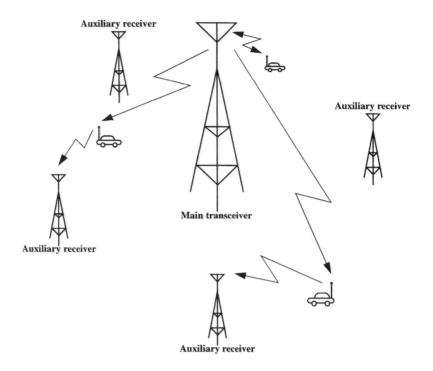

FIGURE 2.2. Mobile Telephone Service (MTS).

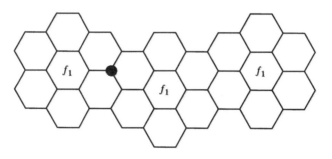

FIGURE 2.3. Illustration of the cellular idea
(1/7 frequency re-use pattern shown).

In Figure 2.3, the cells are idealized as hexagonal in shape. It's important to keep in mind that electromagnetic radiation does not respect such geometric boundaries, so don't take the arrangement in the figure too literally. Nevertheless, together with a central hexagon, we can identify a repeating cluster of seven cells, with each cell operating on a different frequency. This group of seven frequencies may be re-used in the next cluster of cells. This seven-cell re-use pattern is clearly only one possible pattern from among many choices.

It should also be noted that base stations are located at vertices, not at cell centers, as indicated by the big black dot in the figure (only one is shown to reduce clutter). A *tri-sectored* antenna (i.e., a triple antenna, each transmitting at a different frequency and aimed at a different cell) communicates with the three cells the base station touches. By focusing energy only in preferred directions, this arrangement helps reduce interference among analogous cells (i.e., those employing the same frequency) in other clusters.

Before the advent of cellular service as we know it today, the FCC allocated additional spectrum for conventional mobile telephones. In 1956, the Bell System began to offer service in this newly allocated band at 450 MHz. Even with the added spectrum, however, the system could accommodate far fewer users than the number of eager customers. Improvements consequently were incremental and were made in dimensions other than capacity. The IMTS (Improved Mobile Telephone Service) that debuted in 1964 provided full-duplex operation (no push-to-talk button), and callers could dial numbers directly.

2.1.3 FIRST-GENERATION CELLULAR SYSTEMS

Cellular finally made its debut in limited fashion in early 1969, in the form of payphones aboard a train running between New York City and Washington, D.C.[19] The 450-MHz system, limited as it was to this single route, nonetheless possessed the defining features of cellular: frequency re-use and handoff.[20] A few years later, Motorola filed a patent that is often cited as the first expression of the cellular idea as it is practiced today.[21] By 1975 the Bell System had finally received FCC approval to offer trial service – but didn't receive permission to *operate* it until 1977. Trial service finally began in 1978 in Chicago, Illinois, with a transition to full service finally taking place on 12 October 1983. Dubbed AMPS (for Advanced Mobile Phone Service), the analog FM-based system operates in a newly allocated band around 800 MHz (created by reclaiming spectrum previously assigned to upper UHF television channels). See Table 2.1. Just as MTS and IMTS had, AMPS uses frequency-division multiple access (FDMA), in which multiple users may communicate simultaneously by assignment to different frequencies. It also uses frequency-division duplexing (FDD), as had IMTS, to enable a user to talk while listening, just as with an ordinary phone. Recall that, in FDD, different frequencies are used for transmitting and receiving.

It is relevant that cellular was finally made practical by advances in integrated circuit technology. The network gymnastics involved in gracefully implementing

---

[19] C. E. Paul, "Telephones Aboard the Metroliner," *Bell Laboratories Record,* March 1969.

[20] More detail than is found in the tabulated data that follows may be found online at references such as http://www.rfcafe.com/references/electrical/wireless_comm_specs.htm. Because of the possibility of typographical errors, the reader is cautioned to double-check all data as necessary.

[21] Martin Cooper et al., U.S. Patent #3,906,166, filed 17 October 1973, granted 16 September 1975. Bell and Motorola were in a race to realize the cellular concept. Although Bell had been working on the theoretical aspects over a longer period, Motorola was the first to build an actual system-scale prototype and the first to complete a cellular call with a *handheld* mobile phone (on 3 April 1973, according to Cooper, as reported by Dan Gillmor in the 29 March 2003 *San Jose Mercury News*).

Table 2.1. *Some characteristics*
of AMPS

Parameter	Value
Mobile-to-base frequency	824–849 MHz
Base-to-mobile frequency	869–894 MHz
Channel spacing	30 kHz
Multiple access method	FDMA
Duplex method	FDD
Users per channel	1
Modulation	FM

frequency re-use and handoff brings with it a significant computational load. With-out modern digital electronics, the hardware would be impractically cumbersome. In fact, one reason for cellular's relatively long gestation period is that the cost of implementing the required control was simply too high until the late 1970s.

Certainly other countries had been designing similar systems as well. There are too many to name individually, but it is particularly noteworthy that the 450-MHz Nordic Mobile Telephone System (NMT-450, inaugurated in 1981) was the first multinational cellular system, serving Finland, Sweden, Denmark, and Norway. Aside from the fre-quency range, its characteristics are very similar to those of AMPS. Within a decade, the first generation of cellular service had become pervasive.[22]

2.1.4 SECOND-GENERATION CELLULAR SYSTEMS

As cellular systems have evolved, they've naturally become increasingly complex. One area that has become conspicuously abstruse concerns the types of modulation used. To avoid cluttering this overview of wireless systems with modulation minu-tiae, we'll make only passing reference to a number of modulation types in much of what follows and defer more detailed discussion of this topic to a later section.

In tracing the development of second-generation ("2G") cellular services, it's im-portant to recognize that the U.S. and Europe had quite different initial conditions. The U.S. enjoyed a single standard, allowing a user's phone to function in any service area. Europe's political fragmentation was mirrored in its panoply of incompatible first-generation cellular standards. The *Groupe Spéciale Mobile* formed in 1982 with

---

[22] This growth surprised almost everyone. In a famous (notorious?) study commissioned by AT&T around 1982, the total U.S. market for cell phones was projected to saturate at 900,000 well-heeled subscribers by 2000. In fact, there were over 100 million U.S. subscribers in 2000, so the predic-tion was off by over 40 dB. Today, more than a million cell phones are sold worldwide each *day,* and the total number of subscribers exceeds one billion (this is double the installed base of PCs). Acting on the implications of the study, AT&T unfortunately sold its cellular business unit early on, only to pay $11.5 billion to re-enter the market in 1993–1994 when the magnitude of its error had finally become too large to ignore.

the aim of developing a pan-European cellular standard in the 900-MHz band. By mid-1991, the GSM system had begun operation, with the letters now standing for Global System for Mobile Communications. Unlike AMPS, this second-generation system uses digital modulation, which is produced by digitizing the audio signal and then compressing the result. The binary stream is transmitted using FSK, just as in early arc transmitters. A necessary refinement is to low-pass filter the binary stream prior to modulation. A digital stream's bandwidth is constrained only by its rise and fall times and so, without filtering, the high modulation bandwidth would result in an unacceptably broad transmitted spectrum. A Gaussian low-pass filter is used because, of all filter shapes, the Gaussian shape possesses the minimum product of duration and bandwidth.[23] The modulation used by GSM is known as Gaussian *minimum*-shift keying (GMSK) because it happens to choose a frequency difference that is the minimum value necessary to guarantee orthogonality between the two modulation states. As with AMPS, the modulation is pure FM; no amplitude variation is used or desired. This constant-envelope characteristic allows the use of simple, efficient power amplifiers, because their design is not constrained by any particular concern for amplitude linearity; only preservation of the zero crossings matters.

Another feature that distinguishes GSM from AMPS is the former's use of time-division multiple access (TDMA), the modern implementation of Dolbear's century-old proposal. Multiple access is not achieved through time-division alone, however. The available spectrum is divided into numerous 200-kHz–wide sections; within each section, TDMA operates to provide access to eight users (each user is assigned to one of eight time slots). Thus, a combination of TDMA and frequency division provides multiple access.

Because first-generation European systems were incompatible with each other, the *Groupe* was free to design the GSM standard unburdened by any concern for backward compatibility. In the United States, however, the IS-54/IS-136 digital standard (NADC, for North American Digital Cellular) was designed to maintain compatibility with first-generation AMPS while increasing the number of supported users within the allocated spectrum.[24] An IS-54 handset uses the newer standard where supported but otherwise reverts to AMPS. The European success with GSM, however, also encouraged the adoption (in the 1.9-GHz band) of a North American version (usually, but not always, called PCS-1900).[25] A similar deployment of GSM at higher frequency is the United Kingdom's DCS-1800. See Table 2.2. The IS-54 standard (see Table 2.3)

---

[23] Because a Gaussian is its own Fourier transform, *Gaussian shape* applies equally well to the impulse or frequency response. Of course, the acausality of the Gaussian shape means that practical realizations are necessarily only approximations. In addition, one must scale the width of the Gaussian appropriately. For GSM, the product of the filter's −3-dB bandwidth and the bit period is 0.3. This particular value for the *BT* product results from balancing the desire to reduce bandwidth with the need to avoid excessive *intersymbol interference* (too much low-pass filtering would cause smearing, making one bit interfere with the next).

[24] IS-136 and IS-54 and now merged into a single standard.

[25] A pox on marketers for making it virtually impossible to make sense of what's what!

Table 2.2. *Overview of GSM parameters*

Parameter	GSM-900	GSM-1800 (DCS-1800)	PCS-1900
Mobile-to-base frequency	880–915 MHz	1710–1785 MHz	1850–1910 MHz
Base-to-mobile frequency	925–960 MHz	1805–1880 MHz	1930–1990 MHz
Channel spacing	200 kHz	200 kHz	200 kHz
Multiple access method	TDMA/FDM	TDMA/FDM	TDMA/FDM
Duplex method	FDD	FDD	FDD
Users per channel	8	8	8
Modulation	GMSK; $BT = 0.3$	GMSK; $BT = 0.3$	GMSK; $BT = 0.3$
Channel bit rate	270.833 kb/s	270.833 kb/s	270.833 kb/s

Table 2.3. *Overview of IS-54/IS-136*

Parameter	800-MHz band	1900-MHz band
Mobile-to-base frequency	824–849 MHz	1850–1910 MHz
Base-to-mobile frequency	869–894 MHz	1930–1990 MHz
Channel spacing	30 kHz	30 kHz
Number of channels	832	1999
Multiple access method	TDMA/FDM	TDMA/FDM
Duplex method	FDD	FDD
Users per channel	3	3
Modulation	$\pi/4$-DQPSK	$\pi/4$-DQPSK
Channel bit rate	48.6 kb/s	48.6 kb/s

employs TDMA and a form of digital modulation known as quadrature phase-shift keying (QPSK), which we will examine in more detail shortly.[26] For now it suffices to comment that this modulation allows doubling the bit rate for a given bandwidth.

A third second-generation standard was proposed in 1994 by the U.S. company Qualcomm. This standard, IS-95, is based on the use of code-division multiple access (CDMA) *spread-spectrum* technology, which differs in rather significant respects from the wireless standards that preceded it. See Table 2.4. In this technique, each bit of the digital modulation is multiplied by a higher bit-rate digital sequence (the higher rate is called the *chip rate* because it chops up, or chips, the input data stream); see Figure 2.4. If different users are assigned sequences from an orthogonal (or nearly orthogonal) set of codes then they may coexist, with one user's signal appearing somewhat as noise to the others. To distinguish this form of spread spectrum from alternative forms (such as frequency hopping), it is known as DSSS for direct-sequence spread spectrum.

---

[26] Specifically, the modulation is a mutant strain of QPSK known as $\pi/4$-DQPSK. We'll discuss this variant later.

Table 2.4. *Overview of IS-95 CDMA*

Parameter	800 MHz	1900 MHz	Asia
Mobile-to-base frequency	824–849 MHz	1850–1910 MHz	1920–1980 MHz
Base-to-mobile frequency	869–894 MHz	1930–1990 MHz	2110–2170 MHz
Channel spacing	1250 kHz	1250 kHz	1250 kHz
Number of channels	20	48	48
Multiple access method	CDMA/FDM	CDMA/FDM	CDMA/FDM
Duplex method	FDD	FDD	FDD
Users per channel	15–?	15–?	15–?
Modulation	QPSK/OQPSK	QPSK/OQPSK	QPSK/OQPSK
Channel bit rate (chip rate)	1.2288 Mb/s	1.2288 Mb/s	1.2288 Mb/s

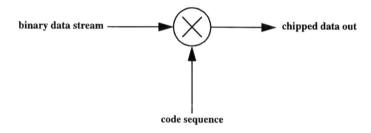

FIGURE 2.4. CDMA modulator.

The chipped output data sequence is then modulated onto the carrier (in this case, through the use of a type of phase modulation called OQPSK, for *offset quadrature phase-shift keying*). CDMA systems suffer from a unique sensitivity known as the *near–far* problem. Because other signals appear noiselike, their aggregate noise power can reduce the desired signal's SNR to the point where demodulation fails. To maximize the number of users per channel per cell requires controlling the power of each user's signal (as received by the base station) to near equality.[27] That is, the base station – in the face of multiple users moving at vehicle speeds, producing rapid and slow fading – must communicate constantly with each mobile unit and make real-time adjustments to the transmit power. This requirement for power control is one of the most difficult aspects of CDMA. Failure to control even a single mobile unit can result in the outage of an entire cell.

This sensitivity produces a particular difficulty during handoff. Suppose a mobile unit is at a cell boundary, so that control is about to shift from one base station to another. There is no guarantee that the power received by the two base stations is the same. Hence, they would send two different power control values to the mobile unit. Resolving the conflict gracefully is difficult. Dropping the call is generally favored

---

[27] Power control benefits just about any wireless system, but it is absolutely essential for CDMA.

over allowing the mobile unit to wipe out an entire cell. Mitigating such problems requires tremendous amounts of computation, and CDMA handsets are consequently the most complex of all in terms of signal processing.

2.1.5 THIRD-GENERATION (3G) CELLULAR SYSTEMS

Both first- and second-generation cellular systems are voice-centric in nature. Although both can support digital data after a fashion (e.g., by using a suitable modem), the data rates are low by modern standards (i.e., up to a couple of kilobytes/s). In contrast, 3G systems are designed from the ground up to support high-speed digital communications. The goals of 3G are to support 144 kb/s in-vehicle data rates and up to 384 kb/s for pedestrians.

An important development is the debut of *packet switching* in 3G systems.[28] Many wireline systems employ packet switching for a simple reason: The channel (network) is a valuable resource whose utility should be maximized. By chopping data into packets and then separately routing the packets dynamically along whatever paths are available, one maximizes the use of the available bandwidth. Many more users can be accommodated than if conventional circuit switching were used, because few users make 100% use of their allotted bandwidth 100% of the time; communication is typically a bursty affair. Packet switching exploits this characteristic. Clearly, the same factors that make packet switching a good strategy in wireline systems apply to wireless ones as well, and 3G systems are the first to acknowledge this truth explicitly.

As an intermediate step prior to full-scale deployment of 3G services, carriers are offering GPRS (*general packet radio service*), with typical data rates similar to fixed dial-up service (theoretical peak value is about 170 kb/s, but typical rates are about 50 kb/s). As its name suggests, this is a packet-switched protocol. The high data rates are achieved by using all eight time slots in a GSM frame. It continues to use the same GMSK as is used by GSM.

Another "2.5G" service is EDGE (*enhanced data rate for global evolution* – the marketers worked overtime on that one), which offers more than double the data rate of GPRS by using a more sophisticated modulation (called 8-PSK, if you really want to know) that encodes 3 bits per symbol. Like GPRS, EDGE is a packet-switched protocol. As can be inferred from its name, EDGE is an overlay on top of existing GSM networks.

2.2 NONCELLULAR WIRELESS APPLICATIONS

2.2.1 IEEE 802.11 (WiFi)

We should not leave the impression that cellular services are the only forms of wireless, especially in view of the rapid deployment of wireless local-area networks

---

[28] Packet switching displaces circuit switching, in which a user has exclusive use of a resource until the communication is complete.

Table 2.5. *Summary of IEEE 802.11b and 802.11a*

Parameter	802.11b	802.11a
Frequency range	2400–2483.5 MHz	5150–5350 MHz 5725–5825 MHz
Channel spacing	FHSS: 1 MHz DSSS: 25 MHz	OFDM: 20 MHz
Number of channels	3 non-overlapping	12 non-overlapping
Multiple access method	CSMA/CA	CSMA/CA
Duplex method	TDD	TDD
Modulation	FHSS: GFSK, $BT = 0.5$ (802.11g: OFDM)	OFDM: 64-QAM for 54 Mb/s
Bit or symbol rate	1, 2, or 11 Mb/s (54 Mb/s for g)	12 MS/s; 5.5–54 Mb/s

(WLANs). The popularity of WLANs speaks to the continuing appeal of tetherless communications.

Many WLAN standards have been proposed, but we will mention only IEEE 802.11 (or "WiFi"[29]). That specification covers several variants, of which we will mention only three – identified by the suffixes a, b, and g. Historically, these have been deployed in the order b, a, and g (think "bag"). See Table 2.5.

Unlike the cellular examples we've seen, these WLAN systems employ *time-division duplexing* (TDD). Transmission and reception thus do not occur simultaneously but may occur fast enough to *appear* simultaneous (to humans, at least). The use of TDD simplifies transceiver design by replacing a somewhat bulky and expensive duplexing filter with a relatively simple switch.[30] See Figure 2.5.

As the table shows, peak bit rates for 802.11b range up to a maximum of 11 Mb/s (several times the data rate of the fastest broadband services currently available to the home). An extension known as 802.11g theoretically increases the peak data rate to 54 Mb/s, propagation and co-channel conditions permitting, while maintaining backward compatibility with 802.11b.

The standard supports several modulation types. One is *frequency-hopped spread-spectrum* (FHSS).[31] Frequency hopping is a method for mitigating interference and poor propagation. As its name suggests, the carrier frequency hops from one value to

[29] *WiFi* ("wireless fidelity") is a marketing term invented as a more accessible alternative to the mouthful "eight-oh-two-dot-eleven." It is vying with *eleven-b* and *eleven-a* for currency.

[30] A duplexer must provide enough isolation so that the receiver is not overwhelmed by transmit power leaking back through the duplexer. Satisfying this requirement at low cost is decidedly challenging, even with the generous ∼5% difference between uplink and downlink frequencies of first- and second-generation cellular systems, for example.

[31] Frequency hopping was evidently first patented by the actress Hedy Lamarr during the Second World War as a technique to avoid jamming of radio-controlled torpedoes and also to enable covert communications. See H. K. Markey et al. (Markey was Lamarr's married name at the time), U.S. Patent #2,292,387, filed 10 June 1941, granted 11 August 1942. Although the system was never

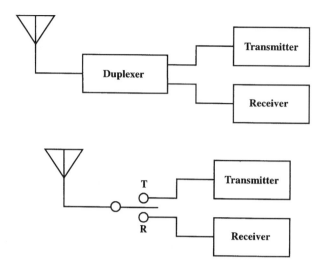

FIGURE 2.5. FDD versus TDD.

another in the course of communicating. Not putting all of one's eggs in one basket reduces the probability of a failed communication.

One may regard frequency hopping's success as due to spreading the spectrum. As we've noted, wideband FM is an early example of how spreading the spectrum beyond the minimum strictly required may improve some aspect of communication. Another spread-spectrum technique, DSSS, is used to support the 11-Mb/s data rate.

For even greater peak data rates, 802.11a offers as high as 54 Mb/s – but at the cost of incompatibility with 802.11b or g. Using a technique known as *orthogonal frequency division multiplexing* (OFDM), 802.11a subdivides a 20-MHz swath of spectrum into a collection of modulated subcarriers. As with DSSS, orthogonality is used to ensure noninterference among the subcarriers. Quadrature amplitude modulation (QAM, a topic we'll take up presently) provides the encoding of multiple bits per symbol. At the maximum bit rate, the use of 64-QAM provides six bits per symbol, allowing 54 Mb/s peak data rates after accounting for overhead.

2.2.2 BLUETOOTH

The arc of history traces a steady evolution of wireless deployment over progressively smaller geographic areas. The Bluetooth standard is a relatively recent example of a transition toward "very local" area networks. Originally envisioned as a low-cost wireless cable replacement technology and named after the Danish Viking King Harald Bluetooth, who "united" Norway and Denmark, Bluetooth is intended to provide basic wireless connectivity over small distances (e.g., order of 10 m) at moderate data rates (e.g., 1 Mb/s). See Table 2.6.

---

used during the war, it nonetheless represents a remarkable conceptual achievement. It was independently re-invented, refined, and finally used over a decade later. For more on the fascinating story, see *Forbes,* 14 May 1990, pp. 136–8.

Table 2.6. *Summary of Bluetooth*

Parameter	North America, most of Europe
Frequency range	2402–2480 MHz
Channel spacing	1 MHz
Number of channels	79
Multiple access method	Frequency hop (1.6k hops/s)
Duplex method	TDD
Users per channel	200 (7 active)
Modulation	GFSK
Symbol rate	1 MS/s

Table 2.7. *Summary of IEEE 802.15.4 ("ZigBee")*

Parameter	North America	North America	Europe
Frequency range	2402–2480 MHz	902–928 MHz	2412–2472 MHz
Channel spacing	5 MHz	5 MHz	5 MHz
Multiple access method	CSMA/CA	CSMA/CS	TDMA
Duplex method	FDD	FDD	FDD
Users per channel	255	255	255
Modulation	OQPSK, $BT = 0.5$	OQPSK, $BT = 0.5$	GFSK, $BT = 0.5$
Peak bit rate	250 kb/s	40 kb/s	250 kb/s

2.2.3 WPAN

Continuing the transition to communication over ever-smaller areas is the wireless personal area network (WPAN). Part of the appeal of Bluetooth has been its promise of low cost. The goal of WPANs is to provide wireless connectivity at still lower cost, enabling the connection of a host of devices that previously would have been precluded from consideration because of expense.

Table 2.7 summarizes key parameters of the recently ratified IEEE 802.15.4 specification for ZigBee. As can be seen, the contemplated data rates are substantially below those of 802.11 but still higher than those of conventional dial-up modem connections. The bandwidths are sufficient for voice and even for highly compressed video. It's expected that the performance will be sufficient for many applications yet modest enough to enable the achievement of low cost. Propagation conditions permitting, communications over a distance of 10–75 meters should be possible. An additional attribute is support for an extremely low standby power consumption (measured in microwatts, or over an order of magnitude lower than Bluetooth, for example), making ZigBee particularly attractive for battery-powered remote monitoring applications.

2.2.4 ULTRAWIDEBAND (UWB)

The wireless age began with impulse-like spark transmissions, and then it moved to carrier-based systems to avoid problems caused by the broadband nature of spark. Recently there has been growing interest in *ultrawideband* (UWB) techniques, many of which rely on impulse-like signals that are reminiscent of spark signals. Acknowledging that the term *UWB* is somewhat ambiguous, an arbitrary working definition has evolved by agreement among many proponents of this technology. If the transmitted signal's fractional bandwidth (i.e., the bandwidth normalized to the center frequency),

$$B = \frac{f_h - f_l}{(f_h + f_l)/2} = \frac{2(f_h - f_l)}{f_h + f_l},$$ (1)

exceeds 0.25, then it is considered ultrawideband. Also, if the mean frequency exceeds 6 GHz, then any signal occupying 1.5 GHz or more is considered UWB even if the fractional bandwidth is smaller than 0.25. The bandwidths are measured at the 10% power points rather than at the more conventional half-power points. Furthermore, these bandwidths are measured after the antenna, because all practical antennas are band-limiting devices whose filtering action must be taken properly into account.

In many popular proposals, the positions of narrow impulse- or doublet-like pulses encode the data. As with CDMA, the modulations are chosen to produce noiselike signals. Because of their broadband nature, statutory limits on emission presently constrain power to low values in order to avoid interference with incumbent narrowband services. There is a fear that these same limits may also constrain UWB's overall performance so seriously as to limit the technology's usefulness. Because of its relative novelty, there is no large base of empirical data from which firm conclusions about this technology may yet be drawn.

2.3 SHANNON, MODULATIONS, AND ALPHABET SOUP

When studying the specialized topic of modulation forms, a dizzying collection of abbreviations and acronyms quickly overwhelms even the experts. Encountering them is unavoidable, however, as we have seen. To understand one important reason for the historical progression from relatively simple modulation forms to rather complex ones, we need to spend a brief moment summarizing a tiny subset of what Shannon says: For the special case of a band-limited channel corrupted by stationary[32] additive white Gaussian noise (AWGN), one may define a *channel capacity,* in bits per second, as

$$C = B[\log_2(\text{SNR} + 1)],$$ (2)

---

[32] By *stationary* we mean that the statistical measures of the noise (e.g., its standard deviation) are time-invariant.

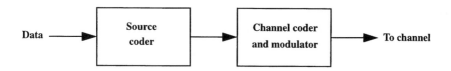

FIGURE 2.6. Conceptual block diagram of modern digital transmitter.

where B is the channel bandwidth (in hertz) and SNR is the signal-to-noise ratio over that bandwidth. Shannon says *much* more than this (see footnote 10), but Eqn. 2 is enough to make our point.

By channel capacity Shannon means that there exists a way to send data with an *arbitrarily low* probability of error at a rate C through a band-limited, noisy channel. Shannon does not show how to do it, he simply proves that one *could* do it. Until his paper, most engineers believed, quite reasonably, that the presence of noise inevitably implies an irreducible error rate. Shannon shows that this belief is wrong.

His equation says that bandwidth and signal-to-noise ratio are both degrees of freedom that one may use to increase the rate at which information is sent through a channel. Trying to devise practical ways to exploit this insight has been an ongoing preoccupation of communications engineers in the fifty or so years since the publication of Shannon's landmark paper. Decades of concerted effort have led to the widespread use of the general structure depicted in Figure 2.6.

The source coder compresses the raw data stream. By effectively removing redundancy, it reduces the average bit rate needed to send the information. Examples of compression algorithms include the various linear predictive coders used routinely for voice in cell phones, as well as algorithms used in JPEG, MPEG (including MP-3), AAC, and so on.[33] The channel coder and modulator handle the task of embedding error-correction capability to accommodate an imperfect channel and then converting the bits into a form suitable for transmission over the channel. These tasks include functions such as equalization (e.g., correction for frequency response).

Note that, in principle, one could combine these functions into a single operation – especially in view of the fact that source coding removes redundancy while channel coding puts some back in.[34] However, any possible improvements are offset by the greatly increased complexity of identifying and achieving those optima and by having to redesign the coding each time the data is to be sent down a different channel. Keeping these operations separate *greatly* simplifies the design and use of systems.

---

[33] Compression algorithms are subdivided into *lossless* and *lossy* types. Lossless algorithms allow the exact reconstruction of the original data sequence, whereas lossy ones discard some information. The latter, not surprisingly, provide greater compression factors but may exhibit objectionable artifacts under certain conditions.

[34] Andrew J. Viterbi, "Wireless Digital Communication: A View Based on Three Lessons Learned," *IEEE Communications Magazine,* September 1991, pp. 33–6.

The widespread success of this strategy suggests that this simplification is achieved at an acceptable cost in performance for many practical cases.

A now-familiar example of this conspicuous success is the ordinary phone-line data modem, which is (allegedly) capable of communicating at 56 kb/s over a 3.2-kHz phone line.[35] From Eqn. 2 we can compute the need for a minimum signal-to-noise ratio in excess of 50 dB. Such a high SNR is rarely seen in practice and accounts for the common experience of data rates below the maximum.[36] Nevertheless, users routinely enjoy data rates well in excess of the line's 3.2-kHz bandwidth (and certainly far in excess of the data rates supported by the earliest modems). This achievement underscores the great progress made in approaching the Shannon limit for practical systems.

As we mentioned, Shannon didn't explain how to achieve these goals, and he certainly did not indicate how robust any given solution would be in the face of departures from the assumptions used in devising the solutions in the first place. Because of the multiple dimensions of the design problem, a family of solutions has evolved, each with particular strengths and weaknesses. The jumbled mess of abbreviations for the different types of modulation reflects this fragmentation.

Although the details themselves can be quite complex, the underlying idea is simple: Shannon says that if the channel bandwidth is fixed then we may still make use of SNR to increase the bit rate. But how? One possibility is to use multiple amplitude levels. For example, suppose we had sufficient SNR to allow resolving 2^N separate amplitude levels. Then we could use those levels to produce a *symbol* that represents N bits. For two bits per symbol, we would need the ability to resolve four amplitude levels, three bits per symbol would require eight levels, and so forth. The exponential growth in the number of amplitude levels implies that reliable demodulation becomes progressively more challenging as we seek to encode even more bits per symbol. In keeping with Shannon's logarithmic factor, linear increases in the number of bits per symbol imply an exponential growth in required SNR. Stated crudely, the channel bandwidth primarily constrains the symbol rate, and the SNR basically constrains the number of bits we may encode per symbol.

The example in Figure 2.7 shows a sequence of symbol values. The amplitude of the carrier, for instance, could be changed on a symbol-by-symbol basis to correspond to the sequence of values shown.

Amplitude and phase represent two separate degrees of freedom, as we have seen with the historical exploitation of each (as in FM and AM).[37] Therefore, once we consider amplitude modulation, we ought to consider phase modulation as well. These

---

[35] This value applies for downloads; for uploads, the data rate ceiling is 33.6 kb/s.

[36] An additional, subtle reason is that the signal power from the phone company is presently (and probably forever) constrained by FCC regulations. Consequently, SNR is insufficient to support 56 kb/s even under the best of conditions.

[37] Keep in mind that phase is the integral of frequency, so when we say "FM" here we are really talking about angle modulations in general.

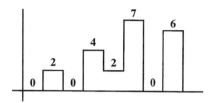

FIGURE 2.7. Example of symbols corresponding to amplitude modulation.

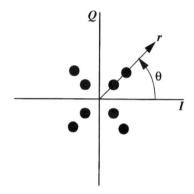

FIGURE 2.8. Symbol constellation for hypothetical polar modulation example (three bits per symbol).

two degrees of freedom in fact represent orthogonal bases for a two-dimensional space; two coordinates suffice to specify any point in a plane. In this particular case, the two coordinates are magnitude and phase. Hence, we may imagine a family of *polar modulations* in which combinations of amplitudes and phases permit us to encode several bits per symbol. As long as the SNR exceeds some minimum, we can reliably separate one symbol from another.

Figure 2.8 conveys some idea of what a polar modulation might look like. Here, each dot represents a possible symbol value in (r, θ) space. The horizontal and vertical axes are labeled I and Q for *in-phase* (with the carrier) and *quadrature* (shifted 90° with respect to the carrier). With eight such symbols in the figure, we can encode three bits per symbol. The eight symbols collectively form a symbol *constellation*. The arrangement shown is but one of many possible ways to organize eight symbols in a plane. In this hypothetical example, there are four possible amplitude values (two on an absolute-value basis) coupled with four possible phases.

Polar modulation may be viewed as a superset of AM and FM; either can be generated by suitably suppressing variation in magnitude or phase. Hence, this example is of more general utility than might appear at first. To underscore this point, consider the conceptual block diagram for a polar modulator illustrated by Figure 2.9. From this block diagram, we see that the phase shifter enables us to provide any necessary

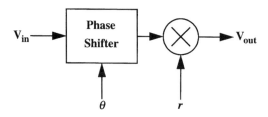

FIGURE 2.9. Polar modulator.

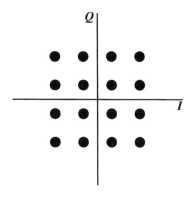

FIGURE 2.10. Constellation for hypothetical Cartesian
modulation (four bits per symbol shown).

rotation and that the multiplier enables any necessary magnitude scaling. Consequently, omission of either the phase shifter or the multiplier leaves us with a pure
amplitude or phase modulator, respectively.

Of course, polar coordinates are not the only possible choice; a rectangular (Cartesian) coordinate representation can also specify any position in 2-space through a
suitable choice of I and Q values. Although both coordinate systems are mathematically equivalent (in being able to represent any possible symbol value with a pair
of coordinates), there may be very different circuit design and system performance
implications involved in choosing one over the other. Just as a polar-to-rectangular
coordinate conversion (or vice versa) is mathematically somewhat involved, a constellation that is readily generated in, say, polar coordinates may be cumbersome to
represent in Cartesian form.

An example of a constellation that is naturally suited to Cartesian coordinates is a
set of symbols arrayed as in Figure 2.10. The constellation in this example consists
of 16 symbols, and thus we may encode 4 bits per symbol. It is left as an exercise for
the reader to verify that representing the constellation in Figure 2.10 with polar coordinates is indeed less compact than the Cartesian representation. By convention, this
type of modulation is known as *quadrature amplitude modulation,* even though the
modulation clearly involves variations in both amplitude and phase. If there are only
two symbols in the set, it's called binary modulation; if four, then *quaternary* is the

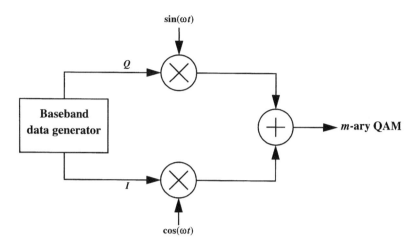

FIGURE 2.11. Quadrature modulator block diagram.

adjective. To keep from having to use and remember ever more cumbersome adjectives, the catchall "*m*-ary QAM" or simply "*m*-QAM" is more frequently employed instead, where *m* is the number of symbols in the complete constellation.[38]

We can imagine generating such a symbol constellation with the type of modulator depicted in Figure 2.11. For *m*-ary QAM, I and Q will take on $\sqrt{m}$ distinct values each.

Compared with polar modulation, QAM has the attribute that the baseband data passes through identical paths (and thus simplifies design) and that the resulting symbols are evenly spaced in the constellation. If noise is also uniformly distributed throughout the symbol space, then the probability of error remains independent of symbol value for QAM.

Coupled with the foregoing theoretical considerations is a large collection of practical ones. For example, even for a given constellation, there is no unique assignment of symbol values. Not all possible assignments may be equally desirable from a practical standpoint. Some arrangements could require cumbersome hardware or aggravate certain system sensitivities. To appreciate some of the constraints that drive the choices made by systems designers, we now briefly survey specific modulation forms used over the history of wireless.

2.3.1. AMPLITUDE MODULATION (AM)

We begin with classic amplitude modulation. Of the several possible ways to express AM mathematically, we'll choose

$$v(t) = [1 + mf(t)] \cos \omega_c t, \tag{3}$$

---

[38] Generally, *QAM* is pronounced as if it were spelled *quam*, as in "I have qualms about 16384-QAM."

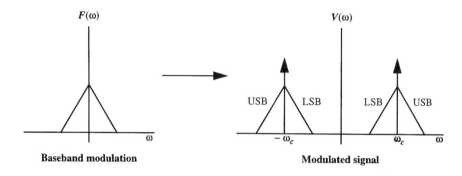

$F(\omega)$

$V(\omega)$

USB   LSB   LSB   USB

$-\omega_c$ ω_c

ω ω

Baseband modulation   **Modulated signal**

FIGURE 2.12. Amplitude modulation (real parts only).

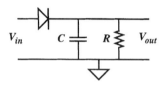

V_{in} C R V_{out}

FIGURE 2.13. Leaky peak detector (envelope detector) for AM demodulation.

where m is the *modulation index* and $f(t)$ is the baseband function to be modulated on a carrier of frequency ω_c. Furthermore, to reduce clutter, we've normalized $v(t)$ to unit carrier amplitude.

The frequency domain picture implied by the equation is illustrated in Figure 2.12. We see the structures identified by Carson in his analysis: the (information-free) remanent carrier, and two redundant sidebands (marked in the figure as LSB and USB for lower and upper sideband, respectively). The carrier will be absent if the baseband modulation directly multiplies the carrier (i.e., if there is no additive factor of unity in Eqn. 3). The carrier's presence is not entirely without value, however, for it permits the use of a very unsophisticated demodulator. As we've seen, the crystal radio is actually little more than a leaky peak detector. Virtually every radio for AM broadcast reception uses the circuit shown in Figure 2.13. This circuit has a chance of working properly only if $[1 + mf(t)]$ never becomes negative. Furthermore, the RC product must be large enough to provide adequate filtering of the carrier "teeth" yet small enough to allow the tracking of the steepest portions of the envelope. In general, this latter consideration forces us to satisfy the following inequality:

$$\frac{1}{f(t)}\left|\frac{d}{dt}f(t)\right| < \frac{1}{RC}. \tag{4}$$

In the special case of a purely sinusoidal modulation, satisfying the inequality is the same as requiring $1/RC$ to be much higher than the modulation frequency.

Ordinary broadcast television also employs AM for the video signal. However, to save spectrum, one of the sidebands is filtered. As a concession to cheap demodulator

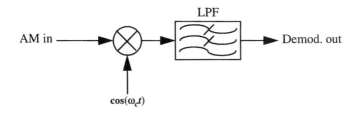

FIGURE 2.14. Conventional AM demodulator.

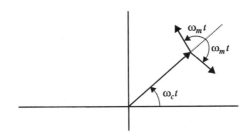

FIGURE 2.15. Phasor representation
of AM with sinusoidal modulation.

hardware, enough of the sideband is permitted to remain to allow the use of the diode envelope detector. Demodulation of this *vestigial sideband* (VSB) AM signal is then followed by a frequency equalizer to compensate for the unequal contributions by the two sidebands.[39]

A more sophisticated demodulator that is free of the peak detector's constraints is a conventional multiplier followed by a low-pass filter, as seen in Figure 2.14. As can be seen from the figure, the symbol for a filter consists of three wavy lines, each representing a frequency band (top for high, middle for center, bottom for low). A slash through any of those wavy lines signifies that the filter blocks that band.

Additional insight into AM may be obtained by examining its phasor representation. For the special case of a (co)sinusoidal modulation, we may write:

$$v(t) = [1 + m \cos \omega_m t] \cos \omega_c t = \cos \omega_c t + (m \cos \omega_m t)(\cos \omega_c t). \qquad (5)$$

After expansion of the product term, the output becomes

$$v(t) = \cos \omega_c t + \frac{m}{2}[\cos(\omega_c + \omega_m)t + \cos(\omega_c - \omega_m)t]. \qquad (6)$$

The three components that comprise the output may be considered as vectors rotating about the origin with an angular velocity equal to the frequency argument (Figure 2.15). The carrier phasor has unit magnitude, and each sideband phasor has a magnitude of $m/2$. We see that the two modulation sideband phasors counterrotate at the modulation rate about the end of the carrier phasor. The vector sum of the

---

[39] Some low-cost televisions omit this filter. The result is a loss of high-frequency detail.

sideband phasors lies in phase with the carrier, so the modulation does not affect the phase of the modulated waveform. The vector sum of all three phasors has a magnitude that varies with time; that's amplitude modulation.

If the baseband signal is binary, the result is often called amplitude-shift keying (ASK). The special case of on–off keying (OOK) is a subset of ASK because the latter can involve shifting between two (or more) nonzero amplitude values.

2.3.2 DSB-SC (SC-AM) AND SSB

We see, as did Carson, that ordinary AM is profligately wasteful. The bulk of the modulated signal's energy is actually in an information-free carrier, so that a transmitter's precious power will be largely wasted. Compounding this suboptimality is the redundancy represented by the two sidebands, which cause AM to consume double the necessary bandwidth.

Carson's recognition of these shortcomings inspired his invention of double-sideband suppressed-carrier AM as well as of single-sideband AM. We now consider how to generate and demodulate these signals.

We could imagine any number of ways of generating SC-AM. For example, we might simply use some sort of filter to remove the carrier component. In some instances, that might even be practical, but a better approach is simply not to generate any carrier component in the first place. We may accomplish this result by ensuring the lack of a DC term in the modulation and then multiplying that DC-free modulation by the carrier term:

$$v(t) = [mf(t)] \cos \omega_c t. \tag{7}$$

With no DC in $f(t)$, there can be no carrier component in the modulated output $v(t)$.[40]

Demodulation of SC-AM is straightforward in principle. All one needs is an oscillator of the frequency ω_c; then multiply its output signal with the SC-AM signal. The result is a re-insertion of the carrier component, converting the problem into the trivially solved one of demodulating a standard AM signal. A diode envelope detector, for example, would work great.

The problem in practical implementations is that if the locally generated version of ω_c is a little off-frequency then DSB exhibits seriously objectionable artifacts, for the envelope will fluctuate at a frequency equal to the magnitude of the frequency difference. For instance, if there is just a 10-Hz error (representing only 10 ppm of a 1-MHz carrier) then the envelope will vary at a 10-Hz rate. With a voice signal, for example, we would experience a flutter of this frequency. Needless to say, such an effect is intolerable.

Solutions include purposefully suppressing the carrier incompletely. Extraction of that small carrier component (e.g., through a narrowband filter) permits the generation of a local oscillator with exactly the right frequency. Another option is to

---

[40] Here we are assuming that the baseband signal's energy is concentrated well below the carrier frequency.

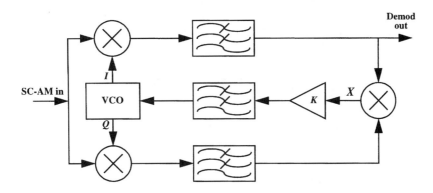

FIGURE 2.16. Costas loop for SC-AM demodulation.

use a circuit known as a phase-locked loop, which is able to infer the correct carrier frequency by effectively computing the average frequency of the symmetrical sidebands. Although we devote a separate chapter to phase-locked loops, for now simply accept that the architecture of Figure 2.16, known as a *Costas loop,* is able to perform this miraculous extraction of a carrier from a carrier-free DSB signal and then use that extracted carrier to effect demodulation.[41]

For a brief and incomplete analysis, assume first that the amplifier gain, K, is very large. Consequently, the voltage at X does not have to be very large to generate the proper control voltage for the VCO (voltage-controlled oscillator). In fact, let us idealize the voltage at X as approximately zero. A zero output from the corresponding multiplier occurs if one of the inputs is zero. This condition occurs when one path is locked in phase and the other is in quadrature. By proper choice of loop signs, we can force the VCO lock condition to be in phase (i.e., its I output is in phase with the carrier that originally modulated the SC-AM signal), allowing us to recover the modulation from the upper path as shown.

That brief explanation aside, we still have two sidebands with DSB, as is readily seen by considering once again a sinusoid for $f(t)$:

$$v(t) = \frac{m}{2}[\cos(\omega_c + \omega_m)t + \cos(\omega_c - \omega_m)t]. \tag{8}$$

To eliminate either the upper or lower sideband, we could consider once again the use of a filter, as Carson originally proposed. However, rather heroic filters are needed in the common case where the modulation frequencies are much smaller than the carrier frequency, because the upper and lower sidebands are then separated by a small amount. It is decidedly nontrivial to construct a filter that provides large attenuation at frequencies not far from where it is also expected to attenuate virtually not at all.

The block diagram depicted in Figure 2.17 shows how Carson's filter method functions to generate SSB (the upper sideband is chosen by the filter in this case).

---

[41] John P. Costas, "Synchronous Communications," *Proc. IRE,* v. 44, no. 12, December 1956, pp. 1713–18. Also see U.S. Patent #3,047,660, filed 6 January 1960, issued 31 July 1962.

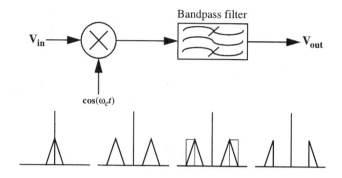

FIGURE 2.17. Carson's filter method of SSB generation
(USB example shown).

Thumbnail sketches of the idealized spectra (only real parts are shown) at various points throughout the system help illustrate how the filter method works.

Aside from the filter design, it's important to note that real mixers are imperfect. Any DC offsets at the input port will produce carrier feedthrough. If a high degree of carrier suppression is required, then some means for mitigating mixer offset will have to be implemented. Sometimes, however, a small DC offset is intentionally introduced specifically for the purpose of generating some carrier feedthrough. This leakage eases demodulation at the receiver because an accurate clone of the carrier is required for proper demodulation. While we're on the subject of carrier feedthrough, it should be noted that there may be transient DC offsets that are signal-dependent (many AC-coupled systems in particular exhibit this phenomenon), causing the appearance of a carrier component in the output that varies with modulation.

In 1924, Carson's colleague Hartley devised a second method for SSB generation that eliminates the need for a sharp bandpass filter.[42] Just as cancellation is the best way to achieve carrier suppression in SC-AM, Hartley's SSB generation method similarly cancels the undesired sideband. To understand how Hartley's method works, recall from the polar modulation and QAM discussions that a two-dimensional representation of a baseband signal provides a complete description. Stated equivalently, a baseband signal can be decomposed into real and imaginary parts.[43]

If we feed such a signal through a quadrature phase shifter, then real parts become imaginary and vice versa (because we are simply rotating the entire space by one quadrant). The imaginary part is an odd function of frequency, so a quadrature

[42] Ralph V. L. Hartley, U.S. Patent #1,666,206, filed 15 January 1925, granted 17 April 1928.

[43] Hartley does not explain his method this way in his patent (his only publication on this architecture). Raymond Heising, in "Production of Single Sideband for Trans-Atlantic Radio Telephony" (*Proc. IRE,* June 1925, pp. 291–312), provides an analysis of SSB that is an exercise in the manipulation of trigonometric identities. Most textbooks still analyze SSB using Heising's approach, leaving as an open question how Hartley and others actually *invented* their architectures in the first place. These folks were very, very smart!

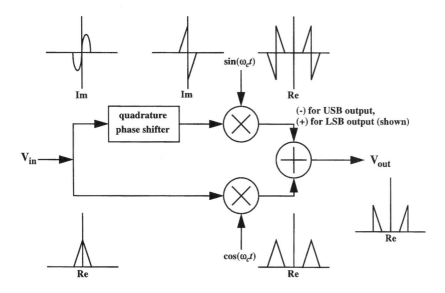

FIGURE 2.18. Hartley's SSB method (the "phasing method").

phase shifter converts the even-symmetric real part into an odd-symmetric imaginary part.[44] This odd symmetry is the key to Hartley's method, for it now provides a means for discriminating the upper and lower sidebands from each other: by sign. The next step is to convert this odd-symmetric component back into a real one (while heterodyning upward at the same time). This conversion is readily achieved by heterodyning with a sinusoid, whose own odd symmetry produces the spectrum shown in Figure 2.18. This upconversion preserves the sign difference between upper and lower sidebands. Hence, simple addition cancels one component (the USB), leaving a pure LSB output. Subtraction of the upper path's contribution from that of the lower leaves us with a pure USB output.

Proper operation of the Hartley modulator requires a precise gain match between the upper and lower paths, for otherwise cancellation will be incomplete. Furthermore, the phase shifter must provide an accurate phase shift over the bandwidth of interest; otherwise, the output of the shifter will contain a superposition of real and imaginary components. If that weren't enough, we also require that the two multipliers be driven by sine and cosine signals that are exactly that: precisely in quadrature, and of equal amplitude. Any imperfections will once again result in incomplete cancellation of the undesired sideband (and possibly produce some distortion of the desired sideband). As with many methods depending on matching, one can generally

---

[44] This phase shifter is also known as a Hilbert transformer. Its transfer function is $-j \operatorname{sgn}(\omega)$, so that it has unit magnitude at all frequencies, a constant phase lage of 90° for positive frequencies, and a phase lead of 90° for negative frequencies. A true Hilbert transformer cannot be realized by a finite network, but useful approximations of the desired behavior may be obtained over any finite frequency interval.

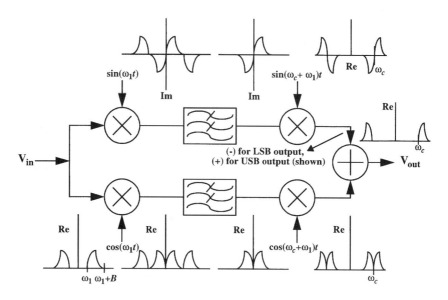

FIGURE 2.19. Weaver's SSB architecture (the "third method").

expect suppressions roughly of the order of 30 dB after moderate work, and perhaps an additional 10–20 dB if more elaborate measures are taken (e.g., autocalibration, etc.). As with the filter method, DC offsets in the mixers will produce carrier feedthrough.

Hartley's "phasing method" is considerably superior to Carson's filtering method, but it still depends on a type of filter that is not quite trivial to realize. Nevertheless, the phasing method dominated until the late 1950s (and lives on in amateur radio gear, along with Carson's filter method). An engineer named Donald K. Weaver had spent a great deal of effort devising ways to design suitable filters for Hartley's method of SSB generation.[45] Perhaps frustrated with the cumbersome design procedures resulting from that effort, Weaver cleverly figured out an alternative. He realized that quadrature heterodyning operation could replace the quadrature phase shifter and thereby obviate the need for tricky filters of any kind.[46]

In Weaver's method (Figure 2.19), the baseband signal is first *downconverted* so that the center of the desired passband is at DC. For reasons that will become clear in a moment, the baseband signal must be DC-free for reasons that have nothing to do with carrier suppression.

If the baseband signal's lower band edge is ω_1, then the pair of downconversion mixers must be driven by oscillators also of frequency ω_1. The mixer outputs are then low-pass filtered to a bandwidth of B (the bandwidth of the baseband signal). Next, the filtered outputs are upconverted about the desired carrier frequency, ω_c, by driving the final pair of mixers with oscillators of frequency $(\omega_c + \omega_1)$.

[45] D. K. Weaver, Jr., "Design of RC Wide-band 90-degree Phase-Difference Network," *Proc. IRE,* v. 42, no. 4, April 1954, pp. 671–6.

[46] D. K. Weaver, Jr., "A Third Method of Generation and Detection of Single-Sideband Signals," *Proc. IRE,* v. 44, no. 12, December 1956, pp. 1703–5.

If there is any DC in the original baseband signal (or, equivalently, if there is any mixer offset), it ultimately converts into a tone in the center of the final modulated output's passband instead of producing carrier feedthrough. In this respect, the Weaver architecture's sensitivity to mixer DC offsets is unique. Other impairments include imperfect quadrature among the oscillators and unequal oscillator amplitudes, as well as gain mismatch. We consider the consequences of these imperfections in Chapter 19.

We will also see in that chapter that the Weaver modulator is extremely versatile because it is effectively a general-purpose complex signal processor. By appropriate choices of the up- and downconversion frequencies and of their associated filter cutoff frequencies, this architecture can be made to perform a host of useful functions. One example is using the architecture to demodulate SSB signals by merely reversing the order of conversions in the Weaver modulator. A choice between upper and lower sidebands may be made simply by selecting the appropriate sign at the output summing node.

Another, older demodulation technique simply re-injects the carrier directly (and blindly). For either of these approaches, there is no method analogous to the Costas loop for SC-AM for regenerating an exact-frequency local carrier from the incoming signal itself. An SSB signal simply lacks the symmetry required to permit such inference of the carrier frequency from the modulation alone. So, to demodulate, the SSB-modulated signal is simply multiplied by a local oscillator whose frequency (hopefully) equals that of the original carrier. This re-insertion of the carrier component gives us a signal whose envelope is the modulation. A standard envelope detector may then be used to complete the process if the Weaver architecture is not used.

Fortunately, frequency offsets are not nearly as serious for SSB as for SC-AM (at least, they manifest themselves in quite different ways). In the case of SSB, a frequency error simply translates all spectral components of the modulation by an amount equal to the error. Thus, any harmonic relationships among the components of the modulation are destroyed, but for signals such as voice, such shifts do not seriously impair intelligibility. To most listeners, voices undergoing upward shifts sound like Donald Duck (really; no kidding).[47] To the extent that Donald Duck speaks intelligibly (and opinions do differ on this point), frequency errors of moderate values (e.g., <50 Hz) are tolerable in SSB.

Finally, there are variants of SSB in which some (or a lot of) carrier energy is allowed to remain to ease demodulation. If the carrier component is small then there is negligible impact on the power efficiency of SSB, but that small carrier component can be extracted (e.g., by a phase-locked loop) to allow perfect demodulation with no Donald Duck effect. In the case where the carrier component is extremely large, the power efficiency advantage disappears but demodulation becomes trivial

---

[47] This phenomenon is distinct from, say, playing a tape recording too fast (which produces sounds like Alvin and the Chipmunks). There, harmonic relationships are preserved; all frequencies are multiplied by the same factor. In SSB, all frequencies are *offset* by the same amount.

(e.g., through the use of a standard diode envelope detector). This latter version of SSB is often dubbed SSB-LC, for "SSB, large carrier."

2.3.3 ANGLE MODULATIONS: FM AND PM

Although *frequency modulation* predates *phase modulation* as an engineering term, we will shortly see that they are closely related. We will thus often refer to them collectively as angle modulations. No matter what we call them, their analysis becomes complex very rapidly if we insist on rigorous and approximation-free derivations throughout, as we'll see shortly. So we won't be so insistent. To facilitate acquisition of useful insights, we begin with narrowband (or *low-index*) FM. We will shortly quantify *narrowband,* but for now we mean that the bandwidth of the modulated signal is not very much greater than that of the original modulation itself.

To begin, consider a general expression for an angle-modulated signal:

$$v(t) = \cos[\omega_c t + \phi(t)]. \tag{9}$$

In turn, let us express the phase function as

$$\phi(t) = mf(t), \tag{10}$$

where once again we refer to m as the modulation index. For the special case of sinusoidal modulation, we have

$$v(t) = \cos[\omega_c t + \phi(t)] = \cos[\omega_c t + m\cos\omega_m t]. \tag{11}$$

This equation, involving the cosine as an argument of a cosine, leads to the appearance of Bessel functions.[48] For the special case of a low modulation index, however, we may obtain a simple, approximate solution. We start by using a standard trigonometric identity to expand the cosine term without approximation:

$$v(t) = \cos[\omega_c t + m\cos\omega_m t]$$
$$= \cos[\omega_c t]\cos[m\cos\omega_m t] - \sin[\omega_c t]\sin[m\cos\omega_m t]. \tag{12}$$

Next, we invoke approximations that hold under the condition $m \ll 1$ to obtain

$$v(t) = \cos[\omega_c t]\cos[m\cos\omega_m t] - \sin[\omega_c t]\sin[m\cos\omega_m t]$$
$$\approx \cos[\omega_c t] - \sin[\omega_c t][m\cos\omega_m t], \tag{13}$$

so that

$$v(t) \approx \cos[\omega_c t] - \sin[\omega_c t][m\cos\omega_m t]$$
$$= \cos[\omega_c t] - \frac{m}{2}\sin[(\omega_c + \omega_m)t] - \frac{m}{2}\sin[(\omega_c - \omega_m)t]. \tag{14}$$

---

[48] For a detailed discussion with Bessel functions galore, see e.g. M. Schwartz, *Information Transmission, Modulation and Noise,* 3rd ed., McGraw-Hill, New York, 1980.

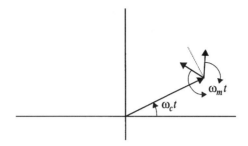

FIGURE 2.20. Phasor representation of narrowband
PM with sinusoidal modulation.

Inspecting this last equation, we see that there are three components of the same frequency and amplitude as in the case of AM. However, there is an important distinction: the phase relationships among the three are different, as is apparent from examining a phasor representation of these three components; see Figure 2.20.

We see that the two sideband phasors continue to counterrotate at the modulation rate, just as in AM. Note, however, the replacement of cosine by sine in the expressions for the sidebands, relative to the AM case. The sideband vectors are thus rotated by 90°. Unlike AM, then, the vector sum of these sideband phasors is in quadrature, rather than in phase, with the carrier. As a consequence, the total vector sum of all three phasors has a nearly constant magnitude. (Had we made no approximations, we would have discovered that the net magnitude is precisely constant.) With or without approximations, we see that the phase angle of the vector sum alternately moves ahead and behind the nominal phase of the carrier. Modulation is encoded as those phase variations.

Not too much additional work is needed to understand frequency modulation, because phase modulation actually subsumes it. Phase is the integral of frequency, so frequency and phase modulations are related by an integration (and possible scaling). More formally, we begin again with

$$v(t) = \cos[\omega_c t + \phi(t)]. \tag{15}$$

Next, by frequency modulation, we mean that

$$\frac{d\phi}{dt} = -k \sin \omega_m t, \tag{16}$$

where we have introduced a minus sign to simplify what is to come. The scale factor k (with dimensions of inverse time, or frequency) is included to make the analysis more general.

If we now integrate Eqn. 16, we obtain

$$\phi(t) = \int_{-\infty}^{t} -k \sin \omega_m t \, dt = \frac{k}{\omega_m} \cos \omega_m t. \tag{17}$$

The fussy reader will note that, strictly speaking, the integral is improper. We don't worry about such minor details; we're engineers. Thus, the integral of sine is cosine, and that's that.[49]

Comparing Eqn. 17 with the corresponding expression for phase modulation (Eqn. 11), we see no functional difference. There isn't even a quantitative difference if we identify the modulation index as

$$|m| = \frac{k}{\omega_m}. \tag{18}$$

Because frequency is the derivative of phase, we see from Eqn. 17 that k is the peak frequency deviation from the carrier. In this context, the modulation index m is also called the *deviation ratio,* because it is the ratio of the peak frequency deviation to the modulation frequency. It is the belief that choosing small k would allow us to shrink the spectrum arbitrarily that Carson so resoundingly disproved. As we've already seen from analysis of the equivalent phase modulation case, the best we can do is to approach the $2\omega_m$ bandwidth of AM in the limit of small modulation indices. A more detailed analysis shows that the spectrum is, in fact, infinitely broad in general. Fortunately, the amplitudes of the various sidebands generally diminish (although not monotonically) as one moves away from the carrier. There is a useful approximation, called Carson's rule, that gives us a measure of bandwidth in which about 98–99% of the sideband power resides:

$$B \approx 2(\omega_m + \Delta\omega) = 2(\omega_m + m\omega_m) = 2\omega_m(1 + m). \tag{19}$$

From this equation we can identify two limiting cases. We've already studied the low-index case. For the very high-index case, we see that the bandwidth is approximately equal to the total peak-to-peak frequency deviation. Carson's rule thus says that a straightforward sum of these two behaviors yields a good approximation to the actual bandwidth.

As an example, consider ordinary broadcast FM radio. There, the modulation is band-limited to 15 kHz. The peak deviation $\Delta\omega$ from the carrier is constrained to a maximum of 75 kHz, which corresponds to a deviation ratio of 5. From Carson's rule, we estimate the maximum bandwidth to be about 180 kHz. This computation explains the 200-kHz channel spacing chosen by the FCC in allocating spectrum for this service.

As a final note, the modulation index m is often represented by the symbol β in other texts.

By analogy with ASK in AM systems, we also have phase-shift keying (PSK) and frequency-shift keying (FSK) in PM/FM systems, in which bits are encoded as discrete values of phase or frequency relative to the carrier. We will consider these digital modulations (and related variants) shortly.

---

[49] Despite the seeming lack of rigor, we have actually introduced no fundamental error. If it makes you feel better, treat the integral as if it described the response of an LTI system whose impulse response is a step. We will implicitly invoke this same maneuver in Chapter 18.

Demodulation may be performed any number of ways. In the earliest systems, several now-obsolete types of demodulators were popular: slope detectors (*discriminators*), ratio detectors, and (Foster–Seeley) discriminators. We will describe these in only the sketchiest terms because they are not particularly appropriate idioms for the integrated circuit era.[50]

The simplest slope detector uses a single bandpass filter tuned off-center, so that FM passing through it converts to AM, after which standard AM demodulation techniques may be applied. The slope of a filter's response is not well approximated as linear except over a very narrow frequency span, so distortion can be unacceptably high. A variation on the slope detector, the Travis discriminator (also known as a double-tuned discriminator), extends the linear range of the slope detector by using two tuned circuits – one tuned above and the other below the nominal center frequency. With proper adjustment (and that's tricky), very linear demodulation is possible.

The Foster–Seeley discriminator also makes use of tuned circuits, but it exploits a filter's phase variation with frequency rather than the amplitude characteristic directly. As with the slope detectors, this variation is converted into an amplitude modulation, which is then subsequently demodulated as usual. In common with those techniques, the output is not solely a function of input frequency, unfortunately. It is also sensitive to input amplitude. To prevent corruption by amplitude variation (which would nullify the chief advantage of wideband FM), these demodulators must be preceded by limiters (essentially comparators) whose job is to present signals of constant amplitude to the discriminators.

The *ratio detector* is a clever modification of the Travis double-tuned circuit that effectively embeds limiter-like action by its inherent insensitivity to amplitude variations. Countless FM receivers in the pre-IC days used ratio detectors because of the cost reductions enabled by eliminating the limiter.

Today, the phase-locked loop reigns supreme for both modulation and demodulation of PM and FM. We take up the detailed consideration of PLLs in a separate chapter dedicated to that subject. For now, merely accept that it is possible to use feedback to control an oscillator in order to produce a frequency equal to the average frequency of an FM or PM input signal. Then the regenerated carrier and the FM or PM input signal feed a phase detector, whose output is the recovered modulation.[51] The same basic architecture may also be used as a modulator.

2.3.4 DIGITAL MODULATIONS

Now that we've taken a detailed look at classical analog modulations, we're ready for the exponential growth in abbreviations and acronyms that comes with a consideration of even a small subset of digital modulations.

---

[50] The interested reader may consult any number of historical references for the full Monty. See e.g. F. E. Terman, *Electronic and Radio Engineering,* 4th ed., McGraw-Hill, New York, 1955.

[51] This brief explanation is necessarily incomplete, but we have no alternative at this point in the textbook.

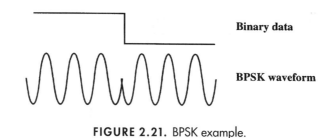

FIGURE 2.21. BPSK example.

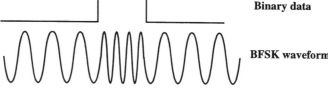

FIGURE 2.22. BFSK example (continuous phase shown).

We've already met phase-shift keying (PSK). In binary PSK (BPSK), the carrier phase shifts between two discrete values as a function of the binary data. By far the most common choice is just to flip polarity and thus maximize the phase difference between the two states (to maximize the ease of demodulation in the presence of various impairments). With this choice also comes a trivial method of generation. Simply multiplying the binary data (offset to switch between values -1 and $+1$) by a carrier produces the desired result. See Figure 2.21.

Closely related to PSK is frequency-shift keying (FSK), just as PM and FM are closely related forms. As you might have guessed, FSK simply involves switching among a set of discrete frequencies. Binary FSK (BFSK) is the special case of switching between just two frequencies. Depending on the particular generation method used, the phase may or may not be continuous across the bit boundaries – unlike BPSK, which offers no option.

A crude and somewhat impractical way of generating BFSK is direct modulation of a VCO. That open-loop method has the disadvantage of being subject to drift with time, temperature, supply voltage, and Coriolis force. Another method is to switch between the outputs of two separate oscillators (in this case, phase is not guaranteed to be continuous across bit boundaries). Again, a phase-locked loop may be used to produce a stable carrier, which is subsequently modulated by the digital data. No matter what method is used, a BFSK waveform might appear as in Figure 2.22.

Both BPSK and BFSK are one-dimensional modulations with a long and honored history. They are simple to generate and simple to detect (BFSK is used in pagers, for example, and was used in the earliest modems, including those for wireless Teletype). As we've already seen, however, they do not use spectrum very efficiently (but as a compensating advantage, they can be detected at very low SNR; BPSK has a

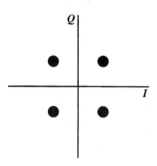

FIGURE 2.23. QPSK (4-PSK)
constellation.

3-dB theoretical advantage over BFSK, but the ease of generating BFSK has made it the more popular choice).

We've already met a version of minimum-shift keying, but without much explanation. We shall not show it, but if one chooses a modulation index of $(n+0.5)$ in FSK, where n is a nonnegative integer, then the two symbol states (frequencies) will be orthogonal (totally uncorrelated). Choosing n equal to zero is the minimum value for which this orthogonality holds. A modulation index of 0.5 thus corresponds to MSK. If, further, the binary data is first filtered through a Gaussian low-pass filter in order to reduce bandwidth, then we obtain the Gaussian minimum-shift keying (GMSK) mentioned in connection with GSM. The term GFSK refers to any FSK produced as the result of Gaussian filtering of the binary modulation regardless of the particular value of modulation index. Thus, GMSK is a subset of GFSK, which is a subset of BFSK, which is a subset of FSK. (Your neurons should be shutting down right about now.)

We now consider quadrature phase modulations, in which only the phase varies; we will keep the amplitude constant for the moment. First, a little advisory note: Over the years, engineers (and, unfortunately, many textbook authors among them) have gotten a bit sloppy about what QPSK stands for. It could stand for either *quadrature* phase-shift keying (which includes any digital phase modulation form that results in a square-grid arrayed symbol constellation) or *quaternary* phase-shift keying (the special case of quadrature PSK in which four symbols comprise the constellation). So keep this in mind as you read some of the literature. We will use QPSK to mean quadrature PSK throughout this text.

The most basic QPSK is quaternary, with the symbols at the corners of a square. Again, there are many possible choices for the specific orientation of the square, so we show the most common one as Figure 2.23. This constellation appears to be the same as for 4-QAM. For this particular case, the two are in fact the same, as far as the constellation is concerned. There is potentially a difference in the details of the resulting output waveform, however. We've seen that prefiltering of the binary data prior to modulation is routinely performed to reduce spectral width. In the case of 4-PSK, the resulting waveform is generally constant-envelope, or very nearly so. In

4-QAM, however, the filtered waveforms amplitude-modulate the two quadrature signals, producing a nonconstant envelope output.

Generation of the modulation is readily achieved by taking the binary data, feeding it to a serial-to-parallel converter with two outputs, and then filtering those binary streams prior to feeding them separately to I and Q modulators.

If hardware were perfect, the palette of modulation options would consist of fewer choices. To understand that statement, consider what would happen if the binary data were to command us to make a transition diagonally in the symbol space. In that case, the phase would make a jump of 180°, just as with BPSK. Unfortunately, many amplifiers will fail to reproduce the transition faithfully. Transmitter power amplifiers (PAs), in particular, exhibit severe design tradeoffs between linearity and efficiency. As a consequence, many PAs may produce objectionable distortion if forced to produce a zero output momentarily. In a concession to that reality, many systems modify QPSK slightly to ease the burden on the hardware. The most popular modification is called *offset* QPSK (OQPSK) or OQAM. In OQPSK, a delay (offset) is inserted in series with either the I or Q path, thus preventing a direct transition between diagonally disposed symbols. The maximum phase shift magnitude is thus constrained to 90°.

There are (infinitely) many ways to orient a square in a plane, and thus many ways to assign the positions of the four symbols for QPSK. Suppose, for example, we were to rotate the entire symbol constellation by $\pi/4$. Then the symbols would lie on the I and Q axes. In a mutant strain of QPSK known as $\pi/4$-QPSK, that rotated constellation is combined with the original one of Figure 2.23. These two constellations are used in alternation, so that the worst-case phase jump is constrained to 135°. This value is not as small as is achieved more simply with OQPSK, but $\pi/4$-QPSK has a slight advantage in permitting *differential encoding* and detection. It is sufficient here to note that differential detection is a near-optimal method (in terms of bit-error rate performance versus SNR), implemented with modest hardware. As a result, $\pi/4$-QPSK is also known as $\pi/4$-DQPSK. This modulation is used in the IS-54 TDMA cellular system.

We finish our digestion of alphabet soup with a brief discussion of two other methods for getting bits from one place to another. We met them earlier, but only in passing. Code-division multiple access (CDMA) uses codes from a (near-) orthogonal code set. Each data bit is chipped by multiplication with the code. Each user is assigned a unique code from that code set, so the resulting chipped signals are also nearly orthogonal. Demodulation is performed by multiplication with the same code set, and it yields a signal only if the particular code matches that used in the original modulation. Data streams originating from multiplication with other codes merely produce noiselike output, thanks to the near-orthogonality.

Recently, attempts to push ever more bits per second per hertz through a channel have motivated wireless engineers to implement orthogonal frequency-division multiplexing (OFDM). In this method, a collection of orthogonal carriers is used instead of the usual single carrier. Each of these carriers is separately modulated, and the collection of modulated carriers is transmitted as an ensemble. Aside from conveying lots of bits per unit time, use of OFDM can facilitate adaptive response to changing channel conditions. For example, if one part of the spectrum drops out (due to fading, for example), the other carriers can compensate. If the carriers are distributed over a sufficiently wide spectrum (where "sufficiently wide" may be a function of factors such as delay spread), then the probability of a failed communication can be made very small. In effect, OFDM embeds a measure of frequency diversity. The highest-speed WLAN standards of today, 802.11a and 802.11g, use OFDM to provide up to 54 Mb/s peak data rates.

Finally, we close this subsection by commenting briefly on detection methods.[52] For the various forms of FSK/PSK, we can identify three classes of detection techniques. One is known as coherent detection, in which the carrier frequency and phase are recovered or extracted (e.g., with a PLL) and then the recovered carrier is used to multiply the received signal. The result of this multiplication is low-pass filtered to complete the demodulation. Because the carrier extraction is often performed with very low-bandwidth (and hence relatively noise-immune) circuitry, coherent detection is almost always the best performer – at least in the presence of Gaussian noise. By "best" we mean that coherent detection usually yields the lowest bit-error rate (BER) for a given SNR.

An alternative method that does almost as well is differential detection, mentioned briefly earlier in connection with $\pi/4$-DQPSK. In differential detection, no explicit extraction of the carrier is performed. Rather, it's performed *implicitly* by making use of correlations produced from bit period to bit period by the carrier. Here, the received signal is multiplied by a version of itself delayed some number of bit periods. That is, the delayed signal is used in place of an explicitly recovered carrier. Because there is no filtering action on the "carrier" analogous to that performed in true coherent detection, differential detection doesn't do quite as well. Nevertheless, it is simple to implement and so has a high performance-to-complexity ratio.

The least capable detection method is one we've already encountered: discriminators, which are a form of noncoherent detector. These have the poorest BER performance for a given SNR. Perhaps because they are already so poor in this respect, they show the least additional degradation in the face of random phase variations in the radio channel.

2.4 PROPAGATION

As we have seen, the Shannon channel capacity theorem states that a high SNR can be used to increase data rate for a fixed channel bandwidth. Fully exploiting this observation in wireless systems is made extremely difficult by the vagaries of signal propagation. Numerous impairments that are inconsequential in wireline communications are first-order considerations in wireless systems. Examples of these impairments include multipath, fading, delay spread, and Doppler shift, just to name a few. Not all of these are independent, but all of them cause grief.

---

[52] For an excellent overview of this topic (and many others), see Donald C. Cox, "Universal Digital Portable Radio Communications," *Proc. IEEE*, v. 75, no. 4, April 1987, pp. 436–77.

To develop at least a superficial appreciation for these factors, we begin by presenting a famous formula due to Harald Friis, known as his free-space transmission formula.[53] A fast derivation begins by considering the power density p measured at a distance r from an isotropic radiator of total power P_t:

$$p(r) = \frac{P_t}{4\pi r^2}. \tag{20}$$

All real radiators are nonisotropic, a fact readily accommodated by introducing a transmit antenna gain factor, G_t:

$$p(r) = G_t \frac{P_t}{4\pi r^2}. \tag{21}$$

The received power is simply the received power density multiplied by the effective area of the receiving antenna:

$$P_r = G_t \frac{P_t}{4\pi r^2} A_r. \tag{22}$$

This formula accurately predicts the performance of free-space (e.g., satellite) links, but it requires considerable modification to describe terrestrial links. Nevertheless, the formula reveals an important dependency on the effective antenna area. For antennas such as parabolic dishes, it's not too difficult to compute at least a good estimate for the area parameter (e.g., as equal to the physical area of the dish). However, the notion of an effective area is a bit hazy for the commonly used dipole antenna.[54] Even so, we can anticipate, on purely dimensional grounds, that the effective area of any antenna (including a dipole) should be proportional to the square of some length parameter. Therefore, we could write

$$P_r = G_x G_t \frac{P_t}{4\pi r^2} l^2, \tag{23}$$

where G_x is whatever it needs to be in order to make the equality hold.

It turns out that the length parameter is wavelength, so that

$$P_r = G_y G_t \frac{P_t}{4\pi r^2} \lambda^2. \tag{24}$$

Expressed in this form, we see that the received power scales as the *square* of the ratio λ/r. Thus, a doubling in frequency cuts by a factor of four the power received at a certain distance. As carrier frequency increases, then, the quality of a link can degrade rapidly. Again, this statement holds for free-space propagation. Hence Eqn. 24 predicts link quality in satellite systems fairly well.

Unfortunately, the assumption of free-space propagation is not at all well satisfied in terrestrial wireless systems. Signals may be absorbed by the medium, couple to

---

[53] Harald T. Friis, "A Note on a Simple Transmission Formula," *Proc. IRE,* v. 41, May 1946, pp. 254–6.

[54] One can still derive a well-defined effective area for a dipole antenna; it's just not something that's obvious by inspection. It turns out to equal $\lambda^2/4\pi$, if you really want to know.

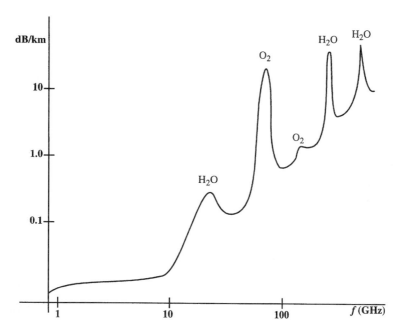

FIGURE 2.24. Approximate atmospheric attenuation versus frequency at sea level, in dry air.

lossy materials during propagation, and suffer reflection, refraction, and diffraction along the way. Compactly and accurately accounting for all of these factors is impossible, but numerous propagation measurements show that the path loss is almost never as favorable as $1/r^2$. In fact, for indoor propagation in particular, a typical path loss dependency is as $1/r^4$ and, in certain cases, the exponent can be as high as *six*.[55] Assuming free-space propagation is therefore almost guaranteed to be uselessly optimistic, and realistic link calculations need to take this observation into account.

Signal attenuation due to absorptive effects in the atmosphere start to become significant in dry air at tens of gigahertz. Below about 40–50 GHz, atmospheric absorption at sea level is typically below 1 dB/km (see Figure 2.24[56]), but heavy rainfall may exacerbate the loss considerably.[57] There are strong absorption peaks centered at around 22 GHz and 63 GHz (give or take a gigahertz here and there). The lower frequency absorption peak is due to water, and the higher frequency one is due to oxygen. The oxygen absorption peak contributes a path loss in excess of 20 dB/km, so it is quite significant. This attenuation, however, may be turned into an attribute if one wishes to permit re-use of spectrum over shorter distances. This property is exploited in various proposals for the deployment of picocells and other short-distance services at 60 GHz.

---

[55] Henry L. Bertoni, *Radio Propagation for Modern Wireless Systems,* Prentice-Hall, Englewood Cliffs, NJ, 2000.

[56] After *Millimeter Wave Propagation: Spectrum Management Implications,* Federal Communications Commission, Bulletin no. 70, July 1997.

[57] These values are in addition to the Friis path loss.

The large amount of spectrum offers high data rates, and the poor propagation is turned to advantage by forcing high frequency re-use ("it's not a bug, it's a feature").

The atmosphere is not the only source of attenuation, of course. Bear in mind that a conventional microwave oven, operating at around 2.45 GHz, depends on the typically high lossiness of many materials in that frequency range.[58] Indoor propagation certainly experiences additional attenuation inevitably contributed by this mechanism. For example, a typical segment of drywall in a home will attenuate a 2.6-GHz signal by about 2 dB on-axis and as much as 10 dB at a 45° angle, and going through a typical floor might attenuate 10–20 dB.[59] Reflection, absorption, and diffraction all conspire to produce these results.

Multipath propagation is another important impairment. A signal almost never proceeds directly from transmitter to receiver in a straight line. Signals radiated from a single source may travel quite different paths to a common receiver. When these signals sum together, they may reinforce or cancel as a function of the paths taken. Depending on the assumptions made, one may obtain the observed $1/r^4$ dependency noted earlier. In mobile applications in particular, multipath is highly variable. Strategies intended to mitigate multipath do not have the luxury of assuming a stationary channel. Fast fading can occur as the result of relative motion between transmitter and receiver or because of the motion of intervening objects. Slow fades are easier to accommodate than fast fades, but both are undesired.

Time-varying multipath produces multiple signals that we will treat as having nearly randomly distributed amplitudes and phases. As with virtually every other engineering approximation involving the word *random,* we will pretend that time-varying multipath is well approximated as a Gaussian process. In turn, a Gaussian process has an envelope that is Rayleigh distributed.[60] As with all approximations, it's important not to rely on its satisfaction for success. Nevertheless, it does provide valuable qualitative insights.

The Rayleigh distribution (see Figure 2.25) is described by

$$p(r) = \frac{\pi r \exp[-\pi r^2/4m^2]}{2m}, \tag{25}$$

where m is the mean. With this distribution (and its corresponding cumulative function), we may estimate that fades exceeding 10 dB (relative to the mean) will occur about 6% of the time.

---

[58] It does not depend on any resonant effects associated with water, as is frequently claimed. The microwave oven mainly relies on simple dielectric loss. The 2.45-GHz frequency is partly a compromise between speed of heating (favoring higher frequencies to increase loss) and delivering power more or less uniformly through the entire volume of food (favoring lower frequencies to increase skin depth). The first microwave oven – the Raytheon *Radarange* – operated at around 915 MHz, and many industrial ovens still do. It is not a coincidence that one of the unlicensed ISM (industrial-scientific-medical) bands is centered around that frequency. Nor is it a coincidence that another ISM band is centered around 2.45 GHz.

[59] Bertoni, op. cit. (see footnote 55).

[60] Cox, op. cit. (see footnote 52).

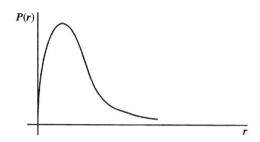

FIGURE 2.25. Rayleigh distribution.

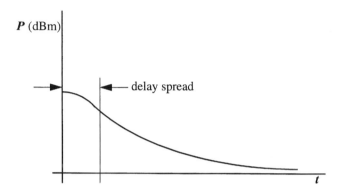

FIGURE 2.26. Idealized delay spread example.

Another way to characterize multipath propagation is by *delay spread,* which is defined formally as the square root of the second moment of the channel's impulse response.[61] Imagine "pinging" the channel and then plotting the power of the returned signal as a function of time (see Figure 2.26). As a very crude rule of thumb, you may estimate the delay spread by inspection of a power–time plot as shown by simply dividing some arbitrary measure of the total width by a value of 3 to 4. This operation is the same as you would perform to estimate the standard deviation of a Gaussian distribution, for example. Again, this is a very crude rule of thumb, to be used only when performing zeroth-order, back-of-the-envelope engineering calculations.

Delay spread is important because the arrival of two otherwise identical signals displaced too much in time will frustrate demodulation of digital signals. It seems reasonable, then, that reliable demodulation would be possible as long as the delay spread were kept at some suitably small fraction of a symbol period. Again, "suitably small" can be taken to mean "about a tenth or twentieth" for *very* crude estimates. As an example, a channel whose delay spread is 30 nanoseconds might support data rates of 1–3 megasymbols per second – contingent upon the satisfaction of many assumptions, and sensitive to many details.[62] Nevertheless, this type of estimation procedure does yield useful order-of-magnitude insights.

---

[61] Cox, ibid.

[62] This value is not atypical for an indoor WLAN, for example.

An alternative way to think about delay spread is in the frequency domain. To zeroth order, the channel will be approximately flat over a bandwidth that is proportional to the reciprocal of the delay spread. A small delay spread implies a relatively wide bandwidth over which the channel characteristics will be constant. In such cases, the fading associated with the delay spread is termed *flat*. If excessive delay spread causes significant variation over the bandwidth of a channel, then the fading is termed *frequency-selective*. Again, no matter what it's called, it's undesired.

Methods for improving link quality in the face of these impairments include: brute-force increases in power; adaptive equalization (to undo frequency-selective fading by flattening the frequency response of the channel); use of directional and adaptive antennas to maximize the power transmitted in a desired direction; diversity; and error correction. We briefly consider each of these in turn.

Increases in power are not nearly as straightforward to implement as one would like because of the nonnegligible probability that fade depths will exceed 10 or even 20 dB. A brute-force power increase by a factor of 10 or 100 to overcome a deep fade is costly in several dimensions and may even violate statutory limits on power. Given how microwave ovens work, one should properly be concerned about effects on human tissue as well.

Rather than increasing power by 10–20 dB indiscriminately, one might use (adaptive) directional antennas to focus the power in a direction that will do the most good. Even though this method is preferable to brute-force increases in total power, the peak power density can still be quite high, so concerns about causing interference or unhealthy biological effects remain, if only over more limited regions of space.

A popular method for mitigating fades is to make use of *diversity*. Diversity can take many forms, and all of them can be quite effective. Antenna diversity involves the use of one or more additional antennas, preferably spaced apart by a reasonable fraction of a wavelength (e.g., a quarter wavelength) to ensure sufficient lack of correlation among the antennas. If a deep fade is caused by an unfortunate cancellation of two strong but out-of-phase signals arriving at one antenna, it is unlikely that an antenna located a quarter of a wavelength away will experience the same unfortunate coincidence. If a deep fade is detected with one antenna then one may simply switch antennas, or combine the signals from the two antennas in some fashion. Both strategies are in use, and both work well. Once multiple antennas are present, "beam forming" to provide adaptive directionality becomes an option as well, at the expense of more complex circuitry.

Frequency diversity is also effective. Again, if a deep fade is caused by two rays arriving with a relative phase shift of 180° at a particular frequency, then switching to a (significantly) different frequency may produce a phase shift of a smaller, less troublesome value. An alternative to switching is simply to occupy a broad bandwidth and dynamically alter the spectral occupation as necessary to avoid transmitting energy where fading is significant. Frequency diversity may be implemented as a supplementary strategy or as an inherent characteristic of the modulation.

Polarization diversity can also be effective. It is common in satellite communications to employ circularly polarized radiations, for example, because the sense of the polarization reverses (e.g., from clockwise to counterclockwise) upon undergoing a reflection. A receiver with a properly polarized antenna will sense the main signal and largely ignore the reflection. Signals undergoing an even number of reflections will have the same polarization sense as the directly arriving signal – but, having suffered attenuations at each reflection, the amplitude of the interference may be small enough to present no serious problem.

The effect of moderate and occasional fades can be mitigated by a combination of modest power increases and redundancy. The latter is more commonly known as *forward error correction* (FEC). If a temporary fade wipes out part of a message, the redundancy enables reconstruction nevertheless. There is a tradeoff between too much redundancy (causing low data rates even when the link quality is high) and too little (causing zero data rates when the link quality is just the slightest bit imperfect). As one might expect, the optimum choice depends on details of propagation, data rate, modulation type, and a host of other factors.

2.5 SOME CLOSING COMMENTS

From examining the history of mobile and portable wireless, it might appear that the band from approximately 500 MHz to about 5 GHz is overrepresented. The appearance of favoritism is not an artifact of selective reporting. That segment of spectrum *is* popular, and for excellent reasons.

First, let's consider what factors might constrain operation at low carrier frequencies. One is simply that there's less spectrum there. More significant, however, is that antennas cannot be too small (relative to a wavelength) if they are to operate efficiently. Efficient low-frequency antennas are thus long antennas. For mobile or portable applications, one must choose a frequency high enough that efficient antennas won't be *too* long. At 500 MHz, where the free-space wavelength is 60 cm, a quarterwave antenna would be about 15 cm long. That value is readily accommodated in a handheld unit.

As frequency increases, we encounter a worsening path loss. One causal factor is the increasing tendency for reflection, refraction, and diffraction, but another is anticipated by the Friis formula. Increasing the frequency by a factor of 10, to 5 GHz, increases the Friis path loss factor by 20 dB. At these frequencies, interaction with biological tissues is nonnegligible, so simply increasing power by a factor of 100 to compensate is out of the question. Operation at higher frequency is accompanied by an ever-decreasing practical radius of communications.

Thus we see that there is an approximate decade span of frequency, ranging from 500 MHz to 5 GHz, that will *forever* remain the sweet spot for large-area mobile wireless. Unlike Moore's law, then, useful spectrum does not expand exponentially over time. In fact, it is essentially fixed. This truth explains why carriers went on a lunatic tear in the late 1990s, bidding hundreds of billions of dollars for 3G spectrum (only to find the debt so burdensome that many carriers have been forced to make "other arrangements"). No doubt, there will be ongoing efforts to maximize the utility of that finite spectrum and also to reclaim spectrum from other services (e.g., UHF television) that arguably use the spectrum less efficiently.

With this high-level overview completed, we now drop down a level or two in the hierarchy. The next chapter begins with coverage of a topic that may be unfamiliar to those trained in the digital age: tuned circuits.

2.6 APPENDIX: CHARACTERISTICS OF OTHER WIRELESS SYSTEMS

It is impossible to list all the services and systems in use, but here we provide a brief sampling of a few others that may be of interest. (A detailed U.S. spectrum allocation chart may be downloaded for free from http://www.ntia.doc.gov/osmhome/allochrt. pdf.) The first of these is the unlicensed ISM (industrial-scientific-medical) band; see Table 2.8. Microwave ovens, transponders, RF ID tags, some cordless phones, WLANs, and a host of other applications and services use these bands. Notice that these bands reside within the "sweet spot" for mobile and portable wireless identified earlier.

Another unlicensed band has been allocated recently in the United States. The *unlicensed national information infrastructure* (UNII) band adds 200 MHz to the existing 5-GHz ISM band and also permits rather high EIRPs in one of the bands. See Table 2.9.

Just to make sure that we don't leave the impression that the mobile and cellular systems discussed in the main part of this chapter are the only uses for wireless, Table 2.10 gives a brief sampling of other (broadcast) wireless systems. Also occasionally useful is Table 2.11, which lists frequency bands and their common (but by no means universal) designations. Not all sources agree on the precise frequency limits of these bands (particularly for the bands below VLF), so it's best to supplement these band designations with actual frequency values whenever it might matter.

In relating wavelength to frequency, just remember that the product of frequency (in hertz) and wavelength (in meters) is the speed of light (very nearly 3×10^8 m/s). Therefore, a 1-MHz signal has a free-space wavelength of almost exactly 300 m; a 1-GHz signal has a 300-mm wavelength. It's useful to note that the frequency (in gigahertz) multiplied by the wavelength (in millimeters) is about 300.

Table 2.8. *ISM band allocations and summary*

Parameter	900 MHz	2.4 GHz	5.8 GHz
Frequency range	902–928 MHz	2400–2483.5 MHz	5725–5850 MHz
Total allocation	26 MHz	83.5 MHz	125 MHz
Maximum power	1 W	1 W	1 W
Maximum EIRP[a]	4 W	4 W (200 W for point-to-point)	200 W

[a] EIRP stands for "effective isotropically radiated power" and equals the product of power radiated and the antenna gain.

Table 2.9. *UNII band allocations and summary*

Parameter	Indoor	Low-power	UNII/ISM
Frequency range	5150–5250 MHz	5250–5350 MHz	5725–5825 MHz
Total allocation	100 MHz	100 MHz	100 MHz
Maximum power	50 mW	250 mW	1 W
Maximum EIRP	200 mW; unit must have integral antenna	1 W	200 W

Table 2.10. *Random sampling of some broadcast systems*

Service/system	Frequency span	Channel spacing
AM radio	535–1605 kHz	10 kHz
TV (ch. 2–4)	54–72 MHz	6 MHz
TV (ch. 5–6)	76–88 MHz	6 MHz
FM radio	88.1–108.1 MHz	200 kHz
TV (ch. 7–13)	174–216 MHz	6 MHz
TV (ch. 14–69)	470–806 MHz	6 MHz

Table 2.11. *Radio frequency band designations*

Band	Frequency range	Wavelength range
Extremely low frequency (ELF)	<30 Hz	>10,000 km
Super low frequency (SLF)	30 Hz to 300 Hz	10,000 km to 1000 km
Ultra low frequency (ULF)	300 Hz to 3 kHz	1000 km to 100 km
Very low frequency (VLF)	3 kHz to 30 kHz	100 km to 10 km
Low frequency (LF)	30 kHz to 300 kHz	10 km to 1 km
Medium frequency (MF)	300 kHz to 3 MHz	1 km to 100 m
High frequency (HF)	3 MHz to 30 MHz	100 m to 10 m
Very high frequency (VHF)	30 MHz to 300 MHz	10 m to 1 m
Ultra high frequency (UHF)	300 MHz to 3 GHz	1 m to 10 cm
Super high frequency (SHF)	3 GHz to 30 GHz	10 cm to 1 cm
Extremely high frequency (EHF)	>30 GHz	<1 cm

Another classification system has its origins in radar work during the Second World War. Based on letters chosen at random to confuse adversaries, a lack of standardization has succeeded in confusing just about everyone.[63] The frequency ranges

---

[63] Wartime secrecy concerns were so great that the radar design community didn't even standardize nomenclature with engineers involved in other communications research elsewhere. We are still living with the resulting legacy of confusion.

Table 2.12. *Microwave band designations (IEEE 521-1984)*

Band	Frequency range
L	1.0 GHz to 2.0 GHz
S	2.0 GHz to 4.0 GHz
C	4.0 GHz to 8.0 GHz
X	8 GHz to 12 GHz
Ku	12 GHz to 18 GHz
K	18 GHz to 27 GHz
Ka	27 GHz to 40 GHz
V	40 GHz to 75 GHz
W	75 GHz to 110 GHz

associated with the letters have changed somewhat over time, and they also vary from country to country (and even within a country). Different companies, also, have used different conventions of their own devising. For these reasons, the letter-based designations are perhaps best avoided (or, at least, supplemented with actual frequency values, as with the previous designations). Nevertheless, they are still used, so we offer here a table of such bands (Table 2.12) as standardized by the IEEE, the only international standard the author could locate. The designations Ku and Ka arose from "under K" and "above K," respectively.

Other systems of letter designations that you may encounter include the waveguide bands as well as those due to organizations and companies such as NASA, Hewlett-Packard, Sperry, Motorola, Narda, Raytheon, and others. These designations (or the frequency ranges associated with a given band) are all a bit different, and they may include bands designated by additional letters and omit others. Bear this in mind as you survey the literature.

CHAPTER THREE

PASSIVE *RLC* NETWORKS

3.1 INTRODUCTION

One characteristic of RF circuits is the relatively large ratio of passive to active components. In stark contrast with digital VLSI circuits (or even with other analog circuits, such as op-amps), many of those passive components may be inductors or even transformers. This chapter hopes to convey some underlying intuition that is useful in the design of *RLC* networks. As we build up that intuition, we'll begin to understand the many good reasons for the preponderance of *RLC* networks in RF circuits. Among the most compelling of these are that they can be used to match or otherwise modify impedances (important for efficient power transfer, for example), cancel transistor parasitics to provide high gain at high frequencies, and filter out unwanted signals.

To understand how *RLC* networks may confer these and other benefits, let's revisit some simple second-order examples from undergraduate introductory network theory. By looking at how these networks behave from a couple of different viewpoints, we'll build up intuition that will prove useful in understanding networks of much higher order.

3.2 PARALLEL *RLC* TANK

Let's just jump right into the study of a parallel *RLC* circuit. As you probably know, this circuit exhibits resonant behavior; we'll see what this implies momentarily. This circuit is also often called a *tank circuit*[1] (or simply *tank*).

We begin by studying its complex impedance, or more directly, its admittance (more convenient for a parallel network); see Figure 3.1. For this network, we know that the admittance is simply

---

[1] So-called either by analogy with acoustic resonators or because it stores energy, as does a tank of water.

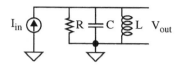

FIGURE 3.1. Parallel *RLC* tank circuit.

$$Y = G + j\omega C + \frac{1}{j\omega L} = G + j\left(\omega C - \frac{1}{\omega L}\right). \tag{1}$$

From inspection of the network (or of the equation), it's easy to see that the admittance goes to infinity both at DC (because the inductor acts as a short there) and at infinitely high frequency (because the capacitor acts like a short there). We may therefore say that, at very low frequencies, the network's admittance is essentially that of the inductor (since its admittance dominates the combination) and is also that of the capacitor at very high frequencies. What divides "low" from "high" is the frequency at which the inductive and capacitive admittances cancel. Known as the *resonant* frequency,[2] this is given by:

$$\left(\omega_0 C - \frac{1}{\omega_0 L}\right) = 0 \implies \omega_0 = \frac{1}{\sqrt{LC}}. \tag{2}$$

A frequently handy rule of thumb is that a 1-nH inductor and a 1-pF capacitor resonate at 5 GHz to an excellent (better than 1%) approximation. Knowing this one datum allows rapid computation of the resonant frequency of any other *LC* combination.[3] In any event, at resonance, the admittance is purely real and equal to *G* by virtue of the cancellation of the reactive terms.

To say that the reactive terms cancel at resonance is certainly correct, but a little glib. As we'll see shortly, the *individual* currents in the inductive and capacitive branches can be surprisingly large, although they cancel each other as far as the external world is concerned. We'll also see that this augmented current is a sign that a downward impedance transformation has taken place, a phenomenon we will often exploit. To explore these behaviors more fully and describe them in the most generally useful way, we need to introduce another parameter, *Q*.

3.2.1   *Q*

Aside from the resonant frequency itself, another important descriptive parameter is the *quality factor,* or simply *Q*. Different (but equivalent) definitions of *Q* abound

---

[2] Some authors use the term "antiresonant" for parallel resonances and reserve the term "resonant" for series circuits. We will use the term "resonant" to indicate *any* cancellation of inductive and capacitive reactances, whether in series or parallel circuits. On those occasions when some distinction is necessary, we'll just say "series resonant" or "parallel resonant."

[3] At lower frequencies, it may be more convenient to remember that a 1-μH inductance and a 1-nF capacitance resonate at 5 MHz.

but, for a system under sinusoidal excitation at a frequency ω, perhaps the most fundamental one is as follows:

$$Q \equiv \omega \frac{\text{energy stored}}{\text{average power dissipated}}. \tag{3}$$

Note that Q is dimensionless and that it is proportional to the ratio of energy stored to the energy lost, per unit time. This definition is fundamental because it says nothing specific about what stores or dissipates the energy. So, as we'll see later on, it applies perfectly well even to distributed systems, such as microwave resonant cavities, where it is not possible to identify individual inductances, capacitances, and resistances. It should also be clear that the notion of Q applies both to resonant and nonresonant systems, so one may talk of the Q of an *RC* circuit. A high-order system may exhibit multiple resonances, each with its own peak Q value. From the fundamental definition, we also see that the value we compute depends on whether or not we include external loading, and perhaps also on how that load connects to the network in question. If we neglect the loading then we refer to the computed value as the unloaded Q, and if we include it then we call it the loaded Q. Whenever the context is ambiguous and the distinction matters, it is important to identify explicitly the type of Q under discussion.

Let's now use this definition to derive expressions for the Q of our parallel *RLC* circuit at resonance.

At the resonant frequency, which we'll denote by ω_0, the voltage across the network is simply $I_{\text{in}}R$. Recall that energy in such a network sloshes back and forth between the inductor and capacitor, with a constant sum at resonance. As a consequence, the peak energy stored in either the capacitor or inductor is equal to the total energy stored in the network at any given time. Since we happen to know the peak capacitor voltage at resonance (it's just $I_{\text{pk}}R$), it's most convenient to use it to compute the network energy:

$$E_{\text{tot}} = \tfrac{1}{2}C(I_{\text{pk}}R)^2. \tag{4}$$

Now we need to compute the average power dissipated. Again, this computation is easy to carry out at resonance, since the network degenerates to a simple resistance there. The average power dissipated in the resistor at resonance is therefore simply

$$P_{\text{avg}} = \tfrac{1}{2}I_{\text{pk}}^2 R. \tag{5}$$

The Q of the network at resonance is then

$$Q = \omega_0 \frac{E_{\text{tot}}}{P_{\text{avg}}} = \frac{1}{\sqrt{LC}} \frac{\tfrac{1}{2}C(I_{\text{pk}}R)^2}{\tfrac{1}{2}I_{\text{pk}}^2 R} = \frac{R}{\sqrt{L/C}}. \tag{6}$$

The quantity $\sqrt{L/C}$ has the dimensions of resistance and is sometimes called the *characteristic impedance* of the network.[4] It is significant because it is equal to the magnitude of the capacitive and inductive reactances at resonance, as is easily shown:

---

[4] This term is usually applied to transmission lines, but it has a certain importance even in lumped networks.

$$|Z_C| = |Z_L| = \omega_0 L = \frac{L}{\sqrt{LC}} = \sqrt{\frac{L}{C}}. \tag{7}$$

We will find that this quantity recurs with some frequency,[5] so keep it in mind.

Before we continue, let's see if our equation for Q makes sense. As the parallel resistance goes to infinity, Q does, too. This behavior seems reasonable since, in the limit of infinite resistance, the network degenerates to a pure LC system. With only purely reactive elements in the network, there is no way for energy to dissipate and Q should go to infinity, just as the equation says it should. Plus, Q also increases as the impedance of the reactive elements decreases (by decreasing L/C), since the pure resistance becomes less significant compared with the reactive impedances.

For completeness, we may derive a couple of additional expressions for the Q of our parallel RLC network at resonance:

$$Q = \frac{R}{|Z_{L,C}|} = \frac{R}{\omega_0 L} = \omega_0 RC. \tag{8}$$

3.2.2 BRANCH CURRENTS AT RESONANCE

As mentioned earlier, the inductive and capacitive branch currents at resonance can differ significantly from the overall network current (which is simply due to the parallel resistance). Let's now compute the magnitude of these currents.

Again, at resonance, the voltage across the network is $I_{in}R$. Since the inductive and capacitive reactances are equal at resonance, the inductive and capacitive branch currents will be equal in magnitude:

$$|I_L| = |I_C| = \frac{|V|}{Z} = \frac{|I_{in}|R}{\omega_0 L} = \frac{|I_{in}|R\sqrt{LC}}{L} = |I_{in}|\frac{R}{\sqrt{L/C}} = Q|I_{in}|. \tag{9}$$

That is, the current flowing in the inductive and capacitive branches is Q times as large as the net current. Hence, if $Q = 1000$ and we drive the network at resonance with a one-ampere current source, then that one ampere will flow through the resistor but *one thousand amperes* will flow through the inductor and capacitor (until they vaporize). From this simple example, you can well appreciate the incompleteness of simply stating that the inductor and capacitor cancel at resonance!

3.2.3 BANDWIDTH AND Q

We've already deduced the behavior of the network at frequencies far removed from resonance, and we've also taken a detailed look at the network's behavior at resonance. Let's now examine the behavior of the tank circuit at frequencies slightly displaced from resonance to round out our analysis.

---

[5] You may recall that the characteristic impedance of a transmission line is given by the same expression, where L and C are interpreted as the inductance and capacitance *per unit length*. We will explore the connection between finite, lumped LC networks and infinite, distributed systems more fully in Chapter 6.

First, let $\omega = \omega_0 + \Delta\omega$. Then we may rewrite our expression for the admittance as

$$Y = G + \frac{j}{\omega L}(\omega^2 LC - 1) = G + \frac{j}{\omega L}[2\Delta\omega\omega_0 + (\Delta\omega)^2]LC. \qquad (10)$$

For suitably small displacements about ω_0 (i.e., for values of $\Delta\omega$ that are small relative to ω_0), this expression simplifies to

$$Y \approx G + j2C\Delta\omega. \qquad (11)$$

This admittance behavior is exactly the same as that of a resistor of value R in parallel with a capacitor of value $2C$, except with $\Delta\omega$ replacing ω. Hence, the shape of the admittance curve for (small) positive displacements about the resonant frequency is the same as that of a parallel RC network, and we may define a half-bandwidth as simply $1/2RC$ by analogy with the RC case. Why "*half*-bandwidth?" Well, from the symmetry of the approximate admittance function derived previously, the shape for displacements below resonance will be the mirror image of that above resonance, so that the *total* -3-dB bandwidth is just $1/RC$.

Something interesting happens when we normalize this bandwidth to the resonant frequency:

$$\frac{\text{BW}}{\omega_0} = \frac{1}{RC\omega_0} = \frac{\sqrt{LC}}{RC} = \frac{\sqrt{L/C}}{R} = \frac{1}{Q}. \qquad (12)$$

Once again, Q has popped into the picture: the fractional bandwidth is simply $1/Q$. For a given resonant frequency, then, higher Q implies narrower bandwidth.

From the foregoing developments you can see that Q is an important parameter, and we shall find in many subsequent examples that significant analytical (and synthetic) advantages often accrue from focusing on it.

3.2.4 RINGING AND Q

Although we have focused on the frequency-domain behavior of *RLC* networks, perhaps it is worthwhile here to mention at least one important time-domain property. The resonant exchange of energy between the inductor and the capacitor leads to an impulse response with the familiar damped oscillatory characteristic. Since Q is a measure of the rate of energy loss, one would expect a higher Q to be associated with more persistent ringing than a lower Q. To place this intuition on a more quantitative basis, recall that the impulse response is a damped sinusoid with an exponential envelope. It is straightforward to show that the time constant of this envelope is $2RC$, so that

$$V(t) \propto V_0 e^{-t/(2RC)}. \qquad (13)$$

The time constant may be rewritten (using Eqn. 12) to yield

$$V(t) \propto V_0 e^{-(t/T)(\pi/Q)}. \qquad (14)$$

Thus, the amplitude decays to $1/e$ of its initial value in Q/π cycles. Because an exponential decays to about 4% of its initial value in π time constants, Q is roughly

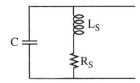

FIGURE 3.2. Not-quite-parallel
RLC tank circuit.

equal to the number of cycles of ringing. This handy rule of thumb is extremely useful for rapidly estimating Q from experimental impulse (or step) response data.

3.3 SERIES *RLC* NETWORKS

We may follow an exactly analogous dual approach to deduce the properties of series *RLC* circuits. The details of the derivations are relatively uninteresting, so here
we simply present the relevant observations and equations.

The resonant condition corresponds again to the frequency where the capacitance
and inductance cancel. Rather than resulting in an admittance minimum, though,
resonance here results in an impedance minimum, with a value of R. The equation
for Q involves the same terms as for the parallel case, but in reciprocal form:

$$Q = \frac{\sqrt{L/C}}{R}. \tag{15}$$

At resonance, the voltage across either the inductor or capacitor is Q times as great
as that across the resistor. Thus, if a series *RLC* network with a Q of 1000 is driven
at resonance with a one-volt source, then the resistor will have that one volt across it
yet a thrilling one thousand volts will appear across the inductor and capacitor.[6]

You are encouraged to explore independently and in more detail the duality of series and parallel resonant circuits.

3.4 OTHER RESONANT *RLC* NETWORKS

Purely parallel or series *RLC* networks rarely exist in practice, so it's important to
take a look at configurations that might be more realistically representative. Consider, for example, the case sketched in Figure 3.2. Because inductors tend to be
significantly lossier than capacitors, the model shown in the figure is often a more
realistic approximation to typical parallel *RLC* circuits.

Since we've already analyzed the purely parallel *RLC* network in detail, it would
be nice if we could re-use as much of this work as possible. So, let's convert the

---

[6] By the way, such resonant voltage magnification is the fundamental basis for Tesla coil operation.
With such techniques, Tesla was able to generate an estimated 5–10 MV in 1899!

circuit of Figure 3.2 to a purely parallel *RLC* network by replacing the series *LR* section with a parallel one. Clearly, such a substitution cannot be valid in general, but over a suitably restricted frequency range (e.g., near resonance) the equivalence is pretty reasonable. To show this formally, let's equate the impedances of the series and parallel *LR* sections:

$$j\omega_0 L_S + R_S = [(j\omega_0 L_P) \parallel R_P] = \frac{(\omega_0 L_P)^2 R_P + j\omega_0 L_P R_P^2}{R_P^2 + (\omega_0 L_P)^2}. \tag{16}$$

If we equate real parts and note that $Q = R_P/\omega_0 L_P = \omega_0 L_S/R_S$,[7] we obtain

$$R_P = R_S(Q^2 + 1). \tag{17}$$

Similarly, equating imaginary parts yields

$$L_P = L_S\left(\frac{Q^2 + 1}{Q^2}\right). \tag{18}$$

We may also derive a similar set of equations for computing series and parallel *RC* equivalents:

$$R_P = R_S(Q^2 + 1), \tag{19}$$

$$C_P = C_S\left(\frac{Q^2}{Q^2 + 1}\right). \tag{20}$$

Let's pause for a moment and look at these transformation formulas. Upon closer examination, it's clear that we may express them in a universal form that applies to both *RC* and *LR* networks:

$$R_P = R_S(Q^2 + 1) \tag{21}$$

and

$$X_P = X_S\left(\frac{Q^2 + 1}{Q^2}\right), \tag{22}$$

where *X* is the imaginary part of the impedance. This way, one need only remember a single pair of "universal" formulas in order to convert any "impure" *RLC* network into a purely parallel (or series) one that is straightforward to analyze. However, one must bear in mind that *the equivalences hold only over a narrow range of frequencies centered about* ω_0.

3.5 *RLC* NETWORKS AS IMPEDANCE TRANSFORMERS

The relative abundance of power gain at low frequencies allows designers to treat it essentially as an infinite resource. Design specifications are thus often expressed simply in terms of a voltage gain, for example, without any explicit reference to or concern for power gain. Hence, circuit design at low frequencies usually proceeds in

[7] If the series and parallel sections are to be equivalent, then their *Q*s certainly must be equivalent.

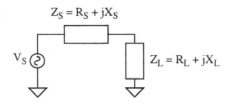

FIGURE 3.3. Network for maximum power
transfer theorem.

blissful ignorance of the maximum power transfer theorem derived in every undergraduate network theory course. In striking contrast with that insouciance, RF circuit design is frequently *preoccupied* with power gain because of its relative scarcity.
Impedance-transforming networks thus play a prominent role in the radio frequency
domain.

3.5.1 THE MAXIMUM POWER TRANSFER THEOREM

To understand more explicitly the value of impedance transformers, we now review
the maximum power transfer theorem (see Figure 3.3). The problem is this: Given
a *fixed source* impedance Z_S, what load impedance Z_L maximizes the power delivered to the load? The power delivered to the load impedance is entirely due to R_L,
since reactive elements do not dissipate power. Hence, the power delivered is simply

$$\frac{|V_R|^2}{R_L} = \frac{R_L|V_S|^2}{(R_L + R_S)^2 + (X_L + X_S)^2}, \tag{23}$$

where V_R and V_S are the rms voltages across the load resistance and source, respectively.

To maximize the power delivered to R_L, it's clear that X_L and X_S should be inverses so that they sum to zero. In addition, maximizing Eqn. 23 under that condition
leads to the result that R_L should equal R_S. Hence, the maximum power transfer from
a fixed source impedance to a load occurs when the load and source impedances are
complex conjugates.

Having established mathematically the condition for maximum power transfer, we
now consider practical methods for achieving it.

3.5.2 THE L-MATCH

The multiplication by Q of voltages or currents in resonant *RLC* networks hints at
their impedance-modifying potential. Indeed, the series–parallel *RC/LR* network
conversion formulas developed in the previous section actually show this property
explicitly. To make this clearer, consider once again the circuit of Figure 3.2, re-
drawn slightly as Figure 3.4. Here we treat R_S as a load resistance for the network.
When this resistance is viewed across the capacitor, it is transformed to an equivalent

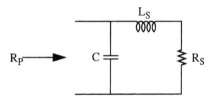

FIGURE 3.4. Upward impedance transformer.

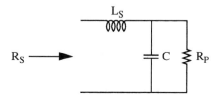

FIGURE 3.5. Downward impedance transformer.

R_P according to the formulas developed in the previous section. From inspection of those "universal" equations, it is clear that R_P will always be larger than R_S, so the network of Figure 3.4 transforms resistances upward. To get a downward impedance conversion, just interchange ports (see Figure 3.5).

There is a nice, intuitive way to keep track of which way the impedance transformation goes. For example, if the circuit in Figure 3.5 is driven by a test voltage source, the result is a parallel RLC network since the Thévenin resistance of the test source is zero. Now, the inductive current in a parallel RLC network at resonance is Q times as large as the current through R_P. This increase in current is seen by the source and may be interpreted as a reduction in resistance.

It should also be clear that interchanging the inductor and capacitor doesn't alter the transformation ratio, although other considerations may dictate whether one chooses a high-pass or low-pass configuration.

This circuit is known as an *L-match* because of its shape (perhaps you have to be lying on your side and dyslexic to see this), and it does have the attribute of simplicity. However, there are only two degrees of freedom (one can choose only L and C). Hence, once the impedance transformation ratio and resonant frequency have been specified, network Q is automatically determined. If you want a different value of Q then you must use a network that offers additional degrees of freedom; we'll study some of these shortly.

As a final note on the L-match, the "universal" equations can be simplified if $Q^2 \gg 1$. If this inequality is satisfied, then the following approximate equations hold:

$$R_P \approx R_S Q^2 = R_S \left(\frac{1}{\omega_0 R_S C} \right)^2 = \frac{1}{R_S} \frac{L_S}{C}, \tag{24}$$

which may be rewritten as

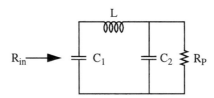

FIGURE 3.6. The π-match.

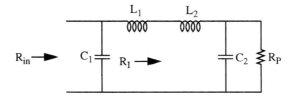

FIGURE 3.7. π-match as cascade of L-matches.

$$R_P R_S \approx \frac{L_S}{C} = Z_0^2, \tag{25}$$

where Z_0 is the characteristic impedance of the network, as discussed in Section 3.2.1.

One may also deduce that Q is approximately the square root of the transformation ratio:

$$Q \approx \sqrt{\frac{R_P}{R_S}}. \tag{26}$$

Finally, the reactances don't vary much in undergoing the transformation:

$$X_P \approx X_S. \tag{27}$$

As long as Q is greater than about 3 or 4, the error incurred will be under about 10%. If Q is greater than 10, the maximum error will be in the neighborhood of 1% or so. Hence, for quick, back-of-the-envelope calculations, these simplified equations are adequate. Final design values can be computed using the full "universal" equations.

3.5.3 THE π-MATCH

As already discussed, one limitation of the L-match is that one can specify only two of center frequency, impedance transformation ratio, and Q. To acquire a third degree of freedom, one can employ the network shown in Figure 3.6.

This circuit is known as a π-match, again because of its shape. The most expedient way to understand this matching network is to view it as two L-matches connected in cascade, one that transforms down and one that transforms up; see Figure 3.7. Here, the load resistance R_P is transformed to a lower resistance (known as the *image* or *intermediate resistance,* here denoted R_I) at the junction of the two inductances. The image resistance is then transformed up to a value R_{in} by a second L-match section.

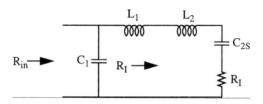

FIGURE 3.8. π-match with transformed right-hand L-section.

Now, it may seem a bit silly to use one L-section to go down, then another to go back up. However, we *have* gained the additional degree of freedom we were seeking. Recall that, for an L-match, Q is fixed at a value roughly equal to the square root of the transformation ratio. Typically, the Q of an L-match isn't particularly high because huge transformation ratios are infrequently required. The π-match decouples Q from the transformation ratio by introducing an intermediate resistance value to transform to, allowing us to achieve much higher Q than is generally available from an L-match, even if the *overall* transformation ratio isn't particularly large.

Because we now have three degrees of freedom (the two capacitances and the sum of the two inductances), we can independently specify center frequency, Q (or bandwidth), and overall impedance transformation ratio. However, as with the L-match (or any other kind of match), impractical or inconvenient component values can result, and some creativity or compromise may be required to generate a sensible design. In many instances, cascading several matching networks may be helpful.

In order to derive the design equations, first transform the parallel RC subnetwork of the right-hand L-section into its series equivalent, as shown in Figure 3.8. When we replace the output parallel RC network with its series equivalent, the series resistance is, of course, R_I. Hence, the Q of the right-hand L-section may be written as

$$\frac{\omega_0 L_2}{R_I} = \sqrt{\frac{R_P}{R_I} - 1} = Q_{\text{right}}. \tag{28}$$

At the same time, recognize that the left-hand L-section also sees a resistance of R_I at the center frequency. Therefore, its Q is given by

$$\frac{\omega_0 L_1}{R_I} = \sqrt{\frac{R_{\text{in}}}{R_I} - 1} = Q_{\text{left}}. \tag{29}$$

The overall network Q is simply

$$Q = \frac{\omega_0 (L_1 + L_2)}{R_I} = \sqrt{\frac{R_{\text{in}}}{R_I} - 1} + \sqrt{\frac{R_P}{R_I} - 1}. \tag{30}$$

Equation 30 allows us to find the image resistance, given Q and the transformation resistances. Once R_I is computed, the total inductance is quickly found:

$$L_1 + L_2 = \frac{Q R_I}{\omega_0}. \tag{31}$$

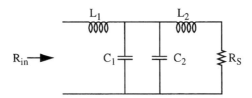

FIGURE 3.9. T-match.

The values of the capacitors are also found readily:

$$C_1 = \frac{Q_{\text{left}}}{\omega_0 R_{\text{in}}}, \tag{32}$$

$$C_2 = \frac{Q_{\text{right}}}{\omega_0 R_P}. \tag{33}$$

As a practical matter, note that finding R_I from Eqn. 30 generally requires iteration. A good starting value can be obtained by assuming that Q is large (even if it isn't). In that case, R_I is approximately given by

$$R_I \approx \frac{\left(\sqrt{R_{\text{in}}} + \sqrt{R_P}\right)^2}{Q^2}. \tag{34}$$

If Q is very large, or if you're just doing some preliminary "cocktail napkin" calculations, then iteration may not even be necessary.

And that's all there is to it.

As a parting note, one final bit of trivia deserves mention. An additional reason that the π-match is popular is that the parasitic capacitances of whatever connects to it can be absorbed into the network design. This property is particularly valuable because capacitance is the dominant parasitic element in many practical cases.

3.5.4 THE T-MATCH

The π-match results from cascading two L-sections in one particular way. Connecting up the L-sections another way leads to the dual of the π-match, as shown in Figure 3.9. Here, what would be a single capacitor in a practical implementation has been decomposed explicitly into two separate ones. The (parallel) image resistance is seen across these capacitors, either looking to the right or looking to the left as in the π-match.

The design equations are readily derived by following an approach analogous to that used for the π-match. The overall network Q is simply

$$Q = \omega_0 R_I (C_1 + C_2) = \sqrt{\frac{R_I}{R_{\text{in}}} - 1} + \sqrt{\frac{R_I}{R_S} - 1}, \tag{35}$$

from which the image resistance may be found. Then:

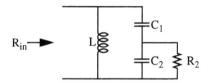

FIGURE 3.10. Tapped capacitor resonator
as a matching network.

$$C_1 + C_2 = \frac{Q}{\omega_0 R_I}, \tag{36}$$

$$L_1 = \frac{Q_{\text{left}} R_{\text{in}}}{\omega_0}, \tag{37}$$

$$L_2 = \frac{Q_{\text{right}} R_S}{\omega_0}. \tag{38}$$

The T-match is particularly useful when the source and termination parasitics are primarily inductive in nature, allowing them to be absorbed into the network.

3.5.5 TAPPED CAPACITOR RESONATOR AS AN IMPEDANCE MATCHING NETWORK

Tapped resonators share with the π- and T-match the ability to set center frequency, Q, and transformation ratio. As a consequence, they are also common idioms in RF design.

An example of such a circuit is the tapped capacitor tank, frequently used in oscillators since it combines a resonator with impedance transformation and so allows one to couple energy out of the tank without degrading Q excessively. See Figure 3.10.

The operation of this impedance transformer is best understood as simply the consequence of the capacitive voltage divider. A voltage reduction in a perfectly lossless network must be accompanied by an impedance reduction proportional to the square of the voltage attenuation if power is to be conserved. This network is not perfectly lossless, but we do expect the impedance transformation ratio to be roughly

$$\frac{R_2}{R_{\text{in}}} \approx \left(\frac{1/sC_2}{1/sC_1 + 1/sC_2} \right)^2 = \left(\frac{C_1}{C_1 + C_2} \right)^2, \tag{39}$$

so that the network transforms a resistance R_{in} downward to a value R_2, or a resistance R_2 upward to a value R_{in}.

To confirm this expectation, let us analyze the resistively loaded capacitive divider in isolation. The admittance of the combination is readily found after a little labor:

$$Y_{\text{in}} = \frac{j\omega C_1 - \omega^2 R_2 C_1 C_2}{j\omega R_2 (C_1 + C_2) + 1}. \tag{40}$$

The real part is

$$G_{\text{in}} = \frac{\omega^2 R_2 C_1^2}{\omega^2 R_2^2 (C_1 + C_2)^2 + 1}. \tag{41}$$

At sufficiently high frequencies, the equivalent shunt conductance simplifies to

$$G_{\text{in}} \approx \frac{\omega^2 R_2 C_1^2}{\omega^2 R_2^2 (C_1 + C_2)^2} = G_2 \cdot \left[\frac{C_1}{C_1 + C_2}\right]^2 = \frac{G_2}{n^2}, \tag{42}$$

as anticipated. Equation 42 also defines a factor, n, which is the turns ratio of an ideal transformer that would yield the same resistance transformation as the capacitive divider. The concept of an equivalent turns ratio will prove particularly useful in unifying the treatment of various oscillators.

For the sake of completeness, we also compute the imaginary part of the admittance:

$$B_{\text{in}} = \frac{\omega C_1 + \omega^3 R_2^2 C_1 C_2 (C_1 + C_2)}{\omega^2 R_2^2 (C_1 + C_2)^2 + 1}, \tag{43}$$

which, at sufficiently high frequencies, approaches a limiting value of

$$B_{\text{in}} \approx \omega \cdot \frac{C_1 C_2}{C_1 + C_2} = \omega \cdot C_{\text{eq}}. \tag{44}$$

Not surprisingly, the resistively loaded tapped capacitor network presents a susceptance equal to that of a series combination of the two capacitances.

The foregoing series of equations serves well for analysis and particularly to develop design intuition. Equations 42 and 44 are also extremely useful for first-cut, back-of-the-envelope designs. However, to carry out a detailed design requires a bit more labor. To derive a series of equations more suitable for design, we now apply parallel–series transformations until we massage the network into something we already know. In broad outline, what we'll do is equivalent to the following procedure. Transform the parallel $R_2 C_2$ combination into its series counterpart. Then combine C_1 with the transformed capacitance to yield a single series RC in parallel with the inductor. Next, transform that series RC into its parallel equivalent. The parallel R thus found is set equal to R_{in}.

Carrying out that strategy, we first determine the required network Q:

$$Q \approx \frac{\omega_0}{\omega_{-3\,\text{dB}}}, \tag{45}$$

where we interpret the bandwidth as the frequency span over which the impedance transformation ratio is to remain roughly constant. We then note that one expression for the network Q is

$$Q = \frac{R_{\text{in}}}{\omega_0 L}, \tag{46}$$

so that

$$L = \frac{R_{in}}{\omega_0 Q}. \tag{47}$$

Next, transform the parallel *RC* section into its series equivalent. The series resistor has a value given by

$$R_{2S} = \frac{R_2}{Q_2^2 + 1} \tag{48}$$

and

$$C_{2S} = C_2 \left[\frac{Q_2^2 + 1}{Q_2^2} \right], \tag{49}$$

where Q_2 is the Q of the parallel *RC* section. We then recognize that the series resistor may also be viewed as the result of transforming R_{in}:

$$R_S = \frac{R_{in}}{Q^2 + 1}. \tag{50}$$

Equating the two expressions for R_S and solving for Q_2 yields

$$Q_2 = \sqrt{\frac{R_2}{R_{in}} (Q^2 + 1) - 1}. \tag{51}$$

The original parallel *RC* has a Q of

$$Q_2 = \omega_0 R_2 C_2, \tag{52}$$

so we may write

$$C_2 = \frac{Q_2}{\omega_0 R_2} = \frac{\sqrt{\frac{R_2}{R_{in}} (Q^2 + 1) - 1}}{\omega_0 R_2}. \tag{53}$$

The only remaining undetermined element is C_1. To derive an equation for its value, first express the series combination of C_1 with C_{2S} as a single capacitance:

$$C_{eq} = \frac{C_1 C_{2S}}{C_1 + C_{2S}}. \tag{54}$$

The network Q then can be expressed as

$$Q = \frac{1}{\omega_0 R_{2S} C_{eq}} = \frac{C_1 + C_{2S}}{\omega_0 R_{2S} C_1 C_{2S}}. \tag{55}$$

Solving for C_1 yields one possible formula for the last unknown capacitor:

$$C_1 = \frac{C_2 (Q_2^2 + 1)}{Q Q_2 - Q_2^2}. \tag{56}$$

The derivations are a little cumbersome, but the ideas behind them are very simple. Once you are well versed in this parallel–series transformation business, figuring out

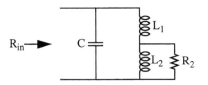

FIGURE 3.11. Tapped inductor resonator.

how tank circuits transform impedances is *conceptually* straightforward; it's just the execution that may be a bit involved.

3.5.6 TAPPED INDUCTOR MATCH

For the sake of completeness, we now consider briefly the tapped *inductor* resonator as a matching network (see Figure 3.11). As you might expect, its behavior is quite similar to that of its tapped capacitor counterpart.

We won't go through a detailed derivation of the design equations since they're essentially the same as for the tapped capacitor case, but we can immediately make the following observation: R_2 must be less than R_{in} because, once again, we have a voltage divider.

As in Section 3.5.5, we may proceed with the following derivations. First, determine the network Q. Then, since

$$Q = \omega_0 R_{in} C, \tag{57}$$

we have

$$C = \frac{Q}{\omega_0 R_{in}}. \tag{58}$$

Next, transform the parallel *RL* section into its series equivalent. The series resistor has a value given by

$$R_S = \frac{R_2}{Q_2^2 + 1}, \tag{59}$$

while the inductor transforms to a value

$$L_{2S} = L_2 \left[\frac{Q_2^2}{Q_2^2 + 1} \right]. \tag{60}$$

We may also consider R_S to be the result of transforming R_{in}:

$$R_S = \frac{R_{in}}{Q^2 + 1}. \tag{61}$$

Equate the two expressions for R_S and solve for Q_2:

$$Q_2 = \sqrt{\frac{R_2}{R_{in}}(Q^2 + 1) - 1}, \tag{62}$$

which is the same expression as for the tapped capacitor network.

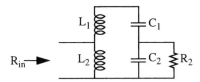

FIGURE 3.12. Double-tapped resonant match.

Having found Q_2, we may write

$$Q_2 = \frac{R_2}{\omega_0 L_2},$$

(63)

so that

$$L_2 = \frac{R_2}{\omega_0 Q_2} = \frac{R_2}{\omega_0 \sqrt{\dfrac{R_2}{R_{in}}(Q^2+1)-1}}.$$

(64)

To solve for the last unknown, note that we may express Q as

$$Q = \frac{\omega_0[L_1 + L_{2S}]}{R_{2S}}.$$

(65)

Solving for L_1 ultimately yields

$$L_1 = L_2 \frac{[QQ_2 - Q_2^2]}{Q_2^2 + 1},$$

(66)

thus completing the design.

3.5.7 DOUBLE-TAPPED RESONATOR

In certain instances, impractical or inconvenient component values may be needed in the various transformation networks considered so far. Quite often, this problem appears in the form of excessive required capacitance. Obtaining additional degrees of freedom to mitigate this problem requires the use of networks with additional elements.

One such network is the double-tapped resonator, displayed in Figure 3.12. This circuit first boosts R_2 to a larger effective parallel resistance across the whole tank than in a standard tapped capacitor network; it then reduces this parallel resistance by the tapped inductors to the desired value R_{in}. This technique therefore increases the required inductance and simultaneously reduces the required capacitance, potentially bringing both closer to comfortably realizable values.

The double-tapped matching network has four components and so provides four degrees of freedom. We may now specify center frequency, impedance transformation ratio, Q (bandwidth), and, say, total inductance (or capacitance).

Naturally, derivation of a design procedure is left as an exercise for the reader. As you might have surmised, focusing on Q invariance and series–parallel transformations is the key to the derivations.

As a final note on this particular matching network, it should be mentioned that – although it solves a thorny practical problem in principle – the finite Q of physically realizable inductors (and, to a lesser extent, of capacitors) imposes a bound on the amount of improvement that one can obtain in practice.

All of the examples considered so far are narrowband matching networks. Following the prescriptions given provides a perfect match at the design frequency, but the mismatch at other frequencies is not controlled; you get what you get. If a good match over a broader frequency range is necessary, one may use a number of networks that approximate a perfect match to arbitrary accuracy over an arbitrarily large bandwidth.[8] The underlying principles are similar to those in filter theory, where approximating functions such as Chebyshev polynomials enable the synthesis of filters that approach ideal magnitude characteristics to any degree desired. The broader the bandwidth, and the better the approximation, the harder we have to work and the more cumbersome the design. Consequently, engineers have devised numerous simplified procedures that nonetheless quite often yield acceptable results. One popular one is as follows. First note the uncorrected impedance (or admittance) at three frequencies, corresponding to the minimum and maximum frequencies as well as the center. Add a lossless network to transform the conductances at the extremes of the band to a value that equals the reciprocal of the conductance at the band center. This maneuver seeks to distribute the error in some reasonably equitable fashion, assuring that the conductance is too high at the edges (for example) and too low at the center. Then finish the design by adding whatever reactance will make the imaginary parts equal at the band extremes. Usually (but not always), this simple procedure will result in a match that stays within reasonable bounds over the specified bandwidth. In cases where somewhat better performance is required, a third (or additional) degree of freedom may provide it.[9]

3.6 EXAMPLES

Let's work out a few examples to make sure that the design procedures are clear. We'll consider L-match, π-match, tapped capacitor, and tapped inductor networks in trying to meet the following specifications: center frequency of 1 GHz, $R_1 = 50\ \Omega$, $R_2 = 5\ \Omega$, and a bandwidth of 25 MHz. Keep in mind that the bandwidth depends on whether the network is single- or doubly-terminated. In the examples that follow, we implicitly assume single terminations.

---

[8] See e.g. the classic work by G. Matthaei, L. Young, and E. M. T. Jones, *Microwave Filters, Impedance-Matching Networks and Coupling Structures,* McGraw-Hill, New York, 1965.

[9] See Matthaei et al., ibid. Also see "Impedance Matching Techniques for Mixers and Detectors," Agilent Applications Note 963.

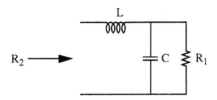

FIGURE 3.13. L-match attempt.

3.6.1 L-MATCH ATTEMPT

More parameters than can generally be set by the L-match are specified in this design. Nevertheless, we'll proceed with an attempt as follows. The network is sketched in Figure 3.13.

(1) The transformation ratio fixes the value of Q as

$$Q = \sqrt{\frac{R_1}{R_2} - 1} = 3. \tag{67}$$

Here, our estimate of bandwidth based on Q would be rather crude, but certainly sufficient to reveal our inability to meet the 25-MHz bandwidth specification. We need a Q of around 40, so we're an order of magnitude away from meeting the requirement.

(2) We've computed Q from the impedance ratio; let's equate it to the Q computed with the inductor impedance to enable a computation of the required inductance:

$$Q = \frac{\omega_0 L}{R_2} \implies L = \frac{QR_2}{\omega_0} \approx 2.39 \text{ nH.} \tag{68}$$

(3) Now let's use yet another, but still equivalent, expression for Q to find the required capacitance value:

$$Q = \omega_0 R_1 C \implies C = \frac{Q}{\omega_0 R_1} = 9.55 \text{ pF.} \tag{69}$$

These values are quite practically realizable in IC form (much more on this subject later), but the design has excessive bandwidth. And that's about all there is for the L-match.

3.6.2 π-MATCH

We follow the procedure outlined in Section 3.5.3.

(1) The required Q is still 40.

(2) A Q of 40 is large enough to calculate the image resistance with the approximate formula

$$R_I \approx \frac{\left(\sqrt{R_{\text{in}}} + \sqrt{R_P}\right)^2}{Q^2} \approx 0.054 \ \Omega. \tag{70}$$

(3) The capacitors are calculated as follows:

$$C_1 = \frac{Q_{\text{left}}}{\omega_0 R_{\text{in}}} \approx 305 \text{ pF}, \tag{71}$$

$$C_2 = \frac{Q_{\text{right}}}{\omega_0 R_P} \approx 96.8 \text{ pF}. \tag{72}$$

(4) Finally, the required inductance is

$$L = \frac{QR_I}{\omega_0} \approx 0.344 \text{ nH}. \tag{73}$$

In any IC technology, these values are highly inconvenient. In particular, the large total capacitance consumes excessive die area. Additionally, the inductance is low enough that it may be difficult to realize accurately and with low loss. Although we've been able to meet the design requirements in principle, we still don't have a practical design if the network is to be completely integrated.

3.6.3 TAPPED CAPACITOR MATCH

Here we proceed as in Section 3.5.5.
(1) The required Q is still 40.
(2) First, we compute L:

$$L = \frac{R_{\text{in}}}{\omega_0 Q} \approx 0.199 \text{ nH}. \tag{74}$$

(3) Next, we find the bottom capacitor, C_2:

$$C_2 = \frac{Q_2}{\omega_0 R_2} = \frac{\sqrt{\dfrac{R_2}{R_{\text{in}}}(Q^2 + 1) - 1}}{\omega_0 R_2} \approx 401 \text{ pF}. \tag{75}$$

(4) We finish by finding the other capacitor's value:

$$C_1 = \frac{C_2(Q_2^2 + 1)}{QQ_2 - Q_2^2} \approx 186 \text{ pF}. \tag{76}$$

As you can see, the tapped capacitor resonator has unfortunately yielded a no less impractical set of values.

3.6.4 TAPPED INDUCTOR MATCH

If we carry out a procedure analogous to our design exercise of Section 3.6.3, we obtain $C \approx 127$ pF, $L_1 \approx 136$ pH, and $L_2 \approx 63$ pH. Although the total capacitance has decreased substantially to a value that would be considered only somewhat onerous, the individual inductances are probably impractically small. Again, even if they could be accurately realized (say, by a suitably short piece of metallization), the typical lossiness of on-chip interconnect would most likely prevent attaining a Q of 40.

There are a few remedies that can be applied to improve the situation. Recognize that the problem fundamentally stems from the high Q-value sought; it is probably easiest to consider reflecting R_2 across the capacitor so that it becomes R_{in}. High Q for this parallel tank implies a small characteristic network impedance (small L/C ratio) relative to R_2, and a small Z_0/R_2 ratio in turn implies small L and large C.

If, for a given transformation ratio, we could boost the effective parallel resistance, then we could employ a larger L/C ratio to achieve the same Q and therefore increase the required inductance to practical values. Hence we need an upward impedance transformer within the transformer. Use of a double-tapped resonator is therefore one possible solution to this difficulty. Another possibility is to cascade one or more additional impedance transformers. The reader is once again invited to pursue these options independently.

PROBLEM SET FOR PASSIVE RLC NETWORKS

PROBLEM 1 An important theorem states that the load impedance should equal the complex conjugate of the source impedance in order to maximize power transfer. This sounds simple enough, but engineers sometimes draw incorrect inferences about how to satisfy this condition. Specifically, consider two approaches to matching a purely resistive 75-Ω load to a source whose impedance at some frequency happens to be $50 + j10$. Downwardly mobile engineer A, after reading about the maximum power transfer theorem, dutifully adds a 25-Ω resistor and a capacitor of $-j10\ \Omega$ in series with the source, whereas engineer B offers a similar solution but replaces the 25-Ω resistor with an appropriately designed L-match.

Quantitatively compare the two approaches by computing explicitly the ratio of powers delivered to the 75-Ω load by these two solutions, assuming equal Thévenin source voltages. Qualitatively explain the reason for the difference, and explain why engineer A can look forward to a long career at SubOptimal Products, Inc.

PROBLEM 2 In addition to maximizing power transfer, impedance-transforming networks are widely used simply to enable a specific amount of power to be delivered to a load. An important example is found in the output stage of a transmitter where, owing to supply voltage limitations, a downward impedance transformation of the antenna resistance is necessary.

A common load impedance is 50 Ω. Suppose we wish to deliver 1 W of power into such a load at 1 GHz, but the power amplifier has a maximum peak-to-peak sinusoidal voltage swing of only 6.33 V because of various losses and transistor breakdown problems. Design the following matching networks to allow that 1 W to be delivered. Use low-pass versions in all cases, and assume that all reactive elements are ideal (if only that were true ...).

(a) L-match.
(b) π-match ($Q = 10$).
(c) T-match ($Q = 10$).
(d) Tapped capacitor ($Q = 10$).
(e) If the maximum allowable on-chip capacitance is 200 pF and the maximum allowable on-chip inductance is 20 nH, are any of your designs amenable to a fully integrated implementation? If so, which one(s)?

PROBLEM 3 As we'll see, resonant circuits are also indispensable for allowing amplifiers to function at high frequencies. To investigate this property in a crude way, consider a standard common-source amplifier whose transistor is modeled with a hybrid-π model that neglects C_{gd} and r_o but includes r_g, C_{gs}, and C_{db} (the junction capacitance from the drain to the substrate, here at source potential).

(a) Assume a source resistance equal to r_g, and call the load resistance R_L. Derive an expression for the voltage gain of the amplifier. At what frequency ω_1 does the magnitude of the voltage gain go to *unity*?
(b) Now assume that we add inductance in series with the source to maximize the voltage developed across C_{gs} at frequency ω_1. What is an expression for this inductance, L_S? What is the new voltage gain at ω_1?
(c) In addition to the input series inductance, suppose we place an inductance from the drain to ground to resonate out the drain capacitance at ω_1. What is an expression for this inductance, L_{out}? Now what is the voltage gain at ω_1?

PROBLEM 4 An ideal dipole receiving antenna that is much shorter than a wavelength may be modeled as a voltage-driven series *RC* network, where the portion of *R* that is due to radiation is given approximately by

$$R_{\text{rad}} \approx 395(l/\lambda)^2, \tag{P3.1}$$

where l is the length of the antenna and λ is the wavelength. This equation works reasonably well up to l/λ ratios of about $1/4$.

(a) First assume that the only resistance in the antenna model is this radiation resistance. Further assume that the value of the equivalent voltage generator is simply the received E-field strength (which you may assume is fixed) multiplied by the antenna length l. Given an "optimum" passive impedance matching network interposed between the antenna and some fixed resistive load R_L, what is the maximum power transfer efficiency, defined as the ratio of power delivered to R_L to the total dissipated in the system, as a function of (normalized) antenna length? Answering this question does not require a numerical value for R_L.
(b) Now suppose that the antenna has some additional loss that is represented by a resistance R_d in series with the radiation resistance term. Does your answer change? How?

PROBLEM 5 Design an L-match to match a 10-Ω source to a 75-Ω load. Assume that the center frequency is 150 MHz.

PROBLEM 6 It was mentioned earlier that an L-match has only two degrees of freedom. Hence, once center frequency and impedance transformation ratio are chosen, the Q (and therefore the bandwidth) are fixed. The π-match adds one degree of freedom, allowing independent choice of all three parameters. Re-do Problem 5 with the additional constraint that the total bandwidth of the match be 15 MHz.

PROBLEM 7 Disappointed by lackluster sales of their low-bandwidth–high-offset op-amps, FromageTech (now a proud subsidiary of SubOptimal Products, Inc.) has decided to hop onto the wireless bandwagon. Their savvy market research department (a guy named Earl) concludes that they should enter the AM radio market.

One of their circuit design problems involves a common-source amplifier with a tuned load that is resonant at the traditional (though suboptimal, of course) intermediate frequency (IF) of 455 kHz. For reasons that are never made clear to you, the drain load resistance at resonance must be five kilohms.[10] At the same time, the output of this stage must ultimately drive a 5-Ω load.

(a) Assume infinite output resistance for the transistor and initially assume infinite Q for all reactive elements. Further assume that the transistor has zero drain–gate capacitance. Devise an L-match to satisfy the design requirements. Make reasonable approximations.
(b) What is the total -3-dB bandwidth of this circuit?
(c) Now suppose that the inductor found in part (a) actually possessed a Q of 100 at the desired center frequency. Find the equivalent series resistance of the inductor and then redesign the matching network, given the assumption that the inductor's series resistance remains constant.
(d) What is the new bandwidth?
(e) What if the transistor were made in a terrible process technology and actually had a 5-pF C_{gd}? Would this affect the performance of this stage very much? Assume that the transistor's gate is driven from essentially zero impedance and that r_g is zero.

PROBLEM 8 The L-match provides only two degrees of freedom. Hence, once the impedance transformation ratio and resonant frequency have been chosen, Q is determined automatically; you get whatever Q you get. Usually, the Q is fairly low, leading to a reasonably broadband match. Although this property is frequently desirable, it is often better to have independent control over all three parameters.

(a) Design a π-match to satisfy the design requirements of Problem 7. Initially assume the same ideal conditions and a desired total -3-dB bandwidth of 10 kHz.

---

[10] Not kilo-ohms. Similarly, megohms, not mega-ohms.

(b) Inevitably, the inductor found in part (a) again has a Q of 100. First find the equivalent series resistance of the inductor; then redesign the matching network to accommodate this additional loss mechanism. Are all the values positive real?

PROBLEM 9 Yet another matching network is the tapped reactance network. It shares with the π-network the ability to set independently the transformation ratio, Q, and center frequency. Repeat Problem 8(a), now using a tapped capacitor matching network. To keep the design procedure as uncomplicated as possible, you may assume large Q from the outset and use the approximate formulas that apply in that regime.

PROBLEM 10 As stated in Problem 4, the impedance of a dipole antenna that is much shorter than a wavelength may be modeled as a voltage-driven series *RC* network, where the portion of R that is due to radiation is given approximately by

$$R_{\text{rad}} \approx 395(l/\lambda)^2, \tag{P3.2}$$

where l is the length of the antenna and λ is the wavelength. Suppose that the antenna capacitance is 15 pF, and assume a l/λ ratio of 0.1.

(a) What value of inductance is required to resonate out the antenna capacitance at a frequency of 30 MHz? What inductor Q is required if its effective series resistance is not to exceed 10% of the antenna resistance?

(b) Now suppose you want to connect this antenna to a receiver whose input impedance is 50 Ω. Design the appropriate L-match. You may unrealistically assume infinite inductor Q. What is the bandwidth, given these assumptions?

PROBLEM 11 Design a π-network to match a source impedance of $5 - j30$ Ω to a 50-Ω resistive load. If the Q of the network is 100, what is the current in each element of the matching network when 1 W is delivered to the load?

PROBLEM 12 Consider the network shown in Figure 3.14. Using the method of series–parallel transformations, simplify the network to find the impedance at 100 MHz when $C = 1$ pF, $L = 10$ nH, $R_S = 15$ Ω, and $R_P = 1$ kΩ.

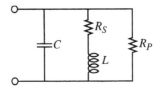

FIGURE 3.14. Lossy *RLC* network.

PROBLEM 13 In the circuit shown in Figure 3.15, calculate the Q and sketch the magnitude and phase response for $R = 1$, 10, and 100 Ω. Also sketch the step response for each of these three cases.

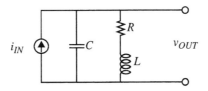

FIGURE 3.15. *RLC* network.

PROBLEM 14 The electrical model for a typical quartz crystal is a series *RLC* circuit in parallel with a shunt capacitor, C_0. Although C_0 does model a physical capacitance, the *RLC* components model electro*mechanical* properties and therefore may take on values not practically realizable with ordinary inductors and capacitors.

A typical 10-MHz crystal has a series resistance of 30 Ω, a Q of 100,000, and a C_0 of 2 pF.

(a) Neglecting C_0, what values for series capacitance and inductance are implied?
(b) By what percentage does the resonant frequency change when C_0 is included, assuming that the main terminals of the crystal are open-circuited? What if the main terminals are shunted by an additional 10 pF of parasitic capacitance?
(c) Plot the impedance (both magnitude and phase) of the crystal model (without the 10-pF parasitic capacitance, but including C_0) from 1 MHz to 100 MHz, as well as a "close-up" in the range of 9 MHz to 11 MHz.

PROBLEM 15

(a) Using the energy definition for Q, show that an alternate formula for Q for a one-port is:

$$Q = \frac{|\text{Im}[Z]|}{\text{Re}[Z]}, \tag{P3.3}$$

where Z is the impedance of the network.
(b) Derive the analogous formula in terms of admittances.
(c) Consider, if you have not already done so, the deficiencies of the definition in part (a). In particular, does the definition make sense for (say) a series *RLC* network? Near resonance? Explain.

PROBLEM 16 This problem considers the issue of matching to a *nonlinear* load.

As seen in numerous examples (particularly in the historical chapters) but not discussed in much detail, a very common AM demodulator is the *envelope detector*. This circuit is evidently nonlinear because of the presence of the diode; see Figure 3.16.

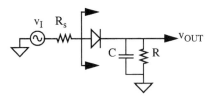

FIGURE 3.16. Envelope detector.

The problem is to determine the effective resistance presented by the envelope detector (i.e., the circuitry to the right of the indicated boundary). This problem is somewhat difficult to solve exactly, but useful approximations can be obtained by invoking several simplifying assumptions. Because different assumptions lead to slightly different answers, please use only the assumptions given, even though some of them may seem dubious. You'll just have to trust that, by some miracle of numerology, the errors introduced cancel so that the final answer is roughly correct.

(a) Let the input voltage v_{in} be $A \cos \omega t$, and initially assume that R_S is zero. Assume further that the voltage dropped across the diode is negligibly small whenever the diode is forward-biased. Note that this assumption is contrary to the "0.6-V" rule that we commonly apply in large-signal analyses. In effect, we are assuming that R is sufficiently large that the current through the diode is quite small. In that case, little voltage is dropped across the diode. Finally, assume that the RC product is much larger than the period of the sinusoidal drive.

With these assumptions, sketch approximately the output voltage v_{out} and the diode current waveforms in steady state. At this point, it is not necessary to identify features quantitatively; approximate shapes suffice. Figure 3.17 shows the input waveform in hashed form for convenience.

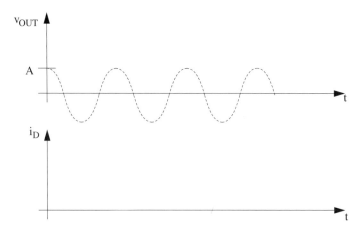

FIGURE 3.17. Approximate output voltage and diode current for envelope detector.

(b) Derive an expression for when the diode turns on (with our reference convention, the peak input occurs at $t = 0$). To simplify the derivation, feel free to assume that $e^{-x} \approx 1 - x$ for small x, and also that $\cos x \approx 1 - x^2/2$ for small x. It's sufficient to tell us when, in the *first period* after $t = 0$, the diode turns on. It may also help to remember that $\cos x = \cos(-x)$.

(c) Assume, somewhat erroneously, that the diode turns off the moment the input reaches its peak. As measured entirely by the capacitor voltage, what is the total energy (in joules) supplied by the input source per cycle?

(d) Invoking energy equivalence is one way to define an effective resistance for a nonlinear load. What we mean is this: A sinusoidal voltage source directly loaded by a resistance R_L delivers a certain amount of energy to that resistive load per cycle. What equivalent value of R_L connected directly to the source would consume the same amount of energy per cycle as calculated in part (c)? Express your answer in terms of R.

Thus, this is the value one should use in designing matching networks to drive an envelope detector, if maximizing power gain is important. As a last trivia note, this consideration is of greatest significance when designing "zero-power" receivers (e.g., crystal radios), since the only source of energy is the incoming wave itself, and maximizing power transfer is therefore critically important.

CHAPTER FOUR

CHARACTERISTICS OF PASSIVE IC COMPONENTS

4.1 INTRODUCTION

We've seen that RF circuits generally have many passive components. Successful design therefore depends critically on a detailed understanding of their characteristics. Since mainstream integrated circuit (IC) processes have evolved largely to satisfy the demands of digital electronics, the RF IC designer has been left with a limited palette of passive devices. For example, inductors larger than about 10 nH consume significant die area and have relatively poor Q (typically below 10) and low self-resonant frequency. Capacitors with high Q and low temperature coefficient are available, but tolerances are relatively loose (e.g., order of 20% or worse). Additionally, the most area-efficient capacitors also tend to have high loss and poor voltage coefficients. Resistors with low self-capacitance and temperature coefficient are hard to come by, and one must also occasionally contend with high voltage coefficients, loose tolerances, and a limited range of values.

In this chapter, we examine IC resistors, capacitors, and inductors (including bondwires, since they are often the best inductors available). Also, given the ubiquity of interconnect, we study its properties in detail since its parasitics at high frequencies can be quite important.

4.2 INTERCONNECT AT RADIO FREQUENCIES: SKIN EFFECT

At low frequencies, the properties of interconnect we care about most are resistivity, current-handling ability, and perhaps capacitance. As frequency increases, we find that inductance might become important. Furthermore, we invariably discover that the resistance increases owing to a phenomenon known as the *skin effect*.

Skin effect is usually described as the tendency of current to flow primarily on the surface (skin) of a conductor as frequency increases. Because the inner regions of the conductor are thus less effective at carrying current than at low frequencies, the useful

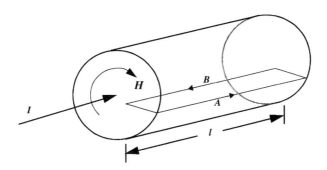

FIGURE 4.1. Illustration of skin effect with isolated
cylindrical conductor.

cross-sectional area of a conductor is reduced, thereby producing a corresponding
increase in resistance.

From that perfunctory and somewhat mysterious description, there is a risk of leaving the impression that all surfaces of a conductor will carry current equally well. To
develop a deeper understanding of the phenomenon, we need to appreciate explicitly the role of the magnetic field in producing the skin effect. To do so qualitatively,
let's consider a solid cylindrical conductor carrying a time-varying current, as shown
in Figure 4.1. Assume for now that the return current (there must always be one in
any real system) is far enough away that its influence may be neglected. A time-
varying current I generates a time-varying magnetic field H. That time-varying field
induces a voltage around the rectangular path shown, in accordance with Faraday's
law. Ohm's law then tells us that the induced voltage in turn produces a current flow
along that same rectangular path, as indicated by the arrows. Now here's the key observation: The direction of the induced current along path A is *opposite* that along
B. The induced current thus adds to the current flowing along one side of the rectangle and subtracts from the other. Taking care to keep track of algebraic signs, we
see that the current along the surface is the one that is augmented whereas the current below the surface is diminished. In other words, current flow is strongest near
the surface; that's the skin effect.

To develop this idea a little more quantitatively, let's apply Kirchhoff's voltage
law (with proper accounting for the induced voltage term, both in magnitude and
sign) around the rectangular path to obtain

$$J_B \rho l - J_A \rho l + \frac{d\phi}{dt} = 0, \tag{1}$$

where J is the current density, ρ is the resistivity, and ϕ, the flux, is perpendicular to
the rectangle shown.

We see that, as deduced earlier, the current density along path A is indeed larger
than along B by an amount that increases as either the depth, frequency, or magnetic
field strength increases and also as the resistivity decreases. Any of these mechanisms
acts to exacerbate the skin effect. Furthermore, the presence of the derivative tells us

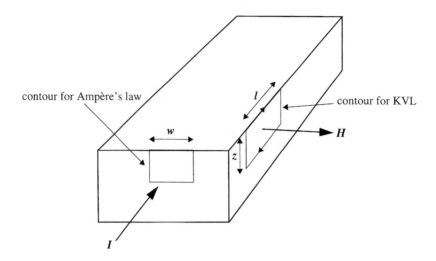

FIGURE 4.2. Subsection of semi-infinite conductive block.

that the current undergoes more than a simple decrease with increasing depth; there is a phase shift as well.

If we now increase the radius of curvature to infinity, we may convert the cylinder into the rectangular structure that is more commonly analyzed to introduce skin effect; see Figure 4.2. We will provide only the barest outline of how to set up the problem, and then simply present the solution.[1]

Computing the voltage induced by H around the rectangular contour proceeds with Kirchhoff's voltage law (KVL) as before:

$$J\rho l - J_0\rho l = \frac{d\phi}{dt} = -\frac{d}{dt}\int_0^z Bl\,dz, \qquad (2)$$

where the subscript 0 denotes the value at the surface of the conducting block.

Now express J and H (and thus B) explicitly as sinusoidally time-varying quantities. For example, let

$$J_0 = J_{s0}e^{j\omega t}, \qquad (3)$$

where the subscript s is chosen arbitrarily to denote the magnitude of these skin-effect variables.

With these substitutions, the KVL equation allows us to write

$$\rho\frac{dJ_s}{dz} = -j\omega B_s = -j\omega\mu H_s. \qquad (4)$$

Here we have used the relation between flux density B and magnetic field strength H,

---

[1] For a detailed derivation, consult any number of excellent texts on electromagnetic theory. See e.g. S. Ramo, T. van Duzer, and J. R. Whinnery, *Fields and Waves in Communications Electronics,* 3rd ed., Wiley, New York, 1994. Also see U. S. Inan and A. S. Inan, *Electromagnetic Waves,* Prentice-Hall, Englewood Cliffs, NJ, 2000.

$$B = \mu H, \tag{5}$$

where μ is the permeability (equal to the free-space value in nearly all integrated circuits).

We need one more equation to finish setting up the differential equation. Ampère's law will give it to us:[2]

$$I_{\text{encl}} = w \int_0^z J \, dz = w H_0 - w H. \tag{6}$$

Making the same substitutions as before now yields

$$\frac{dH_s}{dz} = J_s. \tag{7}$$

Combining Eqn. 4 and Eqn. 7 yields a simple second-order differential equation for the current density:

$$\frac{d^2 J_s}{dz^2} = \frac{j\omega\mu}{\rho} J_s, \tag{8}$$

whose solution is

$$J_s = J_{s0} \exp\left(-\frac{z}{\delta}\right) \exp\left(\frac{-jz}{\delta}\right), \tag{9}$$

and where

$$\delta = \sqrt{\frac{2\rho}{\omega\mu}} = \sqrt{\frac{2}{\omega\mu\sigma}} \tag{10}$$

is known as the skin depth. Notice that the current density decays exponentially from its surface value. Notice also (from the second exponential factor) that there is indeed a phase shift, as argued earlier, with a 1-rad lag at a depth equal to δ.

For this case of an infinitely wide, infinitely long, and infinitely deep conductive block, the skin depth is the distance below the surface at which the current density has dropped by a factor of e. For copper at 1 GHz, the skin depth is approximately 2 μm. For aluminum, that number increases a little bit, to about 2.5 μm. What this exponential decay implies is that making a conductor much thicker than a skin depth provides negligible resistance reduction because the added material carries very little current. Furthermore, we may compute the effective resistance as that of a conductor of thickness δ in which the current density is uniform. This fact is often used to simplify computation of the AC resistance of conductors. To make sure that the result is valid, however, the boundary conditions must match those used in deriving our system of equations: The return currents must be infinitely far away, and the conductor must resemble a semi-infinite block. The latter criterion is satisfied reasonably well if all radii of curvature, and all thicknesses, are at least 3–4 skin depths.

---

[2] Remember what Ampère's law says in words: The integral of the magnetic field around a closed path equals the total current enclosed by that path.

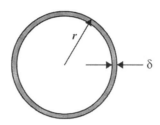

FIGURE 4.3. Application of skin depth concept
to resistance calculation (cross-section shown).

As a specific example, let us estimate the AC resistance of an isolated wire. Assume that the wire's diameter is much larger than the skin depth. In that case, we may estimate the resistance by pretending that all of the current flows in an annulus of depth δ, as seen in Figure 4.3. The resistance is readily computed as

$$R = \frac{\rho l}{A} \approx \frac{\rho l}{2\pi r \delta}, \qquad (11)$$

where l is the length of the wire and we have assumed that the radius $r \gg d$ in constructing the last approximation.

In this case, calculations based on a simple skin-depth assumption yield excellent results. However, in other cases, the results may be grossly in error. In evaluating the risk of making such a mistake, it's helpful to anticipate qualitatively where the currents will be flowing. To do so, recall that – upon introducing skin depth – we invoked a qualitative argument that led us to several important insights. One of these is that skin effect is strongest where the magnetic fields are strongest. So, in determining which surfaces are likely to carry most of the current, we need to identify where the fields are the strongest.

Consider a coaxial system of conductors, as in a cable. There are three surfaces, but not all three exhibit skin effect. The outer cylindrical conductor conveys the return current for the central conductor. The coaxial structure is self-shielding in that both electrical and magnetic fields external to the cable are ideally zero, thanks to cancelling contributions by the two conductors (this attribute is why the coaxial structure is valued in the first place). The magnetic field is therefore the strongest in the space between them, and thus the skin effect is felt most acutely at the surface of the inner conductor and at the inner surface of the outer conductor. In Figure 4.4, the regions of high current density are indicated in black.

The outer surface of the outer conductor carries very little current (again, we're assuming conductor thicknesses that are very much larger than the skin depth). Therefore, computation of its resistance would consider only the black annulus at the inner surface of the outer conductor. *Not all skin exhibits the skin effect.*

To reinforce that last statement, consider another qualitative example. Specifically, consider what happens when we have two parallel cylindrical conductors in proximity, where the current in one serves as the return current for the other. As in

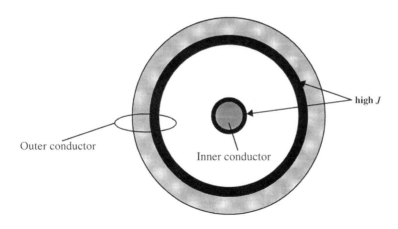

FIGURE 4.4. Coaxial cable cross-section.

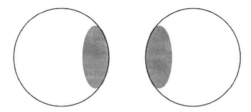

FIGURE 4.5. Cross-section of two-wire line.

the coaxial case, the magnetic field is strongest in the region between the two conductors. Therefore the maximum current densities are found where the wire surfaces face each other. See Figure 4.5. For self-evident reasons, the phenomenon of current crowding at a surface because of current flowing in a nearby conductor is sometimes known as the *proximity effect*.

As one last example, consider a thin, wide conductor (again, with all other conductors very far away). The currents will distribute themselves roughly as shown in Figure 4.6. The current crowds toward the two ends (as a mnemonic aid, imagine this structure to be the central slice of a cylindrical conductor; the ends of this slab correspond to the current-carrying outer surface of the cylinder). Because of the relatively low current density along the long edges, further widening of the conductor produces only modest resistance reductions. Thickening the conductor would have a much stronger effect. The current distribution changes if we bring a conductor near this one.

Computing the effective resistance for these last two structures is clearly not as straightforward as for the isolated conductor. Indeed, accurately computing the effective resistance of such a simple-seeming structure as a single-layer coil is virtually impossible because of all of the interactions among turns. This difficulty highlights the danger of automatically assuming that all surfaces are equally effective at carrying current.

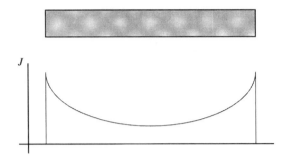

FIGURE 4.6. Thin, wide conductor cross-section
and current density (approximate).

Finally, for a collection of useful skin-effect related formulas, see Harold A.
Wheeler, "Formulas for the Skin Effect" (*Proc. IRE,* v. 30, September 1942, pp.
412–24).

4.3 RESISTORS

There are relatively few good resistor options in standard CMOS (complementary
metal-oxide silicon) processes. One possibility is to use polysilicon ("poly") interconnect material, since it is more resistive than metal. However, most poly these days
is silicided specifically to reduce resistance. Resistivities tend to be in the vicinity
of roughly 5–10 ohms per square (within a factor of about 2–4, usually), so poly is
appropriate mainly for moderately small-valued resistors. Its tolerance is often poor
(e.g., 35%), and the temperature coefficient, defined as

$$\mathrm{TC} \equiv \frac{1}{R}\frac{\partial R}{\partial T}, \tag{12}$$

depends on doping and composition and is typically in the neighborhood of 1000
ppm/°C. Unsilicided poly has a higher resistivity (by approximately an order of
magnitude, depending on doping), and the TC can vary widely (even to zero, in certain cases) as a function of processing details. It is usually not tightly controlled, so
unsilicided poly, if available as an option at all, frequently possesses very loose tolerances (e.g., 50%). Advanced bipolar technologies use self-aligned poly emitters,
so poly resistors are an option there, too.

In addition to their moderate TC, poly resistors have a reasonably low parasitic capacitance per unit area and the lowest voltage coefficient of all the resistor materials
available in a standard CMOS technology.

Resistors made from source–drain diffusions are also an option. The resistivities
and temperature coefficients are generally similar (within a factor of 2, typically) to
those of silicided polysilicon, with lower TC associated with heavier doping. There
is also significant parasitic (junction) capacitance as well as a noticeable voltage coefficient. The former limits the useful frequency range of the resistor, while the latter
limits the dynamic range of voltages that may be applied without introducing significant distortion. Additionally, care must be taken to avoid forward-biasing either

end of the resistor. These characteristics usually limit the use of diffused resistors to noncritical circuits.

In modern VLSI (very large-scale integration) technologies, source–drain "diffusions" are defined by ion implantation. The source–drain regions formed in this way are quite shallow (usually no deeper than about 200–300 nm, scaling roughly with channel length), quite heavily doped, and almost universally silicided, leading to moderately low temperature coefficients (order of 500–1000 ppm/°C).

Wells may be used for high-value resistors, since resistivities are typically in the range of 1–10 kΩ per square. Unfortunately, the parasitic capacitance is substantial because of the large-area junction formed between the well and the substrate; the resulting resistor has poor initial tolerance (±50–80%), large temperature coefficient (typically about 3000–5000 ppm/°C, owing to the light doping), and large voltage coefficient. Well resistors must therefore be used with care.

Sometimes, a MOS transistor is used as a resistor, even a variable one. With a suitable gate-to-source voltage, a compact resistor can be formed. From first-order theory, recall that the incremental resistance of a long-channel MOS transistor in the triode region is

$$r_{ds} \approx \left[\mu C_{\text{ox}} \frac{W}{L} [(V_{GS} - V_T) - V_{DS}] \right]^{-1}. \tag{13}$$

Unfortunately, implicit in this equation is that a MOS resistor has loose tolerance (because it depends on the mobility and threshold), high temperature coefficient (because of mobility and threshold variation with temperature) and is quite nonlinear (because it depends on V_{DS}). These characteristics frequently limit its use to noncritical circuits outside of the signal path. An exception is use of such a resistor in certain gain control applications in which the gate drive is derived from a feedback loop so that variations in device characteristics are automatically compensated.

One other option that is occasionally useful, particularly to prevent thermal runaway in bipolar power stages with paralleled devices, is to use metal interconnect as a small resistor. In most interconnect technologies, metal resistivities are usually on the order of 50 mΩ/square, so resistances up to around 10 Ω are practical.

Aluminum is most commonly used in interconnect and has a temperature coefficient of about 3900 ppm/°C. The TC varies little with temperature and the resistance may be considered PTAT (proportional to absolute temperature) over the military temperature range (−55°C to 125°C) to a reasonable approximation:

$$R(T) \approx R_0 \frac{T}{T_0}, \tag{14}$$

where one data point, the resistance R_0 at temperature T_0, is known.

Some processes offer one or more layers of interconnect made of some silicide (mainly for its superior electromigration properties). The resistivity is about an order of magnitude larger than that of pure aluminum or copper, while the TC is about the same.

A few companies that specialize in analog circuits have modified their processes to provide excellent resistors, such as those made of NiCr (nichrome) or SiCr (sichrome).

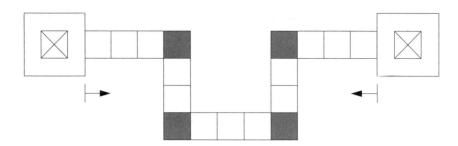

FIGURE 4.7. Example for counting squares.

These resistors possess low TC (order of 100 ppm/°C or less), and thin-film versions are easily trimmed with a laser to absolute accuracies better than a percent. Unfortunately, these processes are not universally available, and the additional process steps increase die cost significantly.

COUNTING SQUARES

For practical reasons, resistors are rarely laid out as one straight, skinny rectangle. Rather, shapes that use the allotted die area most effectively typically have bends in them, raising the question of how one counts squares; see Figure 4.7.

By inspection, it should be clear that a corner piece (shown shaded) doesn't quite count as a full square. A rough approximation treats the shaded regions as equivalent to about half a square. A more rigorous analysis reveals that a corner is in fact equivalent to 0.56 squares.[3] Thus, the resistance between the boundaries marked by the arrows is approximately 15.24 squares.

In general, the resistance of the contact cells (the "dumbbells" at the ends) must also be taken into account. Although the actual value depends on the details of the contact cell layout, the resistance is typically on the order of half a square.

4.4 CAPACITORS

All of the interconnect layers may be used to make traditional parallel plate capacitors (see Figure 4.8). However, ordinary interlevel dielectric tends to be rather thick (order of 0.5–1 μm), precisely to reduce the capacitance between layers, so the capacitance per unit area is small (a typical value is 5×10^{-5} pF/μm^2). Additionally, one must be aware of the capacitance formed by the bottom plate and any conductors (especially the substrate) beneath it. This parasitic bottom plate capacitance is frequently as large as 10–30% (or more) of the main capacitance and often severely limits circuit performance.

---

[3] See e.g. Richard C. Jaeger, *Introduction to Microelectronic Fabrication,* 2nd ed., Prentice-Hall, Englewood Cliffs, NJ, 2001.

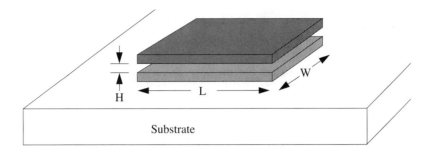

FIGURE 4.8. Parallel plate capacitor.

The standard capacitance formula,

$$C \approx \varepsilon \frac{A}{H} = \varepsilon \frac{W \cdot L}{H}, \tag{15}$$

somewhat underestimates the capacitance because it does not take fringing into account, but it is accurate as long as the plate dimensions are much larger than the plate separation H. In cases where this inequality is not well satisfied, a rough first-order correction for the fringing may be provided by adding between H and $2H$ to each of W and L in computing the area of the plates. Choosing the maximum yields

$$C \approx \varepsilon \frac{(W + 2H) \cdot (L + 2H)}{H} \approx \varepsilon \left[\frac{WL}{H} + 2W + 2L \right]. \tag{16}$$

One of the few bits of good news in IC passive components is that the TC of metal–metal capacitors is quite low, usually in the range of approximately 30–50 ppm/°C, and is dominated by the TC of the oxide's dielectric constant itself, as dimensional variations with temperature are negligible.[4]

The total capacitance per unit area can be increased by using more than one pair of interconnect layers. As of this writing, some process technologies offer five metal layers, so that a quadrupling of the capacitance is possible through the use of a sandwich structure. One can increase the capacitance even further by exploiting *lateral* flux between adjacent metal lines within a given interconnect layer. Allowable adjacent metal spacings have shrunk to values smaller than the spacing between layers, so this lateral coupling is substantial.

A simple structure that illustrates the general idea is shown in Figure 4.9, where the two terminals of the capacitor are distinguished by different shadings. As can be seen, the "top" and "bottom" plates, constructed out of the same metal layer, alternate to exploit the lateral flux. Ordinary vertical flux may also be exploited by arranging the segments of a different metal layer in a complementary pattern, as shown in Figure 4.10.

[4] J. L. McCreary, "Matching Properties, and Voltage and Temperature Dependence of MOS Capacitors," *IEEE J. Solid-State Circuits,* v. 16, no. 6, December 1981, pp. 608–16.

FIGURE 4.9. Example of lateral flux capacitor (top view).

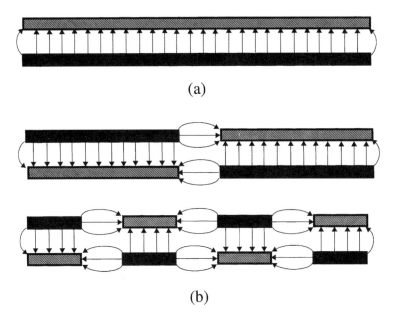

FIGURE 4.10. Evolution of lateral flux capacitor (side views).

An important attribute of a lateral flux capacitor is that the parasitic bottom plate capacitance is much smaller than for an ordinary parallel plate structure, since it consumes less area for a given value of total capacitance. In addition, adjacent plates help steal flux away from the substrate, further reducing bottom plate parasitic capacitance, as seen in Figure 4.11.

Since lateral capacitance depends on the total perimeter, the maximum capacitance is obtained with layout geometries that maximize the amount of perimeter. A

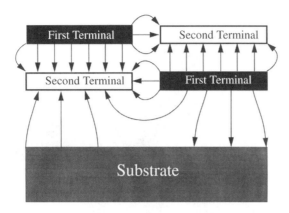

FIGURE 4.11. Illustration of flux stealing.

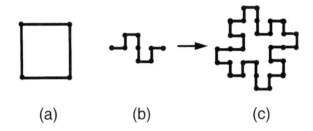

(a)   (b)   (c)

FIGURE 4.12. Construction of a Koch curve.

particularly rich source of useful geometries may be found in the literature on fractals, since they are structures that may enclose a finite area with an infinite perimeter.[5] Because of photolithographic limitations, infinite perimeter is unattainable, but large increases are possible. In certain cases, capacitance increases of a factor of 10 or more can be achieved.

Some ideal fractals have finite area but infinite perimeter. The concept is perhaps best illustrated with an example. *Koch islands* are a family of fractals first introduced as a crude model for the shape of a coastline. The construction of a Koch curve begins with an *initiator,* as shown in the example of Figure 4.12. Here, the initiator is simply a square, so that $M = 4$ sides. The construction continues by replacing each segment of the initiator with a curve called a *generator*; the example shown in the figure has $N = 8$ segments. The size of each segment of the generator is $r = 1/4$ that of each segment of the initiator. By recursively replacing each segment of the resulting curve with the generator, a fractal border is formed. The first step of this process is depicted in Figure 4.12(c). The total area occupied remains constant at each stage in this case because of the particular shape of the generator.

---

[5] A. Shahani et al., "A 12mW, Wide Dynamic Range CMOS Front-End Circuit for a Portable GPS Receiver," *ISSCC Digest of Technical Papers,* Slide Supplement, February 1997.

FIGURE 4.13. Fractal based on Koch islands.

A more complicated Koch island can be seen in Figure 4.13 (a Koch island with $M = 4$, $N = 32$ and $r = 1/8$). The associated initiator of this fractal has four sides, and its generator has 32 segments. It can be noted that the curve is self-similar – that is, each section of it looks like the entire fractal. As we zoom in, more detail becomes visible; this self-similarity at different scales is the essence of a fractal.

The final shape of a fractal can be tailored over wide limits by appropriate choices of initiator and generator. It is also possible to use different generators during successive steps. In practice, though, the palette of options is constrained by the desirability of rectangular shapes for realizing capacitors. Figure 4.14 shows an alternative fractal-based capacitor that does have the attribute of filling a rectangular space more conveniently and fully. The capacitor shown uses only one metal layer with a fractal border, but it may be augmented with complementary layers above it (additional metal layers permitting). For a better visualization of the overall picture, the electrical terminals of this square-shaped capacitor have been identified using two different shadings. Again, additional metal layers may be cross-connected to improve capacitance density further.

In addition to the capacitance density, the quality factor Q is important in RF applications. Here, the degradation in quality factor is minimal because the fractal structure automatically limits the length of the thin metal sections to a few microns,

FIGURE 4.14. Fractal capacitor based on Minkowski sausage.

keeping the series resistance reasonably small. For applications that require low series resistance, lower-dimension fractals may be used. Fractals thus add one more degree of freedom to the design of capacitors, allowing the capacitance density to be traded for a lower series resistance.

In current IC technologies, there is usually tighter control over the lateral spacing of metal layers compared to the vertical thickness of the oxide layers, from wafer to wafer and across the same wafer. Lateral flux capacitors shift the burden of matching away from oxide thickness to lithography. Therefore, using lateral flux can improve matching characteristics. Furthermore, the pseudorandom nature of the structure can compensate, to some extent, for the effects of nonuniformity in the etching process. Still, to achieve accurate ratio matching, multiple copies of a unit cell should be used; this is standard practice in high-precision analog circuit design.

Comparing fractal and conventional interdigitated capacitors, we note that a disadvantage of the latter is its inherent parasitic inductance. Most fractal geometries randomize the direction of the current flow and thus reduce the effective series inductance, whereas for interdigitated capacitors the current flow is in the same direction for all the parallel stubs. In addition, fractals usually have lots of sharp edges whose electric field concentration helps boost capacitance a small amount (generally of the order of 15%). Furthermore, interdigitated structures are more vulnerable to nonuniformity of the etching process. Nevertheless, the simplicity of the interdigitated capacitor favors its use in many situations.

The woven structure shown in Figure 4.15 may also be used to achieve high capacitance density. The vertical lines are in metal 2 and horizontal lines are in metal 1. The two terminals of the capacitor are identified using different shades. Compared to an interdigitated capacitor, a woven structure has much less inherent series inductance,

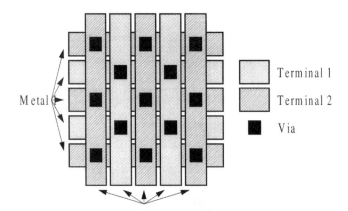

FIGURE 4.15. Woven capacitor.

thanks to current flow in different directions. The resulting higher self-resonant frequency is a welcome attribute of this structure. In addition, the series resistance contributed by vias is smaller than that of an interdigitated capacitor owing to the greater ease of cross-connecting the metal layers. However, the capacitance density of a woven structure is smaller than an interdigitated capacitor with the same metal pitch, because the capacitance contributed by the vertical fields is smaller.

A final way to make full use of the many metal layers available in modern CMOS process technologies is to construct conductive *vertical* plates out of vias in combination with the interconnect metal. This vertical parallel plate (VPP) structure exploits the scaling trends of lateral spacing even more fully than the fractal structures presented.[6]

A standard alternative is to use a MOS capacitor, available in CMOS processes as simply the gate capacitance of an ordinary transistor. Even in some bipolar processes, a MOS capacitor is still available as a special option, where the bottom plate is typically formed by an emitter diffusion, a thinned oxide is the dielectric, and metal or polysilicon is the top plate. Capacitance per unit area depends on the dielectric thickness, but is typically in the range of 1–5 $fF/\mu m^2$, or roughly 20–100 times larger than ordinary interconnect capacitors. Generally, a MOS capacitor will have a small, positive temperature coefficient (on the order of 30 ppm/°C), but only if the semiconductor is doped to degeneracy. Gate capacitors therefore exhibit somewhat larger TC values.

When using a gate capacitor in a CMOS process, it is important to keep the transistor in strong inversion (i.e., keep the gate–source voltage well above the threshold); otherwise, the capacitance will be small, lossy, and highly nonlinear. Occasionally, an accumulation-mode MOSFET, constructed by using n+ source–drain diffusions in an n-well (see Figure 4.16), is made available for use as a capacitor to alleviate

---

[6] R. Aparicio and A. Hajimiri, "Capacity Limits and Matching Properties of Lateral Flux Integrated Capacitors," *Custom Integrated Circuits Digest of Technical Papers,* 2001.

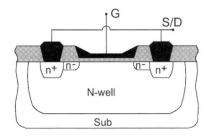

FIGURE 4.16. Accumulation-mode varactor.

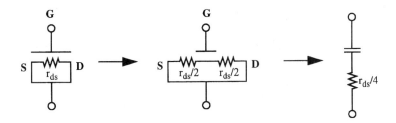

FIGURE 4.17. Evolution of crude gate capacitor model.

this problem. This type of capacitor requires nothing above what is already available in a standard CMOS process, but its characteristics may not be controlled (or even tracked) by the fab.

For both types of gate capacitor, the Q depends on the channel resistance, as defined by Eqn. 13. To develop a crude, first-order model for such a capacitor, consider Figure 4.17. This approximation grossly overestimates the effective series resistance because, in the actual structure, portions of the gate capacitance near the source and drain connect to those terminals through less resistance than do portions near the center of the channel.[7] The model identifies the maximum resistance – that from the center of the channel to the source–drain connection – and puts that worst-case value in series with all of the capacitance. Nevertheless, the model correctly predicts that, to maximize Q, one should use the minimum allowable device length (to minimize r_{ds} for a given bias). Furthermore, one must also exercise care in connecting to the device in order to minimize additional resistive losses.

Another option is to use a junction capacitance, such as that formed by a p+ region in an n-well. Since a junction capacitance depends on the applied bias, such a capacitor is often exploited to make electronically tuned circuits; diodes used this way are known as *varactors* (from variable reactor). Recall that the junction capacitance depends on bias as follows:

---

[7] In fact, it may be shown that a better estimate of the equivalent resistance is actually one third the value shown in Figure 4.17.

$$C_j \approx \frac{C_{j0}}{(1 - V_F/\phi)^n}, \tag{17}$$

where C_{j0} is the incremental capacitance at zero bias, V_F is the amount of forward bias applied across the junction, ϕ is the built-in potential (typically several tenths of a volt), and n is a parameter that depends on the doping profile. For an abruptly doped junction, n has a value of $1/2$, whereas $n = 1/3$ for linearly graded junctions.[8] As you can imagine, junction capacitors are often a poor choice if varactor action is undesired.

The capacitance equation holds well for reverse and weak forward biases, but overestimates the true capacitance for increasing amounts of forward bias. A crude approximation is that the junction capacitance in forward bias is 2–3 times the zero-bias value. Additionally, the parameters ϕ and n should be treated mainly as curve-fitting parameters – they may not always take on physically reasonable values.

Junction capacitors also have relatively large temperature coefficients, varying from about 200 ppm/°C at large reverse bias to perhaps 1000 ppm/°C at zero bias. It is possible to show[9] that the temperature coefficient of a junction capacitance may be expressed approximately as

$$\text{TC} \approx (1 - n)\text{TC}_{\text{Si}} - n\left[\frac{1}{1 - V_F/\phi}\right]\text{TC}_\phi, \tag{18}$$

where n, ϕ, and V_F are as in Eqn. 17, TC_{Si} (the TC of ε_{Si}) is about 250 ppm/°C, and TC_ϕ (the TC of the built-in junction potential) is doping-dependent and typically in the range of about -1000 to -1500 ppm/°C.

Unfortunately, the Q of IC varactors varies in an inverse manner with the tuning range. The asymmetrical doping required to provide a large change in capacitance per unit voltage also guarantees a relatively large series resistance. The largest series resistance occurs when the depletion region is the narrowest and the capacitance is therefore the largest. That is, the Q is lowest when the capacitance is highest. As a consequence, IC varactors must be used with care (e.g., they should not comprise all of a tank's capacitance).

INTERCONNECT CAPACITANCE

At radio frequencies, it becomes especially crucial to obtain accurate values for the parasitic capacitance of interconnect. The parallel plate formula from undergraduate physics often grossly underestimates the capacitance because one dimension is often not much larger than the distance to the next conductor layer. The fringing capacitance is therefore significant, and the simple formula is often unacceptably inaccurate.

---

[8] Values of n in excess of unity may be obtained by "hyperabrupt" junctions but require special processing not available in standard IC technology.

[9] Using information from McCreary (see footnote 4).

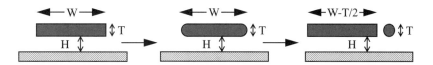

FIGURE 4.18. Yuan's decomposition of parallel plate capacitance into area and fringe components.

We will consider three configurations of conductors: Case 1 is a single wire over a conducting plane of infinite extent; Case 2 is a single wire between two infinite planes; and Case 3 is a wire between two adjacent wires all over a single infinite plane. Those who are uninterested in the somewhat tedious derivations may skip to the summary of the formulas. Finally, keep in mind that, in all cases considered here, a uniform dielectric is assumed. The presence of passivation layers and/or plastic packaging typically increases the capacitance by amounts ranging from about 10% for the topmost layers of interconnect to only 1–2% for the innermost layers.

Case 1: Single Conductor over Ground Plane

The case of a single, isolated wire over a conducting plane is perhaps the easiest one to consider first. One formula for this case, offered by Yuan,[10] has an intuitive appeal because it is physically motivated: it explicitly decomposes the capacitance into area and fringe terms by modeling progressively as shown in Figure 4.18 (we assume that the wire is infinitely long along the direction perpendicular to the plane of this page).

As seen in the figure, the basic idea is to obtain the total capacitance as the sum of two parts. One component is the familiar area (fringeless) term proportional to W/H, while the other (fringe) contribution involves the capacitance due to a wire of diameter T, diminished by the term proportional to $T/2H$. Yuan's formula for the capacitance per unit length is thus

$$C_{\text{Yuan}} \approx \varepsilon \left[\frac{W}{H} + \frac{2\pi}{\ln\{1 + (2H/T)(1 + \sqrt{1 + T/H})\}} - \frac{T}{2H} \right]. \qquad (19)$$

Yuan's approach works well as long as the ratio W/H is not too small, with typical errors in the range of 5% or so.[11] Below W/H values of about 2–3, however, the error grows quite rapidly. Unfortunately, that often can be the regime of interest, particularly when future process technologies are considered. Also, the subexpression for the fringe capacitance is still a bit cumbersome.

[10] C. P. Yuan and T. N. Trick, "A Simple Formula for the Estimation of the Capacitance of Two-Dimensional Interconnects in VLSI Circuits," *IEEE Electron Device Lett.*, v. 3, 1982, pp. 391–3.

[11] E. Barke, "Line-to-Ground Capacitance Calculation for VLSI: A Comparison," *IEEE Trans. Computer-Aided Design*, v. 7, no. 2, February 1988, pp. 295–8.

Another strategy is to abandon any physically motivated approach and directly apply function-fitting techniques to the results of two-dimensional (2D) field-solver simulations. One formula resulting from such an exercise was developed by Sakurai:[12]

$$C_{\text{Sakurai}} \approx \varepsilon \left[\frac{W}{H} + \frac{0.15W}{H} + 2.8 \left(\frac{T}{H} \right)^{0.222} \right], \tag{20}$$

where the area contribution (first term) and fringe contribution (other two terms) are shown separately, as before.

As with Yuan's formula, Sakurai's equation has increasingly poor accuracy at large W/H ratios, but accuracy superior to Yuan's at small W/H, with the accuracy of the equations crossing over in the neighborhood of $W/H = 2$–3, at least for the particular values of conductor and dielectric thicknesses considered in Barke's paper ($T = 1.3\ \mu$m, $H = 0.75\ \mu$m).

Better accuracy can be obtained with a formula that is only marginally more complex (although possibly an increment more computationally efficient) than Sakurai's. Such an equation, developed by v.d. Meijs and Fokkema (hereafter referred to as MF) through function fitting, is:[13]

$$C_{\text{MF}} \approx \varepsilon \left[\frac{W}{H} + 0.77 + 1.06 \left[\left(\frac{W}{H} \right)^{0.25} + \left(\frac{T}{H} \right)^{0.5} \right] \right]. \tag{21}$$

Barke claims that the MF formula typically yields accuracies better than 1% for dimensions appropriate to ICs. The simplicity and accuracy of the MF formula are very attractive (requiring only the availability of a square-root extractor in addition to the usual arithmetic functions). Additionally, as we'll see, the MF formula is a nice basic equation from which formulas for other cases may evolve.

At this point, let's look at the results of some numerical calculations (Table 4.1). For the sake of developing some intuitive feel for the magnitude of fringing, the area term (i.e., the one given by the simple formula from undergraduate physics) is also shown (all dimensions are in μm, all capacitances in fF/μm). For the smaller W/H, all three methods yield essentially equivalent results. For the larger W/H, though, the values exhibit a greater spread. Trusting Barke and taking the MF value as the most accurate, we see that Yuan's formula is perhaps about 10% in error, although this is probably more than good enough for most hand calculations.

It is important to note also that the area component is only a small fraction of the total capacitance when W and H are comparable. For example, at the lower W/H ratio, the area term accounts for less than a fourth of the total, whereas for the higher W/H ratio, the area term is still only about half of the total. These observations have some interesting implications for the dependence on line width of total RC product

---

[12] T. Sakurai and K. Tamaru, "Simple Formulas for Two- and Three-Dimensional Capacitances," *IEEE Trans. Electron Devices*, v. 30, no. 2, February 1983, pp. 183–5.

[13] N. v.d. Meijs and J. T. Fokkema, "VLSI Circuit Reconstruction from Mask Topology," *Integration*, v. 2, no. 2, 1984, pp. 85–119.

Table 4.1. *Capacitance of single wire over single conducting plane*

Method	Capacitance for $W = 1.36$, $H = 1.65$, $T = 0.8$	Capacitance for $W = 2.38$, $H = 0.87$, $T = 0.3$
MF	0.115	0.190
Sakurai	0.115	0.185
Yuan	0.114	0.172
Area term	0.028	0.094

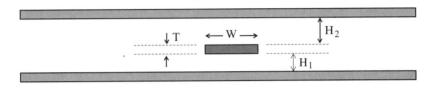

FIGURE 4.19. Conductor arrangement for Case 2.

per length: in the regime where fringing is significant, it is possible to obtain substantial reductions in total wire RC by increasing width, since fringe capacitance grows only slowly with width while R decreases in direct proportion.

Case 2: Wire Sandwiched between Two Conducting Planes

The capacitive load of a single wire between two conducting planes (Figure 4.19) can be calculated using the formula for Case 1 as a starting point, although not necessarily in the way one might think. If one does the obvious and sums the capacitance of the wire to each plane separately, it turns out that the result is a gross overestimate because the fringing term is pessimistically computed.

An important insight is that the addition of a second conducting plane results mainly in a *redistribution* of the fringing field without greatly affecting its magnitude. This observation then suggests that one should sum only the area terms, and then add a *weighted average* of the fringe terms that we would compute for each plane separately.

Ideally, we'd like an expression for the total fringe term that gives us a properly weighted average of the two fringe contributions when the two are comparable, but also converges to just one of these terms in the limit as the spacing to one plate approaches infinity. One relatively simple type of general "averaging" function that has more or less the right kind of behavior (as n approaches infinity) is

$$f(x_1, x_2) = \left[\frac{x_1^n + x_2^n}{2} \right]^{1/n}. \tag{22}$$

Note that a value of $n = 1$ corresponds to the conventional formula for computing an average.

Table 4.2. *Capacitance of single wire between two conducting planes*

Method	Capacitance
Sakurai sum of capitance	0.468 fF/μm
Yuan sum of capitance	0.434 fF/μm
Yuan (with maximum fringe)	0.343 fF/μm
MF (with maximum fringe)	0.375 fF/μm
MF (with weighted fringe)	0.370 fF/μm
2D field-solver value	0.361 fF/μm

As an ad hoc fix, suppose we compute the fringe terms separately, apply such an averaging function, and add that result to the sum of the area terms. With a somewhat arbitrary choice of $n = 4$ (so that the exponents are reasonable), the capacitance of a wire between two planes is

$$
C \approx \varepsilon \left[W \left(\frac{1}{H_1} + \frac{1}{H_2} \right) + 0.77 \right.
$$
$$
\left. + 0.891 \left\{ \left(\frac{W}{H_1} + \frac{W}{H_2} \right)^{0.25} + \left[\left(\frac{T}{H_1} \right)^2 + \left(\frac{T}{H_2} \right)^2 \right]^{0.25} \right\} \right], \quad (23)
$$

which may be rewritten as:

$$
C \approx \varepsilon \left[W \left(\frac{1}{H_1} + \frac{1}{H_2} \right) + 0.77 \right.
$$
$$
\left. + 0.891 \left\{ \left[W \left(\frac{1}{H_1} + \frac{1}{H_2} \right) \right]^{0.25} + T^{0.5} \left(\frac{1}{H_1^2} + \frac{1}{H_2^2} \right)^{0.25} \right\} \right]. \quad (24)
$$

Now let's compare the results obtained with this formula to those calculated by some other methods. Shown in Table 4.2 are values for $W = 2.38\,\mu$m, $H_1 = 0.87\,\mu$m, $H_2 = 0.48\,\mu$m, and $T = 0.3\,\mu$m.

From the table, we can see that the simple sum-of-caps method overestimates the true capacitance by as much as 30%. Yuan's formula fares considerably better for this particular geometry: it yields a value that is about 5% lower than the 2D value if we arbitrarily add only the *larger* fringe term to the sum of area terms. The MF equation (with weighted fringe terms) gives us a value that is only about 2.5% higher than the value computed by a field solver, and it has the advantage that one needn't keep track of which fringe term is larger.

Case 3: Three Adjacent Wires over a Single Plane

For the general case of multiple wires sandwiched between two planes (Figure 4.20), we'd like expressions for the capacitance between adjacent pairs of wires as well as the total capacitance seen by each wire. Unfortunately, simple formulas for adjacent pair capacitances don't seem to exist yet. The best we can offer is another formula of

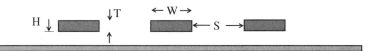

FIGURE 4.20. Conductor arrangement for Case 3.

Table 4.3. *Three adjacent*
lines over ground plane

Method	Total capacitive load on center wire
Sakurai	0.154 fF/μm
2D value	0.155 fF/μm

Sakurai, one that computes the total capacitance of the *middle* wire of three adjacent ones over a *single* conducting plane.

This formula expresses this total capacitance as the sum of two components. The first is simply the ordinary wire-over-ground plane term of Case 1. The second is supposedly the capacitance between the middle wire and its adjacent neighbors. So, it is claimed,

$$C_{\text{total}} = C_{\text{single}} + 2C_{\text{mutual}}, \tag{25}$$

where

$$C_{\text{single}} \approx \varepsilon \left[\frac{1.15W}{H} + 2.8 \left(\frac{T}{H} \right)^{0.222} \right] \tag{26}$$

and

$$C_{\text{mutual}} \approx \varepsilon \left[0.03 \frac{W}{H} + 0.83 \frac{T}{H} - 0.07 \left(\frac{T}{H} \right)^{0.222} \right] \left[\frac{S}{H} \right]^{-1.34}. \tag{27}$$

Note that the mutual capacitance term dies away just a bit faster than the first power of the spacing. Additionally, one may presumably use the MF formula for C_{single} if desired, although Sakurai almost certainly tweaked his coefficients for C_{mutual} so that C_{total} would be more or less correct.

If we let $W = 1.36$ μm, $H = 1.65$ μm, $T = 0.8$ μm, and $S = 1.19$ μm, Sakurai's formula predicts a total capacitive load on the center wire that matches closely the value given by "Maxwell" (a 2D field-solver program); see Table 4.3. This level of agreement is quite good, so it appears that Sakurai's formula is usefully accurate at least for computing the total loading on the center of three wires.

Where one can go awry is in the physical interpretation of the individual terms of his equation. Sakurai implies that the term C_{single} is in fact the capacitance of the middle line to ground when, in reality, the proximity of the two additional conductors alters that capacitance in major ways. As a consequence, the term he calls the coupling capacitance (C_{mutual}) is *not* the capacitance between adjacent pairs of conductors, as one would naturally assume (and as he seems to claim). It is only the *sum*

Table 4.4. *Comparison of individual capacitance terms for Case 3*

Method	Capacitance to ground of center wire (fF/μm)	Capacitance between center and outer wires (fF/μm)	Total capacitance of center wire (fF/μm)
Sakurai	0.115	0.039	0.154
2D value	0.056	0.100	0.156

of his two terms that happens to equal the total capacitive load on the center wire; the partitioning of this total into separate ground and mutual capacitances is another thing altogether.

As a specific example, let's consider the same conductor arrangement for which Table 4.3 was generated and compare calculations from Maxwell with those of Sakurai. This is shown in Table 4.4. As you can see, the sum of the Sakurai components equals that given by 2D simulations, but the individual terms don't agree at all.

4.4   INDUCTORS

From the point of view of RF circuits, the lack of a good inductor is by far the most conspicuous shortcoming of standard IC processes. Although active circuits can sometimes synthesize the equivalent of an inductor, they always have higher noise, distortion, and power consumption than "real" inductors made with some number of turns of wire.

4.5.1   SPIRAL INDUCTORS

The most widely used on-chip inductor is the planar spiral, which can assume many shapes; see Figure 4.21. The choice of shape is more often made on the basis of convenience (e.g., whether the layout tool accommodates non-Manhattan geometries) or habit than anything else. Despite stubborn lore to the contrary, the inductance and Q values attainable are very much second-order functions of shape, so engineers should feel free to use their favorite shape with relative impunity. Octagonal or circular spirals are moderately better than squares (typically on the order of 10%) and hence are favored when layout tools permit their use – or when that modest difference represents the margin between success and failure.

The most common realizations use the topmost metal layer for the main part of the inductor (occasionally with two or more levels strapped together to reduce resistance) and provide a connection to the center of the spiral with a crossunder implemented with some lower level of metal. These conventions arise from quite practical considerations: the topmost metal layers in an integrated circuit are usually the thickest and thus generally the lowest in resistance. Furthermore, maximizing the distance to the substrate minimizes the parasitic capacitance between the inductor and the substrate.

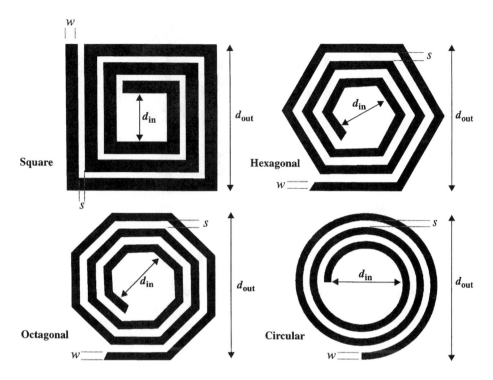

FIGURE 4.21. Planar spiral inductors.

The inductance of an arbitrary spiral is a complicated function of geometry, and accurate computations require the use of field solvers or Greenhouse's method.[14] However, a (very) crude zeroth-order estimate, suitable for quick hand calculations, is

$$L \approx \mu_0 n^2 r = 4\pi \times 10^{-7} n^2 r \approx 1.2 \times 10^{-6} n^2 r, \tag{28}$$

where L is in henries, n is the number of turns, and r is the radius of the spiral in meters. This equation typically yields numbers on the high side, but generally within 30% of the correct value (and often better than that).

For shapes other than square spirals, multiply the value given by the square spiral formula by the square root of the area ratio to obtain a crude estimate of the correct value (see Eqn. 50). Thus, for circular spirals, multiply the square-spiral value by $(\pi/4)^{0.5} \approx 0.89$, and by 0.91 for octagonal spirals.

Perhaps more useful for the approximate *design* of a square spiral inductor is the following equation:

$$n \approx \left[\frac{PL}{\mu_0} \right]^{1/3} \approx \left[\frac{PL}{1.2 \times 10^{-6}} \right]^{1/3}, \tag{29}$$

[14] H. M. Greenhouse, "Design of Planar Rectangular Microelectronic Inductors," *IEEE Trans. Parts, Hybrids, and Packaging,* v. 10, no. 2, June 1974, pp. 101–9. This classic paper describes a readily implemented algorithm for accurately computing the inductance.

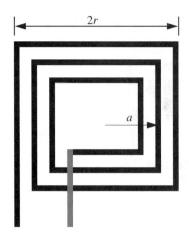

FIGURE 4.22. Hollow spiral inductor.

where P is the winding pitch in turns/meter; we have assumed that the permeability is that of free space.

To get an idea of how area-inefficient such a structure is, consider for example a 120-nH inductor (this amount of inductance is considered small in the context of discrete circuits) made with a spiral pitch P of one turn every five micrometers. Plugging these numbers into our design formula, we find that the number of turns needed is something like 27, corresponding to a required radius of around 140 μm. The area consumed by this inductor is equivalent to that of about eight typical bond pads; it is *huge.* Clearly, the number of such inductors must be kept to a minimum, and it actually may be more economically sensible to consider the use of external inductors in many instances. In general, practical on-chip inductances are in the neighborhood of 10 nH on down.

A great many analytical formulas for the inductance of planar spirals may be found in the literature. These formulas can yield reasonably accurate results for some suitably restricted subset of parameters. Regrettably, there is rarely an explicit statement of those restrictions, or of any error bounds. We offer here a collection of simple formulas that are nonetheless sufficiently accurate to obviate the need for field solvers in nearly all practical cases.

The first of these, which applies to a hollow square spiral inductor (see Figure 4.22), is[15]

$$L \approx \frac{9.375\mu_0 n^2 (d_{avg})^2}{11 d_{out} - 7 d_{avg}},   (30)$$

where d_{out} is the outer diameter and d_{avg} is the arithmetic mean of the inner and outer diameters. Checks with a field solver reveal that this modified Wheeler formula exhibits errors below 5% for typical IC inductors.

---

[15] See H. A. Wheeler, "Simple Inductance Formulas for Radio Coils," *Proc. IRE,* v. 16, no. 10, October 1928, pp. 1398–1400. The original formula applies to circular spirals and yields answers in microhenries with dimensions in inches.

Table 4.5. *Coefficients for current-sheet*
inductance formula

Shape	c_1	c_2	c_3	c_4
Square	1.27	2.07	0.18	0.13
Hexagon	1.09	2.23	0.00	0.17
Octagon	1.07	2.29	0.00	0.19
Circle	1.00	2.46	0.00	0.20

The inductance of planar spirals of all regular shapes can be cast in a simple unified form if we base a derivation on the properties of a uniform current sheet:

$$L \approx \frac{\mu_0 n^2 d_{\mathrm{avg}} c_1}{2} \left[\ln\left(\frac{c_2}{\rho}\right) + c_3 \rho + c_4 \rho^2 \right], \tag{31}$$

where ρ is the *fill factor*, defined as

$$\rho \equiv \frac{d_{\mathrm{out}} - d_{\mathrm{in}}}{d_{\mathrm{out}} + d_{\mathrm{in}}}. \tag{32}$$

From this last equation, you can see why the term "fill factor" is appropriate: ρ approaches unity as the inductor windings fill the entire space, and it approaches zero as the inductor becomes progressively hollower.

The various coefficients c_n are a function of geometry; they are given in Table 4.5 for four representative shapes.[16] The coefficient c_1 is simply the area for a given outer dimension, normalized to the area of the largest circle that can be inscribed within the layout. The factor c_2 is the primary term, while c_3 and c_4 may be considered first- and second-order correction factors, respectively. When all four factors are used, the equations are typically accurate to within a couple of percent (and almost never in error by more than 5%), thus generally obviating the need for a full electromagnetic field solver to evaluate the inductance of such structures. This low maximum error is in contrast with the peak errors in alternative formulas found in the literature. In some cases, those peak errors can exceed 50%.

Figure 4.23 compares the current sheet and modified Wheeler formulas (the *monomial* formula has a mathematical form that is particularly suited for use in optimizations; it is described in the cited reference). These curves are the result of simulating 19,000 inductors and then comparing those simulation results with the formulas. These inductors span 100 pH to 70 nH in value, d_{out} from 100 μm to 400 μm, n from 1 to 20, spacing-to-width ratios s/w from 0.02 to 3, and fill factors from 0.03 to 0.95. Practical inductors do not span anywhere near such a wide range.

---

[16] S. S. Mohan et al., "Simple, Accurate Inductance Formulas," *IEEE J. Solid-State Circuits,* October 1999, pp. 1419–24.

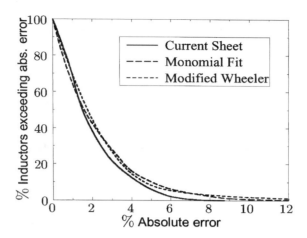

FIGURE 4.23. Comparison of inductor formulas.

On those rare occasions where other regular polygons are of interest, one may use the following analytical formula:

$$L \approx \frac{\mu n^2 d_{\mathrm{avg}} A_{\mathrm{out}}}{\pi d_{\mathrm{out}}^2} \left[\ln\left(\frac{2.46 - 1.56/N}{\rho} \right) + \left(0.20 - \frac{1.12}{N^2} \right) \rho^2 \right], \qquad (33)$$

where A_{out} is the area computed with the outer dimensions and N is the number of sides of the polygon. This formula is simply a restatement of Eqn. 31, with analytical approximations used for the coefficients c_1, c_2 and c_4. The coefficient c_3 is set to zero, which is a good approximation for all regular polygons with more than four sides. This analytical formula is only one or two percent more inaccurate than the tabulated one.

Aside from the large areas potentially consumed, another serious problem with spiral inductors is their relatively large loss. The DC resistive losses are exacerbated by the skin effect, which causes a nonuniform current distribution in a conductor at RF. The consequence is a reduction in the effective cross-section, increasing the series resistance.

In addition to the series resistive loss, capacitance to the substrate is another conspicuous problem of on-chip spirals. In silicon technology, the substrate is close by (typically no more than about 2–5 μm away) and fairly conductive, creating a parallel plate capacitor that resonates with the inductor. The resonant frequency of the LC combination represents the upper useful frequency limit of the inductor and is often so low that the inductor is useless. The proximity of the substrate also degrades Q because of the energy coupled into the lossy substrate.

An additional parasitic element is the shunt capacitance across the inductor that arises from the overlap of the cross-under with the rest of the spiral. The lateral capacitance from turn to turn usually has a negligible overall effect because it is the series connection of these capacitances that ultimately appears across the terminals of the inductor.

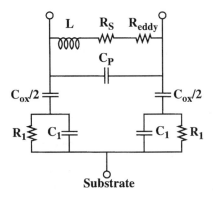

FIGURE 4.24. Model for on-chip spiral inductor.

Figure 4.24 shows a relatively complete model for on-chip spirals.[17] The model is symmetrical, even though actual spirals are not. Fortunately, the error introduced is negligible in most instances.

An estimate for the series resistance may be obtained from the following equation:

$$R_S \approx \frac{l}{w \cdot \sigma \cdot \delta(1 - e^{-t/\delta})}, \qquad (34)$$

where σ is the conductivity of the material, l is the total length of the winding, w and t are the width and thickness of the interconnect, and the skin depth δ is given by

$$\delta = \sqrt{\frac{2}{\omega\mu_0\sigma}}. \qquad (35)$$

Resistance computations based on this modified skin depth formula can yield optimistic estimates because the formula accounts only for skin loss associated with the surface of the conductor that faces the substrate. It thus neglects skin loss associated with the other surfaces (including the proximity effect). It also neglects substrate loss. That latter neglect is generally justified in those cases where the substrate is lightly doped. However, many CMOS processes employ quite heavily doped wafers (e.g., 10 mΩ-cm), and losses associated with currents induced in the substrate (*eddy currents*) may also become important (even dominant) in certain cases. In such technologies, one should therefore expect a purely skin-loss based estimate to be both crude and optimistic. Fortunately, the effect of induced substrate currents on the inductance value is small enough to be ignored (or can be made so) in nearly all cases of practical interest.

Although the analysis is quite complicated, and numerous approximations are cascaded to make the analysis tractable, the resulting crude formula for the eddy-current resistance is not too unwieldy and yields useful design insights:[18]

[17] P. Yue et al., "A Physical Model for Planar Spiral Inductors on Silicon," *IEDM Proceedings,* December 1996.
[18] S. S. Mohan, "Modeling, Design, and Optimization of On-Chip Inductors and Transformers," Ph.D. thesis, Stanford University, 1999.

$$R_{\text{eddy}} \approx \frac{\sigma_{\text{sub}}}{4e} (\mu n f)^2 d_{\text{avg}}^3 \rho^{0.7} z_{n,\text{ins}}^{-0.55} z_{n,\text{sub}}^{0.1}, \tag{36}$$

where σ_{sub} is the substrate conductivity, d_{avg} is the average of the inner and outer diameters, ρ remains the fill factor, and e is our old friend (2.7182818...). The quantity $z_{n,\text{ins}}$ is the total thickness of the insulation between the spiral proper and the heavily doped portion of the substrate, normalized to the average inductor diameter. That insulation is generally a combination of oxide and lightly doped semiconductor but is treated here as a uniform, magnetically transparent material. Similarly, $z_{n,\text{sub}}$ is the substrate skin depth, also normalized to the average inductor diameter. The total series resistance is the sum of R_S and R_{eddy}. The other inductor model elements remain unchanged.

It is worthwhile examining Eqn. 36 to extract some intuition from it. First, note the relatively weak dependence of loss on $z_{n,\text{ins}}$ and the near independence of the loss on $z_{n,\text{sub}}$. Note also that the resistance due to eddy current loss in the substrate is proportional to the square of frequency and to the square of the number of turns. Perhaps more important is its proportionality to the *cube* of the (average) diameter.[19] It is a natural tendency to use wide conductors to reduce skin and DC loss, but we see that – beyond a certain point – eddy current loss dominates and Q actually degrades rapidly with further increases in size. For heavily doped CMOS substrates, then, one must often use inductors with smaller outer diameters (and hence narrower conductors) than is common practice in technologies with semi-insulating substrates (e.g., GaAs). Failure to recognize the existence of this tradeoff has led to a great spread in reported results due to the great spread in inductor layout choices.

Generally, somewhat hollow inductors are best, because the innermost turns tend not to contribute much magnetic flux yet do contribute significant resistance. Hence, removing those turns is a good idea in general. While there is no simply identified universal optimum, a reasonable rule of thumb is to have a 3 : 1 ratio between the outer and inner diameters (corresponding to a fill factor of about 0.5). Fortunately, the optimum conditions are usually relatively flat, so the rule of thumb is satisfactory for most practical cases.

The shunt capacitance C_P is:

$$C_P = n \cdot w^2 \cdot \frac{\varepsilon_{\text{ox}}}{t_{\text{ox}}}, \tag{37}$$

where t_{ox} is the thickness of the oxide between the crossunder and the main spiral.

The capacitance between the spiral and the substrate proper is C_{ox} and is here approximated with a simple parallel plate formula, where the total area is that of the winding:

$$C_{\text{ox}} = w \cdot l \cdot \frac{\varepsilon_{\text{ox}}}{t_{\text{ox}}}. \tag{38}$$

---

[19] One may use Eqn. 36 to deduce that the eddy resistance then grows approximately as the square of the average diameter *for a fixed inductance*.

Resistance R_1 models the substrate's dielectric loss. This loss is simply that associated with current capacitively coupled into the substrate through C_{ox}. The value of R_1 is given by:

$$R_1 \approx \frac{2}{w \cdot l \cdot G_{sub}}, \tag{39}$$

where G_{sub} is a fitting parameter that has the dimensions of conductance per area. It is constant for a given substrate material and distance of the spiral to the substrate, and it has a typical value of about 10^{-7} S/μm^2.

In addition to contributing to loss, the image currents also flow in a direction opposite to those of the main inductor. Hence, the effect of the image is to cancel partially the inductance. That undesirable effect is associated with another: as the temperature increases, the substrate resistivity also increases, reducing the effectiveness of the cancellation. As a consequence, the inductance tends to increase somewhat with temperature. The temperature coefficient may be as high as $\sim$200 ppm/°C, improving as the spiral is moved farther away from the substrate.

The capacitance C_1 reflects the capacitance of the substrate as well as other reactive effects related to the image inductance; it is given by

$$C_1 \approx \frac{w \cdot l \cdot C_{sub}}{2}. \tag{40}$$

As with G_{sub}, C_{sub} is a fitting parameter that is constant for a given substrate and distance of the spiral to the substrate. A typical range for C_{sub} is between 10^{-3} and 10^{-2} fF/μm^2.

With the foregoing set of equations, it is possible to optimize the Q and self-resonant properties of a spiral inductor. One insight gained from optimization exercises is that the innermost turns may be removed to increase Q, since they contribute negligibly to the total flux while contributing measurably to the total loss. Even when this and other optimizations are performed, though, one almost invariably finds that the maximum Q is below 10 (and frequently below 5), so these spirals are unsuitable in many cases.

Some minor refinements can help squeeze a bit more Q out of a given technology. A *patterned ground shield*[20] (PGS; see Figure 4.25) prevents capacitive coupling to the lossy substrate. Slots in the shield avoid a short-circuiting of the magnetic flux by preventing the flow of induced current along the path indicated in the figure. Its presence thus greatly reduces the effect of the dielectric terms C_1 and R_1. At the same time, the shielding greatly reduces coupling of noise from the substrate to the inductor, and vice versa. The penalty paid is a reduction in self-resonant frequency caused by the increased capacitance. If the inductor is constructed out of metal layers sufficiently removed from the shield layer, this penalty can be tolerated in most cases.

[20] P. Yue and S. Wong, "On-Chip Spiral Inductors with Patterned Ground Shields for Si-Based RFIC's," *VLSI Circuits Symposium Digest of Technical Papers,* June 1997.

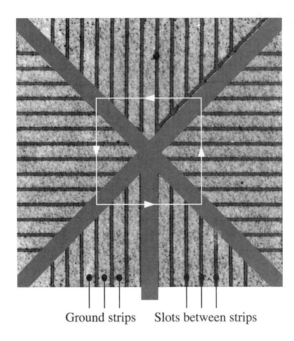

Ground strips   Slots between strips

FIGURE 4.25. Patterned ground shield.

An effective ground shield may even be constructed out of heavily doped polysilicon, allowing one to save the precious metal layers for the inductor itself. One may also place alternating wedges of n-well and substrate underneath the inductor to push image currents deeper into the substrate. Finally, one may employ the hollow spiral layout discussed previously. When used together, these techniques can sometimes increase Q by 50% or more.

As a last comment on the PGS, we should note that not all methods for connecting the PGS to ground are equally good. It's particularly important to reduce the resistance in series with the ground connection. As seen in Figure 4.25, an X-shaped interconnect arrangement reduces the average path length to ground for the PGS strips. The ground connection is made to the center of the X, as shown.

4.5.2 BONDWIRE INDUCTORS

In addition to planar spirals, bondwires are also frequently used to make inductors. Because standard bondwires are 1 mil (that's 0.001 inches, or about 25 μm) in diameter, they have much more surface area per length than planar spirals and hence less resistive loss, and therefore higher Q values. Also, they may be placed well above any conductive planes to reduce parasitic capacitances (thereby increasing the self-resonant frequency) and decrease the loss due to induced image currents. If we may neglect the influence of nearby conductors (i.e., if we assume that the return currents are infinitely far away), then the DC inductance of a bondwire is given by:[21]

$$L \approx \left[\frac{\mu_0 l}{2\pi}\right]\left[\ln\left(\frac{2l}{r}\right) - 0.75\right] \approx 2 \times 10^{-7} l\left[\ln\left(\frac{2l}{r}\right) - 0.75\right]. \quad (41)$$

For a 2-mm–long standard bondwire, this formula yields 2.00 nH, leading to a handy rule of thumb that the inductance is approximately 1 nH/mm. Notice that the inductance does grow faster than linearly with length because there is mutual coupling between parts of the bondwire (i.e., there is a weak transformer action) with a polarity that aids the inductance. From the logarithmic term, however, we see that this effect is minor. For example, going from 5 mm to 10 mm changes the DC inductance per millimeter from 1.19 nH to 1.33 nH (at least according to Eqn. 41). The inductance is similarly insensitive to the wire diameter, so even relatively large wire has an inductance on the order of 1 nH/mm.

The inductance value is not necessarily well controlled (partly because it is weakly frequency-dependent in addition to the obvious geometric dependencies),[22] so circuits using bondwires must be able to accommodate variation in the value of inductance. Despite this limitation, however, bondwire inductors have been used for years in a number of commercially successful amplifiers, and one recent (although somewhat impractical) design[23] uses several bondwires stitched across a die as part of a high-Q resonator for an on-chip voltage-controlled oscillator (VCO).

The Q of a bondwire inductor is relatively easy to estimate. The conductivity of aluminum is about 4×10^7 S/m, and the permeability of free space is $4\pi \times 10^{-7}$ H/m. With these numbers, the skin depth is approximately 2.5 μm at 1 GHz, a handy number to remember. Because the skin depth is small compared with the 25-μm diameter of a typical bond wire, we can readily compute the effective resistance per length as

$$\frac{R}{l} \approx \frac{1}{2\pi r \delta \sigma}. \quad (42)$$

For these numbers, one obtains a resistance of about 125 mΩ/mm at 1 GHz, potentially allowing the synthesis of inductors with Q-values of 50 at that frequency. Because inductive reactance grows linearly with frequency while skin loss grows only as the square root, Q-values approaching 100 might be possible at 5 GHz – although actually achieving such values in practice requires extraordinary care in minimizing all loss, especially at contacts to the inductor.

---

[21] *The ARRL Handbook,* American Radio Relay League, Newington, CT, 1992, pp. 2–18.

[22] With automated die attach and bonding equipment, however, the repeatability can be excellent, with variations held to within 1% or so.

[23] J. Craninckx and M. Steyaert, "A CMOS 1.8GHz Low-Phase-Noise Voltage-Controlled Oscillator with Prescaler," *ISSCC Digest of Technical Papers,* February 1995, pp. 266–7.

The temperature coefficient of a bondwire inductor is due to the combination of two effects. One is simply the linear expansion of the wire with increasing temperature; this component has a TC of approximately 25 ppm/°C. The other is the change in the contribution of the *internal* flux to the total inductance. The resistance goes up with temperature, causing the skin depth to increase, increasing the amount of internal flux (and hence the inductance). The contribution of the internal flux can be determined from the following equation for the internal inductance per unit length of a piece of wire at DC:

$$L_{\text{int}} = \frac{\mu_0}{8\pi}. \tag{43}$$

Equation 43 yields a value of 0.05 nH/mm, so internal inductance evidently accounts for 5% of the total inductance of a typical wire at DC. At 1 GHz, the skin depth is only a tenth the diameter of a bondwire, so the internal inductance decreases substantially (by a factor of 5 to 10 in this case). The change in internal inductance with temperature typically contributes a TC of approximately 20–50 ppm/°C, so one may expect the *total* inductance of a bondwire to possess a TC of roughly 50–70 ppm/°C.

Change in internal inductance also influences the TC of a spiral inductor, exacerbating the positive TC arising from the reduced cancellation by image currents.

Coupled Bondwires

The magnetic fields surrounding bondwires drop off relatively slowly with distance. As a result, there can be substantial magnetic coupling between adjacent (and even more remote) bondwires (as well as other conductors). A measure of this coupling is the *mutual* inductance between them. For two bondwires of equal length, this inductance is given approximately by

$$M \approx \frac{\mu_0 l}{2\pi}\left[\ln\left(\frac{2l}{D}\right) - 1 + \frac{D}{l}\right], \tag{44}$$

where l is the length of the bondwires and D is the distance between them.[24] For a 10-mm length and a spacing of 1 mm, the mutual inductance works out to about 4 nH. Since the inductance of each bondwire in isolation is about 10 nH, the 4-nH mutual inductance represents a coupling coefficient of 40%. The logarithmic dependence of M on spacing means that the coupling decreases rather slowly with distance, so there can be significant interaction even between alternate pins, for example.

4.5.3 MISCELLANEOUS INDUCTANCE FORMULAS

There are a couple of other formulas that are worth knowing, although not all of them have direct relevance to IC inductors. The first is very ancient, and it applies to a

---

[24] This formula is adapted from F. Terman, *Radio Engineer's Handbook*, McGraw-Hill, New York, 1943.

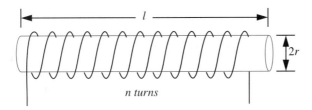

FIGURE 4.26. Single-layer solenoid.

single-layer solenoid[25] (Figure 4.26). For this structure, the inductance in *microhenries* is given by

$$L \approx \frac{n^2 r^2}{9r + 10l},$$

(45)

where r and l are in inches. In SI units, the formula is

$$L \approx \frac{10\pi\mu_0 n^2 r^2}{9r + 10l},$$

(46)

where a free-space permeability is assumed. These formulas provide remarkably good accuracy (typically better than 1%) for single-layer coils as long as the length is greater than the radius.[26]

The equivalent shunt capacitance across the terminals of a single-layer air-core coil is given approximately by

$$C = 1 \times 10^{-12}\left[11.25[D + l] + \frac{D}{4\sqrt{l/D}}\right],$$

(47)

where C is in farads and the diameter D and length l are in meters. This formula is derived from data published by R. G. Medhurst.[27] Regrettably, Medhurst convincingly argues that there is no simple formula for the effective series resistance. The influence of one turn's magnetic field on the current distribution in nearby turns is strong enough that a simple skin-effect formula rarely yields a useful approximation.

Another case of interest is the inductance of a single loop of wire. Despite the simplicity of the structure, there is no exact, closed-form expression for its inductance (elliptic functions arise in the computation of the total flux). However, a useful approximation is given by

$$L \approx \mu_0 \pi r.$$

(48)

[25] Wheeler, op. cit. (see footnote 15).

[26] The best Q is obtained when the windings are spaced approximately by an amount equal to the wire diameter itself.

[27] "H.F. Resistance and Self-Capacitance of Single-Layer Solenoids," *Wireless Engineer,* February 1947, pp. 35–43, and March 1947, pp. 80–92.

This formula tells us that a loop of 1 mm radius has an inductance of approximately 4 nH. A few spot checks with a field solver suggests that this formula typically underestimates the inductance for IC-sized objects, but values are still correct to better than about 25–30%.

In deriving this approximation, the (easily calculated) flux density in the center of the loop is arbitrarily assumed to be one half the average value in the plane of the loop; then the inductance is computed as simply the ratio of total flux to the current. In view of the rather coarse approximation involved, it is remarkable that the formula works as well as it typically does.

Note that, for a single turn and in the limit of zero length, Wheeler's formula (Eqn. 45 and Eqn. 46) converges very nearly to $\mu_0 \pi r$ (within 11%). For fussy folks, a better approximation is provided by[28]

$$L \approx \mu_0 r \left[\ln\left(\frac{8r}{a}\right) - 2 \right], \tag{49}$$

where a is the radius of the wire. With this equation, we see that Eqn. 48 strictly holds only for an r/a ratio of about 20.

To make a crude approximation even more so, Eqn. 48 can be extended to noncircular cases by arguing that all loops with equal area have about the same inductance, regardless of shape. Thus, we may also write

$$L \approx \mu_0 \sqrt{\pi A}, \tag{50}$$

where A is the area of the loop. According to this formula, a closed contour of 1-mm^2 area has an inductance of about 2.2 nH.

We can check the reasonableness of these equations by considering the inductance of a loop of extremely large radius. Since we can treat any suitably short segment of such a loop as if it were straight, we can use the equation for the inductance of a loop to estimate the inductance of a straight piece of wire.

We've already computed that a circular loop of 1-mm radius has an inductance of 4 nH, so we have roughly 4 nH per 6.3 mm length (circumference), which is in the same ballpark as the value given by the more accurate formulas.

4.6 TRANSFORMERS

It used to be that any electrical engineering graduate student would be familiar with the properties of an ideal transformer, at minimum. However, recent classroom evidence reveals that many schools omit material about transformers these days, so perhaps here is as good a place as any to plug that curricular hole (readers not in need of this refresher are invited to skip this section). We'll develop a model for ideal transformers first and then patch it up to model real transformers.

[28] S. Ramo, J. R. Whinnery, and T. Van Duzer, *Fields and Waves in Modern Radio,* Wiley, New York, 1965, p. 311.

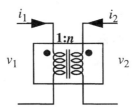

FIGURE 4.27. Ideal 1 : n transformer

A conventional transformer is a magnetically coupled system of inductors. Transformers get their name from their valuable ability to transform voltages, currents, and impedance levels over a relatively broad frequency range. In the simplest case, there are only two inductors, a primary and secondary. Just as the voltage across an isolated inductor is the result of a changing flux, a changing flux produced by the primary of a transformer can induce a voltage in the secondary, and vice versa.

For the ideal 1 : n transformer shown in Figure 4.27, n is the secondary-to-primary turns ratio. A changing magnetic flux common to both inductors thus generates n times the voltage at the secondary as at the primary (the polarity dots in the symbol identify which terminals are in phase). Energy conservation tells us that this voltage boost must be paid for by a corresponding current reduction of precisely the same factor. Because the ratio of voltage to current thus changes by n^2 in going from primary to secondary, an impedance transformation of that factor occurs at the same time. A turns ratio of 3, for example, corresponds to an impedance transformation ratio of 9. The ever-elusive ideal transformer would perform this function over an infinitely wide frequency range (including DC) and with zero loss. Even though such an element is physically unrealizable, it is nonetheless a useful starting point for constructing models of real transformers, as we'll soon see.

In the foregoing ideal example, we have implicitly assumed that all of the magnetic flux produced by, say, the primary couples to the secondary. The aim in most (but not all) transformer design is to approach this ideal as closely as possible. However, as with everything else in life, this aim is imperfectly met in practice, so our model must acknowledges a lack of perfect coupling or otherwise accommodate prescribed values besides unity.

Let L_1 be the inductance of the primary alone (i.e., with the secondary open-circuited) and let L_2 be that of the secondary alone. From the physics of the arrangement, we expect the voltage at any port to be the superposition of a self- and mutual term. The V–I equations for the (still) lossless but imperfectly coupled transformer may therefore be expressed as follows:

$$v_1 = L_1 \frac{di_1}{dt} + M \frac{di_2}{dt} \tag{51}$$

and

$$v_2 = M \frac{di_1}{dt} + L_2 \frac{di_2}{dt}, \tag{52}$$

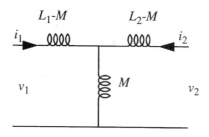

FIGURE 4.28. First-pass lossless transformer
model (T version).

where M, the *mutual inductance* between the windings, enables us to model the degree of coupling between primary and secondary. Reciprocity (another concept emphasized less and less these days) is what permits us to use the same value of M in both primary and secondary voltage equations, even for asymmetrical transformers. Depending on the physical arrangement, mutual inductance may take on positive or negative values – unlike isolated passive inductances. If the coupled flux adds to the self-flux then the mutual inductance is positive. If it opposes the self-flux, it is negative.

Although the total voltage across either the primary or secondary is the superposition of contributions from both the primary and secondary, the individual terms in the equations are isomorphic to that for an ordinary inductance. The corresponding circuit model for a transformer thus contains only inductive elements. See Figure 4.28, where we have implicitly assumed a common connection between ports.

If the primary and secondary are very close to each other, nearly all of the flux from one inductor will couple to the other. If far apart, or if their fields are orthogonally disposed, then the coupling will be negligible and M will be very small. It is useful to describe the continuum of possibilities with a quantitative measure of coupling known, reasonably enough, as the *coupling coefficient*. This is defined as

$$k \equiv \frac{M}{\sqrt{L_1 L_2}}. \qquad (53)$$

Thus, the coupling coefficient is the ratio of the mutual inductance to the geometric mean of the individual inductances. For passive elements, the magnitude of the coupling coefficient may not exceed unity.

Our first-pass model is perfectly respectable, but it suffers from some deficiencies that occasionally motivate the development of alternatives. One specific limitation of the model shown is that it does not explicitly incorporate a turns ratio between primary and secondary; that information is buried inside of the various inductance parameters. A less important limitation is that the primary and secondary share a common terminal. That deficiency is readily repaired simply by cascading the model with an ideal 1:1 transformer.

An alternative model that allows us to separate the ports completely and also explicitly incorporate an arbitrary turns ratio is depicted in Figure 4.29, where

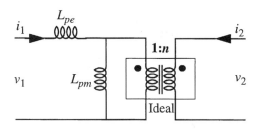

FIGURE 4.29. Alternative lossless transformer model.

$$L_{pe} = L_1(1 - k^2), \tag{54}$$

$$L_{pm} = k^2 L_1, \tag{55}$$

and

$$n = \frac{L_2}{M}. \tag{56}$$

This model contains an ideal transformer at its heart and then uses an isolated (un-coupled) *leakage inductance, L_{pe},* to account for the flux that doesn't participate in primary–secondary coupling. The *magnetizing inductance L_{pm}* models that portion of the primary inductance that does participate in coupling. It is therefore equal to the total primary inductance reduced by an amount equal to the leakage inductance. The magnetizing inductance also properly accounts for a real transformer's failure to function at DC, and it explains why low-frequency transformers are generally bulkier than high-frequency ones.

In cases where the coupling coefficient is close to unity, the magnetizing inductance is generally quite close in value to the primary inductance. For quick calculations of transformer circuits involving tight coupling, they may be treated as equal in most cases.

Having developed a lossless model that accommodates imperfect coupling as well as arbitrary turns ratios, we now need to account for a variety of parasitic elements that are always present. One potential source of significant parasitics is the material around or on which the inductor is wound. Although integrated circuit transformers almost never employ magnetic core materials (the transformers behave essentially as if they were wound on an air core), we'll consider core materials briefly for completeness. All magnetic core materials exhibit loss of at least two types. *Hysteresis loss* arises from the inelasticity of magnetic domain walls. To support magnetic state changes, these walls must move. One may think of a sort of friction as accompanying and inhibiting this wall movement. The energy lost per magnetic state transition is usually well modeled as constant for an excitation of fixed amplitude, so the total power dissipated owing to this mechanism is approximately proportional to frequency. We may account for this loss by adding a frequency-dependent resistance in shunt with the primary winding of our model.

Eddy current loss besets transformers as much as it degrades ordinary inductors. Currents may be induced in any nearby conductor, including electrically conductive core materials, adjacent windings, and conductive substrates. Because the induced voltage is proportional to frequency, eddy current losses are proportional to the square of frequency, as we've seen in the inductor case. The core losses augment those attributable to winding resistance, with due accounting for the skin effect.

In addition to the loss terms, there is also energy storage in the electric field surrounding and suffusing the windings. Hence, a high-frequency model must also include capacitances to account for this additional energy storage mechanism. Further complications arise when attempting to model behavior at frequencies where the dimensions of the transformer are not very small relative to a wavelength. In those cases, a simple lumped description of the transformer will be inadequate.

Finally, to make matters even more complex, all core materials become noticeably nonlinear at sufficiently high flux density, and all parameters are generally functions of temperature as well. These factors, compounded by the incompatibility of most magnetic materials with ordinary integrated circuit processes, explains the conspicuous absence of such materials from commercial RF integrated circuits.

The most important implication of the other nonidealities is that the various parasitics limit both the frequency response and efficiency. The magnetizing inductance shorts out the primary of the ideal transformer at DC, preventing transformer action there, while the winding capacitances perform a similar disservice at high frequencies. Precisely where in the model these parasitics appear – and also their magnitude – are functions of the physical structure, so we now turn to an examination of common ways to build transformers in integrated circuit technologies.

4.6.1 MONOLITHIC TRANSFORMER REALIZATIONS

With that short tutorial out of the way, we're now in a position to consider several ways to realize practical monolithic transformers and to construct corresponding circuit models for them. Figures 4.30–4.33 illustrate common configurations of monolithic transformers. The different realizations offer varying tradeoffs among the self-inductance and series resistance of each port, the mutual coupling coefficient, the port-to-port and port-to-substrate capacitances, resonant frequencies, symmetry, and die area consumed. The models and coupling expressions allow a systematic exploration of these tradeoffs, thereby permitting customization of the transformers for a variety of circuit design requirements.

The characteristics desired of a transformer are application-dependent. As we've suggested, transformers can be configured as three or four (or more) terminal devices. They may be used for narrowband or broadband applications. For example, in a conversion from single-ended to differential, the transformer might be used as a four-terminal narrowband device. In this case, a high mutual coupling coefficient and high self-inductance are desired along with low series resistance. On the other hand, for bandwidth extension applications, the transformer might be used as a broadband three-terminal device. In this case, a small mutual coupling coefficient and high series resistance are acceptable while minimization of all capacitances is of paramount importance.

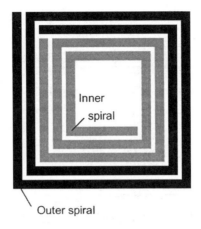

FIGURE 4.30. Tapped transformer.

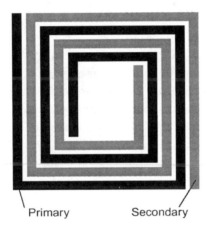

FIGURE 4.31. Interleaved transformer.

The tapped transformer (Figure 4.30) is best suited for three-port applications. It permits the realization of a variety of tapping ratios. This transformer relies only on lateral magnetic coupling. All windings can be implemented with the top metal layer, thereby minimizing port-to-substrate capacitances. Since the two inductors occupy separate regions, the self-inductance is maximized while the port-to-port capacitance is minimized. Unfortunately, this spatial separation also implies low to moderate coupling ($k = 0.3$–0.5) and the consumption of large amounts of chip area.

The interleaved transformer (Figure 4.31) is best suited for four-port applications that demand symmetry. Once again, capacitances can be minimized by implementing the spirals with top-level metal to maximize resonant frequencies. The interleaving of the two inductances permits moderately high coupling ($k = 0.7$) to be achieved at the cost of reduced self-inductance. This coupling may be increased at the cost of higher series resistance by reducing the turn width w and spacing s.

Top View

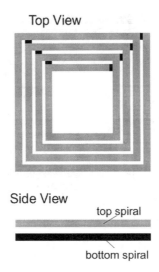

Side View

top spiral

bottom spiral

FIGURE 4.32. Stacked transformer
with completely overlapping spirals.

The stacked transformer (Figure 4.32) uses multiple metal layers and exploits both vertical and lateral magnetic coupling to provide the best area efficiency, the highest self-inductance, and the highest coupling ($k = 0.9$). This configuration is suitable for both three- and four-terminal configurations. The main drawback is the high port-to-port capacitance, or equivalently a low self-resonant frequency. In some cases, such as narrowband impedance transformers, this capacitance may be incorporated as part of the resonant circuit. Also, in multilevel processes, the capacitance can be reduced by increasing the oxide thickness between spirals. For example, in a five-metal process, 50–70% reductions in port-to-port capacitance can be achieved by implementing the spirals on layers five and three instead of on five and four. The increased vertical separation will reduce k by less than 5%. One can also trade off reduced coupling for reduced capacitance by displacing the centers of the stacked inductors (see Figures 4.33). For an offset stacked transformer constructed from two identical spirals, the coupling coefficient is reasonably well approximated by

$$k \approx 0.9 - \frac{\sqrt{x_s^2 + y_s^2}}{d_{\text{avg}}} = 0.9 - \frac{d_s}{d_{\text{avg}}}. \tag{57}$$

As seen from this expression, the coupling coefficient diminishes to zero when the two spirals are offset by an amount about equal to the average diameter. The coupling coefficient then goes negative and ultimately approaches zero asymptotically as the two spirals are offset by increasing amounts.

In a final extension of these ideas, one may exploit the mutual coupling among several spirals stacked atop one another (with or without displaced centers) to produce inductances of larger value than can normally be obtained within a given chip area.

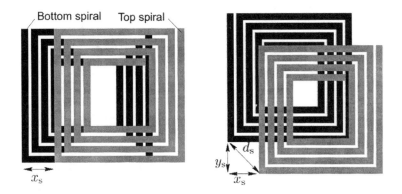

FIGURE 4.33. Stacked transformer with offset top and bottom spirals.

The parasitic capacitance – and the corresponding self-resonant frequency – of the inductor must be carefully evaluated to ensure that the resulting element will function satisfactorily.

4.6.2 ANALYTICAL MODELS FOR PLANAR TRANSFORMERS

Figure 4.34 and Figure 4.35 present the circuit models for tapped and stacked transformers. The corresponding element values for the tapped transformer model are given by the following equations (subscript o refers to the outer spiral, i to the inner spiral, and T to the whole spiral). First,

$$L_\mathrm{T} = \frac{9.375\mu_0 n_\mathrm{T}^2 d_\mathrm{avg,T}^2}{11 d_\mathrm{out,T} - 7 d_\mathrm{avg,T}}, \tag{58}$$

$$L_\mathrm{o} = \frac{9.375\mu_0 n_\mathrm{o}^2 d_\mathrm{avg,o}^2}{11 d_\mathrm{out,o} - 7 d_\mathrm{avg,o}}, \tag{59}$$

and

$$L_\mathrm{i} = \frac{9.375\mu_0 n_\mathrm{i}^2 d_\mathrm{avg,i}^2}{11 d_\mathrm{out,i} - 7 d_\mathrm{avg,i}}, \tag{60}$$

where we have somewhat arbitrarily used the modified Wheeler formula for the inductance of square planar spiral inductors. Here d_out and d_avg are the outer and average diameters, respectively.

Continuing, we have

$$M = \frac{L_\mathrm{T} - L_\mathrm{o} - L_\mathrm{i}}{2\sqrt{L_\mathrm{o} L_\mathrm{i}}}, \tag{61}$$

$$R_\mathrm{s,o} = \frac{\rho l_\mathrm{o}}{\delta w (1 - e^{-t/\delta})}, \tag{62}$$

$$R_\mathrm{s,i} = \frac{\rho l_\mathrm{i}}{\delta w (1 - e^{-t/\delta})}, \tag{63}$$

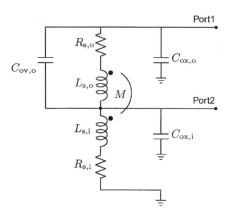

FIGURE 4.34. Tapped transformer model.

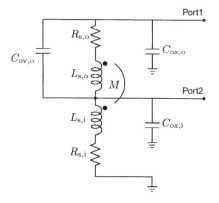

FIGURE 4.35. Stacked transformer model.

$$C_{\text{ov},\text{o}} = \frac{\varepsilon_{\text{ox}}}{t_{\text{ox},\text{t-b}}} \cdot (n_{\text{o}} - 1) w^2, \tag{64}$$

$$C_{\text{ox},\text{o}} = \frac{\varepsilon_{\text{ox}}}{2 t_{\text{ox}}} \cdot l_{\text{o}} w, \tag{65}$$

and

$$C_{\text{ox},\text{i}} = \frac{\varepsilon_{\text{ox}}}{2 t_{\text{ox}}} \cdot (l_{\text{o}} + l_{\text{i}}) w, \tag{66}$$

where L is inductance and R denotes resistance; ρ is the DC metal resistivity; δ is the skin depth; C denotes capacitance; $t_{\text{ox},\text{t-b}}$ is the oxide thickness from top-level metal to bottom metal; n is the number of turns; l is the length of the spiral; w is the turn width; and t is the metal thickness. The subscript ov denotes "crossover."

Expressions for the stacked transformer model are as follows (subscript t refers to the top spiral and b to the bottom spiral):

$$L_{\text{t}} = L_{\text{b}} = \frac{9.375 \mu_0 n^2 d_{\text{avg}}^2}{11 d_{\text{out}} - 7 d_{\text{avg}}}, \tag{67}$$

$$k = 0.9 - \frac{d_s}{d_{\mathrm{avg}}}, \tag{68}$$

$$M = k\sqrt{L_t L_b}, \tag{69}$$

$$R_{s,t} = \frac{\rho_t l}{\delta_t w (1 - e^{-t_t/\delta_t})}, \tag{70}$$

$$R_{s,b} = \frac{\rho_b l}{\delta_b w (1 - e^{-t_b/\delta_b})}, \tag{71}$$

$$C_{ov} = \frac{\varepsilon_{ox}}{2 t_{ox,t\text{-}b}} \cdot l \cdot w \cdot \frac{A_{ov}}{A}, \tag{72}$$

$$C_{ox,t} = \frac{\varepsilon_{ox}}{2 t_{ox,t}} \cdot l \cdot w \cdot \frac{A - A_{ov}}{A}, \tag{73}$$

$$C_{ox,b} = \frac{\varepsilon_{ox}}{2 t_{ox}} \cdot l \cdot w, \tag{74}$$

$$C_{ox,m} = C_{ox,t} + C_{ox,b}. \tag{75}$$

Here k is the coupling coefficient; d_s is the center-to-center spiral distance; $t_{ox,t}$ is the oxide thickness from top metal to the substrate and $t_{ox,b}$ the oxide thickness from bottom metal to substrate; and A_{ov} is the overlap area of the two spirals.

The expressions for the series resistances ($R_{s,o}$, $R_{s,i}$, $R_{s,t}$, and $R_{s,b}$), the port–substrate capacitances ($C_{ox,o}$, $C_{ox,i}$, $C_{ox,t}$, $C_{ox,b}$, and $C_{ox,m}$) and the crossover capacitances ($C_{ov,o}$, $C_{ov,i}$, and C_{ov}) are taken from C. P. Yue, C. Ryu, J. Lau, T. H. Lee, and S. S. Wong, "A Physical Model for Planar Spiral Inductors on Silicon" (*International Electron Devices Meeting Technical Digest,* December 1996, pp. 155–8). Note again that the model only crudely accounts for the increase in series resistance with frequency that is due to skin effect. Patterned ground shields are placed beneath the transformers to isolate them from resistive and capacitive coupling to the substrate.[29] As a result, the substrate parasitics can be neglected for lightly and moderately doped substrates. In cases where heavily doped substrates are used, one must include the eddy current loss terms as well.

The inductance expressions in the foregoing are again based on the modified Wheeler formula. This formula does not take into account the variation in inductance due to conductor thickness and frequency. However, in practical inductor and transformer realizations, the thickness is small compared to the lateral dimensions of the coil and has only a small impact on the inductance; hence it can usually be ignored. For typical conductor thickness ranges (0.5–2.0 μm), the change in inductance is within a few percent for practical inductor geometries. The inductance also changes with frequency owing to changes in current distribution within the conductor. However, over the useful frequency range of a spiral, this variation is again generally negligible (see the work by Yue et al. cited in footnote 29). When compared to field

---

[29] C. P. Yue et al., "On-Chip Spiral Inductors with Patterned Ground Shields for Si-Based RF ICs," *IEEE J. Solid-State Circuits,* v. 33, May 1998, pp. 743–52.

solver simulations, the inductance expression exhibits a maximum error of 8% over a broad design space (outer diameter d_{out} varying from 100 μm to 480 μm, L varying from 0.5 nH to 100 nH, w varying from 2 μm to $0.3d_{out}$, s varying from 2 μm to w, and inner diameter d_{in} varying from $0.2d_{out}$ to $0.8d_{out}$).

For the tapped transformer, the mutual inductance is determined by first calculating the inductance of the whole spiral (L_T), the outer spiral (L_o), and the inner spiral (L_i) and then using the expression $M = (L_T - L_o - L_i)/2$. For the stacked transformer, the spirals have identical lateral geometries and therefore identical inductances. In this case, the mutual inductance is determined by calculating the inductance of one spiral (L_t or L_b) and then using the coupling coefficient (k) in the expression $M = kL_t$. In this last case, the coupling coefficient is given by $k = 0.9 - d_s/d_{avg}$ for $d_s < 0.7d_{avg}$, where again d_s is the center-to-center spiral distance and d_{avg} is the average diameter of the spirals. As d_s increases beyond $0.7d_{avg}$, the mutual coupling coefficient becomes harder to model. Eventually, k crosses zero and reaches a minimum value of approximately -0.1 at $d_s = d_{avg}$. As d_s increases further, k asymptotically approaches zero. At $d_s = 2d_{avg}$ we have $k = -0.02$, indicating that the magnetic coupling between closely spaced spirals is negligible.

The self-inductances, series resistances, and mutual inductances calculated in this way are independent of whether a transformer is used as a three- or four-terminal device. The only elements that require recomputation are the port-to-port and port-to-substrate capacitances. This situation is analogous to that of a spiral inductor used as a single or dual terminal device. As with the inductance formulas, these transformer models obviate the need for full field solutions in all but very rare instances and so enable rapid designs and optimization.

4.7 INTERCONNECT OPTIONS AT HIGH FREQUENCIES

It is certainly true that the silicon substrate's lossiness poses many challenges. One of these concerns the attenuation of interconnect at high frequencies. Recent work, however, shows that it is nonetheless possible to build perfectly serviceable interconnect well into millimeter-wave bands – provided that we select appropriate interconnect structures. In this context the *coplanar waveguide* (or *coplanar transmission line*) is of particular value because its edge-coupled nature helps reduce the coupling of energy into the troublesome substrate.

As can be seen in Figure 4.36, wide lines with wide spacing (as might be used in an attempt to reduce conductor losses) exacerbate substrate coupling and its attendant loss. For technologies with lossy substrates, then, a good choice is to use narrower lines and spacings than would be customary in technologies with low-loss substrates.

As we've already noted, an important degree of freedom is the ever-increasing number of metal layers made available with successive generations of CMOS technology. Associated with those metal layers are additional layers of intermetal dielectric. Thus, it is possible to build coplanar lines far above the substrate by using the topmost metal layer(s), in effect allowing the construction of coplanar lines over a

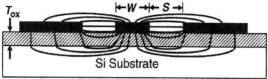

Large *W, S*: Large coupling into substrate

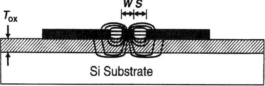

Small *W, S*: Small coupling into substrate

FIGURE 4.36. Coplanar transmission line
(after Kleveland et al.).

Table 4.6. *Coplanar line loss per*
millimeter at 50 GHz

$W = S$ (μm)	t_{ox} (μm)	Loss (dB)
16	40	0.3
8	40	0.55
4	20	0.85

Source: B. Kleveland et al., "50-GHz Interconnect Design in Standard Silicon Technology,"
*IEEE MTT-S International Microwave Sympo-
sium,* 1988.

silicon dioxide substrate. In defiance of Murphy, silicon dioxide is an exceptionally
low-loss material and so interconnect made over and within it will exhibit low loss
up to exceedingly high frequencies.

Use of a combination of these strategies enables the attainment of attenuations that
are as good as the best achieved in other technologies, as seen in Table 4.6. It should
be noted that 50 GHz is not a fundamental limit by any means. Millimeter-wave
operation of CMOS is thus largely a matter of waiting for scaling to deliver transistors fast enough to take advantage of the interconnect.

MOSFETs as Switches

One attractive characteristic of MOSFETs is their ability to act as excellent switches.
Here we highlight some of the applications and limitations of MOSFETs as RF
switches and also some ways to improve performance.

In many transceivers, the transmitter and receiver do not operate simultaneously;
hence it is natural in such systems to use a switch to permit the sharing of a common

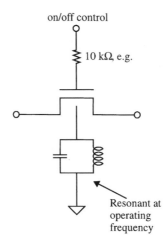

on/off control

$10\ k\Omega$, e.g.

Resonant at
operating
frequency

FIGURE 4.37. High-linearity RF switch
(after Talwalkar).

antenna. In this application the T/R switch needs to exhibit low insertion loss and
high linearity in the "on" state while presenting a high impedance in the "off" state.

It is traditional to drive the gate of a MOSFET switch with a relatively high impedance to improve on-state linearity. This method exploits the device's own capacitances to maintain roughly constant terminal voltage relationships over an RF cycle,
thereby minimizing modulation of device characteristics. Required turn-on/turn-off
times are generally orders of magnitude slower than the RF period, so driving the
gate through a relatively large impedance presents no fundamental difficulty.

The remaining sources of nonlinearity are then the nonlinear drain- and source-to-
bulk capacitances and the change in terminal voltages with respect to the back gate
(substrate) voltage. In process technologies that are amenable (or if PMOS switches
are used, at a $2\times$ penalty in "on" resistance for a given device width), the body terminal of individual transistors may be available as a separate terminal. In those cases, a
high impedance of some sort (e.g., a suitable inductor) connected to the nominal DC
voltage to which the body is normally connected would allow all four terminals of
the MOSFET to translate together. Such a high substrate impedance also mitigates
the nonlinearity due to the drain and source diodes.

It may be impractical using available on-chip elements to provide the necessary
high impedance directly. In such cases, it is advantageous to exploit the boosts in impedance that accompany resonance (see Chapter 5). In that case, the switch circuit
would appear as shown in Figure 4.37.[30] Using these techniques, 1.5-dB insertion
loss, 30-dB isolation, and 28-dBm 1-dB compression point have been demonstrated
with an experimental 5.2-GHz switch built in a standard 0.18-μm digital CMOS

---

[30] See N. Talwalkar, "Integrated CMOS Transmit-Receive Switch Using On-Chip Spiral Inductors,"
Ph.D. dissertation, Stanford University, 2003.

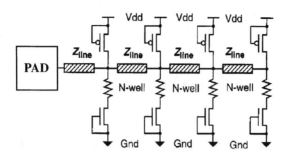

FIGURE 4.38. High-bandwidth distributed ESD
protection circuit (after Kleveland et al.).

process.[31] The triple-well process allows access to the relevant body terminal to permit selective insertion of the resonant circuit. The results achieved compare favorably with discrete RF switches.

Electrostatic Discharge (ESD)

A commonly asked question concerns RF-compatible electrostatic discharge protection. If adequate ESD survivability cannot be provided without intolerable degradation of RF characteristics, then RF CMOS will forever remain a useless academic activity.

The easiest method is simply to ignore ESD altogether in the initial design of RF circuits. Then, after completing the design of the RF circuitry proper, add the maximum amount of ESD protection consistent with preserving the desired RF performance. Whatever ESD tolerance results then becomes the accepted specification. At low frequencies, this method succeeds more than its fair share of the time. In the gigahertz realm, however, more sophisticated methods are called for.

The problem arises from the added capacitance associated with ESD protection devices (e.g., clamp diodes and the like). To frame the problem a bit more quantitatively, the associated capacitance is typically several picofarads (perhaps as high as 10 pF). Needless to say, such a large capacitance is rarely tolerable at high frequencies (consider that the reactance of a 5-pF capacitor is only 32 Ω at 1 GHz). Reducing the capacitance, however, means having to effect a painful tradeoff between ESD survivability and bandwidth (or maximum operating frequency). However, as we shall detail in Chapter 6, it is possible to arrange for these added capacitances to form part of a pure *delay line*. A delay line may still have a large bandwidth, but we can effect a critically important decoupling of delay from bandwidth. The resulting structure, known as a *distributed* ESD protection device, appears as Figure 4.38.

As can be seen in the figure, short segments of interconnect (acting as delay lines) are periodically loaded by the ESD diodes, whose capacitance merely adds to that

---

[31] We'll have more to say later about what "compression point" means. For now, just accept that it's a measure of linearity and that larger values are better.

of the line segments, increasing their delay.[32] The level of ESD protection is a function of the total capacitance (number of segments) and can be extraordinarily large. Using this approach, 12-kV survivability with the human-body model (HBM) as well as 800-V survivability with the charged-device model (CDM) have been achieved.[33]

These levels of HBM and CDM survivability greatly exceed those in many low-frequency commercial designs. The only drawback is the relatively large die area consumed by these circuits. However, it is generally true that only a small fraction of the total pins of a given IC need to operate at full speed, so most of the pins may use small, conventional ESD structures.

4.8 SUMMARY

This chapter has presented numerous models and formulas for on-chip (and off-chip) passive elements. A hierarchy of formulas with roughly constant simplicity–accuracy products allows one to carry out computations at any level of design – ranging from initial exploration to final verification – reducing greatly (even obviating) the need for analysis by electromagnetic field solvers in many instances.

4.9 APPENDIX: SUMMARY OF CAPACITANCE EQUATIONS

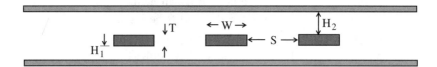

FIGURE 4.39. Conductor arrangement for all cases.

Case 1: Wire over Single Conducting Plane

$$C \approx \varepsilon \left[\frac{W}{H} + 0.77 + 1.06 \left\{ \left(\frac{W}{H} \right)^{0.25} + \left(\frac{T}{H} \right)^{0.5} \right\} \right]. \tag{76}$$

Case 2: Wire between Two Conducting Planes

$$C \approx \varepsilon \left[W \left(\frac{1}{H_1} + \frac{1}{H_2} \right) + 0.77 \right.$$
$$\left. + 0.891 \left\{ \left[W \left(\frac{1}{H_1} + \frac{1}{H_2} \right) \right]^{0.25} + T^{0.5} \left(\frac{1}{H_1^2} + \frac{1}{H_2^2} \right)^{0.25} \right\} \right]. \tag{77}$$

---

[32] In other variations, one could use lumped inductances (such as planar spirals) in place of the line segments.

[33] B. Kleveland et al., "Distributed ESD Protection for High-Speed Integrated Circuits," *IEEE Electron Device Lett.*, v. 21, no. 8, August 2000, pp. 390–2.

Case 3: Three Adjacent Wires over Single Ground Plane

$$C_{\text{total}} = C_{\text{single}} + 2C_{\text{mutual}},\tag{78}$$

$$C_{\text{single}} \approx \varepsilon\left[1.15\frac{W}{H} + 2.8\left(\frac{T}{H}\right)^{0.222} - 1.31\left(\frac{T}{H}\right)^{0.222}\left(\frac{S}{H}\right)^{-1.34}\right],\tag{79}$$

$$C_{\text{mutual}} \approx \varepsilon\left[0.03\frac{W}{H} + 0.83\frac{T}{H} + 0.585\left(\frac{T}{H}\right)^{0.222}\right]\left[\left(\frac{S}{H}\right)^{-1.34}\right],\tag{80}$$

where C_{single} is the capacitance from the middle wire to ground and C_{mutual} is the capacitance between adjacent pairs of wires. The capacitance between the outer pair of wires is generally negligible owing to the Faraday shielding provided by the middle conductor.

Also, the formulas for two wires over ground plane are similar. The mutual capacitance is the same (and, of course, you count it only once when computing the total capacitive load on either wire); the term C_{single} (which then denotes the capacitance of either wire to ground) differs only in that the factor 1.31 changes to 0.655.

PROBLEM SET FOR PASSIVE IC COMPONENTS

PROBLEM 1 Extend to circular capacitors the simple fringe correction method outlined for square capacitors. Compute the ratio of your corrected capacitance to the uncorrected value for the following ratios of plate spacing to diameter: 0.005, 0.01, 0.025, 0.05, 0.1. Compare your result to the actual correction factors shown in Table 4.7.

Table 4.7. *Correction factors for fringing in circular capacitors*

s/D	$C_{\text{corr}}/C_{\text{uncorr}}$
0.005	1.023
0.01	1.042
0.025	1.094
0.05	1.167
0.10	1.286

PROBLEM 2

(a) Design a 10-nH square spiral inductor in which the total length of the interconnect is 3500 μm, the spacing between turns is 2 μm, the metal and oxide are both 1-μm thick, and the metal conductivity is 4×10^7 S/m. The oxide has a relative dielectric constant of 3.9.

(b) Compute the values for the model elements of your design at 1.5 GHz if one terminal is grounded. Initially assume that the substrate is a superconductor.

(c) If this inductor is used as part of a tank in which the external capacitance and resistance are 500 fF and 10 kΩ, respectively, what is the impedance of the combination at resonance? Do not bother recomputing the model parameter values for the new resonant frequency.

(d) Change your model to use the default values for G_{sub} and C_{sub}. Use SPICE to determine the new impedance magnitude at resonance. Again, do not bother recomputing the other model parameter value. Compare with your previous result.

PROBLEM 3 A parallel resonant tank circuit is constructed from a 4-turn, 80-μm$\times$80-μm square spiral inductor, and a 5-μm$\times$5-μm capacitor constructed from two metal layers separated by an atypically thin 0.2-μm oxide dielectric ($\varepsilon = 3.9\varepsilon_0$).

(a) Initially neglecting fringing and *all* other parasitics, what is the nominal resonant frequency of this network?

(b) Now foolishly assume that the substrate is a superconductor and that the bulk of the inductor is built out of a layer that is 3 μm above the substrate. Ignoring the shunt capacitance of the inductor's crossunder, but not the capacitance to the substrate, what now is the approximate parallel resonant frequency of the network if the substrate is connected to one terminal of the tank? The inductor windings are 8 μm wide and 1 μm thick. You may assume a symmetrical inductor model.

(c) What is the value of the inductor's effective series resistance at this new resonant frequency if the interconnect has a conductivity of 5×10^7 S/m? It may help to know that the permeability is about 1.26×10^{-6} H/m.

(d) One measure of how little loss a reactive network has is Q, the *quality factor*. What is the Q of this resonator if it is defined here as $\omega L/R$ and is measured at the resonant frequency?

PROBLEM 4 A common problem, especially in digital systems, is how best to size interconnect. A wider line has more capacitance, but lower resistance, so how wide is wide enough? To put this question on a somewhat quantitative basis, use the Sakurai formula for a single conductor over a ground plane and derive an equation for the RC product of such a line, where R and C are the total resistance and capacitance, respectively. We will use a simplified approach here and ignore the effect of interconnect loading on whatever has to drive it. Instead, we focus entirely on the interconnect itself.

(a) If the propagation delay of signals is proportional to RC, how does the delay increase if the length doubles?

(b) Your formula should show that the RC product asymptotically approaches a minimum value as the width goes to infinity. What width produces a delay just 25% above this minimum value? Express your answer in terms of the thickness T and the height H above the substrate.

PROBLEM 5 Junction capacitors are normally used in reverse bias as varactors. To explore why they are almost never used in (strong) forward bias, assume that the diode behaves as follows in the forward direction:

$$i_D \approx I_S e^{v_j/V_T}. \tag{P4.1}$$

Assume that the thermal voltage V_T is 25 mV at the operating temperature, and assume that the diode is built in such a fashion that the forward current is 1 mA at a junction voltage of 0.5 V.

(a) Calculate the incremental resistance at 1 mA.

(b) If the zero-bias capacitance C_{j0} is 2 pF, what is the capacitance at the forward bias of 0.5 V? Assume an abruptly doped (step) junction and a ϕ of 0.8 V.

(c) Compute the reactance of the capacitance found in part (b) at 1 GHz. Does the varactor appear mainly resistive or capacitive at this frequency?

PROBLEM 6 Design a 7.3-nH inductor. You have at your disposal a total of 6 mm of bondwire and 900 μm^2 of die area. Assume that the bondwire length can be controlled to no better than 10%. Maximize the Q of the resulting inductor, subject to the constraint that the final inductor value be within 5% of the target value. For simplicity's sake, you may assume that the following planar spiral inductance formula is exact:

$$L \approx \mu_0 n^2 R. \tag{P4.2}$$

PROBLEM 7 Derive an expression for the resistance of interconnect as a function of temperature for two cases as follows.

(a) The skin depth is very small compared with the conductor dimensions.

(b) The skin depth is very large compared with the conductor dimensions.

You may assume that the resistivity of the interconnect material is itself PTAT.

(c) Using your result to part (a), how much variation in Q would you expect for a square spiral inductor between $-55°C$ and $+125°C$?

PROBLEM 8

(a) Derive a circuit model for two coupled bondwires, each of which is 7 mm in length, that are separated by 4 mm. You may ignore resistive and capacitive parasitics.

(b) Assume that one bondwire is driven by a voltage source through a resistance of 50 Ω. Also assume that the other bondwire is shunted by a 200-Ω–load resistance. If the voltage source provides a 1-V unit step, use SPICE to plot the voltage across the load resistor.

(c) Now double the separation to 8 mm and repeat. What do you conclude about the effectiveness of separation as a means to reduce parasitic coupling?

PROBLEM 9 A 10-kΩ polysilicon resistor is to be made out of material with a sheet resistivity of 100 Ω per square.

(a) Determine the minimum dimensions of this resistor if the width cannot be controlled to an uncertainty of better than 0.2 μm, and if the variation in resistance due to width variation must be kept below 5%. For simplicity, assume a straightforward linear layout.

(b) Determine the parasitic capacitance to the substrate if the oxide dielectric layer (relative dielectric constant: 3.9) is 1 μm thick. Using a single *RC* model for this structure, what is the approximate maximum frequency above which this resistor ceases to appear predominantly resistive?

PROBLEM 10 Another constraint on conductor dimensions is imposed by *electromigration* effects. At high enough current densities, momentum transfer between electrons and metal atoms can cause physical motion of parts of the interconnect. A narrowing of interconnect causes an increase in the current density, which accelerates the narrowing, and so on until either the resistance increases to unacceptable levels or the interconnect actually open-circuits.

Electromigration rules for most commonly used interconnect metals usually dictate an upper bound within a factor of 2 of about 10^9 A/m^2 DC current (much larger densities are permitted for high-frequency sinusoidal currents because little net migration can occur in such a case).

(a) Assuming a maximum allowable current density of 2×10^9 A/m^2, determine the minimum acceptable interconnect width capable of supporting a current of 100 mA if the conductor is 0.5 μm thick.

(b) Compute the resistance per millimeter of your design if the conductivity is 4×10^7 S/m.

(c) Estimate the parasitic capacitance per millimeter if the oxide is 1 μm thick.

CHAPTER FIVE

A REVIEW OF
MOS DEVICE PHYSICS

5.1 INTRODUCTION

This chapter focuses attention on those aspects of transistor behavior that are of immediate relevance to the RF circuit designer. Separation of first-order from higher-order phenomena is emphasized, so there are many instances when crude approximations are presented in the interest of developing insight. As a consequence, this review is intended as a supplement to – rather than a replacement for – traditional rigorous treatments of the subject. In particular, we must acknowledge that today's deep-submicron MOSFET is so complex a device that simple equations cannot possibly provide anything other than first-order (maybe even zeroth-order) approximations to the truth. The philosophy underlying this chapter is to convey a simple story that will enable first-pass designs, which are then verified by simulators using much more sophisticated models. Qualitative insights developed with the aid of the zeroth-order models enable the designer to react appropriately to bad news from the simulator. We design with a simpler set of models than those used for verification.

With that declaration out of the way, we now turn to some history before launching into a series of derivations.

5.2 A LITTLE HISTORY

Attempts to create field-effect transistors (FETs) actually predate the development of bipolar devices by over twenty years. In fact, the first patent application for a FET-like transistor was filed in 1926 by Julius Lilienfeld, but he never constructed a working device.[1] Before co-inventing the bipolar transistor, William Shockley also tried to modulate the conductivity of a semiconductor to create a field-effect transistor. Like

---

[1] U.S. Patent #1,745,175, filed 8 October 1926, granted 28 January 1930. It's not a MOSFET, as is sometimes reported, for it uses a *vertical* gridlike control electrode disposed within a solid body.

Lilienfeld, problems with his materials system, copper compounds,[2] prevented success. Even after moving on to germanium (a much simpler – and therefore much more easily understood – semiconductor than copper oxide), Shockley was still unable to make a working FET. In the course of trying to understand the reasons for this spectacular lack of success, Shockley's Bell Laboratories colleagues John Bardeen and Walter Brattain stumbled across the point-contact bipolar transistor, the first practical semiconductor amplifier. Unresolved mysteries with that device (such as negative β, among others) led Shockley to invent the junction transistor, and the three eventually won a Nobel Prize in physics for their work.

By 1950, a transistor based on the modulation of a semiconductor's effective cross-sectional area had been successfully demonstrated. This junction FET (JFET) is a useful device, but it's not what Shockley had originally sought to build.

A decade later, Kahng and Atalla of Bell Labs finally succeeded in making a silicon MOSFET, taking advantage of the fortuitous discovery that silicon's own oxide does a superb job of taming the pesky interface states that had frustrated earlier attempts in other materials systems. However, mysterious (and maddening) drifts in device characteristics inhibited commercialization of MOS technology until contamination by sodium ions was identified as the main culprit and remedial protocols put in place. Within a short time, MOSFET technology became the preferred way to make integrated circuits, owing to relatively simple fabrication and the potential for high circuit density.

5.3 FETs: THE SHORT STORY

Although the quantitative details are a bit complicated, the basic idea that underlies the operation of a FET is simple: Start with a resistor and add a third terminal (the gate) that somehow allows modulation of the resistance between the other two terminals (the source and drain). If the power expended in driving the control terminal is less than that delivered to a load, power gain results.

In a junction FET (see Figure 5.1), a reverse-biased p–n junction controls the resistance between the source and drain terminals. Because the width of a depletion layer depends on bias, a gate voltage variation alters the effective cross-sectional area of the device, thereby modulating the drain–source resistance. Because the gate is one end of a reverse-biased diode, the power expended in effecting the control is virtually zero, and the power gain of a junction FET is correspondingly very large.

A junction FET is normally conducting and requires the application of a sufficiently large reverse bias on the gate to shut it off. Because control is effected by altering the extent of the depletion region, such FETs are called depletion-mode devices.

---

[2] Recitifers made of cuprous oxide had been in use since the 1920s, even though the detailed operating principles were not understood. Around 1976, with decades of semiconductor research to support him, Shockley took one last shot at making a copper oxide FET (at Stanford, in fact). Still unsuccessful, he despaired of ever knowing why. (See the July 1976 issue of the *Transactions on Electron Devices* for Shockley's reminiscences and fascinating stories from many other pioneers.)

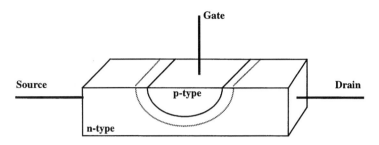

FIGURE 5.1. n-channel junction FET (simplified; most practical devices have two gate diffusions).

While JFETs are not the type of FETs used in mainstream IC technology, the basic idea of conductivity modulation underlies the operation of the ones that are: MOSFETs.

In the most common type of MOSFET, the gate is one plate of a capacitor separated by a thin dielectric from the bulk of a nearly insulating semiconductor. With no voltage applied to the gate, the transistor is essentially nonconductive between the source and drain terminals. When a voltage of sufficient magnitude is applied to the gate, charge of the opposite polarity is induced in the semiconductor, thereby enhancing the conductivity. This type of device is thus known as an enhancement-mode transistor.

As with the JFET, the power gain of a MOSFET is quite large (at least at DC); there is virtually no power expended in driving the gate since it is basically a capacitor (at low frequencies, anyway). We will revisit this issue when discussing MOSFET behavior at high frequencies, where the gate impedance exhibits a resistive component that limits power gain.

5.4 MOSFET PHYSICS: THE LONG-CHANNEL APPROXIMATION

The previous overview leaves out a great many details – we certainly can't write any device equations based on the material presented so far, for example. We now undertake the task of putting this subject on a slightly more quantitative basis. In this section, we will assume that the device has a "long" channel. We will see later that by "long channel" we actually mean "low electric field." The behavior of short-channel devices will still conform reasonably well to the equations derived in this section if the applied voltages are low enough to guarantee small electric fields.

As you well know, a basic n-channel MOSFET (Figure 5.2) consists of two heavily doped n-type regions, the source and drain, which constitute the main terminals of the device. The gate was made of metal in early incarnations but is now made of heavily doped polysilicon, whereas the bulk of the device is p-type and is typically rather lightly doped. In much of what follows, we will assume that the substrate

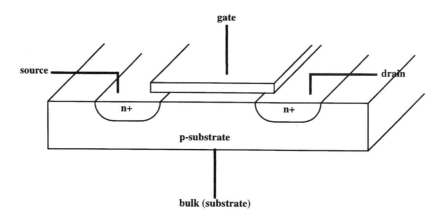

FIGURE 5.2. n-channel MOSFET.

(bulk) terminal is at the same potential as the source. However, it is extremely important to keep in mind that the substrate constitutes a fourth terminal whose influence cannot always be ignored.

As an increasing positive voltage is applied to the gate, holes are progressively repelled away from the surface of the substrate. At some particular value of gate voltage (the threshold voltage V_t), the surface becomes completely depleted of charge. Further increases in gate voltage induce an *inversion layer,* composed of electrons supplied by the source (or drain), that constitutes a conductive path ("channel") between source and drain. When the gate–source voltage is several kT/q above V_t, the device is said to be in strong inversion.

The foregoing discussion implicitly assumes that the potential across the semiconductor surface is a constant (i.e., that there is zero drain-to-source voltage). With this assumption, the induced inversion charge is proportional to the gate voltage above the threshold, and the induced charge density is constant along the channel. However, if we do apply a positive drain voltage V, then the channel potential must increase in some manner from zero at the source end to V at the drain end. The net voltage available to induce an inversion layer therefore decreases as one approaches the drain end of the channel. Hence, we expect the induced channel charge density to vary from a maximum at the source (where V_{gs} minus the channel potential is largest) to a minimum at the drain end of the channel (where V_{gs} minus the channel potential is smallest), as shown by the shaded region representing charge density in Figure 5.3.

Specifically, the channel charge density has the following form:

$$Q_n(y) = -C_{ox}\{[V_{gs} - V(y)] - V_t\}, \tag{1}$$

where $Q_n(y)$ is the charge density at position y, C_{ox} is ε_{ox}/t_{ox}, and $V(y)$ is the channel potential at position y. Note that we follow the convention of defining the y-direction as along the channel. Note also that C_{ox} is a capacitance *per unit area.* The negative sign simply reflects that the charge is made up of electrons in this NMOS example.

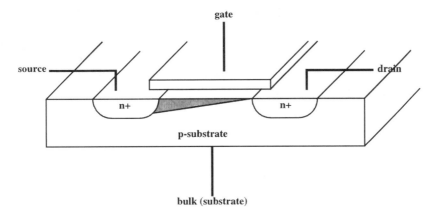

FIGURE 5.3. n-channel MOSFET (shown at boundary between triode and saturation).

Equation 1 is all we really need in order to derive the most important equations governing the terminal characteristics.

5.4.1 DRAIN CURRENT IN THE LINEAR (TRIODE) REGION

The linear or triode region of operation is defined as one in which V_{gs} is large enough (or V_{ds} small enough) to guarantee the formation of an inversion layer for the whole distance from source to drain. From our expression for the channel charge density, we see that it has a zero value when

$$[V_{gs} - V(y)] - V_t = 0. \tag{2}$$

The charge density thus first becomes zero at the drain end at some particular voltage. Hence, the boundary for the triode region is defined by

$$[V_{gs} - V_{ds}] - V_t = 0 \implies V_{ds} = V_{gs} - V_t \equiv V_{d\text{sat}}. \tag{3}$$

As long as $V_{ds} < V_{d\text{sat}}$, the device will be in the linear region of operation.

Having derived an expression for the channel charge and having defined the linear region of operation, we are now in a position to derive an expression for the device current in terms of the terminal variables. Current is proportional to charge times velocity, so we've just about got it:

$$I_D = -WQ_n(y)v(y). \tag{4}$$

The velocity at low fields (remember, this is the "long channel" approximation) is simply the product of mobility and electric field. Hence,

$$I_D = -WQ_n(y)\mu_n E, \tag{5}$$

where W is the width of the device.

Substituting now for the channel charge density, we obtain:

$$I_D = -WC_{\text{ox}}[V_{gs} - V(y) - V_t]\mu_n E. \tag{6}$$

Next, we note that the (y-directed) electric field E is simply (minus) the gradient of the voltage along the channel. Therefore,

$$I_D = \mu_n C_{\text{ox}} W [V_{gs} - V(y) - V_t]\frac{dV}{dy}, \tag{7}$$

so that

$$I_D\,dy = \mu_n C_{\text{ox}} W [V_{gs} - V(y) - V_t]\,dV. \tag{8}$$

Now integrate along the channel and solve for I_D:

$$\int_0^L I_D\,dy = I_D L = \int_0^{V_{ds}} \mu_n C_{\text{ox}} W [V_{gs} - V(y) - V_t]\,dV. \tag{9}$$

At last, we have the following expression for the drain current in the triode region:

$$I_D = \mu_n C_{\text{ox}} \frac{W}{L}\left[(V_{gs} - V_t)V_{ds} - \frac{V_{ds}^2}{2}\right]. \tag{10}$$

Note that the relationship between drain current and drain-to-source voltage is nearly linear for small V_{ds}. Thus, a MOSFET in the triode region behaves as a voltage-controlled resistor.

The strong sensitivity of drain current to drain voltage is qualitatively similar to the behavior of vacuum tube triodes, which lend their name to this region of operation.

5.4.2 DRAIN CURRENT IN SATURATION

When V_{ds} is high enough so that the inversion layer does not extend all the way from source to drain, the channel is said to be "pinched off." In this case, the field felt by the channel charge ceases to increase, causing the total current to remain constant despite increases in V_{ds}.

Calculating the value of this current is easy; all we need do is substitute $V_{d\text{sat}}$ for V_{ds} in our expression for current:

$$I_D = \mu_n C_{\text{ox}} \frac{W}{L}\left[(V_{gs} - V_t)V_{d\text{sat}} - \frac{V_{d\text{sat}}^2}{2}\right], \tag{11}$$

which simplifies to

$$I_D = \frac{\mu_n C_{\text{ox}}}{2} \frac{W}{L}(V_{gs} - V_t)^2. \tag{12}$$

Hence, in saturation, the drain current has a square-law dependence on the gate–source voltage and is (ideally) independent of drain voltage. Because vacuum tube pentodes exhibit a similar insensitivity of plate current to plate voltage, this regime is occasionally called the pentode region of operation.

The transconductance of such a device in saturation is easily found from differentiating our expression for drain current:

$$g_m = \mu_n C_{\text{ox}} \frac{W}{L} (V_{gs} - V_t), \tag{13}$$

which may also be expressed as

$$g_m = \sqrt{2 \mu_n C_{\text{ox}} \frac{W}{L} I_D}. \tag{14}$$

Thus, in contrast with bipolar devices, a long-channel MOSFET's transconductance depends only on the square root of the bias current.

5.4.3 CHANNEL-LENGTH MODULATION

So far, we've assumed that the drain current is independent of drain–source voltage in saturation. However, measurements on real devices always show a disappointing lack of such independence. The primary mechanism responsible for a nonzero output conductance in long-channel devices is channel-length modulation (CLM). Since the drain region forms a junction with the substrate, there is a depletion region surrounding the drain whose extent depends on the drain voltage. As the drain voltage increases, the depletion zone's width increases as well, effectively shortening the channel. Since the effective length thus decreases, the drain current increases.

To account for this effect, the drain current equations for both triode and saturation are modified as follows:

$$I_D = (1 + \lambda V_{ds}) I_{D0}, \tag{15}$$

where I_{D0} is the drain current when channel-length modulation is ignored and where the parameter λ is a semi-empirical constant whose dimensions are those of inverse voltage. The reciprocal of λ is often given the symbol V_A and called the Early voltage, after the fellow who first explained nonzero output conductance in bipolar transistors (where it is caused by an analogous modulation of base width with collector voltage). The graphical significance of the Early voltage is that it is the common extrapolated zero current intercept of the V_{ds}–I_D device curves. Measurements on real devices almost never permit extrapolation to a single point, but engineers are rarely bothered by such trivial details.

5.4.4 DYNAMIC ELEMENTS

So far, we've considered only DC parameters. Let's now take a look at the various capacitances associated with MOSFETs. These capacitances limit the high-frequency performance of circuits, so we need to understand where they come from and how big they are.

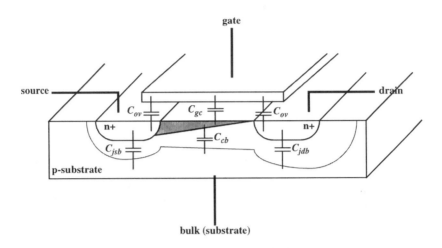

FIGURE 5.4. MOSFET capacitances.

First, since the source and drain regions form reverse-biased junctions with the substrate, one expects the standard junction capacitance from each of those regions to the substrate. These capacitances are denoted C_{jsb} and C_{jdb}, as shown in Figure 5.4, where the extent of the depletion region has been greatly exaggerated.

There are also various parallel plate capacitance terms in addition to the junction capacitances. The capacitors shown as C_{ov} in Figure 5.4 represent gate–source and gate–drain *overlap* capacitances; these are highly undesirable but unavoidable. During manufacture, the source and drain regions may diffuse laterally by an amount similar to the depth that they diffuse. Hence, they bloat out a bit during processing and extend underneath the gate electrode by some amount. As a crude approximation, one may take the amount of overlap, L_D, as 2/3 to 3/4 of the depth of the source–drain diffusions. Hence,

$$C_{ov} \approx \frac{\varepsilon_{ox}}{t_{ox}} W L_D = 0.7 C_{ox} W x_j, \tag{16}$$

where x_j is the depth of the source–drain diffusions, ε_{ox} is the oxide's dielectric constant (about $3.9\varepsilon_0$), and t_{ox} is the oxide thickness.

The parallel plate overlap terms are augmented by fringing and thus the "overlap" capacitance would be nonzero even in the absence of physical overlap. In this context, one should keep in mind that (in modern devices) the gate electrode is actually considerably thicker than the channel is long, so the relative dimensions of Figure 5.4 are misleading. Think of a practical gate electrode as a tall oak tree instead of a thin plate. In addition, the interconnecting wires to the source and drain are hardly of negligible dimensions. See Figure 5.5. Because the thickness of the gate electrode now scales little (if at all), the "overlap" capacitance now changes somewhat slowly from generation to generation.

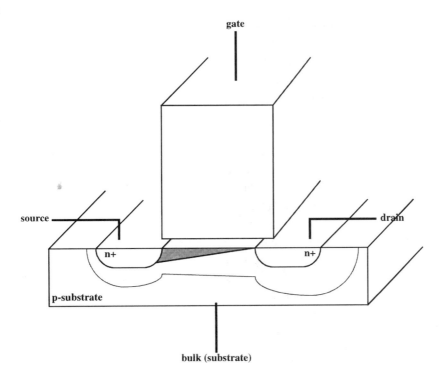

FIGURE 5.5. More accurate depiction of relative gate electrode size.

Another parallel plate capacitance is the gate-to-channel capacitance, C_{gc}. Since both the source and drain regions extend into the region underneath the gate, the effective channel length decreases by twice the bloat, L_D. Hence, the total value of C_{gc} is

$$C_{gc} = C_{ox}W(L - 2L_D). \tag{17}$$

In strong inversion, the charge carriers at the surface and those in the bulk are of the opposite type. In between there is a depletion region. As a result, there is also a capacitance between the channel and the bulk, C_{cb}, that behaves as a junction capacitance. Its value is approximately

$$C_{cb} \approx \frac{\varepsilon_{Si}}{x_d}W(L - 2L_D), \tag{18}$$

where x_d is the depth of the depletion layer, whose value is given by:

$$x_d = \sqrt{\frac{2\varepsilon_{Si}}{qN_{sub}}|\phi_s - \phi_F|}. \tag{19}$$

The quantity within absolute-value bars is the difference between the surface potential and the Fermi level in the substrate. In strong inversion (for both triode and saturation regions), this quantity has a magnitude of twice the Fermi level.

The channel is not an explicitly accessible terminal of the device, so finding how the various capacitive terms contribute to the terminal capacitances requires knowledge of how the channel charge divides between the source and drain. In general, the values of the terminal capacitances depend on the operating regime because bias conditions affect this partitioning of charge. For example, when there is *no* inversion charge (the device is "off"), the gate–source and gate–drain capacitances are just the overlap terms to a good approximation.

When the device is in the linear region there is an inversion layer, and one may assume that the source and drain share the channel charge equally. Hence, half of C_{gc} adds to the overlap terms. Similarly, the C_{jsb} and C_{jdb} junction terms are each augmented by half of C_{cb} in the linear region.

In the saturation region, potential variations at the drain region don't influence the channel charge. Hence, there is no contribution to C_{gd} by C_{gc}; the overlap term is all there is. The gate–source capacitance *is* affected by C_{gc}, but "detailed considerations"[3] show that only about 2/3 of C_{gc} should be added to the overlap term. Similarly, C_{cb} contributes nothing to C_{db} in saturation but does contribute 2/3 of its value to C_{sb}.

The gate–bulk capacitance may be taken as zero in strong inversion (both in triode and in saturation, as the channel charge essentially shields the bulk from what's happening at the gate). When the device is off, however, there is a gate-voltage–dependent capacitance whose value varies in a roughly linear manner between C_{gc} and the series combination of C_{gc} and C_{cb}. Below (but near) threshold, the value is closer to the series combination and approaches a limiting value of C_{gc} in deep accumulation, where the surface majority carrier concentration increases above that of the bulk owing to the positive charge induced by the strong negative gate bias. In deep accumulation, the surface is strongly conducting and may therefore be treated as essentially a metal, leading to a gate–bulk capacitance that is the full parallel plate value.

The variation of this capacitance with bias presents one additional option for realizing varactors. To avoid the need for negative supply voltages, the capacitor may be built in an n-well using n+ source and drain regions to form an accumulation-mode MOSFET capacitor, as described in Chapter 3. The terminal capacitances are summarized in Table 5.1.

5.4.5 HIGH-FREQUENCY FIGURES OF MERIT

It is perhaps natural to attempt to characterize multidimensional quantities with a single number; laziness is universal, after all. In the specific case of high-frequency performance, two figures of merit are particularly popular. These are ω_T and $\omega_{\max}$,

---

[3] The 2/3 factor arises from the calculation of channel charge and inherently comes from integrating the triangular distribution assumed in Figure 5.3 in the square-law regime.

Table 5.1. *Approximate MOSFET terminal capacitances*

	Off	Triode	Saturation
C_{gs}	C_{ov}	$C_{gc}/2 + C_{ov}$	$2C_{gc}/3 + C_{ov}$
C_{gd}	C_{ov}	$C_{gc}/2 + C_{ov}$	C_{ov}
C_{gb}	$C_{gc}C_{cb}/(C_{gc} + C_{cb})$ $< C_{gb} < C_{gc}$	0	0
C_{sb}	C_{jsb}	$C_{jsb} + C_{cb}/2$	$C_{jsb} + 2C_{cb}/3$
C_{db}	C_{jdb}	$C_{jdb} + C_{cb}/2$	C_{jdb}

which are the frequencies at which the current and power gains (respectively) are extrapolated to fall to unity. It is worthwhile to review briefly their derivation since many engineers forget the origins and precise meanings of these quantities, often drawing incorrect inferences as a result.

The most common expression for ω_T assumes that the drain is terminated in an incremental short circuit while the gate is driven by an ideal current source. As a consequence of the shorted termination, ω_T does not include information about drain–bulk capacitance. The current–source drive implies that series gate resistance similarly has no influence on ω_T. Clearly, both r_g and C_{jdb} can have a strong effect on high-frequency performance, but ω_T simply ignores this reality.

Furthermore, the gate-to-drain capacitance is considered only in the computation of the input impedance; its feedforward contribution to output current is neglected. With these assumptions, the ratio of drain current to gate current is

$$\left| \frac{i_d}{i_{\text{in}}} \right| \approx \frac{g_m}{\omega(C_{gs} + C_{gd})}, \tag{20}$$

which has a unity value at a frequency

$$\omega_T = \frac{g_m}{C_{gs} + C_{gd}}. \tag{21}$$

Now, the frequency at which the (extrapolated) current gain goes to unity really has no fundamental importance; it is simply easy to compute. Perhaps more relevant is the frequency at which the maximum power gain is extrapolated to fall to unity. To compute ω_{max} in general is quite difficult, however, so we will invoke several simplifying assumptions to make an approximate derivation possible. Specifically, we compute the input impedance with an incrementally shorted drain and ignore the feedforward current through C_{gd}, just as in the computation of ω_T. We do consider the feed*back* from drain to gate through C_{gd} in computing the output impedance, however, which is important because computation of the maximum power gain requires termination in a conjugate match.

With these assumptions, we can calculate the power delivered to the input by the current source drive as simply

$$P_{\text{in}} = \frac{i_{\text{in}}^2 r_g}{2},$$ (22)

where r_g, the series gate resistance, is the only dissipative element in the input circuit.

The magnitude of the short-circuit current gain at high frequencies is approximately given by the same expression used in the computation of ω_T:

$$\left| \frac{i_D}{i_{\text{in}}} \right| \approx \frac{\omega_T}{\omega}.$$ (23)

It is also straightforward to show that the resistive part of the output impedance is roughly

$$g_{\text{out}} \approx g_m \cdot \frac{C_{gd}}{C_{gd} + C_{gs}} = \omega_T \cdot C_{gd}.$$ (24)

If the conjugate termination has a conductance of this value then the power gain will be maximized, with half of the g_m generator's current going into the conductance of the termination and the balance into the device itself. The total maximum power gain is therefore

$$\frac{P_L}{P_{\text{in}}} \approx \frac{\frac{1}{2} \left(\frac{\omega_T}{\omega} \cdot i_{\text{in}} \cdot \frac{1}{2} \right)^2 \frac{1}{(\omega_T \cdot C_{gd})}}{\frac{i_{\text{in}}^2 r_g}{2}} \approx \frac{\omega_T}{\omega^2 4 r_g C_{gd}},$$ (25)

which has a unity value at a frequency given by

$$\omega_{\text{max}} \approx \frac{1}{2} \sqrt{\frac{\omega_T}{r_g C_{gd}}}.$$ (26)

It is clear that ω_{max} depends on the gate resistance, so it is more comprehensive in this regard than ω_T. Because judicious layout can reduce gate resistance to small values, ω_{max} can be considerably larger than ω_T for many MOSFETs. The output capacitance has no effect on ω_{max}, because it can be tuned out with a pure inductance and therefore does not limit the amount of power that may be delivered to a load.

Measurements of both ω_{max} and ω_T are carried out by increasing the frequency until a noticeable drop occurs in maximum power gain or current gain. A simple extrapolation to unity value then yields ω_{max} and ω_T. Because these are extrapolated values, it is not necessarily a given that one may actually construct practical circuits operating at, say, ω_{max}. These figures of merit should instead be taken as rough indications of high-frequency performance capability.

Transit-Time Effects (Nonquasistatic Behavior)

The lumped models of this chapter clearly cannot apply over an arbitrarily large frequency range. As a rough rule of thumb, one may usually ignore with impunity the true distributed nature of transistors up to roughly a tenth or fifth of ω_T. As frequencies increase, however, crude lumped models become progressively inadequate. The most conspicuous shortcomings may be traced to a neglect of transit-time ("nonquasistatic" or NQS) effects.

To understand qualitatively the most important implications of transit-time effects, consider applying a step in gate-to-source voltage. Charge is induced in the channel and then drifts toward the drain, arriving some time later because of the finite carrier velocity. Hence, the transconductance has a phase delay associated with it.

A side effect of this delayed transconductance is a change in the input impedance: the delayed feedback from the channel back through the gate capacitance necessarily prevents a pure quadrature relationship between gate voltage and gate current. As a consequence, the applied gate voltage performs work on the channel charge. This dissipation must be accounted for in any correct circuit model. Van der Ziel has shown[4] that, at least for long-channel devices, the transit delay causes the gate admittance to have a real part that grows as the square of frequency:

$$g_g = \frac{\omega^2 C_{gs}^2}{5g_{d0}}. \tag{27}$$

To get roughly calibrated on the magnitudes implied by Eqn. 27, assume that g_{d0} is approximately equal to g_m. Then, to a crude approximation,

$$g_g \approx \frac{g_m}{5}\left(\frac{\omega}{\omega_T}\right)^2. \tag{28}$$

Hence, this shunt conductance is negligible as long as operation well below ω_T is maintained. However, the *thermal noise* associated with this conductance may not be ignored in the design of low-noise circuits (the fluctuation–dissipation theorem[5] of physics tells us that "dissipation implies noise"). We shall see later that a proper noise figure calculation must take this noise source into account. Finally, because the derivation of the maximum unity power gain frequency presented earlier neglects nonquasistatic effects, it overestimates the true value of $\omega_{\max}$.

5.4.6 TECHNOLOGY SCALING IN THE LONG-CHANNEL LIMIT

Now that we have examined both the static and dynamic behavior of MOSFETs, we can derive an approximate expression for ω_T in terms of operating point, process parameters, and device geometry. We've already derived an expression for g_m, so all we need is an expression for the requisite capacitances. To simplify the derivation,

---

[4] *Noise in Solid State Devices and Circuits,* Wiley, New York, 1986.

[5] Strictly speaking, the theorem applies only to systems in thermal equilibrium. Although transistor circuits violate this condition, the general qualitative insight regarding the relationship between dissipation and noise remains valuable.

let us assume that C_{gs} dominates the input capacitance and is itself dominated by the parallel plate capacitance. Then, in saturation:

$$\omega_T \approx \frac{g_m}{C_{gs}} \approx \frac{\mu_n C_{\text{ox}} (W/L)(V_{gs} - V_t)}{\frac{2}{3} WLC_{\text{ox}}} = \frac{3}{2} \frac{\mu_n (V_{gs} - V_t)}{L^2}. \tag{29}$$

Hence, ω_T depends on the inverse square of the length and increases with increasing gate–source voltage. Remember, though, that this equation holds only in the long-channel regime.

5.5 OPERATION IN WEAK INVERSION (SUBTHRESHOLD)

In simple MOSFET models (such as the one we've presented so far), the device conducts no current until an inversion layer forms. However, mobile carriers don't abruptly disappear the moment the gate voltage drops below V_t. In fact, exercising a little imagination, one can discern a structure reminiscent of an n–p–n bipolar transistor when the device is in the subthreshold regime, with the source and drain regions functioning as emitter and collector, respectively, and the (noninverted) bulk behaving a bit like a base.

As V_{gs} drops below threshold, the current decreases in an exponential fashion, much like a bipolar transistor. Rather than dropping at the 60 mV/decade rate of such a bipolar, however, the current in all real MOSFETs drops more slowly (e.g., 100 mV/decade) owing to the capacitive voltage division between gate and channel and between channel and source or bulk.

5.5.1 BACKGROUND

Recall that in strong inversion the density of charge carriers induced at the surface is at least as great as the (oppositely charged) mobile carrier density in the bulk. This inversion charge then drifts from source to drain under the influence of a lateral field. In fact, we have implicitly assumed so far that carrier transport is *entirely* by drift. Furthermore, we have simply stated (without explanation) that strong inversion corresponds to values of gate overdrives of at least "several" kT/q. We now consider device behavior in weak inversion (a term we will use interchangeably with subthreshold), where gate overdrives do not satisfy this requirement; we also quantify what is meant by "several."

To extend our analysis into the weak inversion regime, first acknowledge that the mobile carrier density at the surface does not drop abruptly to zero as the gate voltage diminishes. However, because the carrier density will be low, the contribution to drain current by drift will be small. We therefore need to consider the real possibility that current transport by *diffusion* may also be important in this operating regime. This acknowledgment is the key to understanding operation in weak inversion.

Before presenting any equations, it's useful to examine a couple of V–I plots in order to develop a feel for the phenomena we are trying to describe mathematically. For example, a plot of the square root of drain current versus gate–source voltage

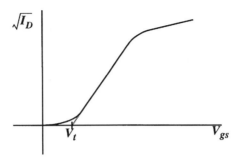

FIGURE 5.6. *V–I* plot for a "more real"
MOSFET ($V_{ds} > V_{dsat}$).

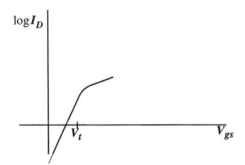

FIGURE 5.7. Semilog *V–I* plot for a
"more real" MOSFET ($V_{ds} > V_{dsat}$).

ought to be a simple straight line – if we believe in the simplest long-channel equations. However, real devices behave differently, as shown in Figure 5.6

One noteworthy feature of this plot is that the current does not drop to zero below threshold. Rather, it tails off in some manner, and the region where this occurs is what we are calling the weak inversion regime.[6] Beyond some amount above threshold, the current is proportional to the square of the overdrive; we call this region the *square-law* regime. At very high overdrives, the dependence of drain current on overdrive becomes subquadratic owing to various high-field effects (both lateral and vertical).

From the plot in Figure 5.6, we see that the threshold voltage is perhaps most naturally defined as the *extrapolated* zero-current intercept of the linear portion of the curve above threshold.

Additional insights result if we present the same data on a semilog plot; see Figure 5.7. Below threshold, we observe that the current depends exponentially on the gate–source voltage – instead of quadratically (or subquadratically).

Keep these plots in mind as we present the equations that follow.

---

[6] It is by no means required that the drain current go to zero at zero V_{gs}, by the way.

5.5.2 SUBTHRESHOLD MODEL EQUATIONS

The equations we present here made their debut in the level-2 SPICE model, the first to accommodate operation in weak inversion. Although level 3 replaced the equations for strong inversion, it subsumed those for weak inversion without modification.

The weak inversion model begins by defining a subthreshold slope parameter, n, as follows:

$$n = 1 + \frac{qN_{\text{FS}}}{C_{\text{ox}}} + \frac{C_B}{C_{\text{ox}}}, \tag{30}$$

where N_{FS} is the (voltage-dependent) fast surface state density, C_{ox} is the gate capacitance per unit area, and C_B is the depletion capacitance (defined in turn as dQ_B/dV_{BS}) per unit area. The parameter N_{FS} is the primary subthreshold model parameter. One adjusts its value until the simulated subthreshold behavior matches experimental data (if only it were that easy …). Note also that the value of n always exceeds unity. A typical value for n is in the range of 3–4, but your mileage may vary.

The subthreshold slope parameter works in tandem with the thermal voltage, kT/q (about 26 mV at room temperature) to define a value of V_{gs} that marks an arbitrary boundary between strong and weak inversion. This boundary, conventionally called V_{ON}, is defined as

$$V_{\text{ON}} = V_t + n\frac{kT}{q}. \tag{31}$$

Hence the gate overdrive corresponding to the onset of strong inversion is simply nkT/q. Therefore, whenever we say that operation in strong inversion requires that one exceed the threshold by "several" kT/q, we mean "at least n" units of kT/q. For typical values of n, this prescription corresponds (at room temperature) to minimum gate overdrives of about 100 mV to guarantee operation in strong inversion.

The drain current at V_{ON} is usually called I_{ON}. Continuity of the current equations across the boundary is assured by requiring that the strong and weak inversion equations predict the same drain current, I_{ON}, at V_{ON}. In practice, I_{ON} is found by evaluating the strong inversion equations at a gate overdrive of nkT/q. However, even though equality of currents at the boundary is then guaranteed, the derivatives of current unfortunately almost never share this continuity, and quantities that are sensitive to the derivatives (such as conductance) will typically take on nonsensical values at or near the boundary between strong and weak inversion. Regrettably, this limitation of the level-2/level-3 model is a fundamental one, and the user has no recourse other than to regard with skepticism any simulation result associated with operation near threshold. The message here is "know when to distrust your models."

The exponential behavior in weak inversion is described quantitatively by an ideal diode law as follows:

$$I_D = I_{\text{ON}}\left[\exp\left(\frac{qV_{\text{od}}}{nkT} - 1\right)\right], \tag{32}$$

where V_{od} is the gate overdrive, which can take on negative values. Note that, at zero overdrive, the drain current is smaller than the current at V_{ON} by a factor of e.

From Eqn. 32 it is easy to see why n is known as the subthreshold slope parameter. If we consider a semilog plot of drain current versus gate voltage, the slope of the resulting curve is proportional to $1/n$. One measure of this slope is the voltage change required to vary the drain current by some specified ratio. At room temperature, a factor-of-10 change in drain current results for every $60n$ mV of gate voltage change. Thus, a lower n implies a steeper change in current for a given change in gate voltage. A small value of n is desirable because a more dramatic difference between "off" and "on" states is then implied. This consideration is particularly important in the modern gigascale era, where small leakage currents in nominally "off" devices can result in significant total chip current flow. For example, a 100-pA subthreshold current per transistor may seem small, but ten billion transistors leaking this amount implies a total "off" current of one ampere!

5.5.3 SUMMARY OF SUBTHRESHOLD MODELS

From the equations presented, the reader can appreciate that N_{FS} is the main parameter that can be adjusted to force agreement between simulation and experiment. Perfect agreement is not possible, but considerable improvement can be obtained with a suitable choice of N_{FS}. That said, level 2 and level 3 offer an alternative way of modeling weak inversion. This alternative separately computes the drift and diffusion contributions to drain current and then adds them together, an approach that avoids any artificial distinction between strong and weak inversion regions of operation. For those interested in studying this alternative further, see for example Antognetti et al., "CAD Model for Threshold and Subthreshold Conduction in MOSFETs" (*IEEE J. Solid-State Circuits,* v. 17, 1982). A brief summary is also found in Massobrio and Antognetti's *Semiconductor Device Modeling with SPICE* (McGraw-Hill, New York, 1993, p. 186).

Finally, many bipolar analog circuits are often translated into MOS form by operating the devices in this regime. However, such circuits typically exhibit poor frequency response because MOSFETs possess small g_m (but good g_m/I) in this region of operation. As devices continue to shrink, the frequency response can nonetheless be good enough for many applications, but careful verification is in order.

5.6 MOS DEVICE PHYSICS IN THE
SHORT-CHANNEL REGIME

The continuing drive to shrink device geometries has resulted in devices so small that various high-field effects become prominent at moderate voltages. The primary high-field effect is that of velocity saturation.

Because of scattering by high-energy ("optical") phonons, carrier velocities eventually cease to increase with increasing electric field. As the electric field approaches about 10^6 V/m in silicon, the electron drift velocity displays a progressively weakening dependence on the field strength and eventually saturates at a value of about 10^5 m/s.

In deriving equations for long-channel devices, the saturation drain current is assumed to correspond to the value of current at which the channel pinches off. In short-channel devices, the current saturates when the carrier velocity does.

To accommodate velocity saturation, begin with the long-channel equation for drain current in saturation:

$$I_D = \frac{\mu_n C_{ox}}{2} \frac{W}{L} (V_{gs} - V_t)^2,\tag{33}$$

which may be rewritten as

$$I_D = \frac{\mu_n C_{ox}}{2} \frac{W}{L} (V_{gs} - V_t) V_{dsat,l}.\tag{34}$$

Here the long-channel V_{dsat} is denoted $V_{dsat,l}$ and is equal to $(V_{gs} - V_t)$.

As stated earlier, the drain current saturates when the velocity does, and the velocity saturates at smaller voltages as the device gets shorter. Hence, we expect V_{dsat} to diminish with channel length.

It can be shown that V_{dsat} may be expressed more generally by the following approximation:[7]

$$V_{dsat} \approx (V_{gs} - V_t) \parallel (LE_{sat}) = \frac{(V_{gs} - V_t)(LE_{sat})}{(V_{gs} - V_t) + (LE_{sat})}.\tag{35}$$

Hence,

$$I_D = \frac{\mu_n C_{ox}}{2} \frac{W}{L} (V_{gs} - V_t)[(V_{gs} - V_t) \parallel (LE_{sat})],\tag{36}$$

where E_{sat} is the field strength at which the carrier velocity has dropped to half the value extrapolated from low-field mobility.

It should be clear from the preceding equations that the prominence of "short channel" effects depends on the ratio of $(V_{gs} - V_t)/L$ to E_{sat}. If this ratio is small, then the device still behaves as a long device; the actual channel length is irrelevant. All that happens as the device shortens is that less gate overdrive $(V_{gs} - V_t)$ is needed to induce the onset of these effects.

Given our definition of E_{sat}, the drain current may be rewritten as

$$I_D = WC_{ox}(V_{gs} - V_t)v_{sat}\left[1 + \frac{LE_{sat}}{V_{gs} - V_t}\right]^{-1}.\tag{37}$$

A typical value for E_{sat} is about 4×10^6 V/m. Although E_{sat} is somewhat process-dependent, we will treat it as constant in all that follows.

For values of $(V_{gs} - V_t)/L$ that are large compared with E_{sat}, the drain current approaches the following limit:

$$I_D = \frac{\mu_n C_{ox}}{2} W(V_{gs} - V_t)E_{sat}.\tag{38}$$

---

[7] Ping K. Ko, "Approaches to Scaling," *VLSI Electronics: Microstructure Science,* v. 18, Academic Press, New York, 1989.

That is, the drain current eventually *ceases to depend on the channel length*. Furthermore, the relationship between drain current and gate–source voltage becomes incrementally *linear* rather than square-law.

Let's do a quick calculation to obtain a rough estimate of the saturation current in the short-channel limit. In modern processes, t_{ox} is less than 2.5 nm (and shrinking all the time), so that C_{ox} is about 0.015 F/m^2. Assuming a mobility of 0.04 m^2/V-s, an E_{sat} of 4×10^6 V/m, and a gate overdrive of 1 V, the drain current under these conditions and assumptions exceeds one milliampere per micron of gate width. Despite the crude nature of this calculation, actual devices do behave similarly, although one should keep in mind that channel lengths must be much shorter than 0.5 μm in order to validate this estimate for this value of overdrive.

Since the gate overdrive is more commonly a couple hundred millivolts in analog applications, a reasonably useful number to keep in mind for rough order-of-magnitude calculations is that the saturation current is of the order of 200 mA for every millimeter of gate width for devices operating in the short-channel limit. Keep in mind that this value does depend on the gate voltage and C_{ox} (among other things) and is thus a function of technology scaling.

In all modern processes, the minimum allowable channel lengths are short enough for these effects to influence device operation in a first-order manner. However, note that there is no *requirement* that the circuit designer use minimum-length devices in all cases; one certainly retains the option to use devices whose lengths are greater than the minimum value. This option is regularly exercised when building current sources to boost output resistance. Furthermore, many process technologies offer devices with different gate oxide thicknesses (and corresponding minimum lengths) to facilitate interfacing with higher-voltage legacy I/O circuits, further expanding the palette of options.

5.6.1 EFFECT OF VELOCITY SATURATION ON TRANSISTOR DYNAMICS

In view of the first-order effect of velocity saturation on the drain current, we ought to revisit the expression for ω_T to see how device scaling affects high-frequency performance in the short-channel regime.

First, let's compute the limiting transconductance of a short-channel MOS device in saturation:

$$g_m \equiv \frac{\partial I_D}{\partial V_{gs}} = \frac{\mu_n C_{ox}}{2} W E_{sat}. \tag{39}$$

Using the same numbers as for the limiting saturation current, we find that the transconductance should be roughly 1 mS per micron of gate width (easy numbers to remember: everything is roughly 1 somethings per μm). Note that the only practical control over this value at the disposal of a device designer is through the choice of t_{ox} to adjust C_{ox} (unless a different dielectric material is used).

To simplify calculation of ω_T, assume (as before) that C_{gs} dominates the input capacitance. Further assume that short-channel effects do not appreciably influence charge sharing, so that C_{gs} still behaves approximately as in the long-channel limit:

$$C_{gs} \approx \tfrac{2}{3} WLC_{ox}. \tag{40}$$

Taking the ratio of g_m to C_{gs} then yields:

$$\omega_T \approx \frac{g_m}{C_{gs}} \approx \frac{(\mu_n C_{ox}/2)WE_{sat}}{\tfrac{2}{3}WLC_{ox}} = \frac{3}{4}\frac{\mu_n E_{sat}}{L}. \tag{41}$$

We see that the ω_T of a short-channel device thus depends on $1/L$ instead of on $1/L^2$. Additionally, note that it does not depend on bias conditions (but keep in mind that this independence holds only in saturation) nor on oxide thickness or composition.

To get a rough feel for the numbers, assume a μ of 0.04 m^2/V-s, an E_{sat} of 4×10^6 V/m, and an effective channel length of 0.18 μm. With these values, f_T works out to nearly 100 GHz (again, this value is very approximate, but it compares reasonably well with typical measured peak values of 70 GHz – particularly if we correct for C_{gd}). In practice, substantially smaller values are measured because smaller gate overdrives are used in actual circuits (so that the device is not operated deep in the short-channel regime), and also because the overlap capacitances are not actually negligible (in fact, they are frequently within a factor of 2 or 3 of C_{gs}). As a consequence, practical values of f_T could be as much as a factor of 3 or so lower than the peak values most often reported.

Minimum effective channel lengths continue to shrink, of course, and process technologies just making the transition from the laboratory to production possess practical f_T values in excess of 100 GHz. At the ultimate limit of MOS scaling (thought to arrive when gate lengths are about 10 nm), the corresponding f_T values should be around 1.5 THz. This range of values is similar to that offered by many high-performance bipolar processes, and it is one reason that MOS devices are increasingly found in applications previously served only by bipolar or GaAs technologies.

5.6.2 THRESHOLD REDUCTION

We've already seen that higher drain voltages cause channel shortening, resulting in a nonzero output conductance. When the channel length is small, the electric field associated with the drain voltage may extend enough toward the source that the effective threshold diminishes. This *drain-induced barrier lowering* (DIBL, pronounced "dibble") can cause dramatic increases in subthreshold current (keep in mind the exponential sensitivity of the subthreshold current). Additionally, it results in a degradation in output conductance beyond that associated with simple channel-length modulation.

A plot of threshold voltage as a function of channel length shows a monotonic decrease in threshold as length decreases. At the 0.5-μm level, the threshold reduction can be 100–200 mV over the value in the long-channel limit, corresponding to potential increases in subthreshold current by factors of 10 to 1000.

To reduce the peak channel field and thereby mitigate high-field effects, a lightly doped drain (LDD) structure is almost always used in modern devices. In such a transistor, the doping in the drain region is arranged to have a spatial variation, progressing from relatively heavy near the drain contact to lighter somewhere in the channel. In some cases, the doping profile results in overcompensation in the sense that higher drain voltages actually *increase* the threshold over some range of drain voltages before ultimately decreasing the threshold. Not all devices exhibit this *reverse short-channel effect,* since its existence depends on the detailed nature of the doping profile. Additionally, PMOS devices do not exhibit high-field effects as readily as do NMOS transistors, since the field strengths necessary to cause hole velocity to saturate are considerably higher than those that cause electron velocity saturation.

5.6.3 SUBSTRATE CURRENT

The electric field near the drain can reach extraordinarily large values with moderate voltages in short-channel devices. As a consequence, carriers can acquire enough energy between scattering events to cause impact ionization upon their next collision. Impact ionization by these "hot" carriers creates hole–electron pairs and, in an NMOS device, the holes are collected by the substrate while the electrons flow to the drain (as usual). The resulting substrate current is a sensitive function of the drain voltage, and this current represents an additional conductance term shunting the drain to ground. This effect is of greatest concern when one is seeking the minimum output conductance at high drain–source voltages.

5.6.4 GATE CURRENT

The same hot electrons responsible for substrate current can actually cause *gate current.* The charge comprising this gate current can become trapped in the oxide, causing upward threshold shifts in NMOS devices and threshold reductions in PMOS devices. Although this effect is useful if one is trying to make nonvolatile memories, it is most objectionable in ordinary circuits because it degrades long-term reliability.

As device scaling continues on its remarkable exponential trajectory, gate oxide becomes thin enough for tunneling current to become an issue. At the 0.13-μm process generation, the physical gate oxide is on the order of 1 nm thick – corresponding to only a few atomic layers of oxide or thereabouts – and is typically controlled to a precision of one atom. Scaling much further simply can't continue, so the industry is currently searching actively for a replacement for our beloved silicon dioxide. A consensus on what that replacement will be has yet to emerge.

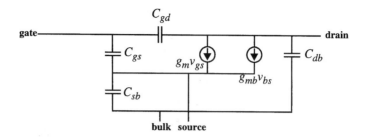

FIGURE 5.8. Incremental MOSFET model including back-gate effect (resistive elements not shown).

5.7 OTHER EFFECTS

5.7.1 BACK-GATE BIAS

Another important effect is that of *back-gate bias* (often called the "body effect"). Every MOSFET is actually a four-terminal device, and one must recognize that variations in the potential of the bulk relative to the other device terminals will influence device characteristics. Although the source and bulk terminals are frequently tied together, there are important instances when they are not. An example that quickly comes to mind is a current–source biased differential pair – the potential of the common source connection is higher than that of the bulk in this case, and it moves around with input common-mode voltage.

As the potential of the bulk terminal of an NMOS transistor becomes increasingly negative with respect to the source, the depletion region formed between the channel and the bulk increases in extent, increasing the amount of fixed negative charge in the channel. This increased charge tends to repel negative charge coming from the source and thus increases the value of V_{gs} required to form and maintain an inversion layer. Therefore, the threshold voltage increases. This back-gate bias effect (so called because the bulk may be considered another gate terminal) thus has both large- and small-signal implications. The influence of this variation is accounted for in level-1 SPICE models using the following equation:

$$V_t = V_{t0} + \gamma \left(\sqrt{2\phi_F - V_{BS}} - \sqrt{2\phi_F} \right). \tag{42}$$

The parameter V_{t0} is the threshold voltage at zero bulk-to-source voltage, ϕ_F is the Fermi level deep in the bulk, and V_{BS} is the bulk-to-source voltage. (A common error is to forget that the SPICE model parameter PHI *already includes the factors of 2* shown in Eqn. 42.)

The small-signal model accommodates back-gate bias effect by adding another dependent current source, this time controlled by the bulk-to-source voltage; see Figure 5.8. Formally, the back-gate transconductance is simply the derivative of drain current with respect to bulk-to-source voltage. In saturation, this transconductance is

$$g_{mb} \equiv \frac{i_d}{v_{bs}} = \left[\mu C_{\mathrm{ox}} \frac{W}{L} (V_{gs} - V_t) \right] \cdot \frac{\partial V_t}{\partial V_{bs}} = g_m \frac{\gamma}{2\sqrt{2\phi_F + V_{sb}}}. \qquad (43)$$

The parameter γ is called, sensibly enough, GAMMA in the level-1 SPICE models. The back-gate transconductance g_{mb} is typically no larger than about 30% of the main transconductance and is frequently about 10% of g_m. However, these are hardly universal or static truths, so one should always check the detailed models and bias conditions before making such assumptions.

Unintended – and much undesired – modulation of the source–bulk potential can also occur owing to static and dynamic signal currents flowing through the substrate. This coupling can cause serious problems in mixed-signal circuits. Extremely careful attention to layout is necessary in order to reduce noise problems arising from this mechanism.

5.7.2 TEMPERATURE VARIATION

There are two primary temperature-dependent effects in MOS devices. The first is a change in threshold. Although its precise behavior depends on the detailed device design, the threshold tends to have a TC (temperature coefficient) similar to that of base-emitter voltage V_{BE}, namely, about -2 mV/°C (within a factor of 2).

The other effect, that of mobility reduction with increasing temperature, tends to dominate because of its exponential nature:

$$\mu(T) \approx \mu(T_0) \left[\frac{T}{T_0} \right]^{-3/2}, \qquad (44)$$

where T_0 is some reference temperature (e.g., 300 K). At a fixed bias, then, the drain current drops as temperature increases.

5.7.3 NORMAL-FIELD MOBILITY DEGRADATION

As the gate potential increases, the electrons in the channel are encouraged to flow closer to the silicon–oxide interface. Remember that the interface is full of dangling bonds, various ionic contaminants, and abandoned cars. As a consequence, there is more scattering of carriers and thus a decrease in mobility. Hence, the available drain current drops below what one would expect if mobility were to stay fixed. Since the vertical field is proportional to the gate overdrive, it is perhaps not surprising that the actual drain current is the value given by the previous equations, multiplied by the following factor:

$$\frac{1}{1 + \theta(V_{gs} - V_t)}, \qquad (45)$$

where θ, the normal-field mobility degradation factor, has a typical value in the range of 0.1–1 V^{-1}. It is technology-dependent, growing as t_{ox} shrinks. In the absence of measured data, an *extremely* crude estimate of θ can be obtained from

$$\theta \approx \frac{2 \times 10^{-9} \text{ m/V}}{t_{\text{ox}}}. \tag{46}$$

Although there are certainly additional effects that influence the behavior of real devices (e.g., variation of threshold voltage along the channel), the foregoing phenomena are of greatest relevance to the designer of analog and RF circuits.

5.8   SUMMARY

We've seen that long- and short-channel devices exhibit different behavior and that the differences are caused by variation in mobility with electric field. What distinguishes "long" from "short" is actually a function of electric field strengths. Because electric field is dependent on length, longer devices do not exhibit these high-field effects as readily as shorter ones do.

5.9   APPENDIX A: 0.5-μm LEVEL-3 SPICE MODELS

The following set of models is fairly typical for a 0.5-μm (drawn) process technology. Because level-3 models are quasiempirical, not all of the parameters may take on physically reasonable values. They have been adjusted here to provide a reasonable fit to measured device V–I characteristics as well as to limited dynamic data inferred primarily from ring oscillator frequency measurements.

It should be mentioned that there are many other SPICE MOSFET model sets in existence, with BSIM4 a currently popular and widely supported one. The newer models provide better accuracy at the expense of an exponential growth in the number of parameters, not all of which are physically based. In the interest of providing reasonable accuracy with the simplest models, we will make extensive use of the relatively primitive level-3 models presented here.

```
*SPICE LEVEL3 PARAMETERS
.MODEL NMOS NMOS LEVEL=3 PHI=0.7 TOX=9.5E-09 XJ=0.2U TPG=1
+ VTO=0.7 DELTA=8.8E-01 LD=5E-08 KP=1.56E-04
+ UO=420 THETA=2.3E-01 RSH=2.0E+00 GAMMA=0.62
+ NSUB=1.40E+17 NFS=7.20E+11 VMAX=1.8E+05 ETA=2.125E-02
+ KAPPA=1E-01 CGDO=3.0E-10 CGSO=3.0E-10
+ CGBO=4.5E-10 CJ=5.50E-04 MJ=0.6 CJSW=3E-10
+ MJSW=0.35 PB=1.1
* Weff=Wdrawn-Delta_W, The suggested Delta_W is 3.80E-07

.MODEL PMOS PMOS LEVEL=3 PHI=0.7 TOX=9.5E-09 XJ=0.2U TPG=-1
+ VTO=-0.950 DELTA=2.5E-01 LD=7E-08 KP=4.8E-05
+ UO=130 THETA=2.0E-01 RSH=2.5E+00 GAMMA=0.52
+ NSUB=1.0E+17 NFS=6.50E+11 VMAX=3.0E+05 ETA=2.5E-02
+ KAPPA=8.0E+00 CGDO=3.5E-10 CGSO=3.5E-10
```

+ CGBO=4.5E$-$10 CJ=9.5E$-$04 MJ=0.5 CJSW=2E$-$10
+ MJSW=0.25 PB=1
* Weff=Wdrawn$-$Delta_W, The suggested Delta_W is 3.66E$-$07

5.10 APPENDIX B: THE LEVEL-3 SPICE MODEL

This brief appendix summarizes the static device equations corresponding to the level-3 SPICE model. Not surprisingly, the conspicuous limitations of the level-1 model led to development of a level-2 model, which improved upon the level-1 model by including subthreshold conduction. Unfortunately, level 2 has serious problems of its own (mostly related to numerical convergence) and has been largely abandoned as a consequence. Level 3 is a quasiempirical model that not only accommodates subthreshold conduction but also attempts to account for both narrow-width and short-channel effects. Although it, too, is not quite free of numerical convergence problems (largely resulting from a discontinuity associated with the equations that use parameter KAPPA), it possesses just enough additional parameters to provide usable fits to data for a useful range of modern devices over a useful range of operating conditions. Such fits generally require adjustment of parameters to values that may appear to conflict with physics. Even so, it may be said that the level-3 model is essentially the last with parameters that are at least somewhat traceable to the underlying physics and whose equation set is sufficiently small to make calculations by hand or by spreadsheet a practical option (however, you may disagree with this assertion once you see the full equation set). For this reason, many engineers use level-3 models in the early stages of a design, both for hand calculations (to develop important insights into circuit operation) and for initial simulations (to obtain answers more quickly when simulating large circuits over a large parameter space). Later, as the design is believed close to complete, more sophisticated models (e.g., BSIM4) can be used essentially as a verification tool or perhaps for final optimization.

In the equations that follow, the reader may often ask, "Why does the equation have *that* form?" Often the answer is simply, "Because it works well enough without incurring a massive computational overhead." The reader will also note that the equations for level 3 differ in some respects from those presented in the main body of the chapter. Welcome to the world of quasiempirical fitting, where one engineer may favor a different approximation than another.

Without further ado, here are the equations. In triode, the equation for drain current is:

$$I_D = \mu_{\text{eff}} C_{\text{ox}} \frac{W}{L - 2L_D} \left[(V_{gs} - V_t)V_{ds} - \frac{(1 + F_B)V_{ds}^2}{2} \right], \qquad (47)$$

where the effective mobility is now a parameter whose value depends on both the lateral and vertical field. Dependence on lateral field is accommodated using a slightly different approach than outlined previously:

$$\mu_{\text{eff}} = \frac{\mu_s}{1 + \dfrac{\mu_s}{v_{\max} L_{\text{eff}}} V_{ds}}, \tag{48}$$

where μ_s is the carrier mobility at the surface. Dependence of this surface mobility on vertical field is modeled exactly as in the main part of this chapter, however:

$$\mu_s = \frac{\mu_0}{1 + \theta(V_{gs} - V_t)}, \tag{49}$$

where μ_0 is the low-field mobility.

The function F_B attempts to capture the dependence of channel charge changes on the full three-dimensional geometry of the transistor. This function is given by

$$F_B = \gamma \frac{F_S}{4\sqrt{2\phi - V_{bs}}} + F_N. \tag{50}$$

Note that a zero F_B corresponds to ordinary long-channel triode behavior. Regrettably, the complete expression for F_B is an unholy mess, involving various subexpressions and containing quite a few empirical quantities. The reader will naturally be tempted to ask many questions about these equations in particular, but will have to suffer largely in silence.

The first subexpression is not too bad:

$$F_N = \Delta \frac{\pi \varepsilon_{\text{Si}}}{2 C_{\text{ox}} W}. \tag{51}$$

This equation accounts for the change in threshold as the width narrows and comes about fundamentally as a result of considering the fringing field at the edges of the gate. This fringing is modeled as a pair of quarter-cylinders, explaining where the $\pi/2$ factor comes from. As the width increases to infinity, this correction factor approaches zero, as it should.

On the other hand, it is simply hopeless to extract much of intuitive value from the second subexpression, that for F_S:

$$F_S = 1 - \frac{x_j}{L_{\text{eff}}} \left[\frac{L_D + W_C}{x_j} \sqrt{1 - \left(\frac{W_P/x_j}{1 + W_P/x_j} \right)^2} - \frac{L_D}{x_j} \right], \tag{52}$$

where

$$W_P = \sqrt{\frac{2\varepsilon_{\text{Si}}(2\phi_F - V_{bs})}{q N_{\text{sub}}}} \tag{53}$$

and

$$\frac{W_C}{x_j} = d_0 + d_1 \frac{W_P}{x_j} + d_2 \left(\frac{W_P}{x_j} \right)^2. \tag{54}$$

The various fitting constants d_n have the following intuitively obvious values:

$$d_0 = 0.0631353, \tag{55}$$

$$d_1 = 0.8013292, \tag{56}$$

$$d_2 = -0.01110777. \tag{57}$$

The best we can do is to note that the function F_S captures changes in threshold arising from back-gate bias changes as well from narrow-width effect.

Changes in drain current arising from DIBL are also treated as due to shifts in threshold:

$$V_t = V_{FB} + 2\phi_F - \sigma V_{ds} + \gamma F_S \sqrt{2\phi_F - V_{bs}} + F_N(2\phi_F - V_{bs}), \qquad (58)$$

where

$$\sigma = \eta \frac{8.15 \times 10^{-22}}{C_{\text{ox}} L_{\text{eff}}^3}. \qquad (59)$$

The drain current equation considered so far describes the triode region of operation. To develop a corresponding equation for the saturation region, we "merely" substitute $V_{d\text{sat}}$ for the drain–source voltage in the triode current equation. Short-channel effects alter $V_{d\text{sat}}$ to a considerably more complicated form, however:

$$V_{d\text{sat}} = \frac{V_{gs} - V_t}{1 + F_B} + \frac{v_{\text{max}} L_{\text{eff}}}{\mu_s} - \sqrt{\left(\frac{V_{gs} - V_t}{1 + F_B}\right)^2 + \left(\frac{v_{\text{max}} L_{\text{eff}}}{\mu_s}\right)^2}. \qquad (60)$$

The effective channel length is the drawn length minus the sum of twice the lateral diffusion and less a drain-voltage–dependent term:

$$L_{\text{eff}} = L_{\text{drawn}} - 2L_D - \Delta L, \qquad (61)$$

where

$$\Delta L = x_d \left[\sqrt{\left(\frac{E_P x_d}{2}\right)^2 + K(V_{ds} - V_{d\text{sat}})} - \frac{E_P x_d}{2} \right]. \qquad (62)$$

In turn, we have

$$x_d = \sqrt{\frac{2\varepsilon_{\text{Si}}}{q N_{\text{sub}}}}. \qquad (63)$$

Finally, the lateral electric field at the nominal pinch-off point is

$$E_P = \frac{\dfrac{v_{\text{max}} L_{\text{eff}}}{\mu_s} \left(\dfrac{v_{\text{max}} L_{\text{eff}}}{\mu_s} + v_{d\text{sat}} \right)}{l_{\text{eff}} v_{d\text{sat}}} \qquad (64)$$

When fitting a level-3 model to data, one adjusts η and K to match the observed output conductance, taking care to fit subthreshold conduction behavior through an appropriate choice of η. Notice that λ appears nowhere as an explicit parameter (unlike in level 1), although one could, in principle, derive an expression for it from the system of equations provided.

Having gone through this arduous documentation exercise, we conclude now by providing extremely brief explanations of the various model parameters in Table 5.2.

Table 5.2. *SPICE level-3 model parameters*

Parameter name	Conventional symbol	Description
PHI	$\lvert 2\phi_F \rvert$	Surface potential in strong inversion
TOX	t_{ox}	Gate oxide thickness
XJ	x_j	Source–drain junction depth
TPG		Gate material polarity: 0 for Al, −1 if same as subtrate, +1 if opposite substrate; this parameter is ignored if VTO is specified
VTO	V_{T0}	Threshold at $V_{bs} = 0$
DELTA	Δ	Models threshold dependence on width
LD	L_D	Source–drain lateral diffusion, for computing L_{eff}; *not* used to calculate overlap capacitances
KP	$k' = \mu_0 C_{\mathrm{ox}}$	Process transconductance coefficient
UO	μ_0	Low-field carrier mobility at surface
THETA	θ	Vertical-field mobility degradation factor
RSH	$R_{\square}$	Source–drain diffusion sheet resistance; multiplied by NRS and NRD to obtain total source and drain ohmic resistance, respectively
GAMMA	γ	Body-effect coefficient
NSUB	N_A or N_D	Equivalent substrate doping
NFS		Fast surface state density (needed for proper subthreshold calculation)
VMAX	v_{max} or v_{sat}	Maximum carrier drift velocity
ETA	η	Models changes in threshold due to V_{ds} variations (e.g., DIBL)
KAPPA	K	Models effects of channel-length modulation
CGDO	C_{gd0}	Gate–drain overlap capacitance per width
CGSO	C_{gs0}	Gate–source overlap capacitance per width
CGBO	C_{gb0}	Gate–bulk overlap capacitance per length
CJ	C_{j0}	Zero-bias bulk bottom capacitance per unit source–drain area; multiplied by AS and AD to obtain total bottom capacitance of source and drain at $V_{sb} = V_{db} = 0$
MJ	m_j	Bottom source–drain junction grading coefficient
CJSW	$C_{j\mathrm{SW}}$	Zero-bias sidewall junction capacitance per unit perimeter of source–drain adjacent to field; multiplied by PS and PD to obtain total sidewall capacitance
MJSW	$M_{j\mathrm{SW}}$	Sidewall junction grading coefficient
PB	ϕ_j	Bulk junction potential barrier used to compute junction capacitance for other than zero bias

As with the level-1 model, parameters NRS, NRD, AS, AD, PS, and PD are all specified in the device description line – not in the model set itself – because they depend on the dimensions of the device.

DEFICIENCIES OF THE LEVEL-3 MODEL

Although the achievable fits are sometimes remarkably good, it should hardly surprise you that various problems are unavoidable when using something as simple as level 3 to model something as complex as a modern short-channel MOSFET.

Models of all types often exhibit strange behavior at operating-regime boundaries, and level 3 is unfortunately no exception to this rule. Subthreshold current is improperly calculated if parameter NFS is not included. Even when it is included, anomalies in drain current and transconductance are still to be expected near threshold because of discontinuous derivatives. Similarly, "strange things" tend to happen to the model's output conductance near $V_{d\text{sat}}$.

Although narrow-width effects are included in level 3, the best fits occur if an adjusted width (rather than the drawn width) is used in the device equations. The effective value of width is typically smaller than the drawn width. In the absence of actual data, a reasonable estimate for the reduction is approximately 10–20% of the minimum drawn channel length. For example, in a 0.18-μm process, one should probably subtract about 0.02–0.04 μm from the drawn width and use the resulting value in all subsequent device equations.

It is the prevailing wisdom that level 3 becomes increasingly inaccurate beyond the 1-μm technology boundary. The prominence of drain engineering (e.g., use of lightly doped drains – LDD), among other factors, in modern processes is one of several reasons. At the 0.13-μm generation currently in production, it is quite difficult to obtain good fits with level-3 models over more than quite a restricted range of operating conditions. Commercial foundries consequently have not bothered to develop level-3 models for several process generations.

5.11 APPENDIX C: LEVEL-1 MOS MODELS

The complexity of MOS device physics is mirrored in the complexity of the models used by commercial simulators. As we've emphasized repeatedly, *all* models are "wrong," and it is up to the circuit designer to use models only where they are sufficiently correct. It is unfortunate that this judgment comes only slowly with experience, but perhaps it is sufficient here merely to raise awareness of this issue – so that you can be on guard against the possibility (rather, probability) of modeling-induced simulation error.

This brief appendix summarizes the static device equations corresponding to level-1 SPICE models. By examining these, perhaps you can begin to appreciate the nature of the assumptions underlying their development and thus also appreciate the origins of their shortcomings.

THE LEVEL-1 STATIC MODEL

The level-1 SPICE model totally neglects phenomena such as subthreshold conduction, narrow-width effects, lateral-field dependent mobility (including velocity saturation), DIBL, and normal-field mobility degradation. Back-gate bias effect (body effect) is included, but its computation is based on very simple physics. The model can also accommodate channel-length modulation, but only in a crude way: the user must provide a different value of λ *for each different channel length.* Repeat: Using level 1, SPICE does not, cannot, and will not automatically recompute λ for different channel lengths. This limitation is serious enough that it is advisable to use only devices of channel lengths for which actual experimental data exists and from which the various values of λ have been obtained. Level-1 equations for static behavior are thus extremely simple and correspond most closely to those outlined in the main development of this chapter.

Recall that, in triode, the drain current of an NMOS transistor with drawn dimensions W/L is given by:

$$I_D = \mu_n C_{\text{ox}} \frac{W}{L - 2L_D} \left[(V_{gs} - V_t)V_{ds} - \frac{V_{ds}^2}{2} \right](1 + \lambda V_{ds}), \tag{65}$$

while the saturation drain current is given by

$$I_D = \mu_n C_{\text{ox}} \frac{W}{2(L - 2L_D)} (V_{gs} - V_t)^2 (1 + \lambda V_{ds}). \tag{66}$$

The model comprises the parameters in Table 5.3, which (like Table 5.2) also lists the corresponding mathematical notation. With level-1 models, SPICE computes the threshold voltage using the following equation:

$$V_t = V_{t0} + \gamma \left(\sqrt{2\phi_F - V_{bs}} - \sqrt{2\phi_F} \right). \tag{67}$$

As emphasized in Section 5.7.1 and as shown in the last two tables, the model parameter PHI already includes the factors of 2 shown in Eqn. 67.

Level-1 models also confer on SPICE the option of computing both the body-effect coefficient and the zero-bias threshold voltage from process variables if γ and VTO are not specified. However, it is simpler and better to obtain these values from actual data, so the list in Table 5.3 omits several parameters that are used only to compute them from scratch. The table also omits some resistances in series with the device terminals and neglects parameters related to device noise, a subject that is treated in greater detail in Chapter 11.

Greatly expanded explanations of various SPICE models are presented well by G. Massobrio and P. Antognetti in *Semiconductor Device Modeling with SPICE* (2nd ed., McGraw-Hill, New York, 1993), although there are maddening omissions in some of the derivations.

Table 5.3. *SPICE level-1 model parameters*

Parameter name	Conventional symbol	Description
PHI	$\lvert 2\phi_F \rvert$	Surface potential in strong inversion
TOX	t_{ox}	Gate oxide thickness
TPG		Gate material polarity: 0 for Al, -1 if same as subtrate, $+1$ if opposite substrate; this parameter is ignored if VTO is specified
VTO	V_{T0}	Threshold at $V_{bs} = 0$
LD	L_D	Source–drain lateral diffusion, for computing L_{eff}; *not* used to calculate overlap capacitances
KP	$k' = \mu_0 C_{\text{ox}}$	Process transconductance coefficient
UO	μ_0	Low-field carrier mobility at surface
RSH	$R_\square$	Source–drain diffusion sheet resistance; multiplied by NRS and NRD to obtain total source and drain ohmic resistance, respectively
LAMBDA	λ	Channel-length modulation factor
GAMMA	γ	Body-effect coefficient
NSUB	N_A or N_D	Equivalent substrate doping
CGDO	C_{gd0}	Gate–drain overlap capacitance per width
CGSO	C_{gs0}	Gate–source overlap capacitance per width
CGBO	C_{gb0}	Gate–bulk overlap capacitance per length
CJ	C_{j0}	Zero-bias bulk bottom capacitance per unit source–drain area; multiplied by AS and AD to obtain total bottom capacitance of source and drain at $V_{sb} = V_{db} = 0$
MJ	m_j	Bottom source–drain junction grading coefficient
CJSW	$C_{j\text{SW}}$	Zero-bias sidewall junction capacitance per unit perimeter of source–drain adjacent to field; multiplied by PS and PD to obtain total sidewall capacitance
MJSW	$M_{j\text{SW}}$	Sidewall junction grading coefficient
PB	ϕ_j	Bulk junction potential barrier used to compute junction capacitance for other than zero bias

5.12 APPENDIX D: SOME EXCEPTIONALLY CRUDE SCALING LAWS

On occasion, actual information about a technology is absent – or you have partial information and you have to guess the rest. What follows here is a crude guide to guessing.

For many generations of CMOS technology, the ratio of minimum gate length to effective oxide thickness has remained roughly constant at a value of 45–50. Hence, a 0.5-μm technology generally has an $\sim$10-nm–thick gate oxide, and the 0.13-μm technology currently in high-volume manufacturing has an *effective* oxide thickness of about 2.5 nm. The qualifier "effective" is necessary because the thickness of the surface depletion layer in polysilicon gate electrodes is no longer negligibly small compared with the physical oxide thickness. Gate oxide has today just about reached its practical limit, however, so the factor-of-50 rule of thumb will certainly change as new gate dielectrics debut.

The nominal supply voltage limits used to be roughly the minimum gate length multiplied by about 10 V/μm. Starting with the 1.2-V, 0.13-μm generation, the supply voltage targets are set to scale approximately as the square root of the channel length.

Finally, a crude estimate for the rms mismatch in device threshold is

$$\Delta V_t \approx \frac{P \cdot t_{\text{ox}}}{\sqrt{WL}}, \tag{68}$$

where W and L are in microns; gate oxide thickness is in nanometers; and P is the Pelgrom coefficient, which has a typical value (within a factor of 2) of 2 mV-μm/nm.[8]

PROBLEM SET FOR MOS DEVICE PHYSICS

PROBLEM 1   In long-channel devices, C_{gs} typically dominates over C_{gd}. This problem investigates whether this remains the case for short-channel devices.

(a) Using the approximation $L_D \approx \frac{2}{3}x_j$, derive an expression for C_{gd}/C_{gs} in terms of channel length L and x_j.
(b) Plot C_{gd}/C_{gs} versus L for $x_j = 50$ nm, 150 nm, and 250 nm, and for L ranging from 0.5 μm to 5 μm. What do you deduce about the scaling of these capacitances with channel length?

PROBLEM 2   The current mirror is a ubiquitous subcircuit. However, merely specifying the current ratio alone does not constrain the design sufficiently to determine individual device dimensions, so other criteria must be considered. In circuits that require a high degree of isolation, one criterion is to minimize the sensitivity to substrate voltage fluctuations. To illustrate the nature of the problem, compare two simple 1 : 1 NMOS current mirrors. The reference current for both is 100 μA, while the device widths in the two mirrors are 10 μm and 100 μm.

(a) Assume that noise coupled from a nearby circuit may be modeled as a 1-MHz, 100-mV amplitude sinusoidal voltage generator connected between ground and the body (bulk) terminal of the NMOS transistors. Assume that we measure the

[8] M. J. M. Pelgrom et al., "Matching Properties of MOS Transistors," *IEEE J. Solid-State Circuits*, v. 24, no. 5, October 1989, pp. 1433–40, and "Transistor Matching in Analog CMOS Applications," *Proc. International Electron Devices Meeting*, December 1998, pp. 915–18.

output current as it flows into a DC source of 2 V. Which mirror will exhibit worse sensitivity to the substrate fluctuation? Explain your answer in terms of parameters of the four-terminal incremental model.

(b) Verify your answer using the level-3 SPICE models of this chapter, and find from simulations the actual incremental component of the output current for the two mirrors.

PROBLEM 3   Derive a more general expression for ω_T that is valid in both the short- and long-channel regimes, rather than only in either limit, by first deriving a general expression for the transconductance. You may neglect back-gate bias effect.

PROBLEM 4   This problem investigates the importance of various correction factors on the *V–I* characteristics of MOSFETs.

(a) First assume operation in the long-channel regime and plot the drain current as a function of drain–source voltage as the gate voltage is varied in 200-mV steps from 0 to 3 V. Assume a mobility of 0.05 m^2/V-s and a t_{ox} of 10 nm, and let V_{ds} range from 0 to 5 V. Assume $W/L = 10$.

(b) Now take velocity saturation into account. Assume that LE_{sat} is 1.75 V, and re-plot as in part (a). By what factor has the maximum drain current diminished?

(c) Now take into account vertical-field mobility degradation. Assume that θ is 0.2 V^{-1} and replot. By what factor has the maximum drain current now diminished, relative to the long-channel case?

PROBLEM 5   If the gate-to-source overdrive voltage of an NMOS transistor is held constant at 1 V, then what is the percentage change in device transconductance as the temperature increases from 300 K to 400 K?

PROBLEM 6   The voltage dependency of gate capacitance can lead to distortion and other errors in analog circuits. To explore this idea further, consider the amplifier circuit shown in Figure 5.9. Here, capacitor C_3 represents the gate capacitance of a transistor inside the op-amp, whose open-loop gain is G. Suppose we wish to use this circuit as an inverting buffer, so we initially select $C_1 = C_2 = C$.

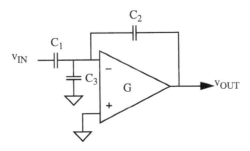

FIGURE 5.9. Amplifier with voltage-sensitive capacitance.

(a) Derive an expression for the gain error, defined as the difference in the magnitudes of the input and output. Let the input voltage be 1 V.

(b) Clearly, the nominal value of gain error can in principle be reduced to zero – for example, by a suitable adjustment of the feedback capacitor. However, such an adjustment cannot compensate for voltage-dependent capacitance variations in C_3. Specifically, suppose C_3 has a voltage sensitivity as follows:

$$C_3 = C_{3,0}(1 + \alpha V), \tag{P5.1}$$

where $C_{3,0}$ is the capacitance value at zero bias and α is a first-order voltage sensitivity coefficient. Derive an expression for the minimum acceptable gain G if the variation in gain error is to be kept below 1 part in 10^5 as the input voltage swings from 0 to 5 V.

PROBLEM 7   Channel-length modulation (CLM) is one effect that can cause the output current to depend on drain voltage in saturation. To investigate CLM in a very approximate way, assume that the region near the drain is in depletion. Hence, the effective length of the channel (the portion in inversion) is smaller than the physical length by an amount equal to the extent of this depletion region. Furthermore, the amount of depletion depends on the drain voltage, thus leading to a nonzero output conductance in saturation.

In order to simplify the development, assume the square-law characteristics

$$i_D = \frac{\mu C_{ox}}{2} \frac{W}{L_{eff}} (V_{gs} - V_t)^2, \tag{P5.2}$$

where $L_{eff} = L - \delta$. In turn, use a simple formula for the depletion layer's extent:

$$\delta = \sqrt{\frac{2\varepsilon_{Si}}{q N_{sub}} (V_{ds} - V_{dsat})}. \tag{P5.3}$$

Derive an expression for the output conductance from these equations.

PROBLEM 8   Current mirrors are often designed to provide a current ratio other than unity. In principle, this ratio can be set by adjusting either width or length but, in practice, the length is almost never varied to provide the desired current ratio. Explain.

PROBLEM 9   As discussed in the text, the magnitude of the threshold voltage increases if the source–bulk junction is increasingly reverse-biased. This effect can be modeled formally as follows:

$$\Delta V_t = \gamma \left(\sqrt{2|\phi_s| + |V_{sb}|} - \sqrt{2|\phi_s|} \right), \tag{P5.4}$$

where the parameter γ is the body-effect coefficient.

(a) In the inverting amplifier circuit shown in Figure 5.10, what is the output voltage if the input is grounded? Assume that the substrate is grounded and take body effect into account.

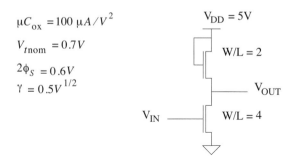

$$\mu C_{ox} = 100 \ \mu A/V^2$$
$$V_{tnom} = 0.7V$$
$$2\phi_S = 0.6V$$
$$\gamma = 0.5V^{1/2}$$

$V_{DD} = 5V$

W/L = 2

V_{OUT}

V_{IN} W/L = 4

FIGURE 5.10. Inverting amplifier.

(b) Calculate the output voltage when the input is connected to the 5-V supply.

PROBLEM 10 One way to model the cumulative effects of DIBL and channel-length modulation is to multiply the standard equations by the following correction factor:

$$1 + \lambda V_{ds}. \tag{P5.5}$$

The parameter λ (and its reciprocal, the Early voltage) thus accounts for the observed increase in drain current in saturation as drain voltage increases. Since MOSFETs are thus imperfect current sources in saturation, one must be careful in choosing device dimensions and biasing conditions.

For a simple square-law device, determine the appropriate length and width to provide a drain current within 1% of 1 mA as the drain voltage ranges from 3 V to 5 V. Assume a mobility of 0.05 m^2/V-s, a C_{ox} of 3.5 mF/m^2, a gate overdrive of 1.5 V, and a λ of 0.1 V^{-1}.

CHAPTER SIX

DISTRIBUTED SYSTEMS

6.1 INTRODUCTION

There are two important regimes of operating frequency, distinguished by whether one may treat circuit elements as "lumped" or distributed. The fuzzy boundary between these two regimes concerns the ratio of the physical dimensions of the circuit relative to the shortest wavelength of interest. At high enough frequencies, the size of the circuit elements becomes comparable to the wavelengths, and one cannot employ with impunity intuition derived from lumped-circuit theory. Wires must then be treated as the transmission lines that they truly are, Kirchhoff's "laws" no longer hold generally, and identification of R, L, and C ceases to be obvious (or even possible).

Thanks to the small dimensions involved in ICs, though, it turns out that we can largely ignore transmission-line effects well into the gigahertz range, at least on-chip. So, in this text, we will focus primarily on lumped-parameter descriptions of our circuits. For the sake of completeness, however, we should be a little less cavalier (we'll still be cavalier, just less so) and spend some time talking about how and where one draws a boundary between lumped and distributed domains. In order to do this properly, we need to revisit (briefly) Maxwell's equations.

MAXWELL AND KIRCHHOFF

Many students (and many practicing engineers, unfortunately) forget that Kirchhoff's voltage and current "laws" are *approximations* that hold only in the lumped regime (which we have yet to define). They are derivable from Maxwell's equations if we assume quasistatic behavior, thereby eliminating the coupling terms that give rise to the wave equation. To understand what all this means, let's review Maxwell's equations (for free space) in differential form:

$$\nabla \cdot \mu_0 \mathbf{H} = 0, \tag{1}$$

$$\nabla \cdot \varepsilon_0 \mathbf{E} = \rho, \tag{2}$$

$$\nabla \times \mathbf{H} = \mathbf{J} + \varepsilon_0 \frac{\partial \mathbf{E}}{\partial t}, \tag{3}$$

$$\nabla \times \mathbf{E} = -\mu_0 \frac{\partial \mathbf{H}}{\partial t}. \tag{4}$$

The first equation says that there is no net magnetic charge (i.e., there are no magnetic monopoles). If there were net magnetic charge, it would cause divergence in the magnetic field. We won't be using that equation at all.

The second equation (Gauss's law) acknowledges that there is net electric charge, and says that electric charge is the source of divergence of the electric field. We won't really use that equation either.

The third equation (Ampere's law, with Maxwell's famous modification) states that both "ordinary" current *and* the time rate of change of the electric field produce the same effect on the magnetic field. The term that involves the derivative of the electric field is the famous displacement (capacitive) current term that Maxwell pulled out of thin air to produce the wave equation.

Finally, the fourth equation (Faraday's law) says that a changing magnetic field causes curl in the electric field.

Wave behavior arises fundamentally because of the coupling terms in the last two equations: A change in $\mathbf{E}$ causes a change in $\mathbf{H}$, which causes a change in $\mathbf{E}$, and so on. If we were to set either μ_0 or ε_0 to zero, the coupling terms would disappear and no wave equation would result; circuit analysis could then proceed on a quasistatic (or even static) basis.

As a specific example, setting μ_0 to zero makes the electric field curl-free, allowing $\mathbf{E}$ to be expressed as the gradient of a potential (within a minus sign here or there). It then follows identically that the line integral of the $\mathbf{E}$-field (which is the voltage) around any closed path is zero:

$$V = \oint \mathbf{E} \cdot dl = \oint (-\nabla \phi) \cdot dl = 0. \tag{5}$$

This is merely the field-theoretical expression of Kirchhoff's voltage law (KVL).

To derive KCL (Kirchhoff's current law), we proceed in the same manner but now set ε_0 equal to zero. Then, the curl of $\mathbf{H}$ depends only on the current density $\mathbf{J}$, allowing us to write:

$$\nabla \cdot \mathbf{J} = \nabla \cdot (\nabla \times \mathbf{H}) = 0. \tag{6}$$

That is, the divergence of $\mathbf{J}$ is identically zero. No divergence means no net current buildup (or loss) at a node.

Of course, neither μ_0 nor ε_0 is actually zero. To show that the foregoing is not hopelessly irrelevant as a consequence, recall that the speed of light can be expressed as[1]

---

[1] Applying the "duck test" version of Occam's razor ("if it walks like a duck and quacks like a duck, it must be a duck"), Maxwell showed that light and electromagnetic waves are the same thing. After all, if it travels at the speed of light and reflects like light, it must be light. Most would agree that the derivation of Maxwell's equations represents the crowning intellectual achievement of the 19th century.

$$c = 1/\sqrt{\mu_0 \varepsilon_0}. \tag{7}$$

Setting μ_0 or ε_0 to zero is therefore equivalent to setting the speed of light to infinity. Hence, KCL and KVL are the result of assuming infinitely fast propagation; we expect them to hold reasonably well as long as the physical dimensions of the circuit elements are small compared with a wavelength so that the finiteness of the speed of light is not noticeable:

$$l \ll \lambda, \tag{8}$$

where l is the length of a circuit element and λ is the shortest wavelength of interest.

To develop a feel for what this constraint means numerically, consider an IC subcircuit whose longest dimension is 1 mm. If we arbitrarily say that "much less than" means "a factor of at least 10 smaller than," then such a subcircuit can be treated as lumped if the highest-frequency signal on the chip has a wavelength greater than roughly 1 cm. In free space, this wavelength corresponds to a frequency of about 30 GHz. On chip, the frequency limit decreases a bit because the relative dielectric constants for silicon and silicon dioxide are well above unity (11.7 and 3.9, resp.), but it should be clear that a full transmission-line treatment of the on-chip design and analysis problem is generally unnecessary until fairly high frequencies are reached.

In summary, the boundary between lumped circuit theory (where KVL and KCL hold, and where one can identify R, L, and C) and distributed systems (where KVL/KCL don't hold, and where R, L, and C can't always be localized) depends on the *size of the circuit element relative to the shortest wavelength of interest.* If the circuit element (and interconnect is certainly a circuit element in this context) is very short compared with a wavelength, we can use traditional lumped concepts and incur little error. If not, then use of lumped ideas is inappropriate. Because the dimensions of CMOS circuit blocks are typically much shorter than a wavelength, we shall glibly ignore transmission-line effects for most of this textbook.

6.2 LINK BETWEEN LUMPED AND DISTRIBUTED REGIMES

We now turn to the problem of extending into the distributed regime the design intuition developed in the lumped regime. The motivation is more than merely pedagogical for, as we shall see, extremely valuable design insights emerge from this exercise. An important example is that *delay,* instead of gain, may be traded for bandwidth.

Interconnect is an example of a system that may be treated successfully at lower frequencies as, say, a simple RC line. With that type of mindset, reduction of RC "parasitics" in order to increase bandwidth becomes a major preoccupation of the circuit and system designer (particularly of the IC designer). Unfortunately, reduction of parasitics below some minimum amount is practically impossible. Intuition from lumped-circuit design would therefore (mis)lead us into thinking that the bandwidth is limited by these irreducible parasitics. Fortunately, a proper treatment of interconnect as a transmission line, rather than as a finite lumped RC network, reveals otherwise. We find that we may still convey signals with exceedingly large bandwidth as long as we acknowledge (indeed exploit) the true, distributed nature of the interconnect. By using, rather than fighting, the distributed capacitance (and inductance), we may therefore effect a decoupling of delay from bandwidth. This new insight is extremely valuable, and applies to active as well as passive networks. Aside from giving us an appreciation for transmission-line phenomena, such an understanding will lead us to amplifier topologies that relax significantly the gain–bandwidth trade-off that characterizes lumped systems of low order.

Given these important reasons, we now undertake a study of distributed systems by extension of lumped-circuit analysis.

6.3 DRIVING-POINT IMPEDANCE OF ITERATED STRUCTURES

We begin by studying the driving-point impedance of uniform, iterated structures. It's important to note that certain nonuniform structures (e.g., exponentially tapered transmission lines[2]) have exceedingly useful properties, but we'll limit the present discussion to a consideration of uniform structures only.

Specifically, consider the infinite ladder network shown in Figure 6.1. Even though resistor symbols are used here, they represent arbitrary impedances.

To find the driving-point impedance of this network without summing an infinite series, note that the impedance to the right of node C is the same as that to the right of B, and the same as that to the right of A.[3] Therefore, we may write

$$Z_{in} = Z + [(1/Y) \parallel Z_{in}], \tag{9}$$

which expands to

$$Z_{in} = Z + \frac{Z_{in}/Y}{1/Y + Z_{in}} \implies (Z_{in} - Z)\left(\frac{1}{Y} + Z_{in}\right) = \frac{Z_{in}}{Y}. \tag{10}$$

Solving for Z_{in} yields:

$$Z_{in} = \frac{Z \pm \sqrt{Z^2 + 4(Z/Y)}}{2} = \frac{Z}{2}\left[1 \pm \sqrt{1 + \frac{4}{ZY}}\right]. \tag{11}$$

In the special case where $Z = 1/Y = R$,

$$Z_{in} = \left(\frac{1 + \sqrt{5}}{2}\right)R \approx 1.618R. \tag{12}$$

---

[2] For those of you who are curious, the exponentially tapered line allows one to achieve a broadband impedance match instead of the narrowband impedance match that a quarter-wave transformer provides. The transformation ratio can be controlled by choice of taper constants.

[3] This is an extremely useful technique for analyzing such structures, but a surprisingly large percentage of engineers have never heard of it, or perhaps don't remember it. In any event, it certainly saves a tremendous amount of labor over a more straightfoward approach, which would require summing various infinite series.

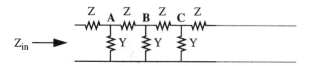

FIGURE 6.1. Ladder network.

This ratio of Z_{in} to R is known as the *golden ratio* (or *golden section*) and shows up in contexts as diverse as the aesthetics of Greek geometers, Renaissance art and architecture, and solutions to several interesting (but largely useless) network theory problems.

IDEAL TRANSMISSION LINE AS INFINITE LADDER NETWORK

Let's now consider the more general case of the input impedance in the limit where $|ZY| \ll 1$, and where we continue to disallow negative values of Z_{in}. In that case, we can simplify the result to

$$Z_{\text{in}} \approx \sqrt{Z/Y}. \tag{13}$$

We see that if Z/Y happens to be frequency-independent, then the input impedance will also be frequency-independent.[4] One important example of a network of this type is the model for an ideal transmission line. In the case of a lossless line, $Z = sL$ and $Y = sC$, where L and C represent differential (in the mathematical sense) circuit elements. The input impedance (called the *characteristic* impedance Z_0) for an ideal, lossless infinite transmission line is therefore

$$Z_{\text{in}} \approx \sqrt{Z/Y} = \sqrt{sL/sC} = \sqrt{L/C}. \tag{14}$$

Because Y, the admittance of an infinitesimal capacitance, approaches zero as the length of the differential element approaches zero, while the reactance of the differential inductance element approaches zero at the same time, the ratio $1/YZ$ approaches infinity and so satisfies the inequality necessary to validate our derivation. The result – that we are left with a purely real input impedance for an infinitely long transmission line – should be a familiar one, but perhaps this particular path to it might not be.

An often-asked question concerns the fate of the energy we launch down a transmission line. If the impedance is purely real, then the line should behave as a resistor and should dissipate energy like a resistor. But the line is composed of purely reactive (and hence dissipationless) elements, so there would appear to be a paradox.

The resolution is that the energy doesn't end up as heat if the line is truly infinite. The energy just keeps traveling down the line forever, and so is lost to the external world just as if it had heated up a resistor and its environs; the line acts like a black hole for energy.

---

[4] Ladder networks with this property are called "constant-k" lines, since $Z/Y = k^2$ for a constant k.

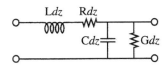

FIGURE 6.2. Lumped *RLC* model of infinitesimal transmission-line segment.

6.4 TRANSMISSION LINES IN MORE DETAIL

The previous section examined the impedance behavior of a lossless infinite line. We now extend our derivation of the characteristic impedance Z_0 to accommodate loss. We also introduce an additional descriptive parameter, the *propagation constant* γ.

6.4.1 LUMPED MODEL FOR LOSSY TRANSMISSION LINE

To derive the relevant parameters of a lossy line, consider an infinitesimally short piece of line, of length dz, as shown in Figure 6.2. Here, the elements L, R, C, and G are all quantities *per unit length* and simply represent a specific example of the more general case considered earlier.

The inductance accounts for the energy stored in the magnetic field around the line, while the series resistance accounts for the inevitable energy loss (such as due to skin effect) that all ordinary conductors exhibit. The shunt capacitance models the energy stored in the electric field surrounding the line, and the shunt conductance accounts for the loss due to mechanisms such as ordinary ohmic leakage as well as loss in the line's dielectric material.

6.4.2 CHARACTERISTIC IMPEDANCE OF A LOSSY TRANSMISSION LINE

To compute the impedance of a lossy line, we follow a method exactly analogous to that in Section 6.3:

$$Z_0 = Z\,dz + [(1/Y\,dz) \parallel Z_0] = Z\,dz + \frac{Z_0}{1 + (Y\,dz)Z_0}. \qquad (15)$$

We will consider the limiting behavior of this expression as dz approaches zero, so we may use the first-order binomial expansion of $1/(1 + x)$:

$$Z_0 = Z\,dz + \frac{Z_0}{1 + (Y\,dz)Z_0}$$
$$\approx Z\,dz + Z_0[1 - (Y\,dz)Z_0] = Z_0 + dz(Z - YZ_0^2). \qquad (16)$$

Cancelling Z_0 from both sides, we see that the final term in parentheses must equal zero. The characteristic impedance is thus

$$Z_0 = \sqrt{\frac{Z}{Y}} = \sqrt{\frac{R + j\omega L}{G + j\omega C}}. \tag{17}$$

If the resistive terms are negligible (or if RC just happens to equal GL), the equation for Z_0 collapses to the result we derived earlier:

$$Z_0 = \sqrt{L/C}. \tag{18}$$

Because the impedance approaches $\sqrt{L/C}$ at sufficiently high frequency, independent of R or G, it is sometimes known as the *transient* or *pulse* impedance.

6.4.3 THE PROPAGATION CONSTANT

In addition to the characteristic impedance, one other important descriptive parameter is the *propagation constant,* usually denoted by γ. Whereas the characteristic impedance tells us the ratio of voltage to current at any one point on an infinitely long line, the propagation constant enables us to say something about the ratio of voltages (or currents) between any two points on such a line. That is, γ quantifies the line's attenuation properties.

Consider the voltages at the two ports of a given subsection. The ratio of these voltages is readily computed from the ordinary voltage divider relationship:

$$V_{n+1} = V_n \left\{ \frac{Z_0 \parallel (1/Y\,dz)}{Z\,dz + [Z_0 \parallel (1/Y\,dz)]} \right\}. \tag{19}$$

Thus,

$$\frac{V_{n+1}}{V_n} = \frac{Z_0 \parallel (1/Y\,dz)}{Z\,dz + [Z_0 \parallel (1/Y\,dz)]} = \frac{Z_0}{Z_0 Z Y (dz)^2 + Z_0 + Z\,dz}. \tag{20}$$

Because we will use this expression in the limit of very small dz, we may discard the term proportional to $(dz)^2$ and again use the binomial expansion of $1/(1+x)$ to preserve only the first-order dependence on dz (remember, we're engineers – the whole universe is first-order to us!). This yields

$$\frac{V_{n+1}}{V_n} \approx \frac{Z_0}{Z_0 + Z\,dz} = \frac{1}{1 + (Z/Z_0)\,dz} \approx 1 - \frac{Z}{Z_0}\,dz = 1 - \sqrt{ZY}\,dz. \tag{21}$$

Despite our glibness, the net error in these approximations actually does converge to zero in the limit of zero dz.

Let us rewrite the previous equation as a difference equation:

$$V_{n+1} = V_n \left(1 - \sqrt{ZY}\,dz\right) \implies \frac{V_{n+1} - V_n}{dz} = -\sqrt{ZY}\,V_n. \tag{22}$$

In the limit of zero dz, the difference equation becomes a differential equation:

$$\frac{dV}{dz} = -\sqrt{ZY}\,V. \tag{23}$$

The solution to this first-order differential equation should be familiar:

$$V(z) = V_0 e^{-\sqrt{ZY}z}.$$ (24)

That is, the voltage at any position z is simply the voltage V_0 (the voltage at $z = 0$) times an exponential factor. The exponent is conventionally written as $-\gamma z$ so that, at last,

$$\gamma = \sqrt{ZY} = \sqrt{(R + j\omega L)(G + j\omega C)}.$$ (25)

To develop a better feel for the significance of the propagation constant, first note that γ will be complex in general. Hence, we may express γ explicitly as the sum of real and imaginary parts:

$$\gamma = \sqrt{(R + j\omega L)(G + j\omega C)} = \alpha + j\beta.$$ (26)

Then

$$V(z) = V_0 e^{-\gamma z} = V_0 e^{-(\alpha + j\beta)z} = V_0 e^{-\alpha z} e^{-j\beta z}.$$ (27)

The first exponential term becomes smaller as distance increases; it represents the pure attenuation of the line. The second exponential factor has a unit magnitude and contributes only phase.

6.4.4 RELATIONSHIP OF γ TO LINE PARAMETERS

To relate the constants α and β explicitly to transmission-line parameters, we make use of a couple of identities. First, recall that we may express a complex number in both exponential (polar) and rectangular form as follows:

$$M e^{j\phi} = M \cos \phi + jM \sin \phi.$$ (28)

Here, M is the magnitude of the complex number and ϕ is its phase. The polar form allows us to compute the square root of a complex number with ease (thanks to Euler):

$$\sqrt{M e^{j\phi}} = \sqrt{M} e^{j\phi/2} = \sqrt{M} \cos(\phi/2) + j\sqrt{M} \sin(\phi/2).$$ (29)

The last factoid we need to recall from undergraduate math is a pair of half-angle identities:

$$\cos(\phi/2) = \sqrt{\tfrac{1}{2}(1 + \cos \phi)}$$ (30)

and

$$\sin(\phi/2) = \sqrt{\tfrac{1}{2}(1 - \cos \phi)}.$$ (31)

Now, γ is the square root of a complex number:

$$\gamma = \sqrt{ZY} = \sqrt{(R + j\omega L)(G + j\omega C)}$$
$$= \sqrt{(RG - \omega^2 LC) + j\omega(LG + RC)}.$$ (32)

Making use of our identities and turning the crank a few revolutions, we obtain:

$$\alpha = \sqrt{\tfrac{1}{2}\left[\sqrt{\omega^4(LC)^2 + \omega^2[(LG)^2 + (RC)^2] + (RG)^2} + (RG - \omega^2 LC)\right]} \quad (33)$$

and

$$\beta = \sqrt{\tfrac{1}{2}\left[\sqrt{\omega^4(LC)^2 + \omega^2[(LG)^2 + (RC)^2] + (RG)^2} - (RG - \omega^2 LC)\right]}. \quad (34)$$

These last two expressions may appear cumbersome, but that's only because they are. We may simplify them considerably if the product RG is small compared with the other terms. In such a case, the attenuation constant may be written as

$$\alpha \approx \sqrt{\tfrac{1}{2}\left[\sqrt{\omega^4(LC)^2 + \omega^2[(LG)^2 + (RC)^2]} - \omega^2 LC\right]}, \quad (35)$$

which, after a certain amount of bloodletting, further simplifies to

$$\alpha \approx \frac{R}{2}\sqrt{\frac{C}{L}} + \frac{G}{2}\sqrt{\frac{L}{C}}. \quad (36)$$

This, in turn, may be further approximated by

$$\alpha \approx \frac{R}{2}\sqrt{\frac{C}{L}} + \frac{G}{2}\sqrt{\frac{L}{C}} \approx \frac{R}{2Z_0} + \frac{GZ_0}{2}. \quad (37)$$

Thus, the attenuation per length will be small as long as the resistance per length is small compared with Z_0, and if the conductance per length is small compared with Y_0.

Turning our attention now to the equation for β, we have

$$\beta = \mathrm{Im}[\gamma] \approx \omega\sqrt{LC}. \quad (38)$$

In the limit of zero loss (both G and $R = 0$), these expressions simplify to

$$\alpha = \mathrm{Re}[\gamma] = 0 \quad (39)$$

and

$$\beta = \mathrm{Im}[\gamma] = \omega\sqrt{LC}. \quad (40)$$

Hence, a lossless line doesn't attenuate (no big surprise). Since the attenuation is the same (zero) at all frequencies, a lossless line *has no bandwidth limit*. In addition, the propagation constant has an imaginary part that is exactly proportional to frequency. Since the delay of a system is simply (minus) the derivative of phase with frequency, the delay of a lossless line is a constant, independent of frequency:

$$T_{\mathrm{delay}} = -\frac{\partial}{\partial\omega}\Phi(\omega) = -\frac{\partial}{\partial\omega}(-\beta z) = \sqrt{LC}\, z. \quad (41)$$

We can now appreciate the remarkable property of distributed systems alluded to in the introduction: The capacitance and inductance *do not directly cause a bandwidth reduction*. They result only in a propagation delay. If we were to increase the inductance or capacitance per unit length, the delay would increase but bandwidth (ideally) would not change. This behavior is quite different from what one observes with low-order lumped networks.

Also in stark contrast with low-order lumped networks, a transmission line may exhibit a *frequency-independent* delay, as seen here. This property is extremely desirable, for it implies that all Fourier components of a signal will be delayed by precisely the same amount of time; pulse shapes will be preserved. We have just seen that a lossless line has this property of zero dispersion. Since all real lines exhibit nonzero loss, though, must we accept dispersion (nonuniform delays) in practice? Fortunately, as Heaviside[5] first pointed out, the answer is No. If we exercise some control over the line constants, we can still obtain a uniform group delay *even with a lossy line* (at least in principle). In particular, Heaviside discovered that choosing RC equal to GL (or, equivalently, choosing the L/R time constant of the series impedance Z equal to the C/G time constant of the shunt admittance Y) leads to a constant group delay. There is nonzero attenuation, of course (can't get rid of that, unfortunately), but the constant group delay means that pulses only get smaller as they travel down the line; they don't smear out (disperse).

Showing that Heaviside was correct isn't too hard. Setting RC and GL equal in our exact expressions for α and β yields

$$\alpha = \mathrm{Re}[\gamma] = \sqrt{RG} \tag{42}$$

and

$$\beta = \mathrm{Im}[\gamma] = \omega\sqrt{LC}. \tag{43}$$

Note that the expression for β is the same as that for a lossless line and hence also leads to the same frequency-independent delay.

Though the attenuation is no longer zero, it continues to be frequency-independent; the bandwidth is still infinite as long as we choose $L/R = C/G$. Furthermore, the characteristic impedance becomes exactly equal to $\sqrt{L/C}$ at all frequencies, rather than approaching this value asymptotically at high frequencies.

Setting $LG = RC$ is best accomplished by increasing either L or C, rather than by increasing R or G, because the latter strategy increases the attenuation (presumably an undesirable effect). Michael Pupin of Columbia University, following through on the implications of Heaviside's work, suggested the addition of lumped inductances periodically along telephone transmission lines to reduce signal dispersion. Such "Pupin coils" permitted significantly improved telephony in the 1920s and 1930s.[6]

---

[5] By the way, he was the first to use vector calculus to cast Maxwell's equations in modern form and also the one who introduced the use of Laplace transforms to solve circuit problems.

[6] Alas, the use of lumped inductances introduces a bandwidth limitation that true, distributed lines do not have. Since bandwidth and channel capacity are closely related, all of the Pupin coils (which had been installed at great expense) eventually had to be removed (at great expense) to permit an increase in the number of calls carried by each line.

6.5 BEHAVIOR OF FINITE-LENGTH TRANSMISSION LINES

Now that we've deduced a number of important properties of transmission lines of infinite length, it's time to consider what happens when we terminate finite-length lines in arbitrary impedances.

6.5.1 TRANSMISSION LINE WITH MATCHED TERMINATION

The driving-point impedance of an infinitely long line is simply Z_0. Suppose we cut the line somewhere, discard the infinitely long remainder, and replace it with a single lumped impedance of value Z_0. The driving-point impedance must remain Z_0; there's no way for the measurement apparatus to distinguish the lumped impedance from the line it replaces. Hence, a signal applied to the line simply travels down the finite segment of line, eventually gets to the resistor, heats it up, and contributes to global warming.

6.5.2 TRANSMISSION LINE WITH ARBITRARY TERMINATION

In general, a transmission line will not be terminated in precisely its characteristic impedance. A signal traveling down the line maintains a ratio of voltage to current that is equal (of course) to Z_0 until it encounters the load impedance. The termination impedance imposes its own particular ratio of voltage to current, however, and the only way to reconcile the conflict is for some of the signal to reflect back toward the source.

To distinguish forward (incident) quantities from the reflected ones, we will use the subscripts i and r, respectively. If E_i and I_i are the incident voltage and current, then it's clear that

$$Z_0 = \frac{E_i}{I_i}. \tag{44}$$

At the load end of things, the mismatch in impedances gives rise to a reflected voltage and current. We still have a linear system, so the total voltage at any point on the system is the superposition of the incident and reflected voltages. Similarly, the net current is also the superposition of the incident and reflected currents. Because the current components travel in opposite directions, the superposition here results in a subtraction. Thus, we have

$$Z_L = \frac{E_i + E_r}{I_i - I_r}. \tag{45}$$

We may rewrite this last equation to show an explicit proportionality to Z_0 as follows:

$$Z_L = \frac{E_i + E_r}{I_i - I_r} = \frac{E_i}{I_i}\left[\frac{1 + E_r/E_i}{1 - I_r/I_i}\right] = Z_0\left[\frac{1 + E_r/E_i}{1 - I_r/I_i}\right]. \tag{46}$$

The ratio of reflected to incident quantities at the load end of the line is called Γ_L and will generally be complex. Using Γ_L, the expression for Z_L becomes

$$Z_L = Z_0 \left[\frac{1 + E_r/E_i}{1 - I_r/I_i} \right] = Z_0 \left[\frac{1 + \Gamma_L}{1 - \Gamma_L} \right]. \tag{47}$$

Solving for Γ_L yields

$$\Gamma_L = \frac{Z_L - Z_0}{Z_L + Z_0}. \tag{48}$$

If the load impedance equals the characteristic impedance of the line, then the reflection coefficient will be zero. If a line is terminated in either a short or an open, then the reflection coefficient will have a magnitude of unity; this value is the maximum magnitude it can have (for a purely passive system such as this one, anyway).

We may generalize the concept of the reflection coefficient so that it is the ratio of the reflected and incident quantities at any arbitrary point along the line:

$$\Gamma(z) = \frac{E_r e^{\gamma z}}{E_i e^{-\gamma z}} = \frac{E_r}{E_i} e^{2\gamma z} = \Gamma_L e^{2\gamma z}. \tag{49}$$

Here we follow the convention of defining $z = 0$ at the load end of the line and locating the driving source at $z = -l$. With this convention, the voltage and current at any point z along the line may be expressed as:

$$V(z) = V_i e^{-\gamma z} + V_r e^{\gamma z}, \tag{50}$$

$$I(z) = I_i e^{-\gamma z} - I_r e^{\gamma z}. \tag{51}$$

As always, the impedance at any point z is simply the ratio of voltage to current:

$$Z(z) = \frac{V_i e^{-\gamma z} + V_r e^{\gamma z}}{I_i e^{-\gamma z} - I_r e^{\gamma z}} = Z_0 \left[\frac{1 + \Gamma_L e^{2\gamma z}}{1 - \Gamma_L e^{2\gamma z}} \right]. \tag{52}$$

Substituting for Γ_L and doing a whole heck of a lot of crunching yields

$$\frac{Z(z)}{Z_0} = \frac{\dfrac{Z_L}{Z_0}(e^{-\gamma z} + e^{\gamma z}) + (e^{-\gamma z} - e^{\gamma z})}{\dfrac{Z_L}{Z_0}(e^{-\gamma z} - e^{\gamma z}) + (e^{-\gamma z} + e^{\gamma z})}. \tag{53}$$

Writing this expression in a more compact form, we have

$$\frac{Z(z)}{Z_0} = \frac{\dfrac{Z_L}{Z_0} - \tanh \gamma z}{1 - \dfrac{Z_L}{Z_0} \tanh \gamma z}. \tag{54}$$

In the special case where the attenuation is negligible (as is commonly assumed to permit tractable analysis), a considerable simplification results:

$$\frac{Z(z)}{Z_0} = \frac{\dfrac{Z_L}{Z_0} - j\tan\beta z}{1 - j\dfrac{Z_L}{Z_0}\tan\beta z} = \frac{Z_L\cos\beta z - jZ_0\sin\beta z}{Z_0\cos\beta z - jZ_L\sin\beta z}. \tag{55}$$

Here, z is the actual coordinate value and will always be zero or negative.

As a final comment, note that this expression is periodic. Such behavior is strictly observed only in lossless lines, of course, but practical lines will behave similarly as long as the loss is negligible. Periodicity implies that one need consider the impedance behavior only over some finite section (specifically, a *half*-wavelength) of line. This observation is exploited in the construction of the Smith chart, a brief study of which is taken up in Chapter 7.

6.6 SUMMARY OF TRANSMISSION-LINE EQUATIONS

We've seen that both the characteristic impedance and the propagation constant are simple functions of the per-length series impedance and shunt admittance:

$$Z_0 = \sqrt{\frac{Z}{Y}} = \sqrt{\frac{R + j\omega L}{G + j\omega C}}, \tag{56}$$

$$\gamma = \sqrt{ZY} = \sqrt{(R + j\omega L)(G + j\omega C)}. \tag{57}$$

Using these parameters – in conjunction with the definition of reflection coefficient – allows us to develop an equation for the driving-point impedance of a lossy line terminated in an arbitrary impedance. In the case of a lossless (or negligibly lossy) line, the expression for impedance takes on a reasonably simple and periodic form, setting the stage for discussion of the Smith chart.

6.7 ARTIFICIAL LINES

We've just seen that an infinite ladder network of infinitesimally small inductors and capacitors has a purely real input impedance over an infinite bandwidth. Although structures that are infinitely long are somewhat inconvenient to realize, we can always terminate a finite length of line in its characteristic impedance. Energy, being relatively easy to fool, cannot distinguish between real transmission line and a resistor equal to the characteristic impedance, so the driving-point impedance of the properly terminated finite line remains the same as that of the infinite line, and still over an infinite bandwidth.

There are instances when we might wish to approximate a continuous transmission line by a finite lumped network. Motivations for doing so may include convenience of realization or greater control over line constants. However, use of a finite lumped approximation guarantees that the characteristics of such an artificial line cannot match

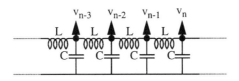

FIGURE 6.3. Lumped delay line.

those of an ideal line over an infinite bandwidth.[7] The design of circuits that employ lumped lines must take this bandwidth limitation into account.

One important use of artificial lines is in the synthesis of delay lines; see Figure 6.3. Here, we use LC L-sections to synthesize our line. As in the continuous case, the driving-point impedance is just

$$Z_{\text{in}} = \sqrt{L/C}, \tag{58}$$

while the delay per section is

$$T_D = \sqrt{LC}. \tag{59}$$

The value of a lumped delay line is that one may obtain large delays without having to use, say, a kilometer of coaxial cable.

6.7.1 CUTOFF FREQUENCY OF LUMPED LINES

Unlike the distributed line, the lumped line presents a real, constant impedance only over a finite bandwidth. Eventually, the input impedance becomes purely reactive,[8] indicating that real power can be delivered neither to the line nor to any load connected to the other end of the line. The frequency at which this occurs is known as the line's *cutoff* frequency, which is readily found by using the formula for the input impedance of an infinite (but lumped) LC line, reprised here from Section 6.3 for convenience:

$$Z_{\text{in}} = \frac{Z}{2}\left[1 \pm \sqrt{1 + \frac{4}{ZY}}\,\right]. \tag{60}$$

Here, let $Y = j\omega C$ and $Z = j\omega L$. Then the input impedance is

$$Z_{\text{in}} = \frac{j\omega L}{2}\left[1 \pm \sqrt{1 - \frac{4}{\omega^2 LC}}\,\right]. \tag{61}$$

---

[7] One easy way to see this is to recognize that a true transmission line, being a delay element, provides unbounded phase shift as the frequency approaches infinity. A lumped line can provide only a finite phase shift because of the finite number of energy storage elements and hence a finite number of poles.

[8] From inspection of the network, it should be clear that driving-point impedance eventually collapses to that of the input inductor, since the capacitors act ultimately like shorts.

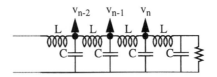

FIGURE 6.4. One choice for
terminating lumped lines.

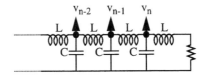

FIGURE 6.5. Alternative choice for
terminating lumped lines.

At sufficiently low frequencies, the term under the radical has a net negative value.
The resulting imaginary term, when multiplied by the $j\omega L/2$ factor, provides the
real component of the input impedance.

As the frequency increases, however, the magnitude of the term under the radical
sign eventually becomes zero. At and above this frequency, the input impedance is
purely imaginary, and no power can be delivered to the line. The cutoff frequency is
therefore given by

$$\omega_h = \frac{2}{\sqrt{LC}}. \tag{62}$$

Since the lumped line's characteristics begin to degrade well below the cutoff frequency, one must usually select a cutoff well above the highest frequency of interest.
Satisfying this requirement is particularly important if good pulse fidelity is necessary.

In designing artificial lines, the L/C ratio is chosen to provide the desired line
impedance, while the LC product is chosen small enough to provide a high enough
cutoff frequency to allow the line to approximate ideal behavior over the desired
bandwidth. If a specified overall time delay is required, the first two requirements
define the minimum number of sections that must be used.

6.7.2 TERMINATING LUMPED LINES

There's always a question as to how one terminates the circuit of Figure 6.4. For example, one choice is to end in a capacitance and simply terminate across it. Another
choice (see Figure 6.5) is to end in an inductance. Although both of these choices
will work after a fashion, a better alternative is to compromise by using a *half-section*
at both ends of the line, as shown in Figure 6.6.

Such a compromise extends the bandwidth over the choices of Figures 6.4 and 6.5.
Each half-section contributes half the delay of a full section, so putting one on each

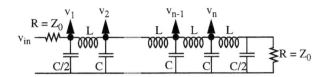

FIGURE 6.6. Half-sections for line termination.

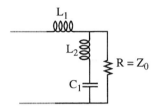

FIGURE 6.7. *m*-derived half-section
for line termination.

end adds the delay of a full section. Furthermore, and more important, a half-section has twice the cutoff frequency of a full section, which is precisely why better bandwidth is obtained.

6.7.3 *m*-DERIVED HALF-SECTIONS

The port impedance of the *LC* half-section begins to increase significantly at about 30–40% of the cutoff frequency owing to the parallel resonance formed by the output capacitance and the rest of the reactance it sees. This behavior can be moderated by the use of half-sections that are only marginally more elaborate than the single *LC* pair. Specifically, if the capacitor is replaced by a series *LC* branch, then the frequency range over which the impedance stays roughly constant can be increased even further because the decreasing impedance of the series resonant branch helps offset the increasing impedance.

A simple network that achieves the desired result is shown in Figure 6.7. The element values are given by the following equations:

$$L_1 = \frac{mL}{2}, \tag{63}$$

$$L_2 = \frac{1 - m^2}{2m}L, \tag{64}$$

$$C_1 = \frac{mC}{2}. \tag{65}$$

A network modified in this manner is called an *m*-derived half-section because, for any value of the parameter *m*, the nominal characteristic impedance remains the same as that of the simple *LC* half-section. This can be verified by direct substitution into Eqn. 60. The impedance stays roughly constant up to about 85% of the cutoff frequency for a value of *m* equal to roughly 0.6. This choice is therefore a common one.

6.8 SUMMARY

We have identified a fuzzy boundary between the lumped and distributed regimes, and found that lumped concepts may be extended into the distributed regime. In carrying out that extension, we have discovered that there are several, perhaps many, ways to trade gain for delay rather than bandwidth. As a final observation on this subject, perhaps it is worthwhile to reiterate that avoiding a straight gain–bandwidth tradeoff requires a gross departure from single-pole dynamics. Hence, all the structures we've seen that trade gain for delay involve many energy storage elements. Another way to look at this issue is to recognize that if we are to trade delay for anything then we must have the ability to provide large delays. But large delays imply a large amount of phase change per unit frequency, and if we are to operate over a large bandwidth then the total phase change required is very large. Again, this need for large amounts of phase shift necessarily implies that many poles (and hence many inductors and capacitors) will be required, resulting in the relatively complicated networks we've seen.[9] If one is pursuing operation over the largest possible bandwidth, however, use of these distributed concepts is all but mandatory. As we shall see in the chapter on high-frequency amplifier design, distributed concepts may be applied to active circuits to allow the realization of amplifiers with exceptionally large bandwidth by trading delay in exchange for the improved bandwidth.

PROBLEM SET FOR DISTRIBUTED SYSTEMS

PROBLEM 1

(a) What is the impedance looking into a lossless quarter-wave transmission line terminated with a resistive load? Let Z_0 be the line's characteristic impedance and R_L the load resistance.

(b) Given your answer to (a), propose a method for matching an 80-Ω load to a 20-Ω source at 500 MHz.

PROBLEM 2

(a) Calculate the input impedance of a lossless transmission line of 0.6 wavelength in extent, and whose characteristic impedance is 50 Ω, if the load impedance is $60.3 + j41.5$ Ω.

(b) Repeat (a) for a lossy line whose attenuation is 3.1 dB per wavelength.

---

[9] An exception is the superregenerative amplifier. There, the sampled nature of the system effectively causes aliases of the single stage to act in a way that is similar to a cascade of such stages.

PROBLEM 3   This problem considers an extremely useful property of transmission lines that are $\lambda/8$ in extent. Derive an expression for the input impedance of a lossless $\lambda/8$ line of characteristic impedance Z_0, terminated in an arbitrary impedance $R + jX$. Comment on potential applications for what you have discovered.

PROBLEM 4   In much discrete RF work, the standing wave ratio (SWR) is used because it is easier to measure than the reflection coefficient. The two quantities are related, however:

$$|\Gamma_L| = \frac{\text{SWR} - 1}{\text{SWR} + 1}. \tag{P6.1}$$

(a) Show that SWR is the ratio of the peak voltage amplitude to the minimum voltage amplitude along a lossless line.
(b) What is the SWR of a properly terminated line? Of a shorted line? An open-circuited line?
(c) What are the reflection coefficient and SWR for a 50-Ω line terminated in 45 Ω?

PROBLEM 5   Design an artificial line with a characteristic impedance of 75 Ω and a cutoff frequency of 1 GHz. Use SPICE to plot the gain and phase of the line when terminated in 75 Ω, as well as the input impedance (both magnitude and phase) from DC to 2 GHz. Would your design be suitable for an on-chip implementation? Explain.

PROBLEM 6   Consider an *RC diffusion line,* a transmission line in which there is no inductance.

(a) Derive expressions for α, β, and the characteristic impedance.
(b) Derive an expression for the input impedance of such a line when it is terminated in an open circuit.
(c) Using your answer to part (b), show that the real part of the input impedance at low frequencies is equal to one third the total resistance. This equality allows one to derive a single-pole lumped RC model to approximate (crudely) the fully distributed case by simply substituting the total capacitance and one third the total resistance into the model.

PROBLEM 7   An ideal lossless transmission line with inductance of 3 nH/mm and of length l is terminated by a 60-Ω resistor. A voltage step is applied at the input at time $t = 0$. What length of line and characteristic impedance do we need to provide a 5-ns delay and a maximum amplitude across the load resistor? What is the capacitance per unit length of your line?

PROBLEM 8

(a) Suppose we construct a simple 3.2-ns lumped delay line out of ten identical low-pass L-sections, where each inductance is 5π nH and each capacitance is $20/\pi$ pF. To two significant digits, what is the characteristic impedance of this line?

(b) What is the cutoff frequency in hertz?

(c) Using SPICE, plot the frequency response (magnitude and phase) of this line from DC to twice the frequency found in (b). Drive the line with a source whose resistance is the value found in part (a), and terminate it with another resistance of this value.

(d) Plot the response to a unit step.

(e) Now use simple half-sections on the ends. Replot as in (c) (up to the same frequency as before) and (d).

(f) Replace the simple half-sections with m-derived half-sections, with m equal to 0.6. Replot as in (c) and (d). Any improvement in bandwidth? Comment on the appearance of the step response relative to that in part (e).

PROBLEM 9   Construct a discrete artificial line with five simple LC sections. The 100-pF capacitors you are allowed to use happen to have a series parasitic inductance of 2 nH, while the 100-μH inductors at your disposal have a parasitic shunt capacitance of 5 pF. Plot the frequency response of this line when it is terminated in its characteristic impedance. Does the line behave as you expect? Explain. Resimulate without the parasitic elements and compare.

PROBLEM 10   Dielectric loss can generally be neglected in the low-gigahertz range for on-chip lines. However, substrate and conductor losses can be quite significant. In order to reduce the former, *microstrip* lines are sometimes used. Such transmission lines use a conductor over a ground plane. Occasionally, even *striplines* are employed, which are shielded structures in which a conductor is sandwiched between two ground planes. Using the formula from Chapter 4 for the capacitance of such lines, estimate the *inductance* per unit length for the two cases if the relative dielectric constant is 3.9. From those values, derive an expression for the characteristic impedance for the two types of line. Using your formulas, approximately what range of characteristic impedances could one expect for on-chip lines?

CHAPTER SEVEN

THE SMITH CHART AND S-PARAMETERS

7.1 INTRODUCTION

The subject of CMOS RF integrated circuit design resides at the convergence of two very different engineering traditions. The design of microwave circuits and systems has its origins in an era where devices and interconnect were usually too large to allow a lumped description. Furthermore, the lack of suitably detailed models and compatible computational tools forced engineers to treat systems as two-port "black boxes" with frequency-domain graphical methods. The IC design community, on the other hand, has relied on the development of detailed device models for use with simulation tools that allow both frequency- and time-domain analysis. As a consequence, engineers who work with traditional RF design techniques and those schooled in conventional IC design often find it difficult to converse. Clearly, a synthesis of these two traditions is required.

Analog IC designers accustomed to working with lower-frequency circuits tend to have, at best, only a passing familiarity with two staples of traditional RF design: Smith charts and S-parameters ("scattering" parameters). Although Smith charts today are less relevant as a computational aid than they once were, RF instrumentation continues to present data in Smith-chart form. Furthermore, these data are often S-parameter characterizations of two-ports, so it is important, even in the "modern" era, to know something about Smith charts and S-parameters. This chapter thus provides a brief derivation of the Smith chart, along with an explanation of why S-parameters won out over other parameter sets (e.g., impedance or admittance) to describe microwave two-ports.

7.2 THE SMITH CHART

Recall from Chapter 6 the expression for the reflection coefficient in terms of the normalized load impedance:

$$\Gamma = \frac{\dfrac{Z_L}{Z_0} - 1}{\dfrac{Z_L}{Z_0} + 1} = \frac{Z_{nL} - 1}{Z_{nL} + 1}. \tag{1}$$

The relationship between the normalized load impedance and Γ is bi-unique: knowing one is equivalent to knowing the other. This observation is important because the familiar curves of the Smith chart are simply a plotting, in the Γ-plane, of contours of constant resistance and reactance.

A natural question to ask is why one should go to the trouble of essentially plotting impedance in a nonrectilinear coordinate system, since it's certainly more straightforward to plot the real and imaginary parts of impedance directly in standard Cartesian coordinates.

There are at least two good reasons for the non-obvious choice. One is that trying to plot an infinite impedance directly poses self-evident practical problems. Plotting Γ instead neatly handles impedances of arbitrary magnitude because $|\Gamma|$ cannot exceed unity for passive loads. The other reason is that, as shown in Chapter 6, Γ repeats every half-wavelength when a lossless transmission line is terminated in a fixed impedance. Hence, plotting Γ is a natural and compact way to encode this periodic behavior. Much of the computational power of the Smith chart derives from this encoding, allowing engineers to determine rapidly the length of line needed to transform an impedance to a particular value, for example.

The relationship between impedance and Γ given in Eqn. 1 may be considered a mapping of one complex number into another. In this case, it is a special type of mapping known as a *bilinear transformation* because it is a ratio of two linear functions. Among the various properties of bilinear transformations, a particularly relevant one is that circles remain circles when mapped. In this context, a line is considered a circle of infinite radius. Hence, circles and lines map into either circles or lines.

With the aid of Eqn. 1, it is straightforward to show that the imaginary axis of the Z-plane maps into the unit circle in the Γ-plane, while other lines of constant resistance in the Z-plane map into circles of varying diameter that are all tangent at the point $\Gamma = 1$; see Figure 7.1.

Lines of constant reactance are orthogonal to lines of constant resistance in the Z-plane, and this orthogonality is preserved in the mapping. Since lines map to lines or circles, we expect constant-reactance lines to transform to the circular arcs shown in Figure 7.2. The Smith chart is just the plotting of both constant-resistance and constant-reactance contours in the Γ-plane without the explicit presence of the Γ-plane axes.

Because, as mentioned earlier, the primary role of the Smith chart these days is as a standard way to present impedance (or reflectance) data, it is worthwhile taking a little time to develop a familiarity with it. The center of the Smith chart corresponds to zero reflection coefficient and, therefore, a resistance equal to the normalizing impedance.

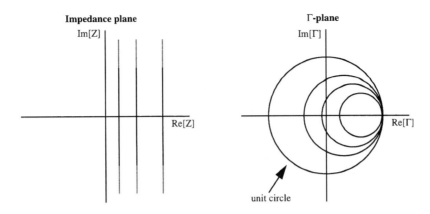

FIGURE 7.1. Mapping of constant-resistance lines in Z-plane to circles in Γ-plane.

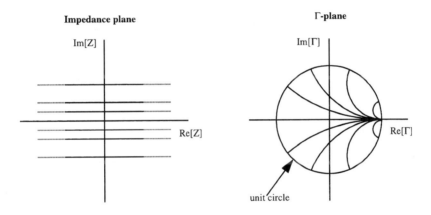

FIGURE 7.2. Mapping of constant-reactance lines in Z-plane to contours in Γ-plane.

The bottom half of the Z-plane maps into the bottom half of the unit circle in the Γ-plane, and thus capacitive impedances are always found there. Similarly, the top half of the Z-plane corresponds to the top half of the unit circle and inductive impedances. Progressively smaller circles of constant resistance correspond to progressively larger resistance values. The point $\Gamma = -1$ corresponds to zero resistance (or reactance), and the point $\Gamma = 1$ corresponds to infinite resistance (or reactance).

As a simple, but specific, example, let us plot the impedance of a series RC network in which the resistance is 100 Ω and the capacitance is 25 pF, all normalized to a 50-Ω system. Since the impedance is the sum of a real part (equal to the resistance) and an imaginary part (equal to the capacitive reactance), the corresponding locus in the Γ-plane must lie along the circle of constant resistance for which $R = 2$. The reactive part varies from minus infinity at DC to zero at infinite frequency. Since

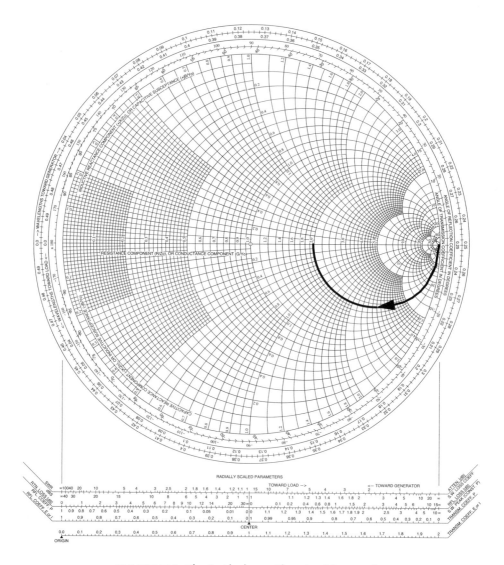

FIGURE 7.3. The Smith chart with series *RC* example.

it is always negative in sign, the locus must be just the bottom half of that constant resistance circle, traversed clockwise from $\Gamma = 1$ as frequency increases, as seen in Figure 7.3.

There are numerous other properties of Smith charts, and the types of computations that may be performed graphically and rapidly with them are truly remarkable. However, since machine computation has largely displaced that role of Smith charts, we direct the interested reader to Smith's papers for further applications.[1]

---

[1] For example, see P. H. Smith, "An Improved Transmission Line Calculator," *Electronics,* v. 17, January 1944, p. 130.

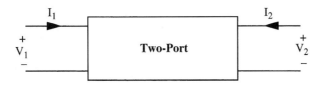

FIGURE 7.4. Port variable definitions.

7.3 S-PARAMETERS

Systems can be characterized in numerous ways. To simplify analysis and perhaps elucidate important design criteria, it is often valuable to use macroscopic descriptions, which preserve input–output behavior but discard details of the internal structure of the system. At lower frequencies, the most common representations use impedance or admittance parameters, or perhaps a mixture of the two (called, sensibly enough, *hybrid* parameters). Impedance parameters allow one to express port voltages in terms of port currents. For the two-port shown in Figure 7.4, the relevant equations are:

$$V_1 = Z_{11}I_1 + Z_{12}I_2, \tag{2}$$

$$V_2 = Z_{21}I_1 + Z_{22}I_2. \tag{3}$$

It is most convenient to open-circuit the ports in succession to determine the various Z parameters experimentally, because various terms then become zero. For instance, determination of Z_{11} is easiest when the output port is open-circuited because the second term in Eqn. 2 is zero under that condition. Driving the input port with a current source and measuring the resulting voltage at the input allows direct computation of Z_{11}. Similarly, open-circuiting the input port, driving the output with a current source, and measuring V_1 allows determination of Z_{12}.

Short-circuit conditions are used to determine admittance parameters, and a combination of open- and short-circuit conditions allows determination of hybrid parameters. The popularity of these representations for characterizing systems at low frequencies traces directly to the ease with which one may determine the parameters experimentally.

At high frequencies, however, it is quite difficult to provide adequate shorts or opens, particularly over a broad frequency range. Furthermore, active high-frequency circuits are frequently rather fussy about the impedances into which they operate, and may oscillate or even expire when terminated in open or short circuits. A different set of two-port parameters is therefore required to evade these experimental problems. Called *scattering* parameters (or simply S-parameters), they exploit the fact that a line terminated in its characteristic impedance gives rise to no reflections.[2]

---

[2] K. Kurokawa, "Power Waves and the Scattering Matrix," *IEEE Trans. Microwave Theory and Tech.*, v. 13, March 1965, pp. 194–202.

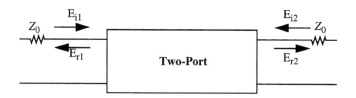

FIGURE 7.5. S-parameter port variable definitions.

Interconnections between the instrumentation and the system under test can therefore be of a comfortable length, since no short or open circuit needs to be provided; this greatly simplifies fixturing.

As implied earlier, terminating ports in open or short circuits is a convenience for lower-frequency two-port descriptions because various terms then become zero, simplifying the math. Scattering parameters retain this desirable property by defining input and output variables in terms of incident and reflected (scattered) *voltage waves,* rather than port voltages or currents (which are difficult to define uniquely at high frequencies, anyway).

As can be seen in Figure 7.5, the source and load terminations are Z_0. With the input and output variables defined as shown, the two-port relations may be written as

$$b_1 = s_{11}a_1 + s_{12}a_2, \tag{4}$$

$$b_2 = s_{21}a_1 + s_{22}a_2, \tag{5}$$

where

$$a_1 = E_{i1}/\sqrt{Z_0}, \tag{6}$$

$$a_2 = E_{i2}/\sqrt{Z_0}, \tag{7}$$

$$b_1 = E_{r1}/\sqrt{Z_0}, \tag{8}$$

$$b_2 = E_{r2}/\sqrt{Z_0}. \tag{9}$$

The normalization by the square root of Z_0 is a convenience that makes the square of the magnitude of the various a_n and b_n equal to the power of the corresponding incident or reflected wave.

Driving the input port with the output port terminated in Z_0 sets a_2 equal to zero, and allows us to determine the following parameters:

$$s_{11} = \frac{b_1}{a_1} = \frac{E_{r1}}{E_{i1}} = \Gamma_1, \tag{10}$$

$$s_{21} = \frac{b_2}{a_1} = \frac{E_{r2}}{E_{i1}}. \tag{11}$$

Thus, s_{11} is simply the input reflection coefficient, while s_{21} is a sort of gain since it relates an output wave to an input wave. Specifically, its magnitude squared is called the forward transducer power gain with Z_0 as source and load impedance.

Similarly, terminating the input port and driving the output port yields

$$s_{22} = \frac{b_2}{a_2} = \frac{E_{r2}}{E_{i2}} = \Gamma_2, \tag{12}$$

$$s_{12} = \frac{b_1}{a_2} = \frac{E_{r1}}{E_{i2}}. \tag{13}$$

Here, we see that s_{22} is the output reflection coefficient; s_{12} is the reverse transmission, whose magnitude squared is the reverse transducer power gain with Z_0 as source and load impedance.

Once a two-port has been characterized with S-parameters, direct design of systems may proceed in principle without knowing anything about the internal workings of the two-port. For example, gain equations and stability criteria can be recast in terms of S-parameters.[3] However, it is important to keep in mind that a macroscopic approach necessarily discards potentially important information, such as sensitivity to parameter or process variation. For this reason, S-parameter measurements are often used to derive element values for models whose topologies have been determined from first principles or physical reasoning.

To summarize, the reasons that S-parameters have become nearly universal in high-frequency work are that "zero"-length fixturing cables are unnecessary, there is no need to synthesize a short or open circuit, and terminating the two-port in Z_0 greatly reduces the potential for oscillation.

7.4 APPENDIX A: A SHORT NOTE ON UNITS

The inability to identify unique voltages and currents in distributed systems, coupled with the RF engineer's preoccupation with power gain, has made power the natural quantity on which to focus in RF circuits and systems. Power levels are expressed in watts, of course, but what can confuse and frustrate the uninitiated are the various decibel versions. For example, "dBm" is quite commonly used. The "m" signifies that the 0-dB reference is one milliwatt, while a "dBW" is referenced to one watt. If the reference impedance level is 50 Ω, then 0 dBm corresponds to a voltage of about 223 mV rms.

As clear as these definitions are, there are some who actively insist on confusing volts with watts, redefining 0 dBm to mean 223 mV rms *regardless of the impedance level*. Not only is this redefinition unnecessary, for one may always define a dBV, it is also dangerous. As we will see, critical performance measures (such as linearity and noise figure) intimately involve true power ratios, particularly in studying cascaded systems. Confusing power with voltage ratios leads to gross errors. Hence, throughout this text, 0 dBm truly means one milliwatt, and 0 dBV means one volt. Always.

---

[3] A representative reference that covers this topic is G. Gonzalez, *Microwave Transistor Amplifiers*, 2nd ed., Prentice-Hall, Englewood Cliffs, NJ, 1997.

With that out of the way, we return to some definitions. Common when discussing noise or distortion products in oscillators or power amplifiers is "dBc," where the "c" here signifies that the 0-dB reference is the power of the carrier.

Most engineers are familiar with the engineering prefixes ranging from that for 10^{-12} (pico) to that for 10^{12} (tera). Supplementing those below pico are femto (f), atto (a), zepto (z) and yocto (y), some of which sound like the names of less well-known Marx brothers. Above tera are peta (P), exa (E), zetta (Z) and yotta (Y). You can see that abbreviations for prefixes associated with positive exponents are capitalized (with the exception of that for kilo to avoid confusion with K, the unit of absolute temperature), and those for negative exponents are rendered in lower case (with μ being used for micro to avoid confusion with the abbreviation for milli). With these additional prefixes, you may express quantities spanning an additional 24 orders of magnitude. Presumably, that should suffice for most purposes.

By international convention, if a unit is named after a person, only the abbreviation is capitalized (W and watt, not Watt; V and volt, not Volt, etc.).[4] This choice assures that "two watts" refers only to two joules per second, not to two members of the Watt family, for example. Absolute temperature is measured in kelvins (not the redundant "degrees kelvin") and is abbreviated K. The liter (or litre) is an exception; its abbreviation may be l or L, the latter to avoid confusion with the numeral 1.

Finally, a good way to start a fistfight among microwave engineers is to argue over the pronunciation of the prefix "giga-." Given the Greek origin, both *g*s should be pronounced as in "giggle," the choice now advocated by both ANSI (American National Standards Institute) and the IEEE (Institute of Electrical and Electronics Engineers). However, there is still a sizable contingent who pronounce the first *g* as in "giant." The depth of emotion felt by some advocates of one or the other pronunciation is all out of proportion to the importance of the issue. Feel free to test this assertion at the next gathering of microwave engineers you attend. Ask them how they pronounce it, tell them they're wrong (even if they aren't), and watch the Smith charts fly.

7.4.1 DEFINITIONS OF POWER GAIN

While we're on the subject of definitions, we ought to discuss "power gain." You might think, quite understandably, that the phrase "power gain" has an unambiguous meaning, but you would be wrong. There are four types of power gain that one frequently encounters in microwave work, and it's important to keep track of which is which.

Plain old *power gain* is defined as you'd expect: it's the power actually delivered to some load divided by the power actually delivered by the source. However, because the impedance that is loading the source may not be known (particularly at

---

[4] The decibel is named after Alexander Graham Bell. Hence dB, not db, is the proper abbreviation.

high frequencies), measuring this quantity can be quite difficult in practice. As a consequence, other power gain definitions have evolved.

Transducer power gain (a term we have already used in this chapter) is the power actually delivered to a load divided by the power *available* from the source. If the source and load impedances are some standardized value then computing the power delivered and the available source power is relatively straightforward, sidestepping the measurement difficulties alluded to in the previous paragraph. From the definition you can also see that transducer gain and power gain will be equal if the input impedance of the system under consideration happens to be the complex conjugate of the source impedance.

Available power gain is the power available at the output of a system divided by the power available from the source. *Insertion power gain* is the power actually delivered to a load with the system under consideration *inserted,* divided by the power delivered to the load with the source connected directly to the load. Depending on context, any of these power gain definitions may be the appropriate one to consider.

Finally, note that if the input and output ports are all matched then the four definitions of power gain converge. In that instance only would it be safe to say "power gain" without being more specific.

7.5 APPENDIX B: WHY 50 (OR 75) Ω?

Most RF instruments and coaxial cables have standardized impedances of either 50 or 75 ohms. It is easy to infer from the ubiquity of these impedances that there is something sacred about these values, and that they should therefore be used in all designs. In this appendix, we explain where these numbers came from in the first place to see when it does and does not make sense to use those impedances.

7.5.1 POWER-HANDLING CAPABILITY

Consider a coaxial cable with an air dielectric. There will be, of course, some voltage at which the dielectric breaks down. For a fixed inner conductor diameter, one could increase the outer diameter to increase this breakdown voltage. However, the characteristic impedance would then increase, which by itself would tend to reduce the power deliverable to a load. Because of these two competing effects, there is a well-defined ratio of conductor diameters that maximizes the power-handling capability of a coaxial cable.

Having established the possibility that a maximum exists, we need to dredge up a couple of equations to find the actual dimensions that lead to this maximum. Specifically, we need one equation for the peak electric field between the conductors and another for the characteristic impedance of a coaxial cable:

$$E_{\max} = \frac{V}{a \ln(b/a)}, \tag{14}$$

and

$$Z_0 = \sqrt{\frac{\mu}{\varepsilon}} \cdot \frac{\ln(b/a)}{2\pi} \approx \frac{60}{\sqrt{\varepsilon_r}} \cdot \ln\left(\frac{b}{a}\right). \tag{15}$$

where a and b are (respectively) the inner and outer radius and ε_r is the relative dielectric constant, which is essentially unity for our air line case.

The next step is to recognize that the maximum power deliverable to a load is proportional to V^2/Z_0. Using our equations, that translates to

$$P \propto \frac{V^2}{Z_0} = \frac{[E_{max} \cdot a^2 \ln(b/a)^2]}{\left(60/\sqrt{\varepsilon_r}\right) \cdot \ln(b/a)} = \frac{\sqrt{\varepsilon_r}[E_{max}^2 \cdot a^2 \ln(b/a)]}{60}. \tag{16}$$

Now take the derivative, set it equal to zero, and pray for a maximum instead of a minimum:

$$\frac{dP}{da} = \frac{d}{da}\left[a^2 \ln\left(\frac{b}{a}\right)\right] = 0 \implies \frac{b}{a} = \sqrt{e}. \tag{17}$$

Plugging this ratio back into our equation for the characteristic impedance gives us a value of 30 Ω. That is, to maximize the power-handling capability of an air dielectric transmission line of a given outer diameter, we want to select the dimensions to give us a Z_0 of 30 Ω.

But wait: 30 does not equal 50, even for relatively large values of 30. So it appears we have not yet answered our original question. We need to consider one more factor: cable attenuation.

7.5.2 ATTENUATION

In Chapter 6 we derived a general expression for the attenuation constant of a transmission line that accounts for both dielectric and conductor losses. It may be shown (but we won't show it) that the attenuation per length due to dielectric loss is practically independent of conductor dimensions. Simplifying our equation to account only for the attenuation due to resistive loss yields

$$\alpha \approx \frac{R}{2Z_0}, \tag{18}$$

where R is the series resistance per unit length. At sufficiently high frequencies (the regime we're concerned with at the moment), R is due mainly to the skin effect. To reduce R we would want to increase the diameter of the inner conductor (to get more "skin"), but that would tend to reduce Z_0 at the same time, and it's not clear how to win. Again, we see a competition between two opposing effects, and we expect the optimum to occur once more at a specific value of b/a and hence at a specific Z_0.

Just as before, we invoke a couple of equations to get to an actual numerical result. The only new one we need here is an expression for the resistance R. If we make the usual assumption that the current flows uniformly in a thin cylinder of a thickness equal to the skin depth δ, we can write:

$$R \approx \frac{1}{2\pi\delta\sigma}\left[\frac{1}{a} + \frac{1}{b}\right], \tag{19}$$

where σ is the conductivity of the wire and δ is the same as always:

$$\delta = \sqrt{\frac{2}{\omega\mu\sigma}}. \tag{20}$$

With these equations, the attenuation constant may be expressed as

$$\alpha = \frac{R}{2Z_0} \approx \frac{\dfrac{1}{2\pi\delta\sigma}\left[\dfrac{1}{a} + \dfrac{1}{b}\right]\sqrt{\varepsilon_r}}{2\left[60\ln\left(\dfrac{b}{a}\right)\right]}. \tag{21}$$

Taking the derivative, setting it equal to zero, and now praying for a minimum instead of a maximum yields

$$\frac{d\alpha}{da} = 0 \implies \frac{d}{da}\frac{\dfrac{1}{a} + \dfrac{1}{b}}{\ln\left(\dfrac{b}{a}\right)} = 0 \implies \ln\left(\frac{b}{a}\right) = 1 + \frac{a}{b}, \tag{22}$$

which, after iteration, yields a value of about 3.6 for b/a. That value corresponds to a Z_0 of about 77 Ω. Now we have all the information we need.

First off, cable TV equipment is based on a 75-Ω world because it corresponds (nearly) to minimum loss. Power levels there are low, so power-handling capability is not an issue. So why is the standard there 75 and not 77 ohms? Simply because engineers like round numbers. This affinity for round numbers is also ultimately the reason for 50 Ω (at last). Since 77 Ω gives us minimum loss and 30 Ω gives us maximum power-handling capability, a reasonable compromise is an average of some kind. Whether you use an arithmetic or geometric mean, 50 Ω results after rounding. And that's it.

7.5.3 SUMMARY

Now that we understand how the macroscopic universe came to choose 50 Ω, it should be clear that one should feel free to use very different impedance levels if performance is limited neither by the power-handling nor the attenuation characteristics of the interconnect. As a result, IC engineers in particular have the luxury of selecting vastly different impedance levels than would be the norm in discrete design. Even in certain discrete designs, it is worth evaluating the tradeoffs associated with using 50 Ω (or some other standard value) rather than simply using the standard values as a reflex.

PROBLEM SET FOR SMITH CHART
AND S-PARAMETERS

PROBLEM 1 As discussed in the text, the Smith chart is nothing more than a mapping of contours of constant resistance and reactance from the impedance plane to the reflectance plane.

(a) Explicitly derive expressions for these mappings.

(b) Show that circles map into circles.

PROBLEM 2   Can one plot a Smith chart locus for a lossy transmission line? Explain. If it is possible, plot an example.

PROBLEM 3   The bilinear mapping that leads to the Smith chart maps an infinitely large domain into a finite, periodic range. Show specifically that it is periodic in a *half*-wavelength.

PROBLEM 4   Since the Smith chart is periodic in a half-wavelength, it is easy to see that quarter-wavelength lines possess an impedance reciprocation property. Demonstrate how one may exploit this property to convert impedances to admittance, and vice versa, with ease on the Smith chart.

PROBLEM 5   The various two-port representations are, of course, equivalent to each other since they ultimately describe the same system. Show this explicitly by converting impedance parameters to S-parameters, and vice versa.

PROBLEM 6   Convert hybrid parameters to S-parameters and vice versa.

PROBLEM 7   Convert admittance parameters to S-parameters and vice versa.

PROBLEM 8   Plot, on a Smith chart, the impedance of a series *RLC* network. Let the normalized resistance be unity. Choose the inductance and capacitance to yield a contour that is neither too large nor too small for comfortable plotting.

PROBLEM 9   It is a rule of thumb that plots on a Smith chart always go clockwise as frequency increases. Defend this rule, if possible. Are there exceptions?

PROBLEM 10   It is observed that networks whose Smith trajectories are far from the origin (of the Γ-plane) are narrowband. Explain.

CHAPTER EIGHT

BANDWIDTH ESTIMATION TECHNIQUES

8.1 INTRODUCTION

Finding the -3-dB bandwidth of an arbitrary linear network can be a difficult problem in general. Consider, for example, the standard recipe for computing bandwidth:

(1) derive the input–output transfer function (using node equations, for example);
(2) set $s = j\omega$;
(3) find the magnitude of the resulting expression;
(4) set the magnitude $= 1/\sqrt{2}$ of the "midband" value; and
(5) solve for ω.

It doesn't take a great deal of insight to recognize that explicit computation (by hand) of the -3-dB bandwidth using this method is generally impractical for all but the simplest systems. In particular, the order of the denominator polynomial obtained in step (1) is equal to the number of poles (natural frequencies), which in turn equals the number of degrees of freedom (measured, say, by the number of initial conditions one may independently specify), which in turn equals the number of independent energy storage elements (e.g., L or C), which in turn can be as large as the number of energy storage elements (phew!). Thus, a network with n capacitors might require the equivalent of finding the roots of an nth-order polynomial. If n exceeds just four, no algebraic closed-form solution exists. Even if $n = 2$, it might be labor-intensive to obtain the final numerical result.

Machine computation is cheap and getting cheaper all the time, so perhaps the analysis of networks doesn't present much of a problem. However, we are interested in developing *design* insight, so that if a simulator tells us that there is a problem then we have some idea of what to do about it. We therefore seek methods that are reasonably simple to apply yet convey the desired insight, even if they yield answers that might be approximate. Simulators can then be used to provide final quantitative verification.

Two such approximate methods are open- and short-circuit time constants. The former method provides an estimate of the high-frequency rolloff point, the latter of the low-frequency point. These methods are valuable because they identify which elements are responsible for the bandwidth limitation. This information alone is often sufficient to suggest what modifications should be tried next.

8.2   THE METHOD OF OPEN-CIRCUIT TIME CONSTANTS

The method of open-circuit time constants (OCτs), also known as "zero value" time constants, was developed in the mid-1960s at MIT. As we shall see, this powerful technique allows us to estimate the bandwidth of a system almost by inspection, and sometimes with surprisingly good accuracy. More important, and unlike typical circuit simulation programs, *OCτs identify which elements are responsible for bandwidth limitations.* The great value of this property in the design of amplifiers hardly needs expression.

To begin development of this method, let us consider all-pole transfer functions only. Such a system function may be written as follows:

$$\frac{V_o(s)}{V_i(s)} = \frac{a_0}{(\tau_1 s + 1)(\tau_2 s + 1) \cdots (\tau_n s + 1)}, \tag{1}$$

where the various time constants may or may not be real.

Multiplying out the terms in the denominator leads to a polynomial that we shall express as

$$b_n s^n + b_{n-1} s^{n-1} + \cdots + b_1 s + 1, \tag{2}$$

where the coefficient b_n is simply the product of all the time constants and b_1 is the sum of all the time constants. (In general, the coefficient of the s^j term is computed by forming all unique products of the n time constants taken j at a time and summing all $n!/j!(n-j)!$ such products.)

We now assert that, near the -3-dB frequency, the first-order term typically dominates over the higher-order terms, so that (perhaps) to a reasonable approximation we have

$$\frac{V_o(s)}{V_i(s)} \approx \frac{a_0}{b_1 s + 1} = \frac{a_0}{\left(\sum_{i=1}^{n} \tau_i\right)s + 1}. \tag{3}$$

The bandwidth of our original system in radian frequency as estimated by this first-order approximation is then simply the reciprocal of the effective time constant:

$$\omega_h \approx \frac{1}{b_1} = \frac{1}{\sum_{i=1}^{n} \tau_i} = \omega_{h,\text{est}}. \tag{4}$$

Before proceeding further, we should consider the conditions under which our neglect of the higher-order terms is justified. Let us examine the denominator of the transfer function near our estimate of ω_h. For the sake of simplicity, we start with a second-order polynomial with purely real roots.

Now, at our estimated -3-dB frequency, the original denominator polynomial is

$$-\tau_1 \tau_2 \omega_{h,\text{est}}^2 + j(\tau_1 + \tau_2)\omega_{h,\text{est}} + 1. \tag{5}$$

Note that the magnitude of the second term is unity (why?). As a consequence, both

$$\tau_1 \omega_{h,\text{est}} \tag{6}$$

and

$$\tau_2 \omega_{h,\text{est}} \tag{7}$$

must have magnitudes no greater than unity. Thus, the product of these terms (which is equal to the magnitude of the leading term of the polynomial) must be small compared to the magnitude of the second (first-order) term. The worst case occurs when the two time constants are equal, and even then the second-order term is only one fourth as large as the first-order term. Extending these arguments to polynomials of higher order reveals that the estimate of the bandwidth based simply on the coefficient b_1 is generally reasonable since the first-order term generally does dominate the denominator. Furthermore, the bandwidth estimate is usually conservative in the sense that *the actual bandwidth will almost always be at least as high as estimated by this method.*

So far, all we've done is show that a first-order estimate of the bandwidth is possible if one is given the sum of the pole time constants ($= b_1$). Alas, such information is almost never available, apparently casting serious doubt on the value of our entire enterprise, since the whole point was to avoid things such as direct computation of the pole locations in the first place.

Fortunately, it is possible to relate the desired time-constant sum, b_1, to (more or less) easily computed network quantities. The new recipe is thus as follows. Consider an arbitrary linear network comprising only resistors, sources (dependent or independent), and m capacitors. Then:

(a) compute the effective resistance R_{jo} facing each jth capacitor with all of the other capacitors removed (open-circuited, hence the method's name);
(b) form the product $\tau_{jo} = R_{jo}C_j$ (the subscript o refers to the open-circuit condition) for each capacitor;
(c) sum all m such "open-circuit" time constants.

Remarkably, the sum of open-circuit time constants formed in step (c) is in fact precisely equal to the sum b_1 of the pole time constants, a result proved by R. B. Adler (see also Section 8.4). Thus, at last, we have

$$\omega_{h,\text{est}} = \frac{1}{\sum_{j=1}^{m} R_{jo}C_j}. \tag{8}$$

8.2.1 OBSERVATIONS AND INTERPRETATIONS

The method of OCτs is relatively simple to apply because each time-constant calculation involves the computation of just a single resistance, although one must be wary of the impedance-modifying potential of dependent sources (such as the transconductance of a transistor model). In any event, the amount of computation required is typically substantially (indeed, often fantastically) less than that needed for an exact solution.

The greatest value of the technique lies in its identification of those elements implicated in bandwidth limitations – that is, those elements whose associated open-circuit time constants dominate the sum. This knowledge can guide the designer to effect appropriate modifications to circuit values or even suggest wholesale topological changes. In contrast, SPICE and other typical simulators only provide a numerical value for the bandwidth while conveying little or nothing about what the designer can do to alter the performance in a desired direction.

The origin of this property of OCτs may be regarded intuitively as follows. *The reciprocal of each* jth *open-circuit time constant is the bandwidth that the circuit would exhibit if that* jth *capacitor were the only capacitor in the system.* Thus, each time constant represents a *local bandwidth degradation term*. The method of OCτs then states that the linear combination of these individual, local limitations yields an estimate of the total bandwidth. The value of OCτs derives directly from the identification and approximate quantification of the local bandwidth bottlenecks.

8.2.2 ACCURACY OF OCτs

One must be careful not to place too much faith in the ability of OCτs to provide accurate estimates of bandwidth in all cases. This situation should hardly be surprising in view of the rather brutal truncation to first order of the denominator polynomial. However, there are numerous conditions under which OCτ estimates are fairly reasonable, as we have seen.

It should be clear that an OCτ bandwidth estimate is in fact exact for a first-order network since *no* truncation of terms is involved there. Not surprisingly, then, the OCτ estimate will be quite accurate if a network of higher order happens to be dominated by one pole (that is, if one pole is much lower in frequency than all of the other poles). There are many systems of practical interest, such as operational amplifiers, that are designed to have a dominant single pole and thus for which OCτ estimates are quite accurate.

Unfortunately, there are so many other conditions under which OCτs give poor estimates that some caveat is necessary. For example, complex poles quite commonly arise (intentionally or otherwise) in the design of wideband multistage amplifiers. Often the physical origin of these complex poles can be traced to the interaction of the primarily capacitive input impedance of one stage (as in a common-source configuration) with the inductive component of the output impedance of source followers.

The reason that the presence of complex poles upsets OCτ estimates is as follows: The coefficient b_1 is the sum of the pole time constants and thus ignores the imaginary parts of complex poles, since they must appear in conjugate pairs. However, the true bandwidth of, say, a two-pole system does depend on both the real and imaginary parts. As a result, gross errors in OCτ estimates are not uncommon if complex poles are present in abundance.

The nature and magnitude of the problem are best illustrated with an example. Consider the simplest possible case, a two-pole transfer function:

$$H(s) = \left[\frac{s^2}{\omega_n^2} + \frac{2\zeta s}{\omega_n} + 1 \right]^{-1}. \tag{9}$$

The OCτ bandwidth estimate is found from the coefficient of the s term:

$$\omega_h \approx \frac{\omega_n}{2\zeta}; \tag{10}$$

it may be shown that the actual bandwidth is

$$\omega_h = \omega_n \left[1 - 2\zeta^2 + \sqrt{2 - 4\zeta^2 + 4\zeta^4} \right]^{1/2}. \tag{11}$$

In this particular case, we see that the OCτ estimate predicts monotonically increasing bandwidth as the damping ratio ζ approaches zero, while the actual bandwidth asymptotically approaches about $1.55\omega_n$. Thus, it is possible for OCτ estimates to be *optimistic* – in this case, wildly so. At a ζ of about 0.35, OCτ estimates are correct; for any higher damping ratio, OCτs are pessimistic. Fortunately, the poles of amplifiers are usually designed to have relatively high damping ratios (to control overshoot and ringing in the step response, and to minimize peaking in the frequency response) and so, for most practical situations, OCτ estimates are pessimistic.

Since it is generally not possible to tell by inspection of a network if complex poles will be an issue, one must always keep in mind that the primary value of OCτs is in identifying those portions of a circuit that control the bandwidth, rather than in providing accurate bandwidth estimates. Circuit simulators will take care of the latter task.

8.2.3 OTHER IMPORTANT CONSIDERATIONS

Although application of open-circuit time constants is reasonably straightforward, there are one or two final issues that deserve consideration. An extremely important idea is that not all capacitors in a network belong in the OCτ calculations. For instance, fairly large coupling capacitors are frequently used in discrete designs to connect the output of one stage to the input of the next without the bias point of one stage upsetting that of the other. Blind application of the OCτ method would lead one to conclude erroneously that the larger this capacitor, the lower the bandwidth (time constants here often correspond to the audio range, suggesting that large bandwidths are not possible). Fortunately, real circuit behavior defies these implications.

The problem stems from the presence of zeros associated with the coupling capacitors. Recall that the assumed form for the system function consists of poles only. Since all zeros are thus assumed at infinitely high frequency, it is hardly surprising that the presence of low-frequency zeros confounds our estimates of bandwidth.

The solution is to *preprocess the network* prior to application of the OCτ method. That is, recognize that the coupling capacitors are effectively short circuits relative to the impedances around them at frequencies near the upper bandwidth limit. Thus, *one may apply OCτs only to models that are appropriate to the high-frequency regime.*

It is usually obvious which capacitors are to be ignored (considered short circuits), but there are occasions when one is not so sure. In these cases, a simple thought experiment usually suffices to decide the issue. Now OCτs are concerned only with those capacitors that limit high-frequency gain. As a consequence, the removal (i.e., the open-circuiting) of a capacitor that belongs in the OCτ calculation should result in an increase in high-frequency gain. The test, therefore, is to consider exciting the network at some high frequency and imagining what would happen to the gain if the capacitor in question were open-circuited. If the gain would go up then the capacitor belongs in the OCτ calculation, since we infer from the thought experiment that the capacitor does indeed limit the high-frequency gain. If the gain would not change (or even decrease, as in the coupling capacitor case) upon removal, that capacitor should probably be short-circuited. The necessary conclusions can usually be reached without taking pencil to paper.

One last issue that deserves some attention concerns the relationship between the individual open-circuit time constants and the time constants of the poles. We have asserted (without formal proof) only that the *sums* of these time constants are equal to each other. *One must therefore resist the temptation to equate an open-circuit time constant with a corresponding pole location.* Indeed, the number of poles may not even equal the number of capacitors (consider the trivial case of two capacitors in parallel). Since the number of open-circuit time constants and the number of poles may be unequal, one clearly cannot expect each OCτ to equal the time constant of a pole in general.

8.2.4 SOME USEFUL FORMULAS

When computing open-circuit time constants for transistor amplifiers, care is required because the feedback action of the g_m generator modifies resistances. As a consequence, one should explicitly apply a test source (choose the type that will most directly allow computation of v_{gs}) to derive expressions for the effective resistances. To illustrate a general method, we will derive formulas for the resistances facing C_{gs} and C_{gd}. To simplify the derivations, we will ignore body effect and output resistance. However, complete formulas including both of those effects are provided at the end of this chapter for reference. Derivations are (surprise!) left as an exercise for the reader.

Consider a comprehensive model for a MOSFET with external resistances added in series with each terminal (except for the substrate, which is our ground reference), as shown in Figure 8.1. Although this model explicitly includes the back-gate transconductance g_{mb} and output resistance r_o, we will not use them in this first set of derivations. In all of the SPICE runs that follow, however, the complete model will be used.

First, let's compute the resistance facing C_{gs}. Applying a test voltage source v_t (since that choice directly fixes the value of v_{gs}), we exploit superposition (once v_{gs}

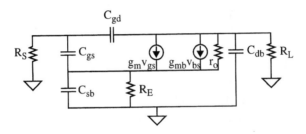

FIGURE 8.1. Incremental model for open-circuit resistance calculations.

is known, we may treat the transconductance generator as an independent current source of value $g_m v_t$) to obtain, when all is said and done,

$$i_t = \frac{v_t}{R_S + R_E} + \frac{g_m R_E v_t}{R_S + R_E}, \tag{12}$$

so that the equivalent resistance facing C_{gs} is given by

$$r_{1o} = \frac{R_S + R_E}{1 + g_m R_E}. \tag{13}$$

Hence, r_{1o} is the sum of the resistors divided by $1 + g_m R_E$.

Now, to compute the resistance facing C_{gd}, use a test current source (you may try a test voltage source, but you'll regret it). The intervening algebra is a little involved, but the resistance may be expressed in the following mnemonic form:

$$r_{2o} = r_{\text{left}} + r_{\text{right}} + g_{m,\text{eff}} r_{\text{left}} r_{\text{right}}, \tag{14}$$

where r_{left} is the resistance between the left terminal and ground, r_{right} is defined between the right terminal and ground, and $g_{m,\text{eff}}$ is the *effective* transconductance (defined as the ratio of current from the dependent current source to the voltage between the left terminal and ground). For our model, we have

$$r_{\text{right}} = R_L, \tag{15}$$

$$r_{\text{left}} = R_S, \tag{16}$$

$$g_{m,\text{eff}} = g_m \cdot \frac{1}{1 + g_m R_E}. \tag{17}$$

8.2.5 MORE USEFUL FORMULAS

To save you the trouble of hunting down (or deriving) the equations you might need to carry out bandwidth calculations, here are many of the relevant equations in one place. Because this is not a textbook on network theory, we provide only the barest sketches of derivations. That said, there should still be enough that you could reconstruct intermediate steps if needed.

Part of what distinguishes an experienced analog designer from others is the ability to apply the simplest approximation that captures the phenomena of interest. Throughout, recognize that many of the equations here simplify considerably if you ignore various factors (e.g., body effect, channel-length modulation, DIBL, etc.). In some cases, we offer those simplifications. In others, they are left for you to finish. In all cases, pause and consider when it may be acceptable to neglect certain of these phenomena – and how that selective and conscious neglect simplifies the equations.

Common-Source Input Resistance

At frequencies where we may neglect quasistatic effects, the input resistance is effectively infinite (not a function of g_{mb} or CLM/DIBL). In this era of superthin gate oxides, leakage is more than measurable and the input resistance is consequently not truly infinite. However, we will pretend that for most purposes it still remains high enough that infinity is a useful approximation.

Common-Source Output Resistance

Use a test current source, i_t. Since $v_{gs} = v_{bs}$, we can simply sum the two transconductances into a single one (let's call it $g_{m,\text{tot}}$). Compute the resistance looking into the drain, and call it r_{out}. The net resistance at the drain will therefore be r_{out} in parallel with any external resistance, R_L. Thus we have

$$v_{gs} = -i_t R_S, \tag{18}$$

$$v_{r0} = (i_t - g_{m,\text{tot}} v_{gs}) r_0 = i_t (1 + g_{m,\text{tot}} R_S) r_0, \tag{19}$$

$$V_{\text{test}} = i_t R_S + i_t (1 + g_{m,\text{tot}} R_S) r_0 = i_t [R_S + r_0 + g_{m,\text{tot}} r_0 R_S]. \tag{20}$$

So, at last, the resistance looking into the drain (and so *not* including R_L) is

$$r_{\text{out}} = R_S + r_0 + g_{m,\text{tot}} r_0 R_S = R_S + r_0 + (g_m + g_{mb}) r_0 R_S. \tag{21}$$

Common-Source Voltage Gain

Multiplication of the total resistance at the drain by the effective transconductance gives us the gain. To find the effective transconductance, short the drain to ground (incrementally speaking) and then measure the ratio of short-circuit output current to input voltage:

$$v_{gs} = v_{\text{in}} - i_{\text{out}} R_S, \tag{22}$$

$$v_{bs} = -i_{\text{out}} R_S; \tag{23}$$

$$i_{\text{out}} = g_m v_{gx} + g_{mb} v_{bs} - i_{\text{out}} \frac{R_S}{r_0}$$

$$= g_m (v_{\text{in}} - i_{\text{out}} R_S) + g_{mb} (-i_{\text{out}} R_S) - i_{\text{out}} \frac{R_S}{r_0}; \tag{24}$$

$$g_{m,\text{eff}} = \frac{i_{\text{out}}}{v_{\text{in}}} = \frac{g_m}{1 + (g_m + g_{mb}) R_S + R_S/r_0} = \frac{g_m}{1 + (g_m + g_{mb} + g_0) R_S}, \tag{25}$$

where we have used $g_0 = 1/r_0$.

The voltage gain is therefore

$$A_v = -g_{m,\text{eff}}(r_{\text{out}} \parallel R_L)$$

$$= -\frac{g_m}{1 + (g_m + g_{mb} + g_0)R_S}([R_S + r_0 + (g_m + g_{mb})r_0 R_S] \parallel R_L), \quad (26)$$

which simplifies initially to

$$A_v = -\frac{g_m R_L}{1 + (g_m + g_{mb} + g_0)R_S} \cdot \frac{R_S + r_0 + (g_m + g_{mb})r_0 R_S}{R_S + r_0 + (g_m + g_{mb})r_0 R_S + R_L}. \quad (27)$$

After a little more work, the equation's complexity reduces further, leading to a quite reasonable expression:

$$A_v = \frac{g_m R_L}{1 + (g_m + g_{mb})R_S + g_0(R_L + R_S)}. \quad (28)$$

The reader should verify that this expression reduces to the expected approximations when the body effect may be neglected, when there is no source degeneration, and so on.

Common-Gate Input Resistance

Use a test voltage source. Note that once again $v_{gs} = v_{bs}$ and so we can merge the two transconductances into a single $g_{m,\text{tot}}$. Compute the resistance looking into the source first; then accommodate R_S at the end if desired or necessary. You may use superposition to simplify the derivation (treat transconductances as independent current sources in this computation, since the control voltage is fixed *throughout this particular experiment*).

We start with

$$v_{gs} = -v_{\text{test}}. \quad (29)$$

With transconductance $g_{m,\text{tot}}$ disabled, we compute one contribution to the test current:

$$i_{t1} = \frac{v_{\text{test}}}{r_0 + R_L}. \quad (30)$$

Next, disable (short out) the test voltage source, taking care not to zero out $g_{m,\text{tot}}$, and compute the other contributions to the test current. The current from the transconductance is added to the current flowing through r_0. Then

$$i_{t2} = -g_{m,\text{tot}}v_{gs} + \left(-i_{t2}\frac{R_L}{r_0}\right) = g_{m,\text{tot}}v_{\text{test}} - i_{t2}\frac{R_L}{r_0}$$

$$\implies i_{t2} = \frac{g_{m,\text{tot}}v_{\text{test}}}{1 + R_L/r_0} = \frac{g_{m,\text{tot}}r_0 v_{\text{test}}}{r_0 + R_L}; \quad (31)$$

$$i_t = i_{t1} + i_{t2} = \frac{v_{\text{test}}}{r_0 + R_L} + \frac{g_{m,\text{tot}}r_0 v_{\text{test}}}{r_0 + R_L} = \frac{v_{\text{test}}}{r_0 + R_L}(1 + g_{m,\text{tot}}r_0). \quad (32)$$

Hence, the resistance looking into the source is

$$r_{\text{in}} = \frac{v_{\text{test}}}{i_t} = \frac{r_0 + R_L}{1 + g_{m,\text{tot}}r_0} = \frac{r_0 + R_L}{1 + (g_m + g_{mb})r_0}. \quad (33)$$

Note that, if $R_L \ll r_0$ and if the transistor's "intrinsic voltage gain" $g_m r_0 \gg 1$, then the resistance looking into the source is approximately

$$r_{in} \approx \frac{1}{g_m + g_{mb}}. \tag{34}$$

If you desire the total resistance between the source node and ground, merely compute the parallel combination of r_{in} and R_S.

Common-Gate Output Resistance

This resistance is precisely the same as the output resistance of the degenerated common-source amplifier:

$$r_{out} = R_S + r_0 + g_{m,tot} r_0 R_S = R_S + r_0 + (g_m + g_{mb}) r_0 R_S. \tag{35}$$

If we drive the source terminal directly with a voltage source, then $R_S = 0$ and the equation for the resistance looking into the drain collapses to simply r_0. That remains a reasonably good approximation if R_S is small compared with r_0 and if $(g_m + g_{mb}) R_S$ is small compared with unity.

Common-Gate Voltage Gain

As before, let's first compute the effective transconductance of the amplifier, defined here as the ratio of short-circuit drain current to source voltage:

$$v_{gs} = -v_{in}, \tag{36}$$

$$v_{bs} = v_{gs}; \tag{37}$$

$$i_{out} = -g_{m,tot} v_{gx} + \frac{V_{in}}{r_0} = v_{in}(g_{m,tot} + g_0)$$

$$\implies g_{m,eff} = g_{m,tot} + g_0 = (g_m + g_{mb} + g_0). \tag{38}$$

The voltage gain from source to drain is therefore

$$A_{v0} = g_{m,eff}(r_{out} \parallel R_L) = (g_m + g_{mb} + g_0)(r_0 \parallel R_L), \tag{39}$$

where we have used the fact that, in this first computation, we are driving the source terminal directly with a voltage source. Consequently, the total drain resistance is just $r_0 \parallel R_L$, so the gain in Eqn. 39 is that from the source to drain, as stated.

The voltage divider formed at the input must be taken into account in order to complete our calculation of the overall gain:

$$A_v = A_{v0} \cdot \frac{r_{in}}{r_{in} + R_S}. \tag{40}$$

Thus,

$$A_v = (g_m + g_{mb} + g_0)(r_0 \parallel R_L) \cdot \frac{\dfrac{r_0 + R_L}{1 + (g_m + g_{mb})r_0}}{\dfrac{r_0 + R_L}{1 + (g_m + g_{mb})r_0} + R_S}, \tag{41}$$

which simplifies a bit to

$$A_v = (g_m + g_{mb} + g_0)(r_0 \parallel R_L) \cdot \frac{r_0 + R_L}{r_0 + R_L + R_S + (g_m + g_{mb})r_0 R_S} \qquad (42)$$

and still further to

$$A_v = \frac{(g_m + g_{mb} + g_0)R_L}{1 + g_0(R_L + R_S) + (g_m + g_{mb})R_S}. \qquad (43)$$

As expected, the gain expression collapses to more familiar forms when body effect is neglected, the source resistance R_S is zero, the transistor's output conductance is zero, ... The reader should verify these statements independently.

Source-Follower Resistance

Again, we may treat the source follower's *input* resistance as infinite if we neglect gate leakage and nonquasistatic effects. This result is independent of body effect and CLM/DIBL.

The source follower's *output* resistance is the same as the input resistance of a common-gate stage:

$$r_{\text{out}} = \frac{r_0 + R_L}{1 + g_{m,\text{tot}}r_0} = \frac{r_0 + R_L}{1 + (g_m + g_{mb})r_0}. \qquad (44)$$

In most source followers, R_L is chosen very small compared with r_0. In such cases, the output resistance simplifies to

$$r_{\text{out}} \approx \frac{r_0}{1 + (g_m + g_{mb})r_0}. \qquad (45)$$

If, in addition, the transistor's intrinsic voltage gain is high, then the unity factor in the denominator can be neglected, simplifying the output resistance expression even further:

$$r_{\text{out}} \approx \frac{1}{g_m + g_{mb}}. \qquad (46)$$

Source-Follower Voltage Gain

Again, compute an effective transconductance and then multiply by the output resistance. That output resistance needs to include any external loading that might appear in parallel with the resistance seen when looking into the transistor's source. Now

$$v_{gs} = v_{\text{in}}, \qquad (47)$$

$$v_{bs} = v_{gs} \qquad (48)$$

(for output resistance only; $v_{bs} = 0$ for effective transconductance calculations). Therefore,

$$i_{\text{out}} = g_m v_{\text{in}} - i_{\text{out}}\frac{R_L}{r_0}, \qquad (49)$$

$$g_{m,\text{eff}} = \frac{i_{\text{out}}}{v_{\text{in}}} = \frac{g_m}{1 + R_L/r_0}. \qquad (50)$$

Overall voltage gain is thus

$$A_{v0} = g_{m,\text{eff}}(r_{\text{out}} \parallel R_S) = \frac{g_m}{1 + R_L/r_0} \cdot \left[\left(\frac{r_0 + R_L}{1 + (g_m + g_{mb})r_0} \right) \parallel R_S \right], \quad (51)$$

which, after some effort, simplifies to

$$A_{v0} = \frac{g_m r_0 R_S}{(g_m + g_{mb})r_0 R_S + (r_0 + R_L + R_S)}. \quad (52)$$

Alternatively, we may write

$$A_{v0} = \frac{g_m R_S}{(g_m + g_{mb})R_S + 1 + g_0(R_L + R_S)} = \frac{1}{1 + \dfrac{g_{mb} + g_0}{g_m} + \dfrac{1 + g_0 R_L}{g_m R_S}}, \quad (53)$$

from which it is perhaps somewhat easier to see that the voltage gain can only *approach* unity, as we would expect from a source follower.

As usual, the reader should verify that the exact equation simplifies to the numerous known approximations in the case where body effect is zero, transistor g_0 may be neglected, R_L is zero, and so forth.

Resistances for OC Time Constant Calculations

It's important to recall that a two-port model is a general and complete description for the terminal behavior of any single-input, single-output amplifier. Hence, the effective resistance facing any capacitor may be evaluated directly from the two-port model if the corresponding parameters are known. We further simplify the situation by assuming that we may safely neglect reverse transmission. This assumption of unilateral behavior is well satisfied in many (but not all) practical amplifiers. Always check assumptions in any case where it matters!

Using a current source in a hybrid model, we find that the effective resistance may be stated in the simple, universal, mnemonic form we've already cited:

$$r_{\text{eq}} = r_{\text{left}} + r_{\text{right}} + g_{m,\text{eff}} r_{\text{left}} r_{\text{right}}, \quad (54)$$

where r_{left} is the resistance seen between the capacitor's left terminal and ground, r_{right} is that seen between the right terminal and ground, and $g_{m,\text{eff}}$ is the effective transconductance – defined as the ratio of short-circuit output current to input voltage.

This equation may also be expressed as

$$r_{\text{eq}} = r_{\text{left}} + r_{\text{right}} - A_{vf} r_{\text{left}}, \quad (55)$$

where we have recognized the product of effective transconductance and r_{right} as (minus) the voltage gain between the two terminals of the capacitor. From this form of the equation we can see directly the effect of voltage gain between the capacitor's terminals, a phenomenon first explained (in the context of vacuum tubes) by John M. Miller of the National Bureau of Standards in 1919.[1]

---

[1] John M. Miller, "Dependence of the Input Impedance of a Three-Electrode Vacuum upon the Load in the Plate Circuit," *Scientific Papers of the Bureau of Standards,* v. 15, 1919–1920, pp. 367–85.

Resistance facing c_{gd}. For the drain–gate capacitance, the effective resistance is

$$R_G + (r_{\text{out}} \parallel R_L) + \frac{g_m}{1 + (g_m + g_{mb} + g_0)R_S}(r_{\text{out}} \parallel R_L)R_G, \qquad (56)$$

where

$$r_{\text{out}} \parallel R_L = [R_S + r_0 + (g_m + g_{mb})r_0 R_S] \parallel R_L. \qquad (57)$$

The expression for the effective resistance simplifies considerably in the case of a common-gate connection, where R_G is often zero (or very small). In that case, Eqn. 56 is all you need.

In a source-follower connection, where R_L is typically zero (or very small), an even greater simplification results. In that case, R_G is perhaps a good approximation to the resistance facing c_{gd}. In other words, the expression for this resistance is somewhat complicated only for the case of a common-source amplifier.

Resistance facing c_{gs}. Here, the same basic two-port model applies, but now the model parameters must be chosen to reflect the fact that the capacitor in question connects between the gate and source terminals. That is, the amplifier under consideration has its input at the gate and provides an output at the source. If we use R_G to denote the resistance to the left, then resistance to the right is simply the total output resistance of a source follower (i.e., the resistance looking into the source terminal in parallel with any external R_S). Similarly, the effective transconductance is also merely that of a source follower. Thus we have actually derived all of the pieces already. We just need to put them together:

$$r_{\text{eq}} = R_G + \left[\left(\frac{r_0 + R_L}{1 + (g_m + g_{mb})r_0}\right) \parallel R_S\right]$$
$$- \left[\frac{g_m}{1 + R_L/r_0}\right]R_G\left[\left(\frac{r_0 + r_L}{1 + (g_m + g_{mb})r_0}\right) \parallel R_S\right], \qquad (58)$$

which we try to simplify as follows:

$$R_{\text{eq}} = R_G + \left[\frac{R_S}{r_0 + R_L + R_S + (g_m + g_{mb})r_0 R_S}\right][(r_0 + R_L) - g_m r_0 R_G]. \qquad (59)$$

After some crunching, we finally get something that does look simpler:

$$r_{\text{eq}} = \frac{R_G(r_0 + R_L + R_S + g_{mb}r_0 R_S) + R_S(r_0 + R_L)}{r_0 + R_L + R_S + (g_m + g_{mb})r_0 R_S}. \qquad (60)$$

Collecting terms in a slightly different way may also prove useful:

$$r_{\text{eq}} = \frac{R_G + R_S + g_{mb}R_G R_S + g_0(R_G R_L + R_G R_S + R_S R_L)}{1 + (g_m + g_{mb})R_S + g_0(R_L + R_S)}. \qquad (61)$$

Note that, in the limit of no body effect and infinite r_0, the equivalent resistance facing c_{gs} indeed simplifies to a result we've seen before:

$$r_{\text{eq}} \approx \frac{R_G + R_S}{1 + g_m R_S}. \qquad (62)$$

Resistance facing c_{db}. In the general case, this is just the resistance looking into the drain of a transistor in parallel with any additional resistance connected to that drain (call that extra resistance R_L to be consistent with our notation so far). We've already found the resistance looking into the drain:

$$r_{out} = R_S + r_0 + g_{m,tot} r_0 R_S = R_S + r_0 + (g_m + g_{mb}) r_0 R_S. \qquad (63)$$

So, just place that value in parallel with any R_L that happens to be present.

Resistance facing c_{sb}. Similarly, this resistance is just that looking into the source of a transistor in parallel with any other resistance connected to it (we've been calling it R_S). Thus, the source-follower output resistance equation is what we need here:

$$r_{out} = \frac{r_0 + R_L}{1 + (g_m + g_{mb}) r_0}. \qquad (64)$$

Just compute the parallel combination of r_{out} and R_S to get the final answer.

After a little practice, these equations will help you to zip through bandwidth calculations.

8.2.6 A DESIGN EXAMPLE

We've seen that the method of open-circuit time constants promises to simplify design while conveying important insight. Let's now carry out an actual design to see if it lives up to this promise.

Suppose we want an amplifier with the following specifications:

voltage gain magnitude: >18 dB (or about a factor of 8);
−3-dB bandwidth: >450 MHz (implies a maximum OCτ sum of ~350 ps).

Furthermore, assume that we must meet these specifications with a 2-kΩ source resistance driving the input and a 1-pF capacitive load on the output. In a truly practical design, there would usually be additional specifications (such as maximum allowed power consumption, dynamic range, etc.), but we'll keep the design space restricted for now.

Further suppose that we are to meet these specifications with transistors from the 0.5-μm (drawn) technology described in Chapter 5. To simplify the process, let us use just one size of device, and just one bias current for all transistors. In a better design, of course, one would generally use different biases and different device sizes, but we need to impose some arbitrary constraints if we are to complete our task in finite space!

Arbitrarily selecting a per-transistor bias current of 3 mA, a 150-μm–wide NMOS transistor in this process technology has the following approximate element values when operating in saturation:

$$C_{gs} = 220 \text{ fF}, \qquad C_{sb} = 130 \text{ fF}, \qquad C_{gd} = 45 \text{ fF}, \qquad C_{db} = 90 \text{ fF};$$

$$r_o = 2 \text{ k}\Omega, \qquad g_m = 12 \text{ mS}, \qquad g_{mb} = 1.8 \text{ mS}.$$

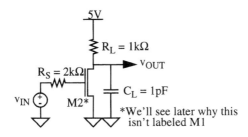

FIGURE 8.2. First-cut design (biasing not shown).

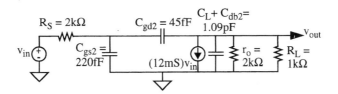

FIGURE 8.3. Incremental model of first-pass design.

Even though some of the capacitances are bias voltage–dependent, we will assume that they are constant at the values shown.

The only way to start a design is, well, to start. Put *something* (almost anything) down. *It's easier to edit than to create,* so virtually any reasonable initial condition is acceptable. A few simple calculations will let you know fairly quickly if you're on the right track, and you can always obsess later about the particulars. So, let's start with the common-source (CS) configuration (after all, it provides voltage gain and has a moderately high input impedance). In all that follows, we'll neglect the details of how biasing is taken care of (since we're focusing on dynamic performance issues), but be aware that any practical design must include careful attention to the bias problem.

Recalling that (neglecting body effect) the voltage gain from gate to drain of a basic CS amplifier is $-g_m R_L$, and being mindful that we do have to worry about gain loss from the additional loading by the transistor's own output resistance, let's shoot for a $g_m R_L$ product that is 50% larger than the gain specification. With the resulting choice of 12 for $g_m R_L$, we find that we must select $R_L = 1\,\text{k}\Omega$. Our circuit then appears as shown in Figure 8.2.

The corresponding incremental model is then as sketched in Figure 8.3. Note that the source–bulk potential is zero, so the back-gate transconductance contributes zero current and the source–bulk capacitance is shorted out.

From the model, it's easy to see that the low-frequency voltage gain just barely meets specifications:

$$A_V = -g_{m2}(R_L \parallel r_o) = -8. \tag{65}$$

Now, let's estimate the bandwidth to see just how bad the news there is:

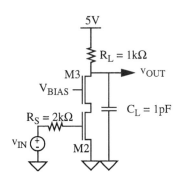

FIGURE 8.4. Second pass: cascode amplifier.

$$\tau_{gs2} = C_{gs2}r_{gs2} = (220\,\text{fF})(R_S) = 440\,\text{ps}, \tag{66}$$

$$\tau_{gd2} = C_{gd2}r_{gd2} = (45\,\text{fF})(r_{\text{left}} + r_{\text{right}} + g_{m2}r_{\text{left}}r_{\text{right}}) \approx 840\,\text{ps}, \tag{67}$$

$$\tau_{db2} = C_{db2}r_{db2} = C_{db2}(R_L \parallel r_o) = 60\,\text{ps}, \tag{68}$$

$$\tau_L = C_L(R_L \parallel r_o) = 670\,\text{ps}; \tag{69}$$

$$\text{BW} \approx \frac{1}{(440\,\text{ps} + 840\,\text{ps} + 60\,\text{ps} + 670\,\text{ps})} \approx 500\,\text{Mrps}. \tag{70}$$

We see that our bandwidth is about 79 MHz (SPICE says 86 MHz), so we're quite a bit shy of our goal of 450 MHz.

Now, who's the big culprit? From our four calculated time constants, we see that there are two similar-sized ones. The larger of these is associated with the drain–gate capacitance, C_{gd2}, even though that capacitance is numerically the smallest, because its effect is Miller-multiplied by the gain. So, if we are to improve bandwidth, we must figure out how to mitigate the Miller effect.

Recall that the Miller effect arises from connecting a capacitance across two nodes that have an inverting voltage gain between them. So, one possible solution would be to distribute the gain among N stages instead of trying to get all of our gain from one stage. You are encouraged to explore this promising option independently.

Another possibility is to isolate (somehow) the offending capacitance so that it no longer appears across a gain stage. We will pursue this approach as we attempt to get all of our gain in one stage. The *cascode* amplifier eliminates the Miller effect precisely by performing this isolation. Consider the circuit shown in Figure 8.4.

As usual, the value of V_{BIAS} is not terribly critical. It just has to be high enough to guarantee that M_2 stays in saturation yet low enough to guarantee that M_3 stays in saturation. A value of 2.3 V satisfies these conditions comfortably in this particular case, and this is the value used in the SPICE simulations.

Before plunging mindlessly ahead into a pile of computations, let's think about how this circuit works. The input voltage is converted into an output current by transistor M_2 (i.e., M_2 is a transconductor). Transistor M_3 merely transfers this current to the output load resistor. Now, the output is at the drain of M_3, while the input is at the gate of M_2, and there is no capacitance directly across the two nodes. Hence, there is very little Miller multiplication, and we expect a significant improvement in bandwidth.[2]

The isolation provided by cascoding also has a beneficial effect on the gain. Voltage changes at the drain of M_3 have hardly any effect on the drain current of M_2. Hence, the output current changes little. An equivalent statement is that the output resistance has increased. In this particular case, the increase is enough to eliminate the effect of r_o for all practical purposes. We would therefore expect a gain very near -12, and SPICE simulations show that it is about -11. If this excess gain holds up as the design evolves, it may be traded off for improved bandwidth, if needed or desired.

Returning to open-circuit time-constant estimates of bandwidth, draw the model corresponding to this cascode connection, and calculate the resistance facing each capacitance. Out of laziness, there will be no more incremental models from here on out, so you're on your own now ("some assembly required"):

$$\tau_{gs2} = C_{gs2}r_{gs2} = (220\,\text{fF})(R_S) = 440\,\text{ps} \quad \text{(unchanged)}; \tag{71}$$

$$\tau_{gd2} = C_{gd2}r_{gd2} \approx (45\,\text{fF})\left(R_S + \frac{1}{g_{m3}} + g_{m2}R_S\frac{1}{g_{m3}}\right)$$

$$\approx 184\,\text{ps} \quad \text{(better!)}. \tag{72}$$

Equation 72 is a bit approximate because we are neglecting the effect of g_{mb3} and r_o in this calculation (as well as in several to follow). Because the body effect degrades transconductance, we are somewhat underestimating the true effective resistance ("r_{right}," specifically). A more accurate calculation shows a 200-ps time constant. The error for this iteration is thus negligible. In any case, the error introduced by neglecting these effects tends to offset the typical pessimism of open-circuit time constants. We have

$$\tau_{gs3} = C_{gs3}r_{gs3} \approx (220\,\text{fF})(1/g_{m3}) \approx 18\,\text{ps} \quad \text{(new)}, \tag{73}$$

$$\tau_{gd3} = C_{gd3}r_{gd3} \approx (45\,\text{fF})(R_L) \approx 45\,\text{ps} \quad \text{(new)}, \tag{74}$$

$$\tau_{sb3} = C_{sb3}r_{sb3} \approx (130\,\text{fF})(1/g_{m3}) \approx 11\,\text{ps} \quad \text{(new)}, \tag{75}$$

$$\tau_{db3} = C_{db3}r_{db3} \approx (90\,\text{fF})(R_L) \approx 90\,\text{ps} \quad \text{(new)}, \tag{76}$$

$$\tau_{db2} = C_{db2}r_{db2} \approx (90\,\text{fF})(1/g_{m3}) \approx 8\,\text{ps} \quad \text{(better)}, \tag{77}$$

$$\tau_L = C_L r_L = C_L R_L = 1000\,\text{ps} \quad \text{(worse!)}; \tag{78}$$

$$\text{BW} \approx \frac{1}{(1750\,\text{ps})} \approx 570\,\text{Mrps}. \tag{79}$$

With a new (estimated) bandwidth of about 90 MHz (SPICE says 109 MHz), we can see that the cascode connection has given us a substantial improvement in bandwidth. But we still have a long way to go.

---

[2] To complete the argument, note that the gain between the gate and drain of M_2 is -1, so that C_{gd2} is not multiplied by very much at all.

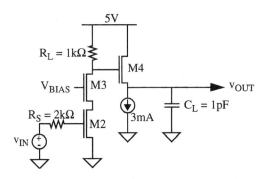

FIGURE 8.5. Third pass: cascode amplifier with output source follower.

Looking at the new big offender, we see that it involves the load capacitance C_L. Driving it with such a high (1-kΩ) resistance is obviously the problem, so we should be able to reduce that time constant to a small value with a source follower; this is shown in Figure 8.5.

Again, we'll ignore biasing details. Just assume that we put a current source (or a plain old resistor) in the source leg of M_4 to bias it to 3 mA. For purposes of time-constant calculations, we'll see that the resistance of the bias network is easily made negligible, so it doesn't really matter what we assume.

The source follower does not quite have unit gain because there is a capacitive voltage division between C_{gs4} and C_{sb4}. A careful calculation, verified by SPICE, reveals that the gain has dropped from -11 to -9.5. Fortunately, this value is still in excess of the desired value.

Calculating the time constants for this iteration yields the following list:

$$\tau_{gs2} = 440 \text{ ps} \quad \text{(unchanged)}, \tag{80}$$

$$\tau_{gd2} = 184 \text{ ps} \quad \text{(unchanged)}, \tag{81}$$

$$\tau_{db2} = 8 \text{ ps} \quad \text{(unchanged)}, \tag{82}$$

$$\tau_{sb3} = 11 \text{ ps} \quad \text{(unchanged)}, \tag{83}$$

$$\tau_{gs3} = 18 \text{ ps} \quad \text{(unchanged)}, \tag{84}$$

$$\tau_{gd3} = 45 \text{ ps} \quad \text{(unchanged)}, \tag{85}$$

$$\tau_{db3} = 90 \text{ ps} \quad \text{(unchanged)}, \tag{86}$$

$$\tau_{gd4} = C_{gd4}r_{gd4} \approx (45 \text{ fF})(R_L) \approx 45 \text{ ps} \quad \text{(new)}, \tag{87}$$

$$\tau_{gs4} = C_{gs4}r_{gs4} \approx (220 \text{ fF})(1/g_{m4}) \approx 18 \text{ ps} \quad \text{(new)}. \tag{88}$$

This last equation is a bit more approximate than usual because of the neglect of g_{mb4} with a 1-kΩ driving resistance (the correct value is about 45 ps). A more careful derivation shows that the resistance in Eqn. 88 should be multiplied by about $(1 + g_{mb4}R_L)$, so one may neglect this factor only if $g_{mb}R_L$ is much smaller than unity.

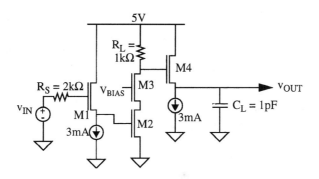

FIGURE 8.6. Fourth pass: cascode amplifier with two source followers.

Fortunately, this particular time constant is not dominant in this case, so the large percentage error in τ_{gs4} has an insignificant effect on the overall time-constant sum.

Continuing:

$$\tau_{sb4} = C_{sb4}r_{sb4} \approx (130\,\text{fF})(1/g_{m4}) \approx 11\,\text{ps} \quad \text{(new)}, \tag{89}$$

$$\tau_L = C_L r_L \approx (1\,\text{pF})(1/g_{m4}) \approx 80\,\text{ps} \quad \text{(better!)}; \tag{90}$$

$$\text{BW} \approx \frac{1}{(906\,\text{ps})} \approx 1.1\,\text{Grps}. \tag{91}$$

So now we're up to about 175 MHz (SPICE says 222 MHz); we "only" need to pick up another factor of about 2 in bandwidth.

Looking over our latest list of time constants, we see that τ_{gs2} dominates by far because of the 2-kΩ source resistance. One obvious remedy is to add an input buffer to reduce the resistance driving the gate–source capacitance of M_2; see Figure 8.6. A recomputation of the gain reveals that the slight attenuation of the added source follower takes us down to a gain of −8, leaving us with no more margin.

With these changes, we expect to get pretty close to the desired bandwidth because τ_{gs2} is about half the total, and we can probably drop it to near zero. Recomputing the time constants, we obtain:

$$\tau_{db2} = 8\,\text{ps} \quad \text{(unchanged)}, \tag{92}$$

$$\tau_{sb3} = 11\,\text{ps} \quad \text{(unchanged)}, \tag{93}$$

$$\tau_{gs3} = 18\,\text{ps} \quad \text{(unchanged)}, \tag{94}$$

$$\tau_{gd3} = 45\,\text{ps} \quad \text{(unchanged)}, \tag{95}$$

$$\tau_{db3} = 90\,\text{ps} \quad \text{(unchanged)}, \tag{96}$$

$$\tau_{gd4} = 45\,\text{ps} \quad \text{(unchanged)}, \tag{97}$$

$$\tau_{gs4} = 18\,\text{ps} \quad \text{(unchanged)}, \tag{98}$$

$$\tau_{sb4} = 11\,\text{ps} \quad \text{(unchanged)}, \tag{99}$$

$$\tau_L = 80\,\text{ps} \quad \text{(unchanged)}, \tag{100}$$

$$\tau_{gs1} = C_{gs1}r_{gs1} \approx (220\,\text{fF})(1/g_{m1}) \approx 18\,\text{ps} \quad \text{(new)}. \tag{101}$$

Again, this last calculation is in error because $g_{mb1}R_S$ is 3.6, so the time constant really ought to be multiplied by $(1 + 3.6) = 4.6$ to yield about 83 ps. A more careful calculation that also takes r_{o1} into account reveals that the time constant here is in fact about 86 ps. We now have

$$\tau_{gd1} = C_{gd1}r_{gd1} \approx (45 \text{ fF})(R_S) \approx 90 \text{ ps} \quad \text{(new)}, \tag{102}$$

$$\tau_{sb1} = C_{sb1}r_{sb1} \approx (130 \text{ fF})(1/g_{m1}) \approx 11 \text{ ps} \quad \text{(new)}, \tag{103}$$

$$\tau_{gs2} = C_{gs2}r_{gs2} \approx (220 \text{ fF})(1/g_{m1}) \approx 18 \text{ ps} \quad \text{(better!)}, \tag{104}$$

$$\tau_{gd2} = C_{gd2}r_{gd2} \approx (45 \text{ fF})\left[\frac{1}{g_{m1}} + \frac{1}{g_{m3}} + g_{m2}\left(\frac{1}{g_{m1}}\right)\left(\frac{1}{g_{m3}}\right)\right]$$
$$\approx 11 \text{ ps} \quad \text{(better!)}; \tag{105}$$

$$\text{BW} \approx \frac{1}{(474 \text{ ps})} \approx 2.1 \text{ Grps}. \tag{106}$$

The estimated bandwidth has now increased to about 340 MHz. Owing to the conservative nature of the estimate, it is reasonable to expect the actual bandwidth to be quite close to our goal. In fact, SPICE simulations show that the bandwidth is about 540 MHz, well in excess of the target value. If desired, some of this excess bandwidth could be exchanged for increased gain.

Suppose, though, that SPICE were to confirm your worst fears, and you find that the amplifier just doesn't quite make it. Are there any other modifications that you could try? The answer, of course, is Yes. One option was passed over earlier: distribute the gain among several stages. Using two or more stages, it would be a trivial matter to beat the bandwidth specification by a handy margin.

These tricks are by no means the only ones, and we will spend a considerable amount of time in Chapter 9 exploring some important alternative methods. However, to whet your appetite and stimulate some thinking, here are some vague allusions to other possibilities.

The method of open-circuit time constants assumes an all-pole transfer function, and it gives more accurate answers if all the poles are real. Consider the effect of purposefully violating these assumptions by allowing zeros and/or complex poles. Careful placement of zeros (antipoles) or complex poles will extend the bandwidth, although the frequency response may no longer be monotonic. One way to form complex poles is through feedback (just think of a two-pole root locus, for example) or through the resonance of inductors (real or synthetic) with capacitors. The surprisingly large bandwidth of the last circuit in the chain of design iterations is largely due to the formation of complex poles arising from the interaction of the gate–source capacitance of M_2 with the *inductive* output impedance of source follower M_1.

Zeros can be formed, for example, by capacitors in parallel with source bypass resistors. As we'll see, a judicious choice of capacitor value can cause this zero to cancel a bandwidth-limiting pole.

Another possibility, made most practical in differential systems, is to use positive feedback to generate negative capacitances. These negative capacitances can cancel positive ones to yield bandwidth increases. Of course, there is a chance of unstable behavior that must be carefully watched, but this method, called *neutralization,* can yield useful bandwidth improvements. We will turn to a detailed examination of these themes very soon.

8.2.7 SUMMARY OF OPEN-CIRCUIT TIME CONSTANTS

We have seen that the method of open-circuit time constants is an extremely valuable tool for designing amplifiers for good dynamic performance, mainly because of its ability to identify the problem areas of the circuit. Because of the tremendous insights gained with very modest effort, we are generally willing to overlook its quantitative limitations, such as the often highly conservative nature of the estimated bandwidth. As long as we take care to use the method only with models that apply to the high-frequency regime, we are assured of reasonable answers.

As a parting remark, it should be noted that the influence of inductances can be incorporated as well into the method, although with generally unsatisfactory results for reasons that will be explained shortly.

The most intuitive way to understand how one incorporates the effect of inductances on bandwidth is to recall that each time-constant term represents an individual, local contribution to the bandwidth limitation; we treat the system at each step of the calculation as if that jth reactive element were the sole one. So, evidently, one treats all of the inductors as *short* circuits when computing the appropriate effective resistances. The L/R time constants are then added to the various RC terms to yield the grand total from which the bandwidth is estimated.

Having said all of this, the presence of explicit inductances and capacitances almost guarantees the formation of troublesome complex pole pairs, often causing the method to yield gross underestimates of bandwidth. This difficulty is exacerbated by the common occurrence of finite zeros. Furthermore, the parasitic inductances in a circuit are often much more difficult to estimate accurately than the capacitances. As a consequence, inductors are rarely taken into consideration. However, one should be aware that while small values of effective resistance minimize the time constant due to capacitances, they tend to maximize the time constant due to the inductances. Hence, if one goes to extremes in reducing the various Rs in the quest for ever-greater bandwidth, there is often a point not only of diminishing returns but even of reversals. Typically, such considerations become important as one pursues bandwidths exceeding, say, 20–50 MHz in discrete designs, where stray inductances of under a few nanohenries are almost impossible to achieve.

In conclusion, the method of open-circuit time constants is an indispensable guide in the design of amplifiers. With it one can design intelligently and confidently to satisfy a given bandwidth specification. To be sure, the method has its quantitative shortcomings, but the valuable intuition provided is more than sufficient compensation.

8.3   THE METHOD OF SHORT-CIRCUIT TIME CONSTANTS

8.3.1   INTRODUCTION

We've already seen how the method of open-circuit time constants allows us to estimate the high-frequency -3-dB point of an arbitrarily complex system by decomposing the bandwidth computation into a succession of first-order calculations. Each of the time constants represents a local bandwidth degradation term, and the sum of these individual degradation terms equals the reciprocal of the overall bandwidth. As we saw, open-circuit time constants are valuable because they identify which elements limit the bandwidth.

Suppose that, instead of estimating the high-frequency -3-dB point, we wanted to find the *low*-frequency -3-dB point of an AC-coupled system. How would we calculate how large the coupling capacitors must be in order to achieve a specified low-frequency breakpoint? Fortunately, we may invoke a procedure that is analogous to the method of open-circuit time constants. This dual technique is known as the method of short-circuit time constants (SCτs).

8.3.2   BACKGROUND

In the method of open-circuit time constants, we assumed that the zeros of the network were all at infinitely high frequency, so that the transfer function consisted only of poles. In the case of short-circuit time constants, we instead assume that all of the zeros are at the *origin,* and that there are as many poles as zeros. Thus, the corresponding system function may be written as follows:

$$\frac{V_o(s)}{V_i(s)} = \frac{ks^n}{(s+s_1)(s+s_2)\cdots(s+s_n)}, \tag{107}$$

where the various pole frequencies may or may not be real and where k is simply a constant to fix up the scale factor.

Multiplying out the terms in the denominator leads to a polynomial that we shall express as

$$s^n + b_1 s^{n-1} + \cdots + b_{n-1}s + b_n, \tag{108}$$

where the coefficient b_1 is the sum of all of the pole frequencies, and b_n is the product of all of the pole frequencies. (In general, the coefficient of the s^j term is computed by forming all unique products of the n frequencies taken j at a time and summing all $n!/j!(n-j)!$ such products.)

We now assert that, near the low-frequency -3-dB breakpoint, the higher-order terms dominate the denominator, so that we obtain

$$\frac{V_o(s)}{V_i(s)} \approx \frac{ks^n}{s^n + b_i s^{n-1}} = \frac{ks}{s + \sum_{i=1}^{n} s_i}. \tag{109}$$

The low-frequency -3-dB point of our original system in radian frequency as estimated by this first-order approximation is then simply the sum of the pole frequencies:

$$\omega_l \approx b_1 = \sum_{i=1}^{n} s_i = \omega_{l,\text{est}}. \tag{110}$$

Before proceeding further, we should consider the conditions under which our neglect of the lower-order terms is justified. Let us examine the denominator of the transfer function near our estimate of ω_l. For the sake of simplicity, we consider a second-order polynomial with purely real roots, s_1 and s_2.

Now, at our estimated -3-dB frequency, the original denominator polynomial is

$$-\omega_{l,\text{est}}^2 + j\omega_{l,\text{est}}(s_1 + s_2) + s_1 s_2. \tag{111}$$

Substituting our expression for the estimated -3-dB point, we obtain

$$-[s_1^2 + s_2^2 + 2s_1 s_2] + j[s_1^2 + s_2^2 + 2s_1 s_2] + s_1 s_2. \tag{112}$$

Clearly, the last term is small compared with the magnitudes of the other terms. Thus, the neglect of all but the two highest-order terms involves little error. The worst case occurs when the two pole frequencies are equal, and even then the error is not terribly large. Extending these arguments to polynomials of higher order reveals that the estimate of the low-frequency cutoff based simply on the coefficient b_1 is generally reasonable, since the higher-order terms do in fact dominate the denominator. Furthermore, the low-frequency cutoff estimate is conservative in the sense that *the actual cutoff frequency will almost always be as low as or lower than estimated by this method.*

So far, all we've done is show that a first-order estimate of the bandwidth is possible if one is given the sum of the pole frequencies ($= b_1$). Of course, if we knew the pole frequencies then we could compute this sum directly. Fortunately, as was the case with open-circuit time constants, it is possible to relate the desired pole-frequency sum, b_1, to (more or less) easily computed network quantities.

The recipe is thus as follows. Consider an arbitrary linear network comprising only resistors, sources (dependent or independent), and m capacitors. Then:

(a) compute the effective resistance R_{js} facing each jth capacitor with all of the other capacitors *short*-circuited (the subscript s refers to the short-circuit condition for each capacitor);
(b) compute the "short-circuit frequency" $1/(R_{js}C_j)$;
(c) sum all m such short-circuit frequencies.

The sum of the reciprocal short-circuit time constants formed in step (c) turns out to be precisely equal to the sum b_1 of the pole frequencies. Thus, at last, we have

$$\omega_{l,\text{est}} = \sum_{j=1}^{m} \frac{1}{R_{js}C_j}. \tag{113}$$

8.3.3 OBSERVATIONS AND INTERPRETATIONS

The method of SCτs is relatively simple to apply for precisely the same reasons that OCτs are easy to apply – namely, each time-constant calculation involves the computation of just a single resistance, although once again we must be wary of the impedance-modifying potential of dependent sources. In any event, the amount of computation required still is typically substantially less than that needed for an exact solution.

Again, the greatest value of the technique lies in its identification of those elements implicated in bandwidth limitations. *The reciprocal of each jth short-circuit time constant is the low-frequency -3-dB breakpoint that the circuit would exhibit if that jth capacitor were the only capacitor in the system.* The method of SCτs then states that the linear combination of these individual, local limitations yields an estimate of the overall -3-dB point. The value of SCτs derives directly from the identification and approximate quantification of the local degradation terms.

Although the development so far has considered only capacitances, inductances also can be incorporated into the method. However, their presence often complicates significantly the decision of which reactive elements really belong in the computation.

The most intuitive way to understand how one incorporates the effect of inductances on bandwidth is to recall that each reciprocal time-constant term represents an individual, local contribution to the low-frequency cutoff; we treat the system at each step of the calculation as if that jth reactive element were the sole limiting one. So, evidently, one treats all of the inductors as *open* circuits when computing the appropriate effective resistances. The R/L frequencies are then added to the various $1/RC$ frequencies to yield the total estimated low-frequency cutoff point.

8.3.4 ACCURACY OF SCτs

As with OCτs, truncation of the denominator polynomial means that one must be careful not to place too much faith in the ability of SCτs to provide accurate estimates of ω_l in all cases. This caveat notwithstanding, it should be clear that an SCτ estimate is in fact exact for a first-order network since *no* truncation of terms is involved there. Not surprisingly, then, the SCτ estimate will be quite accurate if a network of higher order happens to be dominated by one pole (here, that means that one pole is much *higher* in frequency than all the other poles).

8.3.5 OTHER IMPORTANT CONSIDERATIONS

Although application of short-circuit time constants is pretty straightforward, there are one or two fine points that merit discussion. As with OCτs, not all capacitors in a network belong in the SCτ calculations. For instance, the capacitors in a transistor model almost never belong. Blind application of the SCτ method would lead to curious (erroneous) results (and a whole heap of extra calculations).

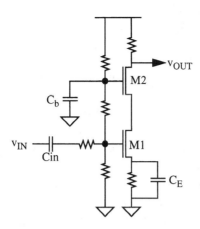

FIGURE 8.7. Cascode amplifier.

The problem is easily understood if you remember that we assumed that all the zeros are at the origin, and that the number of poles equals the number of zeros, so that the gain in the limit of infinitely high frequency is flat, not zero. We violate these assumptions rather severely if we include all the stuff that causes high-frequency rolloff (i.e., all the stuff that $OC\tau$s worry about). The solution is to preprocess the network prior to application of the $SC\tau$ method. That is, recognize that all the capacitors that limit high frequency gain are effectively open circuits relative to the impedances around them at frequencies near ω_l. Thus, *one may apply $SC\tau$s only to models that are appropriate to the low-frequency regime.*

It is usually obvious which capacitors are to be ignored (considered open circuits), but there are occasions when one is not so sure. In these cases, a simple thought experiment usually suffices to decide the issue. Now $SC\tau$s are concerned only with those capacitors that limit low-frequency gain. As a consequence, the removal (i.e., the open-circuiting) of a capacitor that belongs in the $SC\tau$ calculation should result in a decrease in low-frequency gain. The test, therefore, is to consider exciting the network at some low frequency and imagining what would happen to the gain if the capacitor in question were taken out of the circuit (open-circuited). If the gain would decrease then the capacitor belongs in the $SC\tau$ calculation, since we infer from the result of the thought experiment that the capacitor does indeed limit the low-frequency gain. If the gain would not change (or even increase) upon removal, that capacitor should be open-circuited and left out of the computation. Again, as with $OC\tau$s, the necessary conclusions can usually be reached without taking pencil to paper.

To underscore these issues, let's consider a specific example, the cascode amplifier. As seen in the accompanying schematic (Figure 8.7), there are three capacitors. The input coupling capacitor, C_{in}, removes any DC from the input signal to prevent upsetting the bias of the amplifier. Source bypass capacitor C_E is chosen to short the source of M_1 to ground at all signal frequencies to restore the gain lost by the source degeneration resistor. Bias bypass capacitor C_b guarantees that the gate of M_2

is an incremental ground at high frequencies to keep the open-circuit time-constant sum small.

Let's use our thought-experiment technique to deduce which of these three capacitors belongs in the SCτ calculation. If we begin with C_{in}, we note that the low-frequency gain does decrease (to zero, in fact) if we take it out of the circuit. Hence, it belongs in the calculation. Similarly, C_E belongs in the calculation because its removal also reduces the low-frequency gain.

What about C_b? What happens to the low-frequency gain if we take it out of the circuit? The answer can be either trivial or too deep to fathom, depending on how you approach the question. The easiest way to get to the answer is to recognize that M_1 is a device that converts an incoming voltage to an incremental drain current. All M_2 does is take this current and pass it on to the output load resistor. Therefore, whether or not the gate of M_2 is an incremental ground is irrelevant, and the removal of C_b will therefore have essentially no effect on the low-frequency gain. Thus, C_b does *not* belong in the calculation.

The importance of not blindly applying the method cannot be overemphasized. In fact, one of the earliest expositions of the method erroneously includes C_b.[3]

One last issue that deserves some attention concerns the relationship between the reciprocals of the individual short-circuit time constants and the pole frequencies. We have asserted (again without formal proof) only that the *sums* of these frequencies are equal to each other. Therefore, just as with open-circuit time constants, one must resist the temptation to equate each reciprocal short-circuit time constant with a corresponding pole frequency. Since the number of short-circuit time constants and the number of poles may not even be equal, one cannot expect each SCτ to equal the time constant of a pole in general.

8.3.6 SUMMARY AND CONCLUDING REMARKS

We have seen that the method of short-circuit time constants shares with its dual, the method of open-circuit time constants, a number of advantages and disadvantages. It is an invaluable tool for designing amplifiers because of its ability to identify the problem areas of the circuit. Because of the tremendous insights gained with extremely modest effort, we are generally willing to overlook its quantitative limitations, such as the often highly conservative nature of the estimated low-frequency cutoff point. As long as we take care to apply the method only to models that apply to the low-frequency regime, we are assured of reasonable answers.

In conclusion, the method of short-circuit time constants helps one design circuits to satisfy a given low-frequency cutoff specification. Despite the quantitative shortcomings of the method, the valuable intuition provided and the labor saved are more than sufficient compensation.

---

[3] P. E. Gray and C. L. Searle, *Electronic Principles,* Wiley, New York, 1969, pp. 542–6.

8.4 FURTHER READING

For a proof of the equality of the sum of open-circuit and pole time constants, see P. E. Gray and C. L. Searle, *Electronic Principles* (Wiley, New York, 1969, pp. 531–5).

By the way, this work has been extended to allow the exact computation of *all* the poles of a network. It involves the computation of various cross-products of open- and *short*-circuit time constants to obtain the coefficients of all the powers of *s* in the denominator of the transfer function. Originally developed by B. L. Cochrun and A. Grabel, it was simplified by S. Rosenstark, but the method is sufficiently labor-intensive that most engineers often resort to some sort of simulation instead. However, it occasionally proves useful (especially if you choose to automate the procedure by writing your own code). For more information see Cochrun and Grabel's paper, "A Method for the Determination of the Transfer Function of Electronic Circuits" (*IEEE Trans. Circuit Theory,* v. 20, no. 1, January 1973, pp. 16–20), and Rosenstark's book, *Feedback Amplifier Principles* (Macmillan, New York, 1986, pp. 67–77).

8.5 RISETIME, DELAY, AND BANDWIDTH

8.5.1 INTRODUCTION

The method of open-circuit time constants allows one to estimate the overall bandwidth from local *RC* products. In this section, we develop a number of ways to estimate bandwidth from time-domain parameters. In this connection, one occasionally encounters various rules of thumb, such as "bandwidth times risetime equals 2.2," "risetimes add quadratically" or "buy low, sell high." As useful as they are, however, they aren't entirely reliable. To identify when these rules of thumb hold, we now turn to their formal derivation.

We start by deriving a rule that appears trivial, obvious, and irrelevant: The total delay of a cascade of systems is the sum of the individual delays. The reason for starting here is to introduce some analytical techniques and insights that have far broader applicability.

As always, don't worry too much about all the mathematical minutiae; the derivations are provided simply for completeness. Those interested primarily in the application of these relationships may skip the intervening math and simply take note of the results.

8.5.2 DELAY OF SYSTEMS IN CASCADE

We shall see that many analytical advantages accrue from defining delay (and later, risetime) in terms of *moments of the impulse response.* As seen in Figure 8.8, one delay measure is the time it takes for the impulse response to reach its "center of mass," that is, the normalized value of its first moment:

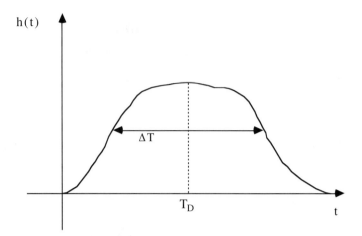

FIGURE 8.8. Illustrative impulse response.

$$T_D \equiv \frac{\int_{-\infty}^{\infty} th(t)\, dt}{\int_{-\infty}^{\infty} h(t)\, dt}. \tag{114}$$

This quantity is also known as the Elmore delay in some of the literature, after the fellow who first used this moment-based approach.[4]

This particular measure of time delay derives much of its utility from the fact that it is readily related to a number of Fourier transform identities, allowing us to exploit the full power of linear system theory. Specifically, the first moment is

$$\int_{-\infty}^{\infty} th(t)\, dt = -\frac{1}{j2\pi} \frac{d}{df} H(f) \Big|_{f=0}. \tag{115}$$

The normalization factor is simply the DC gain,

$$\int_{-\infty}^{\infty} h(t)\, dt = H(0), \tag{116}$$

so that

$$T_D \equiv \frac{\int_{-\infty}^{\infty} th(t)\, dt}{\int_{-\infty}^{\infty} h(t)\, dt} = -\frac{1}{j2\pi H(0)} \frac{d}{df} H(f) \Big|_{f=0}. \tag{117}$$

Using this definition, one finds that the Elmore delay for a single-pole low-pass system is just the pole time constant, τ.

Now that Monsieur Fourier has graciously helped us out by providing a definition of time delay purely in terms of his transforms, the derivation becomes straightforward. Specifically, consider two systems with impulse responses $h_1(t)$ and $h_2(t)$

---

[4] W. C. Elmore, "The Transient Response of Damped Linear Networks with Particular Regard to Wideband Amplifiers," *J. Appl. Phys.*, v. 19, January 1948, pp. 55–63.

with corresponding Fourier transforms $H_1(f)$ and $H_2(f)$. From basic linear system theory, we know that the Fourier transform of these two systems in cascade is just the product of the individual transforms, so that $H_{\text{tot}} = H_1 H_2$. The overall time delay is therefore

$$T_{D,\text{tot}} = -\frac{1}{j2\pi H_1(0) H_2(0)} \frac{d}{df} H_1 H_2 \bigg|_{f=0}, \tag{118}$$

which we may expand to obtain

$$T_{D,\text{tot}} = -\frac{1}{j2\pi H_1(0) H_2(0)} \left[H_2(0) \frac{dH_1}{df} \bigg|_{f=0} + H_1(0) \frac{dH_2}{df} \bigg|_{f=0} \right]. \tag{119}$$

From this we immediately (well okay, maybe not quite *immediately*) note that

$$T_{D,\text{tot}} = T_{D1} + T_{D2}, \tag{120}$$

which was to be shown.

We see that use of this particular definition of time delay has led us to the intuitively satisfying result that the overall delay of a cascade of systems is simply the sum of the individual delays.

8.5.3 RISETIME OF SYSTEMS IN CASCADE

Deriving a risetime addition rule presents a somewhat more significant challenge. In particular, it turns out that developments based on the conventional 10–90% definition of risetime are almost certainly doomed to fail because of the analytical difficulties involved with combining exponentials of differing time constants. Since this measure of risetime is arbitrary anyway, we might as well seek an alternative (but still arbitrary) definition of risetime that permits tractable analysis.

Just as we employed the first moment of the impulse response in defining the time delay, we find the second moment useful in defining the risetime. Referring again to Figure 8.8, note that the quantity ΔT is a measure of the duration of the impulse response and hence also a measure of the risetime of the step response (since the step response is the integral of the impulse response). Specifically, ΔT is twice the "radius of gyration" about the "center of mass" (T_D) of $h(t)$. Recalling some dusty relationships from first-year calculus, we find that

$$\left(\frac{\Delta T}{2} \right)^2 \equiv \left[\frac{\int_{-\infty}^{\infty} t^2 h(t)\, dt}{\int_{-\infty}^{\infty} h(t)\, dt} - (T_D)^2 \right]. \tag{121}$$

Again this definition allows the use of Fourier transform identities. In particular,

$$\int_{-\infty}^{\infty} t^2 h(t)\, dt = -\frac{1}{(2\pi)^2} \frac{d^2}{df^2} H(f) \bigg|_{f=0}, \tag{122}$$

so that

$$t_{\text{rise}}^2 = (\Delta T)^2$$

$$= 4\left[\frac{-\dfrac{1}{(2\pi)^2}\dfrac{d^2}{df^2}H(f)\Big|_{f=0}}{H(0)} - \left(-\frac{1}{j2\pi H(0)}\frac{d}{df}H(f)\Big|_{f=0}\right)^2\right]. \tag{123}$$

Here we have made use of the equation for delay developed in Section 8.5.2.
Simplifying (!), we obtain

$$t_{\text{rise}}^2 = \frac{4}{(2\pi)^2 H(0)}\left[-\frac{d^2}{df^2}H(f)\Big|_{f=0} - \frac{1}{H(0)}\left(\frac{d}{df}H(f)\Big|_{f=0}\right)^2\right]. \tag{124}$$

The Elmore risetime for a single-pole low-pass system is 2τ.

Proceeding as in Section 8.5.2, we now consider two systems, each with its own
risetime. Then

$$t_{\text{rise, tot}}^2 = \frac{4}{(2\pi)^2 H_1(0)H_2(0)}\left[-\frac{d^2}{df^2}H_1H_2\Big|_{f=0}\right.$$

$$\left. - \frac{1}{H_1(0)H_2(0)}\left(\frac{d}{df}H_1H_2\Big|_{f=0}\right)^2\right]. \tag{125}$$

After a small algebraic miracle, this leads to the desired result at last:

$$t_{\text{rise, tot}}^2 = t_{\text{rise1}}^2 + t_{\text{rise2}}^2. \tag{126}$$

Thus we see that the *squares* of the individual risetimes add linearly to yield the
square of the overall risetime. Stated alternatively, the individual risetimes add in
root-sum-squared (rss) fashion to yield the overall risetime:

$$t_{\text{rise, tot}} = \sqrt{t_{\text{rise1}}^2 + t_{\text{rise2}}^2}. \tag{127}$$

Now that we've derived these results, we should spend some time discussing conditions under which the foregoing formulas may yield unsatisfactory estimates of
delay or risetime. In particular, consider what happens to the calculated delay and
risetime if the integral of $h(t)$ is nearly zero. This situation might arise, for example,
if $h(t)$ oscillates more or less evenly about zero. Since the integral of $h(t)$ appears
as a normalizing factor in the denominator of our expressions for delay and risetime,
inappropriate values for these quantities may result.

To handle this difficulty one might propose a modification of our definitions so that
the delay and risetime depend on the moments of the *square* (or perhaps the absolute
value) of $h(t)$. "It is left as an exercise for the reader" to show that such modifications result in exceedingly unpleasant expressions that are cumbersome to use and
interpret. Thus, the simple expressions presented here are understood to apply best
when the impulse response is unipolar (or, equivalently, when the step response is
monotonic). If the individual systems satisfy this requirement then the relationships

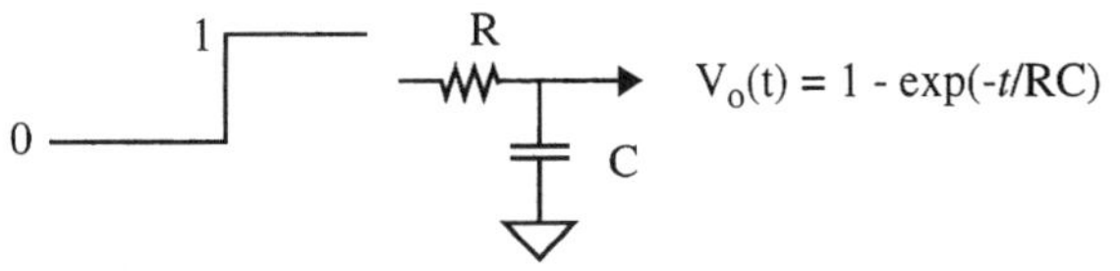

FIGURE 8.9. *RC* low-pass filter and step response.

derived here will hold well. The greater the departure from the step-response monotonicity condition, the less appropriate the use of these formulas.

8.5.4 A (VERY SHORT) APPLICATION OF THE RISETIME ADDITION RULE

Aside from permitting one to predict the risetime of a cascade of systems, the risetime addition rule may be used to extend the limits of instrumentation. Consider, for example, trying to measure the risetime of a system whose bandwidth is about the same as that of the instrumentation. Specifically, assume that an oscilloscope with a known risetime of 5 ns displays a value of 6 ns for the risetime of a system under test. Using the risetime addition rule, we can infer that the true system risetime is about 3.3 ns, saving us the trouble and expense of trying to make this measurement with equipment that is faster still.

8.5.5 BANDWIDTH–RISETIME RELATIONS

We now take up the problem of examining the rule of thumb that led us to this endeavor in the first place:

$$\omega_{-3\,\mathrm{dB}} t_{\mathrm{rise}} \approx 2.2, \tag{128}$$

where $\omega_{-3\,\mathrm{dB}}$ is the -3-dB bandwidth in radians per second and t_{rise} is the 10–90% risetime in response to a step.

Where does this rule come from? Consider our old friend, the simple RC low-pass filter displayed in Figure 8.9. Given the equation for the response to a unit voltage step, it is straightforward to compute the 10–90% risetime:

$$t_{\mathrm{rise}} = RC \ln\left(\frac{0.9}{0.1}\right) \approx 2.2RC. \tag{129}$$

Note that this value is about 10% higher than the Elmore risetime computed earlier.

In addition to the risetime, we already know that the -3-dB bandwidth (in radians per second) is simply $1/RC$. Hence, the bandwidth–risetime product is in fact about 2.2, as the rule states.

Since it was derived for a first-order case, should we expect the rule to hold generally for systems of arbitrary order? Well, let's look at a couple of other cases. Consider, for example, the step response of a two-pole system:

$$V_o(t) = 1 - \frac{1}{\sqrt{1-\zeta^2}} e^{-\zeta\omega_n t} \sin\left(\sqrt{1-\zeta^2}\omega_n t + \Phi\right), \tag{130}$$

where

$$\Phi = \tan^{-1}\left[\frac{\sqrt{1-\zeta^2}}{\zeta}\right]. \tag{131}$$

The -3-dB bandwidth of this system is given by:

$$\omega_h = \omega_n\left(1 - 2\zeta^2 + \sqrt{2 - 4\zeta^2 + 4\zeta^4}\right)^{1/2}. \tag{132}$$

Let's use these formulas to explore what happens as we change ζ. For the extreme case of a damping ratio of zero, the risetime and bandwidth are

$$t_r\big|_{\zeta=0} = \frac{1}{\omega_n}[\sin^{-1}0.9 - \sin^{-1}0.1] \approx \frac{1.02}{\omega_n}, \tag{133}$$

$$\omega_h\big|_{\zeta=0} \approx 1.55\omega_n, \tag{134}$$

so that the corresponding bandwidth–risetime product is

$$\omega_h t_r\big|_{\zeta=0} \approx 1.6, \tag{135}$$

or about 72% of the value obtained for the first-order case.

For a reasonably well-damped system, we might expect closer agreement with the first-order result. As a specific example, if we set $\zeta = 1/\sqrt{2}$ then the risetime and bandwidth are

$$t_r\big|_{\zeta=1/\sqrt{2}} \approx \frac{2.14}{\omega_n} \tag{136}$$

and

$$\omega_h\big|_{\zeta=1/\sqrt{2}} = \omega_n, \tag{137}$$

so that

$$\omega_h t_r\big|_{\zeta=1/\sqrt{2}} \approx 2.14, \tag{138}$$

or a value within a small percentage of the first-order result. Note that the product of bandwidth and Elmore risetime is 2.0 for a single-pole system.

In general, the bandwidth–risetime product will range between 2 and 2.2 if the system is well damped (or more precisely, if the impulse response is unipolar so that the step response is monotonic, for the same reasons that prevailed in the moment-based expressions for risetime and time delay); the product will decrease if the system is not very well damped. However, even in the case of no damping at all, we have seen that the bandwidth–risetime product still does not deviate that much.

Because most systems of practical interest are generally well damped, we can expect their bandwidth–risetime product to be about 2.2. Therefore, measurement of the step response risetime is often an expedient way to obtain a reasonably accurate estimate of the bandwidth: only one experiment has to be performed, and step excitations are often easier to generate than sine waves.[5]

8.5.6 OPEN-CIRCUIT TIME CONSTANTS, RISETIME ADDITION, AND BANDWIDTH SHRINKAGE

As we have seen, bandwidth and risetime have a roughly constant product (at least for systems that are "well behaved"). In addition, the risetimes of cascaded systems increase in root-sum-squared fashion. From these two relationships, we can deduce a bandwidth shrinkage law. It is instructive to compare the results of this exercise with the bandwidth shrinkage law derived in Chapter 9.

Consider a cascade of N identical amplifiers, each of which is single-pole with a time constant τ. Combining the risetime addition rule with the bandwidth–risetime relationship yields

$$\text{BW} \approx \frac{1}{\sqrt{\sum_1^N \tau^2}} = \frac{1}{\tau\sqrt{N}}. \tag{139}$$

Compare that approximate result with the more exact (but still approximate) relationship

$$\text{BW} \approx \frac{\sqrt{\ln 2}}{\tau\sqrt{N}} \approx \frac{0.833}{\tau\sqrt{N}} \tag{140}$$

(see Chapter 9). As can be seen, the functional dependence on N is the same; the equations differ only by a relatively small multiplicative factor.[6]

Note that the method of open-circuit time constants would predict quite a different result. Since the effective time constant is found there by summing all the individual time constants, the OCτ-estimated bandwidth would be

$$\text{BW} \approx \frac{1}{\tau N}. \tag{141}$$

The difference is significant, and underscores yet again how the use of open-circuit time constants can lead to extremely pessimistic estimates of bandwidth if a single pole does not dominate the transfer function.

8.6 SUMMARY

The methods of open- and short-circuit time constants allow us to estimate rapidly the upper and lower -3-dB frequencies, almost by inspection of the network. As

---

[5] At low frequencies, anyway.

[6] It should be mentioned that one consequence of the difference between the Elmore and 10–90% risetimes is that Elmore somewhat underestimates the 10–90% risetime growth. A better estimate for identical stages in cascade is about $1.1\sqrt{n}$.

long as the circuit satisfies the assumptions well, the methods yield reasonably accurate answers. More important, however, is the valuable design insight provided.

Another way to estimate bandwidth is from a measurement of risetime. We've seen that moments of the impulse response allow us to exploit the power of linear system theory to show that delays add linearly and that risetimes add in root-sum-squared fashion. Furthermore, we've seen that the product of bandwidth and risetime is roughly constant and approximately equal to 2.2. For all of these relations, accuracy is greatest when the step response is monotonic. That is, as long as the step response has negligible overshoot and/or ringing, the results derived here will hold well. If these conditions are not well satisfied then all bets are off. Therefore, do not fall into the trap of believing that these rules of thumb are exact and universally applicable.

As long as we keep this caveat in mind, we can use these relationships to extend significantly the boundaries of our instrumentation or to make quantitative inferences about frequency-domain performance from time-domain measurements (or vice versa) when the necessary conditions are well satisfied (as they often, but not always, are).

PROBLEM SET FOR BANDWIDTH ESTIMATION

PROBLEM 1   Consider a truly differential amplifier whose poles are all known to be purely real. The differential gain has a measured bandwidth of ω_h, and the response appears very much like a single-pole system, but direct application of the method of open-circuit time constants yields an estimate that is off by about a factor of 2 in the low direction. Identify the likely source of the problem and describe how to fix it.

PROBLEM 2   It was asserted in the chapter that the method of open-circuit time constants is always conservative for an all-pole system in which all the poles are real. Prove this assertion.

PROBLEM 3   One might wonder how a single-pole RC model could ever adequately describe anything that is truly a higher-order system. In order to explore one aspect of this question, use open-circuit time constants to derive, for arbitrary n, a bandwidth estimate for the RC network shown in Figure 8.10.

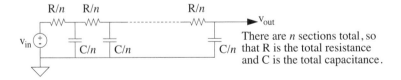

FIGURE 8.10. *RC* ladder network.

(a) What is the estimated bandwidth of the network in the limit of infinite n? *Hint:* It may be helpful to know that the sum of the first n integers is $n(n+1)/2$.

(b) For the specific case of $n = 4$, compare the open-circuit time-constant estimate of bandwidth with the result of SPICE simulations.

PROBLEM 4 Use open-circuit time constants to estimate the bandwidth of a high-pass, single-pole RC network, and compare with an analytical value obtained from inspection of the actual transfer function. The estimate, of course, is grossly in error. Explain why.

PROBLEM 5 Derive the rule that states: The drain–gate capacitance sees an effective open-circuit resistance given mnemonically by "$r_{\text{left}} + r_{\text{right}} + g_{m,\text{eff}} r_{\text{left}} r_{\text{right}}$," where $g_{m,\text{eff}}$ is the ratio of short-circuit drain current to the gate voltage (not gate-to-source voltage). In this case, neither a test voltage nor a test current is a perfect choice. However, fewer algebraic steps are needed if a test current is used.

PROBLEM 6 Derive an expression for the open-circuit resistance facing C_{gs}. Assume the existence of resistances in the gate and source branches, but you may neglect any resistances connecting drain to gate.

PROBLEM 7 Because there is a relationship between bandwidth and risetime, the method of open-circuit time constants seemingly offers a way to estimate risetime. Conversely, one may also estimate the bandwidth if given the risetime.

Specifically, consider a cascade of n amplifiers, each of which has a single pole whose time constant is τ. Assume that each amplifier has an infinite input impedance and a zero output impedance.

(a) Use open-circuit time constants to estimate the overall bandwidth, then estimate the 10–90% risetime from that bandwidth.

(b) Use the risetime addition rule to estimate the overall risetime, then use that estimate to estimate the bandwidth. Compare with your answer to part (a) and comment.

(c) For the specific case of $n = 5$ and $\tau = 1$ s, use SPICE to find the actual risetime and bandwidth. Compare the simulation results with your previous answers. What may you conclude about the accuracy of open-circuit time-constant estimates of risetime? How would your answer to that question change if the time constants were unequal? Specifically, what if one time constant were dominant (i.e., much larger than the rest)?

PROBLEM 8 It was mentioned in the chapter that it is not correct to include in the short-circuit time-constant calculation the gate bypass capacitor for a cascoding device. Show explicitly why by deriving formally the current transfer function of the common-gate amplifier depicted in Figure 8.11. You may neglect all reactive parasitics in the derivation.

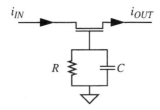

FIGURE 8.11. Common-gate amplifier for
short-circuit time-constant calculation.

(a) What is the actual transfer function? Sketch a Bode plot of magnitude and phase.
(b) From the transfer function, explain why the value of gate bypass capacitance is essentially irrelevant for determining the low-frequency breakpoint.

PROBLEM 9 We've seen that feedback can modify resistances considerably (as with, e.g., the Miller effect), especially within a device. Of course, this impedance-modifying property is hardly a local phenomenon. To underscore this point, calculate the resistance facing the capacitor in the circuit of Figure 8.12. Here, the input signal v_{in} consists of a DC bias and an incremental signal term.

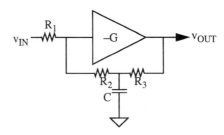

FIGURE 8.12. Feedback-biased amplifier
for resistance calculation.

(a) Derive an expression for the resistance facing the capacitor.
(b) What circuit consequence ensues from a capacitor that is too small?
(c) If all the resistors are 10 kΩ and G is 10^4, what value of capacitor is required to produce a low-frequency breakpoint of 20 Hz? Is this capacitor value consistent with an integrated realization?

PROBLEM 10 The "superbuffer" has been suggested as a means to bootstrap out the gate–drain capacitance of a follower and thereby increase bandwidth. To demonstrate this action explicitly, derive expressions for the open-circuit time constants of the circuit shown in Figure 8.13. You may neglect body effect as well as all resistive parasitics.

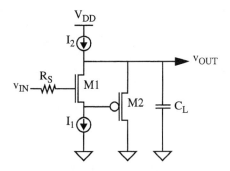

FIGURE 8.13. Superbuffer for open-circuit time-constant calculation.

Compare your derivation with the time-constant sum obtained with M_2 and I_2 removed. Discuss the conditions under which use of the superbuffer is beneficial.

CHAPTER NINE

HIGH-FREQUENCY
AMPLIFIER DESIGN

9.1 INTRODUCTION

The design of amplifiers at high frequencies involves more detailed considerations than at lower frequencies. One simply has to work harder to obtain the requisite performance when approaching the inherent limitations of the devices themselves. Additionally, the effect of ever-present parasitic capacitances and inductances can impose serious constraints on achievable performance.

At lower frequencies, the method of open-circuit time constants is a powerful intuitive aid in the design of high-bandwidth amplifiers. Unfortunately, by focusing on minimizing various RC products, it leads one to a relatively narrow set of options to improve bandwidth. For example, we may choose to distribute the gain among several stages or alter bias points, all in an effort to reduce effective resistances. These actions usually involve an increase in power or complexity, and at extremes of bandwidth such increases may be costly. In other cases, open-circuit time constants may erroneously predict that the desired goals are simply unattainable.

A hint that other options exist can be found by revisiting the assumptions underlying the method of open-circuit time constants. As we've seen, the method provides a good estimate for bandwidth only if the system is dominated by one pole. In other cases, it yields increasingly pessimistic estimates as the number of poles increases or their damping ratio increases. Additionally, open-circuit time constants can grossly underestimate bandwidth if there are zeros in the passband. These observations suggest that alternative ways to increase bandwidth might involve the deliberate violation of the very conditions on which the method of open-circuit time constants relies.

Following up on this idea leads us to a collection of extremely useful bandwidth extension techniques. We'll consider what happens if we *purposefully* introduce a zero or complex pole pair into a transfer function, for example, or if we deliberately construct a system of high order. We'll also consider the design of *narrowband* amplifiers. As we'll see, it is considerably easier to obtain gain if it is to be provided only over a narrow band centered about some nominal operating frequency.

We'll conclude with a more general consideration of the problem of designing amplifiers with large gain–bandwidth products. We'll find that the commonly held belief that gain and bandwidth must trade off linearly is false. Instead, it is possible to construct networks in which gain trades off more with *delay,* allowing much greater flexibility in constructing amplifiers with large gain–bandwidth products.

We begin with the study of a simple example, the shunt-peaked amplifier, whose behavior is not well predicted by open-circuit time constants.

9.2 ZEROS AS BANDWIDTH ENHANCERS

9.2.1 THE SHUNT-PEAKED AMPLIFIER

Back in the 1930s, when television was being developed, one problem of critical importance was that of designing amplifiers with a reasonably flat response over the 4-MHz video bandwidth. Although obtaining this bandwidth seems trivial today, it was challenging with the devices available at the time. Just to make things more difficult, the amplifier also had to be cheap enough for use in a mass-market consumer item, so the number of vacuum tubes had to be kept to an absolute minimum.

A technique that satisfied this requirement of large bandwidth at low cost is known as *shunt peaking,* and it was used in countless television sets at least up to the 1970s. Stripped to its essentials, a shunt-peaked amplifier is sketched in Figure 9.1. This amplifier is a standard common-source configuration, with the addition of the inductance.

If we assume that the transistor is ideal, then the only elements that control the bandwidth are R, L, and C. The capacitance C may be taken to represent all the loading on the output node, including that of a subsequent stage (perhaps arising from the input capacitance of another transistor, for example). The resistance R is the effective load resistance at that node and the inductor provides the bandwidth enhancement, as we now proceed to demonstrate.

Given our assumptions, we may model the amplifier for small signals as shown in Figure 9.2. It's clear from the model that the transfer function v_{out}/i_{in} is just the impedance of the RLC network, so it should be straightforward to analyze. Before launching into a detailed derivation, though, let's think about why adding an inductor this way should give us a bandwidth extension.

First, we know that the gain of a purely resistively loaded common-source amplifier is proportional to $g_m R_L$. We also know that when a capacitive load is added, the gain eventually falls off as frequency increases because the capacitor's impedance diminishes. The addition of an inductance in series with the load resistor provides an impedance component that increases with frequency (i.e., it introduces a zero), which helps offset the decreasing impedance of the capacitance, leaving a net impedance that remains roughly constant over a broader frequency range than that of the original RC network.

An equivalent time-domain interpretation may be provided by considering the step response. The inductor delays current flow through the branch containing the resistor, making more current available for charging the capacitor, reducing the risetime.

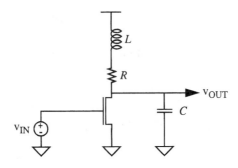

FIGURE 9.1. Shunt-peaked amplifier.

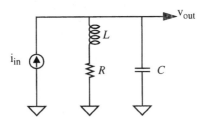

FIGURE 9.2. Model of shunt-peaked amplifier.

To the extent that a faster risetime implies a greater bandwidth, an appropriate choice of inductor therefore increases the bandwidth.

Formally, the impedance of the RLC network may be written as

$$Z(s) = (sL + R) \parallel \frac{1}{sC} = \frac{R[s(L/R) + 1]}{s^2LC + sRC + 1}. \tag{1}$$

In addition to a zero there are two poles (possibly complex), definitely violating the conditions for open-circuit time constants.

Since the gain of the amplifier is the product of g_m and the magnitude of $Z(s)$, let's now compute the latter as a function of frequency:

$$|Z(j\omega)| = R\sqrt{\frac{(\omega L/R)^2 + 1}{(1 - \omega^2 LC)^2 + (\omega RC)^2}}. \tag{2}$$

Notice that, in contrast with the simple RC case, there is a term in the numerator (from the zero) that increases with increasing frequency. Furthermore, the $1 - \omega^2 LC$ term in the denominator contributes to an increase in $|Z|$ for frequencies below the LC resonance as well. Both of these terms extend bandwidth.

Unfortunately, this last equation is not nearly as useful for design as it is for analysis; we don't have an explicit guide that tells us how to select L given R and C, for example. As will be shown, there is no single "optimum" value of inductance, but we can narrow the range of possibilities by imposing one or two arbitrary (but nonetheless sensible) requirements.

To facilitate subsequent derivations, we introduce a factor m, defined as the ratio of the RC and L/R time constants:

$$m = \frac{RC}{L/R}. \tag{3}$$

Then, our transfer function becomes

$$Z(s) = (sL + R) \parallel \frac{1}{sC} = \frac{R(\tau s + 1)}{s^2 \tau^2 m + s \tau m + 1}, \tag{4}$$

where $\tau = L/R$.

The magnitude of the impedance, normalized to the DC value ($= R$) as a function of frequency, is then

$$\frac{|Z(j\omega)|}{R} = \sqrt{\frac{(\omega \tau)^2 + 1}{(1 - \omega^2 \tau^2 m)^2 + (\omega \tau m)^2}}, \tag{5}$$

so that

$$\frac{\omega}{\omega_1} = \sqrt{\left(-\frac{m^2}{2} + m + 1\right) + \sqrt{\left(-\frac{m^2}{2} + m + 1\right)^2 + m^2}}, \tag{6}$$

where ω_1 is the uncompensated -3-dB frequency ($= 1/RC$).

The problem, then, is to choose a value of m that leads to some desired behavior. Maximizing the bandwidth is one obvious possibility. After a certain amount of effort, one finds that this maximum occurs at a value of

$$m = \sqrt{2} \approx 1.41, \tag{7}$$

which extends the bandwidth to a value about 1.85 times as large as the uncompensated bandwidth. Anyone who has labored to meet a tough bandwidth specification can well appreciate the value of nearly doubling bandwidth through the addition of a single inductance at no increase in power.

Unfortunately, however, this choice of m leads to nearly a 20% peak in the frequency response, a value often considered undesirably high. To moderate the peaking, one might seek a bandwidth other than the absolute maximum by increasing m. One specific choice is to set the magnitude of the impedance equal to R at a frequency equal to the uncompensated bandwidth. Solving for this condition yields a value of 2 for m, with a corresponding bandwidth of

$$\omega = \omega_1 \sqrt{1 + \sqrt{5}} \approx 1.8 \omega_1. \tag{8}$$

Hence, the bandwidth in this case is still quite close to the maximum. Further calculation shows that the peaking is substantially reduced, to about 3%.

The arbitrary choice that leads to this result is frequently used because it yields such a significant bandwidth enhancement without excessive frequency response peaking. However, there are many cases where one desires the frequency response to be completely free of peaking. Thus, perhaps one might seek the value of m that maximizes the bandwidth but subject to the constraint of no peaking.

The conditions for such maximal flatness may be found through the following general technique: Form an expression for the frequency response magnitude (or, as is frequently more convenient, the square of the magnitude), and maximize the number of derivatives whose value is zero at DC.

Carrying out this method manually is frequently labor-intensive, but in this particular example, a straightforward calculation reveals that the magic value of m is

$$m = 1 + \sqrt{2} \approx 2.41, \tag{9}$$

which leads to a bandwidth that is about 1.72 times as large as the unpeaked case. Hence, at least for the shunt-peaked amplifier, both a maximally flat response and a substantial bandwidth extension can be obtained simultaneously.

In other situations, there may be a specification on the time response of the amplifier, rather than on its frequency response. One example of practical interest is an oscilloscope deflection amplifier, whose time response (characterized, say, by the step or pulse response) must be "well behaved." That is, not only must we amplify uniformly the various spectral components of the signal over as large a bandwidth as practical, but the *phase relationships* among its Fourier components must be preserved as well. If the spectral components do not experience equal delay (measured in *absolute time,* not degrees), potentially severe distortion of the waveshape can occur. Such "phase distortion" is objectionable for the bit errors it can cause in digital systems or for its obvious negative implications for the fidelity of analog instrumentation such as oscilloscopes.

To quantify this type of distortion, first consider the phase behavior of a pure time delay. If all frequencies are delayed by an equal amount of *time,* then this fixed amount of time delay must represent a linearly increasing amount of *phase shift* as frequency increases. Phase distortion will be minimized if the deviation from this ideal linear phase shift is minimized.

Evidently, then, we wish to examine the delay as a function of frequency. If this delay is the same for all frequencies, we will have no phase distortion (other than the change in shape that results from the ordinary filtering provided by any bandlimited amplifier). Formally, the delay is defined as follows:

$$T_D(\omega) \equiv -\frac{d\phi}{d\omega}, \tag{10}$$

where ϕ is the phase shift of the amplifier at frequency ω.

Unfortunately, it is impossible for a network of finite order to provide a constant time delay over an infinite bandwidth, since infinite phase shift would ultimately be required whereas poles and zeros contribute only bounded amounts of phase shift. All we can do in practice, then, is to provide an approximation to a constant delay over some finite bandwidth.

By analogy with the frequency response case, we see that a maximally flat time delay will result if we maximize the number of derivatives of $T_D(\omega)$ whose value is zero at DC. Again, this method is general.

Table 9.1. *Shunt-peaking summary*

Condition	$m = R^2 C/L$	Normalized bandwidth	Normalized peak frequency response
Maximum bandwidth	~1.41	~1.85	1.19
$|Z| = R$ @ $\omega = 1/RC$	2	~1.8	1.03
Maximally flat frequency response	~2.41	~1.72	1
Best group delay	~3.1	~1.6	1
No shunt peaking	∞	1	1

Because of the involvement of arctangents in expressing the phase shift due to poles and zeros, computing the relevant derivatives is generally quite a bit more unpleasant than in the magnitude case. Even for our shunt-peaked amplifier, which is only second-order, the amount of labor is tremendous. Ultimately, however, one may derive the following cubic equation for m (computational aids are of tremendous benefit here):

$$m^3 - 3m^2 - 1 = 0, \tag{11}$$

whose relevant root is

$$m = 1 + \left[\frac{3 + \sqrt{5}}{2}\right]^{1/3} + \left[\frac{3 - \sqrt{5}}{2}\right]^{1/3} \approx 3.10, \tag{12}$$

corresponding to a bandwidth improvement factor of a bit under 1.6.

Since the conditions for maximally flat frequency response and maximally flat time delay do not coincide, one must compromise (this situation is hardly limited to the example of the shunt-peaked amplifier, of course). We therefore see that, depending on requirements, there is a range of useful inductance values; see Table 9.1. A larger L (smaller m) gives a larger bandwidth extension but poorer pulse fidelity, whereas a smaller L yields less bandwidth improvement but better pulse response.

9.2.2 SHUNT PEAKING: A DESIGN EXAMPLE

Even though shunt peaking traces its origins to 4-MHz video amplifiers from the 1930s, it is a useful trick even in the modern era for the same reasons it was originally valued: it allows one to squeeze the maximum performance from a given technology. This observation is particularly relevant given the acknowledged inferiority of CMOS. Furthermore, it is especially important to note that the technique does not require a high-Q peaking inductor and is therefore quite compatible with IC realizations. To underscore this point, consider the problem of designing a 1.5-GHz common-source broadband amplifier intended to provide gain to a block of phase-modulated channels. In this application, then, phase linearity is important, so we will choose $m = 3.1$ for best group delay uniformity.

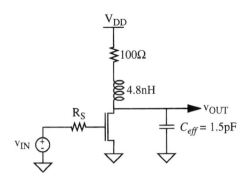

FIGURE 9.3. Shunt-peaked amplifier example.

Let the total capacitive loading on the drain be 1.5 pF (from both the transistor and loading by interconnect and subsequent stages), and assume that the load resistance cannot be made smaller than 100 Ω without increasing by an unacceptable amount the power consumed to keep gain constant. If the bandwidth is entirely controlled by the output node then the bandwidth of the amplifier is just a bit over 1 GHz, somewhat shy of the 1.5-GHz goal.

If we assume that the minimum acceptable resistance is used, then the required shunt-peaking inductor is readily calculated as

$$L = \frac{R^2 C}{3.1} = 4.8 \text{ nH}. \tag{13}$$

A 4.8-nH planar spiral inductor is readily implemented in standard CMOS technologies; the amplifier appears in Figure 9.3. With this inductor, the estimated bandwidth increases to approximately 1.7 GHz, comfortably in excess of the requirement. Again, this improvement is obtained without increasing the power consumed by the stage. Finally, note that the Q of the drain network is approximately 0.5 at 1.7 GHz, so inductors with modest Q (such as IC spiral inductors) suffice.

9.2.3 MORE ON ZEROS AS BANDWIDTH ENHANCERS

We've seen from the shunt-peaked amplifier example that zeros are quite useful, despite their neglect by the method of open-circuit time constants. To illustrate their utility with another simple (but relevant) example, consider an oscilloscope probe. Contrary to what one might think, it is most emphatically *not* just a glorified piece of wire with a tip on one end and a connector on the other. Think about this fact: Most "10 : 1" probes (so-called because they provide a factor-of-10 attenuation) present a 10-MΩ impedance to the circuit under test yet may provide a bandwidth of 200 MHz. But intuition would suggest that the maximum allowable capacitance consistent with this bandwidth is about 80 aF (0.08 fF)! So, how can probes provide such a large bandwidth while presenting a 10-MΩ impedance? The answer is that the combination of a probe and oscilloscope isn't an "ordinary" *RC* network.

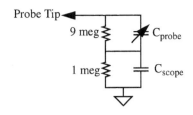

FIGURE 9.4. Simplified oscilloscope/probe model.

A simplified model of the scope/probe combination is shown in Figure 9.4. The 1-MΩ resistor represents the oscilloscope input resistance; C_{scope} represents the scope's input capacitance.

Inside the probe, there is a 9-MΩ resistor to provide the necessary 10:1 attenuation at low frequencies. However, to avoid the tremendous bandwidth degradation that would result from use of only a simple 9-meg resistor, the probe also has a capacitor in parallel with that 9-meg resistor. At high frequencies, the 10:1 attenuation is actually provided by the capacitive voltage divider. It may be shown (and you should show it) that when the top RC equals the bottom RC, the attenuation is exactly a factor of 10, *independent of frequency.* There is a zero that cancels precisely the slow pole, leading to a transfer function that has no bandwidth limitation. This case thus represents the ultimate in bandwidth "enhancement."

Because it is impossible to guarantee exact cancellation with fixed elements, all 10:1 probes have an adjustable capacitor. To appreciate more completely the necessity for such an adjustment, let's evaluate the effect of imperfect pole-zero cancellation with the following transfer function:

$$H(s) = \frac{\alpha \tau s + 1}{\tau s + 1}. \tag{14}$$

For simplicity's sake, note that no attenuation factor is included in this expression. Thus, the ideal value of the constant α is unity, so that $H(s)$ is ideally unity at all frequencies.

Let's now consider the step response of this system (often known as a *pole-zero doublet*). The initial- and final-value theorems tell us that the initial value is α and that the final value is unity. Because state evolves exponentially with the time constant of the pole,[1] we can rapidly sketch a couple of possible step responses, one with $\alpha_1 < 1$ and one with $\alpha_2 > 1$; see Figure 9.5.

We see that the response jumps immediately to α, but it settles down (or up, as the case may be) to the final value with a time constant of the pole. If α happens to equal unity, the response reaches final value in zero time. In all practical circuits, of

---

[1] For some inexplicable reason, there seems to be a fair amount of confusion about this point. The presence of the zero merely alters the initial error, but this error always settles to the final value with a time constant of just the pole.

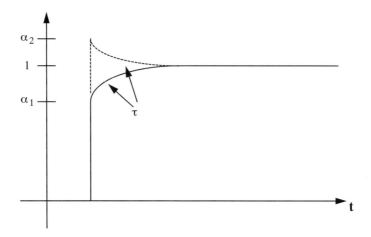

FIGURE 9.5. Possible step responses of pole-zero doublet.

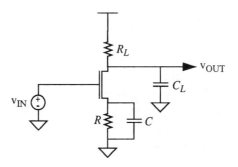

FIGURE 9.6. Zero-peaked
common-source amplifier.

course, additional poles limit the risetime to a nonzero value, but the general idea should be clear from this example.

From Figure 9.5, it is easy to see the importance of adjusting the capacitor to avoid gross measurement errors. This calibration is most easily performed by examining the response to a square wave slow enough to see the slow settling due to the doublet's pole and by adjusting the capacitor for the flattest time response.

This notion of cancelling a pole with a zero can be used to extend the bandwidth of active circuits as well, of course. This type of cancellation may be implemented as in the degenerated common-source amplifier of Figure 9.6. Here, C is *not* chosen large enough to behave as a short at all frequencies of interest. Instead, it is chosen just large enough to begin shorting out R when C_L begins to short out R_L. It is therefore relatively straightforward to understand that ideal compensation should result when $RC \approx R_L C_L$. As with the oscilloscope probe example, proper adjustment is necessary to obtain the best response.

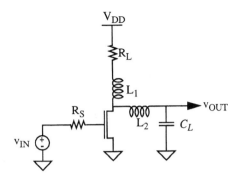

FIGURE 9.7. Amplifier with shunt and
series peaking.

9.2.4 TWO-PORT BANDWIDTH ENHANCEMENT

Shunt peaking is a form of bandwidth enhancement in which the additional elements form a one-port with the original load. Although the near doubling of bandwidth provided is impressive, it is possible to do better still by employing a two-port network between amplifier and load.

One may augment shunt peaking by additionally separating the load capacitance from the output capacitance of the device. If a series inductor is used to perform this separation, the overall result is a combination of shunt and *series* peaking; see Figure 9.7.

It may be shown that, in the absence of L_1, the maximum bandwidth is $\sqrt{2}$ times that of the uncompensated case. Series peaking actually predates shunt peaking, but its bandwidth boost comes entirely from the peaking provided by complex poles; there is no zero to help things along. The smaller bandwidth improvement guaranteed the relative obscurity of series peaking once shunt peaking was invented. For the sake of completeness, the desired inductance if only series peaking is used is given by

$$L_2 = \frac{R^2 C}{m}, \tag{15}$$

where $m = 2$ corresponds to the maximum bandwidth and maximally flat amplitude case. A choice of $m = 3$ leads to maximally flat group delay and a bandwidth boost factor of about 1.36.

We won't spend any more time analyzing this combination because it is an intermediate step on the way to a much better bandwidth extension method. The next evolutionary step is to add an inductance between the device and the rest of the network, as shown in Figure 9.8.

This combination of shunt and series peaking works as follows. Just as in the step response of an ordinary shunt-peaked amplifier, the flow of current into the load resistor continues to be deferred by the action of L_1. This action alone speeds up

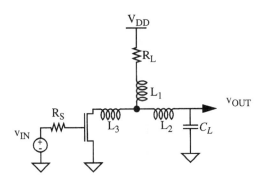

FIGURE 9.8. Shunt and double-series peaking.

the charging of the load capacitance. In addition to that mechanism, the transistor initially has to drive only its own output capacitance for some time, because L_3 delays the diversion of current into the rest of the network. Hence, risetime at the drain improves, which we again interpret as implying an improved bandwidth. Some time after the drain voltage has risen significantly, the voltage across the load capacitance begins to rise as current finally starts to flow through L_2. Hence, such a network charges the capacitances *serially in time,* rather than in parallel. The trade-off is an increased delay in exchange for the improved bandwidth. We will see that this bandwidth–delay tradeoff is a recurrent theme.

To save die area, the combination of three inductors can be realized conveniently as a pair of magnetically coupled inductors (i.e., a transformer), since the equivalent circuit model of such a connection is precisely the arrangement we seek. These may be implemented as a pair of spiral inductors that have been placed on top of each other and offset appropriately to obtain the desired amount of coupling. A further improvement is possible if a small bridging capacitance is added across the inductors to create a parallel resonance. The increased circulating currents associated with the resonance help to push the bandwidth out even further.

After even more agony than suffered in deriving the equations for the shunt-peaking case, one may show that the coupled inductances should each have a value given by

$$L = \frac{R^2 C_L}{2(1+k)}, \tag{16}$$

where L is interpreted as the primary or secondary inductance with the other winding open-circuited. Hence, this is the value of inductance used in designing and laying out each spiral, for example.

The bridging capacitance should have a value of

$$C_c = \frac{C_L}{4}\left[\frac{1-k}{1+k}\right]. \tag{17}$$

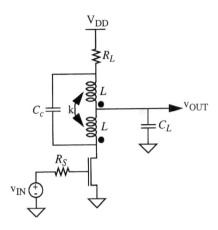

FIGURE 9.9. Amplifier with T-coil
bandwidth enhancement.

It may also be shown that a coupling coefficient of $1/3$ yields a Butterworth-type (maximally flat amplitude) response, while a k of $1/2$ leads to maximally flat group delay. These coupling coefficients are not particularly large and are therefore readily obtained in practice. Two adjacent bondwires typically have coupling coefficients in this range, for example.

Applying these conditions then leads to the amplifier of Figure 9.9. The resulting network is called a "T-coil" (because of the way the schematic is drawn) and has been used for over forty years in oscilloscope circuitry. The T-coil is capable of almost tripling the bandwidth (the theoretical maximum is a $2\sqrt{2}$ improvement, or about $2.83\times$, obtained with the Butterworth condition) if the output capacitance of the device is negligibly small compared with the load capacitance.

It may be shown that the bandwidth is maximized if the junction of the two inductors drives the higher-capacitance node. In Figure 9.9, we have assumed that the load capacitance is larger than the output capacitance of the transistor. The drain and load capacitance connections may be reversed if the output capacitance happens to exceed the load capacitance.

As a final refinement, some additional compensation for the output capacitance of the transistor may be provided by adding more inductance in series with it, effectively providing more series peaking. A nearly equivalent result may be obtained merely by tapping the inductors at other than their midpoint (in this case, closer to the load resistor end).

Modifying our earlier shunt-peaked design example yields the amplifier of Figure 9.10, where – to keep the comparison fair – we have continued to assume that we desire a maximally flat group delay. As can be seen, the total inductance has doubled. However, since the two inductors are on top of each other with a small offset, the additional area is modest (on the order of 50%). In addition, the 125-fF bridging

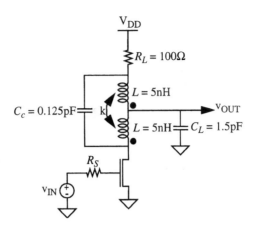

FIGURE 9.10. T-coil bandwidth
enhancement example.

capacitance might be provided as an inherent byproduct of the overlapping inductor layout. The theoretical bandwidth improvement factor provided by this circuit is about 2.7.[2] Hence, roughly a 2.7-GHz bandwidth can be expected, substantially better than the 1.7-GHz bandwidth of the shunt-peaked case. It is important to underscore that this improvement is obtained without an increase in power.

We will later appreciate this structure as an intermediate evolutionary step on the way to a completely "distributed amplifier" (to be discussed shortly), in which parasitic capacitances are absorbed into structures that trade gain for delay rather than for bandwidth. (For example, consider a transmission line – it consists of inductance and capacitance, but these elements impose no limit on bandwidth because the capacitances are charged serially in time.) In the meantime, it may be considered simply as a more sophisticated way to divert current away from the load resistor and into the load capacitance.

9.3 THE SHUNT–SERIES AMPLIFIER

In contrast with the open-loop architectures we've studied so far, an alternative approach to the design of broadband amplifiers is to use negative feedback. One particularly useful broadband circuit that employs negative feedback is the shunt–series amplifier. Its name derives from the use of a combination of shunt and series feedback, and its utility derives from the relative constancy of input and output impedances over a broad frequency range (which makes cascading much less complicated), as well

---

[2] Again, this value assumes that the output capacitance of the transistor is negligibly small compared with the load capacitor. If this inequality is not well satisfied, additional series compensation will be required to achieve bandwidth boosts of this order.

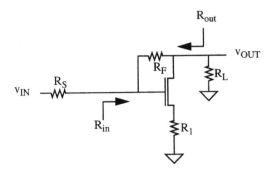

FIGURE 9.11. Shunt–series amplifier
(biasing not shown).

as from its ease of design. In addition, the dual feedback loops confer the usual benefits normally associated with negative feedback – namely, a reduced dependency on device parameters, improved distortion, broader bandwidth and a nicer complexion.

Stripped of biasing details, the shunt–series amplifier is depicted in Figure 9.11, where R_S now denotes the resistance of the input source and R_L is the load resistance. Thus, the amplifier core consists of just R_F, R_1, and the transistor. To understand how this amplifier works, initially assume that R_1 is large enough (relative to the reciprocal of the transistor's g_m) that it degenerates the overall transconductance to approximately $1/R_1$. Since R_1 is in series with the input and output loops, the degeneration by R_1 is the "series" contribution to the name of this amplifier.

To continue the analysis, assume also that R_F is large enough that its loading on the output node may be neglected. With these assumptions, the voltage gain of the amplifier from the gate to the drain is approximately $-R_L/R_1$.

Although we have assumed that R_F has but a minor effect on gain, it has a controlling influence on the input and output resistance. Specifically, it reduces both quantities through the (shunt) feedback it provides. Additionally, the reduction of input and output resistances helps to increase the bandwidth still further by reducing the open-circuit time-constant sum.

To compute the input resistance R_{in}, we use the fact that the gain from gate to drain is approximately $-R_L/R_1$. If, as seems reasonable, we may neglect gate current, then the input resistance is due entirely to current flowing through R_F. Applying a test voltage source at the gate terminal allows us to compute the effective resistance in the usual way. Just as in the classic Miller effect, connecting an impedance across two nodes that have an inverting gain between them results in a reduction of impedance. Formally, R_{in} is given by

$$R_{in} = \frac{R_F}{1 - A_V} \approx \frac{R_F}{1 + R_L/R_1}, \tag{18}$$

where A_V is the voltage gain from gate to drain.

Now, to compute the output resistance, apply a test voltage source to the drain node and again take the ratio of v_{test} to i_{test}:

$$R_{\text{out}} = \frac{R_F + R_S}{1 + R_S/R_1} \approx \frac{R_F}{1 + R_S/R_1}. \tag{19}$$

If the source and load resistances are equal (a particularly common situation in discrete realizations), then the denominators of Eqn. 18 and Eqn. 19 are approximately equal. Since the numerators are also approximately equal, it follows that R_{in} and R_{out} are themselves nearly equal. If $R_S = R_L = R$ then we may write:

$$R_{\text{out}} \approx R_{\text{in}} \approx \frac{R_F}{1 + R/R_1} \approx \frac{R_F}{1 - A_V}. \tag{20}$$

The ease with which this amplifier provides a simultaneous impedance match at both input and output ports accounts in part for its popularity. Once the impedance level and gain are chosen, the required value of the feedback resistor is easily determined. Coupling knowledge of the load resistance with the required gain leads quickly to the necessary value of R_1. To complete the design, a suitable device width and bias point must be chosen. Generally, these choices are made to ensure sufficient g_m to validate the assumptions used in developing the foregoing set of equations.

DETAILED DESIGN OF SHUNT–SERIES AMPLIFIER

The foregoing presentation outlines the first-order behavior of the shunt–series amplifier in order to help the development of design intuition. To carry out a more detailed design, however, we now consider some of the second-order factors neglected in the previous section.

Low-Frequency Gain and Input–Output Resistances

We start by computing the gain from gate to drain since it allows us to find the input and output resistances easily. Once the gate-to-drain gain and input resistance are known, the overall gain is trivially found from the voltage divider relationship.

First, recall that the effective transconductance of a common-source amplifier with source degeneration is

$$g_{m,\text{eff}} = \frac{g_m}{1 + g_m R_1}. \tag{21}$$

Note from Eqn. 21 that the effective transconductance is approximately $1/R_1$ as long as $g_m R_1$ is much larger than unity.

Applying a test voltage from gate to ground causes a drain current to flow through both the load and feedback resistors. Some fraction of the test voltage also feeds forward directly to the output. Superposition allows us to treat each of these contributions to the output voltage separately:

$$v_{\text{out}} = -g_{m,\text{eff}} v_{\text{test}} \frac{R_F R_L}{R_F + R_L} + v_{\text{test}} \frac{R_L}{R_F + R_L}. \tag{22}$$

Solving for the gain yields

$$A_V = \frac{v_{\text{out}}}{v_{\text{test}}} = -\frac{R_L}{R_1} \cdot \left[\frac{1}{1 + 1/g_m R_1}\right] \cdot \left[\frac{1}{1 + R_L/R_F}\right] \cdot \left[1 - \frac{1}{g_{m,\text{eff}} R_F}\right]. \quad (23)$$

Although not the most compact expression, Eqn. 23 shows the gain derived earlier from first-order theory multiplied by three factors (in brackets), each of which is ideally unity.

The first "nonideal" factor reflects the influence of finite g_m on the effective transconductance. While $g_{m,\text{eff}}$ approaches $1/R_1$ in the limit of large $g_m R_1$, this first factor shows quantitatively the effect of noninfinite $g_m R_1$. The second term is the result of the loading by R_F on the output node. As long as R_F is substantially larger than the load resistance R_L, the gain reduction is small.

The final gain reduction factor is due to feedforward of the input signal to the output. This feedforward reduces the gain because the ordinary gain path inverts but the feedforward path does not. Hence, the feedforward term partially cancels the desired output. The transconductance of the feedforward term is $1/R_F$; hence, as long as this parasitic transconductance is small compared with the desired transconductance $g_{m,\text{eff}}$, the gain loss is negligible.

Having examined the complete gain equation term by term, we now present a much more compact (but still exact) expression, useful for calculations to follow:

$$A_V = -\frac{R_L}{R_E} \cdot \left[\frac{R_F - R_E}{R_F + R_L}\right], \quad (24)$$

where R_E is simply the reciprocal of the effective transconductance. The upshot is simply that, in order to obtain the desired gain, one must choose a value of R_1 (or R_E) that is somewhat smaller than anticipated on the basis of the first-order equations.

Now that we have a complete expression (two, even) for the low-frequency gain, we can obtain a more accurate value for the resistance between gate and ground:

$$R_{\text{in}} = \frac{R_F}{1 - A_V}, \quad (25)$$

which, after using Eqn. 24, becomes

$$R_{\text{in}} = \frac{R_F}{1 + \dfrac{R_L}{R_E}\left(\dfrac{R_F - R_E}{R_F + R_L}\right)} = \frac{R_E(R_F + R_L)}{R_E + R_L}. \quad (26)$$

In general, one designs specifically for a particular value of gain. Assuming success at achieving that goal, the value of feedback resistance necessary to produce a desired input resistance is readily found simply from Eqn. 25.

The output resistance (i.e., as seen by R_L) is also simple to find. Again, we apply a test voltage source to the drain node and compute the ratio of test voltage to test current. Performing this exercise yields

$$R_{out} = \frac{v_{test}}{i_{test}} = \frac{R_F + R_S}{1 + g_{m,eff}R_S} = \frac{R_F + R_S}{1 + R_S/R_E} = \frac{R_E(R_F + R_S)}{R_E + R_S}. \tag{27}$$

Comparing the expressions for input and output resistance, we see that if R_S and R_L are equal (as is commonly the case) then R_{in} and R_{out} will also be precisely equal. This happy coincidence is one reason for the tremendous popularity of this topology.[3]

It should be emphasized that, when carrying out a design (as opposed to analysis), the desired gain is known. Hence, if the input and output resistances are to be equal, selection of the feedback resistor is trivial from Eqn. 25. The value of R_1 is then chosen to provide the correct gain, completing the design.

Bandwidth and Input–Output Impedances

Having presented exact expressions for various low-frequency quantities (gain and input–output resistances), we now derive approximate expressions for the bandwidth as well as the input and output *impedances* of this amplifier.

Before plowing through a slew of equations, let's see if we can anticipate the qualitative behavior of these quantities. Because this amplifier is a low-order system, we expect gain and bandwidth to trade off more or less linearly. Furthermore, precisely because it is a low-order system, an open-circuit time-constant estimate of bandwidth should be reasonably accurate.

We also expect the input impedance to possess a capacitive component, partly because of the presence of C_{gs}, but also because of the augmentation of C_{gd} by the Miller effect. The output impedance, on the other hand, could behave differently because the shunt feedback that reduces the output resistance becomes less effective as frequency increases. As a result, the output impedance could actually rise with frequency, leading to an inductive component in the output impedance.

Having made those predictions, let us proceed with a calculation of the open-circuit time-constant sum. To simplify the development, assume that the only device capacitances are C_{gs} and C_{gd}. Furthermore, neglect the series gate resistance. Finally, assume that the source and load resistance both equal a value R.

The effective resistance facing C_{gd} is clearly R_F in parallel with a resistance given by

$$r_{left} + r_{right} + g_{m,eff}r_{left}r_{right}, \tag{28}$$

so that the resistance is:

$$R_F \parallel (R_S + R_L + g_{m,eff}R_S R_L) = R_F \parallel R(2 + g_{m,eff}R). \tag{29}$$

After substitution for R_F, this becomes

---

[3] It should be noted that this topology is also widely used in the bipolar form in which it was first realized. There is a minor difference in that finite β causes the input and output resistance to be somewhat unequal, although the error is small for typical values of β. The input resistance is smaller by a factor of approximately $1 - 1/2\beta$, while the output resistance is higher by a factor of about $1 + 1/2\beta$. The gain is also slightly lower, by a factor of about $1 - 2/\beta$.

$$R(1 - A_V) \parallel R(2 + g_{m,\text{eff}}R). \tag{30}$$

Note that, in the limit of large gain, the resistance facing C_{gd} approaches

$$|A_V|\frac{R}{2}, \tag{31}$$

as might be anticipated from considering the Miller effect.

Computing the resistance facing C_{gs} is somewhat more involved, but ultimately one may derive the following expression:

$$\frac{R(R_F + R + 2R_1) + R_1 R_F}{(2R + R_F)(1 + g_m R_1) + g_m R^2}. \tag{32}$$

In the limit of large gain, Eqn. 32 simplifies to

$$\frac{R}{R_1}\frac{1}{g_m}. \tag{33}$$

Note that the ratio R/R_1 is approximately the magnitude of the gain (from gate to drain). Because both open-circuit resistances are then roughly proportional to gain, the gain–bandwidth product of the shunt–series amplifier is approximately constant.

The estimated bandwidth of the amplifier in this limit is therefore

$$\text{BW} \approx \left[|A_V|\left(\frac{C_{gs}}{g_m} + \frac{RC_{gd}}{2}\right)\right]^{-1}. \tag{34}$$

Having derived an approximate expression for the bandwidth, we now consider the input impedance. As stated earlier, the input impedance should possess a capacitive component because of C_{gs} and the Miller-multiplied C_{gd}. A crude approximation to the total capacitance may be obtained simply by assuming that the impedance at the gate controls the bandwidth of the amplifier. That is, assume that the time constant of the amplifier's pole is the product of the source resistance R_S ($= R$) and the capacitance at that node. With that assumption, the effective input capacitance is just the bracketed term of Eqn. 34 divided by R:

$$C_{\text{in}} \approx \frac{C_{gs}}{g_m R_1} + C_{gd}\frac{|A_V|}{2}. \tag{35}$$

In almost all practical cases, the Miller-augmented C_{gd} dominates.

Note that the presence of this capacitance, which effectively appears between gate and ground, makes it impossible to achieve a perfect input impedance match at all frequencies. Furthermore, as the frequency increases, C_{gs} progressively shorts out and so connects the source degeneration resistance R_1 to the gate node. Hence, even the input resistance tends to degrade as well, diminishing as the frequency increases.

These effects can be mitigated to a certain extent by using some simple techniques. First, an L-match can be used to transform the resistive part up to the desired level, such as 50 Ω, at some nominal frequency (generally a little beyond where the quality of the match has begun to degrade noticeably). Of the possible types of L-matches, the best choice is usually one that places an inductance in series with the gate and a shunt capacitance across the amplifier input; such a network becomes transparent at low frequencies, where no correction is required.

The series inductance of the L-match generally leaves a residual *inductive* component. This inductance is easily compensated by simply augmenting the shunt capacitance of the L-network. With this compensation, the frequency range over which a reasonably good input match is obtained can often be doubled.

To compute the output impedance, apply a test voltage source to the drain and calculate the ratio of the test voltage to the current that it supplies. In the limit of high gain, one finds that the output impedance includes an inductive component whose value is approximately

$$L_{\text{out}} \approx \frac{ARC_{gs}}{g_m},$$
(36)

where C_{gd} has been neglected.

In order to develop a deeper understanding of the origins of this inductance, note that the gate voltage is some fraction of the test voltage applied to the drain. Specifically, the gate voltage is an attenuated and low-pass–filtered version of the applied drain voltage due to the capacitance at the gate. Hence, the gate voltage lags behind the voltage at the drain. The transistor then converts the lagging gate voltage into a lagging drain current. From the viewpoint of the test source, it must supply a current with a component that lags the applied voltage. This phase relationship between voltage and current is characteristic of an inductance.

From this insight, we can assess the effect of neglecting C_{gd}. Since C_{gd} supplies a leading component of voltage at the gate, it tends to offset the inductive effect. As a result, the output inductance actually observed can be considerably smaller than the upper bound estimated by Eqn. 36 if C_{gd} is not negligibly small.

9.4 BANDWIDTH ENHANCEMENT WITH f_T DOUBLERS

While the bandwidth of an amplifier need not be a strong function of f_T, it remains true that a higher f_T increases it, all other things equal. Now, device f_T is bounded within any given technology, so it would seem that once biasing conditions that maximize f_T have been established, the designer has done all that can be done. However, this facile conclusion overlooks the possibility of *topological* routes to increasing f_T.

Recall that the equation for f_T is

$$2\pi f_T = \frac{g_m}{C_{gs} + C_{gd}}.$$
(37)

Loosely speaking, then, f_T is the ratio of transconductance to input capacitance. If a way could be found to, say, decrease input capacitance without decreasing transconductance, f_T would increase.

The ordinary differential pair may be considered an f_T doubler by this definition, for the device capacitances are in series as far as a differential input is concerned.

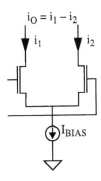

FIGURE 9.12. Differential
pair as f_T doubler.

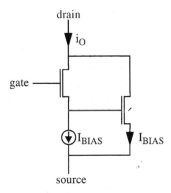

FIGURE 9.13. Darlington
pair as f_T doubler.

Hence, the differential input capacitance is one half that of each transistor. The differential transconductance, on the other hand, is unchanged because, although the input voltage divides equally between the two transistors, the differential output current is twice the current in each device. Hence, the overall stage transconductance is equal to that of each transistor, and a doubling of f_T results; see Figure 9.12.

Because it is not always convenient to arrange for differential signal paths, it is sometimes desirable to synthesize a single-ended f_T doubler. The differential pair can be converted into a single-ended doubler by interchanging the gate and source connections on one device. The attendant polarity reversal allows us to take the output as the sum (rather than difference) of the two drain currents, so merely tying the two drains together completes the transformation, as shown in Figure 9.13.

One may recognize the result as topologically identical to a Darlington pair. An important distinction, however, is that both transistors should be biased to roughly the *same* current to make the device f_Ts approximately equal. There are numerous methods for satisfying this bias requirement, but one particularly simple and convenient arrangement is the CMOS version of a bipolar circuit developed by Carl

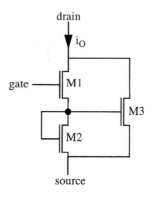

FIGURE 9.14. Battjes f_T doubler.

Battjes of Tektronix;[4] see Figure 9.14. By using mirror M_2–M_3, transistors M_1 and M_3 are guaranteed to operate with substantially equal drain currents. Because the capacitances of M_2 and M_3 are in parallel, though, this circuit does not quite provide a doubling of f_T. The actual increase is about a factor of 1.5.

With f_T doubling circuits it is often possible to obtain 50% increases in bandwidth, although the exact improvement depends on numerous and variable factors. Chief among these are how dependent on f_T the circuit's bandwidth happens to be and how much source–bulk parasitic capacitance there is. Clearly, if the bandwidth is limited by something else (e.g., an external load capacitor interacting with a load resistance), increased f_T will provide little improvement. Nevertheless, f_T doublers are valuable for pushing bandwidth beyond what one would normally believe are the limits of a given technology.

9.5 TUNED AMPLIFIERS

9.5.1 INTRODUCTION

We've already seen that the design of broadband amplifiers can be guided by the method of open-circuit time constants, with a possible assist from bandwidth extension tricks such as shunt peaking. However, it is not always necessary (or even desirable) to provide gain over a large frequency range. Often, all that is needed is gain over a narrow frequency range centered about some high frequency.

Such tuned amplifiers are used extensively in communications circuits to provide selective amplification of wanted signals and a degree of filtering of unwanted signals. As we'll see shortly, eliminating the requirement for broadband operation allows one to obtain substantial gain at relatively high frequencies. That is, to zeroth order, the effort required to get a gain of 100 over a bandwidth of 1 MHz is roughly

---

[4] "Monolithic Wideband Amplifier," U.S. Patent #4,236,119, granted 25 November 1980.

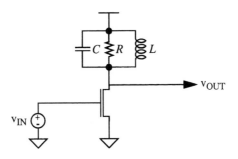

FIGURE 9.15. Amplifier with single tuned load.

independent of the center frequency about which that bandwidth is obtained; the difficulty in obtaining a specified gain–bandwidth product is approximately constant and independent of center frequency (within certain limits). Furthermore, the power required to obtain this gain can be considerably less for a narrowband implementation. This last consideration is particularly important when designing portable equipment, where battery life is a major concern.

9.5.2 COMMON-SOURCE AMPLIFIER WITH SINGLE TUNED LOAD

To understand why the gain–bandwidth product should be roughly independent of center frequency, consider the amplifier shown in Figure 9.15 (biasing details have been omitted). If we drive from a zero-impedance source (as shown) and if we can neglect series gate resistance, then the drain–gate capacitance C_{gd} may be absorbed into the capacitance C. In that case, we can model the circuit as an ideal transconductor driving a parallel RLC tank. At low frequencies the inductor is a short and the incremental gain is zero, whereas at high frequencies the gain goes to zero because the capacitor acts as a short. At the resonant frequency of the tank, the gain becomes simply $g_m R$ since the inductor and capacitor cancel.

For this circuit the total -3-dB bandwidth is, as usual, simply $1/RC$. Hence, the product of gain (measured at resonance) and bandwidth is just

$$G \cdot \text{BW} = g_m R \cdot \frac{1}{RC} = \frac{g_m}{C}. \tag{38}$$

For this example, with all of its simplifying assumptions, we obtain a gain–bandwidth product that is *independent* of center frequency, as advertised.

To underscore the profound implications of this last statement, consider two alternative methods for obtaining a gain of 1000 at 10.7 MHz (e.g., for the IF section of an FM radio). We could attempt a broadband amplifier design, which would require us to achieve a gain–bandwidth product of over 10 GHz (not a trivially accomplished goal). Or we could recognize that, for the FM radio example, we need only obtain this gain over a 200-kHz bandwidth,[5] in which case we only have to achieve something like a 200-MHz gain–bandwidth product, a considerably easier task.

The fundamental difference between these two approaches is, of course, due to the cancellation of the load capacitance by the inductor in the tuned amplifier. As long as we have direct access to the terminals of any parasitic capacitance (and can make them appear across the tank), we can resonate out this capacitance with an appropriate choice of inductance and obtain a constant gain–bandwidth product at any arbitrary center frequency.

Naturally, *real* circuits don't work quite as neatly; we suspect that we probably won't be able to get gain at 100 THz from Jell-O™ transistors, for example, no matter how good our inductor is. But it remains true that, as long as we seek center frequencies that are reasonable,[6] tuned loads allow us to obtain roughly constant gain–bandwidth product.

9.5.3 DETAILED ANALYSIS OF THE TUNED AMPLIFIER

The analysis just performed invokes many simplifying assumptions. In particular, the choice of a zero source resistance and zero gate resistance allowed us to absorb the drain–gate capacitance into the tank network, permitting the inductance to offset its effects. Since C_{gd} might have a more serious effect if it were no longer possible to absorb it directly into the tank, let's consider more realistic models for the circuit and examine what happens.

Specifically, let's now allow for nonzero source resistance and nonzero series gate resistance, as shown in Figure 9.16. The corresponding incremental model is depicted in Figure 9.17.

Using this model, we can compute two important impedances (actually admittances, to be precise). First, we'll find the equivalent admittance seen to the left of the *RLC* tank; then we'll find the admittance seen to the right of the source resistance R_S.

In carrying out this analysis, it is better to apply a test *voltage* source across the tank to find the equivalent admittance seen to its left. Remember, you'll get the same answer whether you use a test voltage or a test current (assuming you make no errors, or at least the same errors), but a test voltage is more convenient here because it most directly fixes the value of v_{gs}, the voltage that determines the value of the controlled source.

The precise details are somewhat messy and essentially unrewarding, but the end result is that the admittance seen by the tank consists of an equivalent resistance (which we'll ignore for now) in parallel with an equivalent capacitance. This capacitance is given by

---

[5] This value applies to commercial broadcast FM radio; your mileage may vary.
[6] We'll quantify this better a little later, but for now pretend that "reasonable" means "reasonably well below ω_T."

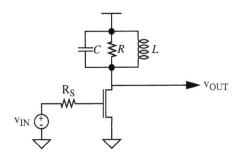

FIGURE 9.16. Amplifier with single tuned load.

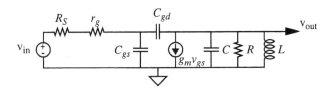

FIGURE 9.17. Incremental model for circuit.

$$C_{eq} = C_{gd}[1 + g_m R_{eq}] = C_{gd}[1 + g_m(R_S + r_g)]. \tag{39}$$

Notice that C_{eq} can be fairly large. This is actually an alternative manifestation of the Miller effect, now viewed from the output port. Some fraction of the voltage applied to the drain appears across v_{gs}, where it excites the g_m generator. The resulting current adds to that through the capacitors and must be supplied by the test source, so the source sees a lower impedance. One component of that current is due to a simple capacitive voltage divider and is thus in phase with the applied voltage. It therefore represents a resistive load on the tank, causing a gain reduction. Another component of the current leads the applied voltage and therefore represents an additional capacitive load on the tank.

The additional capacitive loading by C_{eq} shifts downward the resonant frequency of the output tank. Although this shift can be compensated by a suitable adjustment of the inductance, it is generally inadvisable to operate in a regime where the resonant frequency depends critically on poorly controlled, poorly characterized, and potentially unstable transistor parasitics. It is therefore desirable to select C relatively large compared with the expected variation in parameters, so that the total tank capacitance remains fairly independent of process and operating point. The unfortunate tradeoff is a reduction in the gain–bandwidth product for a given transconductance.

A more serious effect of C_{gd} becomes apparent when we consider the input impedance (or, more directly, the input admittance). Since the intermediate details are again of little use outside of deriving the one bit of trivia we're about to state, we'll simply present the result:

$$y_{\text{in}} = \frac{y_L y_F}{y_L + y_F} + \frac{g_m y_F}{y_L + y_F}, \tag{40}$$

where y_{in} is the admittance seen to the right of C_{gs}, y_F is the admittance of C_{gd}, and y_L is the admittance of the RLC tank.[7]

If, as is often the case, the magnitude of the feedback admittance y_F is small compared to that of y_L, then we may write

$$y_{\text{in}} \approx y_F + \frac{g_m(j\omega C_{gd})}{y_L}. \tag{41}$$

The significance of this result becomes apparent when you observe that y_L has a net negative imaginary part at frequencies where the tank looks inductive (i.e., below resonance), so that the second term on the right-hand side of the equation (and therefore y_{in}) can have a *negative real* part; that is, the input of the circuit can act as if a negative resistor were connected to it. Having negative resistances around can encourage oscillation (which is just fine if this is your intent, but more typically is not). We certainly have all of the necessary ingredients: inductance, capacitance, and negative resistance. If there were no C_{gd}, there would be no such problem.

The difficulty with C_{gd}, then, is that it couples the input and output circuits in potentially deleterious ways. It loads the output tank and decreases gain, detunes the output tank, and can cause instability. This latter problem is particularly severe if one attempts to add a tuned circuit to the input as well. Furthermore, even before true instability sets in, the interaction of tuned circuits at both ports may make it extremely challenging to achieve proper tuning.

Unfortunately, C_{gd} will always be nonzero (in fact, it is typically about 30–50% of the main gate capacitance, so it is hardly negligible). To mitigate its various undesirable effects therefore requires the use of some topological tricks.

9.6 NEUTRALIZATION AND UNILATERALIZATION

One strategy derives naturally from recognizing that the problem stems from coupling the input and output ports. Removing the coupling should therefore be of benefit. This decoupling of output from input should feel familiar – it is precisely what eliminates the Miller effect from common-source amplifiers, and what works there works here as well. See Figure 9.18. By providing isolation between input and output ports with the common-gate stage, we eliminate (or at least greatly suppress) detuning and the potential for instability, thus allowing the attainment of larger gain–bandwidth products.

Another topology that achieves these objectives is the source-coupled amplifier (which may be viewed as a source follower driving a common-gate stage), as shown

---

[7] To avoid obscuring the argument, we neglect explicit mention of the transistor's output admittance in this development. However, it may be considered part of y_L, so the treatment presented is more general than it might appear at first glance.

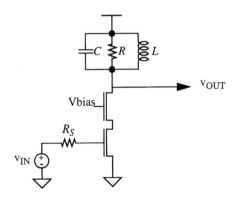

FIGURE 9.18. Cascode amplifier
with single tuned load.

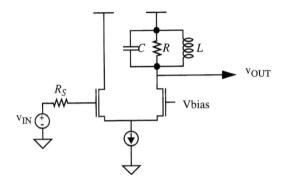

FIGURE 9.19. Source-coupled amplifier
with single tuned load.

in Figure 9.19. Once again, this structure isolates the output from the input and therefore does not suffer as seriously from the instability and detuning problems of the simple common-source stage.

Both the cascode and source-coupled amplifier behave similarly with regard to isolation. The cascode provides roughly twice the gain for a given total current (because all of this current can be used to set g_m), whereas the source-coupled amplifier requires less total supply voltage (since the two transistors aren't stacked as in the cascode). The choice of which topology to use is usually based on such considerations of headroom and gain.

The circuits of Figures 9.18 and 9.19 are examples of nearly "unilateral" amplifiers, that is, ones in which signals can flow only one way over large bandwidths. You can well appreciate the value of unilateralization; aside from conferring the circuit benefits we've already discussed, it makes analysis and design much easier by reducing or eliminating unintended and undesired feedback.

If we cannot (or choose not to) eliminate undesired feedback, another approach is to cancel it to the maximum possible extent. Since this cancellation is rarely perfect

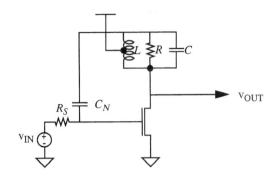

FIGURE 9.20. Neutralized common-source amplifier.

over large bandwidths, this approach is generally called "neutralization"[8] to distinguish it from more broadband unilateralization techniques that do not depend on cancellations.

The classic neutralized amplifier is shown in Figure 9.20. Notice that the inductor has been replaced by something slightly more complex: a tapped inductor, or *autotransformer*. By symmetry, the voltages at the top and bottom of the inductor are exactly 180° out of phase in the connection shown.[9] Therefore, the drain voltage and the voltage at the top of neutralizing capacitor C_N are 180° out of phase. Now, if the undesired coupling from drain to gate is due only to C_{gd} then, by symmetry, selection of C_N equal to C_{gd} guarantees that there is no net feedback from drain to gate! The current through the neutralizing capacitor is equal in magnitude and opposite in sign to that through C_{gd}; we have removed the coupling from output to input by adding more coupling from output to input (it's just out of phase so that the *net* coupling is zero).

Neutralization was originally implemented with tapped transformers, but the poor quality of (and large area consumed by) on-chip transformers makes this particular method unattractive for IC implementation. Observe, however, that the tapped transformer is used simply to obtain a signal inversion. Since inversions are easily obtained other ways, practical neutralized IC amplifiers are still realizable. One topology uses a differential pair to obviate the need for a transformer, as seen in Figure 9.21.

Because perfect neutralization with these techniques depends on feeding back a current that is *precisely* the same as that through C_{gd}, the neutralizing capacitor C_N

---

[8] Neutralization was developed for AM broadcast radios in the 1920s by Harold Wheeler while working for Louis Hazeltine. His invention allowed the attainment of large, stable gains from tuned RF amplifiers, and it thus reduced the number of gain stages (and hence the number of vacuum tubes) required in a typical radio, permitting significant cost reductions over many rival approaches.

[9] Note that autotransformers are not strictly necessary here. They are merely a historically common and convenient means of obtaining two voltages that are precise inverses of each other. Clearly, other ways to provide a signal and its inverse exist (consider the example of Figure 9.21).

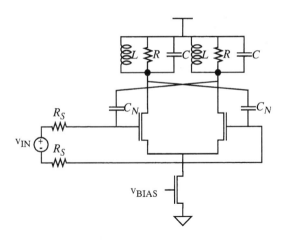

FIGURE 9.21. Neutralized common-source amplifier
(more practical for ICs).

must match C_{gd} precisely. Unfortunately, C_{gd} is somewhat voltage-dependent. Perhaps because of the difficulty of providing precise cancellation in the face of this variability, neutralization has found limited application in semiconductor amplifiers. Vacuum tubes, with their highly linear and relatively constant coupling capacitances, were much better candidates for use of this technique. Nevertheless, with sufficient diligence, it is possible to obtain usefully large gain–bandwidth improvements in semiconductor-based amplifiers by using neutralization.

9.7 CASCADED AMPLIFIERS

So far in our study of high-frequency amplifiers, we've looked at open-circuit time constants, shunt peaking, peaking with zeros, tuned amplifiers, unilateralization, and neutralization – but all mainly in the context of single-stage circuits. However, it is frequently the case that we can't get enough gain out of one stage. The question of how many stages one should use then naturally arises. Furthermore, if each stage has a certain bandwidth, what bandwidth will the overall amplifier have? Finally, is there some optimum number of stages one should use to maximize the overall bandwidth at a given gain in a given technology? To answer these questions, we now consider the properties of cascaded amplifiers.

9.7.1 BANDWIDTH SHRINKAGE

Let's suppose that each amplifier stage has a unit DC gain (to simplify the math marginally) and a single pole. The amplifier's transfer function is then

$$H(s) = \frac{1}{\tau s + 1}. \tag{42}$$

A cascade of n such amplifiers will therefore have an overall transfer function of

$$A(s) = \left(\frac{1}{\tau s + 1}\right)^n. \tag{43}$$

We find the bandwidth in the standard way by computing the magnitude of the transfer function and solving for the -3-dB rolloff frequency:

$$|A(j\omega)| = \left|\left(\frac{1}{j\omega\tau + 1}\right)\right|^n = \frac{1}{\sqrt{2}}, \tag{44}$$

so that

$$\left(\frac{1}{\sqrt{(\omega\tau)^2 + 1}}\right)^n = \frac{1}{\sqrt{2}}. \tag{45}$$

Clearing radicals yields

$$[(\omega\tau)^2 + 1]^n = 2, \tag{46}$$

and solving for the bandwidth at last gives us

$$\omega = \frac{1}{\tau}\sqrt{2^{1/n} - 1}. \tag{47}$$

That is, the bandwidth of the overall amplifier is the bandwidth of each stage, multiplied by some funny factor. As n approaches infinity, the overall bandwidth tends toward zero.

The precise form of the bandwidth shrinkage is perhaps a little hard to see from this formula. For large n, though, we can simplify the term under the radical sign to make the relationship substantially clearer. Mathematicians would suggest using a series expansion of $2^{1/n}$ and then using only the first couple of terms. An equivalent (albeit roundabout) method is to exploit the somewhat better-known expansion for e^x ($\equiv \exp\{x\}$).

We begin by recognizing that

$$2^{1/n} = \exp\{\ln(2^{1/n})\} = \exp\left\{\frac{1}{n}\ln 2\right\}. \tag{48}$$

Then, for large n, we can write

$$\exp\left\{\frac{1}{n}\ln 2\right\} \approx 1 + \frac{1}{n}\ln 2. \tag{49}$$

We thus derive the interesting result that the bandwidth behaves approximately as follows:

$$\omega = \frac{1}{\tau}\sqrt{2^{1/n} - 1} \approx \frac{1}{\tau}\sqrt{\frac{1}{n}\ln 2} \approx \frac{0.833}{\tau\sqrt{n}}. \tag{50}$$

That is, the bandwidth shrinks as the inverse square root of the number of stages, at least in the limit of large n.

Before going further, we might want to get a better feel for how much of an approximation is involved here, especially since we're going to use this result later on.

Table 9.2. *Bandwidth versus n*

n	Actual BW (normalized)	Approximate BW (normalized)	−Error (%)
1	1	0.833	16.7
2	0.643	0.589	9.4
3	0.510	0.481	5.7
4	0.435	0.416	4.4
5	0.386	0.372	3.6
6	0.350	0.340	2.9
7	0.323	0.315	2.5
8	0.301	0.294	2.3

Clearly, the formula is off by about 17% for $n = 1$. We hope that the error decreases rapidly, and it does, as Table 9.2 shows. Observe that the error drops below 5% pretty quickly, so the approximate bandwidth shrinkage formula is reasonably accurate. Note also that the approximation underestimates the true bandwidth by a bit.

We can also deduce from the equation and the table that the method of open-circuit time constants does a pretty rotten job of estimating bandwidth for this cascade of identical amplifiers. In this case, we have n identical poles, so that we would predict from the OCτ method that bandwidth goes directly as $1/n$ when, in fact, it goes as the reciprocal square root.

Now that we've derived this result, we can use it to determine the gain per stage that maximizes overall system bandwidth.

9.7.2 OPTIMUM GAIN PER STAGE

With the bandwidth shrinkage formula, we're in a position to identify the optimum strategy to maximize bandwidth in a cascaded amplifier, given a stated gain requirement and technology constraints.

Again, we'll assume that all the stages are identical (because if one were slower than any other, it would represent the bandwidth bottleneck for the whole amplifier), each with a single pole *whose frequency depends inversely on the stage gain.* That is, each stage has a constant gain–bandwidth product, so that stage gain and bandwidth trade off linearly. Our goal is to find the number of stages that, for a given overall gain requirement, maximizes the bandwidth (and hence the overall gain–bandwidth product).

Assume that the overall gain is to be G, so that each amplifier stage must have a gain of $G^{1/n}$. If each stage has the same gain–bandwidth product ω_T, then the single-stage bandwidth will be

$$\mathrm{BW_{ss}} = \frac{\omega_T}{G^{1/n}}. \tag{51}$$

From the approximate bandwidth shrinkage formula, we may write the following expression for the bandwidth of the total amplifier:

$$\mathrm{BW_{tot}} \approx \frac{\omega_T}{G^{1/n}} \cdot \frac{\sqrt{\ln 2}}{\sqrt{n}}. \tag{52}$$

The reciprocal of the bandwidth (somewhat handier for what we'll do shortly) is thus

$$\frac{1}{\mathrm{BW_{tot}}} \approx \left(\frac{1}{\omega_T \sqrt{\ln 2}} \cdot \sqrt{n} \right) G^{1/n}. \tag{53}$$

We'll now maximize the total bandwidth by minimizing its reciprocal:

$$\frac{d}{dn} \left(\sqrt{n} G^{1/n} \right) = 0. \tag{54}$$

Taking the derivative, cancelling terms, and solving yields

$$\ln(G^{1/n}) = \tfrac{1}{2} \implies G^{1/n} = e^{1/2}. \tag{55}$$

According to this analysis, the gain per stage should therefore be chosen as the square root of e if we want to maximize the overall bandwidth.[10]

The number of stages corresponding to this optimum is

$$n = 2 \ln G, \tag{56}$$

and the overall bandwidth corresponding to this condition is

$$\mathrm{BW_{tot}} = \omega_T \cdot \sqrt{\frac{\ln 2}{2e \cdot \ln G}} \approx \frac{0.357 \omega_T}{\sqrt{\ln G}}. \tag{57}$$

From this last expression, we can see that the overall bandwidth is relatively insensitive to the value of overall gain when this optimum is chosen. In fact, the product of bandwidth and the *square root of the log* of gain is constant. Perhaps this insensitivity is again best illustrated in tabular form; see Table 9.3.

Ignoring the minor practical detail of noninteger values of n, we can see that the bandwidth changes by less than a factor of 2 even though the gain changes by a factor of 100. Clearly, even though *each* amplifier stage has a constant gain–bandwidth product, the overall amplifier does not. In fact, the gain–bandwidth product actually grows without bound as n increases, because a cascade of this type trades off bandwidth for the square root of the log of gain.

It shouldn't take much reflection to conclude that a constant gain–bandwidth product is really only a property of single-pole systems. That many commonly encountered systems are dominated by a small number of poles (say, one) has led to the

---

[10] A similar derivation, but using open-circuit time constants rather than the bandwidth shrinkage formula, leads one to predict an optimum gain per stage of e instead of its square root. The difference is not significant because the optimum conditions are relatively "flat," that is, the overall bandwidth is not overly sensitive to the precise value of gain per stage. Specifically, it is easy to show that, if one uses a gain of e per stage, the bandwidth for large n degrades by a factor of only $\sqrt{2/e}$, or about 0.86.

Table 9.3. *Maximum BW and*
G · BW versus G

G	n	Maximum BW (normalized)	G · BW
10	4.6	0.24	2.4
20	6.0	0.21	4.2
50	7.8	0.18	9.0
100	9.2	0.17	17
200	10.6	0.16	32
500	12.4	0.14	70
1000	13.8	0.14	140

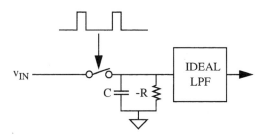

FIGURE 9.22. Superregenerative amplifier.

widely held misconception that gain and bandwidth *must* trade off linearly simply because they often do. However, we've seen that this relationship breaks down dramatically when the order of the system grows to large values. This observation is quite useful, for it suggests that one may purposefully construct systems of high order specifically to decouple gain from bandwidth. We will exploit this observation to construct amplifiers that trade bandwidth for only weak functions of gain.

9.7.3 THE SUPERREGENERATIVE AMPLIFIER

Back in the 1920s, Edwin H. Armstrong developed the *superregenerative* amplifier for radio. It employed enough positive feedback to drive an amplifier into a special intermittent oscillatory condition and, in so doing, enabled the attainment of spectacular amounts of gain from a single stage. It was the first circuit to violate, in a major way, the linear gain–bandwidth tradeoff "law." Although his implementation was actually a bandpass amplifier, we will analyze the corresponding low-pass version; its gain–bandwidth characteristics are the same as its bandpass progenitor.

A greatly simplified superregenerative amplifier is shown in Figure 9.22. Notice that there is a *negative* resistor in this system. We can always synthesize the equivalent of a negative resistance by using active devices, so invoking its existence here is perfectly realistic.

As a consequence of the negative resistance, the RC time constant has a negative value and the pole is in the right-half s-plane. We therefore obtain an exponential *growth* in the capacitor voltage whenever the sampling switch is open. The longer we wait before closing the switch again, the greater the gain, by an exponentially growing factor.

A quantitative analysis of this amplifier is straightforward. Assume that the switch closes for an infinitesimally short time and that the input source is capable of instantly charging up the capacitor. It is not necessary even to approach these conditions in reality, but accepting these assumptions simplifies the analysis without introducing any fundamental errors.

When the switch opens, the capacitor voltage ramps up exponentially from the initial voltage (which is v_{in}):

$$v_{out}(t) = v_{in}e^{t/RC}. \tag{58}$$

This exponential growth is allowed to continue for a period T, and the exponentially growing signal is averaged by an ideal low-pass filter. Hence,

$$\overline{v_{out}(t)} = \frac{1}{T}\int_0^T v_{in}e^{t/RC}\, dt = \frac{RC}{T}(e^{T/RC} - 1)v_{in}. \tag{59}$$

If the time constant RC is short compared with the sampling period T, we get a gain factor that is exponentially related to that ratio.

To involve bandwidth explicitly, note that – since we have a sampled system – we must satisfy the Nyquist sampling criterion. Hence, we must choose $1/T$ higher than twice the highest frequency component of the input signal. That is, we must have

$$BW < \frac{1}{2T}. \tag{60}$$

Therefore, the product of bandwidth and the log of the gain for this type of amplifier is

$$BW \cdot \ln G = \frac{1}{2RC} + \frac{1}{2T}\ln\left(\frac{RC}{T}\right). \tag{61}$$

Thus, we see that the superregenerative amplifier trades off bandwidth for the log of gain, to a reasonable approximation. As in the case of a cascade of amplifiers, the implication is that the bandwidth changes little as one varies the gain over a large range. In fact, if one is willing to endure a sufficiently long regeneration interval, the gain can be made enormously large and so allow the overall gain–bandwidth product to exceed that of the active device(s) involved.

Another difference between the superregenerative amplifier and a cascade of conventional single-pole amplifiers is that it accomplishes this gain–bandwidth tradeoff with just one RC. The periodically time-varying nature of this amplifier endows it with some of the properties of a high-order system with only one energy-storage element.

With a bandpass version of this amplifier, Armstrong was able to obtain so much gain from a single vacuum tube that he could amplify fundamental noise sources (which we'll study very soon) to audible levels. He and RCA quite reasonably assumed that this remarkable property would be extremely useful in radios.

The superregenerative amplifier is not used very much these days, however, for a variety of practical reasons. Chief among them is that it is an oscillator, and RF versions of the superregenerator are actually parasitic transmitters that can cause interference. Additionally, even in the absence of a signal, the superregenerator amplifies noise to audible and annoying levels. These characteristics have limited superregenerative circuits to relatively low-tech applications such as children's walkie-talkies (the most inexpensive ones – they're easy to spot because they have a characteristic, annoying hiss even when no signal is being received). Nevertheless, it is a fascinating amplification principle since it allows one to obtain, in a single stage, an overall gain–bandwidth product that can greatly exceed the ω_T of the device involved.

9.7.4 A CONUNDRUM

We've now studied several types of amplifiers (and there are others) that trade off bandwidth for only weak functions of gain. It should be clear from these examples that the notion of a fixed gain–bandwidth product is seriously flawed. But most engineers recognize that there's no free lunch and start looking for other things that might be trading off with gain. It turns out that the most important of these parameters is *delay.* It is sometimes (but not always) the case that there is a stronger tradeoff between gain and delay than there is between gain and bandwidth. In many applications (such as in TV or optical fiber systems) communication is a one-way affair, so delay is frequently more tolerable than limited bandwidth. It's nice to know that there are practical cases where we can improve a parameter that we do care about by degrading a parameter that we don't.

Having accepted that the coupling between gain and bandwidth is evidently weak at best (as we saw in Chapter 6), we might reasonably ask what gain or bandwidth could be obtained if we were willing to tolerate an arbitrary delay in the response. The surprising answer (in principle, anyway), is that we could obtain *arbitrarily large* gains at a fixed bandwidth if we didn't care about delay.

We've already seen examples of gain–delay tradeoffs. The T-coil compensator, as well as cascaded and superregenerative amplifiers, exhibit larger gains or improved bandwidth at the expense of longer delays. In the case of the cascaded amplifier, adding stages increases gain but also increases delay (although at a slower rate than the gain increase). Similarly, lengthening the regeneration interval in the superregenerative amplifier increases the gain (exponentially) but also increases (obviously) the delay.

To understand how one might synthesize circuits that exhibit a more direct bandwidth–delay tradeoff than we've identified thus far, recall that the reciprocal of risetime is a measure of bandwidth. Now imagine an amplifier that saves all of the energy in an input step for some long period of time without producing any output, then dumps it all at once to the output, yielding a very fast risetime (= high bandwidth). Such an amplifier would indeed trade delay for bandwidth directly. From the foregoing description, we see that such an amplifier must have the ability to provide large delays (to allow this tradeoff in the first place) over a large bandwidth. This requirement in turn suggests that networks of relatively high order are needed, since a pole (or zero) approximates a time delay only over a limited frequency interval. To proceed thus requires a more detailed understanding of networks of high order. As we'll see in the next section, transmission lines and their lumped approximations are particularly valuable in this context, allowing one to construct amplifiers with bandwidths approaching ω_T.

9.7.5 THE DISTRIBUTED AMPLIFIER

Without question, the most elegant exploitation of distributed concepts is the distributed amplifier invented by W. S. Percival of the United Kingdom in 1936. He apparently didn't talk about it very much, though, and widespread awareness of this scheme had to await the publication in 1948 of a landmark paper by Ginzton, Hewlett, Jasberg, and Noe.[11]

In the abstract to their paper, the authors note that "the ordinary concept of 'maximum bandwidth–gain product' does not apply to this distributed amplifier." Let's see how this structure achieves a gain-for-delay tradeoff without affecting bandwidth. As can be seen in Figure 9.23, inputs to the transistors are supplied by a tapped delay line, and the outputs of the transistors are fed into another tapped delay line. Although simple sections are shown, the best performance is obtained when m-derived or T-coil sections are employed, as discussed earlier.

A voltage step applied to the input propagates down the input line, causing a step to appear at each transistor in succession. Each transistor generates a current equal to its g_m multiplied by the value of the input step, and the currents of all the transistors ultimately sum in time coherence if the delays of the input and output lines are matched.

Since each tap on the output line presents an impedance of $Z_0/2$, the overall gain is

$$A_V = \frac{n g_m Z_0}{2}. \tag{62}$$

In contrast with ordinary amplifier cascades, this amplifier has an overall gain that depends linearly on the number of stages and therefore can operate at frequencies where each stage actually has a gain smaller than unity. Consequently, the distributed amplifier can operate at substantially higher frequencies than can conventional amplifiers. Furthermore, since the delay is also proportional to the number of stages,

---

[11] E. L. Ginzton, W. R. Hewlett, J. H. Jasberg, and J. D. Noe, "Distributed Amplification," *Proc. IRE,* August 1948, pp. 956–69.

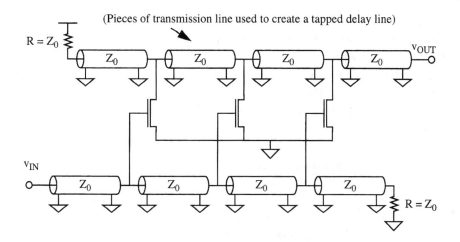

FIGURE 9.23. Distributed amplifier.

this amplifier does trade gain for delay; bandwidth does not factor into the tradeoff in any direct, obvious way.

Another way to look at this amplifier is to recognize that one source of bandwidth limitation in conventional amplifiers is the drop in input impedance with increasing frequency that accompanies input capacitance. Here, however, we absorb the device's input capacitance into the constants of the tapped delay line.[12] Hence, until the cutoff frequency of the line itself is approached, the input impedance remains constant and equal to Z_0.

Similarly, the output capacitance of the devices can be absorbed into the output line. Since input capacitances are usually larger than output capacitances, some significant adjustment of line constants is necessary to guarantee matched delays.

Lest you think that to achieve a practically useful level of balance is possible only in theory, you should know that *vacuum tube* distributed amplifiers were successfully used in many Tektronix® oscilloscopes for many years (their model 513 was the first to use this type of amplifier). The amplifiers were used in the final vertical deflection stage and typically involved six or seven "matched" pairs of vacuum tubes. Bandwidths of roughly $\omega_T/2$ were routinely achieved, so that 100-MHz general-purpose oscilloscopes were available by around 1960.[13]

Given these attributes, one may reasonably ask why this type of amplifier is not ubiquitous today. Part of the reason is that it is rather power hungry, since many stages are required to provide a given gain. Another is that the active devices that

---

[12] We are assuming that the input impedance of the device looks capacitive at high frequencies. However, this assumption is not always satisfied, and the departure from this assumption must be taken into account in practical designs if good results are to be achieved.

[13] The distributed amplifiers in the Tektronix 585A 100-MHz oscilloscope used 6DJ8 duo-triodes, which have f_Ts of roughly 300 MHz. The delay lines were composed of T-coils, which provide better bandwidth than ordinary m-derived lumped approximations.

supplanted vacuum tubes, bipolar transistors, have several characteristics that make them unsuitable for use in distributed amplifiers. The biggest offender is the parasitic base resistance, r_b, which spoils line Q and therefore degrades the line. Bipolar distributed amplifiers consequently acquired an unsavory reputation. Finally, the lumped lines could not be integrated until very recently, when devices improved enough so that frequencies of operation increased to a range where fully integrated lines become practical. Distributed amplifiers all but disappeared as a consequence.

They finally made their reappearance in about 1980, when workers in GaAs technology rediscovered the principle. Since that time, distributed amplifiers have been constructed in a variety of compound semiconductor technologies, with InP versions achieving 100-GHz bandwidths. Few fully integrated CMOS implementations have been reported in the literature. A CMOS distributed amplifier with a bandwidth of approximately 25 GHz using 0.18-μm technology demonstrates that no theoretical reason explains this scarcity.[14] This bandwidth is approximately half of the transistors' f_T.

9.8 AM–PM CONVERSION

Bandwidth and gain are but two of many considerations in RF amplifier design. Another preoccupation (particularly in systems that use complex digital modulations) is the mitigation of *AM-to-PM* (amplitude modulation to phase modulation) *conversion.* It is possible (and altogether too common) for the phase shift of any signal processing block to depend on the amplitude of the input. One common mechanism that can produce an amplitude-dependent phase shift is the change in bias point that generally accompanies nonlinear operation. For example, consider an amplifier with weak even-order nonlinearity. At large amplitudes, the even-order terms can result in the production of DC. This signal-dependent DC in turn adds to the bias already present. Shifts in bias can easily alter bandwidths and slew rates and thereby change the phase shift.

One important reason that we care about AM–PM conversion is that many modulations now in use exploit both amplitude and phase domains to maximize data rate within a given bandwidth. Demodulators assume orthogonality between these two domains, but AM–PM conversion destroys the orthogonality. If sufficiently severe, AM–PM conversion could result in serious increases in bit-error rate and could even render a communications system useless.

The crude picture shown as Figure 9.24 conveys the essentials of this problem. Needless to say, if the demodulator expects symbols arrayed as in the picture on the left – but the symbols are actually as in the right-hand constellation – errors might occur.

Phase errors can be important in other systems as well. For example, both PAL and NTSC color television systems encode hue as a phase relative to that of a reference

[14] B. Kleveland et al., "Monolithic CMOS Distributed Amplifier and Oscillator," *ISSCC Digest of Technical Papers,* February 1999.

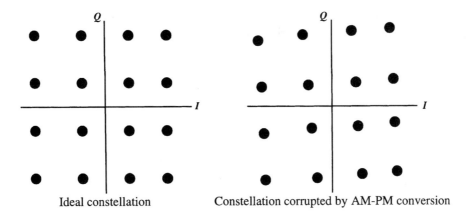

Ideal constellation   Constellation corrupted by AM-PM conversion

FIGURE 9.24. Effect of AM–PM conversion on symbol constellation.

carrier while encoding the degree of color saturation as an amplitude (see Chapter 16). If there is AM–PM conversion, then there can be a visible change in hue as the color deepens. To avoid objectionable visible artifacts, all elements in the signal processing chain must satisfy strict specifications on the allowable phase shifts (*differential phase,* as it's called there) arising from this mechanism.

9.9 SUMMARY

We have seen that purposeful violations of the conditions assumed in the development of open-circuit time constants can lead to significant bandwidth improvements. Only simple networks are used, and the techniques do not require an increase in power. Both shunt peaking and peaking through zeros provide these improvements, albeit at some cost in pulse response fidelity. A tradeoff exists between the amount of bandwidth increase and pulse distortion, but the large improvements obtained for trivial effort make these methods worth considering.

We've seen that use of tuned loads allows the attainment of essentially the same gain bandwidth at high frequencies as at low frequencies by means of exploiting the resonant cancellation (by inductances) of parasitic and explicit capacitances. We've also seen that coupling from output to input can severely limit the practically attainable gain–bandwidth products by loading and detuning the output tank, destabilizing the amplifier, and modifying port impedances that complicate the cascading of stages.

Detuning and destabilization can be suppressed greatly through the use of unilateral topologies (such as the cascode or source-coupled amplifier) that provide isolation between output and input ports over a wide frequency range, or through the use of neutralization to cancel the undesired feedback over some frequency and operating point range.

The notion of a fixed gain–bandwidth product was shown to be false, as demonstrated for example by superregenerative amplifiers and by cascades of ordinary amplifiers, which all exhibit more of a gain–delay than a gain–bandwidth tradeoff. In this point of view, then, we see that series peaking, shunt peaking, and T-coil compensations, which effectively distribute the load, may be thought of as a sequence of ever-better approximations on the way to the distributed amplifier, which distributes the active device as well as the output load.

PROBLEM SET FOR HIGH-FREQUENCY AMPLIFIER DESIGN

PROBLEM 1   Plot gain and phase for the shunt-peaked network shown in Figure 9.25. Let m take on values from 1 to 5 in steps of 0.2.

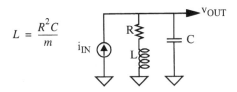

$$L = \frac{R^2 C}{m}$$

FIGURE 9.25. Shunt-peaked *RLC* network.

PROBLEM 2   Derive the equations for series peaking (Figure 9.26). Show formally what conditions maximize bandwidth and yield maximally flat response and time delay. For these three conditions, provide expressions for the bandwidth achieved. Verify your answers with simulations.

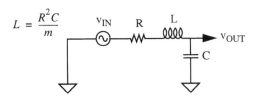

$$L = \frac{R^2 C}{m}$$

FIGURE 9.26. Series-peaked *RLC* network.

PROBLEM 3   Unfortunately, the relatively large parasitic capacitances of typical MOSFETs can prevent f_T doublers from working as well as one would like. Explore this idea further by simulating both the Darlington and Battjes doublers, using any model set (e.g., the level-3 SPICE models provided in Chapter 5). Terminate each in an incremental short circuit (provided by a 2-V DC source), and measure the incremental current gain as frequency increases. Determine f_T with a first-order extrapolation to the unit gain frequency. Compare with the f_T value for a single device at the same bias current. Which parasitic capacitances cause the disappointment (if there is one)?

PROBLEM 4 Following a development parallel to the one shown in the chapter, derive an expression for the gain–bandwidth behavior of an *RLC bandpass* super-regenerative amplifier.

PROBLEM 5 Provide an explicit expression that shows under what conditions y_{in} may have a negative real part in simple common-source amplifiers with tuned load. If you were to build such an amplifier and discover that y_{in} was negative, what remedies could you apply? Offer two specific solutions.

PROBLEM 6 Design a single-stage tuned amplifier to meet the following small-signal specifications:

> |voltage gain|: >50, measured at the center frequency;
> total bandwidth (−3-dB): >1 MHz;
> center frequency: 75 MHz;
> source resistance: 50 Ω;
> load: 10 pF, purely capacitive.

Use the process characteristics from Chapter 5.

Assume that the (external) inductor's self-resonant frequency is high enough to be neglected. You may also assume that the inductor Q is 200 at the center frequency of 75 MHz.

PROBLEM 7 This problem explores a number of impedance transformation issues that are of interest in high-frequency design.

(a) High-speed followers have a tendency to ring or even oscillate when driving capacitive loads. Your task is to identify the conditions that may cause this problem and to propose a solution.

First consider a source follower as shown in Figure 9.27. Assume that the drain–gate capacitance C_{gd} is zero, as is the parasitic gate resistance r_g. Derive an expression for the incremental input impedance as seen by the source, v_I. What is the real part of your answer?

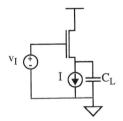

FIGURE 9.27. Capacitively loaded source follower.

(b) Over what range of load capacitance can the real part of the input impedance be negative?

(c) For a load capacitance known to fall within the range of part (b), modify the circuit so that the source v_I always sees an impedance whose real part is positive, keeping in mind that the follower is intended for high-speed operation. Provide formulas to allow computation of the values of any added components.

(d) Now consider a CE or CS circuit with inductive source degeneration. This inductance may arise unintentionally from unavoidable wiring parasitics, or it may be inserted purposefully. In any case, derive an expression for the input impedance of the circuit depicted in Figure 9.28. What is the real part of the input impedance?

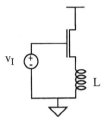

FIGURE 9.28. Inductively loaded source follower.

PROBLEM 8   Consider the zero-peaked amplifier shown in Figure 9.29.

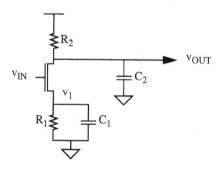

FIGURE 9.29. Zero-peaked amplifier (biasing not shown).

(a) Derive an exact expression for the small-signal *dynamic* transconductance $i_d(s)/v_{in}(s)$. Assume that the (ordinary, static) transconductance of the MOSFET is some value g_m, and that there are *no* transistor parasitics. Assume also that the MOSFET is in saturation.

(b) Choose C_1 so that the zero cancels the output pole. What is the bandwidth of the overall amplifier if $g_m R_1 = 9$? Express your answer in terms of the output pole frequency, $1/R_2 C_2$. Recognize that the output pole frequency is the bandwidth of the unpeaked amplifier.

PROBLEM 9 This problem concerns some practical difficulties in high-frequency amplifier design. Consider a common-gate amplifier (Figure 9.30) in which parasitic inductance is explicitly modeled (assume that the transistor is magically biased into the forward-active region). Since the drain is tied directly to V_{DD}, this is a low-gain circuit, so use an appropriately simplified incremental model (e.g., neglect r_o).

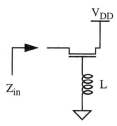

FIGURE 9.30. Common-gate amplifier
with parasitic inductance.

(a) Assuming that one may neglect all *junction* capacitances as well as r_g, derive an expression for the input impedance (as a function of s). Make no other approximations.

(b) Since the incremental model is valid only for frequencies well below ω_T, your answer to part (a) applies only to that restricted frequency range. Simplify your answer to part (a) accordingly, and derive expressions for the real and imaginary parts of the impedance when $s = j\omega$. *For the rest of this problem,* assume that the simplified equations apply.

(c) It should be clear that you can always model any impedance as the series connection of a resistor and a reactive element. For this equivalent input impedance of this particular circuit, what is that reactive element, and what is the expression for its value?

(d) A typical value for the inductance of a straight piece of wire is about 1 nH/mm. As a consequence, it is difficult to construct any real circuits with parasitic inductances much smaller than several nanohenries. Suppose we happen to have 10 nH of total parasitic inductance between the base terminal and ground, and suppose further that the value of C_{gs} is 10 pF (yes, this implies a gargantuan device). Above what frequency f_{crit} would there *potentially* be a stability problem at the input? You need to preserve only two significant digits in your answer. Explain *why* there could be a stability problem above f_{crit}.

(e) Suppose now that we drive the circuit with a small-signal sinusoidal voltage source whose Thévenin resistance is 50 Ω (purely real) at a frequency of $3f_{\text{crit}}$. Suppose further that $1/g_m$ happens to be 5 Ω at our particular bias point. What is the ratio of the voltage at the source to that of the input voltage?

PROBLEM 10 Design a fully integrated amplifier with the following *small-signal* specs:

|voltage gain|: >10, measured at "moderately low" frequency;
bandwidth (-3-dB): >500 MHz;
source resistance: 50 Ω;
load: 1 pF, purely capacitive;
maximum frequency response peak: $<10\%$;
total supply power: <50 mW.

Use device models from Chapter 5. Assume that the supply voltage is 3.3 V.

You may use up to 20 nH of on-chip inductance and 5 nH of bondwire inductance. Assume on-chip spirals have a Q of 5 at 1 GHz, and pretend that the corresponding effective series resistance remains constant at all frequencies. Ignore self-resonance of all inductors. You are also permitted up to 20 pF of on-chip capacitance. You may assume that the capacitor is ideal in all respects.

PROBLEM 11 This problem explores in greater detail the notion of nonuniform group delay as a source of distortion.

(a) Plot the delay versus frequency for a single RC *high*-pass filter. The frequency axis should be normalized to $1/RC$.

(b) Consider a second-order low-pass section whose transfer function is

$$H(s) = \left[\frac{s^2}{\omega_n^2} + \frac{s}{Q\omega_n} + 1 \right]^{-1}. \tag{P9.1}$$

Repeat part (a) for $Q = 0.5$, 1, and 2, now normalizing frequency to ω_n.

(c) For the shunt-peaked amplifier, we noted that there is a tradeoff between maximum bandwidth improvements and pulse fidelity. Recall that its ideal transfer function is

$$A(s) = g_m(sL + R) \parallel \frac{1}{sC} = \frac{g_m R[s(L/R) + 1]}{s^2 LC + sRC + 1}$$

$$= A_0 \frac{s(L/R) + 1}{s^2 LC + sRC + 1}. \tag{P9.2}$$

Repeat part (a) for the following ratios of RC to L/R: 2, 2.5, 3, 3.5. Plot your results as a function of ω/ω_1, where ω_1 is $1/RC$, the uncompensated amplifier's bandwidth. You may find it helpful to make use of the results of parts (a) and (b). Phases of cascaded linear systems add, so their delays also add.

PROBLEM 12 Out of laziness, most engineers choose the common-gate device in a cascode to be of the same size as the common-source device. However, this choice is almost never the optimum. In the questions that follow, you may neglect body effect but do *not* neglect parasitic device capacitances.

(a) Discuss what happens as the cascoding device is made progressively narrower than the common-source device. Assume that the same bias current flows through both (again, this choice is not necessarily optimal).

(b) Discuss what happens as the cascoding device is made progressively wider.

(c) Given your insights from (a) and (b), describe a formal procedure for finding the optimal size of the cascoding device to maximize bandwidth.

CHAPTER TEN

VOLTAGE REFERENCES
AND BIASING

10.1 INTRODUCTION

The previous chapter on amplifier design generally ignored the issue of generating suitable bias voltages or currents. This neglect was by conscious design in order to minimize clutter in the circuit diagrams. In this chapter we finally take up the study of this important topic, focusing on a variety of ways to generate voltages and currents that are relatively independent of supply voltage and temperature. Because CMOS offers relatively limited options for realizing bias circuits, we'll see that some of the most useful biasing idioms are actually those based on bipolar circuits. A parasitic bipolar device exists in every CMOS technology and may be used, for example, in a bandgap voltage reference. Even though the characteristics of parasitic transistors are far from ideal, the performance of bias circuits made with such devices is frequently vastly superior to that of "pure" CMOS bias circuits.

In what follows, it is worthwhile to keep in mind that any voltage we produce must depend on some collection of parameters that ultimately have the dimensions of a voltage (such as kT/q, for example). Similarly, any current we produce must depend on parameters that ultimately have the dimensions of current (such as V/R). Although seemingly obvious and trivial statements, we'll see that they are extremely useful guides for the design of stable references.

10.2 REVIEW OF DIODE BEHAVIOR

Although the voltage across a forward-biased diode is relatively insensitive to current because of the logarithmic dependence of diode current on diode voltage, its variation with temperature is significant. To understand the precise nature of the temperature dependence, recall that the diode voltage may be expressed as

$$V_D = nV_T \ln\left(\frac{I_D}{I_S}\right), \tag{1}$$

where V_T is the thermal voltage kT/q and n, the ideality factor, is typically between 1 and 1.5 in diodes. Transistor V_{BE}s (BE denotes "base-emitter") conform more closely

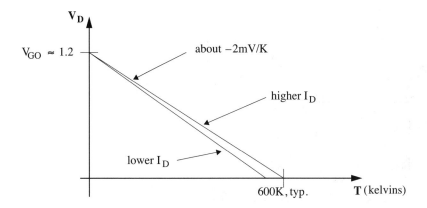

FIGURE 10.1. Approximate behavior of V_D versus temperature.

to the "ideal diode law" than do ordinary diodes, so we will assign n a value of unity in all that follows.

It is frequently (but incorrectly) inferred from Eqn. 1 that V_D has a positive TC because of its proportionality to V_T. The fly in the ointment is that I_S itself has an exponential temperature dependence, and this alters the situation considerably. To clarify matters, consider the following quasiempirical expression for I_S:

$$I_S = I_0 \exp\left(-\frac{V_{G0}}{V_T}\right), \qquad (2)$$

where I_0 is some process- and geometry-dependent current[1] (I_0 is typically around 20 orders of magnitude larger than I_S at room temperature, so I_0 is much larger than typical values of I_D), and where V_{G0} is the bandgap voltage (about 1.2 V) extrapolated to absolute zero.

Using this detailed expression for I_S, we can expand the equation for V_D as[2]

$$V_D = V_{G0} - V_T \ln\left(\frac{I_0}{I_D}\right). \qquad (3)$$

Thus, we see that the junction voltage decreases linearly from a value of V_{G0}, as seen in the plot of V_D versus temperature at constant diode current (Figure 10.1). Note that this equation tells us that V_D *always* equals V_{G0} at absolute zero.[3] Furthermore, it's easy to see that the temperature coefficient at any temperature is simply

---

[1] It also depends weakly on temperature, but we'll defer a detailed discussion about the behavior of I_0 until Section 10.5.

[2] The minus sign is not an error. Just remember that the argument of the log here is typically much larger than unity.

[3] Again, this value is an extrapolated one. It must be stressed that the behavior of real junctions at both extremes of temperature will differ from that shown; the equations presented lose validity at extremely cold temperatures (say, <100 K) because of carrier freeze-out (i.e., the dopants fail to ionize) and bandgap variation with temperature, and at high temperatures (>450–500 K) because the silicon goes intrinsic.

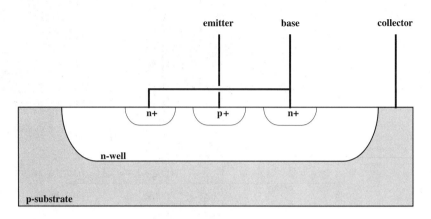

FIGURE 10.2. Parasitic substrate p–n–p in n-well CMOS (not drawn to scale).

$$\frac{dV_D}{dT} = -\frac{V_{G0} - V_D}{T}.$$
(4)

With the assumption of constant I_0, the temperature coefficient is independent of temperature and equal to about -2 mV/K. This linearly decreasing behavior is known as CTAT, for "complementary to absolute temperature." Note that the voltage does depend (logarithmically) on diode current, so the temperature coefficient also depends somewhat on the diode current, with lower currents associated with higher temperature coefficients.

Although a V_D-based reference can provide an output that depends very little on supply voltage, the CTAT behavior may or may not be acceptable, depending on the application. However, we shall see that the CTAT behavior of a V_D is particularly valuable for use in a class of references based on the bandgap voltage V_{G0}. We'll take up the detailed study of bandgap references in Section 10.5.

10.3 DIODES AND BIPOLAR TRANSISTORS IN CMOS TECHNOLOGY

The most flexible option for realizing diodes and bipolar transistors in standard CMOS technology derives from the parasitic substrate p–n–p transistor available in n-well processes. The p+ source–drain diffusions serve as the emitter, the n-well as the base, and the substrate as the collector. To reduce series base resistance, it is advisable to surround completely the emitter with n+ diffusions placed as close to the emitter as the design rules allow, as suggested by Figure 10.2.

As with its counterpart in inexpensive bipolar processes, the substrate p–n–p in CMOS technology can only be used in circuits that allow the collector to be at substrate potential. Fortunately, there are numerous circuits that satisfy this condition. For example, a simple voltage "reference" can be constructed with this device connected as a grounded diode, where the emitter is the anode and the cathode is the base and collector (substrate) tied together.

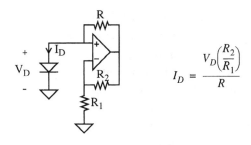

FIGURE 10.3. Self-biased reference.

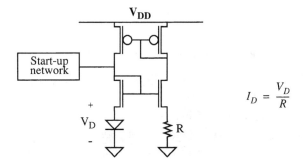

FIGURE 10.4. Alternative self-biased reference.

10.4 SUPPLY-INDEPENDENT BIAS CIRCUITS

In order to minimize sensitivity to power supply variations, it is desirable to derive the bias currents for reference voltages from the reference voltages themselves, rather than directly from the power supply. Although it may seem a violation of some fundamental law (the "no free lunch" principle), it is possible to arrange for this condition. To illustrate how one may accomplish this feat, consider the circuit of Figure 10.3. As you can see, the current through the diode depends on the diode voltage itself, rather than on the supply voltage. This technique thus provides excellent supply voltage independence.

An important practical note is that a start-up network is always necessary in self-biased circuits because there are two states, one which is stable in the conventional sense and another in which all currents are zero.[4] The start-up network guarantees that the circuit gets out of the undesired metastable state.

Most practical implementations of the self-biased circuit dispense with the op-amp, as seen in Figure 10.4. The PMOS mirror[5] enforces equality of the NMOS

---

[4] Even though the all-zero state is metastable, practical circuits are found in this state a maddeningly large portion of the time. A start-up network is therefore mandatory for reliable operation.

[5] Better mirrors would generally be used in practice to avoid supply-dependent mirror ratios; simple ones are shown to reduce schematic clutter.

drain currents, and hence that of the NMOS V_{gs}. Thus, the diode voltage appears across R; the corresponding current is the same in both halves of the mirror and is therefore the bias current of the diode itself. Thus, as in the op-amp version of this circuit, the diode provides its own bias current.

The self-biased circuit of Figure 10.4 is quite versatile. It should be clear that the diode may be replaced by a variety of elements. For example, a diode-connected MOSFET would produce a bias current of V_{gs}/R, or a zener diode (if available) could be used instead. As we'll see in the next section, the self-biased circuit is particularly useful in realizing bandgap voltage references in CMOS technology. Some minor subtleties concerning the operation of the quad of MOSFETs in Figure 10.4 are considered in Problem 12.

10.5 BANDGAP VOLTAGE REFERENCE

Because IC technology directly offers no reference voltages that are inherently constant, the only practical option is to combine two voltages with precisely complementary temperature behavior. Thus, the general recipe for making temperature-independent references is to add a voltage that goes up with temperature to one that goes down with temperature. If the two slopes cancel, the sum will be independent of temperature.

Without question, the most elegant realization of this idea is the bandgap voltage reference. It produces an output voltage that is traceable to fundamental constants and therefore relatively insensitive to variations in process, temperature, and supply.

The first widely used bandgap voltage reference was designed by Bob Widlar in the hugely popular and revolutionary LM309 5-V regulator IC from National Semiconductor. It was the first reference whose initial accuracy was good enough to eliminate the requirement for adjustment by the end user. Thus, only three terminals were needed (allowing use of inexpensive transistor packages), making this part as easy to use as one could hope.

To understand quantitatively how bandgap references work, we need to re-examine the detailed behavior of junction voltage with temperature. Since transistor junctions exhibit more nearly ideal characteristics than ordinary diodes, we will assume bandgap implementations that use transistors. A plot of V_{BE} versus temperature is sketched in Figure 10.5.[6]

Recall that V_{BE} is nearly perfectly CTAT (i.e., it goes down linearly with temperature). Now suppose we add to this CTAT V_{BE} a voltage that is perfectly proportional to absolute temperature (PTAT). If we choose the slope of the PTAT term equal in magnitude to that of the CTAT term, the sum will be independent of temperature (see Figure 10.6). We see that something funny happens above about 600 K, but the fact

---

[6] Again, keep in mind that this plot of V_{BE} is a slight fiction because we have neglected the small curvature caused by the weak temperature dependence of I_0. The correction is second-order, and we will take care of this little detail shortly.

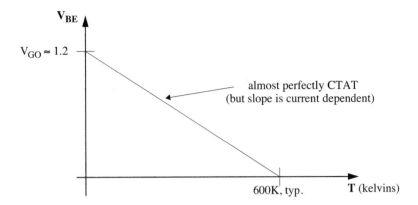

FIGURE 10.5. V_{BE} versus temperature.

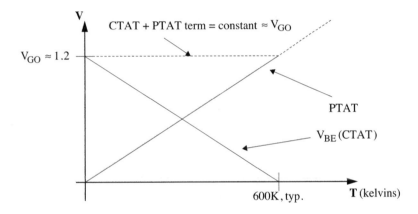

FIGURE 10.6. Illustration of bandgap reference principle.

that the principle fails at temperatures high enough to melt lead is rarely a practical concern.

Note that the addition of a PTAT and CTAT voltage in the proper ratio yields an output equal to the bandgap voltage (extrapolated to 0 K), independent of temperature. Stated another way, if we adjust the PTAT component to make the output voltage equal to V_{G0} at any temperature, then the output voltage will equal V_{G0} at all temperatures – at least in this slightly simplified picture.

At this point, it's natural to consider how one obtains a PTAT voltage, since this whole concept relies on having one around. Let's start with the familiar equation for V_{BE}:

$$V_{BE} = V_T \ln\left(\frac{I_C}{I_S}\right). \tag{5}$$

Using this expression, we can readily compute the *difference* in two V_{BE}s for identical transistors operating at two different values of collector current (or, more generally,

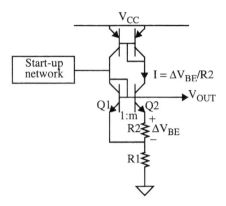

FIGURE 10.7. Classic Brokaw bandgap
reference circuit.

for transistors made in the same process, operating at two different values of collector current density):

$$\Delta V_{\text{BE}} = V_{\text{BE2}} - V_{\text{BE1}} = V_T \ln\left(\frac{J_{C2}}{J_{C1}}\right). \tag{6}$$

The misleading I_S term drops out, so we can conclude confidently that ΔV_{BE} truly is PTAT if the collector current densities are in a fixed ratio. Thus, while each V_{BE} is nearly CTAT, the *difference* between two V_{BE}s is perfectly PTAT.

10.5.1 CLASSIC BANDGAP REFERENCE

Now that we've got all the ingredients, all that remains is to sum the CTAT V_{BE} term with the right amount of PTAT ΔV_{BE}. Although one could imagine a number of methods for doing so, the Brokaw cell is a particularly elegant (and accurate) implementation of the bandgap reference. The classic bipolar implementation is shown in Figure 10.7 (again, basic mirrors are shown for simplicity's sake); we'll modify this circuit shortly for implementation in CMOS technology.

As we shall see, the output voltage is the sum of a PTAT voltage and a V_{BE}. Here, Q_1 and Q_2 operate at a fixed current density ratio of m (>1) set, for example, by ratioing the emitter areas. Now, by KVL, the voltage across R_2 is the difference in V_{BE}s of Q_1 and Q_2, and is therefore PTAT and equal to $V_T \ln m$. Assuming that the TC of R_2 is negligibly small, the current passing through it will also be PTAT. Furthermore, the current through R_1 is simply twice that through R_2, since the two collector currents are equal.[7] Therefore, the voltage drop across the entire resistor string is purely PTAT. Finally, the output voltage is just this PTAT voltage plus the V_{BE} of Q_2, as advertised. With proper choice of R_1 and R_2, the output voltage will have zero TC. As a free bonus, a PTAT voltage is available at the emitters of Q_1 and Q_2, providing thermometer outputs.

---

[7] We are neglecting errors due to mismatch, nonzero base currents, and finite Early voltage.

Design Example

To carry out an actual design, we need some characterization data for our process. As a specific example, suppose we go to the lab and find that $V_{\text{BE}} = 0.65$ V at 300 K and 100 μA for a transistor of Q_2's size. Furthermore, let $m = 8$. This choice[8] of m sets $\Delta V_{\text{BE}} = 53.8$ mV (a number comfortably larger than any offsets that we expect) at 300 K. Since we definitely know the value of V_{BE} at 100 μA, a prudent choice for the collector currents would be this value of 100 μA, and this choice then fixes the value of $R_2 = \Delta V_{\text{BE}}/100 \ \mu\text{A} = 538 \ \Omega$. Now, since we want the output voltage to be 1.2 V, the drop across R_1 must be $1.2 - V_{\text{BE}} - \Delta V_{\text{BE}} = 0.496$ V. Finally, noting that the current through R_1 is twice that through R_2, we conclude that we should choose $R_1 = 0.496 \ \text{V}/200 \ \mu\text{A} = 2.48 \ \text{k}\Omega$, completing the design.

You may have noticed that the collector currents in the Brokaw cell are not constant (in fact, they are PTAT if we assume that the resistors have zero TC). To see why this does not invalidate all we've done so far (in fact, it is beneficial), it is now time to take care of a few details – namely, those involving the temperature dependency of I_0.

A quasiempirical expression for I_0 is

$$I_0 = A_E B T^r, \tag{7}$$

where A_E is the emitter area, B is a process-dependent constant, T is the absolute temperature, and r is a process-dependent quantity we'll call the *curvature coefficient*. For the relatively deep, diffused emitters of older bipolar processes, r typically has a value between 2 and 3, while for the shallow, implanted (and very heavily doped[9]) diffusions that are more common in modern CMOS and high-speed bipolar processes, r typically ranges from 4 to 6.

With this equation for I_0, we can express V_{BE} as follows:

$$V_{\text{BE}} = V_{G0} - V_T \ln\left(\frac{A_E B T^r}{I_C}\right). \tag{8}$$

Plotting in Figure 10.8 as before, we can see why it is reasonable to call the parameter r the curvature coefficient (aside from the euphonious alliteration).

Because the argument of the log is not quite independent of T, the temperature coefficient of V_{BE} is not quite constant, leading to a small departure from CTAT behavior for V_{BE}. Additionally, we've seen at least one implementation of a bandgap reference in which the collector current is also not constant. So, let's compute the actual TC that results if, in addition to the temperature dependence of I_0, we also consider a collector current that varies as the nth power of T:

---

[8] It is important that Q_2 be laid out as eight instances of Q_1 to guarantee that Q_2 behaves as eight parallel devices of Q_1's size. If Q_1 is placed at the center of a common-centroid arrangement, errors due to process variation will be minimized.

[9] Bandgap narrowing and nonlinearity in the heavily doped emitters are probably responsible for the high values of r.

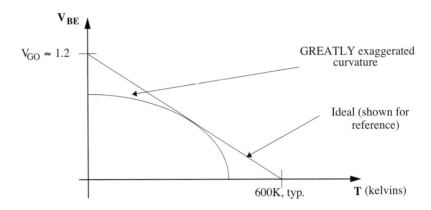

FIGURE 10.8. V_{BE} versus temperature.

$$\frac{dV_{BE}}{dT} = \frac{d}{dT}\left[V_T \ln\left(\frac{CT^n}{A_E BT^r}\right)\right] = \frac{d}{dT}\left[V_T \ln\left(\frac{C}{A_E BT^{r-n}}\right)\right], \tag{9}$$

so that

$$\frac{dV_{BE}}{dT} = \frac{k}{q}\left[\ln\left(\frac{C}{A_E BT^{r-n}}\right) - (r-n)\right], \tag{10}$$

which we may rewrite in a somewhat lower-entropy form, as in Section 10.2:

$$\frac{dV_{BE}}{dT} = -\frac{[V_{G0} - V_{BE} + (r-n)V_T]}{T}. \tag{11}$$

Note that the curvature term disappears if $r = n$, and we're left with the same expression for the temperature coefficient as derived earlier. In the Brokaw cell, $n = 1$, which reduces the effect of yet does not cancel r (remember, r is typically a minimum of 2, and can range up to about 6). Graphically, think of the increasing collector current with temperature as straightening out the V_{BE} curve.

The next question is: How does the curvature term affect the bandgap reference itself? The most expedient answer comes from deriving the condition for net zero TC. Suppose we call GV_T the PTAT component that we add to V_{BE}. Then the TC of the PTAT component may be written as GV_T/T, so the condition for zero TC is

$$\frac{dV_{BE}}{dT} + \frac{GV_T}{T} = 0 \implies G = \frac{[V_{G0} - V_{BE} + (r-n)V_T]}{V_T}, \tag{12}$$

which corresponds to an output voltage of

$$V_{out}\big|_{TC=0} = V_{BE} + GV_T = V_{G0} + (r-n)V_T. \tag{13}$$

This last equation depends on V_T and therefore implies that the output voltage cannot have zero TC at all temperatures; the best we can do is achieve zero TC at one temperature. Furthermore, in order to achieve this zero-TC condition at that one temperature, we need to adjust the output to a voltage *higher* than V_{G0} by an amount equal to $(r-n)V_T$. Fortunately, this correction term is relatively small, typically amounting

Table 10.1. *Output voltage as function of T and r − n*

$r - n$	V_{out} @ $T = -55°C$	V_{out} @ $T = 50°C$	V_{out} @ $T = 150°C$	Maximum error
1	1.226	1.228	1.227	2 mV
2	1.252	1.256	1.253	4 mV
3	1.279	1.284	1.280	5 mV
4	1.305	1.312	1.307	7 mV
5	1.331	1.339	1.333	8 mV

to just tens of millivolts out of a total that is greater than a volt. Hence, the output need only be trimmed to a value a few percent greater than V_{G0} at the temperature where zero TC is desired (generally, the center of the operating temperature range).

At this point, we'd like to quantify the errors that are caused by the curvature. Unfortunately, although the equations we've developed so far are valuable for design, they aren't quite suitable for analysis. To derive one that is, let us choose the factor m so that the output voltage has zero TC at some temperature we'll call T_R (for the reference temperature). Throughout, we'll use the subscript R to denote a variable's value at this reference temperature. With this notational convention, we may express V_{out} as follows:

$$V_{\text{out}}(T) = V_{G0} + \frac{T}{T_R}(r - n)V_{TR} - \frac{T}{T_R}(r - n)V_{TR} \ln\left(\frac{T}{T_R}\right), \qquad (14)$$

or, as some prefer,

$$V_{\text{out}}(T) = V_{G0} + \frac{T}{T_R}(r - n)V_{TR}\left[1 - \ln\left(\frac{T}{T_R}\right)\right]. \qquad (15)$$

Note that this equation has the right limiting behavior: when $T = T_R$, it yields the output voltage corresponding to the zero-TC condition. Also note that, if we were able to arrange for the collector currents to vary as T^n with $n = r$, then the output voltage would have zero TC at all temperatures if V_{out} were adjusted to a value V_{G0} at any temperature. This last observation is at the core of many efforts to synthesize curvature-corrected bandgap references.

Even without elaborate curvature-correction methods, though, the Brokaw cell (where $n = 1$ in the classic implementation) provides outstanding performance; the curvature inherent in bipolars is simply not all that bad, and the Brokaw cell contributes little error of its own. To illustrate this point, let's compute the actual error one could expect, over a temperature range of $-55°C$ to $+150°C$, if T_R is chosen as $+50°C$ and if the quantity $r - n$ ranges from 1 to 5; see Table 10.1.

As is evident, the total change in output voltage is less than a percent over the entire temperature range, even with relatively large values of $r - n$. Furthermore, the output is a maximum at the reference temperature, and drops off above and below this temperature in a quasiparabolic manner. As a consequence, setting T_R equal to

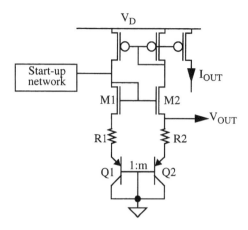

FIGURE 10.9. CMOS bandgap reference.

the center of the desired operating temperature range nearly minimizes the maximum deviation from the value at T_R.

As a final note, the levels of performance in Table 10.1 assume an ideal scenario in which second-order errors (due to device mismatch, nonzero resistor TC, β drift, etc.) are ignored. Actual performance will be somewhat worse in practice owing to the combined effect of these sources. Careful layout of all devices is mandatory in order to minimize errors.

10.5.2 BANDGAP REFERENCES IN CMOS TECHNOLOGY

The classic Brokaw cell uses bipolar transistors in which all device terminals float, so it cannot be implemented directly in this form in CMOS technology. Rearranging to accommodate the restrictions placed on the parasitic substrate p–n–p yields the circuit shown in Figure 10.9.

Transistor Q_2 is designed to have m times the emitter area of Q_1. The quad of CMOS transistors enforces equal emitter currents, so that the *collector* current density ratio is approximately m. Implicit in the last statement is that this circuit has a greater sensitivity to β than the original Brokaw cell. This unfortunate consequence of being forced to use the substrate p–n–p's leads to larger errors than the classic bandgap cell, particularly because β is rarely large enough to be ignored (values of 5–10 are typical). Nevertheless, even a poorly performing bandgap reference is considerably superior to anything that can be built out of pure CMOS components.

Choosing component values for this circuit proceeds in a manner quite similar to that for the classic cell. Begin by specifying a reference temperature T_R at which the TC is to be zero. For illustrative purposes, assume that this temperature is to be 350 K.

The next step is to calculate the target output voltage at this reference temperature. As mentioned previously, the shallow, heavily doped p+ diffusions used to make the emitters lead to relatively large curvature coefficients, with r typically 4 or 5. If no device models are available, a reasonable starting point is to assume a value of 4 for the quantity $r - n$. Hence, the target output voltage should be

$$V_{\text{out}} = V_{G0} + (r - n)V_T \approx 1.32 \text{ V}. \tag{16}$$

Now assume that we have selected 100 μA for the individual emitter currents and that the larger transistor, Q_2, has a V_{BE} of 0.65 V at this current at T_R. Then R_2 is simply

$$R_2 = \frac{V_{\text{out}} - V_{\text{BE2}}}{I_E} = 6.7 \text{ k}\Omega. \tag{17}$$

If we assume an m equal to 8 then V_{BE1} is about 63 mV larger than V_{BE2}. Hence,

$$R_1 = \frac{V_{\text{out}} - V_{\text{BE1}}}{I_E} = 6.07 \text{ k}\Omega, \tag{18}$$

thus completing the design.

As a final comment on this circuit, one usually finds that the bias current is relatively constant with temperature because resistors typically have a positive TC, which offsets the PTAT tendency of the core design. Thus, currents from a mirror slaved to the PMOS mirror will be roughly constant. The precise TC obtained may be adjusted through a suitable choice of resistor values if the ultimate goal is to generate a bias current rather than a reference voltage.

As a final comment on CMOS bandgap circuits in general, it must be noted that the relatively poor matching that CMOS normally exhibits can result in significant errors. Using large devices and operating them at moderately high gate overdrives is beneficial, but it costs power. An alternative technique is to make use of MOSFETs as switches, alternately exchanging the left and right transistors of every pair that is supposed to match.[10] By symmetry, the effect of mismatch should reverse sign at every alternation. If we simply time-average the output with a low-pass RC filter, for example, then the output voltage will be insensitive to offset, to first order. Using this technique, untrimmed 3σ errors of about 1% have been demonstrated in a 0.18-μm CMOS process. Such performance is competitive with many trimmed bipolar implementations. This general technique may be applied wherever symmetry allows it.

10.6 CONSTANT-g_m BIAS

A constant current or constant voltage is often desirable, but this is not always the case. Important examples include situations in which it is the transconductance that must be held constant, such as in the low-noise amplifiers described in Chapter 12.

A circuit whose bias current corresponds to a g_m that is inversely proportional to a reference resistance is a modification of the self-biased CMOS quad of transistors we've already seen; this is shown in Figure 10.10.

---

[10] V. Ceekala et al., "A Method for Reducing the Effects of Random Mismatch in CMOS," *ISSCC Digest of Technical Papers,* February 2002. Also see U.S. Patent #6,535,054, filed 20 December 2001, granted 18 March 2003.

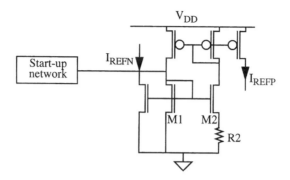

FIGURE 10.10. Basic constant-g_m reference.

To derive an expression for the transconductance of M_1, first use KVL to write

$$V_{gs1} = V_{gs2} + IR_2 \implies V_{od1} = V_{od2} + IR_2, \tag{19}$$

where we have assumed perfect PMOS mirrors and equal threshold voltages for the two long-channel NMOS transistors.

If transistor M_2's width is m times that of M_1, then the two overdrive voltages are related as follows:

$$V_{od2} = \frac{V_{od1}}{\sqrt{m}}. \tag{20}$$

For the long-channel devices we have been considering all along, g_m is simply $2I/V_{od}$. Therefore,

$$V_{od1} = V_{od2} + IR_2 \implies V_{od1} = \frac{V_{od1}}{\sqrt{m}} + IR_2 \implies \frac{2I}{g_{m1}} = \frac{2I}{\sqrt{m}g_{m1}} + IR_2. \tag{21}$$

Solving for the transconductance yields

$$g_{m1} = \frac{2(1 - 1/\sqrt{m})}{R_2}, \tag{22}$$

revealing explicitly that the transconductance is proportional to the reciprocal of reference resistance R_2. If the ratio m is precisely 4, then the proportionality becomes an equality. The currents I_{REFN} and I_{REFP} generated by this cell may be mirrored to slave devices (with or without scaling) to set the transconductance of multiple NMOS devices in other parts of a circuit.

Because this bias generator depends on the quality of the reference, demanding applications may require the use of an external resistance. In noncritical applications, ordinary on-chip resistances (e.g., unsilicided poly) may suffice.

The foregoing development rather optimistically assumes that both NMOS transistors possess the same threshold voltages, despite the unequal source-to-bulk potentials for the devices. In practice, the back-gate bias effect produces unequal threshold shifts and thus introduces an error. To minimize the impact of this back-gate effect, choose relatively high current densities for the devices – within the constraints of maintaining long-channel operation. Doing so assures large overdrives, making the overall

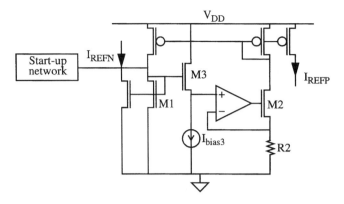

FIGURE 10.11. Improved constant-g_m reference.

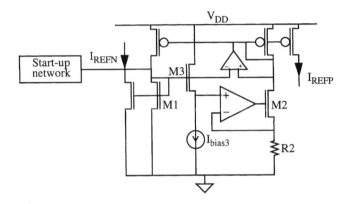

FIGURE 10.12. Minimum-error constant-g_m reference.

circuit's behavior less sensitive to threshold differences. Finally, select a relatively low IR_2 product to minimize the difference in source-to-bulk voltages. Use of both of these strategies together usually permits the attainment of satisfactory performance.

An alternative constant-g_m reference is shown in Figure 10.11. In this circuit, M_2 is largely irrelevant, thanks to the op-amp. As seen in the figure, the voltage across resistor R_2 equals M_1's V_{gs} level-shifted downward by an amount equal to the V_{gs} of M_3. The current I_{bias3} is chosen in conjunction with the dimensions of M_3 to produce a level shift equal to the threshold voltage of M_1. Therefore,

$$V_{gs1} - V_{gs3} \approx V_{gs1} - V_{t1} \approx IR_2 \implies \frac{2I}{g_{m1}} \approx IR_2 \implies g_{m1} \approx \frac{2}{R_2}. \quad (23)$$

Because of the additional degree of freedom represented by I_{bias3}, it is possible for this circuit to exhibit smaller errors than the previous one without requiring small IR_2 products. Still, operation of M_1 with large overdrives remains beneficial.

Mismatched mirrors are an additional error source in a great many analog circuits. Matching may be improved by using large devices biased at reasonably large overdrives. Systematic mismatch is reduced if the drain–source voltages are made as close to equal as possible. The circuit in Figure 10.12 uses another op-amp to force operation of the PMOS mirror transistors with nominally equal drain-source voltages, thereby eliminating systematic errors due to channel-length modulation and DIBL.

With a constant-g_m circuit, one may reduce greatly the variation in g_m-dependent parameters – such as gain, input impedance, and noise figure of LNAs – as temperature, processing, and supply voltage vary.

10.7 SUMMARY

We have seen that the self-biased cell is quite versatile, permitting the generation of currents proportional to the ratio of a voltage in one branch to the resistance in the other. The voltage may be provided by a variety of elements, such as a forward-biased junction. Although a V_{BE} by itself has limited utility as a voltage reference because of its negative TC, its CTAT behavior is valuable in compensating the PTAT ΔV_{BE} in a bandgap reference circuit to yield an output roughly equal to V_{G0} with extremely small temperature variation. Even when parasitic bipolars are used in an otherwise CMOS circuit, the bandgap principle allows the synthesis of more accurate and stable voltages or currents than is possible with ordinary CMOS circuits.

Finally, a constant-g_m bias circuit was presented, allowing the stable biasing of such transconductance-sensitive circuits as filters and LNAs.

PROBLEM SET FOR VOLTAGE REFERENCES AND BIASING

PROBLEM 1   This problem explores in more detail the characteristics of the constant-g_m reference. Refer to Figure 10.13.

Rather than choosing $m = 4$, consider simply making m very large. In the limit, the transconductance approaches a value that is twice what is obtained when $m = 4$. Show this formally by deriving an expression for the transconductance of M_1 if the M_2 is S times as wide as M_1. To simplify the derivation, neglect body effect and assume that the PMOS mirror is an ideal 1:1 mirror.

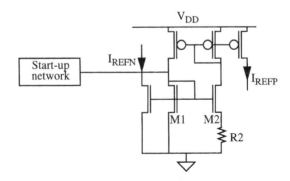

FIGURE 10.13. Basic constant-g_m reference.

PROBLEM 2 In the improved constant-g_m reference of Figure 10.14, investigate the effect of PMOS mirror errors caused by nonzero output conductance. Model the PMOS devices as square-law, and assume that they are ideal except for a channel-length modulation coefficient of $0.1\,\mathrm{V}^{-1}$. Derive a general expression for the transconductance of M_1, no longer assuming that M_3's V_{gs} precisely equals the V_t of M_1.

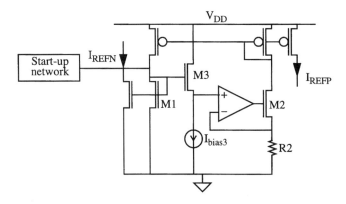

FIGURE 10.14. Improved constant-g_m reference.

PROBLEM 3 Using the simple circuit of the first problem, select device sizes and resistor value to produce an output current sink of $250\,\mu\mathrm{A}$ and a transconductance of $1\,\mathrm{mS}$ at 300 K. Use the level-3 device parameters from Chapter 5 and simulate with SPICE to verify that the design works as desired.

PROBLEM 4 The equations for the CMOS-compatible bandgap reference do not take finite p–n–p β into account. Unfortunately, typical values for β of such transistors are often below 10. Re-derive the expression for the output voltage including β.

PROBLEM 5 In this problem, we consider the settling behavior of a nonideal voltage reference in response to a transient disturbance. Consider the popular circuit shown in Figure 10.15. Assume that each transistor is $10\,\mu\mathrm{m}$ wide and that V_{DD} is 3 V. Use the level-3 device models given in Chapter 5.

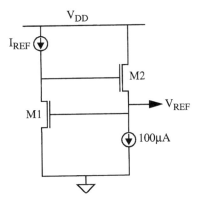

FIGURE 10.15. V_{gs}-based voltage reference.

(a) Choose I_{ref} to make the output voltage 1 V.

(b) What is the low-frequency incremental output resistance?

(c) Now consider what happens if the reference voltage must drive a total load of 3 pF, and if a disturbance happens to bump the output voltage to 1.5 V in 1 ns (e.g., with a fast-acting current source). Calculate the settling time to 1% of the original value of 1 V (you may neglect body effect in your hand calculations) and verify your answer with SPICE. Explain discrepancies quantitatively.

PROBLEM 6   In low-voltage circuits, it becomes difficult or impractical to use ordinary cascode structures to increase the output resistance of current sources. Alternate cascoding techniques can be used to reduce the voltage required, however. An example is sketched in Figure 10.16.

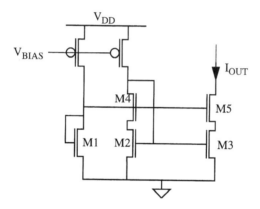

FIGURE 10.16. Low-headroom cascode.

In this circuit, M_1 establishes a bias voltage for the gates of M_4 and M_5. A typical rule of thumb is to make M_1 about 1/4 the width of all the other transistors. However, one can place the design of this circuit on a more rational basis. If we call S the ratio W/L, assume that $S_2 = S_3 = n^2 S_4 = n^2 S_5 = S$, where $n > 1$.

Assume zero output conductance in saturation and neglect body effect. Derive explicitly the condition on S_1 in terms of S and n so that M_2 and M_3 are biased on the edge of saturation. You may assume square-law behavior.

PROBLEM 7   Repeat the previous problem for short-channel devices, expressing your answer in part in terms of E_{sat}. Note in particular that short-channel effects are *helpful* here because the saturation voltage diminishes, allowing operation at lower supply voltages.

PROBLEM 8   Another low-headroom current source is shown in Figure 10.17. Assume for simplicity that all widths are equal. Determine an expression for R that guarantees that both M_1 and M_2 are in saturation. Assume long-channel behavior, and neglect body effect and channel-length modulation.

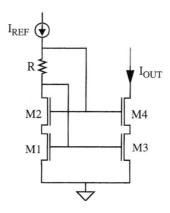

FIGURE 10.17. Alternative low-headroom
cascode current source.

PROBLEM 9 A variation on the constant-g_m bias cell avoids errors due to back-gate
bias (body effect) by forcing the source terminals of both NMOS devices to be at
the same potential; see Figure 10.18. Show that the output current is determined only
by transistor geometry and resistor value. As before, ignore body effect and assume
zero output conductance and identical transistors (except for M_4, which is S times as
wide as the others). Show that the transconductance of M_5 depends only on geometry and the reference resistor R.

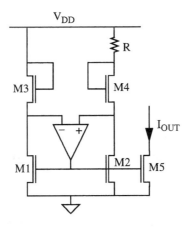

FIGURE 10.18. Alternative constant-g_m reference.

PROBLEM 10

(a) In the circuit of the previous problem, does the polarity of the op-amp matter?
Explain qualitatively why it most certainly does, and how it behaves if the polarity is incorrect.

(b) To answer (a) more quantitatively, derive an explicit expression for the loop transmission of the circuit by breaking the loop at the output of the op-amp, driving

the common gate connection of M_1 and M_2, and observing what comes back from the op-amp output. Watch your signs!

PROBLEM 11   Self-biased circuits abound in this chapter, and we have alluded to the necessity of start-up circuits without actually showing any specific examples. Consider the bandgap reference circuit of Figure 10.19 as an example. Assume that the core of the bandgap has 100 μA of current flowing in each branch.

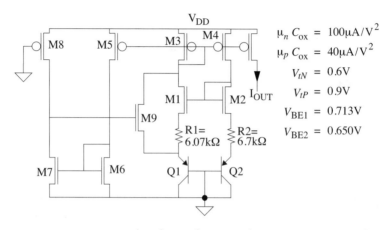

FIGURE 10.19. CMOS bandgap reference with start-up circuit example.

(a) Explain the operation of the start-up network.
(b) When the bandgap is operating, M_9 is supposed to be off. In order to ensure that M_9 does not disturb the operation of the bandgap cell, its V_{gs} should be at least 400 mV below the threshold. What is the smallest ratio of $(W/L)_7$ to $(W/L)_8$ that would satisfy this requirement?
(c) Simulate this circuit by ramping up the supply voltage slowly from zero. At what supply voltage does the circuit "snap" into operation?

PROBLEM 12   Many self-biased circuits in this chapter use a quad of MOSFETs arranged as in the constant-g_m bias circuit of Figure 10.20 (left). It has been argued that the circuit in Figure 10.20 (right) would work just as well.

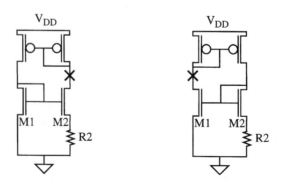

FIGURE 10.20. Alternative connections for constant-g_m reference.

To evaluate this claim, cut each circuit where indicated with an "X." Pull a current out of the PMOS transistors and then see what comes back from the NMOS devices. Evaluate the current ratio at very low currents, remembering that M_2 is the wider NMOS transistor. From this "loop transmission" experiment, explain why both circuits work (or don't work).

CHAPTER ELEVEN

NOISE

11.1 INTRODUCTION

The sensitivity of communications systems is limited by noise. The broadest definition of noise as "everything except the desired signal" is most emphatically *not* what we will use here, however, because it does not separate, say, artificial noise sources (e.g., 60-Hz power-line hum) from more fundamental (and therefore irreducible) sources of noise that we discuss in this chapter.

That these fundamental noise sources exist was widely appreciated only after the invention of the vacuum tube amplifier, when engineers finally had access to enough gain to make these noise sources noticeable. It became obvious that simply cascading more amplifiers eventually produces no further improvement in sensitivity because a mysterious noise exists that is amplified along with the signal. In audio systems, this noise is recognizable as a continuous hiss while, in video, the noise manifests itself as the characteristic "snow" of analog TV systems.

The noise sources remained mysterious until H. Nyquist, J. B. Johnson and W. Schottky[1] published a series of papers that explained where the noise comes from and how much of it to expect. We now turn to an examination of the noise sources they identified.

11.2 THERMAL NOISE

Johnson[2] was the first to report careful measurements of noise in resistors, and his colleague Nyquist[3] explained them as a consequence of Brownian motion: thermally agitated charge carriers in a conductor constitute a randomly varying current that

---

[1] This name is frequently misspelled in English-language publications. It is not all that uncommon to see "Shotkey," "Shottkey," or "Schottkey." While we're at it, "Schmitt," as in the trigger, is also misspelled quite often, with common incorrect renderings being "Shmitt" and "Schmidt."

[2] "Thermal Agitation of Electricity in Conductors," *Phys. Rev.*, v. 32, July 1928, pp. 97–109.

[3] "Thermal Agitation of Electric Charge in Conductors," *Phys. Rev.*, v. 32, July 1928, pp. 110–13.

gives rise to a random voltage (via Ohm's law). In honor of these fellows, thermal noise is often called Johnson noise or, less frequently, Nyquist noise.

Because the noise process is random, one cannot identify a specific value of voltage at a particular time (in fact, the amplitude has a Gaussian distribution), and the only recourse is to characterize the noise with statistical measures, such as the mean-square or root-mean-square values.

Because of the thermal origin, we would expect a dependence on the absolute temperature. It turns out that thermal noise power is exactly proportional to T (the astute might even guess that it is proportional to kT). Specifically, a quantity called the *available noise power* is given by

$$P_{NA} = kT\Delta f, \tag{1}$$

where k is Boltzmann's constant (about 1.38×10^{-23} J/K), T is the absolute temperature in kelvins, and Δf is the noise bandwidth in hertz (equivalent brickwall bandwidth) over which the measurement is made. We will clarify shortly what is meant by the terms "available noise power" and "noise bandwidth," but for now simply note that the noise source is very broadband (infinitely so, in fact, in the simplified picture presented here[4]), so that the total noise power depends on the measurement bandwidth.

Note that this equation tells us that the available noise power has a spectral density that is *independent of frequency,* and that the total power thus grows with bandwidth without limit. This is a bit of a lie, but it is true enough for all bandwidths of interest to the electrical engineer.

With Eqn. 1, we can compute that the available noise power over a 1-Hz bandwidth is about 4×10^{-21} W (or -174 dBm)[5] at room temperature. Further note that the constancy of the noise density implies that the thermal noise power is the same over any given *absolute* bandwidth. Therefore, the noise power in the interval between 1 MHz and 2 MHz is the same as between 1 GHz and 1.001 GHz. Because of this constancy, thermal noise is often described as "white," by analogy with white light. However, the analogy is not exact, since white light consists of constant energy per *wavelength* whereas white noise has constant energy per *hertz*.

The term "available noise power" is simply the maximum power that can be delivered to a load. Recall that the condition for maximum power transfer (for a resistive network) is equality of the load and source resistances. This suggests the use of the network shown in Figure 11.1 to compute the available noise power.

The model of the noisy resistor is enclosed within the dashed box, and is here shown as a noise voltage generator in series with the resistor itself. The power delivered by this noisy resistor to another resistor of equal value is by definition the available noise power:

---

[4] The implication that the available power grows without bound as the bandwidth approaches infinity should hint at the need to modify this formula. We will take care of this detail shortly.

[5] Recall that the reference level for 0 dBm is one milliwatt.

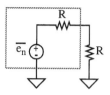

FIGURE 11.1. Network for computing
the thermal noise of a resistor.

$$P_{NA} = kT\Delta f = \frac{\overline{e_n^2}}{4R},$$ (2)

where $\overline{e_n}$ is the open-circuit rms noise voltage generated by the resistor R over the bandwidth Δf at a given temperature. The mean-square open-circuit noise voltage is therefore

$$\overline{e_n^2} = 4kTR\Delta f.$$ (3)

A couple of useful rules of thumb emerge from plugging some numbers into this last equation. At room temperature, a 1-kΩ resistor generates about 4 nV of rms noise over a bandwidth of 1 Hz, while a 50-Ω resistor generates a bit over 0.9 nV (some people just call it 1 nV, since it's easier to remember) of rms noise over that same bandwidth. These values are perhaps among the very few worth committing to memory. Just remember that the rms voltage is proportional to the square root of the bandwidth (and resistance) when scaling these numbers to any bandwidth or resistance.

In many cases, the noise is specified in terms of the spectral density rather than a total value, and is found simply by dividing the mean-square noise value by Δf (or the rms noise value by the square root of Δf). Thus, a 1-kΩ resistor has an rms noise spectral density of about 4 nV/$\sqrt{\text{Hz}}$ (funny units, yes) or a mean-square noise density of approximately 1.6×10^{-17} V^2/Hz. For thermal noise, the spectral density is a constant that depends only on temperature (and Boltzmann's constant), and is independent of frequency.[6] In the more general case, the density may vary with frequency, and one then uses the term "spot noise density" to underscore that the stated density applies only at some specified spot in the spectrum.

We've seen that every physical resistor has a noise source associated with it. In the Thévenin representation of Eqn. 1, there is a voltage source in series with the resistor. Alternatively, we may construct a Norton equivalent model in which a noise current source shunts the resistor. Both models are valid, of course, and the choice of which to use in a given situation is driven by practical considerations, such as computational convenience or intuitive value.

The two noise models for a resistor are displayed in Figure 11.2. Note that the polarity indications on the noise voltage source and the arrow on the noise current

---

[6] See footnote 4, however.

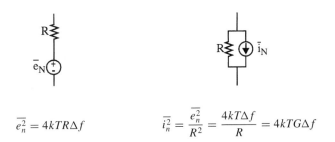

$$e_n^2 = 4kTR\Delta f \qquad \overline{i_n^2} = \frac{\overline{e_n^2}}{R^2} = \frac{4kT\Delta f}{R} = 4kTG\Delta f$$

FIGURE 11.2. Resistor thermal noise models.

source are simply references. They do not imply that the noise has a particular constant polarity (in fact, the noise has a zero mean).

Also note that, since the noise arises from the random thermal agitation of charge in the conductor, the only ways to reduce the noise of a given resistance are to keep the temperature as low as possible and to limit the bandwidth to the minimum useful value as well. Beyond these remedies, there is nothing that can be done about thermal noise.

Now, what about this "brickwall" bandwidth business? The distinction is made to underscore that the noise bandwidth Δf generally is *not* the same as the -3-dB bandwidth. Rather, the noise bandwidth is that of a perfect, brickwall (rectangular) filter that possesses the same area and peak value as the actual power gain-versus-frequency characteristic of the system, including that of the measurement apparatus. The noise bandwidth is therefore

$$\Delta f \equiv \frac{1}{|H_{pk}|^2} \int_0^\infty |H(f)|^2 \, df, \tag{4}$$

where H_{pk} is the peak value of the magnitude of the filter voltage transfer function $H(f)$.

This normalization concept allows comparisons to be made on a standard basis. As a specific example, consider a single-pole RC low-pass filter. We know that the -3-dB bandwidth (in hertz) is simply $1/2\pi RC$, but the equivalent noise bandwidth is computed as

$$\Delta f \equiv \frac{1}{|1|^2} \int_0^\infty \left[\frac{1}{(2\pi fRC)^2 + 1} \right] df = \frac{1}{2\pi RC} \arctan 2\pi fRC \Big|_0^\infty$$
$$= \frac{\pi}{2} f_{3\,dB} = \frac{1}{4RC}. \tag{5}$$

We see that a single-pole low-pass filter (LPF) has a noise bandwidth that is about 1.57 times the -3-dB bandwidth. That the noise bandwidth exceeds the -3-dB bandwidth makes sense, since the lazy rolloff of a single-pole filter allows spectral components of noise beyond the filter's -3-dB frequency to contribute significantly to the output energy of the filter.

A similar calculation shows that a critically damped, second-order low-pass filter has a noise bandwidth that is approximately 1.22 times the -3-dB bandwidth. In general, as the rolloff becomes steeper, the -3-dB and noise bandwidths tend to converge. This behavior should seem reasonable, since steeper rolloffs imply that the filter characteristics more closely approximate those of an ideal brickwall filter.[7]

Now it's time to tie up a few loose ends. It turns out that the spectral density of thermal noise actually increases with frequency, rather than remaining constant. This result follows from a more detailed derivation that takes into account the actual distribution of carrier energies, modified by considerations related to the Heisenberg uncertainty principle.[8] A more general expression for the thermal noise voltage is as follows:

$$\overline{e_n^2} = \frac{h\omega R \Delta f}{\pi} \coth\left(\frac{h\omega}{4\pi kT}\right), \tag{6}$$

where h is Planck's constant, about 6.62×10^{-34} J-s.

Although it is not immediately obvious, this new equation does not completely invalidate everything we did earlier. In fact, there is hardly any difference between this equation and $4kTR\Delta f$ for frequencies below about 80 THz at room temperature. The correction is thus negligible for all reasonable (non-optical[9]) frequencies, so electrical engineers ordinarily don't need to know about it. However, the birth of quantum theory actually traces back to Planck's resolution of a paradox (known as the "ultraviolet catastrophe") associated with the assumption of a constant spectral density in the context of the spectrum of blackbody radiation. This connection is so strong that one source of thermal noise in a wireless communications system is the noise associated with the blackbody radiation of the object on which the antenna is focused.[10]

Some other issues demand attention as well. First, some good news: purely reactive elements generate no thermal noise. Now, all real capacitors and inductors are lossy to some extent, and this loss implies that the impedance has a real part. This real resistance does generate thermal noise, but the purely capacitive or inductive

---

[7] An additional calculation shows that a second-order LPF with $\zeta = 1/\sqrt{2}$ (the minimum value that avoids peaking in the frequency response) has a noise bandwidth that is only about 11% larger than the -3-dB bandwidth. The more lightly damped pole pair gives a frequency response shape that has a flatter gain characteristic in the passband and faster initial rolloff beyond the -3-dB frequency than its more heavily damped cousin, thus more closely approximating a brickwall filter.

[8] H. Heffner, "The Fundamental Noise Limit of Linear Amplifiers," *Proc. IRE*, July 1962, pp. 1604–8.

[9] For reference, visible light spans about an octave, from about 400 THz to 800 THz.

[10] Arno Penzias and Robert Wilson of Bell Labs understood very well exactly how much noise a microwave receiver should exhibit. While meticulously tracking down a stubborn excess noise source they initially attributed to their equipment, they discovered the background microwave radiation that uniformly suffuses the universe. In an encounter with some cosmologists shortly afterward, they learned that the energy of this radiation agreed well with a prediction (by R. H. Dicke) of the energy of the echoes of the Big Bang. For their work, Penzias and Wilson received the 1978 Nobel Prize in physics. Not a bad reward for understanding noise. Let that be a lesson to us all.

parts of the impedance do not. However, note that noisy currents flowing through any impedance, reactive or real, will give rise to a noisy voltage.

It is also important to recognize that thermal noise is not associated with every element represented schematically with a resistor symbol. An important example is found in bipolar transistor models, where r_π is a fictitious resistance in the sense that it is the result of a linearization of the base-emitter junction's exponential $V–I$ characteristic; it does *not* generate thermal noise. However, the parasitic resistance terms, such as r_b and r_c, do.

THERMAL NOISE IN MOSFETs

Drain Current Noise

Since FETs (both junction and MOS) are essentially voltage-controlled resistors, they exhibit thermal noise. In the triode region of operation particularly, one would expect noise commensurate with the resistance value. Indeed, detailed theoretical considerations[11] lead to the following expression for the drain current noise of FETs:

$$\overline{i_{nd}^2} = 4kT\gamma g_{d0}\Delta f, \tag{7}$$

where g_{d0} is the drain–source conductance at zero V_{DS}. The parameter γ has a value of unity at zero V_{DS} and, in long devices, decreases toward a value of $2/3$ in saturation. Note that the drain current noise at zero V_{DS} is precisely that of an ordinary conductance of value g_{d0}.

Some measurements show that short-channel[12] NMOS devices exhibit noise considerably in excess of values predicted by long-channel theory, sometimes by large factors (e.g., by an order of magnitude in extreme cases). Some of the literature attributes this excess noise to carrier heating by the large electric fields commonly encountered in such devices. In this view, the high fields produce carriers with abnormally high energies. No longer in quasi–thermal equilibrium with the lattice, these hot carriers produce abnormal amounts of noise.

Complicating the picture are seemingly contradictory experimental results, with some in at least qualitative accord with the hot-carrier theory and others not. Recent work suggests that a proper acknowledgment of the role of substrate noise appears to go a long way toward explaining much of the excess noise without invoking a hot-carrier theory.[13] Other, more complete theoretical and experimental work casts doubt on the hot-carrier theory altogether, and it now appears safe to say that such putative high-field effects are not significant.[14]

---

[11] A. van der Ziel, "Thermal Noise in Field Effect Transistors," *Proc. IEEE,* August 1962, pp. 1801–12.

[12] As stated in Chapter 5, "short-channel" should be interpreted as "high electric field."

[13] J.-S. Goo et al., "Impact of Substrate Resistance on Drain Current Noise in MOSFETs," *Simulation of Semiconductor Processes and Devices (SISPAD)*, 2001, pp. 182–5.

[14] A. J. Scholten et al., "Noise Modeling for RF CMOS Circuit Simulation," *IEEE Trans. Electron Devices,* v. 50, no. 3, March 2003, pp. 618–32.

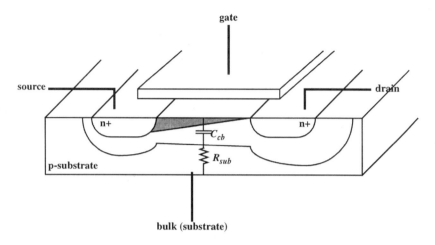

FIGURE 11.3. Simplified illustration of substrate thermal noise.

Figure 11.3 shows a simplified picture of how the thermal noise associated with the substrate resistance can produce measurable effects at the main terminals of the device. At frequencies low enough that we may ignore C_{cb}, the thermal noise of R_{sub} modulates the potential of the back gate, contributing some noisy drain current:

$$\overline{i^2_{nd,\text{sub}}} = 4kTR_{\text{sub}}g^2_{mb}\Delta f. \tag{8}$$

Depending on bias conditions – and also on the magnitude of the effective substrate resistance and size of the back-gate transconductance – the noise generated by this mechanism (often called *epi noise,* whether or not epitaxial layers are present) may actually exceed the thermal noise contribution of the ordinary channel charge. In this regime, layout strategies that reduce substrate resistance (e.g., liberal use of substrate contacts tied together to ground) have a noticeable and beneficial effect on noise.

At frequencies well above the pole formed by C_{cb} and R_{sub}, however, the substrate thermal noise becomes unimportant, as is readily apparent from inspection of the physical structure and the corresponding frequency-dependent expression for the substrate noise contribution:

$$\overline{i^2_{nd,\text{sub}}} = \frac{4kTR_{\text{sub}}g^2_{mb}}{1 + (\omega R_{\text{sub}}C_{cb})^2}\Delta f. \tag{9}$$

The characteristics of many IC processes are such that this pole is often around 1 GHz. Excess noise produced by this mechanism consequently will be most noticeable below about 1 GHz.

At voltages in excess of the nominal supply voltage limits of a technology, it is also possible for avalanche breakdown of drain or source junctions to add considerably to the noise. It is likely that this mechanism is at least partly responsible for the anomalously high levels of noise reported in some of the literature.[15]

---

[15] This confusion is cleared up in the excellent paper by Scholten et al., ibid.

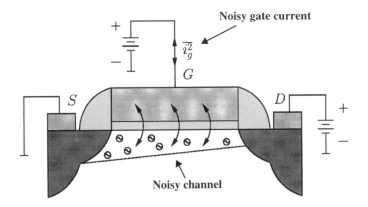

FIGURE 11.4. Induced gate noise.

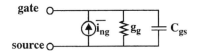

FIGURE 11.5. Gate noise circuit model
(after van der Ziel).

Gate Noise

In addition to drain current noise, the thermal agitation of channel charge has another important consequence: gate noise. The fluctuating channel potential couples
capacitively into the gate terminal, leading to a noisy gate current (see Figure 11.4).
Noisy gate current may also be produced by thermally noisy resistive gate material. Although this noise (whatever its source) is negligible at low frequencies, it can
dominate at radio frequencies. Van der Ziel[16] has shown that the gate noise may be
expressed as

$$\overline{i_{ng}^2} = 4kT\delta g_g \Delta f, \tag{10}$$

where the parameter g_g is

$$g_g = \frac{\omega^2 C_{gs}^2}{5g_{d0}}. \tag{11}$$

Van der Ziel gives a value of 4/3 (twice γ) for the gate noise coefficient, δ, in long-
channel devices.

The circuit model for gate noise that follows directly from Eqn. 8 and Eqn. 9 is a
conductance connected between gate and source, shunted by a noise current source
(see Figure 11.5). The gate noise current clearly has a spectral density that is not
constant. In fact, it increases with frequency, so perhaps it ought to be called "blue
noise" to continue the optical analogy.

For those who prefer not to analyze systems that have blue noise sources, it is
possible to recast the model in a form with a noise *voltage* source that possesses a

[16] A. van der Ziel, *Noise in Solid State Devices and Circuits,* Wiley, New York, 1986.

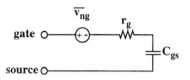

FIGURE 11.6. Alternative gate noise
circuit model.

constant spectral density.[17] To derive this alternative model, first transform the parallel RC network into an equivalent series RC network. If one assumes a reasonably high Q, then the capacitance stays roughly constant during the transformation. The parallel resistance becomes a series resistance whose value is

$$r_g = \frac{1}{g_g} \cdot \frac{1}{Q^2+1} \approx \frac{1}{g_g} \cdot \frac{1}{Q^2} = \frac{1}{5g_{d0}}, \tag{12}$$

which is independent of frequency.

Finally, equate the short-circuit currents of the original network and the transformed version, again with the assumption of high Q. The equivalent series noise voltage source is then found to be

$$\overline{v_{ng}^2} = 4kT\delta r_g \Delta f, \tag{13}$$

which possesses a constant spectral density. Hence, this alternative gate noise model consists of elements whose values are independent of frequency; see Figure 11.6.

Although the noise behavior of long-channel devices is fairly well understood, the precise behavior of δ in the short-channel regime is unknown at present. Given that both the gate noise and drain noise share a common origin, however, it is probably reasonable as a crude approximation to assume that δ continues to be about twice as large as γ. Hence, just as γ is typically 1–2 for short-channel NMOS devices, δ may be taken as 2–4.

Because the two noise sources do share a common origin, they are also correlated. That is, there is a component of the gate noise current that is proportional to the drain current on an instantaneous basis. We will explore the implications of this correlation, both quantitatively and qualitatively, in Chapter 12.

11.3   SHOT NOISE

Another noise mechanism, known as shot noise, was first described and explained by Schottky in 1918.[18] It is therefore occasionally known as Schottky noise in recognition of his achievement. The fundamental basis for shot noise is the granular nature

[17] D. Shaeffer and T. Lee, "A 1.5V, 1.5GHz CMOS Low Noise Amplifier," *IEEE J. Solid-State Circuits,* May 1997.

[18] "Über spontane Stromschwankungen in verschiedenen Electrizitätsleitern" ["On Spontaneous Current Fluctuations in Various Electrical Conductors"], *Annalen der Physik,* v. 57, 1918, pp. 541–67.

of the electronic charge, but how this granularity translates into noise is perhaps not as straightforward as one might think.

Two conditions must be satisfied for shot noise to occur. There must be a direct current flow *and* there must also be a potential barrier over which the charge carriers hop. The second condition tells us that ordinary, linear resistors do not generate shot noise, despite the quantized nature of electronic charge.

The fact that charge comes in discrete bundles means that there are discontinuous pulses of current every time an electron hops an energy barrier. *It is the randomness of the arrival times* that gives rise to the whiteness of shot noise. If all the carriers hopped simultaneously, shot noise would have a much more benign character, as we'll see in a later example.

We would expect the shot noise current to depend on the charge of the electron (since smaller charge would result in less lumpiness and therefore less noise), the total DC current flow (less current also means fewer lumps), and the bandwidth (just as with thermal noise). In fact, shot noise does depend on all of those quantities, as seen in the following equation:

$$\overline{i_n^2} = 2qI_{DC}\Delta f, \tag{14}$$

where $\overline{i_n}$ is the rms noise current, q is the electronic charge (about 1.6×10^{-19} C), I_{DC} is the DC current in amperes, and Δf is again the noise bandwidth in hertz. Note that, like thermal noise, shot noise (ideally) is white and has an amplitude that possesses a Gaussian distribution. As a reference point, the rms current noise density is approximately 18 pA/$\sqrt{\text{Hz}}$ for a 1-mA value of I_{DC}.

The requirement for a potential barrier implies that shot noise will only be associated with nonlinear devices, although not all nonlinear devices necessarily exhibit shot noise. For example, whereas both the base and collector currents *are* sources of shot noise in a bipolar transistor because potential barriers are definitely involved there (two junctions), only the DC gate leakage current of FETs (both MOS and junction types of FETs) contributes shot noise. Because this gate current is normally very small, it is rarely a significant noise source (sadly, though, the same cannot be said of base current).[19]

There are some additional observations worth noting at this point. As with the case of thermal noise, the spectral density does not remain constant to infinite frequency. However, significant departures from simple theory typically do not occur within the useful bandwidth of the devices, and we will consequently assume a constant spectral density in all that follows.

It was also mentioned that the randomness of the arrival times imparts to shot noise its characteristic whiteness. To underscore the importance of the randomness, consider what the shot noise spectrum would look like if all the carriers were to hop the potential barrier simultaneously, and if we could neglect the averaging effects of

---

[19] However, as mentioned previously, the thermally agitated channel charge induces noisy gate currents that can be important at radio frequencies.

nonzero transit time. More specifically, suppose we were to generate a 1.6-mA current by having ten well-trained electrons hop the barrier together every femtosecond. The spectrum of the current would be periodic, with impulses spaced at multiples of an inverse femtosecond, or 1 PHz (that's 10^{15} Hz). Therefore, *no* shot noise would be evident until the first noise component was reached, the frequency of which exceeds that of visible light! Alas, convincing electrons to exhibit this level of cooperation is somewhat difficult, and shot noise appears to be here to stay.

Finally, the term "shot" is not a corruption of "Schottky," as some occasionally assert. It is simply that if you hook up an audio system to a source of shot noise biased at a very low current, the resulting sound is much like that of buckshot (pellets) dropping onto a hard surface.

11.4 FLICKER NOISE

Without question, the most mysterious type of noise is flicker noise (also known as $1/f$ noise or pink[20] noise). No universal mechanism for flicker noise has been identified, yet it is ubiquitous. Phenomena that have no obvious connection, such as cell membrane potentials, the earth's rotation rate, galactic radiation noise, and transistor noise all have fluctuations with a $1/f$ character.

As the term "$1/f$" suggests, the noise is characterized by a spectral density that increases, apparently without limit, as frequency decreases. Measurements have verified this behavior in electronic systems down to a small fraction of a microhertz. One unfortunate implication of the increasing noise with decreasing frequency is the failure of averaging (bandlimiting) to improve measurement accuracy, since the noise power increases just as fast as the averaging interval.

Because of the lack of a unifying theory, mathematical expressions for $1/f$ noise invariably contain various empirical parameters (in contrast with the theoretical cleanliness of the equations for thermal and shot noise), as can be seen in the following equation:

$$\overline{N^2} = \frac{K}{f^n}\Delta f. \tag{15}$$

Here $\overline{N}$ is the rms noise (either voltage or current), K is an empirical parameter that is device-specific (and generally also bias-dependent), and n is an exponent that is usually (but not always) close to unity.

A question that often arises in connection with $1/f$ noise concerns the infinity at DC implied by a $1/f$ functional dependency. It's instructive to carry out a calculation with typical numbers to see why there is no problem, practically speaking.

First, let the parameter n have its commonly occurring value of unity. Then, integrate the density to find the total noise in a frequency band bounded by a lower frequency f_l and an upper frequency f_h:

---

[20] An optical system that accentuates energy at the lower visible frequencies reddens white light, so $1/f$ noise is "pink" by analogy.

$$\overline{N^2} = \int_{f_l}^{f_h} \frac{K}{f}\,df = K \ln\left(\frac{f_h}{f_l}\right). \tag{16}$$

This equation tells us that the total mean-square noise depends on the *log* of the frequency *ratio*, rather than simply on the frequency *difference* (as in thermal and shot noise). Hence, the mean-square value of $1/f$ noise is the same for equal frequency *ratios*; there is thus a certain constant amount of mean-square noise per *decade* of frequency, say, or some specific amount of rms noise per *root* decade of frequency (yes, funny units again for rms quantities).

As a specific numerical example, suppose measurements on an amplifier reveal that its $1/f$ noise has a density of 10 μV rms per root decade. Thus, for the 16-decade frequency interval below 1 Hz, the total $1/f$ noise would be just four times larger, or 40 μV rms. Recognize that 16 decades below 1 hertz is equal to one cycle about every 320 million years,[21] and you have to concede that "DC" infinities are simply not a practical problem. The resolution of the apparent paradox thus lies in recognizing that true DC implies an infinitely long observation interval, and that humans and the electronic age have been around for only a finite time. For any finite observation interval, the infinities simply don't materialize.

11.4.1 FLICKER NOISE IN RESISTORS

Flicker noise also shows up in ordinary resistors, where it is often called "excess noise," since this noise is in addition to what is expected from thermal noise considerations. It is found that a resistor exhibits $1/f$ noise only when there is DC current flowing through it, with the noise increasing with the current. In the discrete world, garden-variety carbon composition resistors are the most conspicuous offenders, while metal–film and wirewound resistors exhibit the smallest amounts of excess noise.

The current-dependent excess noise of carbon composition resistors has been explained by some as the result of the random formation and extinction of "micro-arcs" among neighboring carbon granules.[22] Carbon *film* resistors, which are made differently, exhibit much less excess noise than do carbon composition types. Whatever the explanation, it is certainly true that excess noise increases with the DC bias, so one should minimize the DC drop across a resistor.

The following approximate expression shows explicitly the dependency of this noise on various parameters:

$$\overline{e_n^2} = \frac{K}{f} \cdot \frac{R_\square^2}{A} \cdot V^2 \Delta f, \tag{17}$$

---

[21] Another useful quantity to keep in mind is the number of seconds in a year, which (to an excellent approximation) is the square root of 10^{15} or about 32 million.

[22] C. D. Motchenbacher and F. C. Fitchen, *Low-Noise Electronic Design*, Wiley, New York, 1973, p. 172.

where A is the area of the resistor, $R_\square$ is the sheet resistivity, V is the voltage across the resistor, and K is a material-specific parameter. For diffused and ion-implanted resistors, K has a value of roughly 5×10^{-28} S^2-m^2, whereas for thick-film resistors (not normally available in CMOS processes), K is about an order of magnitude larger.[23]

11.4.2 FLICKER NOISE IN MOSFETs

In electronic devices, $1/f$ noise arises from a number of different mechanisms and is most prominent in devices that are sensitive to surface phenomena. Hence, MOSFETs exhibit significantly more $1/f$ noise than do bipolar devices. One means of comparison is to specify a "corner frequency," where the $1/f$ and thermal or shot noise components are equal. All other things held equal, a lower $1/f$ corner implies less total noise. It is relatively trivial to build bipolar devices whose $1/f$ corners are below tens or hundreds of hertz, and many MOS devices routinely exhibit $1/f$ corners of tens of kilohertz to a megahertz or more.

Charge trapping phenomena are usually invoked to explain $1/f$ noise in transistors. Some types of defects and certain impurities (most plentiful at a surface or interface of some kind) can randomly trap and release charge. The trapping times are distributed in a way that can lead to a $1/f$ noise spectrum in both MOS and bipolar transistors. Since MOSFETs are surface devices (at least in the way that they are conventionally fabricated), they exhibit this type of noise to a much greater degree than bipolar transistors (which are bulk devices). Larger MOSFETs exhibit less $1/f$ noise because their larger gate capacitance smooths the fluctuations in channel charge. Hence, if good $1/f$ noise performance is to be obtained from MOSFETs, the largest practical device sizes must be used (for a given g_m).

The mean-square $1/f$ drain noise current is given by

$$\overline{i_n^2} = \frac{K}{f} \cdot \frac{g_m^2}{WLC_{\text{ox}}^2} \cdot \Delta f \approx \frac{K}{f} \cdot \omega_T^2 \cdot A \cdot \Delta f, \tag{18}$$

where A is the area of the gate $(= WL)$ and K is a device-specific constant. Thus, for a fixed transconductance, a larger gate area and a thinner dielectric reduce this noise term.

For PMOS devices, K is typically about 10^{-28} C^2/m^2, whereas for NMOS devices it is about 50 times larger.[24] One should keep in mind that these constants vary considerably from process to process, and even from run to run, so the values of K given here should be treated as crude estimates. In particular, the superior $1/f$ performance of PMOS devices may be a temporary situation, as it is due to the use of buried channels that may cease to be widely used in the future.

[23] K. Laker and W. Sansen, *Design of Analog Integrated Circuits and Systems,* McGraw-Hill, New York, 1996.
[24] Laker and Sansen, ibid.

11.4.3 FLICKER NOISE IN JUNCTIONS

Forward-biased junctions also exhibit $1/f$ noise. The noise is proportional to the bias current and inversely proportional to the junction area:

$$\overline{i_j^2} = \frac{K}{f} \cdot \frac{I}{A_j} \cdot \Delta f, \tag{19}$$

where the constant K typically has a value of around 10^{-25} A-m^2. Once again, however, considerable variation from process to process is not uncommon.[25]

Flicker noise in bipolar transistors is attributed entirely to the base-emitter junction (since it is the only one in forward bias). It has been established experimentally that only the base current exhibits $1/f$ noise.

11.5 POPCORN NOISE

Another type of noise that can plague semiconductors is known as popcorn noise (also called burst noise, bistable noise, and random telegraph signals, RTS). It is understood even more poorly than $1/f$ noise, and it shares with $1/f$ noise a sensitivity to contamination. Gold-doped[26] bipolar transistors exhibit the highest levels of burst noise, suggesting a particular sensitivity to contamination by metal ions specifically, although not all popcorn noise may be the result of metal ion contamination.

This noise was first observed in point-contact diodes but has also been seen in ordinary junction and tunnel diodes, some types of resistors, and both discrete and integrated circuit junction transistors. Burst noise is characterized by its multimodal (most often bimodal) and hence non-Gaussian amplitude distribution. That is, the noise switches between two or more discrete values, but at random times. The switching intervals tend to be in the audio range (e.g., 10 μs on up), and the popping sound that is heard when a burst noise source is connected to an audio system is why this is known as "popcorn" noise.

As a practical matter, describing popcorn noise mathematically is not a terribly useful exercise,[27] since it is so variable. Some devices exhibit little or no popcorn noise, while others – nominally fabricated the same way – may show large amounts of it. In all cases, meticulous cleanliness in processing is the key to controlling popcorn noise, and describing it with quasiempirical equations therefore has limited practical value. But, for the sake of completeness, here's an equation for it anyway:

$$\overline{N^2} = \frac{K}{1 + (f/f_c)^2} \Delta f. \tag{20}$$

---

[25] Ibid.

[26] The purposeful reduction of carrier lifetime by gold doping is occasionally used in bipolar devices to speed recovery from saturation.

[27] My sincere apologies to the authors of the many excellent dissertations and papers on the phenomenon. But I stand by my statement.

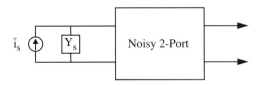

FIGURE 11.7. Noisy two-port driven by noisy source.

Here K is an empirical, device- and fabrication-dependent (and, again, generally bias-dependent) constant, and f_c is a corner frequency below which the burst noise density flattens out.

For frequencies well above f_c, the total mean-square noise between f_l and f_h is

$$Kf_c^2 \left[\frac{1}{f_l} - \frac{1}{f_h} \right]. \tag{21}$$

You will probably never need to use Eqn. 21.

11.6 CLASSICAL TWO-PORT NOISE THEORY

Having developed detailed noise models for individual devices, we now turn to a macroscopic description of noise in two-ports. Focusing on such system noise models can greatly simplify analysis and lead to the acquisition of useful design insight.

11.6.1 NOISE FACTOR

A useful measure of the noise performance of a system is the noise factor, usually denoted F. To define it and understand why it is useful, consider a noisy (but linear) two-port driven by a source that has an admittance Y_s and an equivalent shunt noise current $\overline{i_s}$ (see Figure 11.7).

If we are concerned only with overall input–output behavior, it is an unnecessary complication to keep track of all of the internal noise sources. Fortunately, the net effect of all of those sources can be represented by just one pair of external sources: a noise voltage and a noise current. This huge simplification allows rapid evaluation of how the source admittance affects the overall noise performance. As a consequence, we can identify the criteria one must satisfy for optimum noise performance.

The noise factor is defined as

$$F \equiv \frac{\text{total output noise power}}{\text{output noise due to input source}}, \tag{22}$$

where, by convention, the source is at a temperature of 290 K.[28] The noise factor is a measure of the degradation in signal-to-noise ratio that a system introduces. The

---

[28] You might wonder why a relatively cool 290 K is the reference temperature. The reason is simply that kT is then 4.00×10^{-21} J. Like many practical engineers, Harald Friis of Bell Labs preferred round numbers (see his "Noise Figures of Radio Receivers," *Proc. IRE,* July 1944, pp. 419–22). His suggestion of 290 K as the reference temperature had particular appeal in an era of slide-rule computation, and it was adopted rapidly by engineers and standards committees.

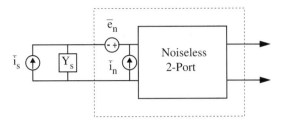

FIGURE 11.8. Equivalent noise model.

larger the degradation, the larger the noise factor. If a system adds no noise of its own then the total output noise is due entirely to the source, and the noise factor is therefore unity.

In the model of Figure 11.8, all of the noise appears as inputs to the noiseless network, so we may compute the noise figure there. A calculation based directly on Eqn. 22 requires the computation of the total power due to all of the sources, and dividing that result by the power due to the input source. An equivalent (and simpler) method is to compute the total short-circuit mean-square noise current and then divide that total by the short-circuit mean-square noise current due to the input source. This alternative method is equivalent because the individual power contributions are proportional to the short-circuit mean-square current, with a proportionality constant (which involves the current division ratio between the source and two-port) that is the same for all of the terms.

In carrying out this computation, one generally encounters the problem of combining noise sources that have varying degrees of correlation with one another. In the special case of zero correlation, the individual *powers* superpose. For example, if we assume, as seems reasonable, that the noise powers of the source and of the two-port are uncorrelated, then the expression for noise figure becomes

$$F = \frac{\overline{i_s^2} + \overline{|i_n + Y_s e_n|^2}}{\overline{i_s^2}}. \tag{23}$$

Note that, although we have assumed that the noise of the source is uncorrelated with the two equivalent noise generators of the two-port, Eqn. 23 does *not* assume that the two-port's generators are also uncorrelated with each other.

In order to accommodate the possibility of correlations between e_n and i_n, express i_n as the sum of two components. One, i_c, is correlated with e_n, and the other, i_u, isn't:

$$i_n = i_c + i_u. \tag{24}$$

Since i_c is correlated with e_n, it may be treated as proportional to it through a constant whose dimensions are those of an admittance:

$$i_c = Y_c e_n; \tag{25}$$

the constant Y_c is known as the *correlation admittance*.

Combining Eqn. 23, Eqn. 24, and Eqn. 25, the noise factor becomes

$$F = \frac{\overline{i_s^2} + \overline{|i_u + (Y_c + Y_s)e_n|^2}}{\overline{i_s^2}} = 1 + \frac{\overline{i_u^2} + |Y_c + Y_s|^2 \overline{e_n^2}}{\overline{i_s^2}}. \tag{26}$$

The expression in Eqn. 26 contains three independent noise sources, each of which may be treated as thermal noise produced by an equivalent resistance or conductance (whether or not such a resistance or conductance actually is the source of the noise):

$$R_n \equiv \frac{\overline{e_n^2}}{4kT\Delta f}, \tag{27}$$

$$G_u \equiv \frac{\overline{i_u^2}}{4kT\Delta f}, \tag{28}$$

$$G_s \equiv \frac{\overline{i_s^2}}{4kT\Delta f}. \tag{29}$$

Using these equivalences, the expression for noise factor can be written purely in terms of impedances and admittances:

$$\begin{aligned} F &= 1 + \frac{G_u + |Y_c + Y_s|^2 R_n}{G_s} \\ &= 1 + \frac{G_u + [(G_c + G_s)^2 + (B_c + B_s)^2]R_n}{G_s}, \end{aligned} \tag{30}$$

where we have explicitly decomposed each admittance into a sum of a conductance G and a susceptance B.

11.6.2 OPTIMUM SOURCE ADMITTANCE

Once a given two-port's noise has been characterized with its four noise parameters (G_c, B_c, R_n, and G_u), Eqn. 30 allows us to identify the general conditions for minimizing the noise factor. Taking the first derivative with respect to the source admittance and setting it equal to zero yields

$$B_s = -B_c = B_{\text{opt}}, \tag{31}$$

$$G_s = \sqrt{\frac{G_u}{R_n} + G_c^2} = G_{\text{opt}}. \tag{32}$$

Hence, to minimize the noise factor, the source susceptance should be made equal to the inverse of the correlation susceptance, while the source conductance should be set equal to the value in Eqn. 32.

The noise factor corresponding to this choice is found by direct substitution of Eqn. 31 and Eqn. 32 into Eqn. 30:

$$F_{\min} = 1 + 2R_n[G_{\text{opt}} + G_c] = 1 + 2R_n\left[\sqrt{\frac{G_u}{R_n} + G_c^2} + G_c\right]. \tag{33}$$

We may also express the noise factor in terms of $F_{\min}$ and the source admittance:

$$F = F_{\min} + \frac{R_n}{G_s}[(G_s - G_{\text{opt}})^2 + (B_s - B_{\text{opt}})^2]. \tag{34}$$

Thus, contours of constant noise factor are non-overlapping circles in the admittance plane.[29]

The ratio R_n/G_s appears as a multiplier in front of the second term of Eqn. 34. For a fixed source conductance, R_n tells us something about the relative sensitivity of the noise figure to departures from the optimum conditions. A large R_n implies a high sensitivity; circuits or devices with high R_n obligate us to work harder to identify, achieve, and maintain optimum conditions. We will shortly see that operation at low bias currents is associated with large R_n, in keeping with the general intuition that achieving high performance only gets more difficult as the power budget tightens.

It is important to recognize that, although minimizing the noise factor has something of the flavor of maximizing power transfer, the source admittances leading to these conditions are generally not the same – as is apparent by inspection of Eqn. 31 and Eqn. 32. For example, there is no reason to expect the correlation susceptance to equal the input susceptance (except by coincidence). As a consequence, one must generally accept less than maximum power gain if noise performance is to be optimized, and vice versa.

11.6.3 LIMITATIONS OF CLASSICAL NOISE OPTIMIZATION

The classical theory just presented implicitly assumes that one is given a device with particular, fixed characteristics, and defines the source admittance that will yield the minimum noise figure given such a device. Although one starts with fixed devices in discrete RF design, the freedom to choose device dimensions in IC realizations points out a serious shortcoming of the classical approach: There are no specific guidelines about what device size will minimize noise. Furthermore, power consumption is frequently a parameter of great interest (even an obsessive one in many portable applications), but power is simply not considered at all in classical noise optimization. We will return to these themes in great detail in the chapter on LNA design, but for now simply be aware of the incompleteness of the classical approach.

11.6.4 NOISE FIGURE AND NOISE TEMPERATURE

In addition to noise factor, other figures of merit that often crop up in the literature are noise figure and noise temperature. The noise figure (NF) is simply the noise factor expressed in decibels.[30]

---

[29] They are also circles when plotted on a Smith chart because the mapping between the two planes is a bilinear transformation, which preserves circles.

[30] Just to complicate matters, the definitions for noise factor and noise figure are switched in some texts.

Table 11.1. *Noise figure, noise factor, and noise temperature*

NF (dB)	F	T_N (K)
0.5	1.122	35.4
0.6	1.148	43.0
0.7	1.175	50.7
0.8	1.202	58.7
0.9	1.230	66.8
1.0	1.259	75.1
1.1	1.288	83.6
1.2	1.318	92.3
1.5	1.413	120
2.0	1.585	170
2.5	1.778	226
3.0	1.995	289
3.5	2.239	359

Noise temperature, T_N, is an alternative way of expressing the effect of an amplifier's noise contribution and is defined as the increase in temperature required of the source resistance for it to account for all of the output noise at the reference temperature T_{ref} (which is 290 K). It is related to the noise factor as follows:

$$F = 1 + \frac{T_N}{T_{\text{ref}}} \implies T_N = T_{\text{ref}} \cdot (F - 1). \tag{35}$$

An amplifier that adds no noise of its own has a noise temperature of 0 K.

Noise temperature is particularly useful for describing the performance of cascaded amplifiers (as discussed further in Chapter 19) and those whose noise factor is quite close to unity (or whose noise figure is very close to 0 dB), since the noise temperature offers a higher-resolution description of noise performance in such cases. This can be seen in Table 11.1. Noise figures in the range of 2–3 dB are generally considered very good, with values around or below 1 dB considered outstanding.

11.7   EXAMPLES OF NOISE CALCULATIONS

Here are some examples of noise calculations to tie up a few loose ends and generally reinforce this material.

Example 1

Quite often, a network will consist of a number of individual noise sources. This example looks at a way to simplify calculations by combining individual noise sources before diving into messy math. Consider the noisy resistive network shown in Figure 11.9. Let's compute the total noise as measured across the output terminals. We'll perform the computation two ways.

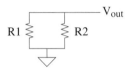

FIGURE 11.9. Noisy resistive network.

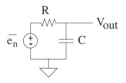

FIGURE 11.10. Capacitively
loaded noisy resistor.

First, we compute the rms noise voltage that each resistor contributes to the output. Then, assuming that each resistor generates noise that is uncorrelated with that of the other (a reasonable assumption here, one would hope), we combine the noise sources in root-sum-squared (rss) fashion to find the rms value of the overall noise. Proceeding in this manner yields the following sequence of computations:

$$\overline{V_{no1}} = \overline{e_{n1}} \frac{R_2}{R_1 + R_2}, \tag{36}$$

$$\overline{V_{no2}} = \overline{e_{n2}} \frac{R_1}{R_1 + R_2}. \tag{37}$$

Combining the individual noise sources yields, after some crunching:

$$\overline{V_{no}^2} = \overline{V_{no1}^2} + \overline{V_{no2}^2} = 4kT(R_1 \parallel R_2)\Delta f. \tag{38}$$

Examination of the final result tells us that one may save a little labor by simply combining the resistances into one equivalent resistance at the outset and then computing the rms noise of that single equivalent resistance. Having illustrated the longer way, you can well appreciate the utility of first combining resistances before plunging into a noise calculation, particularly for more complex networks.

Example 2

Suppose that the only bandwidth limitation in making resistor noise measurements were the ever-present stray capacitance of any physical setup. Derive an expression for the mean-square noise for a network consisting of a resistor R shunted by a capacitance C.

The circuit under consideration is thus as sketched in Figure 11.10, where $\overline{e_n}$ represents the thermal noise of the resistor. Recall that a single-pole RC filter has a noise bandwidth that is $\pi/2$ times the -3-dB bandwidth, and that the -3-dB bandwidth in turn is $1/2\pi RC$ Hz (telling us that the noise bandwidth is just $1/4RC$ Hz). Hence, we may compute the mean-square output voltage noise without too much trouble:

$$\overline{V_{no}^2} = 4kTR\left(\frac{\pi}{2}\frac{1}{2\pi RC}\right) = \frac{kT}{C}. \tag{39}$$

Thus, we see that the mean-square noise voltage is *independent* of the resistance in this case. The reason that the resistance drops out is that a larger resistance has a proportionally larger noise source but also a proportionally smaller bandwidth, so that the total mean-square noise voltage remains constant for a given capacitance.

A deeper insight is that kT represents the maximum available thermal noise energy, while the mean-square energy in a capacitor is simply CV^2. Equating the two and solving for the mean-square voltage leads to the derived expression – namely, kT/C.

Example 3

For a number of years at an advanced technological university in the northeastern United States, an unintentionally cruel hoax was perpetrated on a succession of unsuspecting undergraduates by an equally unsuspecting lecturer. The students were assigned, for their bachelor's thesis, the task of designing and building an amplifier with a -3-dB bandwidth of 1 MHz and a midband voltage gain of 10^6. The source resistance was given as ranging between 100 kΩ and 1 MΩ, and the supply voltage was the old standard ±15 V. Assuming that these hapless students could solve the *enormous* stability problems associated with trying to build an amplifier with a gain–bandwidth product of 1 THz (an extremely difficult problem in its own right), let's see if the thermal noise in the source resistance imposes any significant fundamental limits.

Generously choosing the lower limit of source resistance, we note that the thermal noise of a 100-kΩ resistor has an rms density of 40 nV/$\sqrt{\text{Hz}}$ at room temperature. We expect the noise bandwidth to be greater than the -3-dB bandwidth of 1 MHz, but we'll continue to be generous and assume that the noise bandwidth is also about 1 MHz. The total rms noise contributed by the resistor at the input of the amplifier is therefore about 40 μV, meaning that the noise at the output of this gain-of-10^6 amplifier is calculated to be 40 V rms if we continue to pretend that all of the output noise is due solely to the source resistor. Even with this unrealistically generous assumption, we note that the calculated rms output noise exceeds the total supply voltage by about 33%, so that the amplifier would spend most of its time slamming from rail to rail just from the noise of the input source resistance alone! And naturally, the real situation would be even worse, since the amplifier would necessarily contribute some additional noise of its own. The task, as assigned, is *impossible* unless cryogenic means are made available! This example underscores the importance of first carrying out back-of-the-envelope sanity checks before investing a lot of design effort.

11.8 A HANDY RULE OF THUMB

Making measurements of noise can be rather involved if it is to be done accurately. Typically, a special noise figure instrument (or possibly a spectrum analyzer) is required to determine the noise density as a function of frequency. For very quick assessments of relatively large amounts of noise, a crude measurement is sometimes acceptable. In those cases, an oscilloscope and your eyeball may be the only instruments you need. If we assume that the noise is Gaussian, then the peak-to-peak values very rarely exceed about 5–7 times the rms value. Hence the level-zero eyeball measurement is to connect the noisy DUT to the oscilloscope, make some judgment about what the displayed peak-to-peak value seems to be, and then divide by about 6 to develop an estimate of the rms value.

This method is *very* crude, of course, owing in no small measure to the difficulty of determining what the "true" peak-to-peak value happens to be. The situation is further complicated by the fact that the oscilloscope brightness setting affects what appear to be the peaks; the brighter the trace, the taller the apparent peaks. The same operator may also make significantly different determinations at different times as a function of sleep deprivation, emotional state, and caffeine levels.

A clever extension of the eyeball technique removes much of this uncertainty by converting the measurement into a differential one.[31] Here, the noisy signal simultaneously drives both channels of a dual-trace oscilloscope operating in alternating sweep mode, rather than chop mode (to avoid introducing a correlation between the two sweeps through the oscilloscope's chopping oscillator). With a sufficiently large initial position difference, there will be a dark band between these two traces. Adjust the position controls until the dark band just disappears, with the two traces merging into a single blurry mess with a monotonically decreasing brightness from the center outward. Note that this description implies an independence of the result on the absolute intensity. Remove the noisy signals, and then measure the distance between the two baselines. The resulting value is twice the rms voltage to a good approximation. Absolute accuracies of about 1 dB are possible with this simple method.

The reason this technique works is that a sum of two identical Gaussian distributions has a maximally flat top when the two distributions are separated by exactly twice the rms value.

Because the eye is an imperfect judge of contrast, it is not possible to establish with infinite precision when the dark band disappears. When following the procedure as outlined, most people will perceive the band to have disappeared a little before it actually does. The error resulting from this uncertainty is on the order of 1 dB for most people. Thus, perhaps 0.5 dB should be subtracted from the measurement if you

---

[31] G. Franklin and T. Hatley, "Don't Eyeball Noise," *Electronic Design,* v. 24, November 22, 1973, pp. 184–7.

are very fussy. An alternative is to measure the noise two different ways, one using the procedure given and another with the two traces initially on top of each other. With the latter initial condition, adjust the spacing until the darker area first seems to *appear*. Average the two readings, and also compute the difference between the two readings as a measure of uncertainty. With care and a little practice, sub–1-dB repeatability is readily achievable.

11.9 TYPICAL NOISE PERFORMANCE

To develop an appreciation for what noise performance one may expect in practice, here are a number of examples.

General-purpose circuits tend to be somewhat noisy. For example, the very popular 741 op-amp has poor noise performance, with an equivalent input noise voltage density of around 20 nV/$\sqrt{\text{Hz}}$ (about that of a 20-kΩ resistor) and an equivalent input current noise density of approximately 200 fA/$\sqrt{\text{Hz}}$ (consistent with the 741's input bias current of 100 nA). Both the voltage and current noise typically exhibit $1/f$ corners somewhere between 100 Hz and 1 kHz. The relatively large voltage noise of the 741 can be attributed to the use of active devices as loads, since active devices essentially amplify their own internal noise.

More modern amplifiers have much better noise performance. For example, the OP-27 from Analog Devices exhibits a noise voltage density of around 3 nV/$\sqrt{\text{Hz}}$ and input current noise that is similar to that of a 741, with most of the improvements attributable to the use of ordinary resistive loads in the first stage and a reduction in parasitic base resistance. Process modifications also result in a spectacularly low $1/f$ corner of about 3 Hz for the voltage noise.

Because a major source of input current noise is simply the shot noise associated with the input current itself, one would expect FET-input amplifiers to fare better in this regard. So it is not too surprising that the OP-215 JFET-input op-amp exhibits 10 fA/$\sqrt{\text{Hz}}$ of input current noise density (at room temperature), about a factor of 20 better than the 741.

It is rather challenging to design low-noise amplifiers for 50-Ω RF systems, since the thermal noise associated with the source impedance is so small. In particular, it is critically important to minimize the parasitic base resistance (for bipolar implementations) or gate resistance (for FET circuits). For example, a 50-Ω base resistance already makes it impossible to achieve a noise figure better than 3 dB. To obtain sufficiently small gate or base resistances, it is generally necessary to use a parallel combination of many smaller unit devices.

Despite these difficulties, silicon bipolar RF LNAs (low-noise amplifiers) with noise figures of around 2 dB at 1 GHz are available from a number of sources. This level of noise performance is impressive because a noise figure of 2 dB implies an absolute maximum parasitic base resistance of about 30 Ω. Stated another way, the equivalent input voltage noise of the amplifier must be below 0.7 nV/$\sqrt{\text{Hz}}$ – well below what is commonplace for general-purpose circuits such as op-amps.

Early reports of CMOS LNAs in the gigahertz range also showed great promise, with noise figures in the neighborhood of 3 dB and about 10 mW of power consumption at 1.5 GHz.[32] As we'll see in the next chapter, that modest level of performance by no means represents the best that may be achieved. Further scaling and more careful design have resulted in even better performance.

Note that noise figure is the common way to report noise performance in RF systems since the reference impedance level is known, while equivalent noise voltage and current generators are more commonly used to describe the noise of general-purpose building blocks that might be used with a wide range of source impedances.

11.10 APPENDIX: NOISE MODELS

This appendix summarizes the noise models presented earlier.

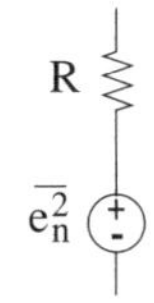

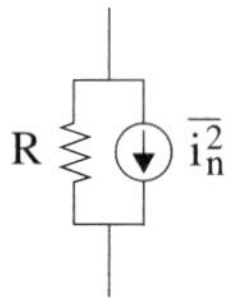

$$\overline{e_n^2} = 4kTR\Delta f \qquad\qquad \overline{i_n^2} = \frac{\overline{e_n^2}}{R^2} = \frac{4kT\Delta f}{R} = 4kTG\Delta f$$

FIGURE 11.11. Resistor thermal noise models.

The following excess mean-square noise voltage is added to the mean-square thermal noise voltage in Figure 11.11 if $1/f$ noise is of interest:

$$\overline{e_n^2} = \frac{K}{f} \cdot \frac{R_\square^2}{A} \cdot V^2\Delta f. \tag{40}$$

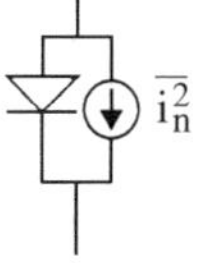

$$\overline{i_n^2} = 2qI_{\mathrm{DC}}\Delta f + \frac{K_1}{f^n}\Delta f$$

FIGURE 11.12. Diode noise model.

For the diode noise model (Figure 11.12), the $1/f$ term may be expanded as follows:

$$\overline{i_j^2} = \frac{K}{f} \cdot \frac{I}{A_j} \cdot \Delta f. \tag{41}$$

---

[32] Shaeffer and Lee, op. cit. (see footnote 17).

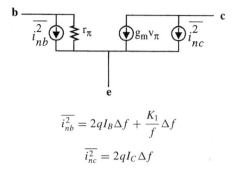

$$\overline{i_{nb}^2} = 2qI_B\Delta f + \frac{K_1}{f}\Delta f$$

$$\overline{i_{nc}^2} = 2qI_C\Delta f$$

FIGURE 11.13. Bipolar transistor
noise model.

The more detailed $1/f$ behavior of Eqn. 41 applies also to the bipolar model of Figure 11.13, which neglects the thermal noise of r_b. In many practical amplifiers, however, this noise is quite important, and may even dominate.

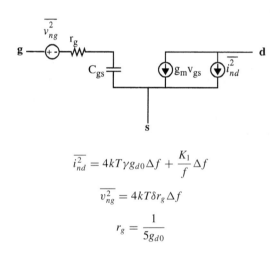

$$\overline{i_{nd}^2} = 4kT\gamma g_{d0}\Delta f + \frac{K_1}{f}\Delta f$$

$$\overline{v_{ng}^2} = 4kT\delta r_g\Delta f$$

$$r_g = \frac{1}{5g_{d0}}$$

FIGURE 11.14. MOS noise model.

For the MOS model of Figure 11.14, the $1/f$ term may be expanded as follows:

$$\overline{i_n^2} = \frac{K}{f} \cdot \frac{g_m^2}{WLC_{ox}^2} \cdot \Delta f \approx \frac{K}{f} \cdot \omega_T^2 \cdot A \cdot \Delta f. \tag{42}$$

PROBLEM SET FOR NOISE

PROBLEM 1 Show formally that the noise figure of a resistive attenuator is equal to its attenuation. In your answer, define carefully what "attenuation" specifically refers to.

PROBLEM 2 Derive a formula for the overall noise figure of a cascade of systems, as shown in Figure 11.15. Here, each noise factor F is computed with respect to the output impedance of the previous stage. Furthermore, each power gain G is the available power gain – that is, the power available under matched conditions.

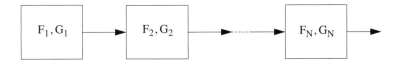

FIGURE 11.15. Cascaded systems for computation of noise figure.

From your formula, what do you deduce about the relative contributions to noise figure of earlier versus later stages?

PROBLEM 3 Derive an expression for the low-frequency noise figure of a *bipolar* transistor. Use the model of Figure 11.13 but with r_b and its thermal noise included. Ignore flicker noise, and assume that all noise sources in the model are uncorrelated. Assume that the source resistance is R_s. *Hint:* Since there are only three noise sources in the device, it is probably not worth deriving a two-port equivalent noise model in this case.

The equation you are about to derive is frequently called Nielsen's equation, after the fellow who first published it.

PROBLEM 4 Using your answer to Problem 3, derive expressions for $F_{\min}$ and R_{opt}.

PROBLEM 5 Because low input current is a highly desirable attribute in many amplifiers, designers have evolved numerous techniques for achieving this goal. The problem is much more challenging for bipolar amplifiers because of the fundamental need for base current. One obvious method is simply to use transistors with large β and then operate the input stage at a low current. Unfortunately, the former trades off with base resistance (among others), and the latter choice degrades g_m and f_T, so gain and speed can suffer. An alternative that is frequently used is to cancel the base current with an internal current mirror, so that the external world does not have to supply it. That way, the input stage can be biased at a relatively large current without causing a large current at the input terminals.

(a) What is the (low-frequency) input shot noise current density if no input current cancellation is used and the base current is 100 nA? We are looking only at the pure shot noise due to the base current; do not worry about reflecting any other noise sources to the input.

(b) Now suppose that we succeed in cancelling nearly all of the input current by using a 99-nA internal current source so that the input current (as determined by external measurements) decreases to 1 nA. What is the input shot noise current density in this case? Comment on the advantages and disadvantages of the current-cancellation method.

PROBLEM 6   A common problem is how to choose an amplifier for best system noise performance. Manufacturers might supply data about the equivalent input noise voltage and current, but engineers sometimes draw incorrect inferences from this information, particularly if they read too much meaning into the term "optimum impedance."

Suppose we have a choice between two amplifiers. They both have $10\,\text{nV}/\sqrt{\text{Hz}}$ input noise voltage density, but amplifier A has $50\,\text{fA}/\sqrt{\text{Hz}}$ input noise current density while amplifier B has twice as much.

(a) What is the optimum source resistance for each amplifier? You may ignore correlation between the noise sources.
(b) If the source resistance happens to be $100\,\text{k}\Omega$, which op-amp should you use? Assume that (ever-elusive) ideal, broadband, arbitrary-ratio, lossless transformers are available.
(c) For your choice in (b), what is the best possible system noise figure?

PROBLEM 7   We've seen that one serious problem with RF ICs is the lack of high-Q inductors. Since low Q is caused by dissipation and since active feedback can compensate for energy loss, there have been many proposals over the years for various active inductor schemes.

One (but by no means the only) problem common to all of these active schemes is that of limited dynamic range. That is, it is not sufficient merely to synthesize an element with inductive small-signal impedance. It is also important to have a large dynamic range, bounded for small signals by the noise floor and for large signals by the linearity ceiling.

This problem explores some noise properties of active inductors. Although we will examine one particular circuit configuration, the outlines of the result apply quite generally to all active inductor circuits, leading to the depressing conclusion that active inductors have limited utility.

(a) First consider, for comparison purposes, a passive parallel RLC tank. What is the equivalent shunt mean-square noise current density? The L and C are ideal lossless elements.
(b) Now consider a magical two-port that reciprocates impedances. Such an element is called a *gyrator,* and its properties were first explored by the Dutch theorist B. D. H. Tellegen of Philips. Impedance reciprocation is potentially relevant because IC processes give us good capacitors. Combining gyrators and good capacitors might then give us good inductors (so goes the classic argument).

One simple way to make an active inductor is with the circuit shown in Figure 11.16. (A quick history note: This circuit derives from the *reactance tube* connection devised in the 1930s for electronic tuning and FM generation.) Note that biasing details are not shown; simply assume that the transistor is somehow biased to produce a specified value of transconductance g_m.

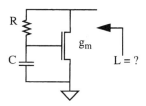

FIGURE 11.16. Active inductor.

Assume that the RC frequency is much lower than the operating frequency of the inductor. Further assume that the resistance R is itself much larger than the reactance of the inductor at all frequencies of interest, and that the transistor model has no reactances and no parasitic resistances. With these assumptions, derive an expression for the inductance.

(c) Derive an expression for the Q of this inductor as a function of frequency if the capacitance is entirely due to the C_{gs} of the transistor. Otherwise, make use of the same assumptions as in part (b). Express your answer in terms of ω_T.

(d) Assume that all of the noise of this circuit comes from just two sources: the resistor R and the drain current noise of the MOSFET. That is, neglect gate current noise. As a consequence, the calculation will underestimate the actual noise somewhat. Even so, we'll see that the news is depressing enough.

What is the most general expression for the short-circuit drain current noise density due solely to the MOSFET itself?

(e) What is the component of short-circuit drain current noise density due to the thermal noise of the resistor?

(f) What is the short-circuit current noise density due directly to the resistor itself? Continue to assume that the RC_{gs} frequency is much lower than the operating frequency of the inductor.

(g) Assume that these noise sources are uncorrelated, and incorporate your expression for Q derived in part (c) to derive a general expression for the total short-circuit noise current density.

Comparing your answers to (a) and (g), you should be able to draw some conclusions about the noise properties of active inductors. Further acknowledging large-signal limitations as well, one must conclude that active inductors have serious problems that are difficult – perhaps even impossible – to evade.

PROBLEM 8 Derive a general expression for the noise bandwidth of a second-order low-pass filter. Using your formula, verify that a critically damped second-order filter has a noise bandwidth that is about 1.22 times the -3-dB bandwidth.

PROBLEM 9 A single-pole amplifier has a voltage gain of 1000 and a -3-dB bandwidth of 1 kHz.

(a) What is the total mean-square output noise voltage if the input noise spectral density is 10^{-15} V^2/Hz and flat? Assume that this noise source completely models all the noise in the system.

(b) Repeat part (a) if the input noise spectral density is not flat and in fact has the following behavior:

$$\frac{\overline{v_{ni}^2}}{\Delta f} = 10^{-15}\left(1 + \frac{10 \text{ Hz}}{f}\right). \tag{P11.1}$$

PROBLEM 10 Explain under what conditions one may add noise sources in root-sum-squared fashion. Derive a more general "addition" law.

PROBLEM 11 Consider the model shown in Figure 11.17 for a sample-and-hold circuit. The resistor models the finite on-resistance of the sampling switch. Since that resistance is thermally noisy, the sampled-and-held voltage will also be noisy. Compute the mean-square value of the output noise.

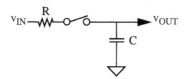

FIGURE 11.17. Sample-and-hold circuit.

PROBLEM 12 Suppose that you are given a single-ended amplifier that possesses a certain equivalent input noise voltage and current. You may assume that these are uncorrelated with each other. Demonstrate that a differential amplifier constructed with two of these single-ended amplifiers achieves the same minimum noise figure but at twice the power consumption. In addition, specify the optimum differential source impedance for the differential amplifier in terms of the optimum single-ended source impedance for the single-ended LNA.

PROBLEM 13 The relationship between reflection coefficient and admittance is given by the following bilinear transformation:

$$\Gamma = \frac{1 - Y}{1 + Y}, \tag{P11.2}$$

where Y is the admittance, normalized to the characteristic impedance of the system (e.g., 50 Ω). Using this relationship, recast the equation for noise factor,

$$F_{\min} = 1 + 2R_n[G_{\text{opt}} + G_c] = 1 + 2R_n\left[\sqrt{\frac{G_u}{R_n} + G_c^2} + G_c\right], \tag{P11.3}$$

in terms of the real and imaginary parts of the reflection coefficient Γ. Are the resulting contours still circles? Are they ever concentric?

PROBLEM 14 For a simple CMOS differential pair biased with a single-transistor MOS current source, find the equivalent mean-square drain current noise of one transistor in the pair in terms of the drain current noise sources of the three transistors, the transconductances of the differential devices, and the output resistance of the current source. You may ignore the output resistance of the differential devices and assume that the gates are driven by a low-resistance source. Do *not* assume that the pair is in the balanced state.

CHAPTER TWELVE

LNA DESIGN

12.1 INTRODUCTION

The first stage of a receiver is typically a low-noise amplifier (LNA), whose main function is to provide enough gain to overcome the noise of subsequent stages (such as a mixer). Aside from providing this gain while adding as little noise as possible, an LNA should accommodate large signals without distortion, and frequently must also present a specific impedance, such as 50 Ω, to the input source. This last consideration is particularly important if a passive filter precedes the LNA, since the transfer characteristics of many filters are quite sensitive to the quality of the termination.

In principle, one can obtain the minimum noise figure from a given device by using the optimum source impedance defined by the four noise parameters: G_c, B_c, R_n, and G_u. This classical approach has important shortcomings, however, as described in the previous chapter. For example, the source impedance that minimizes the noise figure generally differs, perhaps considerably, from that which maximizes the power gain. Hence, it is possible for poor gain and a bad input match to accompany a good noise figure. Additionally, power consumption is an important consideration in many applications, but classical noise optimization simply ignores power consumption altogether. Finally, such an approach presumes that one is given a device with fixed characteristics, and thus offers no explicit guidance on how best to exercise the IC designer's freedom to tailor device geometries.

To develop a design strategy that balances gain, input impedance, noise figure, and power consumption, we will derive analytical expressions for the four noise parameters directly from the device noise model and will then examine several LNA architectures. As we'll see, insights gained from that exercise allow us to design narrowband LNAs with near-minimum noise figure, along with an excellent impedance match and good power gain, all within a power budget that is specified a priori. An important collateral result is a simple formula that yields the optimum device width for a given technology, source impedance, and operating frequency.

12.2   DERIVATION OF INTRINSIC MOSFET TWO-PORT NOISE PARAMETERS

Recall that the MOSFET noise model consists of two sources. The mean-square drain current noise is

$$\overline{i_{nd}^2} = 4kT\gamma g_{d0}\Delta f; \tag{1}$$

the gate current noise is

$$\overline{i_{ng}^2} = 4kT\delta g_g \Delta f, \tag{2}$$

where

$$g_g = \frac{\omega^2 C_{gs}^2}{5g_{d0}}. \tag{3}$$

Further recall that the gate noise is correlated with the drain noise, with a correlation coefficient defined formally as

$$c \equiv \frac{\overline{i_{ng} \cdot i_{nd}^*}}{\sqrt{\overline{i_{ng}^2} \cdot \overline{i_{nd}^2}}}. \tag{4}$$

When the reference direction for the gate noise is from the source to gate (as in Figure 11.5) and that for the drain noise is from the drain to the source (as in Figure 11.14), then the long-channel value of c is theoretically $-j0.395$.[1] Precise measurements of the correlation coefficient are difficult to carry out (especially in the deep submicron regime), but the best published measurements reveal that its magnitude stays within a factor of 2 of this theoretical value, even for devices with drawn channel lengths as small as $0.13\,\mu\text{m}$ (70-nm effective length).[2] For simplicity, we will assume that c remains at its long-channel value in all of the numerical examples that follow. Furthermore, we will neglect the thermal noise due to the resistive gate material, although this source of noise can actually dominate the gate noise when operating devices well below f_T, where nonquasistatic effects (such as induced gate noise) will be less prominent.[3] We will also neglect C_{gd} to simplify the derivation. While the

---

[1] Regrettably, the choice of reference directions here results in a sign that is the opposite of that in van der Ziel's original treatment as well as the treatment in A. J. Scholten et al., "Noise Modeling for RF CMOS Circuit Simulation," *IEEE Trans. Electron Devices*, v. 50, no. 3, March 2003, pp. 618–32.

[2] Ibid. The bulk of the data set is for devices down to $0.18\text{-}\mu\text{m}$ drawn channel length. Corresponding measurements (kindly conveyed by Dr. Scholten) on $0.13\text{-}\mu\text{m}$ devices show correlation coefficient magnitudes of about 0.2 at 5 GHz. Because the resistive part of the gate electrode contributes gate noise current that is fully correlated with its drain current consequences, the overall gate current noise has a net correlation coefficient of magnitude between 0.2 and 1.0.

[3] Ibid. For $0.18\text{-}\mu\text{m}$, 70-GHz f_T transistors operating at 3 GHz, they report that nonquasistatic effects (induced gate noise) account for only about 30% of the gate noise, even after taking extraordinary measures in layout (e.g., double contacts on each gate finger, subdivision into 64 $3\text{-}\mu\text{m}$–wide fingers, etc.) to minimize the contribution by resistive gate material.

achievable noise figure is little affected by C_{gd}, the input impedance can be a strong function of C_{gd}, and this effect must be taken into account when designing the input matching network.

To derive the four equivalent two-port noise parameters, repeated here for convenience,

$$R_n \equiv \frac{\overline{e_n^2}}{4kT\Delta f},$$ (5)

$$G_u \equiv \frac{\overline{i_u^2}}{4kT\Delta f},$$ (6)

$$Y_c \equiv \frac{i_c}{e_n} = G_c + jB_c,$$ (7)

we first reflect the two fundamental MOSFET noise sources back to the input port as a different pair of equivalent input generators (one voltage and one current source).

The equivalent input noise voltage generator accounts for the output noise observed when the input port is short-circuited (incrementally speaking). To determine its value, reflect the drain current noise back to the input as a noise voltage and recognize that the ratio of these quantities is simply g_m. Thus,

$$\overline{e_n^2} = \frac{\overline{i_{nd}^2}}{g_m^2} = \frac{4kT\gamma g_{d0}\Delta f}{g_m^2},$$ (8)

from which it is apparent that the equivalent input noise voltage is completely correlated, and in phase, with the drain current noise. Thus, we can immediately determine that

$$R_n \equiv \frac{\overline{e_n^2}}{4kT\Delta f} = \frac{\gamma g_{d0}}{g_m^2}.$$ (9)

The equivalent input noise voltage generator by itself does not fully account for the drain current noise, however, because a noisy drain current also flows even when the input is *open*-circuited and induced gate current noise is ignored. Under this open-circuit condition, dividing the drain current noise by the transconductance yields an equivalent input voltage which, when multiplied in turn by the input admittance, gives us the value of an equivalent input current noise that completes the modeling of i_{nd}:

$$\overline{i_{n1}^2} = \frac{\overline{i_{nd}^2}(j\omega C_{gs})^2}{g_m^2} = \frac{4kT\gamma g_{d0}\Delta f(j\omega C_{gs})^2}{g_m^2} = \overline{e_n^2}(j\omega C_{gs})^2.$$ (10)

In this step of the derivation, we have assumed that the input admittance of a MOSFET is purely capacitive. This assumption is a good approximation for frequencies well below ω_T, if appropriate high-frequency layout practice is observed to minimize gate resistance. Given this assumption, Eqn. 10 shows that the input noise current i_{n1} is in quadrature, and therefore completely correlated, with the equivalent input noise voltage e_n.

The total equivalent input current noise is the sum of the reflected drain noise contribution of Eqn. 10 and the induced gate current noise. The induced gate noise current itself consists of two terms. One, which we'll denote i_{ngc}, is fully correlated with the drain current noise, while the other, i_{ngu}, is completely uncorrelated with the drain current noise. Hence, we may express the correlation admittance as follows:

$$Y_c = \frac{i_{n1} + i_{ngc}}{e_n} = j\omega C_{gs} + \frac{i_{ngc}}{e_n}$$

$$= j\omega C_{gs} + \frac{g_m}{i_{nd}} \cdot i_{ngc} = j\omega C_{gs} + g_m \cdot \frac{i_{ngc}}{i_{nd}}. \tag{11}$$

To express Y_c in a more useful form, we need to incorporate the gate noise correlation factor explicitly. To do so, we must manipulate the last term of Eqn. 11 in ways that will initially appear mysterious. First, we express it in terms of cross-correlations by multiplying both numerator and denominator by the conjugate of the drain noise current and then averaging each:

$$g_m \cdot \frac{i_{ngc}}{i_{nd}} = g_m \cdot \frac{\overline{i_{ngc} \cdot i_{nd}^*}}{i_{nd} \cdot i_{nd}^*} = g_m \cdot \frac{\overline{i_{ngc} \cdot i_{nd}^*}}{\overline{i_{nd}^2}} = g_m \cdot \frac{\overline{i_{ng} \cdot i_{nd}^*}}{\overline{i_{nd}^2}}. \tag{12}$$

The last equality, in which i_{ng} replaces i_{ngc}, is valid because the uncorrelated portion of the gate noise current necessarily contributes nothing to the cross-correlation.

Using Eqn. 11, we may write the correlation admittance as

$$Y_c = j\omega C_{gs} + g_m \cdot \frac{\overline{i_{ng} \cdot i_{nd}^*}}{\overline{i_{nd}^2}} = j\omega C_{gs} + g_m \cdot \frac{\overline{i_{ng} \cdot i_{nd}^*}}{\sqrt{\overline{i_{nd}^2}}\sqrt{\overline{i_{nd}^2}}}\sqrt{\frac{\overline{i_{ng}^2}}{\overline{i_{ng}^2}}}, \tag{13}$$

which, in turn, may be expressed as

$$Y_c = j\omega C_{gs} + g_m \cdot \frac{\overline{i_{ng} \cdot i_{nd}^*}}{\sqrt{\overline{i_{ng}^2} \cdot \overline{i_{nd}^2}}}\sqrt{\frac{\overline{i_{ng}^2}}{\overline{i_{nd}^2}}} = j\omega C_{gs} + g_m \cdot c\sqrt{\frac{\overline{i_{ng}^2}}{\overline{i_{nd}^2}}}, \tag{14}$$

which explains all of the maneuvering, since the correlation coefficient has finally made an explicit appearance.

Substituting for the term under the radical yields

$$Y_c = j\omega C_{gs} + g_m \cdot c\sqrt{\frac{\delta\omega^2 C_{gs}^2}{5\gamma g_{d0}^2}} = j\omega C_{gs} + \frac{g_m}{g_{d0}} \cdot c\sqrt{\frac{\delta}{5\gamma}} \cdot \omega C_{gs}. \tag{15}$$

If we assume that c continues to be purely imaginary, even in the short-channel regime, we finally obtain a useful expression for the correlation admittance:

$$Y_c = j\omega C_{gs} - j\omega C_{gs}\frac{g_m}{g_{d0}} \cdot |c|\sqrt{\frac{\delta}{5\gamma}} = j\omega C_{gs}\left(1 - \alpha|c|\sqrt{\frac{\delta}{5\gamma}}\right), \tag{16}$$

Table 12.1. *Summary of intrinsic MOSFET two-port noise parameters*

Parameter	Expression
G_c	~ 0
B_c	$\omega C_{gs}\left(1 - \alpha|c|\sqrt{\dfrac{\delta}{5\gamma}}\right)$
R_n	$\dfrac{\gamma g_{d0}}{g_m^2} = \dfrac{\gamma}{\alpha} \cdot \dfrac{1}{g_m}$
G_u	$\dfrac{\delta\omega^2 C_{gs}^2(1 - |c|^2)}{5g_{d0}}$

where we have used the substitution

$$\alpha = \frac{g_m}{g_{d0}}. \tag{17}$$

Since α is unity for long-channel devices and progressively decreases as channel lengths shrink, it is one measure of the departure from the long-channel regime.

We see from Eqn. 16 that the correlation admittance is purely imaginary, so that $G_c = 0.$[4] More significant, however, is the fact that Y_c does not equal the admittance of C_{gs}, although it is some multiple of it. Hence, one cannot maximize power transfer and minimize noise figure simultaneously. To investigate further the important implications of this impossibility, though, we need to derive the last remaining noise parameter, G_u.

Using the definition of the correlation coefficient, we may express the induced gate noise as follows:

$$\overline{i_{ng}^2} = \overline{(i_{ngc} + i_{ngu})^2} = 4kT\Delta f \delta g_g |c|^2 + 4kT\Delta f \delta g_g (1 - |c|^2). \tag{18}$$

The very last term in Eqn. 18 is the uncorrelated portion of the gate noise current, so that, finally,

$$G_u \equiv \frac{\overline{i_u^2}}{4kT\Delta f} = \frac{4kT\Delta f \delta g_g (1 - |c|^2)}{4kT\Delta f} = \frac{\delta\omega^2 C_{gs}^2 (1 - |c|^2)}{5g_{d0}}. \tag{19}$$

A summary of the four noise parameters appears in Table 12.1. With these parameters, we can determine both the source impedance that minimizes the noise figure as well as the minimum noise figure itself:

$$B_{\text{opt}} = -B_c = -\omega C_{gs}\left(1 - \alpha|c|\sqrt{\frac{\delta}{5\gamma}}\right). \tag{20}$$

---

[4] Again, this conclusion is based on a neglect of any resistive term at the input.

Table 12.2. *Estimated*
$F_{\min}$ ($\gamma = 2$, $\delta = 4$)

ω_T/ω	$F_{\min}$ (dB)
20	0.5
15	0.6
10	0.9
5	1.6

From Eqn. 20, we see that the optimum source susceptance is essentially inductive in character, except that it has the wrong frequency behavior. Hence, achieving a broadband noise match is fundamentally difficult.

Continuing, the real part of the optimum source admittance is

$$G_{\text{opt}} = \sqrt{\frac{G_u}{R_n} + G_c^2} = \alpha\omega C_{gs}\sqrt{\frac{\delta}{5\gamma}(1 - |c|^2)}, \qquad (21)$$

and the minimum noise figure is given by

$$F_{\min} = 1 + 2R_n[G_{\text{opt}} + G_c] \approx 1 + \frac{2}{\sqrt{5}}\frac{\omega}{\omega_T}\sqrt{\gamma\delta(1 - |c|^2)}. \qquad (22)$$

In Eqn. 22, the approximation is exact if one treats ω_T as simply the ratio of g_m to C_{gs}. Note that if there were no gate current noise (i.e., if δ were zero), the minimum noise figure would be 0 dB. That unrealistic prediction alone should be enough to suspect that gate noise must indeed exist. Also note that, in principle, increasing the correlation between drain and gate current noise would improve noise figure, although correlation coefficients unrealistically near unity would be required to effect large reductions in noise figure.

Another important observation is that improvements in ω_T that accompany technology scaling also improve the noise figure at any given frequency. To underscore this point, let us assign numerical values to the parameters in Eqn. 22. However, because the detailed behavior of some of these parameters in the short-channel regime is unknown, we will have to make some educated guesses to arrive at estimates of $F_{\min}$. As mentioned in Chapter 11, some measurements of noise in short-channel devices have recently been shown by Scholten et al. (see footnote 1) to be highly suspect, if not grossly in error. The noise parameters for short-channel devices in fact are larger than their long-channel values by less than factors of about 1.5. Nevertheless, in much of what follows we will assume a doubling or tripling of these parameters, out of a sense of conservatism. And as mentioned previously, we will further assume that $|c|$ remains equal to 0.395, even in the short-channel regime.

Table 12.2 shows $F_{\min}$ as a function of normalized frequency if short-channel effects cause a tripling in γ and δ. To the extent that the assumptions made are reasonable, it is encouraging that excellent noise figures are possible, even with increased γ and δ, for frequencies well below ω_T.

The rapid pace of change in IC technology virtually guarantees an incomplete understanding of the behavior of transistors of the most recent generations of technology. If design for low noise is to proceed nonetheless, we must develop an alternative approach to obtaining model parameters. A purely empirical method has the appeal of relying on no theoretical assumptions at all, but its practical implementation requires the design of a suitable, finite experimental suite. Fortunately, the very dimensions of the two-port noise parameters themselves suggest a compact method for obtaining data suitable for use over a wide range of conditions – and suitable also for the extraction of relevant noise parameters for other simulation models.

We've already noted that the two-port noise parameters are impedances or admittances. Even though they do not necessarily represent physical quantities that are directly measurable, they nonetheless obey the same scaling laws. For example, suppose we completely characterize the noise parameters for one MOSFET of some unit size. For constant bias voltages (implying constant current densities) the impedance parameters for two such devices in parallel are half that for a single one, and the admittance parameters are double. Dimensionless parameters remain unchanged. Generalizing, the noise parameters for a device of width W are

$$G_c = \frac{W}{W_0} G_{c0}, \tag{23}$$

$$B_c = \frac{W}{W_0} B_{c0}, \tag{24}$$

$$G_u = \frac{W}{W_0} G_{u0}; \tag{25}$$

$$R_n = \frac{W_0}{W} R_{n0}. \tag{26}$$

The subscript "0" identifies a parameter corresponding to a unit device of width W_0. The noise factor, F, is a dimensionless quantity and is therefore independent of width (though the source admittance that produces a given F scales with width).

These scaling laws mean that one need only characterize the noise parameters for a single device size (which we'll call a unit cell). Although this characterization must include variation of bias over some range, obviating the need to sweep device width as well is of self-evident value. Extrapolations based on these scaling relationships will be accurate as long as one takes care to select a reference device large enough to make negligible any fixturing and layout parasitics and as long as one truly uses multiple instances of this unit cell in the final design, as is well-established practice for good matching in analog design.

It should be mentioned also that using multiple gate fingers and contacting both ends of each one is essential for minimizing the thermal noise contribution from the gate material (and associated vias, contacts, and so on). Although the optimum finger width certainly depends on process details, a reasonable rule of thumb for current technologies is to choose widths on the order of 5 μm.

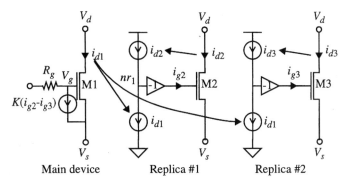

(Arrows point from independent variable to dependent quantity.)

FIGURE 12.1. MOS macromodel to accommodate gate noise (some biasing details omitted for clarity).

A second practical issue concerns the problem of simulation. Most MOS models do not accommodate *induced* gate noise (but they generally do account – at least partially – for the thermal noise of the resistive gate material), and this lack seriously compounds the difficulty of simulating the noise performance of CMOS amplifiers. Fortunately, it is still possible to cobble together a useful substitute: we can build a macromodel out of ordinary SPICE elements and then use it to replace each noise-critical MOSFET in a circuit[5] (see Figure 12.1). For the sake of simplicity, we will choose to set the correlation between gate and drain current noise equal to zero in what follows. We will also not explicitly account for epi noise (substrate thermal noise) here.

Transistor M_1 is the transistor whose noise behavior is to be modeled. Current-controlled current sources transfer copies of its (noisy) drain current to the gate circuits of two replica devices (indirectly, through voltage inverters), each of which is identical to the main device and biased to the same voltages (the drains are not all directly tied together, and neither are the sources; use voltage-controlled voltage sources to enforce equality of terminal voltages). Each replica possesses a provision for feeding back its own noisy drain current to its own gate node (again, through inverters). Summation of the two noisy (and fully uncorrelated) currents there results in a noisy voltage at the gate of M_2, with

$$\overline{v_{g2}^2} = \frac{8kT\Delta f}{g_{d0}} \approx \frac{8kT\Delta f}{g_m}, \tag{27}$$

where the last approximation assumes that g_{d0} is not too different from g_m. The noisy gate voltage causes to flow a noisy gate current whose approximate mean-square value is

[5] D. Shaeffer and T. Lee, *The Design and Implementation of Low-Power CMOS Radio Receivers*, Kluwer, Dordrecht, 1999. To aid in DC convergence, you will probably want to add some small conductance across both dependent sources of value i_{d1}.

$$\overline{i_{g2}^2} \approx \frac{(8kT\Delta f)\gamma[\omega(C_{gs} + C_{gd})]^2}{g_m}. \tag{28}$$

Because half of this gate current noise is the direct result of a contribution by M_1's drain noise, these two noise currents are partially correlated. The same observation holds for the gate noise of M_3. Suppose that our ultimate aim is to produce a gate current noise in M_1 that has zero correlation with its own drain current noise. In that case we must remove the correlation altogether. This decorrelation is readily accomplished by noting that the two replica devices have identical correlated contributions from M_1. Simple subtraction of the gate currents of M_2 and M_3 therefore results in cancellation of the common correlated components. If some calibrated residual correlation is desired (to reflect reality a bit better), then the gate currents should be scaled by different factors prior to subtraction. We do not undertake here the more general goal of controlling the magnitude or phase of this correlated remnant.

After subtraction, the difference current must be scaled appropriately to produce the correct noise current in the gate circuit of M_1,

$$\overline{i_g^2} \approx \frac{(4kT\Delta f)\delta[\omega C_{gs}]^2}{5g_m}, \tag{29}$$

where once again we neglect the effective gate current produced by the resistive gate electrode.

Setting these last two equations equal to each other, we see that the scale factor K should be chosen approximately equal to

$$K \approx \frac{1}{\sqrt{5}(1 + C_{gd}/C_{gs})}, \tag{30}$$

where we have assumed that δ/γ remains equal to about 2, the long-channel value.

For typical process technologies, the drain–gate capacitance is often between one third and one half of the gate–source capacitance. A typical value of K is therefore approximately in the general range of 0.3–0.4. A value closer to unity could be more appropriate in cases where gate resistivity is important.

So many simplifying assumptions are embedded in the foregoing development that you should consider this macromodel approach as largely a stopgap measure – to be used only if you have no alternative. Despite its limitations, the macromodel is nonetheless useful for enabling progress in the absence of detailed device models.

Fortunately, compact and accurate device models for RF noise simulation have recently become widely available. The public-domain MOS11 model from Philips fully accommodates nonquasistatic (NQS) effects by segmenting a device into several pieces, each of which models some portion of the channel and within which the segment behaves quasistatically (the principle is similar to that which allows us to subdivide a transmission line into segments consisting only of lumped elements). Not only does it therefore correctly account for the NQS component of gate noise (and its counterpart, input resistance), it also accounts for the gate noise due to resistive gate electrode material. Similarly, the model accommodates substrate noise and

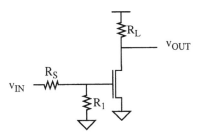

FIGURE 12.2. Common-source amplifier with
shunt input resistor (biasing not shown).

conventional channel thermal noise in all transistor operational regimes.[6] Published measurements show excellent agreement with experiment.

Having presented a crude macromodel – as well as two methods for obtaining noise parameters suitable for use with commercial simulators that support sophisticated models such as MOS11 – we now consider in detail their implications for low-noise amplifier design. Without loss of generality, we use the analytical equations in what follows.

12.3 LNA TOPOLOGIES: POWER MATCH VERSUS NOISE MATCH

The derivations of the previous section show that, for a MOSFET, the source impedance that yields minimum noise factor is inductive in character and generally unrelated to the conditions that maximize power transfer. Furthermore, the input impedance of a MOSFET is inherently capacitive, so providing a good match to a 50-Ω source without degrading noise performance would appear to be difficult. Since presenting a known resistive impedance to the external world is almost always a critical requirement of LNAs, we will first examine a number of circuit topologies that accomplish this feat and then narrow the field of contenders by evaluating their noise properties.

One straightforward approach to providing a reasonably broadband 50-Ω termination is simply to put a 50-Ω resistor across the input terminals of a common-source amplifier; this is shown in Figure 12.2.

Unfortunately, the resistor R_1 adds thermal noise of its own and so attenuates the signal (by a factor of 2) ahead of the transistor. The combination of these two effects generally produces unacceptably high noise figures.[7] More formally, it is straightforward to establish the following lower bound on the noise figure of this circuit:

$$F \geq 2 + \frac{4\gamma}{\alpha} \cdot \frac{1}{g_m R}, \tag{31}$$

---

[6] See http://www.semiconductors.philips.com/Philips_Models/ for documentation on this and other models.

[7] As a specific example, one recently published 800-MHz CMOS amplifier has a noise figure in excess of 11 dB (reported as 6 dB by ignoring the attenuation and noise of R_1).

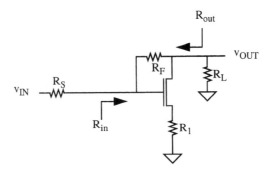

FIGURE 12.3. Shunt–series amplifier (biasing not shown).

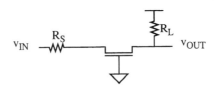

FIGURE 12.4. Common-gate amplifier (biasing not shown).

where $R_S = R_1 = R$. This bound applies only in the low-frequency limit and ignores gate current noise altogether. Naturally, the noise figure is worse at higher frequencies and when gate noise is taken into account.

The shunt–series amplifier, described in Chapter 9, is another circuit that provides a broadband real input impedance. Since it does not reduce the signal with a noisy attenuator before amplifying, we expect its noise figure to be substantially better than that of the circuit of Figure 12.2.

The amplifier sketched in Figure 12.3 suffers from fewer problems than the previous circuit, yet the resistive feedback network continues to generate thermal noise of its own and also fails to present to the transistor an impedance that equals Z_{opt} at all frequencies (perhaps at any frequency). As a consequence, the overall amplifier's noise figure, while usually much better than that of Figure 12.2, still generally exceeds the device F_{min} by a considerable amount (typically a few decibels). Nonetheless, the broadband capability of this circuit is frequently enough of a compensating advantage that the shunt–series amplifier is found in many LNA applications, even though its noise figure is not the minimum possible.[8]

Another method for realizing a resistive input impedance is to use a common-gate configuration. Since the resistance looking into the source terminal is $1/g_m$, a proper selection of device size and bias current can provide the desired 50-Ω resistance; see Figure 12.4.

---

[8] Clearly, the noise figure of this circuit can be low enough to be quite practical, for it is used in an instrument that *measures* noise figure (the 8970A, made by Hewlett-Packard), as one example.

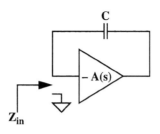

FIGURE 12.5. Impedance
transformation model.

As with the circuit of Figure 12.2, it is straightforward to establish the following lower bound on the noise figure of the common-gate amplifier (again, at low frequencies and neglecting gate current noise):

$$F \geq 1 + \gamma/\alpha. \tag{32}$$

This bound assumes that the resistance looking into the source terminal is adjusted to equal the source resistance, and is about 2.2 dB in the long-channel limit and perhaps as high as 4.8 dB for short devices ($\gamma/\alpha = 2$). The noise figure will be significantly worse at high frequencies and when gate current noise is taken into account.

All three of the preceding topologies suffer noise figure degradation from the presence of noisy resistances in the signal path (including channel resistances, as in the case of the common-gate amplifier). Fortunately, one can provide a resistive input impedance without resistors, contrary to intuition. The first hint of this possibility actually traces back to the vacuum tube era, so a brief historical digression is in order.

Both vacuum tubes and MOSFETs are devices with nominally capacitive input impedances. The key word, however, is "nominally," for if the input impedances actually were purely capacitive then the input could never consume any power, and the power gain would necessarily be infinite even at infinitely high frequency. Such a result defies common sense (and experiment), so one might surmise that the input impedance must possess a resistive component. A MOSFET's gate structure is a parallel plate capacitor, though, and it is perhaps difficult to imagine how a resistive component could arise.[9] The answer is that the bottom plate of the gate capacitor is not at a fixed potential. Rather, the potential varies all along the channel, from source to drain, in a manner that is dependent on the signal on the gate. Furthermore, and most important, finite carrier velocity causes this signal-dependent bottom plate potential to lag somewhat, resulting in both a departure from a pure quadrature relationship between voltage and current and, necessarily, a resistive component in the input impedance.

The essential features of this mechanism may be understood by examining the abstraction of Figure 12.5. The amplifier is ideal in all respects except for a frequency-dependent gain $A(s)$. The input terminal is analogous to the gate (or grid) terminal,

---

[9] Here, we are implicitly assuming that the drain is terminated in a short.

and the amplifier output is connected to the bottom plate of the gate capacitance. The input impedance of this circuit is

$$Z_{\text{in}} = \frac{1}{sC[1 + A(s)]}. \tag{33}$$

Now let $A(s)$ have gain and phase shift:[10]

$$A(s) = A_0 e^{-j\phi}; \tag{34}$$

then

$$Z_{\text{in}} = \frac{1}{j\omega C[1 + A_0 e^{-j\phi}]} = \frac{1}{j\omega C[1 + A_0(\cos\phi - j\sin\phi)]}. \tag{35}$$

Collecting terms and focusing on the denominator, we obtain

$$Y_{\text{in}} = j\omega C[1 + A_0\cos\phi] + A_0\omega C\sin\phi, \tag{36}$$

from which it is apparent that the input admittance indeed possesses a real part whose value depends on the phase lag ϕ. With zero phase lag, the admittance is purely capacitive, as anticipated from quasistatic analyses. If the more realistic scenario of a nonzero phase lag is considered, the equivalent shunt conductance is seen to increase with frequency. Perhaps it is no surprise that measurements show that the phase lag itself grows with frequency, and the equivalent shunt conductance typically increases as the square of frequency, to a good approximation.

Transit-time effects also cause a resistive component of input impedance in vacuum tubes, where the phenomenon was first observed. Because of the finite velocity of charge, then, a real term is an unavoidable reality in charge-controlled devices such as vacuum tubes and FETs.

In the context of low-noise amplifiers, we actually seek to *enhance* this effect, for it can be used to create a resistive input impedance without the noise of real resistors. From the foregoing, it is clear that one possible method is to modify the device (e.g., elongate it) in order to enhance transit time effects directly. However, this approach has the undesirable side effect of degrading high-frequency gain.

A better method is to employ inductive source degeneration.[11] With such an inductance, current flow lags behind an applied gate voltage, behavior that is qualitatively similar to the mechanism described. An important advantage of this method is that one then has control over the value of the real part of the impedance through choice of inductance, as is clear from computing the input resistance of the circuit shown in Figure 12.6.

---

[10] The form shown is chosen simply for convenience. Any function with a phase lag would show the effect as well.

[11] This type of circuit was first investigated in the context of vacuum tube circuits by M. J. O. Strutt and A. van der Ziel, "The Causes for the Increase of the Admittances of Modern High-Frequency Amplifier Tubes on Short Waves," *Proc. IRE,* v. 26, 1936, pp. 1011–32. Aldert van der Ziel would spend the rest of his career studying noise.

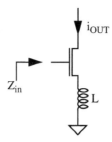

FIGURE 12.6. Inductively degenerated
common-source amplifier.

To simplify the analysis, consider a device model that includes only a transconductance and a gate–source capacitance. In that case, it is not hard to show that the input impedance has the following form:

$$Z_{\text{in}} = sL + \frac{1}{sC_{gs}} + \frac{g_m}{C_{gs}}L \approx sL + \frac{1}{sC_{gs}} + \omega_T L. \tag{37}$$

Hence, the input impedance is that of a series RLC network, with a resistive term that is directly proportional to the inductance value.

More generally, an arbitrary source degeneration impedance Z is modified by a factor equal to $[\beta(j\omega) + 1]$ when reflected to the gate circuit, where $\beta(j\omega)$ is the current gain:

$$\beta(j\omega) = \frac{\omega_T}{j\omega}. \tag{38}$$

The current gain magnitude goes to unity at ω_T (as it should), and it has a capacitive phase angle because of C_{gs}. Hence, for the general case,

$$Z_{\text{in}}(j\omega) = \frac{1}{j\omega C_{gs}} + [\beta(j\omega) + 1]Z = \frac{1}{j\omega C_{gs}} + Z + \left[\frac{\omega_T}{j\omega}\right]Z. \tag{39}$$

Note that capacitive degeneration contributes a *negative* resistance to the input impedance.[12] Hence, any source-to-substrate capacitance offsets the positive resistance from inductive degeneration. It is important to take this effect into account in any actual design.

Whatever the value of this resistive term, it is important to emphasize that it does not bring with it the thermal noise of an ordinary resistor because a pure reactance is noiseless. We may therefore exploit this property to provide a specified input impedance without degrading the noise performance of the amplifier.

The form of Eqn. 37 clearly shows that the input impedance is purely resistive at only one frequency (at resonance), however, so this method can provide only

---

[12] Capacitively loaded source followers are infamous for their poor stability. This negative input resistance is fundamentally responsible, which explains why adding some positive resistance in series with the gate circuit helps solve the problem.

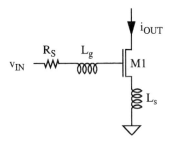

FIGURE 12.7. Narrowband LNA
with inductive source degeneration
(biasing not shown).

a narrowband impedance match. Fortunately, there are numerous instances when narrowband operation is not only acceptable but actually desirable, so inductive degeneration is certainly a valuable technique. The LNA topology we will examine for the rest of this chapter is therefore as shown in Figure 12.7.

The inductance L_s is chosen to provide the desired input resistance (equal to R_s, the source resistance). Since the input impedance is purely resistive only at resonance, an additional degree of freedom, provided by inductance L_g, is needed to guarantee this condition. Now, at resonance, the gate-to-source voltage is Q times as large as the input voltage. The overall stage transconductance G_m under this condition is therefore

$$G_m = g_{m1}Q_{\text{in}} = \frac{g_{m1}}{\omega_0 C_{gs}(R_s + \omega_T L_s)} = \frac{\omega_T}{2\omega_0 R_s}, \tag{40}$$

where we have used the approximation that ω_T is the ratio of g_{m1} to C_{gs}.

Note that the overall transconductance is independent of the device transconductance. This result is the consequence of two competing effects that cancel precisely. Consider narrowing M_1, for example, without changing any bias voltages. The device transconductance would then decrease by the same factor as the width. However, the gate capacitance would also shrink by the same factor, and the inductances would have to increase (again, by the same factor) to maintain resonance. Since the ratio of inductance to capacitance increases, the Q of the input network must increase. The increase in Q cancels precisely the reduction in device transconductance, so that the overall transconductance remains unchanged.

The question remains as to the size of M_1. One might argue that the width of M_1 should be selected with the aid of Eqn. 21, which expresses G_{opt} as a function of gate capacitance. Setting G_{opt} equal to the source conductance then yields the "optimum" value of C_{gs}, which, in turn, allows us to compute the necessary device width.

The foregoing analysis suffices to highlight the first-order behavior of the circuit. A more detailed analysis accommodates effects ultimately traceable to the relatively low intrinsic voltage gain of short-channel MOSFETs. Because scaling trends seem only to worsen this gain reduction, it is worth spending some time discussing how the input impedance changes if we now allow for finite transistor r_0.

Let us continue to neglect C_{gd}, g_{mb}, and C_{sb}. Then it is straightforward to show (see Problem 11) that the input impedance of the circuit in Figure 12.6 is

$$Z_{\text{in}}(j\omega) = \frac{1}{j\omega C_{gs}} + j\omega L + g_m \frac{L}{C_{gs}}\left(\frac{r_0}{r_0 + j\omega L + Z_L}\right). \tag{41}$$

where Z_L is the impedance attached to the drain. Comparing this result with our previous equation, we see that finite output resistance alters the third term in the impedance equation. In particular, we see that the term in parentheses has a unit magnitude only in the limit. In general, the real part of Z_{in} will be reduced and the imaginary part may be altered as well, shifting the resonant frequency of the input loop. If, as is common, the load is a parallel resonant tank, then the quantity $|Z_L + j\omega L|$ might be large enough (relative to r_0) at or near its resonance to cause a significant dip in the real part of the input impedance. Depending on the relative resonant frequencies of the input and output loops, it's possible for the dip to appear below, at, or above the desired center frequency for the overall amplifier. Needless to say, the magnitude and location of the dip are important considerations. If the dip occurs far away from the desired operating frequency, its existence may not pose too great a problem. However, it is common for the dip to occur within a couple of percent of the center frequency (because the resonant frequencies of the input and output circuits are usually designed to be close), resulting in poor input match somewhere in the band of interest.

One possible solution is to employ cascoding. However, this is only partially effective because the same r_0 that causes the problem in the first place also reduces the isolation upon which cascoding depends to work its magic (see Problem 12). In stubborn cases, it may be beneficial to increase the channel length of the cascoding device (lengthening the main device may also help, but doing so involves a more serious tradeoff with other parameters we care about, such as gain and noise figure). Lowering the load resistance of the drain load may also help, but at the cost of reduced gain. Employing some combination of these strategies usually results in a satisfactory design.

Having considered primary and parasitic factors that influence input resistance, we can now complete the design of the LNA.

The best way to illustrate the practical shortcomings of this approach is with a numerical example. Suppose we wish to design an LNA for use in a 50-Ω system at 1.57542 GHz.[13] Using Eqn. 21, we find that the required value of C_{gs} is

$$C_{gs} = \frac{G_{\text{opt}}}{\alpha\omega\sqrt{(\delta/5\gamma)(1 - |c|^2)}} \approx \frac{2G_{\text{opt}}}{\omega} \approx 4 \text{ pF}, \tag{42}$$

---

[13] This frequency corresponds to one used by the Global Positioning System (GPS).

where we have continued to use $\gamma = 2$, $\delta = 4$, and $|c| = 0.395$, and have additionally assumed that α is only a little less than unity. Typical gate overdrives[14] in analog circuits are usually low enough that α is not very much smaller than unity; values in the range of 0.8 to 0.9 are not uncommon. Equation 42 assumes an α of about 0.85, partly in keeping with this observation but mainly to keep the numbers simple.

Because, to a first approximation, device capacitance is about 1 pF per millimeter, a device large enough to produce the required value of C_{gs} would be roughly 4 mm wide. Furthermore, the bias current for such a large device typically would be large also (well over 100 mA, typically). Hence, even though the noise figure would correspond very closely to $F_{\min}$, the power consumed would be unacceptably high for virtually any application. Since power consumption is an important practical constraint, the most generally useful noise optimization technique must consider power a priori. Although amplifiers designed with an explicit power constraint will necessarily exhibit higher noise figures than could be obtained if infinite power consumption were permitted, we should put such tradeoffs on a rational basis to balance gain, noise, power, and input match in a controlled manner.

12.4 POWER-CONSTRAINED NOISE OPTIMIZATION

To develop the desired noise optimization technique, we must express noise figure in a way that takes power consumption explicitly into account. Given a specified bound on power consumption, the method should then yield the optimum device that minimizes noise. Although the detailed derivations are somewhat complex, the end results are remarkably simple. Readers interested primarily in applying the method are invited to skip to the end of this section.

We start with the general expression for noise figure as given by classical noise theory:

$$F = F_{\min} + \frac{R_n}{G_s}[(G_s - G_{\mathrm{opt}})^2 + (B_s - B_{\mathrm{opt}})^2]. \qquad (43)$$

The goal here is ultimately to reformulate the expression for noise figure in terms of power consumption. Once we derive such an equation, we'll minimize it subject to the constraint of fixed power and then solve for the width of the transistor that corresponds to this optimum condition.

To simplify the development, let us assume that the source susceptance B_s is chosen sufficiently close to B_{opt} that we may neglect the difference between the two. We will justify this step formally at a later time. Given this assumption, the expression for noise figure reduces to

$$F = F_{\min} + \frac{R_n}{G_s}(G_s - G_{\mathrm{opt}})^2. \qquad (44)$$

---

[14] Recall that gate overdrive is defined as $(V_{gs} - V_t)$, the gate voltage in excess of the threshold.

Next, rearrange the expression for G_{opt} (Eqn. 21) to define a parameter with the dimensions of a quality factor. This maneuver will help reduce clutter in the equations to come:

$$\frac{G_{\text{opt}}}{\omega C_{gs}} = \alpha \sqrt{\frac{\delta}{5\gamma}(1 - |c|^2)} = Q_{\text{opt}}. \tag{45}$$

To accommodate the possibility of operation with source conductances other than G_{opt}, we also define a similar Q in which G_{opt} is replaced by G_s, the actual source conductance:

$$Q_s \equiv \frac{1}{\omega C_{gs} R_s}. \tag{46}$$

Now re-express Eqn. 44 using Eqn. 21, Eqn. 22, and the noise parameters of Table 12.1:

$$F = F_{\min} + \frac{(\gamma/\alpha)(1/g_m)}{Q_s \omega C_{gs}}(Q_s \omega C_{gs} - Q_{\text{opt}} \omega C_{gs})^2$$

$$= F_{\min} + \left[\frac{\gamma}{\alpha g_m R_s}\right]\left[1 - \frac{Q_{\text{opt}}}{Q_s}\right]^2. \tag{47}$$

The parameters α, g_m, Q_{opt}, and Q_s in Eqn. 47 are linked to power dissipation. We need to make the linkage explicit, however, and rewrite those terms directly in terms of power. To do so, first recall that a simple expression for the drain current is

$$I_D = \frac{\mu_n C_{\text{ox}}}{2}\frac{W}{L}(V_{gs} - V_t)[(V_{gs} - V_t) \parallel (LE_{\text{sat}})], \tag{48}$$

which may be rewritten as

$$I_D = WLC_{\text{ox}} v_{\text{sat}} E_{\text{sat}} \frac{\rho^2}{1 + \rho}, \tag{49}$$

where

$$v_{\text{sat}} = \frac{\mu_n}{2} E_{\text{sat}} \tag{50}$$

and

$$\rho = \frac{V_{gs} - V_t}{LE_{\text{sat}}} = \frac{V_{od}}{LE_{\text{sat}}}. \tag{51}$$

Given Eqn. 49, the power dissipation can be written as follows:

$$P_D = V_{DD}I_D = V_{DD}WLC_{\text{ox}} v_{\text{sat}} E_{\text{sat}} \frac{\rho^2}{1 + \rho}. \tag{52}$$

Furthermore, the transconductance g_m can be found by differentiating Eqn. 49. After a little rearrangement, this may be expressed as

$$g_m = \left[\frac{1 + \rho/2}{(1 + \rho)^2}\right]\left[\mu_n C_{\text{ox}}\frac{W}{L}V_{od}\right] = \alpha\left[\mu_n C_{\text{ox}}\frac{W}{L}V_{od}\right] = \alpha g_{d0}. \tag{53}$$

Another of the parameters of Eqn. 47 linked to power is Q_s. Recall that Q_s is a function of C_{gs}, which in turn is a function of device width. Equation 49 may be solved for W, and the resulting expression substituted into the equation for Q_s, with the following result:

$$Q_s = \frac{P_0}{P_D} \frac{\rho^2}{(1+\rho)},$$ (54)

where

$$P_0 = \frac{3}{2} \frac{V_{DD} v_{\text{sat}} E_{\text{sat}}}{\omega R_s}.$$ (55)

With the aid of these expressions, the noise figure (Eqn. 44) can be written in terms of ρ and P_D.[15] Minimizing the resulting equation is complex enough that it is best solved graphically in the general case if an exact answer is desired. More insight, however, is provided by an approximation that holds if $\rho \ll 1$. Fortunately, that inequality fails to hold only in high-power circuits, a regime in which we are uninterested. Assuming other than high-power operation, then, the minimum noise figure occurs when

$$\rho^2 \approx \frac{P_D}{P_0} \sqrt{\frac{\delta}{5\gamma}(1-|c|^2)} \left[1 + \sqrt{\frac{7}{4}}\right].$$ (56)

Substituting Eqn. 56 into Eqn. 54 yields the value of Q_s that leads to the power-constrained minimum noise figure:

$$Q_{sP} = |c| \sqrt{\frac{5\gamma}{\delta}} \left[1 + \sqrt{1 + \frac{3}{|c|^2}\left(1 + \frac{\delta}{5\gamma}\right)}\right] \approx 4.$$ (57)

Careful inspection of Eqn. 57 reveals that it is relatively insensitive to the particular value of the correlation coefficient and sensitive only to the ratio of δ to γ. As suggested earlier, although the individual values of δ and γ may change owing to hot carrier effects, their *ratio* may vary much less. Hence, the numerical value of 4 for Q_{sP} given in Eqn. 57 may be reasonably invariant. A more exact analysis reveals that the optimum value is typically closer to 4.5, but the achievable noise figure is relatively insensitive to values of Q_{sP} between about 3.5 and 5.5, generally changing by 0.1 dB or less over this range.[16] Lower values lead to circuits that are less sensitive to parameter variations, while less die area is consumed by transistors corresponding to higher values of Q_{sP}.

Once Q_{sP} has been determined, it is a simple matter to provide, at last, an expression for the width of the optimum device:

$$W_{\text{opt }P} = \frac{3}{2} \frac{1}{\omega L C_{\text{ox}} R_s Q_{sP}} \approx \frac{1}{3\omega L C_{\text{ox}} R_s}.$$ (58)

Equation 58 assumes a Q_{sP} of 4.5 and is the key result of this chain of derivations. Because C_{ox} and L tend to scale in a manner that leaves their product approximately

[15] D. K. Shaeffer and T. H. Lee, "A 1.5V, 1.5GHz CMOS Low Noise Amplifier," *IEEE J. Solid State Circuits,* May 1997. Unfortunately, the result is a ratio of sixth-order polynomials.
[16] Shaeffer and Lee, ibid.

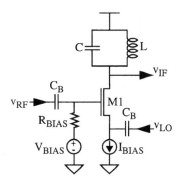

FIGURE 13.5. Square-law MOSFET mixer
(alternative configuration).

$$i_D = \frac{\mu C_{ox} W}{2L} (V_{gs} - V_T)^2. \tag{12}$$

Short-channel (high-field) devices are more linear as a result of velocity saturation, and thus are generally inferior to long devices as mixers.[19]

If the gate–source voltage V_{gs} is the sum of RF, LO, and bias terms, then we may write

$$i_D = \frac{\mu C_{ox} W}{2L} \{V_{BIAS}[v_{RF} \cos(\omega_{RF} t) + v_{LO} \cos(\omega_{LO} t)] - V_T\}^2, \tag{13}$$

from which one may readily find that the conversion gain (here, a transconductance) is simply

$$G_c = \frac{\mu C_{ox} W}{2L} \cdot v_{LO}. \tag{14}$$

This square-law device thus has a conversion transconductance that is independent of bias.[20] It is still dependent on temperature (through mobility variation) and LO drive amplitude, however.

Because perfect square-law behavior is not necessary to obtain mixing action, M_1 can be a bipolar transistor, for example, because the quadratic factor in the series expansion for the exponential i_C–v_{BE} relationship dominates over a limited range of input amplitudes. Precisely because many nonlinearities are well approximated by a square-law shape over some suitably restricted interval, one can estimate the conversion gain for other nonlinear devices used as mixers once the value of the quadratic

---

[19] The reader is reminded once again that "short-channel" actually means "high-field." Hence, even "short" devices may still behave quadratically for suitably small drain–source voltages.

[20] This independence of bias holds only in the square-law regime. Enough bias must therefore be supplied to guarantee this condition. Hence, V_{BIAS} is not permitted to equal zero. In fact, it must be chosen large enough to guarantee that the gate–source voltage always exceeds the threshold voltage, since a MOSFET behaves exponentially in weak inversion.

coefficient (c_2) is found. To underscore this point, let's estimate the conversion gain for one more nonlinear element, a bipolar transistor.

Conversion Gain of a Single Bipolar Transistor Mixer

To simplify the calculation, let us continue to ignore dynamic effects. Then we can use the exponential v_{BE} law:

$$i_C \approx I_S e^{v_{BE}/V_T}. \tag{15}$$

Expansion of this familiar relationship up to the second-order term yields[21]

$$i_C \approx I_C\left[1 + \frac{v_{IN}}{V_T} + \frac{1}{2}\left(\frac{v_{IN}}{V_T}\right)^2\right]. \tag{16}$$

By inspection (well, almost),

$$c_2 = \frac{g_m}{2V_T}, \tag{17}$$

so that an estimate of the conversion gain is:

$$G_c = c_2 v_{LO} = g_m \cdot \frac{v_{LO}}{2V_T}. \tag{18}$$

The conversion gain here is a transconductance proportional both to the standard incremental transconductance and to the ratio of the local oscillator drive amplitude to the thermal voltage. The conversion gain for a bipolar transistor is therefore dependent on bias current, LO amplitude, and temperature.

As in the corresponding derivation for a MOSFET, the foregoing computation ignores parasitic series base and emitter resistances. These resistances can linearize the transistor and therefore weaken mixer action. Thoughtful device layout is thus mandatory to minimize this effect.

13.4 MULTIPLIER-BASED MIXERS

We have seen that nonlinearities produce mixing incidentally through the multiplications they provide. Precisely because the multiplication is only incidental, these nonlinearities usually generate a host of undesired spectral components. Additionally, since two-port mixers have only one input port, the RF and LO signals are generally not well isolated from each other. This lack of isolation can cause the problems mentioned earlier, such as overloading of IF amplifiers, as well as radiation of the LO signal (or its harmonics) back out through the antenna.

Mixers based directly on multiplication generally exhibit superior performance because they ideally generate only the desired intermodulation product. Furthermore, because the inputs to a multiplier enter at separate ports, there can be a high

---

[21] We have implicitly assumed that the base-emitter drive contains a DC component as well as the RF and LO components, so that I_C is nonzero.

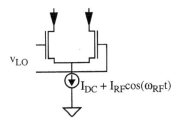

FIGURE 13.6. Single-balanced mixer.

degree of isolation among all three signals (RF, LO, and IF). Finally, CMOS technology provides excellent switches, and one can implement outstanding multipliers with switches.

13.4.1 SINGLE-BALANCED MIXER

One extremely common family of multipliers first converts the incoming RF voltage into a current and then performs a multiplication in the current domain. The simplest multiplier cell of this type is sketched in Figure 13.6.[22] In this mixer, v_{LO} is chosen large enough so that the transistors alternately switch (commutate) all of the tail current from one side to the other at the LO frequency.[23] The tail current is therefore effectively multiplied by a square wave whose frequency is that of the local oscillator:

$$i_{out}(t) = \text{sgn}[\cos \omega_{LO} t]\{I_{BIAS} + I_{RF} \cos \omega_{RF} t\}. \tag{19}$$

Because a square wave consists of odd harmonics of the fundamental, multiplication of the tail current by the square wave results in an output spectrum that appears as shown in Figure 13.7 (ω_{RF} is here chosen atypically low compared with ω_{LO} to reduce clutter in the graph).

The output thus consists of sum and difference components, each the result of an odd harmonic of the LO mixing with the RF signal. In addition, odd harmonics of the LO appear directly in the output as a consequence of the DC bias current multiplying with the LO signal. Because of the presence of the LO in the output spectrum, this type of mixer is known as a *single-balanced* mixer. Double-balanced mixers, which we'll study shortly, exploit symmetry to remove the undesired output LO component through cancellation.

---

[22] Mixers of this general kind are often lumped together and called Gilbert mixers, but only some actually are. True Gilbert multipliers function entirely in the current domain, deferring the problem of *V–I* conversion by assuming that all variables are already available in the form of currents. See Barrie Gilbert's landmark paper, "A Precise Four-Quadrant Multiplier with Subnanosecond Response," *IEEE J. Solid-State Circuits,* December 1968, pp. 365–73.

[23] One may also interchange the roles of LO and RF input, but the resulting mixer has lower conversion gain and worse noise performance, among other deficiencies. A more detailed discussion of this issue is deferred to a later section.

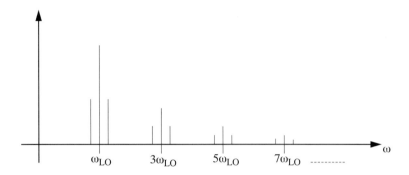

FIGURE 13.7. Representative output spectrum of single-balanced mixer.

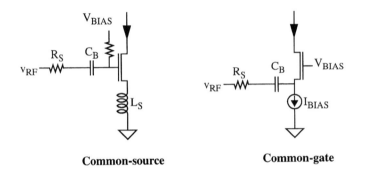

Common-source   **Common-gate**

FIGURE 13.8. RF transconductors for mixers.

Although the current source of Figure 13.6 includes a component that is perfectly proportional to the RF input signal, *V–I* converters of all real mixers are imperfect. Hence, an important design challenge is to maximize the linearity of the RF transconductance. Linearity is most commonly enhanced through some type of source degeneration, in both common-gate and common-source transconductors; see Figure 13.8. The common-gate circuit uses the source resistance R_s to linearize the transfer characteristic. This linearization is most effective if the admittance looking into the source terminal of the transistor is much larger than the conductance of R_s. In that case, the transconductance of the stage approaches $1/R_s$.

Inductive degeneration is usually preferred over resistive degeneration for several reasons.[24] An inductance has neither thermal noise to degrade noise figure nor DC voltage drop to diminish supply headroom. This last consideration is particularly relevant for low-voltage–low-power applications. Finally, the increasing reactance of an inductor with increasing frequency helps to attenuate high frequency harmonic and intermodulation components.

---

[24] Capacitive degeneration has also been tried but is markedly inferior to inductive degeneration because it increases noise and distortion at high frequencies.

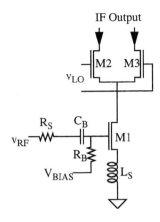

FIGURE 13.9. Single-balanced mixer
with linearized transconductance.

A more complete single-balanced mixer that incorporates a linearized transconductance is shown in Figure 13.9. The value of V_{BIAS} establishes the bias current of the cell, while R_B is chosen large enough not to load down the gate circuit (and also to reduce its noise contribution). The RF signal is applied to the gate through a DC blocking capacitor C_B. In practice, a filter would be used to remove the LO and other undesired spectral components from the output.

The conversion transconductance of this mixer can be estimated by assuming that the LO-driven transistors behave as perfect switches. Then the differential output current may be regarded as the result of multiplying the drain current of M_1 by a unit-amplitude square wave. Since the amplitude of the fundamental component of a square wave is $4/\pi$ times the amplitude of the square wave, we may write:

$$G_c = \frac{2}{\pi} g_m, \tag{20}$$

where g_m is the transconductance of the $V\text{--}I$ converter and G_c is itself a transconductance. The coefficient is $2/\pi$ rather than $4/\pi$ because the IF signal is divided evenly between sum and difference components.

13.4.2 ACTIVE DOUBLE-BALANCED MIXER

To prevent the LO products from getting to the output in the first place, two single-balanced circuits may be combined to produce a double-balanced mixer; see Figure 13.10. We assume once again that the LO drive is large enough to make the differential pairs act like current-steering switches. Note that the two single-balanced mixers are connected in antiparallel as far as the LO is concerned but in parallel for the RF signal. Therefore, the LO terms sum to zero in the output, whereas the converted RF signal is doubled in the output. This mixer thus provides a high degree of

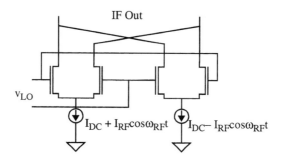

FIGURE 13.10. Active double-balanced mixer.

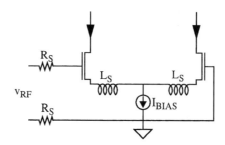

FIGURE 13.11. Linearized differential RF transconductor for double-balanced mixer.

LO–IF isolation, easing filtering requirements at the output. If care is taken in layout, IC realizations of this circuit routinely provide 40 dB of LO–IF isolation, with values in excess of 60 dB possible.

As in the single-balanced active mixer, the dynamic range is limited in part by the linearity of the V–I converter in the RF port of the mixer. So, most of the design effort is spent attempting to find better ways of providing this V–I conversion. The basic linearizing techniques used in the single-balanced mixer may be adapted to the double-balanced case, as shown in Figure 13.11.

In low-voltage applications, the DC current source can be replaced by a parallel LC tank to create a zero-headroom AC current source. The resonant frequency of the tank should be chosen to provide rejection of whatever common-mode component is most objectionable. If several such components exist, one may use series combinations of parallel LC tanks. With such a choice, a complete double-balanced mixer appears as shown in Figure 13.12. The expression for the conversion transconductance is the same as for the single-balanced case.

Noise Figure of Gilbert-Type Mixers

Computing the noise figure of mixers is difficult because of the cyclostationary nature of the noise sources. One technique involves characterization of the time-varying

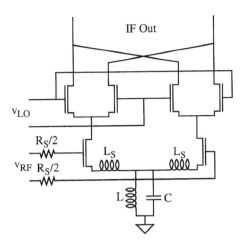

FIGURE 13.12. Minimum supply–headroom
double-balanced mixer.

impulse response, arguing that a mixer is at least linear, if not time-invariant.[25] Although the method is accurate and quite suitable for analysis, its complexity does inhibit acquisition of design insight. Nonetheless, we can identify several important noise sources and make general recommendations about how to minimize noise figure.

One noise source is certainly the transconductor itself, so that its noise figure establishes a lower bound on the mixer noise figure. The same approach used in computing LNA noise figure may be used to compute the transconductor noise figure. The low-headroom mixer of Figure 13.12 may also be modified to act as a low *noise* mixer simply by adding suitable gate inductances to the inductively degenerated pair that receives the RF input. By following a prescription essentially identical to that for stand-alone LNAs (see Chapter 12), it is possible to construct a low-headroom, low-noise mixer that may obviate the need for a separate LNA in some applications. Adjustment of the tuning of the input loop allows a variable tradeoff among conversion gain, noise figure, and distortion.

The differential pair also degrades noise performance in a number of ways. One noise figure contribution arises from imperfect switching, which causes attenuation of the signal current. Hence, one challenge in such mixers is to design the switches (and associated LO drive) to provide as little attenuation as possible.

Another NF contribution of the switching transistors arises from the interval of time in which both transistors conduct current and hence generate noise. Additionally, any noise in the LO is also magnified during this active gain interval. Minimizing the simultaneous conduction interval reduces this degradation, so sufficient LO drive must

[25] C. D. Hull and R. G. Meyer, "A Systematic Approach to the Analysis of Noise in Mixers," *IEEE Trans. Circuits and Systems I,* v. 40, no. 12, December 1993, pp. 909–19.

be supplied to make the differential pair approximate ideal, infinitely fast switches to the maximum practical extent. Finally, the 3-dB attenuation inherent in ignoring either the sum or difference signal automatically degrades noise figure (by 3 dB) since the noise cannot be discarded so readily. As a result, practical current-mode mixers typically exhibit SSB noise figures of at least 10 dB, with values more frequently in the neighborhood of 15 dB.

Linearity of Gilbert-Type Mixers

The IP3 of this type of mixer is bounded by that of the transconductor, so the three-point method used to estimate the IP3 of ordinary amplifiers may also be used here to estimate the IP3 of the transconductor. If the LO-driven transistors act as good switches then the overall mixer IP3 generally differs little from that of the transconductor. To guarantee good switching, it is important to note that – although sufficient LO drive is necessary – excessive LO drive is to be avoided. To understand the reason that excessive LO drive is a liability rather than an asset, consider the effect of ever-present capacitive parasitic loading on the common-source connection of a differential pair. As each gate is driven far beyond what's necessary for good switching, the common-source voltage is similarly overdriven. A spike in current results. In extreme cases, this spike can cause transistors to leave the saturation region. Even if that does not occur, the output spectrum can become dominated by the components arising from the spikes, rather than the downconverted RF. Hence, one should use only enough LO drive to guarantee reliable switching, and no more.

A Short Note on Simulation of Mixer IP3
with Time-Domain Simulators

Just as we noted with simulations of intermodular distortion in amplifiers, common circuit simulators such as SPICE provide accurate mixer simulations only reluctantly, if at all. The problem stems from two fundamental sources: The wide dynamic range of signals in a mixer forces the use of far tighter numerical tolerances than are adequate for "normal" circuit simulations; and the large span of frequencies of important spectral components forces long simulation times. Hence, obtaining an accurate value for IP3 from a transient simulation, for example, is usually quite challenging. Furthermore, a correct noise figure simulation for CMOS mixers is not possible at all with commercially available tools, because the device noise models presently in use are incorrect. The reader is therefore cautioned to treat mixer simulation results with great skepticism.

Because even the "accurate" options available in some simulation tools are orders of magnitude too loose to be useful for IP3 simulations, one specific action that mitigates some of these problems is to tighten tolerances progressively until the simulation results stop changing significantly. In particular, the behavior of the IM3 component in an IP3 simulation is an extremely sensitive indicator of whether the tolerances are sufficiently tight. If the IM3 terms do not exhibit a +3 slope (on a dB scale), chances are high that the tolerances are too loose. One must also make sure that the amplitudes of the two input tones are chosen small enough (i.e., well below either the compression or intercept points) to guarantee quasilinear operation of the mixer; otherwise, higher-order terms in the nonlinearity will contribute significantly to the output and confound the results. In the early phases of design, the three-point method may be applied to the transconductor to estimate its IP3 without having to suffer the agony of a transient simulation.

Another subtle consideration is to guarantee equal time spacing in the transient simulation, since FFT algorithms generally assume uniform sampling. Because some simulators use adaptive time stepping to speed up convergence, significant spectral artifacts can arise when computing the FFT. One may set the time step to a tiny fraction of the fastest time interval of interest to assure convergence without resort to adaptive time stepping. As an example, one might have to use a time step (parameter "delmax" in HSPICE[26]) that is three orders of magnitude smaller than the period of the RF signal. Hence, for a 1-GHz RF input, one might need to use a 1-ps time step. It is this combination of iteration, tight time step, and numerical tolerance problems that causes IP3 simulations to execute so slowly.[27] Again, as with the amplifier case, alternatives to time-domain simulators have evolved in response to these problems.

Additional Linearization Techniques

Because the linearity of these current-mode mixers is controlled primarily by the quality of the transconductance, it is worthwhile to consider additional ways to extend linearity. Philosophically, there are four methods for doing so: predistortion, feedback, feedforward, and piecewise approximation. These techniques can be used alone or in combination. What follows is a representative (but hardly exhaustive) set of examples of these methods.

Predistortion cascades two nonlinearities that are inverses of each other, and it shares with feedforward the need for careful matching. Predistortion is actually nearly ubiquitous, as it is the principle underlying the operation of current mirrors. In a mirror, an input current is converted to a gate-to-source voltage through some nonlinear function that is then undone to produce an output current exactly proportional to the input. Predistortion is also fundamental to the operation of true Gilbert mixers, where a pair of junctions computes the inverse hyperbolic tangent of an input differential current, and a differential pair subsequently undoes that nonlinearity.

Negative feedback computes an estimate of error, inverts it, and adds it back to the input, thereby helping to cancel the errors that distortion represents. The reduction in distortion is large as long as the loop transmission magnitude is large. Because a negative feedback system computes the error a posteriori, the overall closed-loop

---

[26] HSPICE is a trademark of the Meta-Software Corporation.

[27] Measuring these quantities in the laboratory also requires some care. As with the simulation, the amplitudes of the two input tones must be low enough to avoid excitation of higher-order nonlinearities (which would cause a slope of other than +3) yet sufficiently larger than the noise floor.

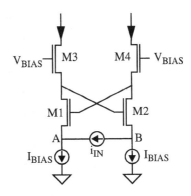

FIGURE 13.13. MOSFET cross-quad.

bandwidth must be kept a small fraction of the inherent bandwidth capabilities of the elements comprising the system; otherwise, the a posteriori estimate will be irrelevant at best and destabilizing at worst. The series feedback examples of this chapter are popular methods for linearizing high-frequency transconductors.

Contrary to common prejudice, positive feedback cannot be precluded as a linearizing technique. Furthermore, since loop transmission magnitudes must be less than unity to guarantee stability, the bandwidth penalty is much less severe than for negative feedback. As an illustrative example, the *cross-quad,* adapted from its bipolar progenitor, uses positive feedback to synthesize a virtual short-circuit; see Figure 13.13.

To show that this connection presents a short circuit to an applied current, i_{in}, consider how the voltages at the sources of M_1 and M_2 change as i_{in} changes. As i_{in} increases, the gate–source voltages of M_2 and M_4 increase by an equal amount, while those of M_1 and M_3 similarly decrease. The voltage at node A is:

$$V_{BIAS} - (V_{gs4} + V_{gs1}). \tag{21}$$

Similarly, the voltage at node B is:

$$V_{BIAS} - (V_{gs3} + V_{gs2}). \tag{22}$$

That is, the voltage at each source terminal is below V_{BIAS} by an amount equal to the sum of a high V_{gs} and a low V_{gs}. Hence, the two source voltages are always equal; the circuit synthesizes a virtual short circuit.[28]

Such a short circuit can be used to shift the burden of linearity away from active elements to a passive element, such as a resistance; this is shown in Figure 13.14. Because nodes A and B are at the same potential, the current injected into A is equal to v_{in}/R_s. This injected current is thus perfectly proportional to the input voltage and is recovered as a differential output current at the drains of M_3 and M_4.

---

[28] This analysis neglects body effect. Practical implementations do not work quite ideally as a result.

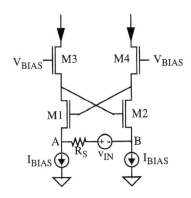

FIGURE 13.14. Cross-quad transconductor.

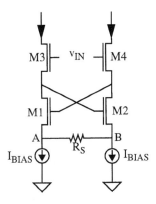

FIGURE 13.15. Alternate connection
of cross-quad transconductor.

A variation of the cross-quad applies the input voltage across the gates of the top pair, as seen in Figure 13.15. The value of the transconductance is still equal to the conductance of R_s.

Feedforward is another linearization technique; it computes an estimate of the error at the same time the system processes the signal, thereby evading the bandwidth and stability problems of negative feedback. However, the error computation and cancellation then depend on matching, so the maximum practical distortion reduction tends to be substantially less than generally attainable with negative feedback. Feedforward is most attractive at high frequencies, where negative feedback becomes less effective owing to the insufficiency of loop transmission.

An example of feedforward correction applied to a transconductor is an adaptation of Pat Quinn's bipolar "cascomp" circuit[29] (Figure 13.16). As can be seen, this transconductor consists of a cascoded differential pair to which an additional

[29] "Feedforward Amplifier," U.S. Patent #4,146,844, issued 27 March 1979, reissued 1984.

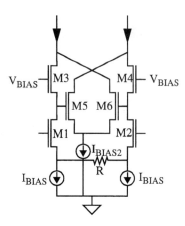

FIGURE 13.16. MOSFET cascomp.

differential pair has been added. Some linearization is provided by the source degeneration resistor R, but significant nonlinearity remains in the transconductance of inner differential pair M_1–M_2. To see this explicitly, consider that the voltage across the resistor is the input voltage minus the difference in gate-to-source voltages of M_1 and M_2:

$$V_R = v_{in} - (v_{gs1} - v_{gs2}) = v_{in} - \Delta v_{gs1}. \qquad (23)$$

The goal is to have a differential output current precisely proportional to v_{in}, so any Δv_{gs} represents an error. The cascoding pair possesses the same Δv_{gs} as the input pair, which is measured by the inner differential pair. A current proportional to this error is subtracted from the main current to linearize the transconductance. The name "cascomp" derives from this combination of a cascode and error compensation. Although the inner pair is shown as an ordinary differential pair for simplicity, it is frequently advantageous to linearize it to increase the error correction range.

Another nonfeedback approach is piecewise approximation, which exploits the observation that virtually any system is linear over some sufficiently small range. It divides responsibility for linearity among several systems, each of which is active only over a small enough range so that the composite exhibits linearity over an extended range.

Gilbert's bipolar "multi-tanh"[30] arrangement is an example of piecewise approximation. In MOS form, it appears as shown in Figure 13.17. Each of the three differential pairs behaves as a reasonably linear transconductance over an input voltage range centered about V_B, 0, and $-V_B$, respectively. For input voltages near zero, the transconductance is provided by the middle pair and is roughly constant for small enough v_{IN}. As the input voltage deviates significantly from zero, the tail current

---

[30] The name derives from the fact that the transfer characteristic of a bipolar differential pair is a hyperbolic tangent.

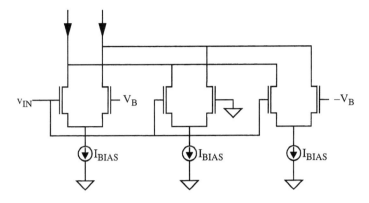

FIGURE 13.17. CMOS g_m cell.

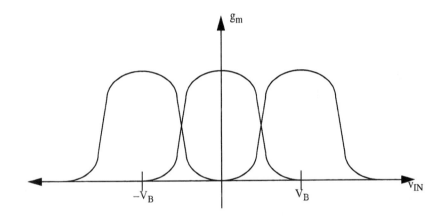

FIGURE 13.18. Illustration of linearization by piecewise approximation.

eventually steers almost completely to one side of the middle pair. However, with an appropriate selection of bias voltage V_B, one of the outer pairs takes over and continues to contribute an increase in output current; see Figure 13.18.

The overall transconductance is the sum of the individual offset transconductances and can be made roughly constant over an almost arbitrarily large range by using a sufficient number of additional differential pairs, each offset appropriately. The trade-off is an increase in power dissipation and input capacitance.

13.4.3 POTENTIOMETRIC MIXERS

Gilbert-type mixers first convert an incoming RF voltage into a current through a transconductor, whose linearity and noise figure set a firm bound on the overall mixer linearity and noise figure. An alternative to using voltage-controlled current sources in V–I converters is to use voltage-controlled resistances. For example, consider varying the resistance of a triode-region MOSFET in a manner inversely proportional to the incoming RF signal. If the voltage between drain and source is maintained at a

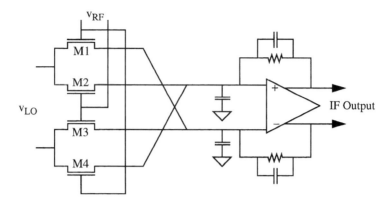

FIGURE 13.19. Potentiometric mixer.

fixed value, the current flowing through the device will be a faithful replica of the
RF voltage, and if v_{ds} varies with the LO then the current will be proportional to
the product of the LO and RF signals. One possible implementation of this idea is
sketched in Figure 13.19.[31] The four MOSFETs perform the mixing, while the capacitors remove the sum frequency component as well as higher-order products.

The RF input drives the gates of the transistors, while the LO drives the sources.
A simplified analysis assumes that the resistances of the transistors are inversely proportional to the RF signal. In that case, the current through the devices is

$$i_{\text{in}} = \frac{v_{\text{LO}}}{r_{ds}} \approx v_{\text{LO}} \cdot \mu C_{\text{ox}} \frac{W}{L} [(v_{\text{RF}} - V_T) - v_{\text{LO}}] \approx K \cdot v_{\text{LO}} \cdot v_{\text{rf}}. \qquad (24)$$

Because the current is then the result of a multiplication of the RF and LO signals,
there are components at the sum and difference frequencies, as desired. This current
flows through the feedback resistors so that the IF signal is available as an output
voltage. The op-amp need only have enough bandwidth to handle the difference frequency component, since the sum component is filtered out by the four capacitances.

Note that, for good linearity, the gate overdrive must greatly exceed v_{LO}. Hence,
v_{RF} must possess a sufficiently large DC component to satisfy this inequality for as
large a value of v_{rf} that must be accommodated.

Practical mixers of this type may exhibit good linearity (e.g., 40 dBm IIP3) but
high noise figures (e.g., 30 dB). The high noise figures are the result of the resistive
thermal noise of the input FETs (which is worst when the signal levels are small) and
the difficulty of providing a good noise match with the broadband op-amp. As a consequence, the overall dynamic range of this type of mixer is typically about the same
as conventional Gilbert-type current-mode mixers.

---

[31] J. Crols and M. Steyaert, "A 1.5GHz Highly Linear CMOS Downconversion Mixer," *IEEE J.
Solid-State Circuits,* v. 30, no. 7, July 1995, pp. 736–42.

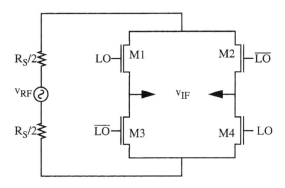

FIGURE 13.20. Simple double-balanced passive CMOS mixer.

13.4.4 PASSIVE DOUBLE-BALANCED MIXER

So far, we've examined active mixers only, with their attendant need for linear transconduction. However, passive mixers have some attractive properties, such as the potential for extremely low-power operation. Considering that CMOS technology offers excellent switches, high-performance multipliers based on switching are naturally realized in CMOS form.

In the active mixers considered so far, representations of the RF signal in the form of currents, rather than the RF voltages themselves, are effectively multiplied by a square-wave version of the local oscillator. An alternative that avoids the *V–I* conversion problem is to switch the RF signal directly in the voltage domain. This option is considerably easier to exercise in CMOS than bipolar form, which is why bipolar mixers are almost exclusively of the active, current-mode type.

The simplest passive commutating CMOS mixer consists of four switches in a bridge configuration (see Figure 13.20). The switches are driven by local oscillator signals in antiphase, so that only one diagonal pair of transistors is conducting at any given time. When M_1 and M_4 are on, v_{IF} equals v_{RF}, and when M_2 and M_3 are conducting, v_{IF} equals $-v_{RF}$. A fully equivalent description is that this mixer multiplies the incoming RF signal by a unit-amplitude square wave whose frequency is that of the local oscillator. Hence, the output contains many mixing products that result from the odd-harmonic Fourier components of the square wave.[32] Luckily, these are often readily filtered out, as discussed previously.

The voltage conversion gain of this basic cell is easy to compute from the foregoing description. Assuming multiplication by a unit-amplitude square wave, we may immediately write

$$G_C = 2/\pi. \tag{25}$$

---

[32] This situation is the same as with the current-mode mixers, however. Also, even harmonics of the LO terms may be nonzero if the duty cycle of the square wave is not exactly 50%.

Here, the $2/\pi$ factor again results from splitting the IF energy evenly between the sum and difference components.[33]

In practice, the actual voltage conversion gain may differ somewhat from $2/\pi$ because real transistors do not switch in zero time. Hence, the incoming RF signal is not multiplied by a pure square-wave signal in general. Perhaps contrary to intuition, however, the effect of this departure from ideal assumptions is usually to *increase* the voltage conversion gain above $2/\pi$.

A more general expression for the voltage conversion gain is somewhat cumbersome to derive, so we will state only the relevant insights here.[34] The output of the mixer may be treated as the product of three time-varying components and a scaling factor:

$$v_{IF}(t) = v_{RF}(t) \cdot \left[\frac{g_T(t)}{g_{T\,max}} \cdot m(t) \right] \cdot \left[\frac{g_{T\,max}}{\overline{g_T}} \right]. \tag{26}$$

The function $g_T(t)$ is the time-varying Thévenin-equivalent conductance as viewed from the IF port, while $g_{T\,max}$ and $\overline{g_T}$ are the maximum and average values, respectively, of $g_T(t)$. The *mixing function,* $m(t)$, is defined by

$$m(t) = \frac{g(t) - g(t - T_{LO}/2)}{g(t) + g(t - T_{LO}/2)}, \tag{27}$$

where $g(t)$ is conductance of each switch and T_{LO} is the period of the LO drive. The mixing function has no DC component, is periodic in T_{LO}, and has only odd harmonic content because of its half-wave symmetry.

The Fourier transform of the first bracketed term in Eqn. 26 has a value of $2/\pi$ at the LO frequency for a square-wave drive (as asserted earlier) and a value of $1/2$ for a sinusoidal drive, so the effective mixing function indeed contributes a higher conversion gain for a square-wave drive. However, the second bracketed term is unity for a square-wave drive (because the peak and average conductances are equal) but $\pi/2$ for a sinusoidal drive. The overall conversion gain is greater with a sinusoidal drive because the second term more than compensates for the smaller contribution by the (effective) mixing function. The difference is not particularly large, however. With a sinusoidal drive, the conversion gain is $\pi/4$ (-2.1 dB), compared with the $2/\pi$ gain (-3.92 dB) obtained with the square-wave drive.

Because of the spectrum of the (effective) mixing function, undesirable products can appear at the IF port of this type of mixer. The subject of filtering therefore deserves careful consideration, especially in connection with the issue of input and

---

[33] If we assume equal source and load terminations, then this gain corresponds to a 3.92-dB voltage and power loss. Many practical implementations, such as the discrete passive mixers discussed in Section 13.6, typically exhibit a somewhat greater conversion loss than this theoretical limit because of additional sources of attenuation (e.g. nonzero switch drop, skin effect loss, etc.). Common conversion losses for mixers of this type are in the neighborhood of 5 dB to 6 dB.

[34] For a detailed derivation, see A. Shahani et. al., "A 12mW Wide Dynamic Range CMOS Front-End for a Portable GPS Receiver," *IEEE J. Solid-State Circuits,* December 1997.

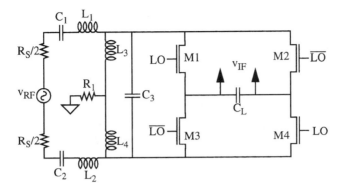

FIGURE 13.21. Low-noise, narrowband passive mixer.

output terminations. In discrete designs, the source and load impedances are usually real and well-defined (50 Ω, for example), but the sources and loads for IC mixers are usually on-chip and not at all standardized. Far from a liability, this lack of standardization is a degree of freedom that the IC engineer can exploit to improve performance. As a specific example, reactive source and load terminations might be preferable because they do not generate noise. Because it is difficult to obtain broadband operation with reactances, narrowband operation is implied for most practical mixers with reactive terminations. Fortunately, there are many applications for which this restriction is not a serious limitation.

In CMOS implementations, the load at the IF port of the mixer is frequently capacitive to an excellent approximation. In such cases, the loading is easily accommodated as forming a simple low-pass filter in conjunction with the resistance of the switches. A detailed analysis[35] reveals that the transfer function of this filter is simply

$$H(s) = \left[s \frac{C_L}{g_T} + 1 \right]^{-1}. \tag{28}$$

We see that the pole frequency is simply the ratio of the average conductance (again, as viewed from the IF port, back through the switches) to the load capacitance. This inherent filtering action may be exploited to provide a much desired attenuation of unwanted mixer products.

A somewhat more elaborate passive mixer that further exploits the freedom to select source and load terminations appears as Figure 13.21.[36] Note that this mixer assumes a capacitive load, represented as C_L in the schematic. This assumption reflects the typical situation in fully integrated CMOS circuits, and it stands in contrast with the resistive terminations common in discrete designs. A capacitive load

---

[35] Shahani et al., ibid.

[36] This example is adapted from A. Shahani et al., "A 12mW Wide Dynamic Range CMOS Front-End for a Portable GPS Receiver," *ISSCC Digest of Technical Papers,* February 1997, pp. 368–9.

generates no thermal noise of its own, and also helps filter out high-frequency noise and distortion.

The input network consists of an L-match in cascade with a parallel tank. The L-match, comprising L_1 and part of the tank capacitance, provides an impedance transformation that moderately boosts the RF signal voltage to help reduce the voltage conversion loss. The parallel tank, formed by L_3 and $C_3 + C_L$, filters out-of-band noise and distortion components present at the input and generated by the mixer itself. Resistor R_1 sets the common-mode potential for the input circuit. Because any nonlinearity in the tank capacitance reduces IP3, C_3 is best implemented as a metal–metal capacitor. To reduce the area consumed, a good choice is to use a lateral flux or fractal capacitor.

Because of the small voltage boost provided by the L-match, the voltage conversion loss can be somewhat better than the 3.92 dB that a simple switch bridge would exhibit ideally. As an example, one implementation in a 0.35-μm technology exhibits a 3.6-dB voltage conversion loss with a 1.6-GHz RF and 1.4-GHz LO.[37]

Both noise figure and IP3 are strong functions of the LO drive, since the resistance of the switches in the "on" state must be kept low and constant to optimize both parameters. The IP3 is also a function of the amount of voltage boost provided by the L-match. This boost may be adjusted downward to trade conversion gain for improved IP3 and, in some cases, it may be appropriate to remove the L-match altogether. Typical SSB noise figures of 10 dB and input IP3 of 10 dBm are readily achievable with an LO drive amplitude of 300 mV.[38] As a crude estimate, the SSB noise figure of this type of mixer is approximately equal to the power conversion loss.

The first edition of this book contained an assertion that the absence of DC bias current in a passive mixer implied the absence of $1/f$ noise. That assertion is not quite correct, because a mixer is a periodically time-varying system. As such, noise centered at integer multiples of the local oscillator can fold down to DC, for example. Thus, $1/f$ noise may still appear at the output of the mixer without requiring any DC bias in the mixer itself.

In cases where it is important to minimize $1/f$ noise in the mixer output, it is generally helpful to (a) reduce the LO drive to the minimum value consistent with acceptable mixing action and (b) design the local oscillator carefully to minimize its close-in phase noise in particular (a topic we take up in detail in Chapter 18). These considerations are particularly important in the design of receivers that are sensitive to $1/f$ noise, which include the direct-conversion (also known as the homodyne or zero-IF) receiver and the low-IF receiver.

To reduce the power consumed by the LO drivers, the gate capacitance of the switches may be resonated with an inductor (for narrowband applications), resulting in a power reduction by a factor of Q^2. It is trivial to reduce the power to the order of a milliwatt or less, even at gigahertz frequencies.

---

[37] A. Shahani et al. (February 1997), ibid.
[38] This value applies to a sine-wave LO.

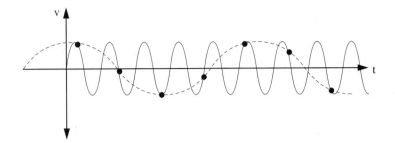

FIGURE 13.22. Illustration of subsampling.

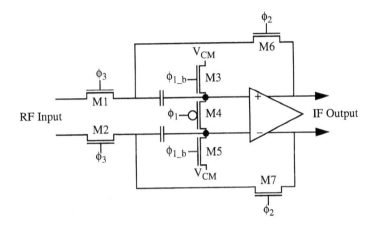

FIGURE 13.23. Track-and-hold subsampling mixer (simplified).

13.5 SUBSAMPLING MIXERS

The high quality of CMOS switches has also been exploited to realize what are some-times called *subsampling* mixers. This type of mixer exploits the observation that the information bandwidth of modulation is necessarily lower than the carrier frequency. Hence, one may satisfy the Nyquist criterion with a sampling rate that is also lower than the carrier frequency, effecting downconversion in the process.

As can be seen in Figure 13.22, the higher-frequency signal is sampled at the instants indicated by the dots, while the downconverted signal is shown as the lower-frequency reconstruction. The theoretical advantage of this approach is that it may be easier to realize samplers that operate at a frequency well below that of the incoming RF signal. From the figure, it should be clear that a properly designed track-and-hold circuit (Figure 13.23) serves as a subsampling mixer.[39]

---

[39] This example is adapted from P. Chan et al., "A Highly Linear 1-GHz CMOS Downconversion Mixer," *IEEE J. Solid-State Circuits,* December 1993.

In the sample (track) mode, transistors M_1 through M_5 are turned on while transistors M_6 and M_7 are placed in the "off" state. Devices M_3, M_4, and M_5 put a voltage equal to the common-mode voltage level V_{CM} on the right-hand terminals of the sampling capacitors, while input switches M_1 and M_2 connect the capacitors to the RF input signal. Because M_6 and M_7 are open, the op-amp is irrelevant in this tracking mode, and the tracking bandwidth is simply set by the RC time constant formed by the total switch resistance and sampling (and parasitic) capacitance. Because the system operates open-loop in this mode, it is easy to obtain tracking bandwidths far in excess of what can be achieved with a feedback structure. For example, it is trivial to obtain tracking bandwidths greater than 1 GHz in a 1-μm technology.

In the hold mode, all switch states are reversed, so that the only conducting transistors are the two feedback devices M_6 and M_7. In this mode, the circuit degenerates to a pair of charged capacitors feeding back around the op-amp. The settling time of this system need only be fast relative to the (slow) sampling period, rather than to the RF signal period. Thus, the bandwidth penalty associated with feedback is not serious.

Although a subsampler is clocked at a relatively low frequency, the sampler must still possess good time resolution or else sampling errors result. Therefore, beyond an adequate tracking bandwidth, one must also have low aperture jitter (i.e., low uncertainty in the sampling instants), and this requirement places extraordinary demands on the phase noise of the sampling clock. Hence, even though the frequency of the sampling clock need only satisfy the Nyquist criterion applied to the *modulation* bandwidth, its absolute time jitter must be a tiny fraction of the *carrier* period.

Another problem is that the sampling operation converts more than just the signal. Noise at the input to the sampler undergoes folding into the IF band, resulting in an unfortunate noise boost roughly equal to the ratio of RF and IF bandwidths. Because the RF bandwidth typically exceeds the IF bandwidth by large amounts, subsampling mixers can exhibit large noise figures (e.g., 25-dB SSB NF). The large linearity implied by the high third-order intercepts often exhibited by these types of mixers is offset by their poor noise performance, so that the dynamic range of the mixer is frequently no better (or even worse) than what one may achieve with conventional architectures. In fact, the noise and IP3 performance of many subsampling mixers can be replicated by preceding a conventional mixer with a resistive divider. In principle, an LNA with sufficient gain may be used to overcome the mixer's noise, but it is difficult in practice to realize LNAs that provide simultaneously high gain and high linearity, so again overall (system) dynamic range may actually suffer. As a result of these problems, one must take great care in applying subsampling.

13.6 APPENDIX: DIODE-RING MIXERS

This appendix considers a number of passive mixers that are common in discrete implementations. The four-diode double-balanced mixer has particularly good characteristics and is nearly ubiquitous in high-performance discrete equipment.

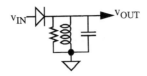

FIGURE 13.24. Simple diode mixer.

13.6.1 SINGLE-DIODE MIXER

The simplest and oldest passive mixer uses a single diode, as seen in Figure 13.24. In this circuit, the output RLC tank is tuned to the desired IF, and v_{IN} is the sum of RF, LO, and DC bias components. The nonlinear V–I characteristic of the diode provides diode currents at a number of harmonic and intermodulation frequencies, and the tank selects only those at the IF.

It is tempting to reject this circuit as hopelessly unsophisticated. It does not provide any isolation, and it doesn't provide any conversion gain, for example. However, at the highest frequencies, it may be difficult to exploit other types of nonlinearities, and such simple mixers may be suitable. In fact, all of the detectors[40] for radar sets developed in WWII were single-diode circuits.[41] Additionally, many early UHF television tuners also used mixers of this type. Much of the modern work in the millimeter-wave bands simply would not be possible without such mixers.

As another note on this circuit, it can be used as a crude demodulator for AM signals if the input signal is the AM signal (at either RF or IF). When used in this manner, the output inductor is removed entirely, no LO is used, and a simple RC network provides the output filtering. Millions of "crystal" radio sets used this type of detector (known in this context as an envelope detector), and even most AM superheterodyne radios built today use a single-diode demodulator.

13.6.2 TWO-DIODE MIXERS

There are several other ways to use diodes as mixers. As we'll see, it will appear that a diode bridge can be used as just about anything, depending on which terminals are defined as input and output and which way the diodes point.[42]

---

[40] We will use the terms "detector" and "demodulator" interchangeably.

[41] The birth of modern semiconductor technology can be traced directly to the development of microwave diodes for radar. By the end of WWII, point-contact microwave diodes capable of operation well into the gigahertz range became widely available.

[42] Diodes can even be used to provide *gain* by exploiting the nonlinear junction capacitance to make a thing known as a *parametric amplifier.* The nonlinearity can be used to transfer energy from a local oscillator (known as the *pump* in par-amp parlance) to the signal, instead of the more conventional transfer of power from a DC source to the signal frequency. Parametric amplifiers can be extremely low-noise devices, since only pure reactances are needed to make them work.

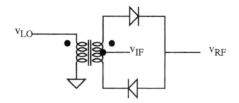

FIGURE 13.25. Single-balanced diode mixer.

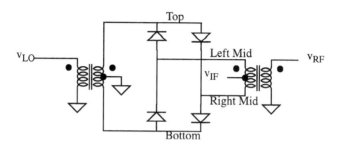

FIGURE 13.26. Double-balanced diode mixer.

With two diodes, it's possible to construct a single-balanced mixer. In this case, one may obtain isolation between LO and IF, but there is poor RF–IF isolation; see Figure 13.25. Assume that the LO drive is sufficient to make the diodes act as switches, regardless of the magnitude of the RF input. With a positive value for v_{LO}, both diodes will be on (note the reference dots on the transformer windings), effectively connecting v_{RF} to the IF output. When v_{LO} goes negative, the diodes open-circuit and disconnect v_{RF}. Hence, this mixer acts the same as the active commutating mixer studied previously.

The poor RF–IF isolation should be self-evident from the comment that the diodes connect the RF and IF ports together whenever the diodes are on. Similarly, it should be evident that symmetry guarantees excellent RF–LO isolation. Whenever the diodes are on, the RF voltage can only develop a common-mode voltage across the transformer windings, so no voltage can be induced at the LO port.

13.6.3 DOUBLE-BALANCED DIODE MIXER

By adding two more diodes and one more transformer, we can construct a double-balanced mixer to provide isolation among all ports (see Figure 13.26). Once again, assume that the LO drive is sufficient to cause the diodes to act as switches. In the circuit shown, the left pair of diodes is on whenever the LO drive is negative, whereas the right pair of diodes is on whenever the LO drive is positive.

With the LO drive positive, the voltage at "Right Mid" must be zero by symmetry, since the center tap of the input transformer is tied to ground. Thus, v_{IF} equals

v_{RF} (again, note the polarity dots). With the LO drive negative, it is "Left Mid" that has a zero potential, and v_{IF} equals $-v_{\text{RF}}$. Hence, this mixer effectively multiplies v_{RF} by a unit-amplitude square wave whose frequency is that of the LO.

Isolation is guaranteed by the symmetry of the circuit. The LO drive forces a zero potential at either the top or bottom terminal of the output transformer, as noted previously. If the RF input is zero, there will be no IF output. Hence, this configuration provides LO–IF isolation. Similarly, we can show LO–RF isolation by considering a zero IF input. Since, again, there is a zero potential at either the top or bottom terminal of the output transformer, there will be no primary voltage and therefore no secondary voltage.

These passive mixers are available in discrete form, and perform exceptionally well. The upper limit on the dynamic range is typically constrained by diode breakdown, and isolation is a function of the matching levels achieved.

With a single quad of diodes, typical double-balanced mixers routinely achieve conversion losses in the neighborhood of 6 dB and isolation of at least 30 dB, and they can accommodate RF inputs of up to 1 dBm at the 1-dB compression point while requiring an LO drive of 7 dBm. Higher RF levels can be accommodated if series connections of diodes are used in place of each diode of Figure 13.26, the drawback being an increased LO drive requirement to guarantee switching operation of the diodes. Using a total of sixteen diodes, for example, extends the RF input range to around 9 dBm but also requires a whopping 13 dBm of LO drive.

13.6.4 FINAL NOTE ON DIODE MIXERS

When actually using such mixers, one should be aware that it is critically important to terminate all ports in the proper characteristic impedance – not only at the RF, IF, and desired LO frequencies, but at the image frequencies as well. If only narrowband terminations are used, it is possible for reflections of various intermodulation products to degrade performance seriously. Hence, it is generally insufficient merely to use a standard RLC tank as an output bandpass filter without an intermediate buffering stage to guarantee a broadband resistive termination. Failure to satisfy this condition can be the source of many perplexing phenomena.

PROBLEM SET FOR MIXERS

PROBLEM 1 Using the device models from Chapter 5, design a single-balanced mixer with inductive source degeneration to achieve an IIP3 of +6 dBm. What is the conversion transconductance?

PROBLEM 2 We've seen the utility of synthesizing a virtual short-circuit for linearizing transconductances. Suppose that someone were to propose the alternative circuit of Figure 13.27.

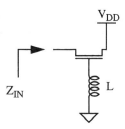

FIGURE 13.27. Common-gate
transconductor with gate
inductance (biasing not shown).

(a) First derive an expression for the incremental impedance looking into the source terminal.

(b) Select a value of inductance that makes the real part of the input impedance zero. Are additional elements needed at the input port to make the imaginary part also zero? If so, then what are they, and what are the expressions for their values?

(c) Having synthesized the virtual short, is this current recoverable as a drain current signal? If so, sketch how this circuit would be used to make a mixer.

PROBLEM 3 Simulate the version of the cross-quad transconductor in which the input signal drives the gates. Use minimum-length devices with a width of 100 μm, a total bias current of 4 mA (supplied from ideal current sources), and a resistance of 200 Ω. Use the model parameters from Chapter 5 to describe the process.

(a) Initially ignore body effect, and measure the low-frequency transconductance with the output terminated in 4-V DC sources. Is it what you expect? Explain.

(b) Now include body effect and repeat. Comment on any differences, and also determine the input common-mode voltage range over which the transconductance is roughly constant.

(c) Find the input-referred third-order intercept from simulations using two equal-amplitude tones, at 95 MHz and 105 MHz, that possess a common-mode value in the center of the range found in (b). Note that this value must be reported as a voltage, since the input impedance is not specified. Be sure to observe the simulation caveats discussed in the chapter to avoid incorrect values.

(d) Compare your answer to (c) with an estimate obtained from the three-point method.

PROBLEM 4 In this problem, we assess the relative merits of square-wave and sinusoidal local oscillator waveforms. Consider the specific case of an ideal multiplier used as a mixer in which the RF signal is $A \sin(\omega_{RF}t)$.

(a) Which type of LO yields a higher conversion gain if they both have the same amplitude?

(b) Compare the relative requirements on postmixing filters for the two LO signals.

PROBLEM 5 Consider the double-balanced passive mixer shown in Figure 13.28.

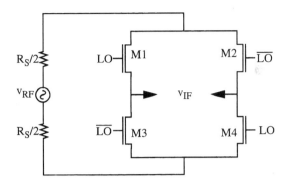

FIGURE 13.28. Simple double-balanced passive CMOS mixer.

(a) If the IF port is terminated in a resistance equal to R_s, what is the conversion gain? Assume infinitely fast switching, and neglect switch resistance.

(b) Because of the effective multiplication by a square wave, the IF output contains components at frequencies in addition to the desired one. Sketch the approximate output spectrum if the RF input is a single frequency sinusoid, and discuss filtering requirements.

(c) How would your answer to (b) change if the LO drive did not possess a perfect 50% duty cycle? Put your answer on a more quantitative basis by explicitly deriving an expression for the spectrum of a non-50% duty cycle square wave. Express your answer in terms of D, the duty ratio (here, ideally 0.5). What does your answer tell you about the need for symmetry in device switching?

PROBLEM 6 In the previous problem, switch resistance was neglected, a deficiency we now repair.

(a) Derive an expression for the maximum acceptable switch resistance if the degradation in conversion gain (relative to the ideal case) is not to exceed 1 dB.

(b) Provide an expression for device width corresponding to your answer in (a). For simplicity, you may assume square-law behavior. Express your formula in terms of the LO gate overdrive.

(c) Provide an expression for the total power consumed by the LO in driving the capacitance of all the switches as computed from your answer to part (b).

PROBLEM 7 The conversion gain equations and three-point method for estimating mixer IIP3 from transconductor IIP3 assume that the current-mode, LO-driven switches are perfect. However, as mentioned in the text, it is possible (perhaps even easy) for improper LO drive to cause degradation of conversion gain as well as significant distortion.

To explore the issue of imperfect switching in more detail, consider the simple differential MOS current switch used in a single-balanced mixer (see Figure 13.29). Simulate this circuit with an ideal current source drive, as shown in the figure. Let the DC bias current be 1 mA, and select an RF current amplitude of 100 μA. With devices (use the level-3 models from Chapter 5) that are each 100 μm wide, plot the amplitude of the 10-MHz IF component (measured as a differential output current into DC voltage sources of a value larger than the common mode of the LO drive) as the LO drive amplitude increases from 0 V to 1 V in 100-mV steps. How does the conversion gain vary? What does this experiment tell you about how one should select the LO amplitude?

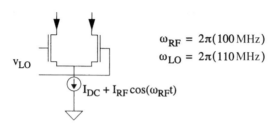

FIGURE 13.29. Single-balanced mixer.

PROBLEM 8 Repeat the previous problem, but instead of focusing on conversion gain, look at the output spectrum at other than the IF. What conclusions can you draw? In particular, explore the effect of the source–bulk capacitance of the transistors. Set them to zero and resimulate. Compare and comment.

PROBLEM 9 Design a three-pair "multi-tanh" transconductor using the device models from Chapter 5. The specifications are a total allowed bias current of 5 mA and a low-frequency transconductance of 20 mS, with a maximum ripple of $\pm10\%$. Verify your design with simulations.

CHAPTER FOURTEEN

FEEDBACK SYSTEMS

14.1 INTRODUCTION

A solid understanding of feedback is critical to good circuit design, yet many practicing engineers have at best a tenuous grasp of the subject. This chapter is an overview of the foundations of classical control theory – that is, the study of feedback in single-input, single-output, time-invariant, linear continuous-time systems. We'll see how to apply this knowledge to the design of oscillators, highly linear broadband amplifiers, and phase-locked loops, among other examples. We'll also see how to extend our design intuition to include many nonlinear systems of practical interest.

As usual, we'll start with a little history to put this subject in its proper context.

14.2 A BRIEF HISTORY OF MODERN FEEDBACK

Although application of feedback concepts is very ancient (Og annoy tiger, tiger eat Og), mathematical treatments of the subject are a recent development. Maxwell himself offered the first detailed stability analyses, in a paper on the stability of the rings of Saturn (for which he won his first mathematical prize), and a later one on the stability of speed-controlled steam engines.

The first conscious application of feedback principles in electronics was apparently by rocket pioneer Robert Goddard in 1912, in a vacuum tube oscillator that employed positive feedback.[1] As far as is known, however, his patent application was his only writing on the subject (he was sort of preoccupied with that rocketry thing, after all), and his contemporaries were largely ignorant of his work in this field.

14.2.1 ARMSTRONG AND THE REGENERATIVE AMPLIFIER

Armstrong's 1915 paper[2] on vacuum tubes contained the first published explanation of how positive feedback (regeneration) could be used to greatly increase the voltage

---

[1] U.S. Patent #1,159,209, filed 1 August 1912, granted 2 November 1915.
[2] "Some Recent Developments in the Audion Receiver," *Proc. IRE,* v. 3, 1915, pp. 215–47.

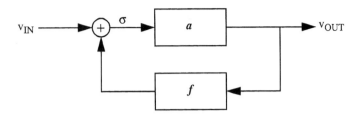

FIGURE 14.1. Positive feedback amplifier block diagram.

gain of amplifiers. Although engineers today have a prejudice against positive feedback, progress in electronics in those early years was largely made possible by Armstrong's regenerative amplifier, since there was no other economical way to obtain large amounts of gain from the primitive (and expensive) vacuum tubes of the day.[3]

We can appreciate the essential features of Armstrong's amplifier by examining the block diagram of Figure 14.1, where the quantity a is known as the forward gain and f is the feedback gain. In our particular example, a represents the gain of an ordinary (i.e., open-loop) single vacuum tube amplifier, while f represents the fraction of the output voltage that is fed back to the amplifier input.

Since we have the block diagram, it's straightforward to derive an expression for the overall gain of this amplifier. First, recognize that

$$\sigma = v_{IN} + f \cdot v_{OUT}. \tag{1}$$

Next, note that

$$v_{OUT} = a \cdot \sigma = a \cdot (v_{IN} + f \cdot v_{OUT}). \tag{2}$$

Solving for the input–output transfer function yields

$$A = \frac{a}{1 - af}. \tag{3}$$

It is evident that any positive value of af smaller than unity gives us an overall gain A that exceeds a, the "ordinary" gain of the vacuum tube amplifier. If we make af equal to 0.9 then the overall gain is increased to ten times the ordinary gain, while an af product of 0.99 gives us a hundredfold gain increase, and so on. In this way, Armstrong was able to get gain from a single stage that others could obtain only by cascading several. This achievement allowed the construction of relatively inexpensive, high-gain receivers and therefore also enabled dramatic reductions in transmitter power because of the enhanced sensitivity provided by this increased gain. In short order, the positive feedback (regenerative) amplifier became a nearly universal idiom, and Westinghouse (to whom Armstrong had assigned patent rights) kept its legal staff quite busy trying to make sure that only licensees were using this revolutionary technology.

---

[3] The effective internal "$g_m r_o$" of vacuum tubes back then was only on the order of 5. Hence the gain per stage was typically quite low, requiring many stages if conventional topologies were used.

14.2.2 HAROLD BLACK AND THE
FEEDFORWARD AMPLIFIER

Although Armstrong's regenerative amplifier pretty much solved the problem of obtaining large amounts of gain from vacuum tube amplifiers, a different problem preoccupied the telephone industry. In trying to extend communications distances, amplifiers were needed to compensate for transmission-line attenuation. Using amplifiers available in those early days, distances of a few hundred miles were routinely achievable and, with great care, perhaps 1000–2000 miles was possible, but the quality was poor. After a tremendous amount of work, a crude transcontinental telephone service was inaugurated in 1915, with a 68-year-old Alexander Graham Bell making the first call to his former assistant, Thomas Watson, but this feat was more of a stunt than a practical achievement.

The problem wasn't one of insufficient amplification; it was trivial to make the signal at the end of the line quite loud. Rather, the problem was *distortion*. Each amplifier contributed some small (say, 1%) distortion. Cascading a hundred of these things guaranteed that what came out didn't very much resemble what went in.

The main "solution" at the time was to (try to) guarantee "small-signal" operation of the amplifiers. That is, by restricting the dynamic range of the signals to a tiny fraction of the amplifier's overall capability, more linear operation could be achieved. Unfortunately, this strategy is quite inefficient since it requires the construction of, say, 100-W amplifiers to process milliwatt signals. Because of the arbitrary distance between a signal source and an amplifier (or possibly between amplifiers), though, it was difficult to guarantee that the input signals were always sufficiently small to satisfy linearity.

And thus was the situation in 1921, when a fresh graduate of Worcester Polytechnic named Harold S. Black joined the forerunner of Bell Laboratories. He became aware of this distortion problem and devoted much of his spare time to figuring out a way to solve it.[4] And solve it he did. Twice.

His first solution involves what is now known as *feedforward correction*.[5] The basic idea is to build two identical amplifiers and use one amplifier to subtract out the distortion of the first. To see how this can be accomplished, consider the block diagram (Figure 14.2) of his feedforward amplifier. Notice that there is no feedback at all; signals move only forward from input to output, as suggested by the name of this amplification technique. The two blocks labeled τ are time-delay elements. They compensate for the group delay τ of each amplifier, assuring that the subtractions operate on signals that have experienced equal delays in traveling through the overall system.

---

[4] The only one of his class of new hires to be passed over for a 10% pay raise three months after starting, Black nearly quit to pursue a career in business. He reconsidered at the last minute, and decided instead to make his mark by solving this critical problem.

[5] U.S. Patent #1,686,792, filed 3 February 1925, granted 9 October 1928.

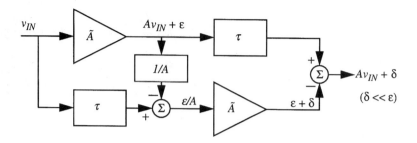

FIGURE 14.2. Feedforward amplifier.

Each amplifier has a nominal gain of A, but may be nonlinear to some degree. To distinguish a nonlinear gain from a perfectly linear one, we use the symbol $\tilde{A}$. The first amplifier takes the input signal and provides a nominal gain of A, but it produces some distortion in the process. Hence, its output is Av_{IN} plus an error voltage denoted by ε. We assume that the amplifier is linear enough that ε is small compared with the desired output Av_{IN}.

The output of the first amplifier also feeds a perfectly linear attenuator, whose gain is $1/A$. The attenuator output is then subtracted from the input to yield a voltage that is a perfectly scaled version of the distortion. This pure distortion signal feeds another amplifier identical to the first one. Because we have assumed that the distortion is small in the first place, we expect the second amplifier to act quite linearly and thus produce an excellent approximation to the original distortion (i.e., $\delta \ll \varepsilon$). That is, we assume that the error in computing the error is itself small.

The distortion signal from the second amplifier is subtracted from the distorted signal of the first amplifier to yield a final output that has greatly reduced distortion. Another feature is that of redundancy, for even if one amplifier fails there still remains some output (just with more distortion).

Black built several such amplifiers, but they proved impractical with the technology of his day. He was encouraged by the positive results he obtained when everything was adjusted right, but it was virtually impossible to maintain the tight levels of matching he needed to make a feedforward amplifier work well all the time. A goal of 0.1% distortion, for example, requires matching to similar levels, and discrete vacuum tube technology simply could not offer this level of matching on a sustained basis.

14.2.3 THE NEGATIVE FEEDBACK AMPLIFIER

While understandably disappointed with the practical barriers he faced with the feedforward amplifier, the basic notion of measuring and cancelling out the offending error terms in the output seemed worthwhile. The practical problem with feedforward was in using two separate amplifiers to accomplish this cancellation. Black began to wonder if one could perform the necessary cancellation with just *one* amplifier. That way, he reasoned, the issue of matching would disappear. It just wasn't clear how to do it.

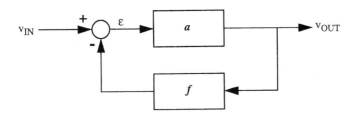

FIGURE 14.3. Negative feedback amplifier block diagram.

Then came the fateful day. On August 2, 1927, while taking the Lackawanna Ferry on the way to work as usual, the idea of the negative feedback amplifier came to him "in a flash."[6] He excitedly sketched his idea on that morning's edition of the *New York Times* and, shortly after arriving at his office twenty minutes later, had it witnessed and signed by a co-worker.

Figure 14.3 shows what he sketched, translated to its simplest form. By following a method exactly analogous to that used in analyzing the positive feedback amplifier, we obtain the following expression for the overall gain of this system:

$$A = \frac{a}{1 + af}. \tag{4}$$

Now make the af product very much larger than unity. In this case,

$$A \approx \frac{1}{f}. \tag{5}$$

As Black observed, the feedback factor f can be implemented with perfectly linear elements, such as resistive voltage dividers, so that the overall closed-loop behavior is linear even though the amplifier in the block a is not. That is, it doesn't matter that a exhibits all sorts of nonlinear behavior as long as $af \gg 1$ under all conditions of interest. The only tradeoff is that the overall, closed-loop gain A is much smaller than the forward gain a. However, if gain is cheap but low distortion isn't, then negative feedback is a marvelous solution to a very difficult problem.

As obviously wonderful the idea of negative feedback is to us today, it was not at all obvious to Black's contemporaries. It was difficult to convince others that it made sense to work hard to design a high-gain amplifier, only to reduce the gain with feedback.

The negative feedback amplifier represented so great a departure from prevailing practice (remember, the Armstrong *positive* feedback amplifier was then the dominant architecture) that it took a dozen years for the British patent office to issue the patent. In the intervening time, they argued that it could not work, and cited a lot of prior art to "prove" their point. Black (and AT&T) finally won in the end, but it did take some doing.

---

[6] H. S. Black, "Inventing the Negative Feedback Amplifier," *IEEE Spectrum,* December 1977, pp. 55–60.

14.3   A PUZZLE

If you've been paying attention, you should be a bit confused. Suppose one makes $af \gg 1$ in the *positive* feedback amplifier. Then the math gives us the following result:

$$A \approx -\frac{1}{f}. \tag{6}$$

It would appear that *either* sign of feedback gives us a linear closed-loop amplifier. So why do we prefer negative feedback?

The math is absolutely, unassailably correct, by the way, and the paradox cannot be resolved within the framework established so far. The problem lies in the implicit assumptions that lead to the math; they are not satisfied by physical systems.

For the resolution to this paradox, we now must consider what happens if a and f are not scalar quantities – that is, if they have some frequency-dependent magnitude and phase. As we shall see shortly, it turns out that the positive feedback amplifier with $af \gg 1$ cannot be made stable because all real systems eventually exhibit increasing negative phase shift with frequency. If nothing else, the finite speed of light guarantees that all physical systems have unconstrained negative phase shift as the frequency increases to infinity. Because of the extremely important role that phase shift plays in determining stability, we will spend a fair amount of time studying it. Before doing so, however, let us examine a number of commonly held misconceptions about negative feedback.

14.4   DESENSITIVITY OF NEGATIVE FEEDBACK SYSTEMS

All sorts of wild claims about negative feedback exist. "It increases bandwidth"; "it decreases distortion"; "it reduces noise, and removes unsightly facial blemishes." Some of these claims can be true but aren't necessarily *fundamental* to negative feedback. As we'll see, there is actually only one absolutely fundamental (but extraordinarily important) benefit of negative feedback systems, and that is the *desensitivity* provided. That is, the overall amplifier possesses an attenuated sensitivity to changes in the forward gain a if $af \gg 1$.

In order to quantify this notion of desensitivity, let's calculate the differential change in A that results from a differential change in a:

$$\frac{dA}{da} = \frac{d}{da}\left(\frac{a}{1+af}\right) = \frac{1}{(1+af)^2} = \frac{A}{a}\left(\frac{1}{1+af}\right). \tag{7}$$

We may rearrange this expression as follows:[7]

$$\frac{dA}{A} = \frac{da}{a}\left(\frac{1}{1+af}\right). \tag{8}$$

---

[7] Mathematicians cringe whenever engineers are this cavalier; we don't worry about such things.

This last equation tells us that a given fractional change in A equals the fractional change in a, *attenuated* ("desensitized") by a factor of $1 + af$. For this reason, the quantity $1 + af$ is often called the *desensitivity* of a feedback system.[8] Thus, if the forward gain varies with time, temperature, or input amplitude, then the overall closed-loop gain exhibits smaller variations since they are attenuated by the desensitivity factor. If the factor af is made extremely large, then the desensitivity will be large and variations in A due to changes in a will be greatly suppressed.

Let's perform a similar analysis to deduce how variations in the feedback factor affect the closed-loop system:

$$\frac{dA}{df} = \frac{d}{df}\left(\frac{a}{1+af}\right) = -\frac{a^2}{(1+af)^2} = \frac{A}{f}\left(-\frac{af}{1+af}\right), \tag{9}$$

so that, on a normalized fractional basis,

$$\frac{dA}{A} = \frac{df}{f}\left(-\frac{af}{1+af}\right). \tag{10}$$

Here, we see that large desensitivity factors do *not* help us as far as variations in feedback are concerned. In fact, in the limit of infinite desensitivity, the fractional change in A has the *same* magnitude as the fractional change in f. This result underscores the importance of having linear feedback networks if overall closed-loop linear operation is the goal (as often, but not always, happens to be the case). For this reason, the feedback block is usually made of passive elements (commonly resistors and capacitors) rather than other amplifiers.

FACIAL BLEMISHES

But how about all of those other claims that are so commonly made about the benefits of negative feedback? Let's examine them, one at a time.

BENT CONCEPTION 1: *Negative feedback extends bandwidth.*

This can be true (but consider an important counterexample, such as the Miller effect), but it's not nearly as magical as it sounds. If negative feedback were to accomplish this bandwidth extension by giving us more gain at high frequencies, then there'd be something to write home about. But, as we'll see in a moment, negative feedback extends bandwidth by *selectively throwing away gain at lower frequencies*. We will demonstrate that one may accomplish precisely the same thing through the use of purely open-loop means.

To see how negative feedback may extend bandwidth, let us suppose that the forward gain is now not a purely scalar quantity but is instead some $a(s)$ that rolls off

---

[8] "Return difference" is another term for this quantity. This name derives from the observation that, if we cut the loop, squirt in a unit signal in one end, and see what dribbles out the other, the difference is $1 + af$.

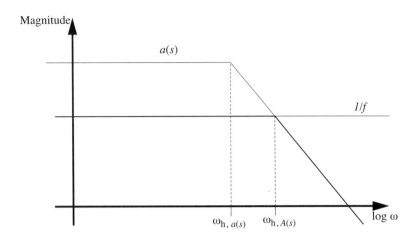

FIGURE 14.4. Bandwidth: $A(s)$ versus $a(s)$.

with single-pole behavior. Now, we said that as long as af had a magnitude large compared with unity, the closed-loop gain was approximately equal to the reciprocal of the feedback gain. We can also see that, in the limit of very small af, the closed-loop and forward gains converge.

An entirely equivalent description is: Plot $|a(s)|$ and $|1/f|$ on the same graph. A good approximation to $|A(s)|$ can be pieced together by choosing the lower of the two curves. *Explanation:* If $|1/f|$ is much lower than $|a(s)|$ then this implies that $a(s)f$ has a large magnitude, and therefore the closed-loop behavior is approximately $1/f$ (the lower curve). If $|1/f|$ is much *higher* than $|a(s)|$, it means that $a(s)f$ has a small magnitude and the closed-loop behavior converges to $a(s)$ (still the lower curve). In the region where $a(s)$ and $1/f$ have similar magnitudes, we can't be sure of what happens precisely, but we can guess and claim that some sort of reasonable approximation might be obtained by continuing to choose the lower of the two curves.

Applying this procedure to our single-pole example generates Figure 14.4 (we have used a straight-line Bode approximation to the actual single-pole curve). As you can see, the response formed by concatenating the lower of the two curves does indeed have a higher corner frequency than that of $a(s)$, but negative feedback has accomplished this extension of bandwidth by reducing gain at lower frequencies, *not* by giving us any more gain at higher frequencies.

Finally, to see that there is nothing special about negative feedback in the context of bandwidth extension, consider that a capacitively loaded resistive divider can have its bandwidth extended simply by placing another resistor in parallel with the capacitor. The bandwidth goes up, but the gain goes down.   QED

MISGUIDED NOTION 2: *Negative feedback reduces noise.*

Actually, negative feedback cannot reduce the *input-referred* noise of a system; that is, we cannot call upon negative feedback to improve the noise-limited sensitivity

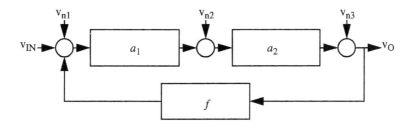

FIGURE 14.5. Feedback system with additive noise sources.

of an amplifier. In fact, the best it can do is preserve the input-referred noise of the open-loop amplifier. Moreover, in most practical cases (e.g., with resistive elements in the loop), feedback typically *increases* the input-referred noise. What feedback can (and generally does) do is reduce the *output* noise.

The idea that negative feedback magically reduces noise stems from an incomplete understanding of the noise properties of the type of system shown in Figure 14.5. For such a system, the individual transfer functions are:

$$\frac{v_{\text{out}}}{v_{\text{in}}} = \frac{a_1 a_2}{1 + a_1 a_2 f}, \tag{11}$$

$$\frac{v_{\text{out}}}{v_{n1}} = \frac{a_1 a_2}{1 + a_1 a_2 f}, \tag{12}$$

$$\frac{v_{\text{out}}}{v_{n2}} = \frac{a_2}{1 + a_1 a_2 f}, \tag{13}$$

$$\frac{v_{\text{out}}}{v_{n3}} = \frac{1}{1 + a_1 a_2 f}. \tag{14}$$

From these equations, we see that the gain from noise source v_{n1} to the output is the same as that from the input to the output. This result should be no surprise: the amplifier cannot distinguish between the input signal and v_{n1}, as they happen to enter the system at the same point.

But the gains to the other two noise sources are smaller, so one might think that there's a benefit after all. In fact, all this observation proves is that noise entering before a gain stage contributes more to the output than noise entering after a gain stage. This yawn-inducing result has nothing to do with negative feedback but only with the fact that we happen to have gain between two nodes where noise signals could enter the system.

To underscore the idea that negative feedback has nothing to do with this result, consider the open-loop structure of Figure 14.6. Note that, with the particular choice of K shown, the input–output transfer functions of the feedback and open-loop amplifiers are exactly the same for every input. Thus we see that feedback offers no magical noise reduction beyond what open-loop systems can provide.

Again, although these properties are not fundamental to negative feedback systems, they may be conveniently obtained through negative feedback. Impedance

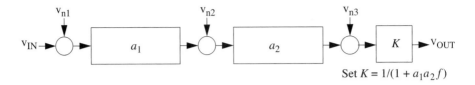

Set $K = 1/(1 + a_1 a_2 f)$

FIGURE 14.6. Open-loop system with additive noise sources.

transformation, for example, can be provided by open- and closed-loop systems, but feedback implementations might be easier to construct or adjust in many instances.

In summary, desensitivity to the forward gain is the *only* inherent benefit conferred by negative feedback systems. Negative feedback may also provide other benefits (perhaps more, even much more, than practically obtainable through open-loop means), but desensitivity is the only *fundamental* one.

14.5 STABILITY OF FEEDBACK SYSTEMS

We have seen that use of negative feedback allows the closed-loop transfer function $A(s)$ to approach the reciprocal of the feedback gain f as (minus) the "loop transmission"[9] $a(s) f(s)$ increases, thereby conferring a benefit if f is less subject to the vagaries of distortion and parameter variation than the forward gain $a(s)$, as is often the case. As argued earlier, this reduction in sensitivity to $a(s)$ is actually the *only* fundamental benefit of negative feedback; all others can be obtained (although perhaps less conveniently) through open-loop means.

Now large gains are trivially achieved, so it would appear that we could obtain arbitrarily large desensitivities without trouble. Unfortunately, we invariably discover that systems become unstable when some loop transmission magnitude is exceeded. And, as luck would have it, the onset of instability frequently occurs with values of loop transmission that are not particularly large. Thus *it is instability* – rather than the insufficiency of available gain – *that usually limits the performance of feedback systems.*

Up to this point, we have discussed instability in rather vague terms. People certainly have some intuitive notions about what is meant, but we need something a bit more concrete to work with if we are to go further. As it happens, there are 2.6 zillion[10] definitions of stability, each with its own subtle nuances. We shall use the bounded-input, bounded-output (BIBO) definition of stability, which states that a system is stable if every bounded input produces a bounded output. Although we shall not prove it here, a system $H(s)$ is BIBO stable if all of the poles of $H(s)$ are in the open left half-plane.

---

[9] To find the loop transmission, break the loop (after setting all independent sources to their zero values), inject a signal into the break, and take the ratio of what comes back to what you put in. For our canonical negative feedback system block diagram, the loop transmission is $-af$.

[10] At last count, plus or minus, as reported by the Bureau of Obscure and Generally Useless Statistics (BOGUS).

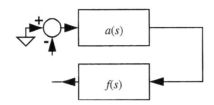

FIGURE 14.7. Disconnected negative
feedback system.

In order to apply this test to our feedback system, we must find the poles of $A(s)$, that is, the roots of $P(s) = 1 + a(s)f(s)$. A direct attack using, say, a root finder is certainly an option, but we're after the development of deeper design insight than this direct approach usually offers. Furthermore, explicit polynomial representations for $a(s)$ and $f(s)$ may not always be available, so we seek alternative methods of determining stability.

All of the alternative methods we will examine focus on the behavior of the loop transmission. *The vast simplification that results cannot be overemphasized.* Determination of the loop transmission is usually straightforward, whereas that of the closed-loop transfer function requires identification of the forward path (not always trivial, contrary to one's initial impression) and an additional mathematical step (i.e., taking the ratio of $a(s)$ to $1 + a(s)f(s)$). Hence, any method that can determine stability from examination of the loop transmission offers a tremendous saving of labor.

14.6 GAIN AND PHASE MARGIN AS
STABILITY MEASURES

Consider cutting open our feedback system (see Figure 14.7). Now imagine supplying a sine wave of some frequency to the inverting terminal of the summing junction. The sine wave inverts there, then gets multiplied by the magnitude of $a(s)f(s)$ and shifted in phase by the net phase angle of $a(s)f(s)$. If the magnitude of $a(s)f(s)$ happens to be unity at this frequency while the net phase of $a(s)f(s)$ happens to be 180°, then the output of the $f(s)$ block is a sine wave of the same phase and amplitude as the signal we originally supplied. It is conceivable, then, that we could dispense with the original input – a sine wave of this frequency *might* be able to persist if we re-close the loop. If the sine wave does survive, it means that we have an output without an input. That is, the system is unstable.

To determine conclusively whether such a persistent sine wave actually exists requires use of the Nyquist stability test. However, derivation of the Nyquist stability criterion is somewhat involved (although reasonably straightforward to apply), and this complication is enough to discourage many from using it.

Instead of the Nyquist test, perhaps the stability measures most often actually used by practicing engineers are a subset of the Nyquist test, *gain margin* and *phase margin*. These quantities are easily computed as follows.

(1) *Gain margin* – Find the frequency at which the phase shift of $a(j\omega)f(j\omega)$ is $-180°$; call this frequency ω_π. Then the gain margin is simply

$$\text{gain margin} = \frac{1}{|a(j\omega_\pi)f(j\omega_\pi)|}. \tag{15}$$

(2) *Phase margin* – Find the frequency at which the magnitude of $a(j\omega)f(j\omega)$ is unity. Call this frequency ω_c, the *crossover frequency*. Then the phase margin (in degrees) is simply

$$\text{phase margin} = 180° + \angle[a(j\omega_c)f(j\omega_c)]. \tag{16}$$

As can be inferred from these definitions, gain and phase margin are measures of how closely $a(j\omega)f(j\omega)$ approaches a magnitude of unity and a phase shift of 180°, the conditions that *could* allow a persistent oscillation. Evidently, these quantities allow us to speak about the *relative* degree of stability of a system since, the larger the margins, the further away the system is from unstable behavior.

Because of the ease with which gain and phase margin are calculated (or obtained from actual frequency response measurements), they are often used in lieu of performing an actual Nyquist test. In fact, most engineers often dispense with a calculation of gain margin altogether and compute only the phase margin. However, it should strike you as remarkable that the stability of a feedback system could be determined by the behavior of the loop transmission at just one or two frequencies, so perhaps you won't be surprised to learn that gain and phase margin are not perfectly reliable guides. In fact, there are many pathological cases (encountered mainly during Ph.D. qualifying exams) in which gain and phase margin fail spectacularly. However, it is true that for *most* commonly encountered systems, stability can be determined rather well by these measures. If there is a question as to the applicability of gain and phase margin then one must use the Nyquist test, which considers information about $a(j\omega)f(j\omega)$ at *all* frequencies. Hence, it can handle the pathological situations that occasionally arise and that cannot be adequately examined using only the gain and phase margin. It is important to remember, then, that the *gain and phase margin tests are only a subset of the more general Nyquist test*. This important point is frequently overlooked by practicing engineers, who are often unaware of the limited nature of gain and phase margin as stability measures. This confusion persists because the stability of the commonest systems happens to be well determined by gain and phase margin, and this success encourages many to make inappropriate generalizations.

Having provided that all-important public service announcement, we can return to gain and phase margin. It is worthwhile to point out that they are easily read off of Bode diagrams (in fact, it is precisely this ease that encourages many designers to use gain and phase margin as stability measures); see Figure 14.8. That is, experimentally derived data may be used to compute gain and phase margin; no explicit modeling step is needed, and no transfer functions have to be determined.

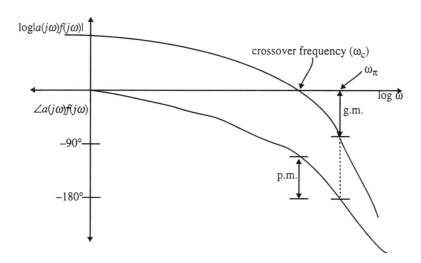

FIGURE 14.8. Gain and phase margin from Bode plots.

Okay, now that we've derived a new set of stability criteria that enable us to quantify degrees of relative stability, what values are acceptable in design? Unfortunately, there are no universally correct answers, but we can offer a few guidelines: One must choose a gain margin large enough to accommodate all anticipated variations in the magnitude of the loop transmission without endangering stability. The more variation in $a_0 f_0$ that one anticipates, the greater the required gain margin. In most cases, a minimum gain margin of about 3–5 is satisfactory.

Similarly, one must choose a phase margin large enough to accommodate all anticipated variations in the phase shift of the loop transmission. Typically, a minimum phase margin of 30°–60° is acceptable, with the lower end of the range generally associated with substantial overshoot and ringing in the step response as well as significant peaking in the frequency response. Note that this range is quite approximate and will vary according to the details of system composition. For example, overshoot might be tolerable in an amplifier but unacceptable in the landing controls of an aircraft.

14.7 ROOT-LOCUS TECHNIQUES

Gain and phase margin use the behavior of the loop transmission to determine the stability of the closed-loop system. We have already noted that a rational transfer function (or even an analytical expression of any type) for the loop transmission is not needed to apply the test. However, if one *is* given a rational transfer function, additional paths to answering the stability question become available.

As noted earlier, one obvious method is simply to compute explicitly the closed-loop transfer function and solve for the roots of the denominator polynomial. Unfortunately, important insights rarely emerge from such an exercise (other than that one should seek an alternative). Fortunately, there exists a method (rather, a collection of techniques) that allows one to sketch rapidly how the poles of the closed-loop system move as some loop-transmission parameter (such as DC gain) varies, given only the loop transmission as starting information.

To understand how one sketches a root locus, recall that the goal is to find roots of the polynomial $1 + a(s)f(s)$. That is, we want to find the values of s that satisfy

$$P(s) = 1 + a(s)f(s) = 0 \implies a(s)f(s) = -1. \tag{17}$$

We may decompose this complex equation into separate constraints on the magnitude and phase angle of the loop transmission:

$$|a(s)f(s)| = 1 \tag{18}$$

and

$$\angle[a(s)f(s)] = (2n + 1) \cdot 180°. \tag{19}$$

Despite their simplicity, we can deduce an amazing amount of information from these last two equations, as we shall now see.

RULE 1 : *The locus starts at the poles of the loop transmission, and it terminates on the zeros (finite or infinite) of the loop transmission.*

This rule derives from the magnitude condition. Suppose we express $a(s)f(s)$ as $kg(s)$, where k represents the gain factor that we are varying and $g(s)$ represents all the rest of $a(s)f(s)$. In this case, the magnitude condition may be stated as

$$|g(s)| = \frac{1}{k}. \tag{20}$$

From this equation, it is apparent that the magnitude of $g(s)$ must be extremely large for very small values of k (that is, for the start of the locus). Hence, values of s that satisfy this magnitude condition are evidently near the poles of $g(s)$. Similarly, for very large values of k (corresponding to the end of the locus), $|g(s)|$ is quite small, indicating values of s near the zeros of $g(s)$.

RULE 2: *If a root-locus branch lies on the real axis, it resides to the left of an odd number of left half-plane poles + zeros and to the right of an odd number of right half-plane poles + zeros.*

Since we are implicitly assuming that k is a positive number in sketching loci corresponding to negative feedback, the only way to satisfy the equation $1 + kg(s) = 0$ is for $g(s)$ to be a negative number. Now, since $g(s)$ is rational by postulate, we may express it in the following manner:

$$g(s) = \frac{\prod_{i=1}^{Z}(\tau_{zi}s + 1)}{\prod_{k=1}^{P}(\tau_{pk}s + 1)}. \tag{21}$$

Each $(\tau s + 1)$ term contributes a minus sign if τs is more negative than -1. Therefore the sign of $g(s)$ will be negative for values of s more negative than (to the left of) an odd number of left half-plane poles + zeros (where τ is positive), including those at the origin. Similarly, $g(s)$ will be negative for s more *positive* than (to the right of) an odd number of right half-plane poles + zeros (where τ is negative). Note that, since complex poles or zeros appear in conjugate pairs, their net contribution to the phase at a test point on the real axis is zero, and thus they have no effect here.

RULE 3: *If the number of poles exceeds the number of zeros by two or more, then the average distance of the poles to the imaginary axis is independent of k.*

This rule derives from a property of polynomials. Specifically, consider that

$$C \cdot \prod_{j=1}^{n} (s + s_j) = C \cdot \left[s^n + s^{n-1} \sum_{j=1}^{n} s_j + \cdots + \prod_{j=1}^{n} s_j \right] = L(s). \qquad (22)$$

We see that the ratio of the two leading coefficients is the sum of the roots of $L(s)$. Note also that the average distance of the roots to the imaginary axis is simply this sum divided by the order of the polynomial (i.e., the number of roots).

To use this observation to derive the rule, let $g(s) = p(s)/q(s)$. Then the characteristic equation (after clearing fractions) becomes

$$P(s) = q(s) + kp(s). \qquad (23)$$

From this we observe that, if the order of $q(s)$ exceeds that of $p(s)$ by 2 or more, then the two leading coefficients are independent of $p(s)$ (and therefore of k) and hence the average distance of the poles to the imaginary axis is a constant.

RULE 4: *As $k \to \infty$, $P - Z$ branches of the locus head off to infinity, asymptotically at angles (with respect to the real axis) given by*

$$\theta_n = \frac{(2n + 1) \cdot 180°}{P - Z}. \qquad (24)$$

In Eqn. 24, n ranges from 0 to $P - Z - 1$, P is the number of finite poles of $g(s)$, and Z is the number of finite zeros of $g(s)$. That the angle condition leads to this rule is easily understood from Figure 14.9.

In the figure, we assume that s_{test} is so far away from the poles and zeros of the loop transmission that each pole or zero of $g(s)$ forms approximately the same angle θ with respect to s_{test}. Each pole then contributes a phase angle of $-\theta$, while each zero contributes a phase angle of θ at s_{test}. Hence, the total phase angle at s_{test} is $Z\theta - P\theta$. Now, the angle condition requires that this total phase angle be an odd multiple of 180° if s_{test} is to be a pole of the closed-loop system (i.e., to lie on the locus). Therefore,

$$Z\theta - P\theta = (2n + 1) \cdot 180° \implies \theta_n = \frac{(2n + 1) \cdot 180°}{Z - P}. \qquad (25)$$

Recognizing that 180° is the same as −180°, we see that this last equation is in fact equivalent to that stated before.

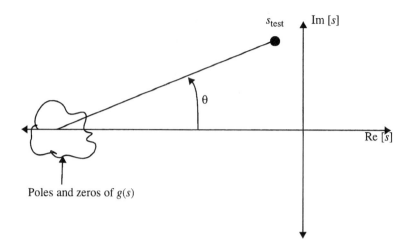

FIGURE 14.9. Asymptotic angle rule.

RULE 5: *The asymptotes of Rule 4 all intersect the real axis at a point given by*

$$\sigma = \frac{\sum \mathrm{Re(poles)} - \sum \mathrm{Re(zeros)}}{P - Z}. \tag{26}$$

To derive this rule, we use an observation of Rule 3 – namely, that the ratio of the two leading coefficients of a polynomial is equal to the sum of the roots. Further note that, since complex roots appear in conjugate pairs, the imaginary parts cancel when forming the sum. Hence, we can write our characteristic equation as

$$P(s) = 1 + \frac{C_1 \cdot \left[s^Z + s^{Z-1} \sum_{i=1}^{Z} \mathrm{Re(zeros)} + \cdots \right]}{s^P + s^{P-1} \sum_{i=1}^{P} \mathrm{Re(poles)} + \cdots} = 0. \tag{27}$$

Because we are interested in the behavior of $P(s)$ for large s, we can approximate $P(s)$ by preserving only the first two terms of the numerator and denominator polynomials. Performing these truncations and dividing through by the numerator polynomial yields

$$0 \approx 1 + \frac{C_1}{s^{P-Z} + s^{P-Z-1} \left[\sum \mathrm{Re(poles)} - \sum \mathrm{Re(zeros)} \right] + \cdots}. \tag{28}$$

Clearing fractions, we see that the ratio of the two leading coefficients (which is the average distance of the asymptotes to the real axis) is a constant, leading to this rule.

As an aside, it should be noted that it is all right for the locus to cross an asymptote. As an additional note, the locus will lie exactly along an asymptote if the pole-zero pattern happens to be perfectly symmetric about the asymptote extended through the point σ (as given by this rule).

RULE 6: *If a real-axis branch of the locus lies between a pair of poles then the locus breaks away from the real axis somewhere between the poles. Similarly, if a real-axis*

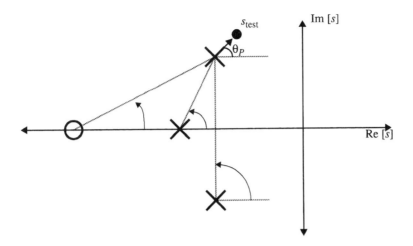

FIGURE 14.10. Angle near complex pole.

branch of the locus lies between a pair of zeros then there will be an entry point be-tween that pair of zeros.

This rule is actually a consequence of Rule 1, since the locus starts at the poles of $g(s)$ and terminates at the zeros of $g(s)$. Hence, if a real-axis branch lies between two poles then the poles must eventually head off to zeros, which lie elsewhere by postu-late. And if a real-axis branch lies between two zeros (including one at infinity), then poles from somewhere will enter the real axis, eventually to terminate at the zeros.

A breakaway point is found by computing the value of s between the poles that minimizes $|g(s)|$; similarly, an entry point is the value of s between the zeros that maximizes the value of $|g(s)|$. In general, iterated function evaluation (a fancy term for trial and error) to find the minima/maxima is more expedient than setting $dg(s)/ds$ equal to zero and solving for the roots of the resulting polynomial,[11] since the range of possible values of s is bounded and known.

RULE 7: *The locus forms an initial angle θ_P with respect to a complex pole, or an angle θ_Z with respect to a complex zero:*

$$\theta_P = 180° - \sum \angle[\text{poles}] + \sum \angle[\text{zeros}] \qquad (29)$$

and

$$\theta_Z = 180° + \sum \angle[\text{poles}] - \sum \angle[\text{zeros}]. \qquad (30)$$

In Eqn. 30, the sums refer to the angles drawn from all of the poles and zeros to the complex pole (zero) in question.

Again, the angle condition is used to derive this rule, as can be seen in Figure 14.10. Here we assume that s_{test} is extremely close to the pole in question, so that the phase

---

[11] In fact, use of this formal method requires finding the roots of a polynomial of degree only one less than that of the original and so is rarely worthwhile.

angle contribution by the other poles and zeros to s_{test} is essentially the same as that to the complex pole. The phase angle at s_{test} is just

$$-\theta_P + \sum \angle[\text{zeros}] - \sum \angle[\text{poles}], \tag{31}$$

and this sum must equal $\pm 180°$, leading to the rule. Similarly, the phase angle contribution to s_{test} near a complex zero is simply

$$\theta_Z + \sum \angle[\text{zeros}] - \sum \angle[\text{poles}], \tag{32}$$

thus completing the derivation.

RULE 8: *If a particular value of s is known to lie on the locus, then the value of k necessary to make that value of s be a closed-loop pole location is given by*

$$k = \frac{1}{|g(s)|}. \tag{33}$$

This rule is simply a restatement of the magnitude condition.

The foregoing eight rules are by no means the only ones that can be deduced from the magnitude and phase conditions, but they should suffice for most purposes.

14.7.1 ROOT-LOCUS RULES FOR POSITIVE FEEDBACK SYSTEMS

The rules we have developed so far apply to the case of negative feedback systems in which $k > 0$. The same basic construction ideas apply even in the case of positive feedback, provided we modify all the rules derived from the phase angle condition. For $k < 0$, the appropriate phase angle condition becomes

$$\angle g(s) = n \cdot 360°. \tag{34}$$

Therefore, all occurrences of $(2n + 1) \cdot 180°$ should be replaced with $n \cdot 360°$, with all associated rules modified accordingly. It is left as an "exercise for the reader" to perform these modifications.

14.7.2 ZEROS OF A(s)

The root locus tells us only how the *poles* of $A(s)$ behave, given the poles and zeros of $g(s)$ as starting information. Since the locus therefore does not necessarily tell us anything about the zeros of $A(s)$, we need to do a little extra work if finding those closed-loop zeros is important. Fortunately, this task is relatively easy, since

$$A(s) = \frac{a(s)}{1 + a(s)f(s)}. \tag{35}$$

Hence the zeros of $A(s)$ are evidently the zeros of $a(s)$ and the poles of $f(s)$.

14.8 SUMMARY OF STABILITY CRITERIA

We have presented several techniques for determining the stability of closed-loop systems through examination of loop transmission behavior. Gain and phase margin are most commonly used in practice, but are actually a subset of the more general Nyquist test. Gain and phase margin (and the Nyquist test) may operate on measured frequency response data; root-locus techniques require a rational transfer function for $a(s)f(s)$. All of these tests are relatively simple to apply and allow one to evaluate rapidly the stability of a feedback system, as well as to assess the efficacy of proposed compensation techniques, without having to determine directly the actual roots of $P(s)$.

14.9 MODELING FEEDBACK SYSTEMS

In our overview of feedback systems so far, we've identified desensitivity as the only fundamental (but extremely important) benefit conferred by negative feedback. We've seen that the larger the desensitivity, the greater the improvement in linearity – that is, the better the reduction in error.

We haven't quantified the notion of error, though, so we need to take care of that little detail now. After all, stability considerations constrain the magnitudes of desensitivity we can obtain, so we ought to learn how to calculate how big the inevitable errors will be. This knowledge will be a useful guide in our efforts at modifying architectures to reduce errors.

Before we can evaluate errors in feedback systems, however, we need to be able to model real systems in a fashion that allows tractable analysis. As we'll see, this task is often difficult because, contrary to intuition, it is not always possible to provide a $1:1$ mapping between the blocks in a model and the circuitry of a real system.

We'll also develop a set of performance measures that allow us to relate various second-order parameters to frequency- and time-domain response parameters. Because many feedback systems are dominated by first- or second-order dynamics by design (owing to stability considerations), second-order performance measures have greater general utility than one might initially recognize.

14.9.1 THE TROUBLE WITH MODELING FEEDBACK SYSTEMS

The noninverting op-amp connection is one of the few examples for which a $1:1$ mapping to our feedback model *does* exist (see Figure 14.11). Suppose we choose the forward gain equal to the amplifier gain,

$$a = G, \tag{36}$$

and choose the feedback factor equal to the resistive attenuation factor:

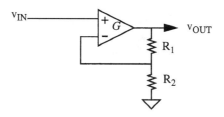

FIGURE 14.11. Noninverting amplifier.

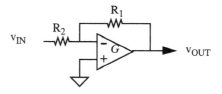

FIGURE 14.12. Inverting amplifier.

$$f = \frac{R_2}{R_1 + R_2}. \tag{37}$$

For this set of model values, we find that the closed-loop gain is indeed $1/f$ in the limit of infinite loop transmission magnitude:

$$A \rightarrow \frac{1}{f} = \frac{R_1 + R_2}{R_2}. \tag{38}$$

For that matter, we find the same loop transmission in both the block diagram and the actual amplifier, so it appears that the model parameters we chose are correct.

However, as suggested earlier, this situation is atypical. One simple case that makes this obvious is the inverting connection shown in Figure 14.12. If we insist on equating the op-amp gain G with the forward gain a of our block diagram, then we must choose the same feedback factor f as for the noninverting case if the loop transmissions are to be equal. However, with that choice, the closed-loop gain does not approach the correct value as the loop transmission magnitude approaches infinity, since we know that inverting amplifiers ideally have a gain given by

$$A = -\frac{R_1}{R_2} \tag{39}$$

whereas our choices lead to

$$A \rightarrow \frac{1}{f} = \frac{R_1 + R_2}{R_2}, \tag{40}$$

the same as for the noninverting case.

Part of the problem is simply that our "natural" choice of $a = G$ is wrong. The other is that we need one more degree of freedom than our two-parameter block diagram provides (consider, for example, the problem of getting a minus sign out of our block diagram).

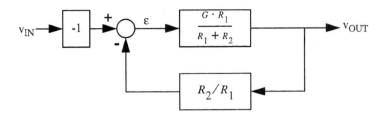

FIGURE 14.13. Inverting amplifier block diagram.

It turns out that there is not necessarily one correct model in general; that is, there are potentially many equivalent models. Operationally speaking, it doesn't matter which of these we use since, by definition, equivalent models all yield the same answer. A procedure for generating one such model is as follows:

(1) select f equal to the (magnitude of the) reciprocal of the ideal closed-loop transfer function;
(2) select a to give the proper loop transmission with the choice made in step (1); and
(3) add a third block in front of (or after) the rest of the model to take care of any sign reversal that might be needed.

There are many other equivalent procedures, of course, but this one makes use of quantities that are usually easy to discover. For example, the simplification that results from letting the loop transmission magnitude go to infinity usually makes finding the ideal closed-loop transfer function a fairly straightforward affair. Additionally, since the loop transmission itself is found from cutting the loop, discovering it is also generally simple.

Let's apply this recipe to the inverting amplifier example. First, we select the feedback factor f equal to R_2/R_1, since the ideal closed-loop transfer function is $-R_1/R_2$. Then, we need to choose a to give us the correct loop transmission:

$$a \equiv -\frac{L(s)}{f} = \left(G \cdot \frac{R_2}{R_1 + R_2} \right) \Big/ \frac{R_2}{R_1} = G \cdot \frac{R_1}{R_1 + R_2}. \qquad (41)$$

Finally, we do need to provide one final sign change, so our complete model for the inverting amplifier finally looks like that shown in Figure 14.13. Again, this model is not necessarily the only correct possibility (e.g., consider making both a and f negative quantities; one may then remove the input negation). But we're usually happy just to find one that works.

14.9.2 CLUTCHES AND LOOP TRANSMISSIONS

We've already seen that the loop transmission is an extremely important quantity since it determines stability and desensitivity. In addition, identifying the loop transmission is usually much easier than figuring out the closed-loop transfer function, making it even more valuable.

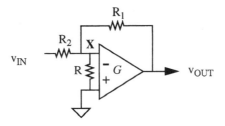

FIGURE 14.14. Nonideal inverting amplifier.

Although finding the loop transmission is a trivial matter if we happen to have a correct block diagram for the system, it may be a bit trickier to find in real systems. The usual problem is how to take loading effects into account.

To see where we might have a problem, let's consider an inverting amplifier that is built with a nonideal op-amp. In this particular case, assume that the nonideality involves some resistance that is connected between the input terminals of the op-amp. The circuit then appears as shown in Figure 14.14. In order to find the loop transmission we suppress all independent sources, so we set the input voltage to zero. Then we have to cut the loop, inject a signal at the cut point, and see what comes back. The ratio of the return signal to the input signal is the loop transmission.

If we cut the branch marked X to the *left* of the resistor R, we will effectively (and incorrectly) eliminate R from the loop transmission when we apply a test voltage to X. To take the loading effect of R properly into account, we need to cut the loop to the *right* of R. Another good choice would be the output of the op-amp.

The general principle is to find a point (if possible) that is driven by a zero impedance, or that drives into an infinite one. That way, there are no loading-effect issues to confound us. Although not all circuits will automatically have such points, it is always possible to generate models that do. For example, consider an emitter follower. We can always model it as an ideal one with input and output resistances added to account for nonideal effects found in the original circuit. By using an ideal follower inside the model, we generate a node whose properties allow us to find the loop transmission of the feedback system of which the follower may be a part.

14.10   ERRORS IN FEEDBACK SYSTEMS

We have already concluded that the only fundamental property of negative feedback systems is the desensitivity to forward gain variations that they provide, with greater loop transmission leading to greater desensitivity. Unfortunately, we've also found that as the loop transmission magnitude increases, we eventually encounter instability, thus imposing serious constraints on the attainable desensitivity.

As a result of having to compromise desensitivity, we expect the loop to produce errors. As a typical example, consider a simple voltage follower made with an op-amp (Figure 14.15). Assume that the amplifier transfer function is single-pole, and then consider what happens if we apply a step input. The output will rise with a first-order

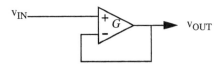

FIGURE 14.15. Op-amp follower.

shape, eventually reaching a value close to the input. How close? It depends on the DC gain of the op-amp. If the input voltage is one volt and the op-amp's DC gain happens to be 1000, then the output will be one volt to within about a part in a thousand. Viewed another way, a follower is supposed to, well, *follow*. So, if the output is to be near a volt and the gain is 1000, the voltage difference at the input has to be 1/1000, or 1 mV. Hence, the error between input and output will be about 1 mV. To reduce this error we need more gain, but we can't always get it owing to stability problems.

Although we can't reduce the error to zero at all frequencies because of the stability issue, we can often obtain zero *steady-state* error in response to a step input. To reduce the steady-state step response error to zero, then, we need infinite gain only at DC, not at all frequencies. As a result, most op-amps are designed to have huge amounts of DC gain (like a million or more) in order to reduce steady-state error to tiny values. To solve the stability problem, they are also generally designed to have single-pole behavior over many decades of frequency. That is, op-amps are generally designed to approximate integrators.

Another way of deducing the necessity for an integration is to recognize that, if we want a nonzero steady-state output for a zero input difference between the op-amp terminals then we need an integrator, since an integrator can provide any DC output with no input.

We can use this observation to deduce other useful and interesting trivia. Suppose, for example, we were interested in a slightly different problem – that of tracking, with zero steady-state error, a *ramp* input (a constant velocity, if you will). If we now assume that the output is a ramp while we have zero input difference, then we need to have *two* integrations in the op-amp's transfer function (the input to the second integrator must be a step, and we already know that another integration is needed to give us zero steady-state error to a step).

You can easily see that *three* integrations allow us to have zero steady-state error in response to a quadratic ramp (a constant acceleration), and so on. Is any of this stuff useful? Absolutely. For example, consider voltage to be proportional to the position of an object we wish to track. Zero step response error corresponds to zero position error, zero ramp response corresponds to zero velocity error, and zero quadratic response corresponds to zero acceleration error. So, if the object we're tracking has a constant acceleration, we can still track it with zero steady-state error if we have three integrations in $G(s)$.

Yes, we do have to worry about stability with more than one integration, but we can fix that up quite easily just by adding enough zeros to cancel the phase shift. If P

is the number of poles, we simply have to provide $P - 1$ zeros well below crossover and we're all right.

When we study phase-locked loops (PLLs), we'll see that this tracking problem reappears in the context of FM modulation–demodulation (as just one example), so we still have to worry about these issues even if we're not building antiaircraft weaponry.

ERROR SERIES

We've seen that it is possible, in principle, to eliminate steady-state errors if we employ enough integrations. However, perfect integrators are difficult to realize in practice. In any case, we'd like some method for quantifying errors in general.

One way is to use an *error series,* that is, to express the error as a power series:

$$\varepsilon(t) = \varepsilon_0 v_i(t) + \varepsilon_1 \frac{dv_i(t)}{dt} + \varepsilon_2 \frac{d^2 v_i(t)}{dt^2} + \cdots . \tag{42}$$

If the series converges quickly, we may truncate after a few terms. Fortunately, if the series does *not* converge quickly, it implies that the system is doing a lousy job of tracking, and we might as well remove the system. Hence, for nearly all practical cases, the error series does converge rapidly.

The various error coefficients may be found from the following equation:[12]

$$\varepsilon_k = \frac{1}{k!} \frac{d^k}{ds^k} \left[\frac{V_e(s)}{V_i(s)} \right] \Bigg|_{s=0}, \tag{43}$$

where $V_e(s)/V_i(s)$ is the input-to-error transfer function.

If you happen to forget this formula, there is another way to obtain the error coefficients. First, find the input-to-error transfer function. Then, divide the numerator polynomial by that of the denominator to obtain a transfer function in *ascending* powers of s. The coefficient of the s^k term in this series turns out to be equal to ε_k.

The various error coefficients have the following physical interpretations. The zeroth-order coefficient is the steady-state step response error, while the first-order term is the steady-state *delay* in response to a ramp input, and so on. Knowledge of the error series thus allows one to estimate rapidly the error for arbitrary input signals.

The more rigorously minded among you may point out that derivatives of the input signal may not exist (or be bounded) if we allow things like step inputs, and therefore we cannot really use our error series. Strictly speaking, that is true, but if we restrict our questions about error to times that are sufficiently removed from such discontinuities then we can still use our decomposition. What is "sufficiently removed?" A reasonable criterion is to wait a few time constants.[13]

---

[12] See e.g. G. C. Newton, Jr., L. A. Gould, and J. F. Kaiser, *Analytical Design of Linear Feedback Controls,* Wiley, New York, 1957, Appendix C.

[13] Because high-order systems have more than one time constant, a more generally useful criterion is to wait a settling time.

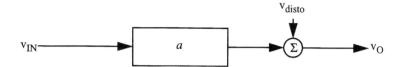

FIGURE 14.16. Model of open-loop system with distortion.

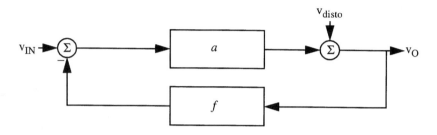

FIGURE 14.17. Feedback system with additive noise sources.

A useful notion is that distortion is a type of error, one that is usually considered from a frequency-domain viewpoint. It is worthwhile spending a little bit of time examining how feedback may affect distortion. For example, how much feedback does one need in order to reduce an amplifier's second harmonic distortion by some factor R? To answer that question, let us model a nonlinear system as a linear system that has been corrupted by additive distortion products; see Figure 14.16. Without loss of generality, suppose that we may express the additive distortion as consisting of just two harmonic distortion products (assuming a sinusoidal excitation):

$$v_{\text{disto}} = v_O[D_2 \cos 2\omega t + D_3 \cos 3\omega t], \qquad (44)$$

where the factors D_2 and D_3 are the amplitudes of the distortion products, normalized to the output fundamental.

Now wrap a feedback loop around that model, as shown in Figure 14.17. For this feedback system, we have

$$\frac{v_O}{v_{\text{in}}} = \frac{a}{1 + af}, \qquad (45)$$

$$\frac{v_O}{v_{\text{disto}}} = \frac{1}{1 + af}. \qquad (46)$$

An open-loop model that is described by the same equations is depicted in Figure 14.18. We see from the open-loop equivalent that, for the same output fundamental amplitude, the distortion products are smaller by a factor equal to the return difference $(1 + af)$. Thus, if we seek to reduce distortion by a factor of (say) 100, then we need to make sure that the loop transmission magnitude is at least about 100 (99, to be exact, but let's call it 100) over the bandwidth where this reduction is needed.

Even though our example happens to consider only the second and third harmonic distortion products, our calculation is in fact applicable much more generally. For

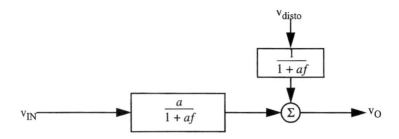

FIGURE 14.18. Open-loop equivalent of feedback system with distortion.

example, it applies as well to intermodulation distortion products, and of any order. So, if we want to reduce errors (described in either the time or frequency domain) by some factor R, we need to choose $1 + af$ equal to R.

14.11 FREQUENCY- AND TIME-DOMAIN CHARACTERISTICS OF FIRST- AND SECOND-ORDER SYSTEMS

An error series is but one of many ways to characterize feedback systems. We could imagine using other measures, such as step response overshoot, settling time, or frequency response peaking. Depending on the context, some or all of these parameters could be of interest.

In this section, we'll simply present a number of exceedingly useful formulas without detailed derivations. In most instances it should be obvious how to derive them, but the tedium involved is too great to merit presentation here. In those cases where the derivation might not be obvious, a comment or two might be added to help point the way.

We have already asserted that it should be possible to characterize most feedback systems as systems of second order at most. This claim derives from the observation that any stable amplifier cannot have more than two (net) poles dominate the loop transmission below crossover. Hence, for feedback systems at least, intimate knowledge of first- and second-order characteristics turns out to be sufficient for most situations of practical interest.

The following formulas all assume that the systems are low-pass with unit DC gain. Therefore, not all of them apply to systems with zeros, for example (you can't have everything, after all). With that little warning out of the way, here are formulas for first- and second-order systems.

14.11.1 FORMULAS FOR FIRST-ORDER LOW-PASS SYSTEMS

Assume that the system transfer function is

$$H(s) = \frac{1}{\tau s + 1}. \tag{47}$$

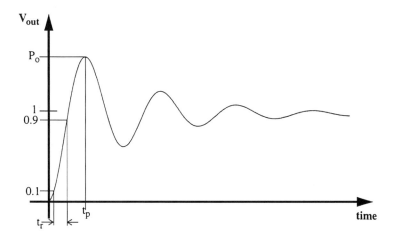

a) Step Response Parameters

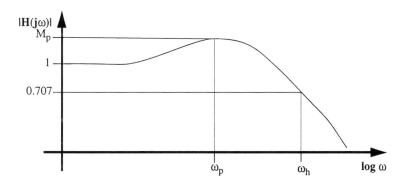

b) Frequency Response Parameters

FIGURE 14.19. First- and second-order parameters.

For this first-order, low-pass system, we have:

$$t_r = \tau \ln 9 \approx 2.2\tau = 2.2/\omega_h, \tag{48}$$

$$P_o = M_p = 1, \tag{49}$$

$$t_p = \infty, \tag{50}$$

$$t_s\big|_{2\%} \approx 4\tau, \tag{51}$$

$$\varepsilon_1 = \tau, \tag{52}$$

$$\omega_p = 0, \tag{53}$$

where the various quantities have the meanings shown in Figure 14.19.

Commentary and Explanations

Equation 48 – The risetime definition used here is the 10% to 90% risetime.

Equation 49 – Both the step and frequency responses are monotonic in a single-pole system.

Equation 50 – Because the step response is monotonic and asymptotically approaches its final value, there is an infinite wait to see the peak.

Equation 51 – An exponential settles to within about 2% of final value in four time constants.

Equation 52 – The steady-state delay in response to a ramp input is equal to the pole time constant.

Equation 53 – The frequency response of a first-order system rolls off monotonically from its DC value. Hence, the peak of the frequency response occurs at zero frequency.

14.11.2 FORMULAS FOR SECOND-ORDER LOW-PASS SYSTEMS

Here, assume a transfer function of the form

$$H(s) = \left[\frac{s^2}{\omega_n^2} + \frac{2\zeta s}{\omega_n} + 1 \right]^{-1}. \tag{54}$$

Then the following relationships hold:

$$t_r \approx 2.2\tau = \frac{2.2}{\omega_h}, \tag{55}$$

$$P_o = 1 + \exp\left(\frac{-\pi\zeta}{\sqrt{1-\zeta^2}} \right), \tag{56}$$

$$t_p = \frac{T_{\text{osc}}}{2} = \frac{\pi}{\omega_n\sqrt{1-\zeta^2}}, \tag{57}$$

$$t_s\big|_{2\%} \approx 4\tau_{\text{env}} = \frac{4}{\zeta\omega_n}, \tag{58}$$

$$\varepsilon_1 = \frac{2\zeta}{\omega_n}, \tag{59}$$

$$M_p = \frac{1}{2\zeta\sqrt{1-\zeta^2}}, \quad \zeta < \frac{1}{\sqrt{2}}, \tag{60}$$

$$\omega_p = \omega_n\sqrt{1-2\zeta^2}, \quad \zeta < \frac{1}{\sqrt{2}}, \tag{61}$$

$$\omega_h = \omega_n\left[1 - 2\zeta^2 + \sqrt{2 - 4\zeta^2 + 4\zeta^4} \right]^{1/2} = \omega_n\big|_{\zeta=1/\sqrt{2}}. \tag{62}$$

Commentary and Explanations

Equation 55 – The risetime of a second-order low-pass system is somewhat dependent on the damping ratio. In the limit of zero damping, the product of bandwidth and risetime can be as small as about 1.6. However, for any reasonably well-damped system, the product will be closer to 2.2.

Equation 56 – The peak of the step response overshoot cannot exceed 100% for a second-order low-pass system.

Equation 57 – The time at which the step response peak overshoot occurs is simply one half the ringing period. Recall that the ringing frequency is equal to the imaginary part of the complex pole pair. The formula for t_p follows directly from these two facts.

Equation 58 – Just as the imaginary part of the pole frequency controls the oscillatory part of the response, the real part controls the decay. As in the first-order case, it takes about four time constants for the envelope to settle to 2% of final value. With the information from Eqn. 57 and Eqn. 58, we can also express the equation for P_o as follows:

$$P_o = 1 + \exp\left(\frac{-\pi\zeta}{\sqrt{1-\zeta^2}}\right) = 1 + \exp\left(-\frac{T_{\text{osc}}/2}{\tau_{\text{env}}}\right). \tag{63}$$

Equation 59 – The steady-state time delay in response to a ramp input is the same as for the first-order case if the damping ratio equals 0.5 and decreases as the damping ratio decreases, approaching zero delay in the limit of zero damping.

Equations 60 and 61 – The frequency response can exhibit a peak at other than zero frequency if the damping ratio is less than 0.707. For greater damping ratios, the response is monotonic and thus exhibits a peak at DC. For smaller damping ratios, the peak magnitude asymptotically approaches infinity in the limit of zero damping.

Equation 62 – The -3-dB frequency equals ω_n at a damping ratio of $1/\sqrt{2}$. The bandwidth is a maximum of about $1.55\omega_n$ in the limit of zero damping.

14.12 USEFUL RULES OF THUMB

Notice that phase margin is conspicuously absent from the set of equations presented in the previous section. To bring phase margin explicitly into the discussion requires making a number of limiting assumptions, because in general there is no unique relationship between, say, phase margin and damping ratio. However, out of necessity, stable systems must behave as first- or second-order systems near crossover, so we may derive a number of relationships for a second-order system and apply them to a much broader class of systems, even though they strictly apply only to the second-order system for which they were derived.

Specifically, assume in all of the following that we have a two-pole system with purely scalar feedback. Further assume that the two loop transmission poles are widely spaced. With these assumptions, one may derive the following relationship between damping ratio and phase margin:

$$\zeta \approx \left[4\left(\{2[\tan(90° - \phi_m)]^2 + 1\}^2 - 1\right)\right]^{-1/4}. \tag{64}$$

This cumbersome equation may be replaced by a remarkably simple approximation that holds over a restricted (but useful) range of phase margins:

$$\zeta \approx \phi_m/100, \tag{65}$$

where ϕ_m is the phase margin in degrees in both Eqn. 64 and Eqn. 65. This relationship is accurate to within about 15% for phase margins less than approximately 70°. Furthermore, it is accurate to better than 10% from about 35° to a bit less than 70°, a range that fortuitously spans the phase margin targets most often encountered in practice. The damping ratio as estimated by Eqn. 65 may also be used to estimate the step response overshoot:

$$P_o = 1 + \exp\left(\frac{-\pi\zeta}{\sqrt{1 - \zeta^2}}\right) \approx 1 + \exp\left(\frac{-\pi\phi_m}{\sqrt{10^4 - \phi_m^2}}\right), \tag{66}$$

where the phase margin is again expressed in degrees. As with the expression for damping ratio, this equation provides reasonable accuracy for phase margins below about 70°.

Another relationship of considerable utility is that between phase margin and frequency response peaking:

$$M_p \approx \frac{1}{\sin \phi_m}. \tag{67}$$

For our prototype second-order system, this equation is accurate to within 1% up to a phase margin of about 55°.

With these approximate equations, it is a simple matter to estimate what phase margin is needed to satisfy an overshoot or peaking specification, or to estimate phase margin from measurements of step response or frequency response. Again, because these equations apply strictly to a two-pole system with widely spaced poles and scalar feedback, they will provide good estimates for systems that are well approximated by such a two-pole system.

14.13 ROOT-LOCUS EXAMPLES AND COMPENSATION

The various root-locus construction rules we've developed are certainly not exhaustive, but they are more than sufficient for the vast majority of loci encountered in practice. We'll now present a few root-locus examples to gain some practice with the method. Then we will examine the subject of compensation.

In the pages that follow are a number of examples, roughly in order of increasing complexity. In all cases, loci for only positive values of k (corresponding to the more common case of negative feedback) are shown.

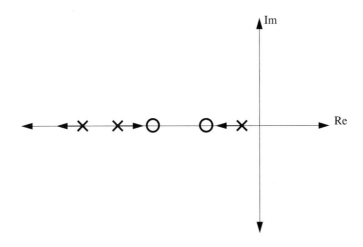

FIGURE 14.20. Root locus with branches on real axis only.

As a general rule, it may be useful to employ something like the procedure below.

(1) First, identify the segments of the real axis that are part of the locus.
(2) Compute asymptotic angles and the real-axis intercept of those asymptotes.
(3) Estimate or calculate breakaway or entry points, if any.
(4) Begin to sketch the locus using additional information, such as the initial angle near a complex pole or zero (if appropriate) or the constancy of the average distance to the imaginary axis (if $P > Z + 2$). Remember that every zero is a terminus for the locus.

Of course, this procedure is hardly unique, but it often suffices. After constructing numerous loci, certain patterns (dare I call them macros?) will often recur, and drawing root loci will become progressively easier.

14.13.1 EXAMPLE: PURELY REAL POLES AND ZEROS THAT STAY REAL

We start with the simple example sketched in Figure 14.20. The rules and facts used here are: (1) real-axis rule; (2) termini of locus are zeros; (3) asymptote rule (left-most pole heads to minus infinity).

14.13.2 EXAMPLE: TWO POLES THAT BECOME COMPLEX

This example is a bit more interesting than the last; see Figure 14.21. The real-axis rule tells us that the real-axis segment between the poles is part of the locus, while the asymptote rules tell us that the asymptotes are at $\pm 90°$ and intersect the real axis exactly halfway between the poles. From the breakaway rule, we can infer that the

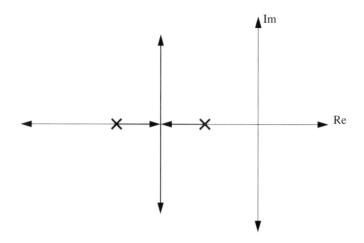

FIGURE 14.21. Two-pole root locus.

poles must leave the real axis somewhere in the middle (exactly in the middle, in this case) of the real-axis segment. Invoking the symmetry argument, the asymptotes are actually exactly a part of the locus.

Note that, strictly speaking, a two-pole negative feedback system is never unstable in the BIBO sense, yet its damping gets progressively smaller as k increases. For such a system, the real part stays constant once the poles become complex, implying that the exponential envelope of the impulse response has a constant shape. However, the imaginary part increases as k increases, so that more oscillations per unit time occur within the exponentially decaying envelope.

Since the exponential envelope doesn't change shape as k increases, the settling time does not change, either. Just because a system becomes less stable does not necessarily imply a degradation in settling time.

Observations and conclusions such as these simply don't emerge from gain and phase margin calculations, illustrating the importance of having several ways to assess system stability. Additional viewpoints accelerate the development of intuition.

14.13.3 EXAMPLE: TWO POLES AND A ZERO

To stabilize the previous system, we may consider the addition of a zero. From a phase margin viewpoint, we regard the improvement in stability as a consequence of the positive phase contributed by the zero. An alternative view, informed by root-locus constructions, is that zeros are attractors of poles, so that properly placed zeros can bend poles away from the imaginary axis and toward more highly damped configurations.

A specific (and frequently occurring) example is sketched in Figure 14.22. The locus appears to include a circle centered on the zero. It turns out to be the correct shape, not just an artifact of the author's laziness.

In constructing this locus, make use of the real-axis rule, the breakaway/entry rule (here, we have an entry point between the finite zero shown and an infinite one), and

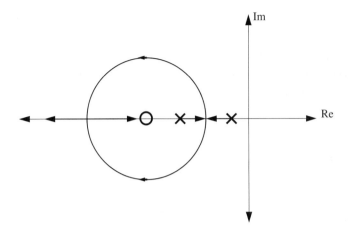

FIGURE 14.22. Root locus of two poles plus zero.

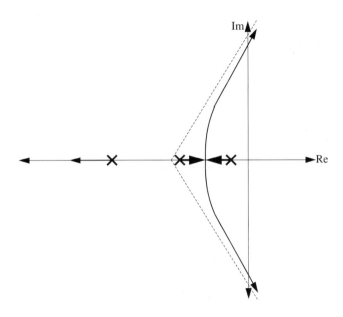

FIGURE 14.23. Three-pole root locus.

the asymptote rule. Those rules are not sufficient to deduce the circular shape shown but are enough to conclude that the poles eventually become purely real again beyond some value of k.

14.13.4 EXAMPLE: SYSTEMS THAT GO UNSTABLE

A simple example of a system that eventually goes unstable beyond some critical value of k is illustrated in Figure 14.23. Use the real-axis, asymptote, and intercept rules to deduce this locus. The asymptotes form angles of 180° and ±60° with respect to the real axis.

As an aside, one can estimate the value of gain at which the poles cross the imaginary axis – as well as the corresponding oscillation frequency – by using the asymptotes. A simple trigonometric calculation yields the intersection of the asymptotes with the imaginary axis; this value is a crude estimate for the oscillation frequency. Then, plug this value into the expression for $g(s) = a(s)f(s)$, compute the magnitude, then take the reciprocal to find the value of k that corresponds to this onset of instability.

If you need exact answers to these questions, simply take $g(s)$ and find the value of $j\omega$ where the phase becomes exactly $-180°$; then compute the magnitude of $g(s)$ at that frequency. That frequency is the oscillation frequency, and the reciprocal of the magnitude of $g(s)$ there is the value of k that just results in instability. This procedure is equivalent to determining the conditions that lead to a phase margin of zero and a gain margin of unity.

Another potentially useful observation is that we can determine the value of k that corresponds to a specified damping ratio. Simply plot rays of constant damping ratio, and compute the intersection of the locus with the rays. Then, plug that value of s into the expression for $g(s)$, compute the magnitude, and take the reciprocal. The result is the required value of k.

14.13.5 EXAMPLE: LOCUS WITH COMPLEX POLES IN $L(s)$

All of our examples so far have had loop transmission poles and zeros that were purely real. Let's do something a little different and exercise the complex-pole construction rule.

Consider the three-pole locus shown in Figure 14.24. Here, we compute the initial angle that the locus makes with the poles by drawing vectors from each pole to the other two poles and then adding up the angle contributions of each. The real-axis, asymptote, and intercept rules tell us what eventually happens overall. Again, the asymptotes themselves may be used to estimate both the value of gain that causes oscillation as well as the oscillation frequency. As with the previous locus, the asymptotes are at $180°$ and $\pm 60°$.

14.13.6 EXAMPLE: LOCUS WITH RIGHT HALF-PLANE ZERO IN $L(s)$

In many operational amplifiers (particularly those implemented in MOS form), there is a significant (i.e., relatively low-frequency) right half-plane zero. It is easy to get tripped up in constructing the locus for such an amplifier if we're asleep at the wheel. Remember the full statement of the real-axis rule: The locus lies to the left of an odd number of left half-plane poles + zeros, and to the right of an odd number of right half-plane poles + zeros.

Consider Figure 14.25. Notice what happens as we increase the loop transmission magnitude: the pole moves to higher and higher frequencies. Then, at some critical

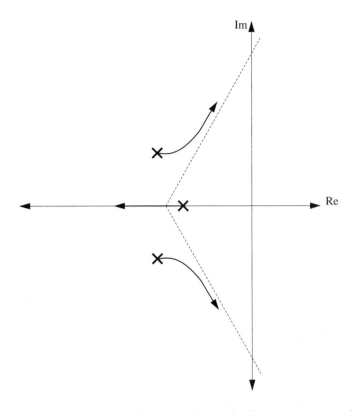

FIGURE 14.24. Root locus of system with complex loop transmission poles.

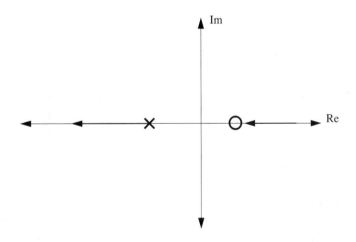

FIGURE 14.25. Root locus of system with right half-plane zero in
loop transmission.

value of gain, the system becomes unstable as the pole moves from minus infinity to
plus infinity (yes, this can, and does, happen at a finite value of k). Then, as the gain
increases still further, the pole asymptotically approaches the zero (from the right
side), as poles always do.

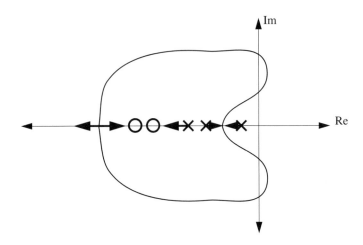

FIGURE 14.26. Root locus of conditionally stable system.

From this locus, you can see why right half-plane zeros are bad news from a stability viewpoint. Again, since zeros are terminal locations for loci, poles will eventually move to them, and if the zeros happen to be in the right half-plane, the system will eventually become unstable.

It is perhaps initially difficult to understand how a zero can ever encourage instability. After all, we use zeros in lead compensators to *improve* stability. What's different here? It is left as an exercise for the reader to show that a right half-plane zero has the same magnitude behavior as a left half-plane zero, but that the phase response has the same general trend as that of a left half-plane *pole*.

14.13.7 EXAMPLE: CONDITIONALLY STABLE SYSTEM

In all of the examples so far, a system may exhibit instability above some critical value of gain. This behavior is certainly common, but it is not the only possible behavior. There are systems of practical interest that are unstable both above and below some range of gain. In this example, we consider the locus of such a *conditionally stable* system (see Figure 14.26).

If we consider the locus at relatively small values of k, we expect to see something like the simple three-pole locus. Assume that the pole-zero constellation is such that the complex pole pair that forms does cross over to the right half-plane. As the gain continues to increase, the attractive influence of the zeros becomes more prominent and the complex poles bend back over to the left half-plane. Hence, the system is unstable only for some finite range of k.

This type of locus can result when we attempt to build a system that possesses a small steady-state error in tracking an input that possesses a constant acceleration. The ideal of three poles at the origin can't be realized in general, so they might end up as shown. To stabilize such a loop, two zeros must be added. If the crossover frequency is too low, however, the positive phase shift of the zeros doesn't do any good, and the system can be unstable. As the gain increases above some critical value, the crossover frequency increases enough for the zeros to work their magic and the system becomes stable again.

We'll later see that such systems may exhibit instability in practice even when our linear analyses tell us that we have plenty of phase margin. The key word is "linear." All real systems have finite dynamic range. If we ever saturate an element in a conditionally stable loop then we effectively reduce the gain, since something that saturates doesn't provide much of an output change for a given input change. From the shape of our locus, we see that there is a real danger of entering the region of conditional instability if a gain reduction should occur. Furthermore, any oscillation associated with this instability might be of sufficient amplitude to sustain this saturated condition, and the system may never recover – a transient overload might be enough to send it into an oscillatory mode from which it cannot escape.

14.14 SUMMARY OF ROOT-LOCUS TECHNIQUES

We have seen that just a handful of simple construction rules are sufficient to guess how closed-loop poles vary as the loop transmission magnitude varies.

Remember that these rules all derive from a magnitude and angle condition on the loop transmission – namely, that a value of s can be a closed-loop pole location only if the phase of $a(s)g(s)$ is an odd multiple of $180°$, and only if the magnitude of $a(s)g(s)$ is unity at the value of s.

Finally, don't forget that the root locus only tells us where the closed-loop *poles* are. It is easy to lose track of this fundamental truth because we employ poles *and* zeros in constructing loci. However, the loop transmission zeros we use in drawing loci are not necessarily the same as closed-loop zeros. If it is necessary to find closed-loop zeros, it is easy to do so by finding the zeros of $a(s)$ and uncancelled poles of $f(s)$.

As a parting note, it must be stated that not all of the given rules gracefully take into account right half-plane poles and zeros, because the phase behavior of poles and zeros in the right half-plane is the opposite of that in the left half-plane. The real-axis rule is an exception. The complex-angle rule must be modified for right half-plane singularities, among others, but we will not make those modifications here. Just be a bit wary when attempting to draw loci if your loop transmission contains right half-plane stuff.

14.15 COMPENSATION

We have developed a number of methods for evaluating the stability of feedback systems. For example, root-locus techniques allow us to sketch rapidly how the closed-loop poles vary as we change some parameter of the loop, but constructing root loci requires a rational expression for the loop transmission in order to find the loop transmission poles and zeros. Gain and phase margin, as well as the Nyquist

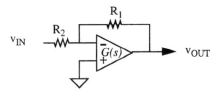

FIGURE 14.27. Inverting amplifier.

test, require knowledge only of loop transmission gain and phase behavior, and this information may be obtained experimentally.

We now shift our focus away from *analyzing* stability to *changing* it. As you might expect, we will draw heavily from insights that are implicit in the various analytical tools developed so far. We shall see that, as usual, the various stability compensation techniques involve tradeoffs among stability, desensitivity, design complexity, and time-domain response quality.

14.16 COMPENSATION THROUGH GAIN REDUCTION

Let us consider one implication of phase margin as a stability measure. If we assume that the loop transmission of our uncompensated system has an increasingly negative phase shift as frequency increases (e.g., because of the presence of poles), then stability could be improved simply by reducing the crossover frequency to a value such that the associated phase shift is less negative. One "low-tech" way to effect such a reduction in crossover frequency is to reduce the loop transmission magnitude by a fixed factor at all frequencies. Since using such an attenuator does not affect phase behavior, it is trivial to calculate the attenuation factor required to satisfy a given phase margin specification.

As a specific example, consider using an op-amp that requires truly zero input current and possesses zero output impedance, but has the following transfer function:

$$G(s) = \frac{10^7}{(s+1)(10^{-3}s+1)}. \tag{68}$$

Suppose we take this op-amp and connect it in an inverting configuration with an ideal closed-loop gain of -99, as shown in Figure 14.27; we have $R_1/R_2 = 99$. With this information, we can readily derive an expression for the loop transmission:

$$-L(s) = \frac{R_2}{R_1+R_2} \cdot G(s) = 10^{-2} \cdot G(s) = \frac{10^5}{(s+1)(10^{-3}s+1)}. \tag{69}$$

Let's now compute the phase margin for this connection. First, we find the crossover frequency. In general, it's most convenient to use simple function evaluation (fancy name for trial and error, guided by a rough Bode plot). In this case, we can pin down the crossover frequency quite accurately without much computation by exploiting a few observations.

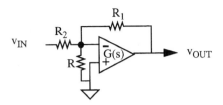

FIGURE 14.28. Inverting amplifier with
reduced-gain compensation.

First, the dominant pole at 1 rps causes a −20-dB/decade rolloff until the second
pole at 1 krps is reached. Without that second pole, the loop transmission would have
a magnitude of 100 at 1 krps.

The second pole accelerates the rolloff ultimately to −40 dB/decade, so only another decade beyond 1 krps takes us from a magnitude of 100 to an extrapolated
magnitude of unity. Since crossover is a decade beyond the second pole, we may
assume that, to a good approximation, the rolloff is −40 dB/decade there, so that
crossover indeed does have the following approximate value:

$$\omega_c \approx 10^4 \text{ rps.} \tag{70}$$

Computation of phase margin is similarly trivial, as crossover occurs well above
both loop transmission poles. The pole at 1 rps may be considered to contribute
−90° at 10^4 rps (the actual phase shift is only 0.0057° shy of −90°, so little error is
involved), while the second pole also contributes nearly −90°. However, a decade
beyond a pole we have a residual error of 5.7°, so here we find that the phase margin
isn't quite zero (but it is small). With such a small phase margin, we would expect
large overshoot in the step response, large peaking in the frequency response, and
extreme sensitivity to any additional negative phase shift from unmodeled poles that
may be lurking in the shadows. In short, the stability is very unsatisfactory.

Suppose we wanted to achieve a phase margin of at least 45° by using gain reduction. How could we do it? From inspection of the expression for loop transmission,
it would appear that changing the ratio of the feedback resistors would be one possibility. If we made R_2 smaller, we would reduce the loop transmission magnitude at
all frequencies. Unfortunately, we would also change the ideal closed-loop gain and,
presumably, we're not permitted that particular degree of freedom.

The solution is to add another resistor, this time *across the input terminals of the
op-amp* (see Figure 14.28). This connection may be puzzling to those who have been
taught to treat an op-amp as an ideal element whose input voltage difference is zero
(the "virtual ground" concept and all that). It is easy to jump to the incorrect conclusion that the addition of resistance in such a place cannot have any effect because
the voltage across it is ideally zero. The key word is "ideally," since we do *not* have
an ideal op-amp. Consider, for example, placing a short circuit across the input terminals of the op-amp. The loop transmission must go to very small (zero) values in
that situation.

Looking at the situation more analytically, note that such an additional resistor appears in parallel with R_2 as far as the loop transmission is concerned, but disappears as far as the ideal closed-loop transfer function is concerned (*this* is where we invoke the virtual ground idea). Hence, we can effect changes in stability without disturbing the ideal closed-loop transfer function.

Let's now compute how much gain reduction we need. Since the phase margin goal is 45°, we need to find the frequency at which $-L(s)$ gives us a phase shift of $-135°$, since that will become the new crossover frequency.

Again, from inspection of the expression for the loop transmission, it should be apparent that the new crossover frequency should be the frequency of the second pole, that is, 10^3 rps, since at that frequency the first pole has contributed essentially $-90°$ and the second pole another $-45°$.

At the new desired crossover frequency, the uncompensated loop transmission has a magnitude of approximately

$$|L(j10^3)| = \frac{10^5}{\left|\sqrt{10^6 + 1}\right| \cdot \left|\sqrt{2}\right|} \approx 70.7. \tag{71}$$

Therefore, this is the factor by which we need to reduce the loop transmission gain.

The old loop transmission may be expressed as

$$-L(s) = \frac{R_2}{R_1 + R_2} \cdot G(s) = \left(\frac{R_1}{R_2} + 1\right)^{-1} \cdot G(s). \tag{72}$$

The new loop transmission is

$$-L(s) = \left(\frac{R_1}{R_2 \parallel R} + 1\right)^{-1} \cdot G(s), \tag{73}$$

which may be rearranged as

$$-L(s) = \left[\frac{R}{R + (R_1 \parallel R_2)}\right] \cdot \frac{R_2}{R_1 + R_2} \cdot G(s), \tag{74}$$

where the term in brackets may be considered the compensator's transfer function $C(s)$.

Since we need to provide a whopping factor-of-70.7 gain reduction, we need $C(s)$ to have a value of $1/70.7$. Therefore, we need to choose R small enough to give us this attenuation. Because of the large attenuation factor, we would expect R to be so small compared with R_1 and R_2 that it should be about 70.7 times smaller than R_2, to a good approximation. A slightly more rigorous calculation yields a value quite close to that estimate:

$$R \approx \frac{R_2}{70.4}. \tag{75}$$

In summary, the reduced-gain compensator has taken the system from a phase margin of about 5.7° to a phase margin of 45° via reducing the loop transmission by a factor of about 70.7 at all frequencies. At the same time, crossover has decreased

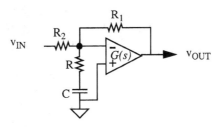

FIGURE 14.29. Inverting amplifier
with lag compensator.

by a factor of 10, from a frequency of 10^4 rps to 10^3 rps. The ideal closed-loop gain
remains -99.

The tradeoffs, of course, are a reduction in bandwidth and desensitivity. Furthermore, the reduction in desensitivity occurs at all frequencies, whereas stability
is determined by just the behavior near crossover (so says phase margin, anyway).
Hence, DC and low-frequency desensitivity are apparently needlessly compromised
by such a simple-minded compensation scheme.

14.17 LAG COMPENSATION

If we could reduce crossover by attenuating the loop transmission only at higher
frequencies, we would leave low-frequency desensitivity untouched while improving stability. Such a selective loop transmission reduction can be accomplished by
adding a capacitor in series with the gain reduction resistor of the previous topology,
as shown in Figure 14.29.

The capacitor prevents the compensation resistor from having any effect at DC
and low frequencies, while the network degenerates to a simple reduced-gain compensator at frequencies high enough for the capacitor to appear as a short. Such a
compensator is known as a *lag compensator* for reasons that will become clear shortly.
Once again, the compensation network has no effect on the *ideal* closed-loop transfer
function because it is connected across two terminals that have no voltage difference
in the ideal limit of infinite op-amp gain.

To discover the *real* effect of this compensator, though, let's derive an expression
for the loop transmission:

$$-L(s) = \left[\frac{R_1}{R_2 \parallel (R + 1/sC)} + 1 \right]^{-1} \cdot G(s), \qquad (76)$$

which, after some manipulation, "simplifies" to

$$-L(s) = \left\{ \frac{sRC + 1}{sC[R(1 + R_1/R_2) + R_1] + (1 + R_1/R_2)} \right\} \cdot G(s). \qquad (77)$$

After further manipulation, this can be expressed in a somewhat more intuitively useful form:

$$-L(s) = \left\{ \frac{sRC + 1}{sC[R + (R_1 \parallel R_2)] + 1} \right\} \cdot \frac{R_2}{R_1 + R_2} \cdot G(s). \tag{78}$$

The term in braces may be considered the transfer function of the compensator, while the rest of the equation is (minus) the loop transmission of the uncompensated system.

At DC, the compensator transfer function is unity, and the system behaves as in the uncompensated case. At very high frequencies, the compensator asymptotically approaches a value of

$$C(s) \rightarrow \frac{R}{R + (R_1 \parallel R_2)}, \tag{79}$$

just as in the reduced-gain compensator case, as expected.

Note that the compensator $C(s)$ contains one zero and one pole. As can be seen from the full expression for $C(s)$, the pole is always at a lower frequency than the zero. It is the pole that causes the loop transmission magnitude to decrease (since the magnitude of $C(s)$ decreases beyond the pole frequency). Unfortunately, an unavoidable side effect is the negative phase shift that is associated with the pole. It is this phase lag that gives this compensator its name.

Clearly, one important design criterion is to make sure that this lagging phase shift has been cancelled by the zero's positive phase shift well below crossover. Otherwise, phase margin will actually degrade rather than improve.

A simple (but not necessarily optimum) design procedure is to begin with the reduced-gain compensator to discover the value of R necessary to force crossover to a low enough frequency to achieve the specified phase margin. For reasons that will become clear shortly, it may be advisable to aim for a phase margin about five or six degrees larger than you ultimately want. Then, place the zero a decade below the new desired crossover by choosing

$$RC = \frac{10}{\omega_{c,\text{new}}} \implies C = \frac{10}{R \cdot \omega_{c,\text{new}}}. \tag{80}$$

With this choice of zero location, the positive phase shift of the zero will be about $5.7°$ shy of its maximum, while the pole (with its lower frequency) has contributed just about all of its $-90°$ of phase shift. Hence, if the reduced-gain compensator is used as a starting point for lag compensator design, the phase margin goal should be augmented by five or six degrees, as stated earlier.

A more thoughtful design might require iteration to complete, since the pole and zero location are both adjustable parameters. Hence, there is no unique set of R and C that provides a given phase margin. The simplified procedure presented here usually suffices, however, to provide either a final design or a reasonable initial design from which further optimization may develop.

The lag compensator provides roughly the same crossover (and hence roughly the same closed-loop bandwidth) as the reduced-gain compensator, but it leaves the low-frequency loop transmission untouched. Hence, it doesn't degrade desensitivity unnecessarily and so allows us to obtain all of the associated benefits, such as reduced steady-state step response error.

However, there is one drawback to the lag compensator that deserves mention. The compensator employs a zero that is well below crossover. Furthermore, since the ideal closed-loop transfer function does not contain a zero, our modeling procedure tells us that the zero must come from the forward path and that the zero thus appears in the closed-loop transfer function. Additionally, from root-locus construction rules, we know that zeros are the terminal locations for loci; they attract poles. Hence, we expect a *closed-loop* pole close to this closed-loop low-frequency zero. Therefore, poles being the natural frequencies of a network, there will be a slow-settling component to transient responses.[14] The problem is equivalent to considering the effect of imperfect pole-zero cancellation.

To explore this idea in more detail, let us examine an isolated pole-zero *doublet*:

$$D(s) = \frac{\alpha \tau s + 1}{\tau s + 1}. \tag{81}$$

To go further, we'll need to recall the initial- and final-value theorems from Laplace transform theory:

$$f(\infty) = \lim_{s \to 0} sF(s), \tag{82}$$

$$f(0) = \lim_{s \to \infty} sF(s). \tag{83}$$

With these formulas, it is straightforward to show that the initial value of the step response of a doublet is just α, while the final value is unity. Note that the initial value is not zero because the high-frequency gain does not go to zero (in fact, it goes to α).

Now that we know that the initial and final values are different, we next have to find how we get from the initial to the final value. Formally, one would use the inverse Laplace transform to discover this information rigorously. However, we can avoid a little labor by reflecting once again on the meaning of the term "natural frequency." Evidently, then, the step response evolves exponentially from its initial to final value with a time constant equal to that of the pole.[15]

Returning to our specific case of the lag compensator, there is a pole-zero doublet formed by the compensating zero and its associated closed-loop pole. Because the zero is well below crossover, the doublet's pole has a much slower time constant than the inverse loop bandwidth. Hence, settling to fine accuracy can be much slower than suggested by the loop bandwidth when a lag compensator is used.

---

[14] Here, "slow" means with respect to the crossover frequency.

[15] This key fact is evidently poorly understood by many. The pole-zero separation determines the *ratio* of initial to final values, while *only the pole* determines the *rate* at which the response settles to the final value from the initial one. The zero thus has *nothing* to do with the time constant that describes the settling.

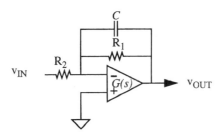

FIGURE 14.30. Inverting amplifier
with lead compensator.

14.18 LEAD COMPENSATION

We have seen that phase margin can be improved by reducing the magnitude of the loop transmission in order to lower the crossover frequency. The tradeoffs associated with such an approach include a loss of desensitivity and the possibility of low-frequency doublet formation.

An alternative compensation method is to alter the phase of the loop transmission, rather than its magnitude. That is, we wish to add a positive or *leading* phase shift near crossover to improve phase margin. One method for doing so in our op-amp example is illustrated in Figure 14.30.

Note that we are no longer maintaining the same ideal closed-loop transfer function. However, as we'll see, the overall closed-loop behavior will generally approach the desired ideal more closely than the reduced-gain or lag-compensated systems.

First, without writing any equations, let's see how the addition of this capacitor should give us a loop transmission zero. As frequency increases, the transmission through the capacitor increases. That's what a zero does, so we get a zero, as advertised. If we choose the capacitor value correctly, we can use the associated zero to bend the phase shift to more positive values and thereby increase phase margin.

There is one danger, however. A zero provides an increasing magnitude characteristic in addition to its positive phase shift. Hence, it also pushes out crossover. Therefore, there is the unfortunate possibility that a poorly placed zero will increase crossover so much that the positive phase shift of the zero will not offset the increased negative phase shift of the uncompensated system. The net effect could actually be a *reduction* in phase margin, so beware of this possibility.

Okay, now it's time for an equation or two. Let's derive an expression for the loop transmission for our lead-compensated system:

$$-L(s) = \frac{R_2}{R_2 + [R_1 \parallel (1/sC)]} \cdot G(s), \tag{84}$$

which may be expressed as

$$-L(s) = \left(\frac{R_2 + R_1}{R_2}\right)\left(\frac{sR_1C + 1}{sR_1C + 1 + R_1/R_2}\right) \cdot \frac{R_2}{R_2 + R_1} \cdot G(s). \tag{85}$$

Here, we see that the loop transmission zero is at a lower frequency than the associated pole, the opposite of the lag compensator.

Designing a lead compensator almost always involves a fair amount of iteration. A few hints may help constrain the search space, however. A reasonable starting point is to place the zero at the uncompensated system's crossover frequency. Vary the zero location about this frequency and find the maximum. If the phase margin specification can be met, you're finished.

Not infrequently, however, you find that the phase margin specification cannot be reached for any value of zero location. In such cases, a combination of gain reduction and lead compensation usually suffices. Unfortunately, with two varying parameters (gain reduction and zero location), finding an optimum can be a bit involved, and machine computation is definitely a tremendous help. Don't turn off your brain, though – you should *always* have a rough idea of what the answer should be, just as a sanity check on the computer's results.

If the necessary gain reduction is too large, then convert the gain reduction into a lag network. The resulting *lead–lag* compensator then gives you maximum desensitivity at DC and at high frequencies.

At this point, you may be wondering how we can get away from the doublet problem that afflicts the lag compensator. The answer is twofold. First, recognize that the lead zero is located near crossover, not well below it. Hence, any closed-loop pole that would be associated with it would have a time constant consistent with the loop bandwidth. That is, any doublet "tail" would settle out at about the same rate as the risetime and hence would be invisible. This observation applies to *all* lead-compensated systems.

A second reason that applies specifically to the particular op-amp connection shown here is that the lead zero does not appear in the forward path. Again, to conclude that this must be the case, recognize that the ideal closed-loop transfer function involves a pole. Hence, the feedback block in our model must supply the zero. The forward gain block does not have a zero. Since the zero appears in the feedback path, it does not show up in the closed-loop transfer function, and thus there is no closed-loop doublet.

A question that often arises at this point is why anyone would ever use anything but a lead compensator. After all, it can actually provide *greater* bandwidth than the uncompensated system and it's free of this doublet problem. The answer is that bandwidth costs power, and sometimes the price is too high. This consideration is particularly significant in mechanical systems, where power requirements are roughly proportional to the cube of bandwidth.[16] In large, industrial machinery, such a relationship between power and bandwidth favors the minimum bandwidth consistent with getting the job done.

---

[16] Here's a quick handwaving "derivation" (put on your windbreakers, 'cause we're going to do a *lot* of handwaving): Power = work/time = (k_1)(inertia)(angular acceleration/time) = $(k_2) \times$ (inertia)(ω^2)/time = $k_3\omega^3$. I warned you.

Even in electronic systems, larger bandwidths are not always desirable. Noise is always present, and larger bandwidth can mean additional noise. If the bandwidth is in excess of what is actually needed, then there is generally an unnecessary degradation of signal-to-noise ratio. In many instances, such a degradation is not tolerable.

14.19 SLOW ROLLOFF COMPENSATION

Accommodating an additional pole in the loop transmission is difficult if the pole's location is unknown or highly variable. Standard compensation methods involve a dominant pole (e.g., at the origin), so any additional poles potentially degrade phase margin (perhaps to zero). Use of a compensating zero is practical only in cases where the location of a troublesome pole is known to a fair degree of accuracy. There are many cases, however, where the uncertainty of pole location is so large that an added zero could very well endanger stability (by pushing bandwidth so high that a whole pile of parasitic poles control crossover) rather than improve it. In such cases, *slow rolloff* compensation is often a valuable option.[17]

The idea behind slow rolloff is simple. Suppose that, instead of a dominant pole, we could implement a dominant *half*-pole (again, at the origin) whose magnitude rolls off as the square root of frequency and whose phase lag is 45°. A half-pole would still be able to guarantee an upper bound on crossover frequency and thus bound the contribution by parasitic poles. A parasitic pole somewhere in-band would contribute at most a 90° lag, so that the worst-case phase margin would still be a healthy 45° (provided there were no other poles).

Of course, half-poles do not exist, but we may approximate them to any desired degree by alternating poles with zeros. If we distribute alternating poles and zeros with a constant frequency ratio α, then the average phase lag will be 45°. The peak deviation from that average will be a function of the frequency ratio. The closer to unity the ratio, the smaller the phase ripple. For values of α ranging from 2 to 100, the following approximation yields the worst-case phase lag in degrees:

$$\phi_{\max} \approx 36 + 22 \log \alpha, \qquad (86)$$

where the logarithmic base is 10. The peak approximation error is less than 3 degrees. See Figure 14.31.

In most applications, it is possible to bound the parasitic pole location to within one or two orders of magnitude. In those cases, a few pole–zero pairs suffice to provide an acceptable approximation to a half-pole. Overall, slow rolloff compensation is extremely useful in those all-too-common cases of considerable modeling uncertainty. The primary drawback is that the large number of pole–zero pairs causes the

---

[17] For an excellent description of slow rolloff (and just about every analog concept in general), see J. K. Roberge, *Operational Amplifiers: Theory and Practice,* Wiley, New York, 1975. No analog designer should be without this superb text.

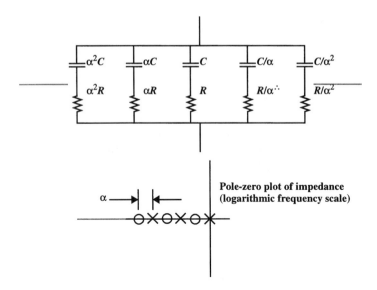

FIGURE 14.31. Slow rolloff network.

step response to suffer from slow settling to fine accuracy. It is the well-known tail problem, multiplied by large factors.

14.20 SUMMARY OF COMPENSATION

We've seen three basic compensation techniques that may be used individually or in combination. Both the reduced-gain and lag compensators seek to improve phase margin by reducing crossover to a value where the corresponding phase shift is less negative than in the uncompensated case, thereby increasing phase margin.

The lag compensator improves on the simple reduced-gain compensator by leaving untouched the low-frequency loop transmission, but it introduces a potentially bothersome doublet that causes slow settling to high accuracy.

The lead compensator improves phase margin by directly improving the phase shift of the loop transmission. As a result, the bandwidth actually increases. Furthermore, the doublet problem disappears because the pole associated with it is just as fast as the overall amplifier, so any "tail" in the response is effectively masked during the risetime. The lead compensator frequently must be combined with either a reduced-gain or lag compensator to provide sufficient degrees of freedom to satisfy a given phase margin specification.

As a final note on compensation, it should be stated that the types discussed here do not comprise an exhaustive list. Additionally, even though these compensators were illustrated with specific op-amp circuits, the fundamental notions apply to all other feedback systems as well. Hence, any method that reduces the loop transmission magnitude uniformly is a reduced-gain compensator, anything that introduces a loop transmission pole–zero pair in which the pole is at the lower frequency is a lag compensator, and so on. Finally, we presented slow rolloff compensation as a means for accommodating large uncertainties in parasitic pole locations – at the cost of slow settling tails.

PROBLEM SET FOR FEEDBACK

PROBLEM 1 One important benefit of negative feedback is the reduction of distortion. This problem explores this property in more quantitative detail. For simplicity, assume that the forward path of a unit feedback system has an input–output transfer characteristic that one may express as the following cubic polynomial:

$$v_{OUT} = a_0[g_1 v_{IN} + g_2 v_{IN}^2 + g_3 v_{IN}^3]. \tag{P14.1}$$

Assume that the system is only weakly nonlinear, so that the quadratic and cubic terms are small compared with the first-order (desired) term.

(a) Derive a cubic polynomial approximation for the overall input–output transfer characteristic. Verify that your equation collapses to $a_0 g_1/(1 + a_0 g_1)$ in the linear limit.
(b) By approximately what factor do the quadratic and cubic terms decrease as the linear loop transmission magnitude increases? Based on your answer, how do higher-order terms vary as the loop transmission magnitude changes?

PROBLEM 2 It was noted in the chapter that one may not always find a 1 : 1 mapping between physical elements in a feedback system and the canonical feedback model. Furthermore, the assignment of transfer functions to each block in the model is not necessarily unique.

Show that representing a system in terms of a feedback model can actually be a philosophical choice by modeling the input–output voltage transfer characteristic of an ordinary resistive voltage divider with a feedback system.

PROBLEM 3 Common errors when cutting the loop to determine loop transmission include failing to account for various loading effects and ignoring the need to establish correct DC operating points. This problem explores these important practical issues in more detail.

Consider the loop sketched in Figure 14.32. The combination of the ideal op-amp and resistor R together model a real op-amp that possesses a nonzero output resistance.

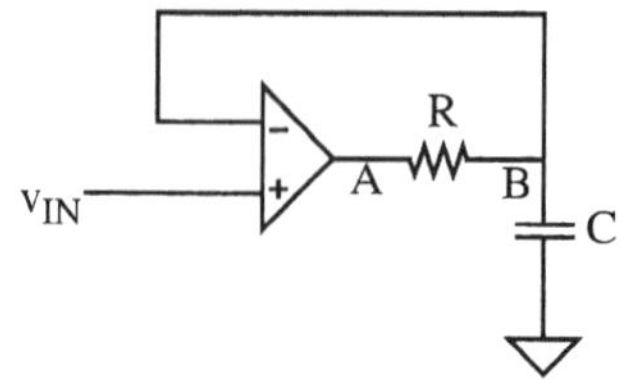

FIGURE 14.32. Feedback system for loop transmission problem.

(a) Shown are two possible locations for breaking the loop, points A and B. Explain why breaking the loop at B and driving that point with a test voltage source yields the incorrect loop transmission.

(b) Suppose that the op-amp also has an extremely high DC gain, as is typical of many general-purpose op-amps. Assume that, when the system is operating as a normal closed-loop feedback system, the output common-mode voltage is near zero. If v_{in} is set to zero value when determining the loop transmission, describe how you would determine what DC value the test generator should have. Should it be zero, or would a different value be a better choice? Explain.

PROBLEM 4 It was asserted in the chapter that the root locus of a lead-compensated two-pole system contains a perfect circle. Show this property formally for the case of a system whose loop transmission consists of two integrators and a single zero whose time constant is τ_z.

PROBLEM 5 Linear analyses are useful approximations, but physical systems exhibit strong nonlinearities if driven hard enough. Aside from obvious side effects such as increased distortion, there can be consequences for stability if the nonlinearity is imbedded within a feedback loop. One common nonlinear phenomenon is that of saturation, in which the output changes little (if at all) beyond a given input amplitude. To the extent that one may thus interpret saturation as a reduction in gain, explain the following observation.

A system whose loop transmission consists of three integrators (implemented with standard op-amp circuits) and two coincident zeros is adjusted to provide a healthy phase margin of nearly 60°. The step response for small amplitudes is well behaved, as would be expected for this phase margin. However, the step response exhibits a distinctly frightening behavior once the excitation is large enough to saturate one of the op-amps. Would you expect the step response ringing in this regime to be of a lower or higher frequency than in the small-signal case? Explain your answers in terms of both a root locus and a Bode plot or phase margin diagram.

PROBLEM 6 Suppose that you are told that the response of a black box appears to be linear and second-order, with a unit DC gain. Measurements reveal that the unit step response has a peak value of 1.38 and that the time for the step response to first pass through unity is 500 ns.

(a) Assuming that it *is* second-order, determine the second-order parameters that can be used to model the system.

(b) Using your model of part (a), estimate the peak value of the output that would result if the system were excited with a unit impulse.

(c) Compute the time required for the impulse response to first return to zero.

(d) Estimate the frequency response peak (M_p) as well as the −3-dB bandwidth for this system.

(e) Estimate the time it takes for the step response to settle to within 2% of its final value.

(f) Describe a test that you would perform in order to confirm that the system is in fact linear to a good approximation.

PROBLEM 7 Most feedback systems we encounter are low-pass in nature, and it easy to get the impression that stability problems are solely a property of such systems. To break us out of this narrow thinking, this problem investigates the stability of *high*-pass systems. For example, AC-coupled feedback amplifiers can sometimes exhibit an interesting low-frequency oscillation known as "motorboating," first observed in vacuum tube audio amplifiers. Suppose that the loop transmission for such an amplifier is

$$L(s) = -\frac{a_0 s^3}{(s+1)(0.1s+1)^2}.$$
(P14.2)

(a) Sketch the root locus for this amplifier for positive values of a_0.
(b) For what range of values of a_0 is the amplifier stable?
(c) At what frequency (or frequencies) can this amplifier oscillate? Your answer to this question should suggest why the term "motorboating" is appropriate.

PROBLEM 8 Phase-locked loops are of great utility in communications systems, among others. Although we will take up the detailed study of PLLs in a later chapter, it is sufficient for now simply to accept that a linear second-order model is a good approximation for a certain class of PLLs. Suppose that such a PLL has been determined to have the following (negative) loop transmission to an excellent approximation:

$$-L(s) = \frac{K(\tau s + 1)}{s^2}.$$
(P14.3)

(a) Determine values for K and τ that yield a closed-loop pole pair with damping ratio of 0.707 and an ω_n of 10 Mrps.
(b) Sketch the root locus for $K > 0$ with the value of τ found in part (a).
(c) What value of K gives us a critically damped loop?
(d) Assume that the zero does appear in the forward path, and assume a unit closed-loop DC gain. Sketch the frequency response of the closed-loop system for the pole location(s) found in part (c).

PROBLEM 9 It is important to develop a facility for modeling physical circuits as feedback systems. In that spirit, model a conventional textbook source follower with a degeneration resistor as a feedback system. Use a simple, low-frequency, small-signal model for the MOSFET, and give expressions for the forward and feedback gains.

PROBLEM 10 Consider a feedback system in which the forward-path transfer function is given by

$$K(s^2 - 3s + 5)$$
(P14.4)

and the feedback transfer function is

$$\frac{1}{s(s^2 + 2s + 4)}.\tag{P14.5}$$

The parameter K is a pure scalar quantity that may be varied.

(a) Sketch the root locus for this system.
(b) Identify the values of K, if any, for which the system can become unstable.
(c) What value of K will give a damping factor of 0.707 to the dominant pole pair?

PROBLEM 11 You are given an operational amplifier that was designed by Sub-Optimal Products, and careful measurements reveal that it has a transfer function given by

$$a(s) = \frac{5 \cdot 10^4}{(s + 1)(10^{-3}s + 1)(10^{-4}s + 1)}.\tag{P14.6}$$

Suppose it is your misfortune to be given the task of using this amplifier in an inverting configuration with an input resistor of 22 kΩ and a feedback resistor of 220 kΩ, so that the ideal closed-loop gain is -10.

(a) Derive an expression for the loop transmission. What is the phase margin? Sketch Bode plots of the loop transmission for this and subsequent parts.
(b) Suppose you are now told that we must have a phase margin of 60°. The first compensation technique that comes to mind is a reduced-gain compensator. By what factor is it necessary to reduce the loop transmission to achieve a 60° phase margin? What value of R_{comp}, placed across the op-amp input terminals, gives us this phase margin? What is the crossover frequency?
(c) The compensator in part (b) throws away loop transmission at all frequencies, but stability is determined mainly by the behavior of the loop near crossover. Recognizing this fact, now amend your compensator to recover the lost gain at low frequencies by designing a lag compensator. Feel free to use this rule of thumb: Place the compensator zero a decade below the crossover frequency of part (b), and the compensator pole a decade below that. What is the new phase margin in this case?
(d) The system is stable, and the low-frequency desensitivity is restored, but you're unhappy with the bandwidth. You therefore consider a lead compensator, after remembering that it works by adding positive phase shift to the loop transmission where you need it, instead of forcing a lower crossover frequency. So you return to the original, uncompensated system and add a capacitor in parallel with the feedback resistor. What value of capacitor gives us the maximum phase margin? What *is* the maximum phase margin?
(e) Suppose it is absolutely imperative to achieve 60° of phase margin. What strategies might you employ to modify the compensator of part (d) to obtain the desired stability? A qualitative answer will suffice. Offer at least two suggestions, however.

PROBLEM 12 Consider two feedback systems, both with unit f. The forward-path transfer function for one system is

$$a(s) = \frac{a_0(0.2s + 1)}{(s + 1)(0.01s + 1)(0.21s + 1)}, \quad \text{(P14.7)}$$

while the other has

$$a(s) = \frac{a_0}{(s + 1)(0.01s + 1)}. \quad \text{(P14.8)}$$

Sketch the root-locus diagrams for both systems. Explain why these two systems have similar closed-loop responses.

PROBLEM 13 Suppose that the open-loop transfer function of an automatic gain control (AGC) loop is found to be

$$a(s) = \frac{10^5}{(0.1s + 1)(10^{-6}s + 1)^2}. \quad \text{(P14.9)}$$

(a) Determine the gain margin, phase margin, and crossover frequency for this system when used with unit feedback.
(b) What value of f results in a phase margin of $45°$?

PROBLEM 14 An op-amp has an open-loop transfer function given by

$$a(s) = \frac{2 \times 10^5}{(0.1s + 1)(10^{-5}s + 1)^2}. \quad \text{(P14.10)}$$

The op-amp is used in a conventional inverting configuration with an ideal gain of -10. Determine the loop transmission gain reduction factor required to provide a phase margin of $45°$. Sketch the magnitude and phase of the loop transmission before compensation.

CHAPTER FIFTEEN

RF POWER AMPLIFIERS

15.1 INTRODUCTION

In this chapter, we study the problem of delivering RF power efficiently to a load. As we'll discover very quickly, scaled-up versions of the small-signal amplifiers we've studied so far are fundamentally incapable of high efficiency, and other approaches must be considered. As usual, tradeoffs are involved, this time among linearity, power gain, output power, and efficiency.

In a continuing quest for increased channel capacity, more and more communications systems employ amplitude and phase modulation together. This trend brings with it an increased demand for much higher linearity (possibly in both amplitude and phase domains). The variety of power amplifier topologies reflects the inability of any single circuit to satisfy all requirements.

15.2 GENERAL CONSIDERATIONS

Contrary to what one's intuition might suggest, the maximum power transfer theorem is largely useless in the design of power amplifiers. One minor reason is that it isn't entirely clear how to define impedances in a large-signal, nonlinear system. A more important reason is that even if we were able to solve that little problem and subsequently arrange for a conjugate match, the efficiency would be only 50% because equal amounts of power are then dissipated in the source and load. In many cases, this value is unacceptably low. As an extreme (but realistic) example, consider the problem of delivering 50 kW into an antenna if the amplifier is only 50% efficient. The circuit dissipation would be 50 kW as well, presenting a rather challenging thermal management problem. Even in the low-power domain of portable communications devices such as cellular phones, high efficiency is extremely desirable to extend battery life or reduce battery weight.

Hence, instead of limiting efficiency to 50% by maximizing power transfer, one generally designs a PA to deliver a specified amount of power into a load with the highest possible efficiency consistent with acceptable power gain and linearity. To

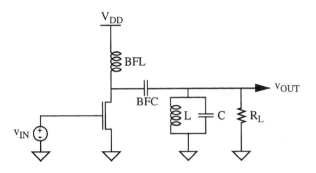

FIGURE 15.1. General power amplifier model.

see how one may achieve these goals by blithely ignoring the maximum power transfer theorem, we now consider a classic power amplifier topology.

15.3 CLASS A, AB, B, AND C POWER AMPLIFIERS

There are four types of power amplifiers, distinguished primarily by bias conditions, that may be termed "classic" because of their historical precedence. These are labeled Class A, AB, B, and C, and all four may be understood by studying the single model sketched in Figure 15.1.[1]

In this general model, the resistor R_L represents the load into which we are to deliver the output power. A "big, fat" inductance, BFL, feeds DC power to the drain, and is assumed large enough so that the current through it is substantially constant. The drain is connected to a tank circuit through capacitor BFC to prevent any DC dissipation in the load. One advantage of this particular configuration is that the transistor's output capacitance can be absorbed into the tank, as in a conventional small-signal amplifier. Another is that the filtering provided by the tank cuts down on out-of-band emissions caused by ever-present nonlinearities. This consideration is particularly important here because we are no longer restricting ourselves to small-signal operation and must therefore expect some distortion. To simplify analysis, we assume that the tank has a high enough Q that the voltage across the tank is well approximated by a sinusoid, even if it is fed by nonsinusoidal currents. This assumption necessarily implies narrowband operation. Although broadband power amplifiers are certainly also of interest, we will limit the present discussion to the narrowband case.

15.3.1 CLASS A AMPLIFIERS

The Class A power amplifier is just a standard, textbook, small-signal amplifier on steroids. The assumption in Class A design (indeed, its defining characteristic) is

---

[1] Many variations on this theme exist, but the operating features of all of them may still be understood with this model.

that bias levels are chosen so that the transistor operates (quasi)linearly. For a bipolar realization, this condition is satisfied by avoiding cutoff and saturation; for MOS implementations, the transistor is kept in the pentode (saturation[2]) region of operation.

The primary distinction between Class A power amplifiers and small-signal amplifiers is that the signal currents in a PA are a substantial fraction of the bias level, and one would therefore expect potentially serious distortion. In narrowband operation, as implied by the general circuit model, a tank circuit solves the distortion problem potentially associated with such large swings so that, overall, linear operation prevails.

Although linearity is certainly desirable, the Class A amplifier provides it at the expense of efficiency because there is always dissipation due to the bias current, even when there is no signal. To understand quantitatively why the efficiency is poor, assume that the drain current is reasonably well approximated by:

$$i_D = I_{DC} + i_{rf} \sin \omega_0 t, \tag{1}$$

where I_{DC} is the bias current, i_{rf} is the amplitude of the signal component of the drain current, and ω_0 is the signal frequency (and also the resonant frequency of the tank). Although we have glibly ignored distortion, the errors introduced are not serious enough to invalidate what follows.

The output voltage is simply the product of a signal current and the load resistance. Since the big, fat inductor BFL forces a substantially constant current through it, KCL tells us that the signal current is none other than the signal component of the drain current. Therefore,

$$v_o = -i_{rf} R \sin \omega_0 t. \tag{2}$$

Finally, the drain voltage is the sum of the DC drain voltage and the signal voltage. The big, fat inductor BFL presents a DC short, so the drain voltage swings symmetrically about V_{DD}.[3] The drain voltage and current are therefore offset sinusoids that are 180° out of phase with each other, as shown in Figure 15.2.

If it isn't clear from the equations, it should be clear from the figure that the transistor always dissipates power because the product of drain current and drain voltage is always positive. To evaluate this dissipation quantitatively, compute the efficiency by first calculating the signal power delivered to the resistor R:

$$P_{rf} = \frac{i_{rf}^2 R}{2}. \tag{3}$$

---

[2] It is unfortunate indeed that the word "saturation" has opposing meanings for MOS and bipolar devices.

[3] This is not a typographical error. The drain actually swings *above* the positive supply. One way to argue that this must be the case is to recognize that an ideal inductor cannot have any DC voltage across it (otherwise, infinite currents would eventually flow). Therefore, if the drain voltage swings below the supply, it must also swing above it. This kind of thinking is particularly helpful in deducing the characteristics of various types of switched-mode power converters.

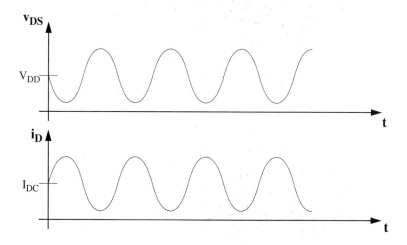

FIGURE 15.2. Drain voltage and current for ideal Class A amplifier.

Next, compute the DC power supplied to the amplifier. Let us assume that the quiescent drain current, I_{DC}, is made just large enough to guarantee that the transistor does not ever cut off. That is,

$$I_{DC} = i_{rf}, \tag{4}$$

so that the input DC power is:

$$P_{DC} = I_{DC}V_{DD} = i_{rf}V_{DD}. \tag{5}$$

The ratio of RF output power to DC input power is a measure of efficiency (usually called the drain efficiency) and is given by

$$\eta \equiv \frac{P_{rf}}{P_{DC}} = \frac{i_{rf}^2(R/2)}{i_{rf}V_{DD}} = \frac{i_{rf}R}{2V_{DD}}. \tag{6}$$

Now, the absolute maximum that the product $i_{rf}R$ can have is V_{DD}. Therefore, the maximum theoretical drain efficiency is just 50%. If one makes due allowance for nonzero minimum v_{DS}, variation in bias conditions, nonideal drive amplitude, and inevitable losses in the filter and interconnect, values substantially smaller than 50% often result – particularly at lower supply voltages, where $V_{DS,\,on}$ represents a larger fraction of V_{DD}. Consequently, drain efficiencies of 30–35% are not at all unusual for practical Class A amplifiers.

Aside from efficiency, another important consideration is the stress on the output transistor. In a Class A amplifier, the maximum drain-to-source voltage is $2V_{DD}$, while the peak drain current has a value of $2V_{DD}/R$. Hence, the device must be able to withstand peak voltages and currents of these magnitudes, even though both maxima do not occur simultaneously. Since scaling trends in IC process technology force reductions in breakdown voltage, the design of PAs becomes more difficult with each passing generation.

One common way to quantify the relative stress on the devices is to define another type of efficiency, called the "normalized power output capability," which is simply the ratio of the actual output power to the product of the maximum device voltage and current. For this type of amplifier, the maximum value of this dimensionless figure of merit is

$$P_N \equiv \frac{P_{\mathrm{rf}}}{v_{DS,\mathrm{pk}}\, i_{D,\mathrm{max}}} = \frac{V_{DD}^2/(2R)}{(2V_{DD})(2V_{DD}/R)} = \frac{1}{8}. \tag{7}$$

The Class A amplifier thus provides linearity at the cost of low efficiency and relatively large device stresses. For this reason, Class A amplifiers are rare in RF power applications[4] and relatively rare in audio power applications (particularly so at the higher power levels) for the reasons cited.[5]

It is important to underscore once again that the 50% efficiency value represents an upper limit. If the drain swing is less than the maximum assumed in the foregoing and if there are additional losses anywhere else, the efficiency drops. As the swing approaches zero, the drain efficiency also approaches zero because the signal power delivered to the load goes to zero while the transistor continues to burn DC power.

15.3.2 CLASS B AMPLIFIERS

A clue to how one might achieve higher efficiency than a Class A amplifier is actually implicit in the waveforms of Figure 15.2. It should be clear that if the bias were arranged to reduce the fraction of a cycle over which drain current and drain voltage are simultaneously nonzero, transistor dissipation would diminish.

In the Class B amplifier, the bias is arranged to shut off the output device half of every cycle. An exact 50% conduction duty cycle is a mathematical point, of course, so true Class B amplifiers do not actually exist. Nevertheless, the concept is useful in organizing a taxonomy. In any case, with intermittent conduction, we expect a gross departure from linear operation. However, we must distinguish between distortion in the output (an earmark of nonlinearity) and proportionality, or lack thereof, between input and output powers (evaluated at the fundamental). A single-ended Class B amplifier may produce a nonsinusoidal output but still act linearly in this sense of input–output power proportionality. We still care about out-of-band spectral components, of course, and a high-Q resonator is absolutely mandatory in order to obtain an acceptable approximation to a sinusoidal output voltage.

Although the single-transistor version of a Class B amplifier is what we'll analyze here, it should be mentioned that most practical Class B amplifiers are push–pull configurations of two transistors (more on this topic later).

For this amplifier, then, we assume that the drain current is sinusoidal for one half-cycle and zero for the other half-cycle:

---

[4] Except, perhaps, in low-level applications, or in the early stages of a cascade.

[5] An exception is the high-end audio crowd, of course, for whom power consumption is often not a constraint.

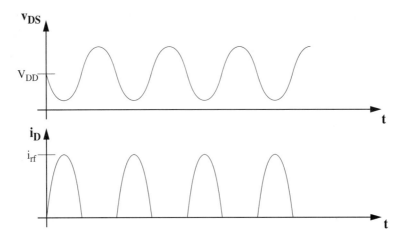

FIGURE 15.3. Drain voltage and current for ideal Class B amplifier.

$$i_D = i_{\text{rf}} \sin \omega_0 t \quad \text{for } i_D > 0. \tag{8}$$

The output tank filters out the harmonics of this current, leaving a sinusoidal drain voltage as in the Class A amplifier. The drain current and drain voltage therefore appear approximately as shown in Figure 15.3.

To compute the output voltage, we first find the fundamental component of the drain current and then multiply this current by the load resistance:

$$i_{\text{fund}} = \frac{2}{T} \int_0^{T/2} i_{\text{rf}}(\sin \omega_0 t)(\sin \omega_0 t)\, dt = \frac{i_{\text{rf}}}{2}, \tag{9}$$

$$v_{\text{out}} \approx \frac{i_{\text{rf}}}{2} R \sin \omega_0 t. \tag{10}$$

Since the maximum possible value of v_{out} is V_{DD}, it is clear from Eqn. 10 that the maximum value of i_{rf} is

$$i_{\text{rf,max}} = \frac{2 V_{DD}}{R}. \tag{11}$$

The peak drain current and maximum output voltage are therefore the same as for the Class A amplifier.[6]

Computing the drain efficiency as before, we first calculate the output power as:

$$P_o = \frac{v_o^2}{2R}, \tag{12}$$

where v_o is the amplitude of the signal across the load resistor. The maximum value of the amplitude remains V_{DD}, so the maximum output power is

---

[6] The assumption of half-sinusoidal current pulses is, necessarily, an approximation. The drain current in practical circuits differs mainly in that the transition to and from zero current is not abrupt. Hence the true device dissipation is somewhat greater, and the efficiency somewhat lower, than predicted by ideal theory.

$$P_{o,\max} = \frac{V_{DD}^2}{2R}. \tag{13}$$

Computing the DC input power requires computation of the average drain current:

$$\overline{i_D} = \frac{1}{T} \int_0^{T/2} \frac{2V_{DD}}{R} \sin \omega_0 t \, dt = \frac{2V_{DD}}{\pi R}, \tag{14}$$

so that the DC power supplied is

$$P_{DC} = \frac{2V_{DD}^2}{\pi R}. \tag{15}$$

Finally, the maximum drain efficiency for a Class B amplifier is

$$\eta = \frac{P_{o,\max}}{P_{DC}} = \frac{\pi}{4} \approx 0.785. \tag{16}$$

The drain efficiency is thus considerably higher than for the Class A PA. Continuing with our hypothetical example of a 50-kW transmitter, the device dissipation would diminish to less than one third of its previous value, from 50 kW to under 14 kW. However, as with the Class A amplifier, the actual efficiency of any practical implementation will be somewhat lower than given by this analysis owing to effects that we have neglected. Nonetheless, it remains true that, all other things held equal, the Class B amplifier offers substantially higher efficiency than its Class A cousin.

The normalized power capability of this amplifier is $1/8$, the same as for the Class A, since the output power, maximum drain voltage, and maximum drain current are the same.[7]

With the Class B amplifier, we have accepted distortion in exchange for a significant improvement in efficiency. Since this tradeoff is effected by reducing the fraction of a period that the transistor conducts current, it is natural to ask whether further improvements might be possible by reducing the conduction angle even more. Exploration of this idea leads to the Class C amplifier.

15.3.3 THE CLASS C AMPLIFIER

In a Class C PA, the gate bias is arranged to cause the transistor to conduct less than half the time. Consequently, the drain current consists of a periodic train of pulses. It is traditional to approximate these pulses by the top pieces of sinusoids to facilitate a direct analysis.[8] Specifically, one assumes that the drain current is of the following form:

$$i_D = I_{DC} + i_{rf} \sin \omega_0 t, \quad i_D > 0, \tag{17}$$

---

[7] A two-transistor push–pull Class B amplifier has a normalized power capability that is twice as large.

[8] See e.g. Krauss, Bostian, and Raab, *Solid-State Radio Engineering,* Wiley, New York, 1981.

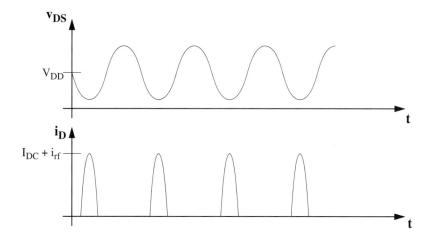

FIGURE 15.4. Drain voltage and current for ideal Class C amplifier.

where the offset I_{DC}, which is analogous to the bias current in a linear amplifier, is actually negative for a Class C amplifier. Of course, the *overall* drain current i_D is always positive or zero. That is, the drain current is a piece of a sine wave when the transistor is active and zero when the transistor is in cutoff. We continue to assume that the transistor behaves at all times as a current source (high output impedance).[9]

Because we still have a high-Q output tank, the voltage across the load remains substantially sinusoidal. The drain voltage and drain current therefore appear as depicted in Figure 15.4. In the derivations that follow, don't worry about being able to replicate all of the details. As we'll see, how one designs such amplifiers differs considerably from what the equations imply, so focus instead on the general conclusions reached rather than the minutiae.

We begin by solving for the total angle over which the drain current is nonzero. In order to reduce the number of steps needed to arrive at the answer, we first rewrite the expression for the drain current in terms of a cosine rather than a sine:

$$i_D = I_{DC} + i_{rf} \cos \omega_0 t, \quad i_D > 0. \tag{18}$$

Clearly, such a maneuver changes nothing since the time origin is arbitrary anyway. With this modification, the current pulses appear as shown in Figure 15.5.

Setting the current equal to zero and solving for the total conduction angle 2Φ yields

$$2\Phi = 2 \cdot \cos^{-1}\left(-\frac{I_{DC}}{i_{rf}}\right), \tag{19}$$

---

[9] Violation of this assumption leads to an exceedingly complex situation. Maximum efficiency is typically obtained when the output power is nearly saturated. Under such conditions, Class C amplifiers force bipolar transistors into saturation (and MOSFETs into triode) for some fraction of a period, making exact analysis exceedingly difficult.

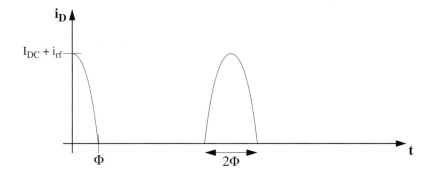

FIGURE 15.5. Detail of drain current waveform.

which may be solved for the "bias current" as follows:

$$I_{DC} = -i_{rf} \cos \Phi. \tag{20}$$

We are now in a position to compute the average drain current:

$$\overline{i_D} = \frac{1}{2\pi} \int_{-\Phi}^{\Phi} (I_{DC} + i_{rf} \cos \phi) \, d\phi = \frac{1}{2\pi} 2\phi I_{DC} + \frac{1}{2\pi} [i_{rf} \sin \phi] \Big|_{-\Phi}^{\Phi}. \tag{21}$$

After substitution with the expression for I_{DC}, this yields:

$$I_{DC} = \frac{i_{rf}}{\pi} [\sin \Phi - \Phi \cos \Phi]. \tag{22}$$

We will use this expression shortly in deriving an equation for the efficiency as a function of conduction angle.

The other quantity we need is a general expression for the power delivered to the load. As with the Class B case, this derivation is simplified because of the high-Q tank circuit, so we need to compute only the fundamental term in the Fourier series:

$$i_{\text{fund}} = \frac{2}{T} \int_0^T i_D \cos \omega_0 t \, dt = \frac{1}{2\pi} (4I_{DC} \sin \Phi + 2i_{rf} \Phi + i_{rf} \sin 2\Phi). \tag{23}$$

Substituting for I_{DC}, we obtain

$$i_{\text{fund}} = \frac{i_{rf}}{2\pi} (2\Phi - \sin 2\Phi). \tag{24}$$

With our expression for the fundamental current through the load, we can easily derive an equation for the maximum output voltage swing:

$$V_{DD} = i_{rf} \frac{R}{2\pi} (2\Phi - \sin 2\Phi), \tag{25}$$

allowing us to solve for the current i_{rf} in terms of V_{DD}:

$$i_{rf} = \frac{2\pi V_{DD}}{R(2\Phi - \sin 2\Phi)}. \tag{26}$$

The peak drain current is the sum of i_{rf} and the bias term:

$$i_{D,pk} = \frac{i_{rf}}{\pi}[\sin \Phi - \Phi \cos \Phi] + \frac{2\pi V_{DD}}{R(2\Phi - \sin 2\Phi)}, \tag{27}$$

which simplifies to

$$i_{D,pk} = \frac{2\pi V_{DD}}{R(2\Phi - \sin 2\Phi)}\left[1 + \frac{(\sin \Phi - \Phi \cos \Phi)}{\pi}\right]. \tag{28}$$

For a fixed output voltage, the peak drain current approaches infinity as the pulsewidth decreases toward zero.

The drain efficiency is readily calculated with the equations we've just derived:

$$\eta_{max} = \frac{2\Phi - \sin 2\Phi}{4(\sin \Phi - \Phi \cos \Phi)}. \tag{29}$$

As the conduction angle shrinks toward zero, the efficiency approaches 100%. While this sounds promising, the gain and output power unfortunately also tend toward zero at the same time, since the fundamental component in the ever-narrowing slivers of drain current shrinks as well. Furthermore, it is clear from the equation for peak drain current that the normalized power-handling capability of the Class C amplifier approaches zero as the conduction angle approaches zero. All of these tradeoffs force the attainment of less than 100% efficiency in practice, since we generally want a reasonable amount of output power as well as high efficiency.

Having endured the foregoing derivations, one might be disappointed that they are typically not used very much in the actual process of designing a Class C PA. One reason is that there are few convenient choices for the gate bias, with zero volts a particularly convenient one. The signal component of the gate drive is then chosen sufficiently large to produce the desired output power. The conduction angle and efficiency therefore are usually not explicit design parameters; they are simply the *consequences* of the choice of zero bias and output power.

Another reason is that the assumptions made (e.g., sinusoidal current spikes, current source behavior of the transistor) are not always satisfied well enough to trust the equations quantitatively. Again, the primary virtue in carrying out the exercise is to develop some general intuition useful for design – mainly, that the efficiency can be large, but at the cost of reduced power-handling capability, gain, and linearity.

15.3.4 THE CLASS AB AMPLIFIER

We have seen that Class A amplifiers conduct 100% of the time, Class B amplifiers 50% of the time, and Class C PAs somewhere between 0 and 50% of the time. The Class AB amplifier, as its name suggests, conducts somewhere between 50% and 100% of a cycle, depending on the bias levels chosen. As a result, its efficiency and linearity are intermediate between those of a Class A and Class B amplifier. This compromise is frequently satisfactory, as one may infer from the popularity of this PA.

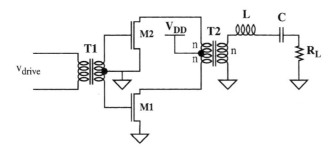

FIGURE 15.6. Class D amplifier.

We do not need to undertake a separate derivation of equations for this amplifier because the equations for the Class C case also apply here (they also include the Class A and Class B case). The only difference is that the bias current is positive, rather than negative.

15.4 CLASS D AMPLIFIERS

The PAs presented so far use the active device as a controlled current source. Another approach is to use the device as a switch, the reasoning being that a switch ideally dissipates no power, for there is either zero voltage across it or zero current through it. Since the switch's *V–I* product is therefore always zero, the transistor dissipates no power and the efficiency must be 100%.

One type of amplifier that exploits this observation is the Class D amplifier. At first glance (see Figure 15.6), it looks the same as a push–pull, transformer-coupled version of a Class B amplifier. In contrast with the parallel tanks we've typically seen, a series *RLC* network is used in the output of this amplifier, since switch-mode amplifiers are the duals of the current-mode amplifiers studied earlier. As a consequence, the output filters are also duals of each other.

The input connection guarantees that only one transistor is driven on at a given time, with one transistor handling the positive half-cycles and the other the negative half-cycles, just as in a push–pull Class B. The difference here is that the transistors are driven hard enough to make them act like switches, rather than as linear (or quasilinear) amplifiers.

Because of the switching action, each primary terminal of the output transformer T_2 is alternately driven to ground, yielding a square-wave voltage across the primary (and therefore across the secondary) winding. When one drain goes to zero volts, transformer action forces the other drain to a voltage of $2V_{DD}$. The output filter allows only the fundamental component of this square wave to flow into the load.

Since only fundamental currents flow in the secondary circuit, the primary current is sinusoidal as well. As a consequence, each switch sees a sinusoid for the half-cycle that it is on, and the transformer current and voltage therefore appear as in Figures 15.7 and 15.8. Because the transistors act like switches, the theoretical efficiency of the Class D amplifier is 100%.

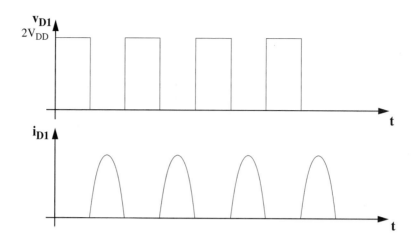

FIGURE 15.7. M_1 drain voltage and current for ideal Class D amplifier.

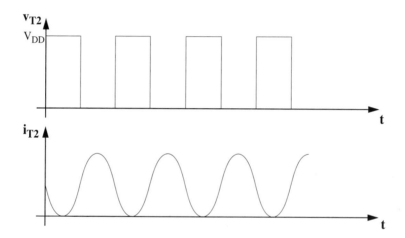

FIGURE 15.8. T_2 secondary voltage and current for ideal Class D amplifier.

The normalized power handling of this amplifier happens to be[10]

$$\frac{P_o}{v_{DS,\text{on}} \cdot i_{D,\text{pk}}} = \frac{1}{\pi} \approx 0.32, \tag{30}$$

which is better than a Class B push–pull and much better than a Class A amplifier. Of course, the Class D amplifier cannot normally provide linear modulation, but it does provide potentially high efficiency and does not stress the devices very much.

One practical problem with this (or any other switching) PA is that there is no such thing as a perfect switch. Nonzero saturation voltage guarantees static dissipation

---

[10] It may help to keep in mind that the amplitude of the fundamental component of a square wave is $4/\pi$ times the amplitude of the square wave.

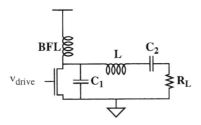

FIGURE 15.9. Class E amplifier.

in the switches, while finite switching speeds imply that the switch *V–I* product is nonzero during the transitions. Hence, switch-mode PAs function well only at frequencies substantially below f_T. Furthermore, a particularly serious reduction in efficiency can result in bipolar implementations if, due to charge storage in saturation, one transistor fails to turn completely off before the other turns on. Transformer action then attempts to apply the full supply voltage across the device that is not yet off, and the *V–I* product can be quite large.

15.5 CLASS E AMPLIFIERS

As we've seen, using transistors as switches has the potential for providing greatly improved efficiency, but it's not always trivial to realize that potential in practice due to imperfections in real switches. The associated dissipation degrades efficiency. To prevent gross losses, the switches must be quite fast relative to the frequency of operation. At high carrier frequencies, it becomes increasingly difficult to satisfy this requirement.

If there were a way to modify the circuit to force a zero switch voltage for a nonzero interval of time about the instant of switching, the dissipation would decrease. The Class E amplifier uses a high-order reactive network that provides enough degrees of freedom to shape the switch voltage to have both zero value *and* zero slope at switch turn-on, thus reducing switch losses. Unfortunately, it does nothing for the turn-off transition, which is often the more troublesome edge, at least in bipolar designs. Another issue, as we'll see later, is that the Class E amplifier has rather poor normalized power-handling capability (worse, in fact, than a Class A amplifier), requiring the use of rather oversized devices to deliver a given amount of power to a load, despite the high potential efficiency (theoretically 100% with ideal switches) of this topology.

The primary virtue of the Class E amplifier is that it is straightforward to design. Unlike typical Class C amplifiers, practical implementations require little postdesign tweaking to obtain satisfactory operation.

With that preamble out of the way, let's take a look at the Class E topology sketched in Figure 15.9. As in earlier examples, the *BFL* simply provides a DC path to the supply and approximates an open circuit at RF. Note additionally that the capacitor C_1 is conveniently positioned, for any device output capacitance can be absorbed into it.

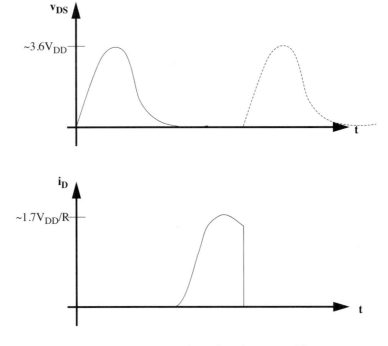

FIGURE 15.10. Waveforms for Class E amplifier.

Derivation of the design equations is more involved than it's worth to do here, but eager readers are directed to the classic paper by the Sokals for details.[11] The design equations are as follows:

$$L = \frac{QR}{\omega}, \tag{31}$$

$$C_1 = \frac{1}{\omega R(\pi^2/4 + 1)(\pi/2)} \approx \frac{1}{\omega(R \cdot 5.447)}, \tag{32}$$

$$C_2 \approx C_1\left(\frac{5.447}{Q}\right)\left(1 + \frac{1.42}{Q - 2.08}\right). \tag{33}$$

For maximum efficiency, one desires the maximum Q consistent with the desired bandwidth. In practice, the achievable Q will often be substantially lower than the value that would limit bandwidth significantly. Once the Q is chosen, design of the Class E PA proceeds in a straightforward manner, using the equations given.

Unfortunately, computation of drain current and voltage waveforms is difficult. However, they look something like the graphs in Figure 15.10 when everything is

---

[11] Invention of the Class E amplifier is usually traced to N. O. Sokal and A. D. Sokal, "Class E, a New Class of High-Efficiency Tuned Single-Ended Power Amplifiers," *IEEE J. Solid-State Circuits*, v. 10, June 1975, pp. 168–76. I am grateful to Prof. David Rutledge of Caltech for calling to my attention G. Ewing's doctoral thesis, "High-Efficiency Radio-Frequency Power Amplifiers," Oregon State University, Corvallis, Oregon, 1964, as the earliest exposition of the concept.

tuned up. Note that the drain voltage has zero slope at turn-on, although the current is nearly a maximum when the switch turns off. Hence, switch dissipation can be significant during that transition if the switch isn't infinitely fast (as is the case with most switches you're likely to encounter). This dissipation can offset much of the improvement obtained by reducing the dissipation during the transition to the "on" state.

Additionally, note that each of the waveforms has a rather dramatic peak-to-average ratio. In fact, a detailed analysis shows that the peak drain voltage is approximately $3.6V_{DD}$, while the peak drain current is roughly $1.7V_{DD}/R$.

The maximum output power delivered to the load is

$$P_o = \frac{2}{1 + \pi^2/4} \cdot \frac{V_{DD}^2}{R} \approx 0.577 \cdot \frac{V_{DD}^2}{R}. \tag{34}$$

The normalized power output capability is therefore

$$\frac{P_o}{v_{DS,\text{on}} \cdot i_{D,\text{pk}}} \approx 0.098. \tag{35}$$

As you can see, the Class E is more demanding of its switch specifications than even a Class A amplifier.

Because of the poor power capability and the reduced efficiency due to switch turn-off losses,[12] practical implementations of the Class E amplifier do not exhibit significantly superior efficiency to well-executed designs of other types (e.g., the Class F amplifier to be described next). Additionally, because of the relatively large switch stresses, Class E amplifiers do not scale gracefully with the trend toward lower-power (and hence lower–breakdown voltage) technologies. For these reasons, Class E amplifiers have not found wide application in CMOS form. However, discrete implementations do not suffer from the severe breakdown voltage constraints of deep-submicron CMOS, and consequently there are countless discrete Class E amplifiers with excellent performance.

15.6 CLASS F AMPLIFIERS

Implicit in the design of Class E amplifiers is the concept of exploiting the properties of reactive terminations to shape the switch voltage and current waveforms to advantage. Perhaps the most elegant expression of this concept is found in the Class F amplifier; see Figure 15.11. Here, the output tank is tuned to resonance at the carrier frequency and is assumed to have a high enough Q to act as a short circuit at all frequencies outside of the desired bandwidth.

The length of the transmission line is chosen to be precisely a quarter-wavelength at the carrier frequency. Recall that a quarter-wavelength piece of line has an "impedance reciprocation" property. That is, the input impedance of such a line is proportional to the reciprocal of the termination impedance:

---

[12] The rather large peak drain current also degrades efficiency in practice, since all real switches have nonzero "on" voltages.

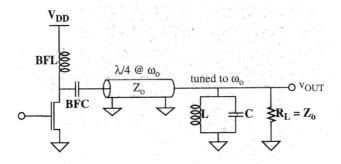

FIGURE 15.11. Class F amplifier.

$$Z_{\text{in}} = \frac{Z_o^2}{Z_L}. \qquad (36)$$

We may deduce from this equation that a *half*-wavelength piece of line presents an input impedance equal to the load impedance, since two quarter-wave sections give us two reciprocations that undo each other.

With that quick review out of the way, we can figure out the nature of the impedance seen by the drain. At the carrier frequency, the drain sees a pure resistance of $R_L = Z_o$, since the tank is an open circuit there, and the transmission line is therefore terminated in its characteristic impedance.

At the second harmonic of the carrier, the drain sees a short, because the tank is a short at all frequencies away from the carrier (and its modulation sidebands), so the transmission line now appears as a half-wavelength piece of line. Clearly, the drain sees a short at *all* even harmonics of the carrier, since the transmission line appears as some integer multiple of a half-wavelength at all even harmonics. Conversely, the drain sees an open circuit at all *odd* harmonics of the carrier, because the tank still appears as a short circuit; the transmission line appears as an odd multiple of a quarter-wavelength and therefore provides a net reciprocation of the load impedance.

Now, if the transistor is assumed to act as a switch, the reactive terminations guarantee that all of the odd harmonics of the drain voltage will see no load (other than that associated with the transistor's own output impedance), and hence a square-wave voltage ideally results at the drain (recall that a square wave with 50% duty ratio has only odd harmonics).

Because of the open-circuit condition imposed by the transmission line at all odd harmonics above the fundamental, the only current that flows into the line is at the fundamental frequency. Hence, the drain current is a sinusoid when the transistor is on. And, of course, the tank guarantees that the output voltage is a sinusoid even though the transistor is on for only a half-cycle (as in a Class B amplifier).

By cleverly arranging for the square-wave voltage to see no load at all frequencies above the fundamental, the switch current is ideally zero both at switch turn-on and turn-off times. The high efficiencies possible are suggested by the waveforms depicted in Figure 15.12. The total peak-to-peak drain voltage is seen to be twice the

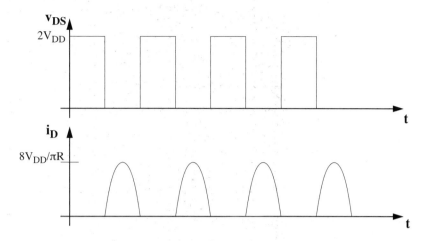

FIGURE 15.12. Drain voltage and current for ideal Class F amplifier.

supply voltage. Therefore, the peak-to-peak voltage of the fundamental component of v_{DS} is

$$(4/\pi)2V_{DD}. \tag{37}$$

Note that the fundamental has a peak-to-peak value that actually exceeds the total v_{DS} swing, thanks to the magic of Fourier transforms.

Now, since only the fundamental component survives to drive the load, the output power delivered is

$$P_o = \frac{[(4/\pi)V_{DD}]^2}{2R}. \tag{38}$$

Since the switch dissipates no power, we can conclude that the Class F amplifier is capable of 100% efficiency in principle. In practice, one can obtain efficiency superior to that of Class E amplifiers. Additionally, the Class F PA has substantially better normalized power-handling capability, since the maximum voltage is just twice the supply while the peak drain current is

$$i_{D,\text{pk}} = \frac{2V_{DD}}{R} \cdot \frac{4}{\pi} = \frac{8}{\pi} \cdot \frac{V_{DD}}{R}. \tag{39}$$

The normalized power handling capability is therefore

$$\frac{P_o}{v_{DS,\text{on}} \cdot i_{D,\text{pk}}} = \frac{\dfrac{[(4/\pi)V_{DD}]^2}{2R}}{2V_{DD} \cdot \left(\dfrac{8}{\pi} \cdot \dfrac{V_{DD}}{R}\right)} = \frac{1}{2\pi} \approx 0.16, \tag{40}$$

or exactly half that of the Class D amplifier. In some respects, the Class F amplifier may be considered equivalent to a single-ended Class D amplifier.

It should be emphasized that Class C, D, E, and F amplifiers are essentially *constant-envelope* amplifiers. That is, they do not normally provide an output that is

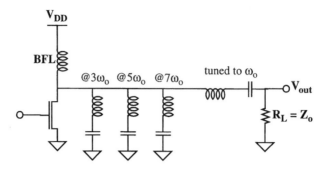

FIGURE 15.13. Inverse Class F amplifier (three-resonator lumped element example shown).

proportional to the input and thus tend to perform best when all we ask of them is a constant-amplitude output (as would be suitable for FM, for example). Nonetheless, we will see later that it is still possible to use these amplifiers in applications requiring linear operation. This capability is important because the modulations used by many modern communications systems involve amplitude modulation (e.g., QAM) to improve spectrum utilization, for which linear operation is necessary. At present, this requirement has frequently forced the use of Class AB amplifiers, with a corresponding reduction in efficiency relative to constant-envelope PA topologies. A general method for providing linear operation at constant-envelope efficiencies remains elusive. In Section 15.7 we will consider in more detail the problem of modulating power amplifiers.

15.6.1 INVERSE CLASS F (F^{-1})

The dual of the Class F is itself a power amplifier with the same theoretical bounds on efficiency as its cousin.[13] Whereas the Class F amplifier's termination appears as an open circuit at odd harmonics of the carrier beyond the fundamental, and as a short circuit at even harmonics, the *inverse Class F* (often denoted by the shorthand F^{-1}) employs a termination that appears as an open circuit at even harmonics and as a short circuit at the odd harmonics; see Figure 15.13.

Again, a transmission line may replace the lumped resonators when it is advantageous or otherwise practical to do so. Here, a piece of line whose length is $\lambda/2$ at the fundamental frequency replaces the paralleled series resonators and is interposed between the drain and the output series *LC* tank.

For an infinite number of series resonators, the drain voltage waveform consequently appears ideally as a (half) sinusoid and the current waveform as a square

---

[13] See e.g. S. Kee et al., "The Class E/F Family of ZVS Switching Amplifiers," *IEEE Trans. Microwave Theory and Tech.*, v. 51, May 2003.

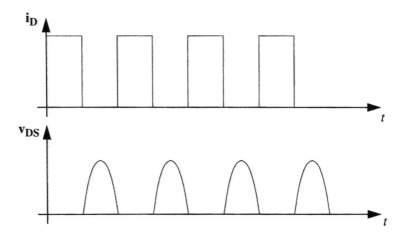

FIGURE 15.14. Drain voltage and current
for ideal inverse Class F amplifier.

wave, as seen in Figure 15.14. Once again, the lack of *V–I* overlap at the switching transitions accounts for the high theoretical efficiency of this architecture, just as with the Class E and standard Class F amplifiers.

15.6.2 ALTERNATIVE CLASS F TOPOLOGY

The topology shown in Figure 15.11 is elegant, but the transmission line may be inconveniently long in many applications. Furthermore, the benefits of an infinite (or nearly infinite) impedance at odd harmonics other than the fundamental are somewhat undermined in practice by the transistor's own output capacitance. Hence, a lumped approximation frequently performs nearly as well as the transmission-line version.

To create such a lumped approximation, replace the transmission line with a number of parallel resonant filters connected in series. Each of these resonators is tuned to a different odd harmonic of the carrier frequency. Quite often, simply one tank tuned to $3\omega_0$ is sufficient. Significant improvement in efficiency is rarely noted beyond the use of the two tanks shown in Figure 15.15. For example, use of one tank tuned to the third harmonic boosts the drain efficiency maximum to about 88%, compared to Class B's maximum of about 78%. Addition of tanks tuned to the fifth and seventh harmonics increases the Class F efficiency limit to 92% and 94%, respectively.[14] Given that practical tank elements are not lossless, the law of diminishing returns rapidly makes the use of additional resonators worse than futile.

---

[14] F. H. Raab, "Class-F Power Amplifiers with Maximally Flat Waveforms," *IEEE Trans. Microwave Theory and Tech.* v. 45, no. 11, November 1997, pp. 2007–12.

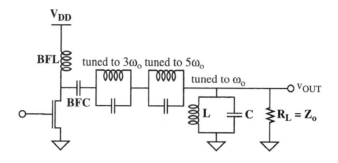

FIGURE 15.15. Alternative Class F amplifier.

15.7 MODULATION OF POWER AMPLIFIERS

15.7.1 CLASS A, AB, B, C, E, F

Modulating a Class A or B amplifier is straightforward in principle because the output voltage is ideally proportional to the amplitude of the signal component of the drain current, i_{rf}. Hence, if i_{rf} is itself proportional to the input drive, linear modulation results. A good approximation to this proportionality is readily achieved with short-channel MOS devices, which possess constant transconductance with sufficient gate voltage. Bipolar devices can provide reasonable linearity as a result of series base resistance, either externally provided or simply that of the device itself. Nevertheless, linearity requirements have become increasingly severe as wireless systems have evolved. A crude (but almost universally used) linearization "method" is power *backoff,* meaning that we ask only for, say, 1 W out of an amplifier capable of 10 W.[15] The rationale for backoff is readily understood by considering the same sort of weakly nonlinear amplifier model we invoked in Chapter 12 to define IP3. Because third-order IM terms drop 3 dB for every 1-dB drop in input power, the ratio between the fundamental and third-order components improves 2 dB for each 1-dB reduction in input power (similarly, the corresponding ratios for fifth-order and seventh-order IM terms theoretically improve by 4 dB and 6 dB, respectively, per 1-dB input power drop). If this trend holds, then there is some input power level below which the output IM3 (and other) distortion products are acceptably low in power relative to the carrier. Because Class A efficiency and output power both diminish as the amount of backoff increases, one should use the minimum value consistent with achieving the distortion objectives. Typical backoff values once were generally below 6–8 dB (and sometimes even as low as 1–3 dB) relative to the 1-dB compression point. These days, it is not unusual to find that backoff values must be as high as 10–20 dB in order to satisfy the stringent linearity requirements of some systems. Since it can be easy to lose track of perspective when expressing quantities in decibels, let's re-examine that

---

[15] If someone elses uses a cheesy technique, it's a hack. If *you* use it, it's a method.

last interval of values: It says that, after you beat your brains out to design a 10-watt RF amplifier, you might find that it meets the specifications for spectral purity only if output powers are kept below a few hundred milliwatts.

Compared to Class A amplifiers, the output IM products of a Class AB amplifier exhibit weaker dependencies on input power (e.g., 2-dB drop in IM3 power per decibel reduction of input power) owing to the latter topology's greater inherent nonlinearity. Worse, it is unfortunately not unusual to encounter cases in which *no* amount of backoff will result in acceptable distortion. Finally, as with the Class A amplifier, backoff frequently degrades efficiency to unacceptably low levels (e.g., to below 5–10% in some cases). We will shortly discuss appropriate linearization alternatives that one might use to relax some of these tradeoffs.

The Class C amplifier poses a more significant challenge, as may be appreciated by studying the equation for the output current derived earlier:

$$i_{\text{fund}} = \frac{i_{\text{rf}}}{2\pi}(2\Phi - \sin 2\Phi). \tag{41}$$

Despite appearances, the fundamental component of the current through the resistive load is generally *not* linearly proportional to i_{rf}, because the trigonometric term in parentheses is also a function of i_{rf}.[16] Hence, Class C amplifiers do not normally provide linear modulation capability and are therefore generally unsuitable for amplitude modulation, at least when a modulated carrier drives the gate circuit.

To obtain linear amplitude modulation from a nonlinear amplifier (e.g., Class C, D, E, or F), it is often advantageous to consider the *power supply* terminal (drain circuit) as an input port. The general idea is simple: Varying the supply voltage varies the output power. The control there can actually be more linear than at the standard input (e.g. the gate). The first to act on this insight was apparently Raymond Heising of AT&T, around 1919, for vacuum tube amplifiers (of course).[17] The Heising modulator (also known as the constant current modulator because of its use of a choke) is shown as Figure 15.16 in its simplest CMOS incarnation.

The modulation amplifier M_2 is loaded by a choke ("Mod. choke") that is chosen large enough to have a high reactance at the lowest modulation frequencies. The voltage V_x is the sum of V_{DD} and the modulation voltage developed across that choke. That sum in turn is the effective supply voltage for M_1 (biased to operate as a Class C amplifier) fed to the drain through the RF choke, as in our standard model for all of the classic PA topologies. Since the two transistors share a common DC supply and since the voltage V_x only approaches ground, transistor M_1's output can never quite go to zero. Consequently, the basic Heising modulator is inherently incapable of

---

[16] That is, with the exception of Class A or B operation. For the other cases, proportionality between drive and response does not occur so that linear modulation is not an inherent property.

[17] See E. B. Craft and E. H. Colpitts, "Radio Telephony," *AIEE Trans.*, v. 38, 1919, p. 328. See also R. A. Heising, "Modulation in Radio Telephony," *Proc. IRE*, v. 9, August 1921, pp. 305–22, and *Radio Review*, February 1922, p. 110.

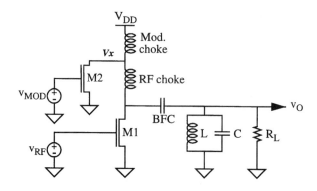

FIGURE 15.16. CMOS Heising modulator
with Class C RF stage (simplified).

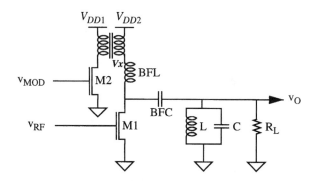

FIGURE 15.17. Alternative drain modulation example.

modulation depths of 100% (this property is a virtue in some instances, because over-modulation – and its attendant gross distortion – is inherently impossible). Typical maximum modulation percentages of 60–80% are not uncommon among commercial examples. In applications where the full modulation range is required, a quick fix is to place a capacitively bypassed resistor in series with the RF choke. The DC drop across the resistor causes M_1 to operate with a lower supply voltage than does M_2. The disadvantage is that the improvement in modulation depth comes at the cost of degraded efficiency, owing to the dissipation in this added resistor.

There are alternative forms of drain modulation that do not suffer this painful trade in efficiency for improved modulation depth. A popular example is shown in Figure 15.17. Here one can find many combinations of supply voltages, transformer turns ratio, and modulation amplitude that will force the drain voltage of M_1 to go to zero (not merely approach it) at an extremum of the modulation. For example, assume that we choose a 1 : 1 transformer. Because we want the secondary voltage to be able to go to zero, the swing at the primary must have an amplitude V_{DD2}, too. In turn, that requirement forces us to choose V_{DD1} somewhat larger than V_{DD2} (to accommodate nonzero drops across M_2) as well as an appropriate gate drive level to produce the desired modulation swing.

More commonly, a single drain supply voltage is used, leading to the need for other than a 1 : 1 transformer turns ratio. By choosing a suitable voltage step-up ratio, 100% modulation can be achieved.

It is important to recognize here that the modulator is itself a power amplifier. As such, these high-level modulators suffer from essentially the same tradeoffs between efficiency and linearity as do the RF stages they modulate. Without care, the power dissipated by the modulator could exceed that of the main RF power amplifier. In a popular variation intended to address this problem, M_2 is replaced by a push–pull Class B stage for improved efficiency. In that case, the output drains of the Class B stage connect to a transformer's primary, whose center tap provides the connection point for the DC supply. An even higher-efficiency alternative is to generate the voltage V_x with a stage operating as a switch-mode (e.g., Class D) amplifier. A challenge is to filter the switching noise sufficiently to meet stringent spectral purity requirements, but the high potential efficiency often justifies the engineering effort. Delta-sigma modulation is occasionally used in switching modulators to shape the noise spectrum in a way that eases filtering requirements.

These few examples show that there are many ways to effect drain (high-level) modulation.[18] However, even though drain modulation permits the nominally linear modulation of nonlinear amplifiers, the degree of linearity may still be insufficient to satisfy exacting requirements on spectral purity. That shortcoming motivates our consideration of a variety of enhancements and alternatives.

15.7.2 LINEARIZATION TECHNIQUES

Envelope Feedback

It's probably a bit generous to refer to power backoff and drain modulation as linearization *techniques*. Backoff pays for linearity with efficiency, and efficiency is too precious a currency to squander. Drain modulation, although superior to gate modulation for certain topologies, still ultimately relies on open-loop characteristics and so the distortion is not under the direct control of the designer. In this subsection we consider a number of ways to improve the linearity of RF power amplifiers at a minimal cost in efficiency.

When faced with the general problem of linearizing an amplifier, negative feedback naturally comes to mind. Closing a classic feedback loop around an RF power amplifier is fraught with peril, however. If you use resistive feedback, the dissipation in the feedback network can actually be large enough in high-power amplifiers to present a thermal problem, to say nothing of the drop in efficiency that always attends dissipation. Reactive feedback doesn't have this problem, but then one must take care to avoid spurious resonances that such reactances may produce. Next, there is the matter of loop transmission magnitude sufficiency, an issue that applies to all

---

[18] Although much of the literature makes no distinction between drain and Heising modulation, we point out that the latter is a subset of the former.

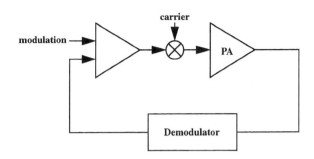

FIGURE 15.18. Negative feedback for
improving modulation linearity.

amplifiers. As shown in Chapter 14, nonlinearities are suppressed by a factor equal to the magnitude of the loop transmission (actually, the return difference, but for large loop transmission magnitudes these quantities are approximately equal), at the cost of an equal reduction in closed-loop gain. A tenfold reduction in closed-loop gain will accompany a tenfold improvement in third-order IM distortion (normalized to the fundamental). One therefore needs an ample supply of excess gain to enable large improvements in linearity. At radio frequencies, it is regrettably often true that available gain is already difficult enough to come by on an open-loop basis. Consequently, it may be hard to obtain significant improvements in linearity without reducing the closed-loop gain to the point that one wins only by losing.

Attempts at other than a Pyrrhic victory by means of cascading a number of gain stages will simply cause the classic stability problem to appear. This problem increases in severity as we seek larger bandwidths owing to the greater likelihood that parasitic poles will fall in band and degrade stability margins.

The astute reader will note that the linearization need only be effective over a bandwidth equal to that of the modulation, and that this bandwidth need not be centered about the carrier. As a specific exploitation of this observation, suppose that we feed back a signal corresponding to the envelope of the output signal (with a demodulator, for example, which can be as crude as a diode envelope detector in noncritical applications) and then use this demodulated signal to close the loop; see Figure 15.18.[19]

Closing the loop at baseband frequencies is potentially advantageous because it is then considerably easier to obtain the requisite excess loop gain over the bandwidth of interest. Still, meeting all of the relevant requirements is not necessarily trivial, particularly if one seeks large improvements in linearity over a large bandwidth. A brief numerical example should suffice to highlight the relevant issues. Suppose that we want to reduce distortion by 40 dB over a bandwidth of 1 MHz; then we must have 40 dB of excess gain at 1 MHz. If the feedback loop is well modeled as single pole then the corresponding loop crossover frequency will be 100 MHz, implying the need for a stable closed-loop bandwidth of 100 MHz as well. Assuring that the loop indeed

---

[19] See, e.g., F. E. Terman and R. R. Buss, "Some Notes on Linear and Grid-Modulated Radio Frequency Amplifiers," *Proc. IRE,* v. 29, 1941, pp. 104–7.

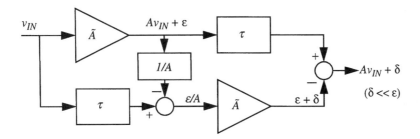

FIGURE 15.19. Feedforward amplifier.

behaves as a single-pole system over this bandwidth is not impossible, but neither is it trivial.[20] From the numbers, it's readily apparent that the difficulty increases rapidly if one seeks greater linearity improvements over a broader bandwidth.

Even if one needs only relatively modest improvements in amplitude linearity, constraints on the phase performance (referenced to baseband) could still present design difficulties. Recall that the phase lag of a single-pole system is 45° at the −3-dB frequency. If there is a tight specification on the permissible phase shift over the passband (e.g., to constrain group delay variation), then the only recourse for a single-pole system is to increase the bandwidth. If the allowable phase error is 5.7°, then the bandwidth must be chosen a decade above the baseband bandwidth. If that error budget shrinks to 0.57°, then the required bandwidth increases another order of magnitude to one hundred times the baseband bandwidth.[21]

These calculations all presume optimistically that the only error source is in the forward path; the feedback is assumed perfect in all respects. In the case of Figure 15.18 this requirement translates into the need for an exceptionally linear demodulator over a wide dynamic range, because a negative feedback system is desensitized only to imperfections in the *forward* path. The overall system's performance is limited by the quality of the feedback, so any nonlinearities and phase shifts in the demodulator will bound the effectiveness of the loop.

These difficulties are sufficiently daunting that a collection of other techniques have evolved as alternatives or supplements to classical negative feedback. Some of these are purely open-loop techniques and hence are not constrained by stability concerns, as discussed in Chapter 14. Furthermore, the linearization techniques we'll present may be used singly or in combination with other techniques, depending on the particular design objectives.

Feedforward

We've already met one open-loop linearization technique: feedforward, devised by Black before inventing the negative feedback amplifier. We reprise it in Figure 15.19.

---

[20] Relaxing the single-pole restriction can help moderate the excess bandwidth requirement but at the risk of creating a conditionally stable feedback system.

[21] One may reduce the demand for excess bandwidths by employing suitable phase-compensating all-pass filters.

As discussed in Chapter 14, the bandwidth over which feedforward provides significant linearity improvements depends in part on the bandwidth over which the group delay of the individual amplifiers may be tracked accurately by realizable time-delay elements.[22] This tracking must remain accurate over time and in the presence of variations in temperature and supply voltage. In many commercial examples, such as some GSM base-station power amplifiers, the delay elements are largely realized with suitable lengths of low-loss coaxial cable. As with most techniques that rely on matching, one might expect improvements of perhaps 30 dB in practice (maybe above 40 dB with great care). In certain cases it may be possible to implement automated trimming techniques, some of which rely on pilot signals sent through the amplifier. Automatic calibration of this nature can enable feedforward to provide excellent linearity with great consistency. If linearity must be improved further still, then one retains the option of combining this technique with others.

Despite the relatively high bandwidth achievable with feedforward, the low efficiency that results from consuming power in two identical amplifiers is certainly a drawback. Although the partial redundancy provided by having two separate gain paths is sometimes a compelling and compensating asset, the efficiency is low enough (typically below 10%) that the general trend is away from feedforward RF power amplifiers for most applications.

Pre- and Postdistortion

Another approach to open-loop linearization exploits the fact that cascading a nonlinear element with its mathematical inverse results in an overall transfer characteristic that is linear. Logically enough, the compensating element is called a *predistorter* if it precedes the nonlinear amplifier and a *postdistorter* if it follows it. Predistortion is by far the more common of the two (because the power levels are lower at the input to the PA proper), and it may be applied either at baseband or at RF. Baseband predistortion is extremely popular because the frequencies are lower and because practical options include both analog and digital techniques (with the latter enjoying increasing popularity because of digital's characteristic flexibility). Another attribute of baseband predistortion is that it may also correct for nonlinearities suffered during upconversion to RF; see Figure 15.20.

Because the principal nonlinearity in an amplifier is associated with gain compression, predistorters succeed only to the extent to which they are able to undo the compression accurately by providing an increasing gain as the input increases. However, it's important to keep in mind that a predistorter cannot increase the saturated output power of an amplifier, and consequently we should expect little or no improvement in the 1-dB compression point. Because IP3 for "well-behaved" nonlinearities is at least somewhat related to compression point, it shouldn't surprise you that predistortion rarely succeeds in reducing IM3 products by much more than about a dozen

---

[22] Because real amplifiers generally do no exhibit constant group delay, designing the compensating delay elements to track this real behavior is quite difficult.

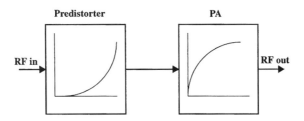

FIGURE 15.20. Illustration of RF predistortion.

decibels. If much greater reductions are needed, predistortion alone is unlikely to succeed.[23]

Corrections for phase errors (including those that may be the result of AM-to-PM conversion) may be provided by a phase shifter placed in series with the input. Most amplifiers tend to exhibit larger phase lag for small-amplitude inputs, so control for the phase shifter must be arranged to provide a compensating phase shift. Constraints here are usually less severe than for amplitude correction, but devising an analog control circuit is complex enough that digital control has become popular. Then, once you go to the trouble of building a digital controller, you might as well use it to control both the gain and phase correctors. It's just a short hop from there to closing a true feedback loop around both the amplitude and phase paths, at which point you've implemented *polar feedback,* a topic about which we'll have more to say shortly.

To achieve even the modest dozen-decibel improvement alluded to earlier – and regardless of whether the predistortion is implemented as a purely analog circuit or digitally controlled element – one must solve the problem of accurately producing the desired inverse transfer characteristic and subsequently assuring that this inverse remains correct over time (and with potentially varying load) in the face of the usual variations in process, voltage, and temperature.[24]

Because a fixed predistorter may prove inadequate to accommodate such drifts, it is natural to consider adaptive predistortion as an alternative. Such a predistorter uses real-time measurements of voltage and temperature, for example, in computing and updating the inverse function periodically. Successful implementation therefore requires a model for the system as well as sensors to measure the relevant input variables. Sadly, system modeling is quite a difficult task, particularly if some of the important variables (such as output load, which can vary wildly in portable applications) cannot

---

[23] Claims of much larger values are occasionally encountered in some of the literature. Upon close examination, however, it turns out that many of these claims are for systems that combine backoff (or some other method) with predistortion, whether or not backoff is explicitly acknowledged.

[24] Predistortion as a method for providing overall linear operation should be familiar to you. A current mirror, for example, actually relies on a pair of inverse, nonlinear conversions (first from current to voltage, then back to current again) to provide truly linear behavior in the current domain. A more sophisticated example that relies on the same basic cascade of nonlinear transductions (I to V, then back again) is a true Gilbert gain cell.

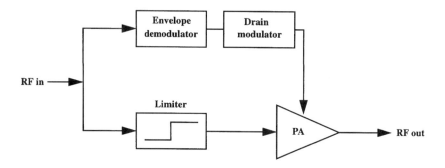

FIGURE 15.21. Kahn EER system.

be measured conveniently (if at all). Compounding the difficulty is that the non-linearity may be hysteretic (have memory) owing to energy storage. In such cases, the present value of the output is a function not solely of the input but also of the past values of the input. These limitations do not imply that predistortion is value-less (quite the contrary, in fact, as many broadcast television transmitters rely on this technique), but they do explain why it's difficult for predistortion to provide large linearity improvements on a sustained basis. As with the other techniques, predistortion may be applied in combination with other methods to achieve overall linearity objectives.

Envelope Elimination and Restoration

Originally developed by Leonard Kahn to improve single-sideband (SSB) transmission systems, envelope elimination and restoration (EER) is not a linearization technique per se but rather a system for enabling linear amplification from nonlinear (constant-envelope) amplifiers through drain modulation.[25] In EER, a modulated RF signal to be linearly amplified is split into two paths; see Figure 15.21. One feeds a *limiting amplifier* (essentially a comparator) to produce a constant-envelope RF signal that is subsequently amplified at high efficiency by a constant-envelope (e.g., Class C) amplifier. The other path feeds an envelope detector (demodulator). The extracted modulation is reapplied to the constant-envelope amplifier using drain modulation. Because EER is not itself a linearization method (it's better to regard it as an efficiency-boosting technique), achievement of acceptable spectral purity may require supplementing EER with true linearization techniques.[26]

Now, it is difficult to build an ideal element of any kind, particularly at RF. Hence it is worthwhile examining what we actually require of the limiter. The role of the limiter in Kahn's EER system is simply to provide adequate drive to the PA stage to assure efficient operation. It so happens, however, that it may actually be advantageous for

---

[25] L. R. Kahn, "Single Sideband Transmissions by Envelope Elimination and Restoration," *Proc. IRE,* v. 40, 1952, pp. 803–6. It really is *Kahn* and not *Khan,* by the way; the latter was Capt. Kirk's nemesis.

[26] D. Su and W. McFarland, "An IC for Linearizing RF Power Amplifiers Using Envelope Elimination and Restoration," *IEEE J. Solid-State Circuits,* v. 33, December 1998, pp. 2252–8.

the input to the PA to follow the envelope of the RF input (at least coarsely), rather than remaining fixed, in order to avoid unnecessarily large (and wasteful) PA drives when the envelope is small. Design of a practical limiter may be considerably easier as a result, because the problem essentially reduces to one of building an amplifier instead of a hard limiter – but without much concern for amplitude linearity.[27] Depending on the characteristics of a particular PA, the limiter may have to provide relatively high gain when the input amplitude is low in order to guarantee that the PA stage is always driven hard enough to assure high efficiency and keep the noise floor low.[28] It may also be necessary to insert a compensating delay (generally in the RF path) to assure that the drain modulation is properly time-aligned with the PA drive. Failure in alignment may affect the ability of EER to function well with low power inputs, thereby resulting in a reduction in the usable dynamic range of output power. Still, it is challenging in practice to achieve dynamic range values much larger than about 30 dB.

Chireix Outphasing (RCA *Ampliphase*) and LINC

A general term for techniques that may obtain linear modulation by combining the outputs of nonlinear amplifiers has come to be called LINC (*li*near amplification with *n*onlinear *c*omponents).[29] The first expression of a LINC idea in the literature is *outphasing modulation,* developed by Henri Chireix (pronounced a bit like "she wrecks") around 1935. Outphasing produces amplitude modulation through the vector addition of two constant-amplitude signals of differing phase.[30] The constant-amplitude characteristic allows the use of highly efficient constant-envelope RF amplifiers, while vector addition obviates the need for drain modulation and thus avoids its associated dissipation.

Outphasing modulation enjoyed intermittent and modest commercial success in the two decades following its invention, but its popularity soared when RCA chose this technology for use in their famous line of broadcast AM transmitters, starting with the 50-kW BTA-50G in 1956.[31] The *Ampliphase,* as RCA's marketing literature called it, would dominate broadcast AM radio transmitter technology for the next fifteen years.

To implement the outphasing method, first perform a single-ended to differential conversion of a baseband signal (if it isn't already available in differential form)

---

[27] We must still be conscious of careful design to avoid AM-to-PM conversion in those communications systems where such conversion may be objectionable. This comment applies to the entire amplifier and thus is independent of whether or not one seeks to implement a classic limiter.

[28] F. Raab, "Drive Modulation in Kahn-Technique Transmitters," *IEEE MTT-S Digest,* v. 2, June 1999, pp. 811–14.

[29] D. C. Cox, "Linear Amplification with Nonlinear Components," *IEEE Trans. Communications,* December 1974, pp. 1942–5.

[30] H. Chireix, "High-Power Outphasing Modulation," *Proc. IRE,* v. 23, 1935, pp. 1370–92.

[31] D. R. Musson, "Ampliphase ... for Economical Super-Power AM Transmitters," *Broadcast News,* v. 119, February 1964, pp. 24–9; also see *Broadcast News,* v. 111, 1961, pp. 36–9. Outphasing boosts efficiency enough to enable the construction of practical transmitters with at least 250-kW output power.

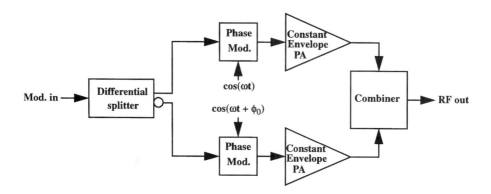

FIGURE 15.22. Block diagram of outphasing modulator.

and then use the outputs to phase-modulate a pair of isochronous – but not synchronous – RF carriers. Amplify the phase-modulated RF signals with highly efficient constant-envelope amplifiers; then sum the two amplified phase-modulated signals together using a simple passive network. The amplitude-modulated RF signal appears across the output of the combiner, as seen in Figure 15.22.

In typical implementations, the quiescent phase shift ϕ_0 between the two amplifier outputs is chosen equal to 135°. The two signal paths are designed to produce maximum phase deviations of 45° and −45° each, so that the total phase difference between the two inputs to the combiner swings between 90° and 180°. When the phase difference is the former, the two signals add to produce the maximum output. When the phase difference is 180°, the two signals cancel, producing zero output. These two extremes correspond to the peaks of a modulated signal with 100% depth.

Much of the design effort in an outphasing system concerns (a) obtaining linear phase modulation and (b) realizing a combiner that has low loss yet prevents the pulling of one amplifier by the other from degrading hard-won efficiency, linearity, and stability. In the *Ampliphase* system, the phase modulator exploits the change in phase one obtains from varying the Q of a tank, whose center frequency is offset from the carrier frequency by some amount; see Figure 15.23.

The output resistance of transistor M_1 acts as a variable resistor, whose value varies with the modulation. As the modulation voltage goes up, M_1's output resistance goes down and so increases the Q of the output tank. Transistor M_2 is simply a transconductor, converting the RF voltage into a current. The phase angle of the output voltage relative to that of the RF current in the drain of M_1 is thus the same (within a sign here or there) as the phase angle of the tank impedance. That angle, in turn, is a function of the tank's Q and is therefore a function of the modulation. Note that a linear dependence of phase shift on modulation voltage is hardly guaranteed with this circuit. Predistortion is used to obtain nominally linear modulation.

Another source of design difficulty is the output-power combining network. In the *Ampliphase* transmitter, the combiner is basically a pair of *CLC* π-networks (necessary for impedance transformation anyway) whose outputs are tied together (each

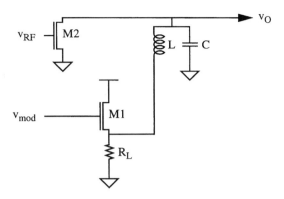

FIGURE 15.23. *Ampliphase* phase modulator (simplified CMOS version; bias details omitted).

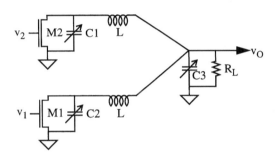

FIGURE 15.24. *Ampliphase* output combiner (simplified CMOS version; bias details omitted).

π-network acts as a lumped approximation to a quarter-wave line). This is shown in Figure 15.24.

This combiner appears simple, but looks are deceiving.[32] Although shown to be workable in a highly successful commercial design, it also illustrates a basic problem with outphasing. The effective impedance seen by the drain of each transistor depends not only on the load connected to the final output but also on the relative phase of the signal at the other transistor's drain. To avoid having to accommodate wild variations in drain load impedance, capacitors C_1 and C_2 must vary (in opposite directions) as a function of instantaneous phase angle of the drive. Needless to say, this requirement for a linear, controllable capacitance serves only to increase the level of design difficulty. In fact, the quest for a practical, low-loss, linear combiner that also provides a high degree of isolation remains unfulfilled. It is principally for this reason that LINC does not dominate today, despite its architectural appeal.

---

[32] The appearance of simplicity is enhanced if we replace the π-networks with suitable pieces of transmission line, although the variable compensating capacitances would still be required.

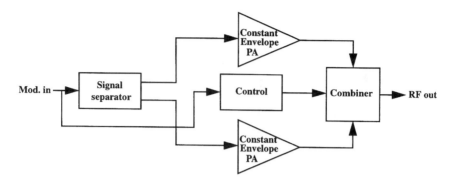

FIGURE 15.25. Block diagram of generalized LINC system.

The difficulties of obtaining linear phase modulation and arranging for the correct quiescent phase – to say nothing of maintaining proper operation over time, temperature and supply voltage – are great enough that broadcast engineers, with a mixture of affection and derision, occasionally referred to these transmitters as *Amplifuzz*. By the mid-1970s, the Ampliphase line had been, well, phased out.[33]

Since that time, engineers have hardly given up on LINC. The availability of sophisticated digital signal processing capability has inspired many to apply that computational power to overcome some of LINC's impairments. A very general block diagram for the resulting LINC appears as Figure 15.25. Regrettably, signal processing gives us only some of what is needed. Design of the combiner in particular remains, for the most part, an ongoing exercise in futility.

Polar Feedback

Because of the general need to correct both phase and amplitude nonlinearities in any signal path, it seems logical to employ a feedback loop around each separately, as we hinted in the discussion of predistortion. The *polar feedback* loop directly implements this idea.[34] Polar feedback is often used in tandem with EER (for the amplitude component), supplemented by a phase detector and phase shifter. See Figure 15.26.

The two control loops are readily identifiable in the figure. The amplitude control loop compares the envelope of the output with that of the input, and the difference drives a drain modulator, just as in EER. The gain function $H_r(s)$ shapes control over the dynamics of the amplitude feedback loop. For example, if it contains an integration then the steady-state amplitude error can be driven to zero. Because the bulk of the loop transmission gain may be obtained at baseband through gain block $H_r(s)$ (rather than at RF), it is possible in principle to suppress nonlinearities by large factors.

The phase control loop examines the phase difference between amplitude-limited versions of the input and output signals. A limiter is a practical necessity because

---

[33] Sorry.

[34] V. Petrovic and W. Goslin, "Polar Loop Transmitter," *Electronics Letters*, v. 15, 1979, pp. 1706–12.

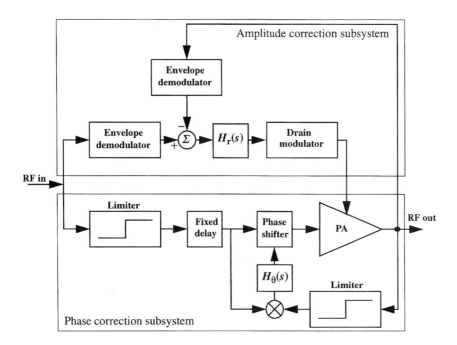

FIGURE 15.26. Example of power amplifier linearized with polar feedback.

most phase detectors are sensitive to both amplitude and phase (or otherwise require some minimum amplitude in order to function properly). Their presence is convenient as well, considering that the basic architecture for EER requires one anyway. The phase error signal drives a phase shifter placed in series with the input to the PA stage and adjusts it accordingly, with dynamics again controlled by a gain block, this time of transfer function $H_\theta(s)$.

One important consideration is to assure that the amplitude and phase corrections line up properly in time. Because the phase and amplitude control subsystems are generally realized with rather different elements, however, there is no guarantee that their delays (or any other relevant characteristics) will match. For example, the bandwidth of the amplitude control loop is a function of the drain modulator's bandwidth. As mentioned earlier, a switch-mode modulator is often used to keep efficiency high. High bandwidth in turn demands exceptionally high switching frequencies in the modulation amplifier. As a result, bandwidths much in excess of 1 MHz are difficult with current technology.

The phase shift loop typically suffers from far fewer constraints and is consequently much higher in bandwidth, with a correspondingly smaller delay. It's therefore usually necessary to insert a fixed compensating delay to assure that the delays through the two control paths match. (In principle, the phase shifter could bear the burden of delay compensation, but making it do so unnecessarily complicates the design of the shifter in most cases.)

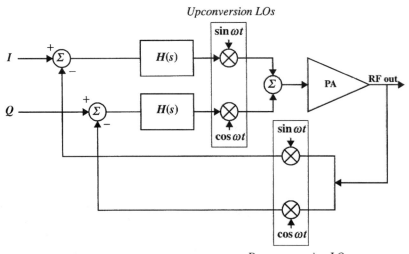

FIGURE 15.27. Transmitter linearized with Cartesian feedback.

Because proper operation of polar feedback requires matching the delays of two very different types of control loops, it is decidedly challenging to obtain high performance consistently. Another complication arises from ever-present AM-to-PM conversion, which couples the two loops in ways that degrade stability. The stability challenge is compounded by the amplitude dependency of AM-to-PM conversion. Polar feedback remains a topic of active research. So far, however, the difficulties of achieving the necessary levels of matching (to say nothing of maintaining same over time in the face of variations in supply, process, and temperature), as well as assuring stability over the full bandwidth and dynamic range of inputs, have proven large enough to prevent large-scale commercialization at the frequencies and bandwidths of greatest interest for mobile communications.

Note that one important idea behind polar feedback is that of decomposing an RF signal into two orthogonal components and then closing a feedback loop around each separately. Given that the polar variables of magnitude and phase represent only one possible choice, perhaps it is a worthwhile exercise to consider another.

Cartesian Feedback

Polar- and rectangular-coordinate representations of a signal are equivalent, so instead of a decomposition into magnitude and phase, we can decompose a signal into in-phase I and quadrature Q components, for example. This rectangular (*Cartesian*) representation has favorable practical implications, and consequently Cartesian feedback has received considerable attention.[35] The block diagram of Figure 15.27 shows

[35] V. Petrovic and C. N. Smith, "The Design of VHF Polar Loop Transmitters," *IEE Comms. 82 Conference,* 1982, pp. 148–55. Also see D. Cox, "Linear Amplification by Sampling Techniques: A New Application for Delta Coders," *IEEE Trans. Communications,* August 1975, pp. 793–8.

that, in contrast with a polar loop, a Cartesian feedback loop consists of two electrically identical paths.

In this setup, the output undergoes a pair of orthogonal downconversions. Baseband symbols I and Q are compared with their corresponding demodulated counterparts.[36] The baseband error signals are computed separately, amplified, upconverted back to RF, and finally summed at the input to the PA stage. Most of the loop gain is obtained at baseband from $H(s)$, rather than at RF, greatly easing loop design.

The fact that both feedback paths are identical in architecture means that Cartesian feedback is free of the matching problem that vexes polar feedback. However, there remain difficult design problems that, once again, have inhibited widespread use of the architecture.

The most significant problem arises from a lack of strict orthogonality between the two loops. Only if orthogonality holds will the two loops act independently and allow design to proceed with relatively few concerns. If the two loops are coupled, the dynamics may change in complex (Murphy-degraded) ways. Worse, the amount by which the loops are not orthogonal may change with time, temperature, and voltage – and also as the RF carrier is tuned over some range. This problem is hardly unique to Cartesian feedback; it's potentially a concern in any system that possesses multiple feedback paths (such as the polar feedback loop).

To evaluate the system-level consequences of this problem, consider a phase misalignment of ϕ between the upconversion and downconversion LOs (we still assume that each pair of LOs consists of orthogonal signals). As with any feedback loop, we may evaluate the loop transmission by breaking the loop, injecting a test signal, and observing what comes back. Doing so, we obtain

$$L_{\text{eff}}(s, \phi) = L_{\text{one}}(s) \cos \phi + \frac{[L_{\text{one}}(s) \sin \phi]^2}{1 + L_{\text{one}}(s) \cos \phi}, \tag{42}$$

where $L_{\text{one}}(s)$ is the transmission around each individual loop.[37]

The effective loop transmission expression helps us understand why Cartesian feedback loops can exhibit "odd behaviors." Depending on the amount of phase misalignment, the overall loop transmission can range from that of a single loop (when the misalignment is zero) all the way to a *cascade of two* single loops (when the misalignment is $\pi/2$). As is true for many control loops, $H(s)$ would be designed to contain an integration (to drive steady-state error to zero, for example). If the misalignment remains zero, that choice presents no problem. However, as the misalignment grows, $H(s)$ now contributes two integration poles to the loop transmission, leading to zero phase margin at best. Any negative phase shift from other sources drives phase margin to negative values.

---

[36] We assume that we have either performed quadrature downconversion in order to obtain the I and Q signals or that we are in fact performing an upconversion of baseband symbols that we have generated digitally. Thus the figure describes a transmitter or an amplifier.

[37] J. Dawson and T. Lee, "Automatic Phase Alignment for a Fully Integrated CMOS Cartesian Feedback PA System," *ISSCC Digest of Technical Papers,* February 2003.

Identification of this mechanism as one significant source of problems with Cartesian feedback is relatively recent. Solutions include automatic phase alignment (to eliminate the source of the stability problem) and carefully crafting $H(s)$ to tolerate the wide variation in loop dynamics as the misalignment varies.[38] Slow rolloff compensation (see Chapter 14) is one possible choice for $H(s)$ if the latter strategy is pursued, which may also be used in tandem with the former. Implementation of these corrective measures enables Cartesian feedback to provide exceptionally large improvements in linearity over extremely wide bandwidths.

15.7.3 EFFICIENCY-BOOSTING TECHNIQUES

Having presented a number of linearization methods, we now focus attention on efficiency-boosting techniques and architectures.

Adaptive Bias

The efficiency of any amplifier with nonzero DC bias current degrades as the RF input power decreases (the Class A is worse than others in this regard). There are many RF PAs, such as those in cell phones, that operate at less than maximum output power a considerable fraction of the time and thus for which the average efficiency is terrible. To improve efficiency at lower power levels, a time-honored technique is to employ *adaptive bias* strategies.[39] Varying the bias current and supply voltage dynamically in accordance with the instantaneous demands on the amplifier can moderate considerably the degradation in efficiency (at least in principle). At high modulation values, the PA stage would operate with a relatively high supply voltage (with correspondingly higher gate bias, for example). At low modulations, the drain supply and gate bias voltages would drop as well. This strategy makes the efficiency a much weaker function of signal amplitude, as desired. Thanks to the advent of flexible, inexpensive, digital control circuitry, it is now considerably easier to implement adaptive bias than it once was.

The controllable drain supply is essentially identical to the drain modulation amplifier in an EER system, with all of the same design challenges. An additional challenge of adaptive biasing overall is that varying so many significant parameters more or less simultaneously is hardly a prescription for linear behavior. Nevertheless, adaptive bias presents an additional degree of freedom to be exercised in the never-ending series of tradeoffs between efficiency and linearity.

The Doherty and Terman–Woodyard Composite Amplifiers

Another efficiency-boosting strategy is to use multiple amplifiers, each responsible for amplification over some subset of the overall power range. By using only the minimum number of amplifiers necessary to provide the desired output power, it's

---

[38] Dawson and Lee, ibid.
[39] F. E. Terman and F. A. Everest, "Dynamic Grid Bias Modulation," *Radio,* July 1936, p. 22.

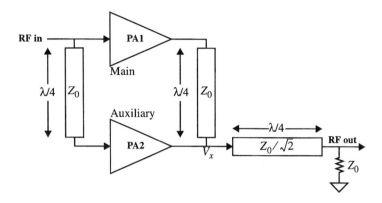

FIGURE 15.28. Doherty amplifier.

possible to reduce unnecessary dissipation. In effect, we implement the electronic equivalent of a turbocharger. The earliest realization of this idea uses two amplifiers and is due to Doherty;[40] see Figure 15.28.

Amplifiers PA1 and PA2 are the main and auxiliary amplifiers, respectively. The auxiliary amplifier is arranged to be cut off for low amplitude inputs. Assuming that PA2's output is an open circuit in this mode, it is straightforward to deduce that the impedance seen by the output of PA1 is then $2Z_0$.[41]

At some predetermined threshold, the auxiliary amplifier turns on and begins to contribute its power to the output. The $\lambda/4$ delay in feeding PA2 matches the $\lambda/4$ delay coupling the output of PA1 to the output of PA2. The contribution of PA2 to V_x is thus in phase with that of PA1. The fact that V_x is larger when PA2 is active implies an increase in the impedance seen by the main amplifier's output delay line; PA2 bootstraps PA1. When reflected back through the $\lambda/4$ line to PA2's output, this increased impedance at V_x is seen as a reduction in impedance. In turn, this reduction in PA2's load resistance increases the power supplied by that amplifier. When both amplifiers are contributing their maximum power, each amplifier sees a load impedance of Z_0 and contributes equally to the output.

After some reflection, one recognizes that this composite amplifier shares a key characteristic with the push–pull Class B amplifier: half of the power is handled by half of the circuit. In fact, the limiting peak efficiencies are identical, at about 78%, when the two amplifiers are partitioned as in Doherty's original design. Average efficiency is theoretically somewhat less than half this value.

The question of how best to select the threshold at which the auxiliary amplifier begins to contribute to the output is answered in large part by examining the envelope probability density function (PDF) of the signals to be processed by the

---

[40] W. H. Doherty, "A New High-Efficiency Power Amplifier for Modulated Waves," *Proc. IRE,* v. 24, 1936, pp. 1136–82.

[41] It should be mentioned that the quarter-wave lines may be approximated by *CLC* π-networks.

amplifier. This consideration is particularly important in view of the trend toward complex modulation methods. For example, a 16-QAM signal theoretically exhibits a 17-dB peak-to-average ratio. A 16-QAM signal with a 16-dBm (40-mW) average output power thus may have occasional peaks as high as 33 dBm (2 W). Designing a conventional 2-W amplifier and operating it with 40-mW average power virtually assures terrible average efficiency. A Doherty-like technique would seem well suited to accommodate modulations with such characteristics, since a highly efficient, low-power main amplifier would be bearing the burden most of the time – with the auxiliary amplifier activated only intermittently to handle the relatively rare high-power peaks. We see that a PDF heavily weighted toward lower powers implies that we should lower the threshold. If we were using some other modulation whose PDF were heavily weighted toward higher power, then we would wish to raise the threshold. Implementing arbitrary power division ratios may be accomplished a number of ways, including adjustment of the coupling impedances and operating the two amplifiers with different supply voltages.[42]

Further improvements in efficiency are possible by subdivision into more than two power ranges. Although the complexity of the load structure that effects the power combining rapidly increases in magnitude, the theoretical boosts in efficiency can be substantial. A doubling in average efficiency is not out of the question, for example.[43]

The similarity with Class B doesn't end with efficiency calculations, unfortunately. A problem akin to crossover distortion afflicts the Doherty amplifier as well. It is perhaps not surprising that efficiency is obtained at the expense of distortion, and one must expend a great deal of engineering effort to suppress nonlinearities (e.g., by embedding a Doherty amplifier within a feedback loop) in both amplitude and phase domains. And as with the outphasing system, the impedance seen at the output of one amplifier is a function of the other amplifier's output. Hence there is ample opportunity for a host of misbehaviors arising from unanticipated interactions.

An extension of the Doherty amplifier is the modulator–amplifier combination of Terman and Woodyard.[44] It is similar to the Doherty amplifier in its use of two amplifiers (driven by $\lambda/4$-delayed RF carrier signals) and of the same output combiner. The difference lies in the modulation capability, provided by injecting modulation in phase to the gate circuit of both amplifiers simultaneously. Because modulation is thus the result of nonlinearities inherent in the device transfer characteristic, less than faithful modulation results. However, by wrapping a feedback loop around the envelope signal, for example, large improvements in linearity may be obtained at low cost in efficiency.

[42] M. Iwamoto et al., "An Extended Doherty Amplifier with High Efficiency over a Wide Power Range," *IEEE MTT-S Digest,* May 2001, pp. 931–4.

[43] F. H. Raab, "Efficiency of Doherty RF Power Amplifier Systems," *IEEE Trans. Broadcast.,* v. 33, September 1987, pp. 77–83.

[44] F. E. Terman and J. R. Woodyard, "A High-Efficiency Grid-Modulated Amplifier," *Proc. IRE,* v. 26, 1938, pp. 929–45.

15.7.4 PULSEWIDTH MODULATION

Another technique for obtaining nearly linear modulation is through the use of pulsewidth modulation (PWM). Amplifiers using this technique are occasionally known as Class S amplifiers, although this terminology is by no means universal.

Such amplifiers do not perform modulation through variation of drive amplitude. Rather, the modulation is accomplished by controlling the duty cycle of constant-amplitude drive pulses. The pulses are filtered so that the output power is proportional to the input duty cycle, and the goal of linear operation at high efficiency is achieved in principle.

Although PWM works well at relatively low frequencies (e.g., for switching power converters up to the low-megahertz range), it is fairly useless at the gigahertz carrier frequencies of cellular telephones. The reason is not terribly profound. Consider, for example, the problem of achieving modulation over a 10 : 1 range at a carrier of 1 GHz. With a half-period of 500 ps, modulation to 10% of the maximum value requires the generation of 50-ps pulses. Even if we were able to generate such narrow pulses (*very* difficult), it is unlikely that the switch would actually turn on completely, and this would lead to large dissipation. Therefore, operation of PWM amplifiers over a large dynamic range of output power is essentially hopeless at high frequencies. Stated another way, the switch (and its drive circuitry) has to be n times faster than in a non-PWM amplifier, where n is the desired dynamic range. As a result, it becomes increasingly difficult to use pulsewidth modulation once carrier frequencies exceed roughly 10 MHz.

15.7.5 OTHER TECHNIQUES

Gain or Power Boost by Cascading

Cascading is so obvious a method for increasing gain and power levels that even mentioning it invites scorn. However there are enough subtleties in cascading PA stages that it's worth the risk, justifying this brief discussion.

Power levels generally scale upward as we proceed from stage to stage in a cascade of amplifiers. If the power consumed by the early stages is low enough then it may be prudent to design those stages with a focus on linearity, deferring to a later stage (or stages) the problem of obtaining high efficiency. In practice, then, the earliest stages may be implemented as Class A amplifiers, with a transition to (say) Class B or C for the last stage, for example.

When using drain modulation with a cascade of stages (e.g., Class C), the level of drain modulation should scale with the power level so that the drive for each stage in the cascade also scales. Without this scaling, there is a risk of overdriving one or more stages, leading to overload-related effects such as slow recovery from peaks, excessive AM-to-PM conversion, and poor linearity.

Finally, cascading always involves a risk of instability, particularly when the stages are tuned. The risk is greatest with Class A amplifiers and is mitigated by using the same general collection of techniques that are effective for low-level amplifiers. In stubborn cases, the *losser* method offers relief when all others fail. As its name implies, the method (if one could so dignify it) employs a resistor placed somewhere in the circuit to throw away gain and Q. Resistors in series with the gate, or across the gate–source terminals, are particularly effective.

Gain Boost by Injection Locking

We have much more to say about injection locking in Chapter 16, so we offer only a brief description here. For now, just accept that it's frequently possible to lock an oscillator's phase to that of a signal injected into an appropriate point in the oscillator circuit – provided certain conditions are met (e.g., frequency of injected signal close enough to the free-running oscillator frequency, amplitude of injected signal within a certain window, and so on). In fact, unwanted injection locking (perhaps caused by signals coupled through the substrate) is a very real problem in RF integrated circuits. As with other parasitic phenomena, virtually every effect that is unwanted in one context can be turned into an advantage in another.

Whereas cascading is an obvious means to increase gain, the relatively low inherent gain of CMOS devices generally implies the necessity for more stages than would be the case in other technologies, such as bipolar. Aside from the increased complexity, it's quite likely that the power consumed would be greater as well. To evade these limits, it's sometimes advantageous to consider building an *oscillator* and somehow influencing its phase or frequency. After all, an oscillator provides an RF output signal with *no* RF input; the "gain" is infinite. *It's easier to influence a signal than to produce it.* Because the power required to effect locking can be small, the apparent gain can be quite large.

Injection locking as an amplification technique is normally limited to constant-envelope modulations because, as its very name implies, the input signal primarily affects the phase. In principle, amplitude modulation could be provided as well (e.g., by varying the bias that controls the oscillation amplitude), but AM-to-PM conversion is usually so serious that this combination is practical only for shallow amplitude modulation depths.

An alternative possibility would be to combine injection locking with outphasing to produce amplitude modulation. Such a combination would theoretically exhibit high gain as well as high efficiency and, as a free bonus, present many subtle design problems. The experimental verification of this statement is left as an exercise for the reader.

Power Boost by Combining

A problem with CMOS IC processes in particular is the low (and ever-diminishing) supply voltage. For recent technologies down to the 0.13-μm process generation, the nominal supply voltage has varied approximately linearly with minimum linewidth:

$$V_{DD} \approx \left(10\frac{V}{\mu m}\right)L_{\text{drawn}}. \tag{43}$$

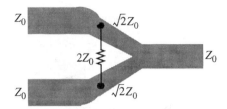

FIGURE 15.29. Wilkinson power combiner.

Perhaps out of necessity, technology roadmaps project a somewhat more moderate rate of decrease for subsequent process generations:

$$V_{DD} \approx 1.2 \sqrt{\frac{L_{\text{drawn}}}{0.13 \, \mu\text{m}}}. \tag{44}$$

For current CMOS process technologies, the supply voltage is already close to 1 V and is projected to dip below it in the near future. Even if breakdown characteristics were to allow periodic excursions close to twice the nominal V_{DD} value (and they don't), delivery of 1 W with a 2-V peak-to-peak swing would require a load resistance of 0.5 Ω. This value is absurdly lower than commonly encountered antenna impedances, for example, so some sort of impedance transformation is inevitably required. Unfortunately, because it is exceptionally challenging to achieve impedance transformation losses that are small compared to 0.5 Ω, efficiency is generally quite poor.

One possible option is simply to use an older process technology to take advantage of its higher breakdown voltage. This strategy works as long as one may tolerate the corresponding reduction in available power gain. Some possibly relevant data points may help frame the problem a bit more quantitatively. Credible power amplifiers up to perhaps 2–3 GHz have been reported using 0.35-μm CMOS technology, and 5-GHz power amplifiers for wireless LANs (IEEE 802.11a) have been built out of 0.25-μm technology.[45] The saturated output power of these examples is below a few watts (and the average output power is typically much, much less).

In discrete implementations, power combiners commonly allow the attainment of higher output power than would otherwise be practical without an impedance transformer. One popular combiner is the *Wilkinson power combiner* (which may also be used in reverse as a power splitter), which theoretically allows lossless power combining when operating into a matched load;[46] see Figure 15.29. Hence, the Wilkinson combiner enables an ensemble of low-power amplifiers to provide high output power.

If larger than 2:1 boosts in output power are needed, several stages of combining may be used in a structure known as a *corporate combiner*. This is depicted

---

[45] D. Su et al., "A 5GHz CMOS Transceiver for IEEE 802.11a Wireless LAN," *ISSCC Digest of Technical Papers*, February 2002.

[46] E. J. Wilkinson, "N-Way Hybrid Power Combiner," *IRE Transactions MTT*, 1960. Unequal power splitting factors are also possible with an asymmetrical structure.

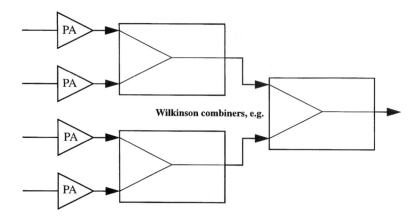

FIGURE 15.30. Corporate power combiner.

schematically in Figure 15.30. These techniques routinely succeed in discrete form, but integration presents numerous challenges. First, the transmission line segments of the combiner are a quarter-wavelength long. As a result, the die area consumed by combiners for power amplifiers in the low-gigahertz frequency range would be impractically large. As a specific example, consider that an on-chip quarter-wave line would be about 4 cm long at 1 GHz. Another factor is the loss of on-chip transmission lines. Again, a 1-dB attenuation is a 21% power loss, and it is difficult to keep losses even this low. Furthermore, the Wilkinson combiner won't be lossless if imperfectly terminated. Reflections resulting from any mistermination are absorbed by the bridging resistor, with a consequent loss in efficiency.

The limitations of this power combining approach (at least for on-chip realizations at centimeter-wave frequencies) are serious enough that no CMOS implementations have appeared in the literature to date.

Another approach is to attack the voltage limitation more directly. Device breakdown certainly forces us to constrain the voltages appearing across the terminals of any individual device, but it should be possible in principle to develop higher output voltages by summation of contributions from individual devices. That is, supply inputs in parallel and then take the outputs in series; the voltage boost reduces the need for absurd impedance transformation ratios. One very simple implementation of this idea is to build a differential power amplifier. In the ideal case, the differential output voltage swing is twice that of a single-ended stage, permitting a quadrupling of output power for a given supply voltage as well as a moderation of any additional impedance transformations that might be required.

A particularly elegant structure that extends this idea is the *distributed active transformer* (DAT).[47] The name is perhaps a bit of a misnomer in that this amplifier is not a distributed system in the same way that a transmission line is, for example;

---

[47] I. Aoki et al., "Distributed Active Transformer – A New Power-Combining and Impedance-Transformation Technique," *IEEE Trans. Microwave Theory and Tech.*, v. 50, January 2002, pp. 316–31.

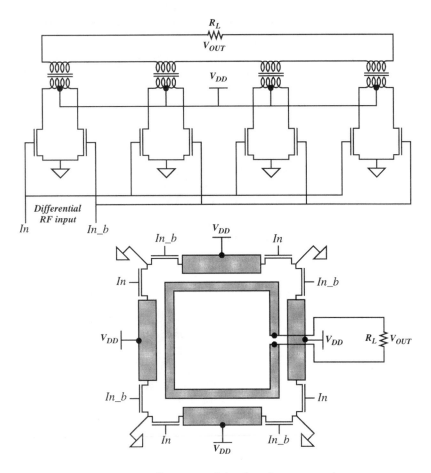

FIGURE 15.31. Illustration of distributed active transformer
with simplified layout (after Aoki et al.).

no distributed parameters are needed to describe it. Rather, the architecture gets its
name from distributing the total power burden among a number of devices, a concrete expression of "divide and conquer."

Suppose that we require more than a doubling of voltage swing in order to achieve
our power goal (or moderate any additional impedance transformation ratio). We
could quadruple the swing (and boost power by a factor of 16) by summing the
contributions of two differential amplifiers. The summation may be accomplished
conveniently by driving the primaries of transformers with the differential stages and
then simply connecting the secondaries in series. Figure 15.31 illustrates this principle, extended to four differential stages, for a theoretical eightfold boost in voltage
swing (relative to that of a single device), corresponding to a power boost of a factor
of 64. (For simplicity, capacitances for producing a resonant load are not shown in
the schematic or layout.)

In the simplified layout shown, each center-tapped drain load is the primary of
an on-chip transformer that has been realized out of coupled lines. The secondary

here is a one-turn square inductor, each arm of which is coupled to its corresponding center-tapped primary. Because the four arms are connected in series, the voltage contributions add as desired and so produce a boosted output voltage across the load R_L. The attendant impedance transformation implies that the current flowing in the secondary is smaller than in the primary by a factor of N. This permits the use of narrower lines than in the primary, as suggested by the relative line widths (shaded areas) in the figure.

Generalizing to N differential pairs with N output transformers, we see that the maximum boost in voltage is $2N$ (again, relative to that of a single transistor) with a corresponding power boost of $4N^2$. Losses in real circuits certainly diminish performance below those maximum limits, but the DAT remains a practical alternative nonetheless. As a benchmark, an early realization of this concept delivers 2.2 W of saturated output power with 35% drain efficiency (31% power-added efficiency) at 2.4 GHz, using a 0.35-μm CMOS process (see footnote 47). Gain is around 8.5 dB for input and output impedances of 50 Ω. This level of performance demonstrates that "CMOS RF power amplifier" is not quite as oxymoronic as it might sound.

15.7.6 PERFORMANCE METRICS

Prior to the advent of complex modulation schemes, it was largely sufficient to frame the design of transmit chains in terms of the specifications we've already presented: saturated output power, third-order intercept, 1-dB compression point, power-added efficiency, and the like. Engineers could rapidly estimate the level of backoff necessary to achieve a certain level of distortion, for example, and that was often enough to construct systems that functioned well. The relatively loose statutory constraints on out-of-band emissions reflected a concession to the crude state of the art. That situation has changed as a result of continuing efforts to reduce waste of precious spectrum through the use of more sophisticated modulation methods. That increased sophistication brings with it an increased sensitivity to certain impairments and thus an obligation to specify and control performance more tightly.

Adjacent channel power ratio (ACPR) is one example of such a performance metric. Developed in response to the difficulty of using a conventional two-tone test to predict adequately the interference generated by a transmitter using complex digital modulations (e.g., CDMA systems), ACPR characterizes interference potential by using representative modulations and then directly measuring the out-of-band power at frequency offsets corresponding to the location of adjacent channels.[48] See Figure 15.32.

Even though quantitatively relating ACPR to IP3 may not be possible in all cases, it remains typically true that ACPR improves with backoff in much the same way IP3

---

[48] Analytical approaches relating two-tone measurements to ACPR can still yield useful insights, however. For a comprehensive discussion of this approach, see Q. Wu et al., "Linear and RF Power Amplifier Design for CDMA Signals: A Spectrum Analysis Approach," *Microwave Journal,* December 1998, pp. 22–40.

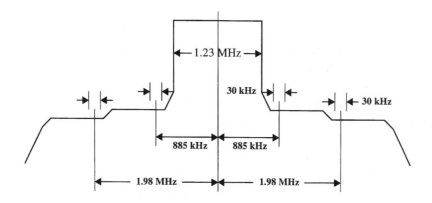

FIGURE 15.32. Example of ACPR specification.

does. That is, for every 1 dB of power backoff, you can expect perhaps 2 dB of ACPR improvement, with the out-of-band skirts moving as an ensemble. This statement holds if third-order nonlinearity dominates, as is often (but not always) the case.

Methods for measuring ACPR are not quite standardized, so in order to interpret reported values properly it's important to know the measurement methods to which the numbers correspond. As an example, handset ACPR for IS-95 CDMA needs to be better than −42 dBc when measured at an 885-kHz offset frequency.[49] A subtlety is that some techniques measure the ratio of integrated power densities and others the ratio of the densities themselves. Further differences involve the choice of integration bandwidths. For example, we could integrate the power density at an offset of 885 kHz over the 30-kHz bandwidth indicated in the figure, then divide by the integral of the power density over the 1.23-MHz bandwidth of the central lobe; strictly speaking, it's *that* ratio that needs to be −42 dBc or better.

In other (much more common) measurement methods, the power density is integrated over a 30-kHz bandwidth centered about both measurement frequencies; then a correction factor is applied to extrapolate the measured ratio to correspond to measurements made using the previous method. That is, given certain assumptions, the two raw ratios will differ by a correction factor

$$\Delta \text{ACPR} = 10 \log \frac{1.23 \text{ MHz}}{30 \text{ kHz}} \approx 16.13 \text{ dB}. \qquad (45)$$

Thus, about 16.1 dB needs to be subtracted from the ACPR measured by the second method in order to correspond to the first.[50] This second method assumes that the average power density in the 30-kHz window about the carrier is the same as in the rest of the 1.23-MHz band. It is important in this context to note that an IS-95 signal typically exhibits 2 dB or so of ripple over that band.

---

[49] Strictly speaking, IS-95 defines the air interface, IS-97 specifies performance of the base station, and IS-98 that of the mobile units.

[50] For more detailed information on measurement methods and interpretation of data, see *Testing CDMA Base Station Amplifiers,* Agilent Applications Note AN 1307.

Another measurement subtlety concerns the nature of signals used in ACPR evaluations. For CDMA systems, the modulations are "noiselike" in nature. As a result, it's tempting to use suitably band-limited noise as a signal for ACPR tests. However, it's important to understand that *noise* and *noiselike* are two different things, much as it might be important to keep track of the difference between *food* and *foodlike*. ACPR is a measure of distortion and, as such, is sensitive to average power level and envelope details like peak-to-average ratio. Those, in turn, are functions of the code set used in generating the modulations. Different noiselike waveforms with identical average power can cause a given amplifier to exhibit rather different ACPR. It is not unusual to see values vary over a 15-dB range as a function of stimulus.

Along with new ways to characterize the effects of distortion comes new terminology. The organic-sounding term *spectral regrowth* refers to the broadening in spectrum that results from distortion. Because distortion increases with power level and also as signals progress through the various stages in a transmitter, it is important to accommodate spectral regrowth in allocating a distortion budget for the various elements in the chain. Thus, to meet (say) a -42-dBc ACPR specification for the entire transmitter, it would be prudent to design the PA stage proper to have a worst-case ACPR at least a few (e.g., 3) decibels better than strictly needed to meet the ACPR specifications.

The philosophical underpinnings of ACPR can be summed up as "be a good neighbor." Specifications are chosen with the hope that compliance will assure that one transmitter minimally interferes with unintended receivers within reception range. In general, however, specifying the out-of-band power at a few discrete frequencies may be insufficient. An ACPR specification by itself, for example, does not preclude relatively strong narrowband emissions. In such cases, it may be necessary to specify a *spectral mask* instead. As its name implies, a spectral mask defines a continuum of limits on emission. Three representative examples are shown in Figure 15.33. One is for GSM (from version 05.05 of the standard), another is for indoor ultrawideband (UWB) systems (as defined in the *FCC Report and Order* of February 2003), and the third is for 802.11b wireless LAN.

The UWB mask in particular is notable for its complexity. The notch between 0.96 GHz and 1.61 GHz exists primarily in order to prevent interference with the global positioning system (GPS). The mask is least constraining from 3.1 GHz to 10.6 GHz and so this slice of spectrum is frequently cited as "the" UWB spectrum allocation. Also notice that the mask specifications are in terms of power spectral density. The absence of a carrier in UWB systems makes the familiar "dBc" and "dBc/Hz" units inapplicable.

Satisfying the "good neighbor" dictum by conforming to such masks is necessary, but not sufficient. We must also ensure that our transmitter produces modulations that *intended* receivers can demodulate successfully. The *error-vector magnitude* (EVM) is particularly well suited for quantifying impairments in many digitally modulated transmitters. The error vector concept applies naturally to systems employing vector modulations of some type (e.g., QAM).

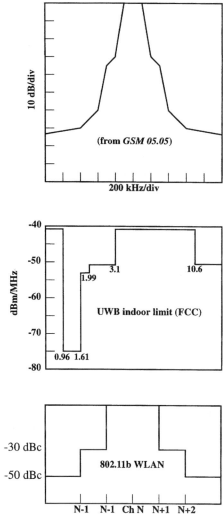

FIGURE 15.33. Examples of transmit spectral masks.

The error vector magnitude is simply the length of the error vector, as seen in Figure 15.34. Every symbol or chip has its own error vector. For 802.11b WLAN at 11 Mb/s, EVM is defined as the rms value over 1000 chips and must be less than a very generous 35%. For 802.11a operating at 54 Mb/s (the maximum specified in the standard), the allowable EVM is 5.6% – a considerably tighter specification than for 802.11b.

When measuring EVM, instrumentation produces plots that look like the one shown in Figure 15.35. In the ideal case, the constellation would appear as a perfect 8 × 8 square array of dots; the smudges would be points. The normalized rms smearing out is the EVM. In the particular case shown, the EVM is approximately 2% – well within specifications.

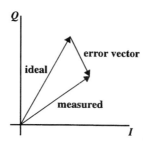

FIGURE 15.34. Illustration
of error vector.

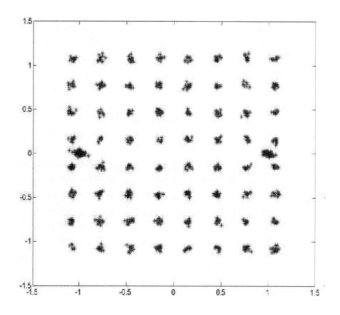

FIGURE 15.35. Measurement of EVM (802.11a example;
64-QAM constellation shown).

15.8   SUMMARY OF PA CHARACTERISTICS

We have seen that Class A amplifiers offer good linearity but poor power-handling capability (0.125 on a normalized basis) and low efficiency (50% at absolute maximum). Class B amplifiers improve on the efficiency (to ~78.5% at best) by reducing the fraction of a period during which the transistor is active, while maintaining the potential for linear modulation.

Class C amplifiers offer efficiencies approaching 100%, but both normalized power-handling capability and power gain approach zero at the same time. Also, they sacrifice linear operation to obtain the improvements in efficiency. Additionally, bipolar Class C amplifiers actually do not satisfy many of the assumptions used in the derivations, and they are difficult to design and construct as a consequence; MOS and vacuum tube implementations tend to be less troublesome in this regard.

Amplifiers based on switching concepts do not readily provide linear modulation capability either, but theoretically offer 100% efficiency at nonzero power-handling capability. Although such perfection is unrealizable, at least the limitation is not an inherent property of the topology.

Class D amplifiers offer a normalized power-handling capability of approximately 0.16, but suffer from potentially large "crowbar" dissipation due to noninfinite switching speeds. Class E PAs solve the dissipation problem for the turn-on transition but impose rather large stresses on the switch. The poor normalized power-handling capability of approximately 0.1, combined with the relatively low breakdown voltage of fineline CMOS, has made the Class E PA a rarity in the CMOS RF realm. Both Class E and Class F employ reactive loading to shape the voltage and current waveforms in ways that reduce switch dissipation, highlighting a powerful tool for improving efficiency.

Having considered numerous architectures and techniques for improving the linearity and efficiency of PAs, and having presented numerous ways to characterize PA performance, we now examine a few simple design examples to highlight classic zeroth-order PA design. The sophistication of modern communications systems calls for equally sophisticated design methods, but these involve rather complex considerations, powerful simulation tools, and large amounts of iteration. We will therefore provide only the barest outlines of an initial design, which would require considerable further refinement to satisfy the demands of modern systems. With expectations suitably lowered, we press onward.

15.9 RF PA DESIGN EXAMPLES

Suppose we want to design a linear amplifier for use in a 1-GHz communications system. The requirements are to supply 1 W into 50 Ω. Assume that a 3.3-V DC power supply is available. We must specify important device parameters, compute all component values, and estimate drain efficiency.

Important note: In the examples that follow, we will make arbitrary assumptions about the minimum values of v_{DS} in the "on" state to simplify design procedures. These assumptions must always be checked and modified in real designs.

15.9.1 CLASS A AMPLIFIER DESIGN EXAMPLE

First, see if the supply voltage is large enough to allow 1 W to be delivered to the load without an impedance transformation:

$$P_{\max} = \frac{V_{DD}^2}{2R} = \frac{(3.3)^2}{2 \cdot 50} \approx 0.1 \text{ W}. \tag{46}$$

Clearly, the supply voltage is not nearly sufficient, so an impedance transformation is required. The maximum value of the transformed resistance is

$$R_{\max} = \frac{V_{DD}^2}{2P_{\max}} = \frac{(3.3)^2}{2 \cdot 1} \approx 5.4 \ \Omega. \tag{47}$$

In practice, the 50-Ω load resistance would have to be transformed to an even lower value owing to such unavoidable dissipative mechanisms as voltage drops in the transistor and power loss in the filter and interconnect. Let us arbitrarily assume that designing for an effective load of 4 Ω compensates sufficiently for these losses. This assumption must be checked in any actual design, of course, and the transformed resistance reduced further if the output power is lower than desired. Alternatively, if the output power is more than sufficient, the transformed resistance may be increased to improve efficiency.

With a 4-Ω load, the peak RF current will not exceed $V_{DD}/R = 825$ mA, and the DC drain current bias must be set to approximately this value. Since the peak drain current is the sum of the bias and peak RF current, the transistor must be designed to supply about 1.65 A with minimum voltage drop. In this case, we can tolerate a reduction in effective V_{DD} of only a couple hundred millivolts or thereabouts, so the "on" resistance of the transistor must be kept below roughly 200 mΩ. In a typical 0.5-μm CMOS technology, device widths of several *millimeters* would therefore be required.

If we assume that 1 W is in fact ultimately delivered to the load after all losses have been taken into account, the drain efficiency would then be

$$\eta = \frac{P_o}{P_{DC}} = \frac{1}{0.825 \text{ A} \cdot 3.3 \text{ V}} \approx 37\%. \tag{48}$$

Therefore, when the amplifier is supplying that 1 W to the load, the transistor will be dissipating about 1.7 W, so the packaging and heat sinking must be designed to keep die temperatures acceptably low with this dissipation.

However, the news is actually worse than suggested by the foregoing computation. The Class A amplifier exhibits worse efficiency as the output swing decreases, since there is always dissipation due to the DC bias current even when the amplifier is delivering zero RF output power. Hence, if a power level below the 1-W target is delivered, the dissipation of the transistor can be substantially greater than 1.7 W. In the worst case, with no RF input, the transistor dissipates the power associated with its DC bias – for this particular example, a value of about 2.7 W. Hence, if the input drive is ever permitted to disappear, the packaging must be made capable of dissipating this greater figure, rather than the 1.7-W value computed earlier. For a Class A amplifier, then, the worst thermal problems are associated with zero input signal.

As a side note, it should be mentioned that significant improvements could be obtained by dynamically varying the bias as a function of the required power level. In that case, the transistor dissipation can be reduced significantly, leading to large improvements in efficiency at lower power levels. This type of adaptive Class A amplifier has been constructed, and efficiency similar to that obtained with Class B amplifiers was obtained.[51] The small complication associated with performing the adaptation has apparently been enough of a barrier to prevent its widespread adoption, but this topology becomes more attractive at higher frequencies because of the higher gain offered by Class A over Class B amplifiers.[52]

To round out the design, we need to specify component values for the output filter and matching network. Assume that the output filter is a simple parallel LC and that the required Q is about 10. The corresponding 100-MHz bandwidth at a center frequency of 1 GHz means that the reactance of the L and C must be 5 Ω in order to give us the required Q of 10. Hence we choose

$$X_L = 5 \implies L = \frac{5}{2\pi \cdot 1\,\text{GHz}} = 0.80\,\text{nH} \tag{49}$$

and

$$X_C = 5 \implies C = \frac{1}{5 \cdot 2\pi \cdot 1\,\text{GHz}} = 31.8\,\text{pF}. \tag{50}$$

Recall that, at resonance, each reactive element in the parallel tank circulates an RF current that is Q times as large as that flowing in the load. Hence, the L and C (*and associated interconnect*) must be able to withstand 2 A of peak current flow in this particular design.[53]

Next, we need to choose the size of the inductor, *BFL* (known as an RF *choke* because it "chokes off" the flow of RF current through it), so that its reactance is large enough. If we arbitrarily choose a factor of 10 as "large enough," we can readily compute the required value.[54] For this RF choke, we wish its reactance to be at least about ten times as large as the 4-Ω resistance of the tank at resonance:

$$X_{BFL} \geq 10 \cdot 4\,\Omega \implies BFL \geq 6.4\,\text{nH}. \tag{51}$$

---

[51] A. Saleh and D. Cox, "Improving the Power-Added Efficiency of FET Amplifiers Operating with Varying-Envelope Signals," *IEEE Trans. Microwave Theory and Tech.*, v. 31, no. 1, January 1983, pp. 51–6. This work is a direct descendant of that by F. E. Terman and F. A. Everest, cited in footnote 39.

[52] Since the transistor is in cutoff half the time in a Class B amplifier, it provides roughly half the gain of a Class A amplifier, all other things being equal. At high frequencies, it becomes increasingly painful to accept a halving of gain.

[53] Such large currents can lead to significant voltage drops along interconnect. If these voltage drops are unconstrained in magnitude and allowed to occur at uncontrolled locations, unanticipated circuit operation can easily occur, including mysterious oscillations (perhaps arising from the creation of parasitic feedback loops, since "ground" may no longer be zero volts everywhere) or other deviations from expected behavior. Again, the message is to be aware that what you construct may not always map one-to-one with your simulations. It is critically important to model *everything* at RF, and this includes seemingly mundane things like wire.

[54] Using factors significantly larger than 10 is not always wise. For example, using a choke that is very much larger than computed on this basis could result in an inductor with a self-resonant frequency that is too low (in which case it appears as a capacitance at the operating frequency instead of as an inductor as desired), or with excessive resistance (which would lead to decreased efficiency and possibly a heat problem).

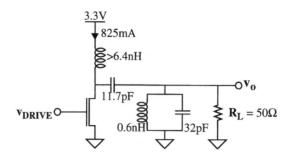

FIGURE 15.36. 1-W Class A amplifier.

This value is low enough that it is readily provided by parasitic bondwire and lead-frame inductance, in rare defiance of Murphy's law.

We must now provide a DC blocking capacitor and an impedance-transforming network. These functions may be combined into one circuit if a high-pass L-match is used, for example. With that choice, the transformation ratio sets the Q to 3.4, and the L-match element values become:

$$L_1 = \frac{R_L}{\omega_0 Q} \approx \frac{50}{2\pi 10^9 \cdot 3.4} \approx 2.3 \text{ nH}, \tag{52}$$

$$C_1 = \frac{1}{\omega_0 Q R_S} \approx \frac{1}{2\pi 10^9 \cdot 3.4 \cdot 4 \; \Omega} \approx 11.7 \text{ pF}. \tag{53}$$

The inductor of the L-match can be combined with the tank inductor to yield the completed circuit shown in Figure 15.36. All of the component values are consistent with realization in IC form, although some effort is required to build an inductor as small as 0.6 nH with low loss, good accuracy, and repeatability. In practice, one or both reactive elements of the tank would be made adjustable to tune the amplifier exactly to the desired center frequency.

An additional practical consideration is that we need a way to establish the proper bias conditions. This may be accomplished with the use of a current mirror. If we consider the output transistor as the output half of a mirror then we can supply, say, 1% of the bias current into another transistor that is 1% the size of the output transistor. Such a biasing method eliminates thermal drift problems associated with fixed-voltage gate bias. If implemented in discrete form, it is critically important to guarantee intimate thermal coupling between the two transistors to ensure that they are at the same temperature.

The two transistors can be connected together directly, or through another RF choke, and the signal to be amplified coupled to the circuit through another DC blocking capacitor; see Figure 15.37. The power consumed by the biasing circuit can be made suitably small by choosing a sufficiently large value of n. Furthermore, the amplitude of the signal at the common gate connection can be augmented by using an inductor in series with the gate drive to provide a resonant boost in amplitude. That

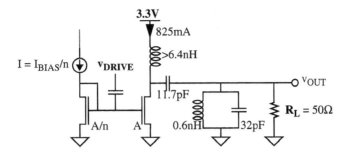

FIGURE 15.37. More complete 1-W Class A amplifier.

is, one may provide some impedance-matching network at the input. In general, practical implementations of power amplifiers would require matching networks to increase power gain.

15.9.2 CLASS AB, B, AND C AMPLIFIER DESIGN EXAMPLE

If we consider only single-ended implementations, then Class AB, B, and C amplifiers look extremely similar, with conduction angle (and therefore biasing details) being the only difference. Hence, the output network is exactly the same for all three, including the choke and output filter values, so a single design example will suffice.

For the Class AB amplifier, the reference bias would result in conduction through less than 360° but more than 180°. Therefore, the drain current of the output transistor would have a quiescent value of less than 825 mA. Since the input drive amplitude must then increase to give us the same output amplitude as in the Class A case, the gain is obviously smaller than for the Class A amplifier.

In the Class B amplifier, the bias is supposed to provide a 180° conduction angle. In practice, it is impossible to achieve a conduction angle of precisely this value, so one may well argue that Class B amplifiers exist only in the theoretical world of academia. Hence, all practical "Class B" amplifiers are actually either Class AB or C PAs. Again, the input drive must increase to maintain the same output amplitude, and the gain is therefore smaller (by a factor of about 2 relative to the Class A amplifier, as alluded to previously).

In the Class C amplifier, the common practice in most semiconductor designs is to use a zero gate bias, crank up the drive amplitude to get the desired output power, and accept whatever conduction angle, gain, and efficiency result. In this case, the reference bias transistor and associated choke (if used) would be removed, and a resistor (or choke) would be tied between the gate terminal and ground (see Figure 15.38). Although a MOS resistor is shown here, an ordinary resistor may also be used. Either way, it is chosen to present a reasonably high resistance to the coupling capacitor.

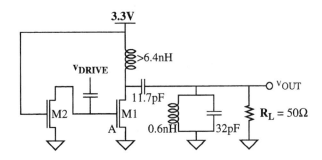

FIGURE 15.38. 1-W Class C amplifier.

15.9.3 CLASS E AMPLIFIER DESIGN EXAMPLE

In this (or in any switching) amplifier, we merely want to drive the transistor hard enough to act as a switch. Any drive in excess of this amount is not only wasted power – in a bipolar implementation, it may also seriously degrade the efficiency by driving the transistor into deep saturation. Hence, we compute the maximum drain current required and then adjust drive conditions on the gate to provide this current, and not much more.

Again, our goal is to supply 1 W into 50 Ω with a 3.3-V DC supply. Recalling that the maximum output power is

$$P_o = 0.577 \cdot \frac{V_{DD}^2}{R}, \tag{54}$$

we compute that the desired load resistance is actually about 6.3 Ω, so we need to transform downward by a bit less than for the previous designs. Initially, we'll assume that we will transform to a 5-Ω load to accommodate various losses; then we'll finish off the design by adding the required impedance transformer.

First, recall the basic topology (shown in Figure 15.39) and the associated equations:

$$L = \frac{QR}{\omega}, \tag{55}$$

$$C_1 = \frac{1}{\omega R (\pi^2/4 + 1)(\pi/2)} \approx \frac{1}{\omega (R \cdot 5.447)}, \tag{56}$$

$$C_2 \approx C_1 \left(\frac{5.447}{Q} \right) \left(1 + \frac{1.42}{Q - 2.08} \right), \tag{57}$$

where R is 5 Ω.

Choosing $Q = 10$ leads us to the following values:

$$L = 8.0 \text{ nH}, \tag{58}$$

$$C_1 = 5.8 \text{ pF}, \tag{59}$$

$$C_2 = 3.8 \text{ pF}. \tag{60}$$

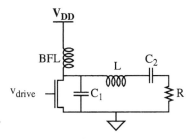

FIGURE 15.39. Class E amplifier.

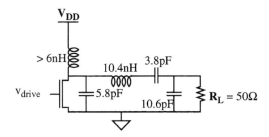

FIGURE 15.40. More complete Class E amplifier.

The required impedance transformation may be provided by a simple low-pass L-match whose element values are given by:

$$L_m \approx 2.4 \text{ nH}, \tag{61}$$

$$C_m \approx 10.6 \text{ pF}. \tag{62}$$

Combining the two inductors into one leads to the final design shown in Figure 15.40. In practice, the drain capacitance of the transistor itself would form part of the 5.8 pF capacitance.

15.10 ADDITIONAL DESIGN CONSIDERATIONS

15.10.1 POWER-ADDED EFFICIENCY

In the foregoing examples, drain efficiency is used to characterize the PAs. However, the definition of drain efficiency involves only the RF output power and the DC input power, so it can assign a high efficiency to a PA that has no power gain. Another measure of efficiency has therefore been developed to yield a figure of merit that takes power gain into account. Power-added efficiency (PAE) simply replaces RF output power with the difference between output and input power in the drain efficiency equation:

$$\text{PAE} \equiv \frac{P_{\text{out}} - P_{\text{in}}}{P_{\text{DC}}}. \tag{63}$$

Clearly, power-added efficiency will always be less than the drain efficiency.

15.10.2 PA INSTABILITY

Amplifiers of any kind can be unstable with certain combinations of load and source impedances, and power amplifiers are no exception. One extremely important problem results from drain-to-gate coupling (or collector-to-base coupling). As noted in the chapter on high-frequency amplifier design, this coupling can cause the input impedance to have a negative real part. In small-signal amplifiers, this problem can be reduced or eliminated entirely by using the various unilateralization techniques described earlier. Unfortunately, these tricks are generally inappropriate for power amplifiers because the requirement for high efficiency precludes the use of any technique (such as cascoding) that diminishes supply headroom. In general, the problem is usually solved through the brute-force means of degrading the input impedance (e.g., through the use of a simple resistor across the input terminals) to make the feedback less significant. Unfortunately, this action has the side effect of reducing gain. In general, MOSFETs – with their larger inherent input impedances – exhibit this stability problem to a greater degree than bipolar devices. In any case, there is usually a significant stability–gain tradeoff due to the feedback capacitance. And of course, thoughtful layout is required to avoid augmenting the inherent device feedback capacitance from an unfortunate juxtaposition of input and output wires.

15.10.3 BREAKDOWN PHENOMENA

MOS Devices

In all of the design examples, downward impedance transformations were required to deliver the desired amount of power into the output load. Clearly, the transformation ratio could be reduced if a higher power supply voltage were permitted, and the reader may reasonably ask why one could not simply demand a higher voltage be made available. The reason is that devices have finite breakdown voltages. Furthermore, as IC technology scales to ever-smaller dimensions, breakdown voltages tend to diminish as well. Thus, increasing transformation ratios are required as devices scale if one wishes to deliver a certain fixed amount of power to the load.

In MOS devices, one may identify four primary limits to allowable applied voltages in PAs. These are drain (or source) diode zener breakdown, drain–source punchthrough, time-dependent dielectric breakdown (TDDB), and gate oxide rupture.

The drain and source regions are quite heavily doped to reduce their resistivity. As a consequence, the diodes they form with the substrate have a relative low breakdown voltage, with typical values of the order of 10–12 V for 0.5-μm technologies.

Drain–source punchthrough is analogous to base punchthrough in bipolar devices and occurs when the drain voltage is high enough to cause the depletion zone around the drain to extend all the way to the source, effectively eliminating the channel. Current flow then ceases to be controlled by the gate voltage. Punchthrough problems can be mitigated by using larger channel lengths at the cost of degraded device transconductance; this, in turn, forces the use of a wider device to maintain output power.

Time-dependent dielectric breakdown is a consequence of gate oxide damage by energetic carriers. With the high fields typical of modern short-channel devices, it is possible to accelerate carriers (primarily electrons) to energies sufficient for them to cause the formation of traps in the oxide. Any charge that gets trapped there then shifts the device threshold. In NMOS transistors, the threshold increases so that the current obtained for a given gate voltage decreases; in PMOS devices, the opposite happens.

Time-dependent dielectric breakdown is cumulative, so it places a limitation on device lifetime. Typically, TDDB rules are designed with a goal of no more than 10% degradation in drive current after 10 years. As an extremely crude rule of thumb, the ratio of gate voltage to oxide thickness must be kept under approximately 0.5 V/nm to satisfy this requirement.

Recent work suggests that TDDB becomes much less of a problem in extremely thin oxides because any trapped charge is close enough to the gate electrode (or channel) not to stay trapped. It appears that the 5-nm and thinner oxides that are now becoming commonplace are relatively free of this problem. As a consequence, the primary limit on allowed gate voltage is imposed by catastrophic oxide rupture, which generally results in an irreversible gate-to-channel short. As another crude rule of thumb, oxide rupture occurs for gate fields exceeding about 1 V/nm. The portion of the gate oxide near the drain frequently ruptures first in a PA, when the gate is at the minimum potential and the drain is at $2V_{DD}$ (or more, depending on topology and load conditions). In future technologies, it is likely that gate oxide rupture will determine the maximum allowable supply voltage in a PA.

Bipolar Devices

Bipolar transistors have no gate oxide to rupture, but junction breakdown and base punchthrough impose important limits on allowable supply voltages. The collector–base junction can undergo avalanche breakdown in which fields are sufficiently high to cause significant hole–electron pair generation and multiplication. In well-designed devices, this mechanism imposes the more serious constraint, although the extremely thin bases that are characteristic of high-f_T devices can often cause base punchthrough to be important as well.

Another, more subtle, problem that can plague bipolar devices is associated with irreducible terminal inductances that act in concert with large di/dt values. When turning off the device, significant base current can flow in the reverse direction until the base charge is pumped out. When the base charge is gone, the base current abruptly ceases to flow, and the large di/dt can cause large reverse voltage spikes across base to emitter. Recall that the base–emitter junction has a relatively low reverse breakdown voltage (e.g. 6–7 V, although some power devices exhibit significantly larger values), and that the damage from breakdown depends on the energy and is cumulative. Specifically, β degrades (and the device also gets noisier). Hence, gain decreases, possibly causing incorrect bias, and the spectrum of the output can show an increase in distortion products as well as a steadily worsening noise floor. In performing simulations of power amplifiers, it is therefore important to look specifically for this effect and take corrective action if necessary.[55] Options include clamping diodes connected across the device (perhaps integral with the device itself to reduce inductances between the clamp diode and the output transistor), or simply reducing $L\, di/dt$ through improved layout or better drive control.

It is possible (but rare) for a similar phenomenon to occur in MOS implementations. As the gate drive diminishes during turn-off, the gate capacitance drops abruptly once the gate voltage goes below the threshold. Again, the $L\, di/dt$ spike may be large enough to harm the device.

15.10.4 THERMAL RUNAWAY

Another problem concerns thermal effects. To achieve high power operation, it is common to use paralleled devices. In bipolars, the base-emitter voltage for a constant collector current has a temperature coefficient of about -2 mV/°C. Therefore, as a device gets hotter, it requires less drive to maintain a specified collector current. Thus, for a fixed drive, the collector current increases dramatically as temperature increases.

Now consider what happens in a parallel connection of bipolars if one device happens to get a little hotter than the others. As its temperature increases, the collector current increases. The device gets hotter still, steals more current, and so on. This thermoelectric positive feedback loop can run out of control if the loop transmission exceeds unity, resulting in rapid device destruction. To solve the problem, some small resistive degeneration in each transistor's emitter leg is extremely helpful. This way, as the collector current tries to increase in any one device, its base-emitter voltage decreases – offsetting the negative TC – and thermal runaway is avoided. Many manufacturers integrate such degeneration (often known as *ballasting*) into the device structure so that none has to be added externally. Even so, it is not uncommon to observe temperature differences of 10°C or more in high-power amplifiers because of this positive feedback mechanism.

Thermal runaway is normally not a problem in MOS implementations because mobility degradation with increasing temperature causes drain current to diminish, rather than increase, with a fixed gate–source drive. A subtle exception can occur if feedback control is used to force the gate–source voltage to increase with temperature to maintain a constant drive current. In that case, device losses increase with temperature, reviving the possibility of a thermal runaway scenario.

For either bipolar or MOS PAs, it is often prudent to include some form of thermal protection to guard against overload. Fortunately, it is trivial in IC implementations to measure temperature with an on-chip thermometer and arrange to reduce device drive accordingly.

---

[55] Again, it is important to have a good model that is based on the actual physical structure.

15.10.5 LARGE-SIGNAL IMPEDANCE MATCHING

The maximum power transfer theorem is useless in designing the output circuit of a power amplifier, but it does have a role in designing the input circuit. Although we have considered MOS implementations almost exclusively, it is worth mentioning that driving bipolar transistors at the large signal levels found in PAs presents a serious challenge if a decent impedance match to the driving source is to be obtained. Since the base–emitter junction is, after all, a diode, the input impedance is highly nonlinear. Recognizing this difficulty, manufacturers of bipolar transistors often specify the input impedance at a specified power level and frequency. However, since there is generally no reliable guide as to how it might change with power level or other operating conditions, and since it is not even guaranteed at any set of conditions, the designer is left with limited design choices.[56] The traditional way of solving this problem is to swamp out the nonlinearity with a small-valued resistor connected from base to emitter. If the resistor is small enough, its resistance dominates the input impedance. In high-power designs, this resistor can be as small as a few ohms or less, so this gives you an idea of how big the problem is. In general, bipolar power amplifiers are "twitchier" than other types because of this input nonlinearity as well as output saturation effects. MOSFETs present different challenges, but they are usually easier to manage.

The bottom line: If you want to optimize a design, expect to do a fair amount of cut-and-try. Bipolar Class C amplifiers generally require the most iterations; other types, fewer.

Finally, statutory requirements on spectral purity cannot always be satisfied with simple output structures such as the single tanks used in these examples. Additional filter sections usually have to be cascaded to guarantee acceptably low distortion. Unfortunately, every filter inevitably adds some loss. In this context, it is important to keep in mind that just 1 dB of attenuation represents a whopping 21% loss. Assiduous attention to managing all sources of loss is therefore required to keep efficiency high.

15.10.6 LOAD-PULL CHARACTERIZATION OF PAs

All of the examples we've considered so far assumed a purely resistive 50-Ω load. Unfortunately, real loads are rarely purely resistive, except perhaps by accident. Antennas in particular hardly ever present their nominal load to a power amplifier, because their impedance is influenced by such uncontrolled variables as proximity to other objects (e.g., a human head in cell-phone applications).

To explore the effect of a variable load impedance on power delivered, one may systematically vary the real and imaginary parts of the load impedance and plot contours

---

[56] This despite helpful applications notes from device manufacturers with optimistic titles such as "Systematic methods make Class C amplifier design a snap."

of constant output power in the impedance plane (or, equivalently, on a Smith chart). The resulting contours are collectively referred to as a *load-pull* diagram.

The approximate shape of a load-pull diagram may be derived by continuing to assume that the output transistor behaves as a perfect controlled-current source throughout its swing. The derivation that follows is adapted from the classic paper by S. L. Cripps,[57] who first applied it to GaAs PAs.

Assume that the amplifier operates in Class A mode. Then the load resistance is related to the supply voltage and peak drain current as follows:

$$R_{\text{opt}} \equiv \frac{2V_{DD}}{I_{D,\text{pk}}}, \tag{64}$$

with an associated output power of

$$P_{\text{opt}} \equiv \left[\tfrac{1}{2}I_{D,\text{pk}}\right]^2 R_{\text{opt}}. \tag{65}$$

Now, if the magnitude of the load impedance is less than this value of resistance, the output power is limited by the current $I_{D,\text{pk}}$. The power delivered to a load in this current-limited regime is therefore simply:

$$P_L = \left[\tfrac{1}{2}I_{D,\text{pk}}\right]^2 R_L, \tag{66}$$

where R_L is the resistive component of the load impedance.

The peak drain voltage is the product of the peak current and the magnitude of the load impedance:

$$V_{\text{pk}} = I_{D,\text{pk}} \cdot \sqrt{R_L^2 + X_L^2}. \tag{67}$$

Substituting for the peak drain current from Eqn. 64 yields

$$V_{\text{pk}} = \frac{2V_{DD}}{R_{\text{opt}}} \cdot \sqrt{R_L^2 + X_L^2}. \tag{68}$$

To maintain linear operation, the value of V_{pk} must not exceed $2V_{DD}$. This requirement constrains the magnitude of the reactive load component:

$$|X_L|^2 \leq (R_{\text{opt}}^2 - R_L^2). \tag{69}$$

The interpretation of the preceding sequence of equations is as follows: For load impedance magnitudes smaller than R_{opt}, the peak output current limits the power; contours of constant output power are lines of constant resistance R_L in the impedance plane, up to the reactance limit in Eqn. 69.

If the load impedance magnitude exceeds R_{opt} then the power delivered is constrained by the supply voltage. In this voltage-swing–limited regime, it is more convenient to consider a load admittance, rather than load impedance, so that the power delivered is

[57] "A Theory for the Prediction of GaAs FET Load-Pull Power Contours," *IEEE MTT-S Digest*, 1983, pp. 221–3.

$$P_L = \left[\frac{V_{DD}}{2}\right]^2 G_L, \tag{70}$$

where G_L is the conductance term of the output load admittance.

Following a method analogous to the previous case, we compute the drain current as

$$i_D = 2V_{DD}\sqrt{G_L^2 + B_L^2}, \tag{71}$$

where B_L is the susceptance term of the output load admittance. The maximum value that the drain current in Eqn. 71 may have is

$$i_{D,\mathrm{pk}} = 2V_{DD}G_{\mathrm{opt}}. \tag{72}$$

Substituting Eqn. 72 into Eqn. 71 and solving the inequality yields

$$|B_L|^2 \leq (G_{\mathrm{opt}}^2 - G_L^2). \tag{73}$$

The interpretation of the foregoing equations is that, for load impedance magnitudes larger than R_{opt}, contours of constant power are lines of constant conductance G_L, up to the susceptance value given by Eqn. 73. The contours for the two impedance regimes together comprise the load-pull diagram.

Load-pull contours are valuable for assessing the sensitivity of a PA to load variations, identifying optimum operating conditions, and possibly revealing hidden vulnerabilities. Experienced PA designers often acquire the ability to diagnose a number of pathologies by inspection of a load-pull contour.

15.10.7 LOAD-PULL CONTOUR EXAMPLE

To illustrate the procedure, let's construct a load-pull diagram for the earlier Class A amplifier example, in which the peak voltage is 6.6 V and the peak current is 1.65 A, leading to a 4-Ω R_{opt}. In order to find the locus of all load admittances (impedances) that allow us to deliver power within, say, 1 dB of the optimum design value, we first compute that a 1-dB deviation from 4 Ω corresponds to about 3.2 Ω and 5.0 Ω. The former value is used in the current-limited regime, and the latter in the voltage-swing–limited regime.

In the current-limited regime we follow the 3.2-Ω constant-resistance line up to the maximum allowable reactance magnitude of about 2.6 Ω, whereas in the swing-limited regime we follow the constant-conductance line of 0.2 S up to the maximum allowable susceptance magnitude of 0.15 S.

Rather than plotting the contours in the impedance and admittance planes, it is customary to plot the diagram in Smith-chart form. Since circles in the impedance or admittance plane remain circles in the Smith chart (and lines are considered to be circles of infinite radius), the finite-length lines of these contours become circular arcs in the Smith chart. The corresponding diagram appears as Figure 15.41, here normalized to 5 Ω and 0.2 S (instead of 50 Ω and 0.02 S) in order to make the contour big enough to see clearly.

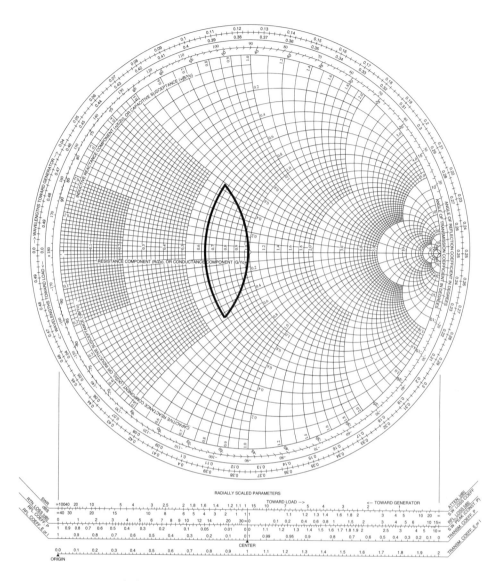

FIGURE 15.41. 1-dB load-pull contour (normalized to 5 Ω) for Class A amplifier example.

The power delivered to a load will therefore be within 1 dB of the maximum value for all load impedances lying inside the intersection of two circles: one of constant resistance (whose value is 1 dB less than the optimum load resistance) and the other of constant conductance (whose value is 1 dB less than the optimum load conductance). Note from this description that one need not compute the reactance or susceptance magnitude limits; the intersection of the two circles automatically takes care of this computation graphically. Hence, construction of theoretical load-pull diagrams is considerably easier than the detailed derivations might imply.

It should be emphasized that the foregoing development assumes that the transistor behaves as an ideal, parasitic-free, controlled current source. Device and packaging parasitics, combined with an external load impedance, comprise the total effective load for the diagram. In constructing practical load-pull diagrams, however, one has knowledge only of the externally imposed impedance. Hence, load-pull contours based on the external impedance values will generally be translated and rotated relative to the parasitic-free case. This minor inconvenience may be exploited to help extract the precise value of these parasitics, since one knows that the center of the contour (after correction for the parasitics) must pass through the real axis of the Smith chart.

15.11 DESIGN SUMMARY

We have seen numerous examples that trade efficiency for linearity and other characteristics. As a consequence, one type of power amplifier cannot satisfy all possible requirements.

The various amplifier topologies that have evolved to accommodate the broad range of requirements span the gamut from poor efficiency and high linearity to high efficiency and low linearity (alas, the highly efficient–highly linear amplifier has yet to be invented). Class A amplifiers provide the best linearity and worst efficiency, whereas switching amplifiers offer the best efficiency and worst linearity.

The maximum power transfer theorem was seen to be largely useless in the design of the output circuit of a PA, although useful in designing the input circuit. Instead, one typically designs to supply a specified amount of power to the load – evaluating later whether efficiency, gain, linearity, and robustness are acceptable (and iterating as necessary). Finally, the need to determine sensitivity to real-world load impedance variation led to the development of the load-pull diagram, which allows one to assess rapidly how the power delivered degrades as the real and imaginary parts of the load vary.

PROBLEM SET FOR POWER AMPLIFIERS

PROBLEM 1 Consider the problem of determining the required device width for a power amp stage. Specifically, suppose that you are to design a device for use in the 1-W Class A amplifier example in this chapter. Recall that the device's maximum allowed resistance in the conducting state was estimated at around 200 mΩ.

(a) Derive a general expression for the required width, given an arbitrary resistance specification. Use the analytical MOSFET models from Chapter 5 that take short-channel effects into account.

(b) With your answer to (a) as a guide, estimate the specific width necessary to satisfy the 200-mΩ constraint if the maximum gate drive is 3.3 V and the threshold voltage is 0.7 V. Assume an effective channel length of 0.35 μm, a C_{ox} of

3.85 mF/m^2, a mobility of 0.05 m^2/V-s, and an E_{sat} of 4×10^6 V/m. Verify your answer with the level-3 NMOS device models from Chapter 5 and iterate if necessary to find an accurate value.

(c) If the previous stage is inductively loaded, the peak drive voltage could potentially double. What device width would be sufficient in such a case?

(d) Approximately speaking, how does the required device width scale with supply voltage if output power and losses due to transistor resistance are to remain constant? Assume a fixed device technology. Comment qualitatively on how your answer would change if the effective channel length and oxide thickness both scaled with supply voltage.

PROBLEM 2   Consider a 500-μm×0.5-μm (drawn) transistor used as a power amplifier in which the drain is allowed to swing from ground to 5 V. Using the level-3 models from Chapter 5, plot C_{gd} and C_{db} as a function of drain voltage over this range for a gate voltage of 0 V, 2.5 V, and 5 V. Explain how these variations might affect the performance of a power amplifier.

PROBLEM 3

(a) Complete the "more complete" 1-W Class A amplifier example of Figure 15.37. Assume that the mirror ratio n is 10, so that the reference bias current is 82.5 mA (neglecting mirror errors). Estimate the specific width necessary to satisfy the 200-mΩ constraint if the maximum gate drive is 6.6 V and the threshold voltage is 0.7 V. Assume an effective channel length of 0.35 μm, a C_{ox} of 3.85 mF/m^2, a mobility of 0.05 m^2/V-s, and an E_{sat} of 4×10^6 V/m. Finally, choose the input coupling capacitor so that the voltage attenuation across it is less than 0.3 dB.

(b) Simulate the design using the level-3 models from Chapter 5. What input signal amplitude is necessary to deliver 1 W to the load?

(c) To what *power* does the voltage in (b) correspond? What is the power gain of this amplifier?

PROBLEM 4 Load-pull experiments are one important way to characterize real power amplifiers. The derivation given in the chapter ignores parasitics of the device and of the packaging, so the sample contour shown applies to the impedance of the external load *combined* with these parasitics. Since real, packaged power amplifiers have nonzero parasitics, load-pull contours based on the *external* load impedance do not look quite like the example given.

(a) Explore this concept further by re-deriving the load-pull contour construction rules if the device and packaging parasitics may be modeled as a shunt admittance $Y = G + jB$. *Hint:* Converting to an equivalent series impedance may be helpful for half of the contour.

(b) As a specific example, suppose that – for the 1-W, 1-GHz Class A PA example – the combination of device and package parasitics may be modeled as shown in Figure 15.42. In this model, the capacitance represents the output capacitance of the device (which, in turn, is modeled as the current source shown), while the inductance represents bondwire parasitics between the package pin and the device itself.

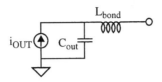

FIGURE 15.42. Simplified
output model.

Suppose that C_{out} is 2 pF and the bondwire inductance is 2 nH. Re-draw the load-pull contour. As with the original example, normalize to 5 Ω to keep the contour comfortably large.

PROBLEM 5   In the Class C PA example, the zero bias voltage is established with a triode-connected FET. Here we consider varying the bias to values other than zero.

(a) Re-design the bias network to allow an external bias source to vary the gate bias from 0 V to increasingly negative values.
(b) Plot the conduction angle, efficiency, *voltage* gain, and drain supply power versus this bias voltage. Comment on how these parameters trade off among each other as the conduction angle varies.

PROBLEM 6   Low distortion can be extremely important in certain applications, depending on the type of modulation used. Class A or AB amplifiers are usually used in such cases. For the 1-W Class A PA example, assume that the output device is 4 mm wide and is built out of the technology modeled with the level-3 parameters given in Chapter 5.

(a) Plot the output power as a function of the amplitude of the gate drive voltage. To simplify the simulations you may, if you so desire, drive the transistor with an ideal sinusoidal voltage source with an appropriate DC offset to set the bias correctly.
(b) What is the 1-dB output-referred compression point?
(c) What is the voltage gain under the conditions in (b)? Compare with the maximum gain and comment.

PROBLEM 7   Another measure of amplifier linearity is the harmonic content of the output.

(a) For the circuit of Problem 6, what are the second and third harmonic components of the output at the 1-dB output compression point?
(b) Repeat part (a), but at half the output power.
(c) Repeat parts (a) and (b) if the output filter inductance has a Q of 10, modeled as a resistance in series with the inductance.

PROBLEM 8　In FM applications, gain linearity is relatively unimportant, but phase linearity is extremely critical because information is encoded as the timing of the zero crossings. If the phase response is highly nonlinear, distortion of the modulation results. Hence, aside from the usual preoccupations with gain and efficiency, the designer of FM systems must also worry about phase distortion.

(a) Re-design the filter in the 1-W Class C amplifier example for maximally flat time delay (you may wish to review the material on this topic from Chapter 9). What Q for the output tank yields this condition?

(b) Complete the design by selecting the width of the output device sufficient to guarantee a maximum resistance of 200 mΩ in the conducting state (with peak voltage of 6.6 on the gate). Use the level-3 models from Chapter 5.

(c) Simulate your design with SPICE, driving the output transistor with an ideal square-wave voltage source with an amplitude of 6.6 V. Over what frequency range does this amplifier provide a time-delay constant to within 10%?

PROBLEM 9　As supply voltages diminish, it becomes increasingly difficult to deliver reasonable amounts of power to a load efficiently, because the load impedance must be transformed to ever-lower values and so tolerable parasitic resistances thus scale at the same time. If an ordinary single-ended stage is used, the required device width can quickly grow to absurd values.

One way to forestall the day of reckoning is to employ a bridged (differential) output stage in which the load is connected between two amplifiers driven out of phase. That way, the voltage swing across the load doubles, and the power delivered to the load may thus quadruple for a given supply voltage.

(a) If the width of the output transistor in a single-ended PA is W, how wide should each device in a differential PA be made in order to maintain equal total on-state losses?

(b) Suppose you must deliver 1 W of power to a 50-Ω load with a 1.5-V supply. What device width would you need to use in a single-ended Class A design to keep the on-state resistance equal to 5% of the load impedance seen by the transistor? Use the level-3 device models from Chapter 5.

(c) Combine your answers to (a) and (b) to estimate the width of each device for a differential design.

PROBLEM 10　Crossover distortion is a problem of push–pull amplifiers, many of which are biased as Class AB stages. This problem explores some approximate ways to predict the crossover distortion of such amplifiers.

Assume that an amplifier with crossover distortion produces no output until the magnitude of the input signal exceeds some critical threshold. Beyond that threshold, the output follows the input, but with an offset. Specifically, assume that the amplifier may be modeled as a simple black box whose output, given a sinusoidal input drive, behaves as follows: $v_{out} = v_{in} - \varepsilon$ for $v_{in} > \varepsilon$; $v_{out} = 0$ for $|v_{in}| < \varepsilon$; and $v_{out} = v_{in} + \varepsilon$ for $v_{in} < -\varepsilon$. Assume that $0 < \varepsilon < 1$.

What values of ε yield a third harmonic component whose amplitude is 0.1%, 1%, and 10% of the fundamental?

PROBLEM 11 In all of the examples given in the chapter, only single-transistor output stages are considered. In general, however, one must cascade several stages in order to obtain sufficient overall power gain. If linearity is important, it is customary for the early stages to be operated as Class A amplifiers and the last to operate perhaps as a Class AB stage (for efficiency).

Expand on the "more complete" 1-W, 1-GHz Class A amplifier of Figure 15.37 by preceding it with enough additional gain stages so that 1 W is delivered to the output load with a 1-mW overall input. Perhaps the first step in carrying out the design, then, is to size the output transistor. Design the interstage coupling networks so as to maximize gain (another choice would be to maximize linearity, but this is an extremely involved, iterative process). Use the level-3 device models from Chapter 5.

What is the overall efficiency of the amplifier – measured here as the ratio of the power delivered to the load – to the total power supplied by the DC source to all stages?

CHAPTER SIXTEEN

PHASE-LOCKED LOOPS

16.1 INTRODUCTION

Phase-locked loops (PLLs) have become ubiquitous in modern communications systems because of their remarkable versatility. As one important example, a PLL may be used to generate an output signal whose frequency is a programmable, rational multiple of a fixed input frequency. The output of such *frequency synthesizers* may be used as the local oscillator signal in superheterodyne transceivers. Phase-locked loops may also be used to perform frequency modulation and demodulation, as well as to regenerate the carrier from an input signal in which the carrier has been suppressed. Their versatility extends to purely digital systems as well, where PLLs are indispensable in skew compensation, clock recovery, and the generation of clock signals.

To understand in detail how PLLs may perform such a vast array of functions, we will need to develop linearized models of these feedback systems. But first, of course, we begin with a little history to put this subject in its proper context.

16.2 A SHORT HISTORY OF PLLs

The earliest description of what is now known as a PLL was provided by H. de Bellescize in 1932.[1] This early work offered an alternative architecture for receiving and demodulating AM signals, using the degenerate case of a superheterodyne receiver in which the intermediate frequency is zero. With this choice, there is no image to reject, and all processing downstream of the frequency conversion takes place in the audio range.

To function correctly, however, the *homodyne* or *direct-conversion* receiver requires a local oscillator (LO) whose frequency is *precisely* the same as that of the incoming carrier. Furthermore, the local oscillator must be in phase with the incoming carrier for maximum output. If the phase relationship is uncontrolled, the gain

---

[1] "La Réception Synchrone," *L'Onde Électrique,* v. 11, June 1932, pp. 230–40.

could be as small as zero (as in the case where the LO happens to be in quadrature with the carrier), or vary in some irritating manner. De Bellescize described a way to solve this problem by providing a local oscillator whose phase is locked to that of the carrier.

For various reasons, the homodyne receiver did not displace the ordinary super-heterodyne receiver, which had come to dominate the radio market by about 1930. However, there has recently been a renewal of interest in the homodyne architecture because its relaxed filtering requirements possibly improve amenability to integration.[2]

The next PLL-like circuit to appear was used in televisions for over three decades. In standard broadcast television, two sawtooth generators provide the vertical and horizontal deflection ("sweep") signals. To allow the receiver to synchronize the sweep signals with those at the studio, timing pulses are transmitted along with the audio and video signals.

To perform synchronization in older sets, the TV's sweep oscillators were adjusted to free-run at a somewhat lower frequency than the actual transmitted sweep rate. In a technique known as *injection locking,*[3] the timing pulses caused the sawtooth oscillators to terminate each cycle prematurely, thereby effecting synchronization. As long as the received signal had relatively little noise, the synchronization worked well. However, as signal-to-noise ratio degraded, synchronization suffered either as timing pulses disappeared or as noise was misinterpreted as timing pulses. In the days when such circuits were the norm, every TV set had to have vertical and horizontal "hold" controls to allow the consumer to fiddle with the free-running frequency and, therefore, the quality of the lock achieved. Improper adjustment caused vertical rolling or horizontal "tearing" of the picture. In modern TVs, true PLLs are used to extract the synchronizing information robustly even when the signal-to-noise ratio has degraded severely. As a result, vertical and horizontal hold adjustments thankfully have all but disappeared.

The next wide application of a PLL-like circuit was also in televisions. When various color television systems were being considered in the late 1940s and early 1950s, the Federal Communications Commission (FCC) imposed a requirement of compatibility with the existing black-and-white standard, and further decreed that

---

[2] However, the homodyne requires exceptional front-end linearity and is intolerant of DC offsets. Furthermore, since the RF and LO frequencies are the same, LO leakage back out of the antenna is a problem. Additionally, this LO leakage can sneak back into the front end, where it mixes with the LO with some random phase, resulting in a varying DC offset that can be several orders of magnitude larger than the RF signal. These problems are perhaps as difficult to solve as the filtering problem, and are considered in greater detail in Chapter 19.

[3] See Balth. van der Pol, "Forced Oscillations in a Circuit with Nonlinear Resistance (Reception with Reactive Triode)," *Philosophical Magazine,* v. 3, January 1927, pp. 65–80, and also see R. B. Adler, "A Study of Locking Phenomena in Oscillators," *Proc. IRE,* v. 34, June 1946, pp. 351–7. The circadian rhythms of humans provide another example of injection locking. In the absence of a synchronizing signal from the sun, a "day" for most people exceeds 24 hours. Note that the free-running frequency is again somewhat lower than the locked frequency.

the color television signal could not require any additional bandwidth. Since mono-chrome television had been developed without looking forward to a colorful future, it was decidedly nontrivial to satisfy these constraining requirements. In particu-lar, it seemed impossible to squeeze a color TV signal into the same spectrum as a monochrome signal without degrading something. The breakthrough was in rec-ognizing that the 30-Hz frame rate of television results in a comblike – rather than continuous – spectrum, with peaks spaced 30 Hz apart. Color information could then be shoehorned in between these peaks without requiring additional bandwidth. To accomplish this remarkable feat, the added color information is modulated on a *sub-carrier* of approximately 3.58 MHz.[4] The subcarrier frequency is carefully chosen so that the sidebands of the chroma signal fall precisely midway between the spec-tral peaks of the monochrome signal. The combined monochrome (also known as the "luminance" or "brightness" signal) and chroma signals subsequently modulate the final carrier that is ultimately transmitted. The U.S. version of this scheme is known as NTSC (for National Television Systems Committee).

Color information is encoded as a vector whose phase with respect to the subcar-rier determines the hue and whose magnitude determines the amplitude ("saturation") of the color. The receiver must therefore extract or regenerate the subcarrier quite accurately to preserve the 0° phase reference; otherwise, the reproduced colors will not match those transmitted.

To enable this phase locking, the video signal includes a "burst" of a number of cycles (NTSC specifications dictate a minimum of 8) of a 3.58-MHz reference oscil-lation transmitted during the retrace of the CRT's electron beam as it returns to the left side of the screen. This burst signal feeds a circuit inside the receiver whose job is to regenerate a continuous 3.58-MHz subcarrier that is phase-locked to this burst. Since the burst is not applied continuously, the receiver's oscillator must free-run during the scan across a line. To prevent color shifts, the phase of this regenerated subcarrier must not drift. Early implementations did not always accomplish this goal successfully, leading some wags to dub NTSC "never twice the same color."[5]

Europe (with the exception of France[6]) chose to adopt a similar chroma scheme, but addressed the phase drift problem by alternating the polarity of the reference every line. This way, phase drifts tend to average out to zero over two successive lines, re-ducing or eliminating perceived color shifts. Thus was born the phase-alternating line (PAL) system.

---

[4] If you really want to know, the exact frequency is 3.579545 MHz, derived from the 4.5-MHz spac-ing between the video and audio carrier frequencies multiplied by 455/572.
[5] In fact, the very earliest such circuits dispensed with an oscillator altogether. Instead, the burst signal merely excited a high-Q resonator (a quartz crystal), and the resulting ringing was used as the regenerated subcarrier. The ringing had to persist for over 200 cycles without excessive decay. Cheesy!
[6] The French color television system is known as SECAM, for Séquentiel Couleur avec Mémoire. In this system, luminance and chrominance information are sent serially in time and reconstructed in the receiver.

Another early application of PLL-like circuits was in stereo FM radio. Again, to preserve backward compatibility, the stereo information is encoded on a subcarrier, this time at 38 kHz. Treating the monaural signal as the sum of a left and right channel (and bandlimited to 15 kHz), stereo broadcast is enabled by modulating the subcarrier with the *difference* between the left and right channels. This L−R difference signal is encoded as a double-sideband, suppressed-carrier (DSB-SC) signal. The receiver then regenerates the 38-kHz subcarrier, and recovers the individual left and right signals through simple addition and subtraction of the L+R monaural and L−R difference signals. To simplify receiver design, the transmitted signal includes a low-amplitude *pilot* signal at precisely half the subcarrier frequency, which is then doubled at the receiver and used to demodulate the L−R signal. As we'll see shortly, a PLL can easily perform this frequency-doubling function even without a pilot, but for the circuits of 1960, it was a tremendous help.

Early PLLs were mainly of the injection-locked variety because the cost of a complete, textbook PLL was too great for most consumer applications. Except for a few exotic situations, such as satellite communications and scientific instrumentation, such "pure" PLLs didn't exist in significant numbers until the 1970s, when IC technology had advanced enough to provide a true PLL for stereo FM demodulation. Since then, the PLL has become commonplace, found in systems ranging from the mundane to the highly specialized.

From the foregoing, it should be clear that phase locking enables a rich variety of applications. With that background as motivation, we begin with a brief look at injection locking in advance of a description of "textbook" PLLs.

INJECTION LOCKING

Even though injection locking is ancient history, don't take that to mean that it is no longer useful. As we'll see, injection-locked systems can actually provide many of the same benefits as complete phase-locked loops, but with much lower complexity and at lower power in many cases.[7] For this reason, it is worthwhile spending a little bit of time examining them in at least some detail.

A simple model that captures the essential features of injection locking consists of a nonlinear element in cascade with a *mode selector* (filter), all in a feedback loop; see Figure 16.1. Note that this model is fairly general; with appropriate choices for the type of nonlinearity and the characteristics of the mode selector, this model could describe an amplifier, a free-running oscillator, a frequency multiplier/divider, or a synchronized (injection-locked) oscillator. This versatility is the reason for the longevity of injection locking.

Assume that, in the absence of an input, the system nonetheless oscillates (not all systems satisfy, or need to satisfy, this criterion; it's just a convenient assumption

---

[7] Unwanted injection locking can also occur (e.g., from coupling through the substrate), so understanding the underlying theory is important for many reasons.

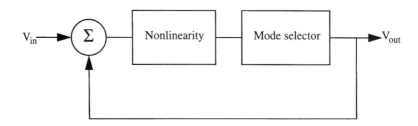

FIGURE 16.1. Model for injection-locked oscillators.

here). If it is to continue oscillating once an input is supplied, the loop must continue to satisfy the Barkhausen oscillation criteria of unit loop transmission magnitude and net zero phase shift around the loop.

The mode selection filter acts on the spectrally rich output of the nonlinearity and selects only the desired term. In turn, that surviving mode interacts nonlinearly with the applied input, thus closing the loop. Synchronization occurs if a self-consistent solution to the Barkhausen criteria exists.

At this point, an example may help to clarify things a bit. Suppose our goal is to design an injection locked oscillator (ILO) intended to act as a frequency divider. Specifically, suppose we wish to design a circuit that divides frequency by a factor of 2. In the language of ILOs, we seek to design a *second (super)harmonic* ILO.[8] If the output frequency is f_{out}, then the mode filter should be a bandpass filter tuned to this frequency. Next, we need a nonlinear element that will produce a spectral component at f_{out}, when the input frequencies are f_{out} and $f_{in} = 2f_{out}$. That is, we want a nonlinearity that produces a difference frequency component. We know from our study of mixers that a second-order nonlinearity will provide mixing action (imperfectly, but that's of no consequence here), so if we can arrange to inject the synchronizing input signal at a point in the circuit that excites a second-order nonlinearity, then injection locking might be achievable – provided that we satisfy the gain and phase criteria around the loop.

An example of a circuit that acts on all of these observations is depicted in Figure 16.2.[9] Here, transistors M_1 and M_2 form the core of a free-running oscillator. (We say much more about this type of circuit in the separate chapter on oscillators.) The synchronizing input signal is applied to node V_x through a differential amplifier (M_3–M_4). Because the oscillator's output signal appears at the gate of M_1 and M_2, the gate–source voltage of M_1 (say) is the sum of the synchronizing voltage and the

---

[8] By convention, nomenclature in ILOs is based on the integer ratio of input frequency to output frequency (precisely backwards from the general characterization by ratios of outputs to inputs). Thus, when the input frequency is lower than the output frequency, the ILO is called a *subharmonic* ILO. When the input frequency is higher than the output frequency, the ILO is called a *superharmonic* (or just plain *harmonic*) ILO.

[9] H. R. Rategh and T. H. Lee, *Multi-GHz Frequency Synthesis and Division,* Kluwer, Dordrecht, 2001.

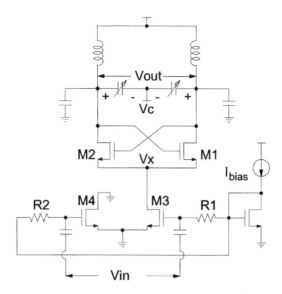

FIGURE 16.2. Injection-locked frequency divider
(after Rategh).

oscillator voltage. The square-law nature of MOSFETs then contributes the desired
second-order nonlinearity naturally.

One desirable property of injection locking is made clear from examination of the
performance of this circuit. In a 0.25-μm CMOS technology, the divider consumes
under 500 μW when operating with an input frequency of 5 GHz. This power consumption is considerably less than that of a conventional flip-flop–based divider in
the same technology and is the result of using resonant circuits. Energy is thus re-
circulated to a certain extent – instead of being thrown away every cycle, as in a
conventional flip-flop. The tradeoffs are an increased die area (because of the inductors) and narrower operational frequency range (because the ILO is a tuned circuit).

Another valuable attribute is that injection locking does not merely produce synchronization of frequencies; it can actually produce *phase locking*. Thus, if the synchronizing signal has lower *phase noise* (a topic we cover in detail in Chapter 18)
than the free-running oscillator, then it's possible to transfer that quality to the output of the locked oscillator. This is shown in the plot of Figure 16.3. The horizontal
axis is the offset frequency from the carrier at which the noise measurement in the
vertical axis is obtained.[10]

The curve in the upper right corner is the phase noise of the oscillator in the absence of a synchronizing input. The high noise (resulting from operation at such low
power) in this unlocked state is evident in the plot. When locked, the phase noise is

---

[10] Again, in a later chapter we'll say much more about precisely what phase noise is and what these
plots mean. For now, all you really need to know is that the vertical axis is a measure of noise, so
lower is better.

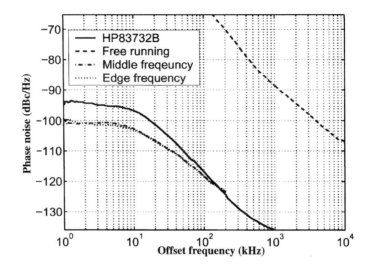

FIGURE 16.3. Phase noise plot of ILO in locked
and unlocked modes.

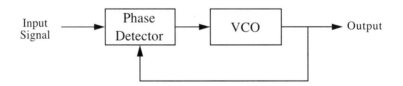

FIGURE 16.4. Phase-locked loop architecture.

as in the lowest curves, with approximately −100 dBc/Hz phase noise at low offset
frequency from the carrier, representing improvements of 60 dB or more. The synchronizing signal in this case is an HP83732B signal generator, whose noise is 6 dB
(a factor of 2) higher than the locked oscillator (at low offset frequencies) because of
the 2 : 1 frequency division; dividing frequency by 2 divides phase by 2 as well.

Having taken a brief look at some of the properties and uses of injection locking,
we now turn to the task of modeling conventional phase-locked loops.

16.3 LINEARIZED PLL MODELS

The basic PLL architecture is shown in Figure 16.4, and is seen to consist of a phase
detector and a voltage controlled oscillator (VCO). The phase detector compares the
phase of an incoming reference signal with that of the VCO, and produces an output that is some function of the phase difference. The VCO simply generates a signal
whose frequency is some function of the control voltage.

The general idea is that the output of the phase detector drives the VCO frequency
in a direction that reduces the phase difference; that is, it's a negative feedback system.

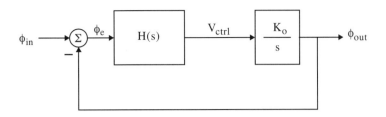

FIGURE 16.5. Linearized PLL model.

Once the loop achieves lock, the phase of the input reference and VCO output signals ideally have a fixed phase relationship (most commonly 0° or 90°, depending on the nature of the phase detector).

Although both the phase detector and VCO may be highly nonlinear in practice, it is customary to assume linearity when analyzing loops that have achieved lock. We will eventually consider a more general case (including the acquisition process), but we have to begin somewhere, and it's best to start simple and add complexity as we go along.

Let us begin with a linearized PLL model, as shown in Figure 16.5. Because we are generally interested in the phase relationship between the input and output signals, the input and output variables are phases in this model, rather than the time waveforms of the actual inputs and outputs. Hence, if you are accustomed to thinking of signals as voltages in a block diagram, the input and output voltages are now proportional to phases.

Another consequence of choosing phase as the input–output variable is that the VCO, whose output frequency depends on a control voltage, is modeled as an integrator, since phase is the integral of frequency. The VCO gain constant K_O has units of radians per second per volt, and merely describes what change in output frequency results from a specified change in control voltage. Also note that, unlike ordinary amplifiers whose outputs are bounded, the VCO is a true integrator. The longer we wait, the more phase we accumulate (unless someone turns off the oscillator).

The phase detector is modeled as a simple subtractor that generates a phase error output ϕ_e that is the difference between the input and output phases. To accommodate gain scaling factors and the option of additional filtering in the loop, a block with transfer function $H(s)$ is included in the model as well.

16.3.1 FIRST-ORDER PLL

The simplest PLL is one in which the function $H(s)$ is simply a scalar gain (call it K_D, with units of volts per radian). Because the loop transmission then possesses just a single pole, this type of loop is known as a first-order PLL. Aside from simplicity, its main attribute is the ease with which large phase margins are obtained.

Offsetting those positive attributes is an important shortcoming, however: bandwidth and steady-state phase error are strongly coupled in this type of loop. Because one generally wants the steady-state phase error to be zero, independent of bandwidth, first-order loops are infrequently used.

We may use our linear PLL model to evaluate quantitatively the limitations of a first-order loop. Specifically, the input–output phase transfer function is readily derived:

$$\frac{\phi_{\text{out}}(s)}{\phi_{\text{in}}(s)} = \frac{K_O K_D}{s + K_O K_D}. \tag{1}$$

The closed-loop bandwidth is therefore

$$\omega_h = K_O K_D. \tag{2}$$

To verify that the bandwidth and phase error are linked, let's now derive the input-to-error transfer function:

$$\frac{\phi_e(s)}{\phi_{\text{in}}(s)} = \frac{s}{s + K_O K_D}. \tag{3}$$

If we assume that the input signal is a constant-frequency sinusoid of frequency ω_i, then the phase ramps linearly with time at a rate of ω_i radians per second. Thus, the Laplace-domain representation of the input signal is

$$\phi_{\text{in}}(s) = \frac{\omega_i}{s^2}, \tag{4}$$

so that

$$\phi_e(s) = \frac{\omega_i}{s(s + K_O K_D)}. \tag{5}$$

The steady-state error with a constant frequency input is therefore

$$\lim_{s \to 0} s\phi_e(s) = \frac{\omega_i}{K_O K_D} = \frac{\omega_i}{\omega_h}. \tag{6}$$

The steady-state phase error is thus simply the ratio of the input frequency to the loop bandwidth; a one-radian phase error results when the loop bandwidth equals the input frequency. A small steady-state phase error therefore requires a large loop bandwidth; the two parameters are tightly linked, as asserted earlier.

An intuitive way to arrive qualitatively at this result is to recognize that, in general, a nonzero voltage is required to drive the VCO to the correct frequency. Since the control voltage derives from the output of the phase detector, there must be a nonzero phase error. To produce a given control voltage with a smaller phase error requires an increase in the gain that relates the control voltage to the phase detector output. Because an increase in gain raises the loop transmission uniformly at all frequencies, a bandwidth increase necessarily accompanies a reduction in phase error.

To produce zero phase error, we require an element that can generate an arbitrary VCO control voltage from a zero phase detector output, implying the need for an infinite gain. To decouple the steady-state error from the bandwidth, however, this element needs to have infinite gain only at DC, rather than at all frequencies. An integrator has the prescribed characteristics, and its use leads to a second-order loop.

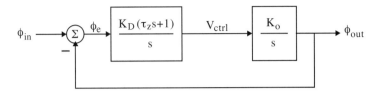

FIGURE 16.6. Model of second-order PLL.

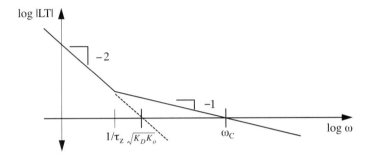

FIGURE 16.7. Loop transmission of second-order PLL.

16.3.2 SECOND-ORDER PLL

The model for a second-order PLL is shown in Figure 16.6. The 90° negative phase shift contributed by the added integrator has to be offset by the positive phase shift of a loop-stabilizing zero. As with any other feedback system compensated in this manner, the zero should be placed well below the crossover frequency to obtain acceptable phase margin.

In this model, the constant K_D has the units of volts per second because of the extra integration. Also thanks to the added integration, the loop bandwidth may be adjusted independently of the steady-state phase error (which is zero here), as is clear from studying the loop transmission magnitude behavior graphed in Figure 16.7. The stability of this loop can be explored with the root-locus diagram of Figure 16.8. As the loop transmission magnitude increases (by increasing $K_D K_O$), the loop become progressively better damped because an increase in crossover frequency allows more of the zero's positive phase shift to offset the negative phase shift of the poles.

For very large loop transmissions, one closed-loop pole ends up at nearly the frequency of the zero, while the other pole heads for infinitely large frequency. In this PLL implementation, the loop-stabilizing zero comes from the forward path; hence, this zero also shows up in the closed-loop transfer function.

It is straightforward to show that the phase transfer function is

$$\frac{\phi_{\text{out}}}{\phi_{\text{in}}} = \frac{\tau_z s + 1}{(s^2/K_D K_O) + \tau_z s + 1},\tag{7}$$

from which we determine that

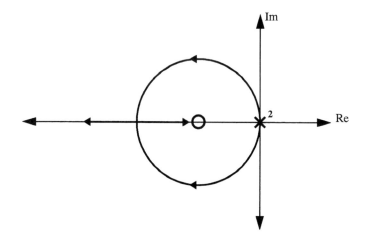

FIGURE 16.8. Root locus of second-order PLL.

$$\omega_n = \sqrt{K_D K_O} \tag{8}$$

and

$$\zeta = \frac{\omega_n \tau_z}{2} = \frac{\tau_z \sqrt{K_D K_O}}{2}. \tag{9}$$

Furthermore, the crossover frequency for the loop may be expressed as

$$\omega_c = \left[\frac{\omega_n^4}{2\omega_z^2} + \omega_n^2 \sqrt{\frac{1}{4}\left(\frac{\omega_n}{\omega_z}\right)^4 + 1} \right]^{1/2}, \tag{10}$$

which simplifies considerably if the crossover frequency is well above the zero frequency, as it often is:

$$\omega_c \approx \frac{\omega_n^2}{\omega_z}. \tag{11}$$

Both Figure 16.7 and Eqn. 10 show that the crossover frequency always exceeds ω_n, which – from Figure 16.7 and Eqn. 8 – is the extrapolated crossover frequency of the loop with no zero. Finally, it should be clear that increasing the zero's time constant improves the damping, given a fixed ω_n. Thus, the bandwidth and stability of a second-order loop may be adjusted as desired while preserving a zero steady-state phase error.

Jitter Peaking in Second-Order PLLs

From the root locus for this loop, we see that the zero is to the right of its associated (closed-loop) pole at larger damping ratios. Hence, the closed-loop frequency response initially exceeds unity until the zero's effect is cancelled by the poles (see Figure 16.9).

We see that there is a rise with frequency, starting at the zero location, then a flattening caused by the first pole. As seen in the figure, the phase transfer function has

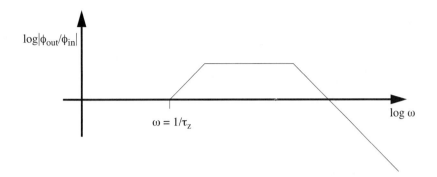

FIGURE 16.9. Closed-loop phase transfer function for PLL (Bode approximation).

a magnitude greater than unity above the zero frequency until the second pole introduces a sufficient rolloff. There is thus a band of frequencies, bounded roughly by the zero location and the second pole, where the magnitude of the transfer function exceeds unity.

The implication of this peaking is that if there is any modulation (intended or otherwise) on the input with spectral components within that certain frequency band, the output modulation will have a phase excursion that exceeds the excursion on the input. Unfortunately, we can see from the locus that such peaking is an inherent property of such loops as long as the zero is contributed by the forward path. Therefore, if this peaking is to be kept to a minimum, we require large loop transmissions to keep the first pole as close to the zero as possible.[11] Although this jitter peaking may be totally eliminated by using a voltage-controlled delay element to provide the loop zero in the feedback path, it is satisfactory in most RF applications simply to choose sufficiently large damping ratios.[12]

16.4 SOME NOISE PROPERTIES OF PLLs

16.4.1 REJECTION OF VCO DISTURBANCES

Aside from the response to the desired input, it is also important in practical systems to evaluate the response to noise (particularly in ICs, where noise generated by other parts of the chip can couple into the PLL). We therefore now examine how the classic PLL behaves in response to noise on the input of the PLL and on the control line of the VCO.

---

[11] The peaking problem is compounded, of course, if there are PLLs in cascade. The overall peaking of a cascade can actually be large enough to cause downstream PLLs to lose lock. While this issue is almost never of concern to RF designers, it can be a significant problem in some digital networks (e.g., token rings).

[12] T. Lee and J. Bulzacchelli, "A 155MHz Clock Recovery Delay- and Phase-Locked Loop," *IEEE J. Solid-State Circuits,* December 1992.

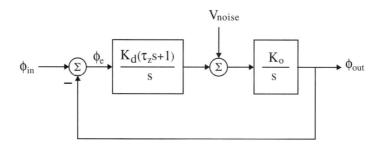

FIGURE 16.10. Linearized PLL model with noise input.

The linear model of Figure 16.10 considers noise as additive at the control port of the VCO. It is a relatively straightforward exercise to show that the noise-to-phase-error transfer function for this system is

$$-\frac{\phi_\varepsilon}{V_N} = \frac{sK_O}{s^2 + s\tau_z K_D K_O + K_D K_O}. \tag{12}$$

Assuming a unit step–function noise input, inverse Laplace-transforming gives us the resulting phase error as a function of time:

$$-\phi_\varepsilon(t) = \frac{\Delta\omega_i}{\omega_n\sqrt{\zeta^2 - 1}} \exp(-\zeta\omega_n t) \sinh\left(\omega_n\sqrt{\zeta^2 - 1}\,t\right), \tag{13}$$

where $\Delta\omega_i$ is the initial frequency error due to the step-function disturbance V_N. The damping ratio ζ of the closed-loop poles is still given by

$$\zeta = \frac{\omega_n \tau_z}{2}, \tag{14}$$

and the natural frequency ω_n remains

$$\omega_n = \sqrt{K_D K_O}. \tag{15}$$

The maximum phase error is given by the following intuitively obvious equation:

$$-\phi_{\varepsilon,\max} = \frac{\Delta\omega_i}{\omega_n\sqrt{\zeta^2 - 1}} \exp\left(-\frac{\zeta}{\sqrt{\zeta^2 - 1}} \tanh^{-1} \frac{\sqrt{\zeta^2 - 1}}{\zeta}\right)$$
$$\cdot \sinh\left(\tanh^{-1} \frac{\sqrt{\zeta^2 - 1}}{\zeta}\right), \tag{16}$$

and occurs at a time

$$t_{\max} = \frac{1}{\omega_n\sqrt{\zeta^2 - 1}} \tanh^{-1} \frac{\sqrt{\zeta^2 - 1}}{\zeta}. \tag{17}$$

Now, while the utility of the foregoing equations may be somewhat elusive owing to their cumbersome nature, these expressions simplify considerably in the limit of high damping:

$$-\phi_{\varepsilon,\max} \approx \frac{\Delta\omega_i}{\omega_c}, \qquad (18)$$

$$t_{\max} \approx \frac{2\ln 2\zeta}{\omega_c}, \qquad (19)$$

where ω_c is the crossover frequency of the loop.

We now have a chance of developing some insight from these equations. We see that the maximum phase error is approximately the ratio of the initial VCO frequency shift to the loop crossover frequency (which is approximately the closed-loop bandwidth). This relationship may be understood intuitively by recalling that phase is the integral of frequency, so that any departure from the correct frequency causes some integration of phase error, and that this accumulation of phase error persists for a time on the order of the reciprocal loop bandwidth.

Depending on the context, this varying phase error is called *jitter* or *phase noise*. From the foregoing equations, it is clear that minimizing jitter or phase noise caused by power supply (or whatever) noise requires maximizing loop bandwidth and minimizing the initial shift in the VCO frequency. Unfortunately, arbitrarily large loop bandwidths are not possible because all practical feedback systems ultimately suffer phase margin degradation from a variety of sources, such as (possibly) poorly modeled parasitic elements. Additionally, many PLLs are sampled-data systems (phase error measurements are made at discrete intervals), and this nature imposes further bounds on the crossover frequency if stability is not to be compromised. These considerations, in conjunction with the need to absorb component tolerances and drifts with temperature and supply voltage, typically force the use of loop bandwidths that are only a small fraction (e.g., <10%) of the clock frequency to guarantee acceptable worst-case phase margins.

At this point, a numerical example may be useful in underscoring the magnitude of the problem. Suppose that, for some value of step disturbance on the power supplies, $\Delta\omega_i$ is 2% of ω_{carrier}. Further assume that the crossover frequency ω_c of the loop is also 2% of ω_{carrier}. In this case, the maximum phase error is one radian, or about 630 ps, with a carrier input of 250 MHz (a period of 4 ns). This magnitude of jitter (over 15%) is generally intolerable.

Viewed from an IC communications systems perspective, the sensitivity of PLLs to external and internal noise sources makes it extremely difficult to merge them with digital circuitry (which tends to generate large amounts of noise with all that switching going on) without compromising the spectral purity of the PLL output. From the equations, we see that the sensitivity to external noise is minimized if we minimize the VCO gain and maximize the loop bandwidth. Minimizing this noise sensitivity remains one of the most significant challenges in the quest for further integration of RF circuitry with digital elements.

16.4.2 REJECTION OF NOISE ON INPUT

We've just seen that maximizing the bandwidth of the PLL helps to minimize the influence of disturbances that alter the VCO frequency. This insight is not too deep –

making a system faster means that it recovers more quickly from errors, whatever the source.

However, there is a potential drawback to maximizing the bandwidth, above and beyond the stability issue. As the loop bandwidth increases, the loop gets better at tracking the input. If the input is noise-free (or at least less noisy than the PLL's own VCO), then there is a net improvement overall. However, if the input signal is *noisier* than the PLL's VCO, then the high-bandwidth loop will faithfully reproduce this input noise at the output. Hence, there is a tradeoff between sensitivity to noise on the input to the loop (a consideration that favors smaller loop bandwidths) and sensitivity to noise that disturbs the VCO frequency (a consideration that we've seen favors larger loop bandwidths).

In general, tuned oscillators (e.g., LC or crystal-based) are inherently less (often much less) noisy, at a given power level, than relaxation oscillators (such as ring or RC phase-shift oscillators). Hence, if the reference input to the PLL is supplied from a tuned oscillator when the VCO is based on a relaxation oscillator topology, larger bandwidths are favored. If, instead, the situation is reversed (a rarer occurrence) and a relaxation oscillator supplies the reference to a crystal oscillator–based PLL, then smaller loop bandwidths will generally be favored.

16.5   PHASE DETECTORS

We've taken a look at the classical phase-locked loop at the block diagram level, with a particular focus on the linear behavior of a second-order loop in lock. We now consider a few implementation details to see how real PLLs are built and how they behave.

In this section, we'll examine several representative phase detectors. In subsequent sections, we will examine one or two types of voltage-controlled oscillators to help develop a feel for how practical loops look, although we will defer a detailed discussion of oscillators to yet a later chapter.

16.5.1   THE ANALOG MULTIPLIER AS A PHASE DETECTOR

In PLLs that have sine-wave inputs and sine-wave VCOs, the most common phase detector by far is the multiplier, often implemented with a Gilbert-type topology. For an ideal multiplier, it isn't too difficult to derive the input–output relationship, as shown in Figure 16.11.

Using some trigonometric identities, we find that the output of the multiplier may be expressed as

$$AB \cos \omega t \cos(\omega t + \phi) = \frac{AB}{2}[\cos(\phi) - \cos(2\omega t + \phi)]. \qquad (20)$$

Note that the output of the multiplier consists of a DC term and a double-frequency term. For phase detector operation, we are interested only in the DC term. Hence, the average output of the phase detector is

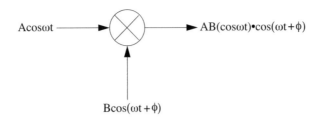

FIGURE 16.11. Multiplier as phase detector.

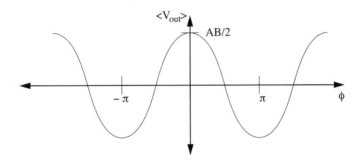

FIGURE 16.12. Multiplier phase detector
output versus phase difference.

$$\langle AB \cos \omega t \cos(\omega t + \phi) \rangle = \frac{AB}{2}[\cos \phi]. \tag{21}$$

We see that the phase detector gain "constant" is a function of the phase angle and is given by

$$K_D = \frac{d}{d\phi}\langle V_{\text{out}} \rangle = -\frac{AB}{2}[\sin(\phi)]. \tag{22}$$

If we plot the average output as a function of phase angle, we get something that looks roughly as shown in Figure 16.12. Notice that the output is periodic. Further note that the phase detector gain constant is zero when the phase difference is zero, and is greatest when the input phase difference is 90°. Hence, to maximize the useful phase detection range, the loop should be arranged to lock to a phase difference of 90°. For this reason, a multiplier is often called a *quadrature* phase detector.

When the loop is locked in quadrature, the phase detector has an incremental gain constant given by:

$$K_D\big|_{\phi=\pi/2} = \frac{d}{d\phi}\langle V_{\text{out}} \rangle\bigg|_{\phi=\pi/2} = -\frac{AB}{2}. \tag{23}$$

In what follows, we will glibly ignore minus signs. The reason for this neglect is that a loop may servo to either a 90° or −90° phase difference (but not to both), depending on the net number of inversions provided by the rest of the loop elements.

Because there are two phase angles (within any given 2π interval) that result in a zero output from the phase detector, there would seem to be two equilibrium points

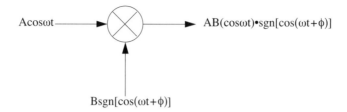

FIGURE 16.13. Multiplier with one square-wave input.

to which the loop could lock. However, one of these points is a stable equilibrium, whereas the other is a *metastable* point from which the loop must eventually diverge. That is, only one of these lock points corresponds to negative feedback.

When speaking of phase errors for a quadrature loop, we calculate the departure from the equilibrium condition of a 90° phase difference. Thus, although the phase difference is 90° in an ideal quadrature loop, the phase *error* is considered to be zero.

16.5.2 THE COMMUTATING MULTIPLIER AS A PHASE DETECTOR

In the previous section, we assumed that both inputs to the loop were sinusoidal. However, one or both of these inputs may be well approximated by a square wave in many cases of practical interest, so let us now modify our results to accommodate a single square-wave input. In this case, we have the situation depicted in Figure 16.13, where "sgn" is the signum function defined as

$$\text{sgn}(x) = 1 \quad \text{if } x > 0, \tag{24}$$

$$\text{sgn}(x) = -1 \quad \text{if } x < 0. \tag{25}$$

Now, recall that a square wave of amplitude B has a fundamental component whose amplitude is $4B/\pi$. If we assume that we care about only the fundamental component of the square wave, then the average output of the multiplier is

$$\langle V_{\text{out}} \rangle = \frac{4}{\pi} \frac{AB}{2} [\cos(\phi)] = \frac{2}{\pi} AB[\cos(\phi)]. \tag{26}$$

The corresponding phase detector gain is similarly just $4/\pi$ times as large as in the purely sinusoidal case:

$$K_D \big|_{\phi=\pi/2} = \frac{d}{d\phi} \langle V_{\text{out}} \rangle \bigg|_{\phi=\pi/2} = -\frac{2AB}{\pi}. \tag{27}$$

Although the expressions for the phase detector output and gain are quite similar to those for the purely sinusoidal case, there is an important qualitative difference between these two detectors. Because the square wave consists of more than just the fundamental component, the loop can actually lock onto harmonics or subharmonics of the input frequency. Consider, for example, the case where the B square-wave input is at precisely one third the frequency of the sinusoidal input frequency. Now, square waves[13] consist of odd harmonics, and the third harmonic will then be at the same frequency as the input sine wave. Those two signals will provide a DC output from the multiplier.

Because the spectrum of a square wave drops off as $1/f$,[14] the average output gets progressively smaller as we attempt to lock to higher and higher harmonics. The attendant reduction in phase detector gain constant thus makes it more difficult to achieve or maintain lock at the higher harmonics, but this issue must be addressed in all practical loops that use this type of detector. Sometimes harmonic locking is desirable, and sometimes it isn't. If it isn't, then the VCO frequency range usually has to be restricted (or acquisition carefully managed) in order to prevent harmonic locking.

Another observation worth making is that multiplication of a signal by a periodic signum function is equivalent to inverting the phase of the signal periodically. Hence, a multiplier used this way can be replaced by switches (also known as "commutators," by analogy with a component of rotating machines). Because switches are easier to implement in some technologies (such as CMOS) than are Gilbert multipliers, this observation can lead directly to simplified circuitry. Even if Gilbert-type multipliers are used, they are often driven with large enough signals on one port that they behave as polarity switches to a good approximation.

16.5.3 THE EXCLUSIVE-OR GATE AS A PHASE DETECTOR

If we now drive an analog multiplier with square waves on *both* inputs, we could analyze the situation by using the Fourier series for each of the inputs, multiplying them together, and so forth. However, it turns out that analyzing this particular situation in the *time* domain is much easier, so that's what we'll do. The reader is welcome (indeed, encouraged) to explore the alternative method and perform the analysis in the frequency domain as a recreational exercise.

In this case, the two square-wave inputs produce the output shown in Figure 16.14. As we change the input phase difference, the output takes the form of a square wave of varying duty cycle, with a 50% duty cycle corresponding to a quadrature relationship between the inputs. Since the duty cycle is in fact proportional to the input phase difference, we can readily produce a plot (Figure 16.15) of the average output as a function of the input phase difference.

---

[13] We are implicitly assuming that the square waves are of 50% duty cycle. Asymmetrical square waves will also contain even as well as odd harmonic components, providing an "opportunity" to lock to even multiples of the incoming reference in addition to odd multiples.

[14] Here's another fun piece of trivia with which to amaze party guests: In general, the spectrum of a signal will decay as $1/f^n$, where n is the number of derivatives of the signal required to yield an impulse. Hence, the spectrum of an ideal sine wave has an infinitely fast rolloff (since no number of derivatives ever yields an impulse), that of an impulse doesn't roll off (since $n = 0$), that of a square wave rolls off as $1/f$, that of a triangle wave as $1/f^2$, and so on.

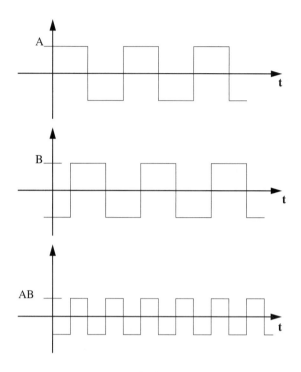

FIGURE 16.14. Multiplier inputs and output.

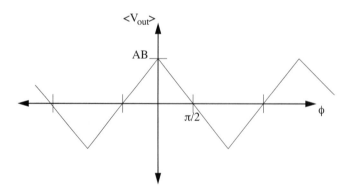

FIGURE 16.15. Multiplier characteristic
with two square-wave inputs.

The phase detector constant *is* a constant in this instance, and is equal to

$$K_D = \frac{2}{\pi} AB. \tag{28}$$

We see that, within a scale factor, this phase detector has the same essential behavior as an analog multiplier with sinusoidal inputs, again interpreting phase errors relative to quadrature.

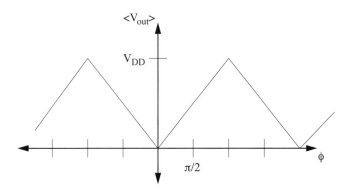

FIGURE 16.16. Characteristic of XOR
as quadrature phase detector.

As in the case with one square-wave input, this phase detector allows the loop to lock to various harmonics of the input. Again, depending on the application, this property may or may not be desirable.

If we examine the waveforms for this detector more closely, we see that they have precisely the same shape as would be obtained from using a digital exclusive-OR gate, the only difference being DC offsets on the inputs and outputs, as well as an inversion here or there. Hence, an XOR may be considered an overdriven analog multiplier. For the special case where the inputs and output are logic levels that swing between ground and some supply voltage V_{DD} (as in CMOS), the phase detector output has an average value that behaves as graphed in Figure 16.16.

The corresponding phase detector gain is then

$$K_D = \frac{V_{DD}}{\pi}. \tag{29}$$

Because of the ease with which they are implemented, and because of their compatibility with other digital circuitry, XOR phase detectors are frequently found in simple IC PLLs.

16.6 SEQUENTIAL PHASE DETECTORS

Loops that use multiplier-based phase detectors lock to a quadrature phase relationship between the inputs to the phase detector. However, there are many practical instances (de Bellescize's homodyne AM detector is one example) where a *zero* phase difference is the desired condition in lock. Additionally, the phase detector constants at the metastable and desired equilibrium points have the same magnitude, resulting in potentially long residence times in the metastable state, perhaps delaying the acquisition of lock.

Sequential phase detectors can provide a zero (or perhaps 180°) phase difference in lock, and they also have vastly different gain constants for the metastable and stable equilibrium points. Additionally, some sequential phase detectors have an output that is proportional to the phase error over a span that exceeds 2π radians.

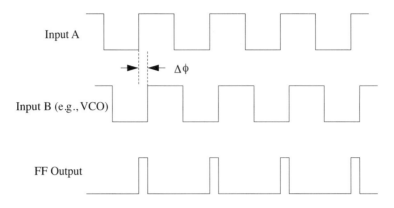

FIGURE 16.17. SR flip-flop phase detector waveforms.

Sequential detectors do possess some disadvantages. Since they operate only on transitions, they tend to be quite sensitive to missing edges (although there are modifications that can reduce this sensitivity); this is in contrast to multipliers, which look at the whole waveform. Furthermore, another consequence of their edge-triggered nature is that they introduce a sampling operation into the loop. As we will see later, sampling inherently adds something similar to a time delay into the loop transmission. The associated increasing negative phase shift with increasing frequency imposes an upper bound on the allowable crossover frequencies that is often substantially more restrictive than if a different phase detector were used.

16.6.1 THE SR FLIP-FLOP AS A PHASE DETECTOR

The simplest sequential phase detector is the set–reset (SR) flip-flop. Here, a transition (say, a positive-going one) on one input sets the flip-flop, while a transition on the other resets it. The waveforms for such a phase detector are sketched in Figure 16.17.

By considering how the output varies as the phase difference varies, we can readily generate a plot of the average output as a function of the phase difference, as shown in Figure 16.18. The gain constant for this detector is:

$$K_D = \frac{V_{DD}}{2\pi};\tag{30}$$

we have assumed the typical case for a CMOS implementation, where the output swings from rail to rail.

From examination of Figure 16.18, we can see that maximizing the phase detector range here requires choosing an equilibrium phase difference of 180°. Furthermore, the gain at the metastable point is extremely high (ideally, infinite), so that the likelihood of the loop residing there is much smaller than for, say, the XOR detector.

The foregoing analysis assumes implicitly that the flip-flop responds equally fast to the set and reset inputs. Any speed difference results in a *static phase error,* as

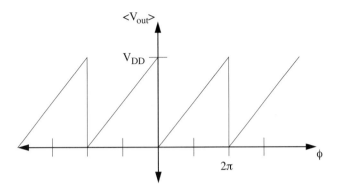

FIGURE 16.18. Characteristic of SR
flip-flop as phase detector.

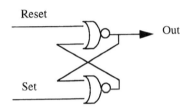

FIGURE 16.19. Textbook SR
flip-flop implementation.

something other than a 180° phase difference is necessary to yield an average output of
$V_{DD}/2$ if the set and reset operations take place at different speeds. Consider, for example, the classic textbook SR flip-flop with cross-coupled NOR gates (Figure 16.19).
Note that this circuit has faster response to the reset than to the set input and hence is
not the preferred implementation for those applications in which a small static phase
error is important.

16.6.2 SEQUENTIAL DETECTORS WITH EXTENDED RANGE

Sometimes, a 0- (rather than 180-) degree phase difference in lock is absolutely necessary. In such cases, the SR flip-flop is usually not a suitable phase detector.[15]
Furthermore, it is often desirable to extend the phase detection range to span more
than one period (say, two).

---

[15] Of course, an inverter may be added to one of the inputs to cancel nominally the 180° phase relationship. However, the inverter delay now adds directly to the phase difference. In some applications,
the associated phase error is of no consequence but in many it is a serious problem, particularly at
high frequencies.

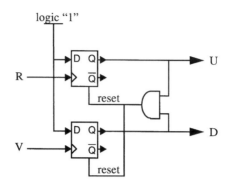

FIGURE 16.20. Phase detector
with extended range.

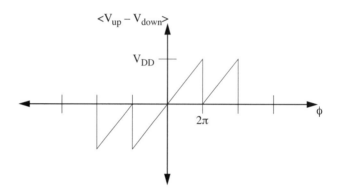

FIGURE 16.21. Characteristic of
extended-range phase detector.

A widely used circuit that possesses both of these attributes consists of two D flip-flops and a reset gate; see Figure 16.20. The designations R and V stand for "reference" and "VCO," while U and D stand for "up" and "down," terms that will mean something shortly.

For this circuit, the up and down outputs have an average *difference* that behaves as shown in Figure 16.21. Note that the input range now spans 4π radians, with a constant phase detector gain of

$$K_D = \frac{V_{DD}}{2\pi}, \tag{31}$$

which is the same as for the SR flip-flop. It should be clear from Figure 16.21 that a 0° lock point should be chosen to maximize the lock range.

One characteristic that occasionally causes trouble is the potential for the generation of runt pulses. If the reset path in Figure 16.20 acts too fast, then the minimum pulsewidth generated at the U and D outputs may be too narrow for the next stage to function reliably. This problem occurs when the R and V inputs are very close to each other, and thus it degrades behavior near the locking point. The degradation typically takes the form of an inability to resolve phase errors reliably near lock. This "dead zone" problem is readily solved by simply slowing the reset path. The insertion of some appropriate number of inverters after the AND gate will guarantee that the U and D outputs will be of a width that is consistent with proper operation of subsequent stages. In lock, both U and D outputs are asserted simultaneously for identical amounts of time.

For those who are interested in building circuits with this type of phase detector, it should be noted that this phase detector is functionally equivalent to that used in the 4044, except for some logical inversions.

16.6.3 PHASE DETECTORS VERSUS FREQUENCY DETECTORS

In many applications, it is important (or at least useful) to have some information about the magnitude of any *frequency* difference between the two inputs. Such information could be used to aid acquisition, for example.

Whereas multiplier-based phase detectors cannot provide such information, sequential phase detectors can. Consider the extended-range phase detector of the previous section. If the frequency of the VCO exceeds that of the reference then the U output will have a high duty cycle, because it is set by a rising edge of the higher-frequency VCO but isn't cleared until there is another rising edge on the lower-frequency reference. Hence, not only does this type of phase detector provide a large and linear phase detection range, it also provides a signal that is indicative of the sign and magnitude of the frequency error. These attributes account for this detector's enormous popularity. Detectors with this frequency discrimination property are known collectively as *phase-frequency* detectors.

It should be mentioned that this detector does have some problems, however. Being a sequential detector, it is sensitive to missing edges. Here, it would misinterpret a missing edge as a frequency error and the loop would be driven to "correct" this error. Additionally, the shape of the phase detector characteristic near zero phase error may actually be somewhat different from what is shown in Figure 16.21, because both the U and D outputs are narrow slivers in the vicinity of the lock point. Since all real circuits have limited speed, the nonzero risetimes will cause a departure from the ideal linear shape shown, as the areas of the slivers will no longer have a linear relationship to the input time (phase) differences.

In some systems, this problem is solved by intentionally introducing a DC offset into the loop so that the phase detector output must be nonzero to achieve lock. By biasing the balanced condition away from the detector's center, the nonlinearities can be greatly suppressed. Unfortunately, this remedy is inappropriate for applications that require small error, since the added offset translates into a static phase error.

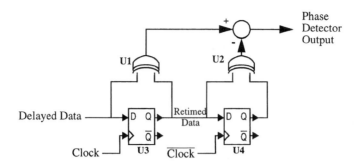

FIGURE 16.22. Hogge's phase detector.

16.6.4 OTHER SEQUENTIAL PHASE DETECTORS

As we've seen, sequential phase detectors can exhibit a great sensitivity to missing pulses. Of course, this behavior can be modified significantly, and a great deal of creative effort has been directed at devising ways to mitigate this problem.

One simple strategy is to have the VCO output cause flip-flops in the phase detector to toggle rather than reset. This way, missing input pulses cause no error (on average) and undesirable loop behavior is greatly minimized, as long as the loop filter can remove the ripple on the control line that results from the toggling.

Another strategy is to recognize that we'd like to implement a "do-nothing" state in the event of missing input pulses. It turns out that it is possible to provide just such a state. Furthermore, many phase detectors of this type can be used to recover the carrier (clock) from certain types of digital data streams.

The first of these "tristate" detectors we'll consider is due to Hogge[16] (even though it doesn't quite solve the problem); see Figure 16.22. This circuit directly compares the phases of the delayed data and the clock in the following manner. After a change in the state of the delayed data, the D input and Q output of D-type flip-flop U_3 are no longer equal, causing the output of XOR gate U_1 to go high. The output of U_1 remains high until the next rising edge of the clock, at which time the delayed data's new state is clocked through U_3, eliminating the inequality between the D and Q lines of U_3. At the same time, XOR gate U_2 raises its output high because the D and Q lines of U_4 are now unequal. The output of U_2 remains high until the next falling edge of the clock, at which time the delayed data's new state is clocked through U_4.

If we assume that the clock has a 50% duty cycle, then U_2's output is a positive pulse with a width equal to half the clock period for each data transition. The output of U_1 is also a positive pulse for each data transition, but its width depends on the phase error between the delayed data and the clock; its width equals half a clock period when the delayed data and the clock are optimally aligned. Hence, the phase error can be obtained by comparing the widths of the pulses out of U_1 and U_2.

[16] C. R. Hogge, "A Self-Correcting Clock Recovery Circuit," *J. Lightwave Technology,* v. 3, no. 6, 1985, pp. 1312–14.

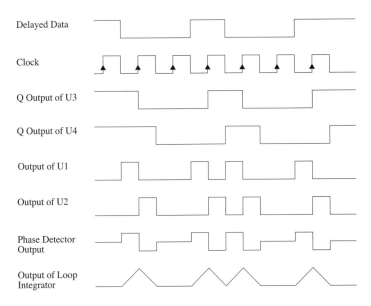

FIGURE 16.23. Waveforms for Hogge's detector in lock.

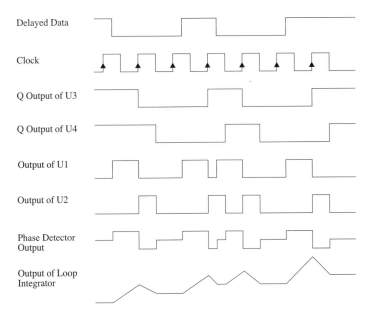

FIGURE 16.24. Waveforms for Hogge's detector with data input ahead of clock.

Figure 16.23 and Figure 16.24 are timing diagrams for this detector with the delayed data and clock optimally aligned (in this case, with the falling edge of the clock) and with data ahead of the clock, respectively. In the former case, the output of the phase detector has zero average value, and there is no net change in the loop integrator's output; in the latter case, the output of the phase detector has a positive average value.

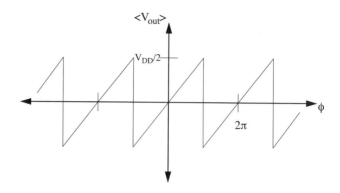

FIGURE 16.25. Characteristic of Hogge's phase detector.

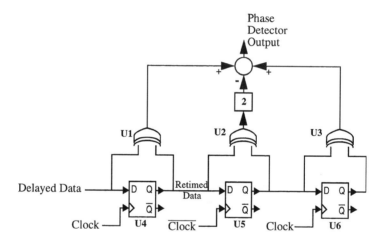

FIGURE 16.26. Triwave phase detector.

As a result, the loop integrator's output exhibits a net increase. Conversely, if the delayed data were behind the clock, the phase detector's output would have a negative average value, and the loop integrator's output would exhibit a net decrease.

Plotting the phase detector's average output (assuming maximum data transition density) as a function of phase error yields the familiar sawtooth characteristic exhibited in Figure 16.25. Consistent with Figure 16.23, the phase detector's average output equals zero when the phase error between the delayed data and the clock is zero.

One noteworthy feature of this phase detector is that the decision-making circuit is an integral component of the phase detector (for the output of flip-flop U_3 is the retimed data). However, this detector does suffer from a sensitivity to the data transition density. Since each triangular pulse on the output of the loop integrator has positive net area (see Figure 16.23), the presence or absence of such a pulse affects the average output of the loop integrator. The (data-dependent) jitter thus introduced is often large enough to be objectionable.

The phase detector shown in Figure 16.26 greatly reduces this problem by replacing the triangular correction pulses (which have net area, even when the delayed data

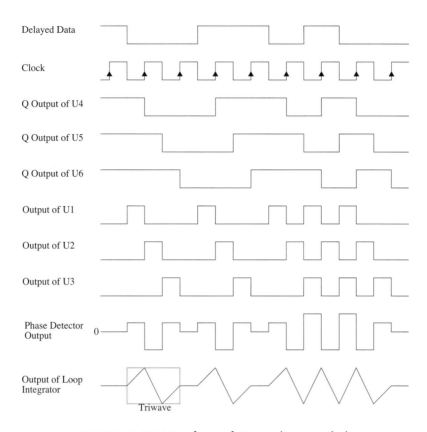

FIGURE 16.27. Waveforms of triwave detector in lock.

and clock are properly aligned) with "triwaves," whose net area is zero when clock and data are aligned.

As in Hogge's detector, the width of U_1's output is dependent on the phase error between the delayed data and the clock, while the output of U_2 and U_3 are always half a clock cycle wide (assuming that the clock possesses a 50% duty cycle). The phase error can thus be obtained by comparing the variable-width pulse from U_1 with the fixed-width pulses from U_2 and U_3. Note that the pulses out of U_1 and U_3 are weighted by 1, while the pulse out of U_2 is weighted by -2.

Figure 16.27 is the timing diagram for the triwave detector with the delayed data and the clock optimally aligned. Note that each data transition initiates a three-sectioned transient (the triwave) on the output of the loop integrator, and that this triwave has zero area. Therefore, its presence or absence does not change the average output of the loop integrator. Hence, the triwave detector exhibits a much reduced sensitivity to data transition density.

However, the triwave detector is somewhat more sensitive to duty-cycle distortion in the clock signal than is Hogge's implementation, owing to the unequal weightings used. This sensitivity to duty cycle can be restored to that of Hogge's implementation with the simple modification shown in Figure 16.28. The modified triwave detector uses two distinct down-integration intervals clocked on opposite edges of the clock,

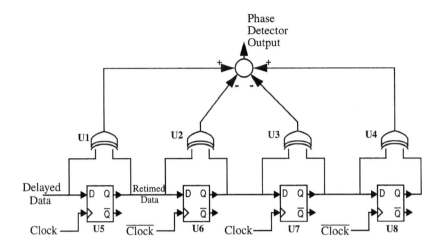

FIGURE 16.28. Modified triwave detector.

rather than a single down-integration of twice the strength clocked on a single clock edge. As a consequence, duty-cycle effects are attenuated.

16.7 LOOP FILTERS AND CHARGE PUMPS

So far, we've examined the behavior of PLLs using a linear model, as well as a number of ways to implement phase detectors. We now consider how to implement the rest of the loop. We'll take a look at various types of loop filters and survey a couple of common techniques for realizing VCOs. We'll wrap up by going through an actual example to illustrate a typical design procedure.

16.7.1 LOOP FILTERS

Recall that we generally want to have zero phase error in lock. Now, the VCO requires some control voltage to produce an output of the desired frequency. To provide this control voltage with a zero output from the phase detector (and hence zero phase error), the loop filter must provide an integration. Then, to ensure loop stability, the loop filter must also provide a zero.

A classic architecture that satisfies these requirements appears as Figure 16.29. It should be easy to deduce the general properties of the loop filter without resorting to equations. (Okay, maybe it's easier after you've done it once or twice so that you already know the answer, but work with me here.)

At very low frequencies, the capacitor's impedance dominates the op-amp's feedback, so the loop filter behaves as an integrator. As the frequency increases, though, the capacitive reactance decreases and eventually equals the series resistance R_2. Beyond that frequency, the capacitive reactance becomes increasingly negligible compared with R_2, and the gain ultimately flattens out to simply $-R_2/R_1$.

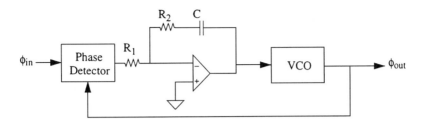

FIGURE 16.29. PLL with typical loop filter.

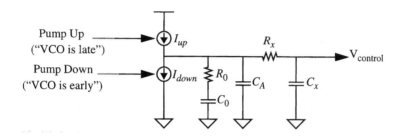

FIGURE 16.30 Idealized PLL charge pump with third-order loop filter.

Stating these observations another way, we have a pole at the origin and a zero whose time constant is R_2C. Furthermore, the value of R_1 can be adjusted to provide whatever loop transmission magnitude we'd like, so the op-amp circuit provides us with the desired loop filter transfer function.

Before going further, it should be mentioned that PLLs need not include an active loop filter of the type shown. In the simplest case, a passive RC network could be used to connect the phase detector with the VCO. However, the static phase error will then not be zero, and the loop bandwidth will be coupled (inversely) with the static phase error. Because of these limitations, such a simple loop filter is used only in noncritical applications.

The circuit of the figure is commonly used in discrete implementations, but a different (although functionally equivalent) approach is used in most ICs. The reason is that it is not necessary to build an entire op-amp to obtain the desired loop filter transfer function. A considerable reduction in complexity and area (not to mention power consumption) can be obtained by using an element that is less general-purpose than an op-amp.

A popular alternative to the op-amp loop filter is the use of a *charge pump,* working in tandem with an RC network. Here, the phase detector controls one or more current sources, and the RC network provides the necessary loop dynamics.

Figure 16.30 shows how a charge pump provides the necessary loop filter action. Here, the phase detector is assumed to provide a digital "pump up" or "pump down" signal. If the phase detector determines that the VCO output is lagging the input reference, it activates the top current source, depositing charge onto the capacitor

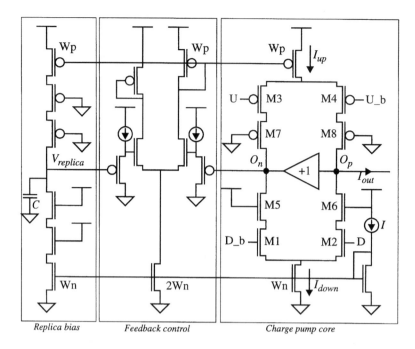

FIGURE 16.31. Example of PLL charge pump.

(pumping up). If the VCO is ahead, the bottom current source is activated, withdrawing charge from the capacitor (pumping down).

If there were no resistor R_0 then we would have a pure integration. As usual, the series resistor provides the necessary loop-stabilizing zero by forcing the high-frequency asymptotic impedance to a nonzero value. Elements C_A, R_x, and C_x provide additional filtering, as described in the next section.

Since switched current sources are easily implemented with a very small number of transistors, the charge pump approach allows the synthesis of the desired loop filter without the complexity, area, and power consumption of a textbook op-amp. The nature of the control also meshes nicely with the many existing digital phase detectors (e.g., sequential phase detectors), such as the one shown in Figure 16.20. When that detector is used with the charge pump of Figure 16.30, the net pump current is given by

$$I = I_{\text{pump}} \frac{\Delta\phi}{2\pi}, \tag{32}$$

where $I_{\text{pump}} = I_{\text{up}} = I_{\text{down}}$. This current, multiplied by the impedance of the filter network connected to the current sources, gives the output voltage.

A typical charge pump appears in Figure 16.31. Analysis of this circuit highlights some of the more important design considerations associated with charge pump design. Transistors M_1 through M_4 are differential switches operated by the up and down commands from the phase detector. Depending on the state of those commands,

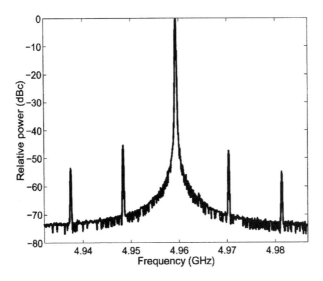

FIGURE 16.32. Output spectrum of synthesizer
with somewhat leaky charge pump.

either source current I_{up} or sink current I_{down} is steered to the output node O_p. Thus, I_{out} equals I_{up} or I_{down}, depending on the phase detector state.

The switches are cascoded by transistors M_5 to M_8 for high output impedance because any leakage increases spur power. To understand why, consider the locked condition. With low leakage, very little net charge needs to be delivered by the charge pump per cycle. There is thus very little ripple on the control line and hence very little modulation of the VCO. As leakage increases, however, the charge pump must make up for an increasing amount of lost charge, implying the necessity for an increasing static phase error. For example, if the leakage is such that the control voltage droops between phase measurements, then the phase error must increase until the net charge deposited as a result of up pulses is just enough greater than that deposited from the down pulses to compensate for the leakage. Cascoding helps reduce control-line ripple by reducing leakage, and it therefore reduces the spur energy (and static phase error). Because the voltage droops between corrections, which occur with a frequency equal to that of the reference input, the control-line ripple also has a fundamental periodicity equal to that of the reference. The spurs are therefore displaced from the carrier by an amount equal to the reference frequency. The existence of large reference frequency spurs is usually a sign of poor charge pump design; see Figure 16.32. Clearly visible are the reference spurs spaced 11 MHz away from the 4.96-GHz carrier. There are additional spurs (spaced integer multiples of 11 MHz away), which correspond to the Fourier components of the ripple on the control line.

For similar reasons it is also important to have equal up and down currents. If one is stronger than the other then a compensating static phase error must again appear, with its attendant consequences for control-line ripple. To mitigate this problem, the charge pump design here uses relatively large devices (to reduce threshold

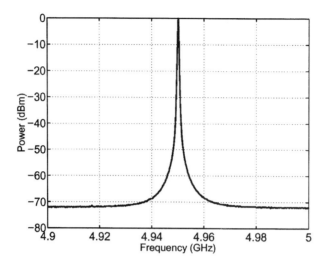

FIGURE 16.33. Spectrum of improved synthesizer.

mismatch) and operates them at moderately large overdrive. In addition, a simple unity-gain buffer forces the unused charge pump output to have the same common-mode voltage as the main output, thus removing systematic mismatch that would arise from operation with unequal drain–source voltages. Supplementing that strategy is a replica bias loop, whose output voltage is compared with the voltage at the unused output of the charge pump. A simple op-amp drives these two voltages to equality (the compensation capacitor C is for loop stability) and thus ensures that all conducting devices in the main core have the same bias voltages as in the replica. The resulting up and down tail currents are then equal within the limits of random mismatch.

Attention to these sorts of details enables the suppression of the reference spurs by large factors, as is apparent from Figure 16.33. Spurs are invisible in this plot and are thus below the noise floor of −70 dBc. The greater than 25-dB reduction in reference spur power represents an improvement by a factor of more than 300.

Control-Line Ripple and Higher-Order Poles

Even when the charge pump is well designed, we must assume nonzero ripple on the control line. As a consequence of the loop-stabilizing zero, there can be significant high-frequency content on the control line that drives the VCO. This "hash" can come from the higher-order mixing products in a multiplier-type detector (i.e., essentially the double frequency term) or from the multiple-order products from a charge-pump–detector combination. If these components are periodic then they produce stationary sidebands (spurs). One obsession of synthesizer designers is the systematic eradication of spurs. Unfortunately, spurs arise very easily from noise injected into the control line – including noise from the supply, from substrate, or even from external fields coupling into the chip. A typical RF VCO may possess tuning sensitivities of tens or hundreds of megahertz per volt, so even a few millivolts of noise will generate noticeable spectral artifacts. The resultant modulation of the VCO frequency may be unacceptable in many applications.

The loop filter's purpose is to remove the "teeth" produced by the phase detection process (which, if you recall, is fundamentally a sampled system in digital implementations) as well as other noise that may couple there, thereby suppressing noise and spurs. For a given loop bandwidth, a higher-order filter provides more attenuation of out-of-band components. However, the higher the order, the harder it is to make the loop stable. For this reason, many simple synthesizer loops are second order, but these rarely provide competitive performance.

Remembering that the VCO adds another pole (at the origin), we see that choosing a three-pole loop filter results in the creation of a fourth-order loop. In the past, no simple closed-form design method existed, so designing such a filter involved staring at lots of plots before giving up and going back to a second- or third-order loop. Luckily this situation has changed quite recently, and we can offer a simple cookbook recipe that is close enough to optimum for most purposes.[17]

Step 1. *Specify a phase margin.* Once this value is chosen, it sets a constraint on capacitor values. Specifically,

$$\text{PM} \approx \text{atan}\left(\sqrt{b+1}\right) - \text{atan}\left(\frac{1}{\sqrt{b+1}}\right), \tag{33}$$

where "atan" is shorthand for "arctangent" and

$$b = \frac{C_0}{C_A + C_X}. \tag{34}$$

It's probably prudent to choose a phase margin a few degrees above the target value to absorb the inevitable negative phase contributions by the sampled nature of the loop and by unmodeled poles and other destabilizing sources. For example, suppose the specified phase margin target is 45°. If we therefore design for 50°, we find (through iteration, for example) that b should be about 6.5.

Step 2. *Select loop crossover frequency,* based on specifications on tracking bandwidth, for example. Combined with the results of Step 1, we find the location of the loop stabilizing zero as follows.

We know that maximizing the loop bandwidth maximizes the frequency range over which the presumably superior phase noise characteristics of the reference oscillator are conferred on the output. Unfortunately, the loop is a sampled data system, and we can only push up the crossover frequency to about a tenth of the phase comparison frequency before the phase lag inherent in a discrete-time phase detector starts to degrade phase margin seriously. As a specific example, assume that the reference frequency (and hence phase comparison frequency) is 2 MHz. Choosing a crossover

---

[17] Rategh and Lee, op. cit. (see footnote 9).

frequency of 100 kHz is more than a decade below the reference frequency, so let's use that value in what follows (you are free to choose some other value, within limits).

For the crossover frequency we have

$$\omega_c \approx \frac{\sqrt{b+1}}{\tau_z} = \frac{\sqrt{b+1}}{R_0 C_0}. \tag{35}$$

Step 3. *Calculate C_0, the value of the zero-making capacitor.* Thus,

$$C_0 = \frac{I_P}{2\pi} \frac{K_0}{N} \frac{b}{\sqrt{b+1}} \frac{1}{\omega_c^2}, \tag{36}$$

where I_P is the charge pump current, N is the divide modulus, and K_0 is the VCO gain constant in radians per second per volt.

Step 4. *Calculate $R_0 = \tau_z/C_0$.* This completes the design of the main part of the loop filter.

Step 5. *Select $\tau_x = R_X C_X$ within the following range:*

$$0.01 < \tau_x/\tau_z < 0.1. \tag{37}$$

Within these wide limits is considerable freedom of choice. You can choose to design for the arithmetic mean, or the geometric mean, or some other kind of mean. Typically, one selects τ_x to be $1/30$ to $1/20$ of τ_z. A bigger time constant results in somewhat better filtering action but tends to be associated with lower stability. Since loop constants aren't constant, it is prudent to design for some margin.

Step 6. *Complete the remaining calculations.* Back in Step 1, we developed a constraint on the capacitance ratios. Having found one of the capacitances, we now know the sum of C_A and C_X. You are free to select the individual values over a quite wide range, as long as they sum to the correct value. Arbitrarily setting them equal is a common choice.[18] Having done so then allows us to determine their absolute values, which subsequently allows us to determine the value of R_X.

This completes the design of the loop filter.

16.7.2 VCOs

Although we will take up the detailed study of oscillators at a later time, we examine here one common architecture for realizing VCOs in integrated circuits: the

---

[18] The noise generated by the resistors in the filter will produce broadband modulation of the VCO, resulting in phase noise. Minimizing the phase noise would impose additional constraints on the loop filter design, but it complicates the situation enough that the cookbook procedure offered here is all we'll consider. Another consideration is to select values that make the overall realization less dependent on parasitics. Generally speaking, using the largest capacitors consistent with the required time constants will help reduce broadband noise modulation of the control voltage.

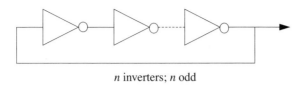

n inverters; *n* odd

FIGURE 16.34. Ring oscillator.

current-starved ring oscillator, in which the effective propagation delay of each inverter in the ring is controlled by a current source.

Ring Oscillators

Ring oscillators are an extremely popular idiom, since they derive from digital-like building blocks. We'll see later that, compared to tuned oscillators (i.e., those that explicitly use high-Q resonators), they have substantially inferior phase noise performance for a given level of power consumption. For many applications, however, their relatively large tuning range and simplicity are strong enough attributes to make them attractive.

The controllable ring oscillator derives from the uncontrolled ring oscillator and consists simply of *n* inverters in a ring, where *n* is odd.[19] In its simplest form, it appears as shown in Figure 16.34.

In the simplest analysis of such an oscillator, it is assumed that each inverter can be characterized by a propagation delay T_{pd}. No stable DC point exists, and a logic level propagates around the loop, experiencing one net inversion each traversal. The oscillation period is therefore simply twice the total propagation delay:

$$f_{osc} = \frac{1}{2n \cdot T_{pd}}. \tag{38}$$

Now, to convert this thing into a controllable oscillator, the propagation delay seems the most natural quantity to adjust.

One can imagine a great many specific methods for adjusting the delay, but they all boil down to either changing the load (e.g., by varying the effective amount of capacitance seen by each inverter output) or varying the current drive of the inverters. One cheesy way to accomplish the latter is shown in Figure 16.35, where a PMOS current mirror provides a limited, variable pull-up current to the CMOS inverter.[20] By adjusting this current, the effective propagation delay of the inverter can be adjusted, altering the oscillation frequency in the process.

---

[19] Ring oscillators in which the individual stages are differential allow the use of an even number of stages. The necessary inversion can be obtained simply by reversing one differential pair of signals.

[20] Obviously, an NMOS mirror could also be used to constrain the pull-down current, but this would reduce the cheese level.

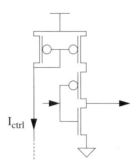

FIGURE 16.35. Simple current-starved CMOS inverter.

The ring oscillator (and its many variants) requires only elements that are normally available in ordinary digital process technology. This attribute, combined with its fundamental simplicity, has made the ring oscillator a near-ubiquitous topology in ICs.

16.8 PLL DESIGN EXAMPLES

Now that we've studied the basics of phase-locked loops, it's time to consider a few design examples. The particular examples we'll study use a commercially available PLL chip, the 4046. It is an inexpensive (~$0.25–$1) CMOS device that contains two phase detectors (one XOR and one sequential phase detector) and a VCO. We will consider the design of a PLL with each of the phase detectors and a couple of loop filters.

The 4046 is a relatively slow device, with a maximum oscillation frequency of only about 1 MHz or so. Still, the design procedure we'll use is generally applicable to PLLs whose output frequency is much higher, so what follows isn't a purely academic exercise. In any event, the device remains useful for many applications even today, and it is certainly an exceptionally inexpensive PLL tutorial lab-on-a-chip.

16.8.1 CHARACTERISTICS OF THE 4046 CMOS PLL

Phase Detector I

The chip contains two phase detectors (PDs). One, known as "phase detector I," is a simple XOR gate. Recall from the section on phase detectors that an XOR has a gain constant given by

$$K_D = \frac{V_{DD}}{\pi} \text{ V/rad.} \tag{39}$$

Throughout these design examples, we will use a power supply voltage of 5 V, so the specific numerical value for our designs will be

$$K_D = \frac{V_{DD}}{\pi} \approx 1.59 \text{ V/rad.} \tag{40}$$

Phase Detector II

The chip's other phase detector (or "comparator") is a sequential phase detector that operates only on the positive edges of the input signals. It has two distinct regions of behavior depending on which input is ahead.

If the signal input edge precedes the VCO feedback edge by up to one period, then the output of the phase detector is set high (that is, to V_{DD}) by the signal edge and sent into a *high-impedance* state by the feedback edge. (We'll see momentarily why it can be advantageous to have this high-impedance state.)

If the signal input edge lags the VCO output by up to one period, then the output is set low (to ground) by the VCO edge and sent into a high-impedance state by the signal input edge. And that's all there is to this phase detector.

The high-impedance state allows one to reduce the amount of ripple on the control line when in the locked state. Hence, the amount of unintended phase and frequency modulation of the VCO during lock can be much smaller than when other detectors are used. It should also be clear that a PLL using this sequential phase detector forces a zero phase difference in lock, in contrast with the quadrature condition that results with an XOR detector.

The other bit of information we need in order to carry out a design is the phase detector gain constant. Unfortunately, this particular detector does not have a particularly well-defined K_D because the output voltage in the high-impedance state depends on external elements, rather than on the phase error alone. A good solution to this problem is to remove the uncertainty by forcing the output voltage to $V_{DD}/2$ during the high-impedance condition (e.g., with a simple resistive divider). With this modification, K_D can be determined.

For phase errors of less than one period (signal input leading), the average output voltage will be linearly proportional to the phase error. The minimum output is $V_{DD}/2$ for zero phase error, and is a maximum value of V_{DD} with a 2π phase error. The minimum output is determined by the added resistive divider, while the maximum output is simply controlled by the supply voltage.

Similarly, in the case of a lagging input signal, the average output voltage will be $V_{DD}/2$ for zero phase error and zero volts for a 2π phase error. Hence, the phase detector characteristic looks as shown in Figure 16.36. After solving Schrödinger's equation with appropriate boundary conditions, we find that the slope of the line is

$$K_D = \frac{V_{DD}}{4\pi} \text{ V/rad.} \tag{41}$$

For our assumed V_{DD} of 5 V, the phase detector gain is approximately 0.40 V/rad.

VCO Characteristics

The VCO used in the 4046 is reminiscent of the emitter-coupled multivibrator used in many bipolar VCOs. Here, an external capacitor is alternately charged one way, then the other, by a current source. A simple differential comparator switches the polarity of the current source when the capacitor voltage exceeds some trip point. The feedback polarity is chosen to keep the circuit in oscillation.

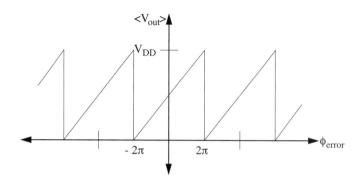

FIGURE 16.36. Characteristic of phase detector II.

The main VCO output is a square wave, derived from one output of the differential comparator. An approximation to a triangle wave is also available across the capacitor terminals. The triangle-wave signal is useful if a sine-wave output is desired, since either a filtering network or a nonlinear waveshaper can be used to convert the triangle wave into some semblance of a sine wave.

Frequency control is provided through adjustment of the capacitive charging current. Both the center frequency and VCO gain can be adjusted independently by choosing two external resistors. One resistor, R_2, sets the charging current (and hence the VCO frequency) in the absence of an input, thus biasing the output frequency-versus-control voltage curve. The other resistor, R_1, sets the transconductance of a common-source stage and therefore adjusts the VCO gain.

Conspicuously absent from the data sheets, however, is an explicit formula for relating the VCO frequency to the various external component values. A quasiempirical (and highly approximate) formula that provides this crucial bit of information is as follows:[21]

$$\omega_{osc} \approx \frac{2\left(\dfrac{V_C - 1}{R_1} + \dfrac{4}{R_2}\right)}{C}. \tag{42}$$

From this formula, the VCO gain constant is easily determined by taking the derivative with respect to control voltage:

$$K_O \approx \frac{2}{R_1 C} \ \text{rad/s/V}. \tag{43}$$

Miscellany

Notice that the phase detector gains are functions of the supply voltage. Additionally, the VCO frequency is a function of V_{DD} as well. Hence, if the supply voltage varies then so will the loop dynamics, for example. If power supply variations (including

---

[21] This formula is the result of measurements on only one particular device with a 5-V power supply. Your mileage may vary, especially if you use resistance values below about 50–100 kΩ (the VCO control function gets quite nonlinear at higher currents). *Caveat nerdus.*

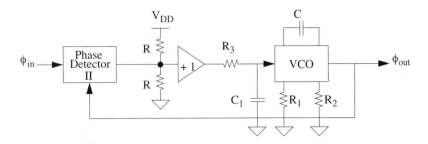

FIGURE 16.37. PLL with phase detector II.

noise) are not to influence loop behavior, it is necessary to provide a well-regulated and well-filtered supply. As a convenience, the 4046 includes a 5.2-V (plus or minus about 15%) zener diode that may be used for this purpose.

The 4046 also includes a simple source follower to buffer the control voltage. This feature is handy for those applications in which the PLL is being used as an FM demodulator, for example. The demodulated signal is equal to the VCO control voltage, so that a buffered version of this control signal is convenient for driving external circuitry.

Finally, the chip includes an "inhibit" control signal that shuts off the oscillator and source follower to reduce chip dissipation to the 100-μW range (even less if the signal input is a constant logic level).

16.8.2 DESIGN EXAMPLES AT LAST

Second-Order PLL with Passive *RC* Loop Filter and PD II

We know that active filters can provide superior performance, particularly with regard to steady-state error. However, there are some applications for which fully passive filters are adequate, and thus for which active filters would simply consume additional area and power.

Suppose we use phase detector II and a simple *RC* low-pass loop filter (without a loop-stabilizing zero). Design a circuit to meet the following specifications:

crossover frequency: 1 krad/sec;
phase margin: 45°;
center frequency: 20 kHz.

Solution: First, we recognize that the high-impedance characteristic of this phase detector requires the use of the resistive divider, as mentioned earlier. Then, to provide the ability to drive an arbitrary *RC* network, it is advisable to add a buffer. Hence, the PLL appears as shown in Figure 16.37.

The value of *R* is not particularly critical but should be large enough to avoid excessive loading of the phase detector's wimpy outputs. Values on the order of tens of kilohms are acceptable. Note that the loop transmission may be written as

$$-L(s) = K_D H_f(s) \frac{K_O}{s} = \frac{V_{DD}}{4\pi} \cdot \frac{1}{sR_3 C_1 + 1} \cdot \frac{K_O}{s}. \tag{44}$$

The phase margin specification *requires* us to choose the pole frequency of the loop filter equal to the desired crossover frequency, since we do not have a loop-stabilizing zero. Having made that choice, we adjust the VCO gain through selection of $R_1 C$. Finally, we choose R_2 to satisfy the center frequency specification.

Carrying out these steps, while being mindful that resistance values should be no lower than about 50 kΩ to validate the quasiempirical VCO equation, yields the following sequence of computations, half-truths, and outright lies.

(1) As stated earlier, the phase margin specification requires a loop filter time constant of 1 ms. Somewhat arbitrarily choose $R_3 = 100$ kΩ, so that $C_1 = 0.01$ μF. Both values happen to correspond to standard component values.
(2) Because the crossover frequency must be 1 krps while $R_3 C_1$ and the phase detector gain constant are both known, K_O must be chosen to yield the desired crossover frequency:

$$|L(j\omega_c)| = K_D \cdot \frac{1}{\sqrt{2}} \cdot \frac{K_O}{10^3 \text{ rps}} = 1 \implies R_1 C = 0.582 \text{ ms}. \tag{45}$$

Arbitrarily choose the capacitor equal to a standard value, 0.001 μF, so that the required resistance is 582 kΩ (not quite a standard value, but close to 560 kΩ, which is). Just for reference, the corresponding VCO gain constant is about 3.56 krps/V.
(3) Now select R_2 to yield the desired center frequency (here defined as the VCO frequency that results with a control voltage of $V_{DD}/2$) with the VCO capacitor chosen in step (2). From the semiempirical VCO formula, we find that R_2 should be approximately 67.3 kΩ (the closest standard value is 68 kΩ). Because of variability from device to device, it is advisable to make R_2 variable over some range if the VCO center frequency must be accurately set.

That completes the design.

With the parameters as chosen, let us compute the VCO tuning range, the steady-state phase error throughout this range, and the lock range (something we haven't explicitly discussed before). The lock range is defined here as the range over which we may vary the input frequency before the loop loses lock.

For the frequency tuning range, we again use the VCO formula. With the values we've chosen, the VCO can tune about 1 kHz above and below the center frequency. This range sets an upper bound on the overall PLL frequency range.

Because of the passive loop filter, the static phase error will not be zero in general since a nonzero phase detector output is required to provide a nonzero VCO control voltage.[22] Now, if we assume that the VCO gain constant is, well, *constant,* we can

---

[22] Here, zero control voltage is interpreted as a deviation from the center value of $V_{DD}/2$.

compute precisely how much control voltage change is required to adjust the frequency over the range computed in step (1). If the corresponding phase error exceeds the $\pm 2\pi$ span of the phase detector, then the loop will be unable to maintain lock over the entire ± 1-kHz frequency range.

The voltage necessary to move the output frequency is found from K_O and is related to the phase detector gain constant and the phase error as follows:

$$\Delta V_{\text{ctrl}} = \frac{\Delta \omega}{K_O} = K_D \phi_{\text{error}}. \tag{46}$$

Using our component values, the phase error is predicted to be about 4.4 rad at 1 kHz off of center frequency. Actual measurements reveal that, at the lower frequency limit (1 kHz below center), the phase error is 4.3 rad. Theory triumphs (here, anyway).

At 1 kHz *above* center, though, the measured phase error is actually about 5.9 rad. The reason for this rather significant discrepancy is that the VCO frequency isn't quite linearly related to the control voltage at higher control voltages. It turns out that a larger-than-expected control voltage is required to reach the upper frequency limit. Hence, a larger phase detector output is required and so a larger corresponding phase error results. Since angles of both 4.3 rad and 5.9 rad are still within the phase detector's linear range, however, it is the VCO's limited tuning range – rather than the phase detector's characteristics – that determines the overall PLL's lock range in this particular case.

Second-Order PLL with Passive *RC* Loop Filter and PD I

It is instructive to re-do the previous design with the XOR phase detector replacing the sequential phase detector. Because the XOR has four times the gain of PD II, the value of K_O must be adjusted downward by this factor to maintain the crossover frequency. We may adjust K_O by increasing R_1 to four times its previous value. In order to maintain a 20-kHz center frequency, R_2 must be adjusted as well (downward). Because the XOR does not have a high-impedance output state, the resistive divider and buffer may be eliminated.

Once these changes have been made, the locked loop displays dynamics that are similar to those observed with the previous design. However, the VCO modifications alter the VCO tuning range and, therefore, the corresponding phase error:

$$\Delta V_{\text{ctrl}} = \frac{\Delta \omega}{K_O} = K_D \phi_{\text{error}} \implies \phi_{\text{error}} = \frac{\Delta \omega}{K_O K_D}. \tag{47}$$

Because R_1 has been changed upward, the VCO tuning range has decreased to a fourth of its previous value, while the product of phase detector gain and VCO gain remains unchanged. Now, the XOR is linear over only a fourth of the phase error span of the sequential phase detector. Hence, for a given crossover frequency and damping, use of the XOR phase detector can cause the loop to possess a narrower lock range.

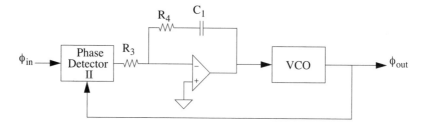

FIGURE 16.38. PLL with active loop filter (defective).

It is left as an exercise for the reader to carry out actual numerical calculations to verify these assertions. (In this case, it turns out that the VCO tuning range is still the limiting factor, but just barely.)

As a few final notes on the use of the XOR, it should be mentioned that this type of detector is sensitive to the duty cycle of the input signals. The ideal triangular characteristic of the XOR phase detector is obtained only when both inputs possess a 50% duty cycle. If there are any asymmetries, the average output will no longer reach both supply rails at the extremes of phase error. The sequential phase detector is an edge-triggered device and so does not suffer this duty-cycle sensitivity.

Another important note is to reiterate that the XOR phase detector allows locking onto harmonics of the input, since the action of the XOR is equivalent to multiplying two sine waves together. The rich harmonic content of square waves provides many opportunities for a coincidence in frequency between components of the input and VCO output, permitting lock to occur. If harmonic locking is undesirable, use of an XOR phase detector may cause some problems.

Second-Order PLL with Active *RC* Loop Filter and PD II

Now let's consider replacing the simple passive *RC* loop filter with an active filter. Let this filter provide a pole at the origin to drive the steady-state phase error to zero. Additionally, assume that we want to achieve precisely the same crossover frequency and phase margin as in the earlier design, but with the additional requirement that the loop maintain lock at least ± 10 kHz away from the center frequency.

To satisfy the phase margin requirement, we need to provide a loop-stabilizing zero to offset the negative phase contribution of our loop filter's integrator. Our first-pass PLL then should look something like Figure 16.38 (VCO components not shown).

Why "first-pass?" The circuit has a small embarrassment: if the input is ahead of the VCO, the phase detector provides a positive output. The inverting loop filter then drives the VCO toward a lower frequency, exacerbating the phase error; we have a positive feedback loop. To fix this problem, we must provide an additional inversion in the control line.

There is another problem with the circuit: The op-amp's noninverting terminal is grounded. The implication is that the output of the loop filter can never integrate up, since the minimum output of the phase detector is ground. To fix this last (known)

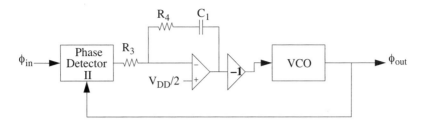

FIGURE 16.39. PLL with active loop filter (fixed).

problem, we need to connect the noninverting terminal to $V_{DD}/2$, as shown in Figure 16.39. Now we can set about determining the various component values.

First, note that our loop transmission is

$$-L(s) = K_D H_f(s) \frac{K_O}{s} = \frac{V_{DD}}{4\pi} \cdot \frac{sR_4C_1 + 1}{sR_3C_1} \cdot \frac{K_O}{s}. \qquad (48)$$

To achieve a 45° phase margin, the zero must be placed at crossover, since the two poles at the origin contribute a total phase shift of $-180°$. Hence, R_4C_1 must equal 1 ms. Choosing values with the same moderately constrained arbitrariness as in the passive filter case, we let $R_4 = 100$ kΩ, so that the value of C_1 is 0.01 μF.

Next, note that the loop transmission magnitude is controlled by both R_3 and K_O, so we would have an underconstrained problem if achieving a specified crossover frequency were the only consideration. Since there is a requirement on the lock range of the loop, however, there is an additional constraint that allows us to fix both R_3 and K_O. Specifically, the control voltage has an effect on VCO frequency only from about 1.2 V to 5 V, according to the empirical formula.[23] The center of this voltage range is 3.1 V, not the 2.5 V implicitly assumed. If we continue to use 2.5 V as our definition of center, though, the lock range will not be symmetrical about 20 kHz. As there is no specification about a symmetrical lock range, we will remain consistent in our use of 2.5 V as the control voltage that corresponds to the center frequency of the VCO.

With that choice, the lower frequency limit is smaller than the higher one. To satisfy our 10-kHz specification, we must be able to change the VCO frequency by 10 kHz (or more) with the control voltage at its minimum value of 1.2 V, corresponding to a deviation of 1.3 V from the center. Hence, we require

$$K_O > \frac{2\pi \cdot 10 \text{ kHz}}{1.3 \text{ V}} \approx 4.8 \times 10^4 \text{ rps/V}. \qquad (49)$$

Maintaining a center frequency of 20 kHz with this VCO gain constant leads to the following choices for the three VCO components:

$$C = 0.001 \ \mu\text{F}, \quad R_1 = 42 \text{ k}\Omega, \quad R_2 = 130 \text{ k}\Omega.$$

---

[23] The control voltage term is not allowed to take on a negative value in the formula.

Here, the closest standard (10% tolerance) resistors for R_1 and R_2 are 39 kΩ and 120 kΩ, respectively.

Finally, having determined everything else, the crossover frequency requirement fixes the value of the op-amp input resistor:

$$R_3C_1 = \frac{K_D K_O}{\omega_c^2} \cdot \sqrt{2} \approx 27.7 \text{ ms.} \qquad (50)$$

Therefore, $R_3 = 2.8$ MΩ (2.7 meg is the closest standard value), and the design is complete.

Note that, for this design, it is definitely the VCO tuning range and not the phase detector characteristics that determines the lock range. With a loop filter that provides an integration, any steady-state VCO control voltage can be obtained with zero phase error. Therefore, the phase detector characteristics are irrelevant with respect to the steady-state lock range.

16.9 SUMMARY

The design examples presented are representative of typical practice, although they are a very tiny subclass of a vast universe of possible PLL applications. In Section 17.7 we will encounter one more application of PLLs, in frequency synthesis, since synthesizers are exceedingly important building blocks for modern RF communications gear.

PROBLEM SET FOR PHASE-LOCKED LOOPS

PROBLEM 1   Consider the PLL shown in Figure 16.40. Assume that the phase detector is a simple CMOS XOR whose logic levels are ground and V_{DD}. Further assume that both the input to the loop and the VCO output are square waves that swing between ground and V_{DD}. Finally, assume that the VCO has a perfectly linear relationship between control voltage and output frequency of 10 MHz per volt. Polarities are such that an increase in control voltage causes an increase in VCO frequency.

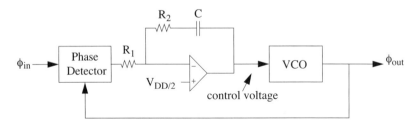

FIGURE 16.40. Second-order PLL.

(a) First suppose that the loop has been in lock forever, and that the input signal frequency has remained constant for that time. Sketch the input signals to the XOR (as functions of time).

(b) Derive expressions for the loop transmission and $\phi_{\text{out}}/\phi_{\text{in}}$.

(c) Initially assume $R_2 = 0$ and $R_1 = 100\ \Omega$. What value of C gives us a loop crossover frequency of 100 kHz? What is the phase margin? Assume that the op-amp is ideal.

(d) With the value of C from part (c), allow nonzero values of R_2 to provide a phase margin of 45° while preserving a 100-kHz crossover frequency.

(e) Now suppose that a frequency divider of factor N is inserted into the feedback path. With the component values of part (d), what is the largest N that can be tolerated without shrinking phase margin below 14°? Do not consider "divider delay" in the phase margin calculations.

PROBLEM 2   Derive the transfer function for a charge pump phase detector. Assume perfect switches, a pump current of I, and a pump capacitor of C. How does your answer change when a resistor is inserted in series with the capacitor?

PROBLEM 3   When an exclusive-OR gate is used as a phase detector, harmonic locking becomes possible. If both the input and VCO signals are square waves with precisely 50% duty cycle, then only odd harmonic locking is allowed.

(a) Show that, if the input signal has a duty ratio D other than 50%, locking to even harmonics is possible. *Hint:* Find the Fourier series representation for a non–50% duty cycle square wave.

(b) Provide an explicit expression for the loop transmission of such a first-order PLL as a function of harmonic number. Normalize your answer in terms of the loop transmission for the fundamental.

(c) Based on your answer to (b), what conclusions may you draw about the ease or difficulty of locking onto harmonics?

PROBLEM 4   Statements such as "PLLs using XORs lock in quadrature," or "PLLs using (certain) sequential phase detectors lock in phase" are incomplete because the lock point is actually a function of more than simply the type of phase detector used. Offsets, either random or purposefully introduced, may also change the lock point (see Figure 16.41). Here, assume that the loop filter contains an integration with zero offset.

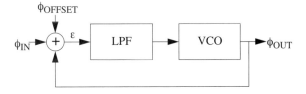

FIGURE 16.41. PLL with offset.

(a) What is the lock point of this loop if the phase detector, here represented as the three-input summing node, is an XOR?

(b) Suppose that the ϕ_{offset} is related to an offset voltage through the phase detector gain constant, K_D. What offset voltage corresponds to a phase error (offset) of 0.1 rad if the phase detector is a simple CMOS XOR with a V_{DD} of 3.3 V?

(c) One way for offsets to be introduced is through variation in bias voltages, such as in the circuit of Figure 16.42. If the phase detector itself has no offset, express the lock point of this PLL as a function of V_{BIAS}. For simplicity, you may assume that the phase detector characteristic is symmetrical about zero, as with the extended-range phase-and-frequency detector discussed in the chapter.

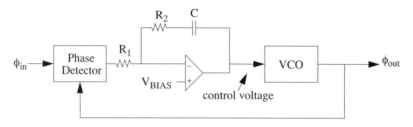

FIGURE 16.42. Second-order PLL with offset.

PROBLEM 5   In real PLL circuits, of course, the output of the VCO is used by something else. As a consequence, it is frequently necessary to buffer the VCO signal. However, all real buffers are noisy, so even if the VCO output proper is quite clean (spectrally speaking), it is possible for buffer noise to negate the considerable design effort expended in making the VCO itself relatively noise-free.

There are two choices for how to close the loop if a buffer is explicitly considered. One is to close the loop around the buffer, and the other is to close the loop without the buffer. To understand the noise consequences of these two choices, model the buffer as an ideal unit-gain block whose output is corrupted by additive noise. For simplicity, you may assume that the added noise is white.

(a) First consider the case depicted in Figure 16.43. For simplicity, assume that the input signal to the PLL is completely noise-free. Furthermore, assume that the VCO is also noise-free. Given these two assumptions, sketch the expected output spectrum.

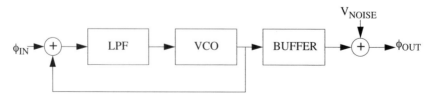

FIGURE 16.43. PLL with noisy buffer outside of loop.

(b) Now consider enclosing the buffer (and its noise source) inside the PLL, as shown in Figure 16.44. With the same assumptions as in part (a), what is the spectrum of the output now? Take into account the effect of the loop filter on the resulting spectrum.

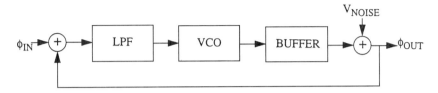

FIGURE 16.44. PLL with noisy buffer inside of loop.

(c) Given your answers to the previous parts of this question, which architecture leads to a cleaner output spectrum? Would your answer change if the input signal to the PLL were not perfectly noiseless? If so, how?

PROBLEM 6 In all ordinary PLLs, the voltage driving the VCO consists of a DC value (corresponding to the correct average frequency of the PLL output) on which some ripple component rides.

(a) Assume that an exclusive-OR gate is used as the phase detector. Sketch the output of the phase detector as a function of time, assuming square-wave inputs when the PLL is locked.

(b) If you additionally assume that the VCO has a perfectly linear frequency-versus-control voltage characteristic, then the sketch you drew in part (a) is in fact a plot of frequency versus time. Clearly, the output spectrum cannot be spectrally pure. Using whatever simulation tools are at your disposal, plot the output spectrum of a perfect sinusoidal VCO driven by such a control voltage waveform.

(c) Comment on the effectiveness of filtering the control voltage to reduce the ripple. Can you provide arbitrarily large attenuation of this ripple? If not, why not?

PROBLEM 7 Assume that, for a particular first-order loop, the VCO has a gain constant K_O of 200π Mrps/V, K_D is 0.8 V/rad, and the oscillation frequency $f_{osc} = 500$ MHz. Sketch the control voltage at the output of the phase detector if the input frequency jumps from 500 MHz to 650 MHz.

PROBLEM 8 In many PLL-based frequency synthesizers, a frequency divider is used in the feedback path, allowing the generation of output frequencies that are multiples of an input reference frequency. These dividers are almost always constructed out of digital logic elements that can introduce noise into the loop, as shown in Figure 16.45.

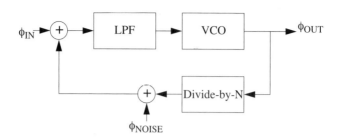

FIGURE 16.45. PLL with noisy divide-by-*N* in feedback path.

Sketch a Bode plot of the transfer function $\phi_{\text{out}}/\phi_{\text{noise}}$ and discuss how the bandwidth of the loop can alter the effect of the divider's noise. You may assume that the loop filter contains an integration and a loop-stabilizing zero; the VCO is modeled as a perfect integrator.

PROBLEM 9   Consider the second-order phase-locked loop used to generate a 1-GHz output signal (Figure 16.46). Noise coupled into the PLL by other circuitry is modeled as additive at the control port of the VCO. The VCO has a gain constant K_O of 400 Mrps/V.

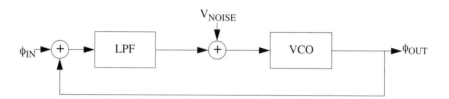

FIGURE 16.46. Second-order PLL with noise injection.

The input reference signal is supplied by a crystal oscillator whose output spectrum is exceptionally free of noise. The design problem is to select loop constants that guarantee reasonable immunity to V_{noise}.

(a) Determine the effective phase detector gain K_D and loop-stabilizing zero time constant τ_z to produce a maximum output phase error of one radian in response to a 100-mV step on V_{noise}.

(b) How would your answer to part (a) change if the maximum tolerable phase error were 0.1 rad?

PROBLEM 10   Practical VCOs generate imperfect signals. Suppose we model one such imperfection, noise, as a voltage added to the output of the VCO; see Figure 16.47.

In this model, then, the VCO is assumed to produce a perfectly pure output signal that is corrupted by the additive noise voltage shown. Assume further that the input

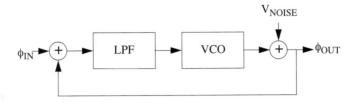

FIGURE 16.47. PLL with noisy VCO.

signal to the loop is also noiseless. Derive an expression for the output spectrum in terms of the transfer characteristics of the loop elements. You may assume that the spectrum of V_{noise} is white.

What does your equation say about how one should select loop bandwidth?

CHAPTER SEVENTEEN

OSCILLATORS AND SYNTHESIZERS

17.1 INTRODUCTION

Given the effort expended in avoiding instability in most feedback systems, it would seem trivial to construct oscillators. However, simply generating some periodic output is not sufficient for modern high-performance RF receivers and transmitters. Issues of spectral purity and amplitude stability must be addressed.

In this chapter, we consider several aspects of oscillator design. First, we show why purely linear oscillators are a practical impossibility. We then present a linearization technique that uses *describing functions* to develop insight into how nonlinearities affect oscillator performance, with a particular emphasis on predicting the amplitude of oscillation.

A survey of resonator technologies is included, and we also revisit PLLs, this time in the context of frequency synthesizers. We conclude this chapter with a survey of oscillator architectures. The important issue of phase noise is considered in detail in Chapter 18.

17.2 THE PROBLEM WITH PURELY LINEAR OSCILLATORS

In negative feedback systems, we aim for large positive phase margins to avoid instability. To make an oscillator, then, it might seem that all we have to do is shoot for zero or negative phase margins. Let's examine this notion more carefully, using the root locus for *positive* feedback sketched in Figure 17.1.

This locus recurs frequently in oscillator design because it applies to a two-pole bandpass resonator with feedback. As seen in the locus, the closed-loop poles lie exactly on the imaginary axis for some particular value of loop transmission magnitude. The corresponding impulse response is therefore a sinusoid that neither decays nor grows with time, and it would seem that we have an oscillator.

There are a couple of practical difficulties with this scenario, however. First, the amplitude of the oscillation depends on the magnitude of the impulse (it is a linear

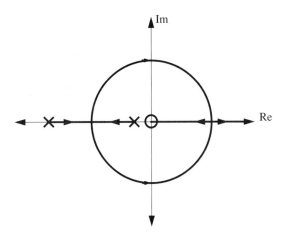

FIGURE 17.1. Root locus for oscillator example.

system, after all). This behavior is generally undesirable; in nearly all cases, we want the oscillator to produce a constant-amplitude output that is independent of initial conditions. Another problem is that if the closed-loop poles don't lie *precisely* on the imaginary axis, then oscillations will either grow or decay exponentially with time.

These problems are inherent in any purely linear approach to oscillator design. The solution to these problems therefore lies in a purposeful exploitation of non-linear effects; *all practical oscillators depend on nonlinearities.* To understand just how nonlinearities can be beneficial in this context, and to develop intuition useful for both analysis and design, we now consider the subject of describing functions.

17.3 DESCRIBING FUNCTIONS

We have seen that linear descriptions of systems often suffice, even if those systems are nonlinear. For example, the incremental model of a bipolar transistor arises from a linearization of the device's inherent exponential transfer characteristic. As long as excitations are "sufficiently small," the assumption of linear behavior is well satisfied.

An alternative to linearizing an input–output transfer characteristic is to perform the linearization in the *frequency domain.* Specifically, consider exciting a nonlinear system with a sinusoid of some particular frequency and amplitude. The output will generally consist of a number of sinusoids of various frequencies and amplitudes. A linear description of the system can be obtained by discarding all output components except the one whose frequency matches that of the input. The collection of all possible input–output phase shifts and amplitude ratios for the surviving component comprises the describing function for the nonlinearity. If the output spectrum is dominated by the fundamental component, results obtained with a describing function approximation will be reasonably accurate.

To validate further our subsequent analyses, we will also impose the following re-striction on the nonlinearities: they must generate no subharmonics of the input (DC is a subharmonic). The reason for this restriction will become clear momentarily. For RF systems, this requirement is perhaps not as restrictive as it initially appears, because bandpass filters can often be used to eliminate subharmonic and harmonic components.

As a specific example of generating a describing function, consider an ideal comparator whose output depends on the input as follows:

$$V_{\text{out}} = B \, \text{sgn} \, V_{\text{in}}. \tag{1}$$

If we drive such a comparator with a sine wave of some frequency ω and amplitude E, then the output will be a square wave of the same frequency but of a constant amplitude B, independent of the input amplitude. Furthermore, the zero crossings of the input and output will coincide (so there is no phase shift). Hence, the output can be expressed as the following Fourier series:

$$V_{\text{out}} = \frac{4B}{\pi} \sum_{1}^{\infty} \frac{\sin \omega n t}{n}, \quad n \text{ odd.} \tag{2}$$

Preserving only the fundamental term ($n = 1$) and taking the ratio of output to input yields the describing function for the comparator:

$$G_D(E) = \frac{4B}{\pi E}. \tag{3}$$

Since there is no phase shift or frequency dependence in this particular case, the describing function depends only on the input amplitude.

Note that the describing function for the comparator shows that the effective gain is *inversely proportional* to the drive amplitude, in contrast with a purely linear system in which the gain is *independent* of drive amplitude. We shall soon see that this inverse gain behavior can be extremely useful in providing negative feedback to stabilize the amplitude.

17.3.1 A BRIEF CATALOG OF DESCRIBING FUNCTIONS

Having shown how one goes about generating describing functions, we now present a small list, without derivation, of describing functions for some commonly encountered nonlinearities.[1]

For a saturating amplifier (Figure 17.2) we have

$$G_D(E) = \begin{cases} K & \text{if } E < E_M, \\ (2K/\pi)\left(\sin^{-1} R + R\sqrt{1 - R^2}\right) & \text{if } E > E_M, \end{cases}$$

where $R = E_M/E$. For an amplifier with crossover distortion (Figure 17.3), the describing function is

$$G_D(E) = \begin{cases} 0 & \text{if } E < E_M, \\ K\left[1 - (2/\pi)\left(\sin^{-1} R + R\sqrt{1 - R^2}\right)\right] & \text{if } E > E_M. \end{cases}$$

---

[1] See e.g. the excellent book by J. K. Roberge, *Operational Amplifiers,* Wiley, New York, 1975.

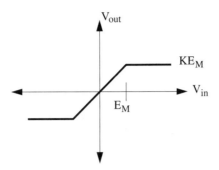

FIGURE 17.2. Transfer characteristic for saturating amplifier.

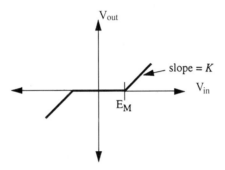

FIGURE 17.3. Transfer characteristic for amplifier with crossover distortion.

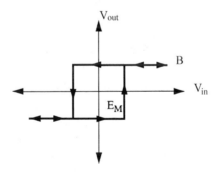

FIGURE 17.4. Transfer characteristic for Schmitt trigger.

Finally, for a Schmitt trigger we have

$$G_D(E) = 4B/\pi E, \quad \angle[-\sin^{-1} R]$$

for $R < 1$ (see Figure 17.4). In this last example, the value of R must be less than unity because otherwise the Schmitt never triggers, in which case the comparator output would be only a DC value of either B or $-B$.

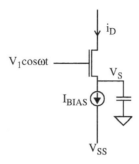

FIGURE 17.5. Large-signal transconductor.

It is important to note that describing functions themselves are linear even though the functions that they describe may be nonlinear (got that?). Hence, superposition holds; the describing function for a sum of nonlinearities is equal to the sum of the individual describing functions. This property is extremely useful for deriving describing functions for nonlinearities not included in the short catalog presented here.

17.3.2 DESCRIBING FUNCTIONS FOR MOS AND BIPOLAR TRANSISTORS

Although the foregoing collection of describing functions is extremely useful, perhaps more relevant to the RF oscillator design problem are describing functions for one- and two-transistor circuits, since the high frequencies that characterize RF operation are difficult to generate with many transistors in a loop.

To illustrate a general approach, consider the circuit in Figure 17.5. The capacitor is assumed large enough to behave as a short at frequency ω, and the transistor is ideal. We will be using this circuit in tuned oscillators, so the bandpass action provided by the tank guarantees that describing function analysis will yield accurate results.

Before embarking on a detailed derivation of the large-signal (i.e., describing function) transconductance, let's anticipate the qualitative outlines of the result. As the amplitude V_1 increases, the source voltage V_s is pulled to higher values, reaching a maximum roughly when the input does. Soon after the gate drive heads back downward from the peak, the transistor cuts off as the input voltage falls faster than the current source can discharge the capacitor. Because the current source discharges the capacitor between cycles, the gate–source junction again forward-biases when the input returns to near its peak value, resulting in a pulse of drain current. The cycle repeats, so the drain current consists of periodic pulses.

Remarkably, we do not need to know any more about the detailed shape of drain current in order to derive quantitatively the large-signal transconductance in the limit of large drive amplitudes. The only relevant fact is that the current pulses consist

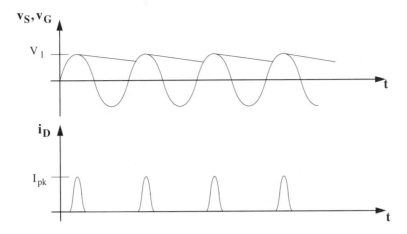

FIGURE 17.6. Hypothetical source and gate voltage, and drain current, for large input voltage.

of relatively narrow slivers in that limit, as in the hypothetical plots of gate voltage, source voltage, and drain current shown in Figure 17.6.[2]

Whatever the current waveform, KCL demands that its average value equal I_{BIAS}. That is,

$$\langle i_D \rangle = \frac{1}{T} \int_0^T i_D(t)\, dt = I_{\text{BIAS}}. \tag{4}$$

Now, the fundamental component of the drain current has an amplitude given by

$$I_1 = \frac{2}{T} \int_0^T i_D(t) \cos \omega t\, dt. \tag{5}$$

Although we may not know the detailed functional form of $i_D(t)$, we do know that it consists of narrow pulses in the limit of large drive amplitudes. Furthermore, these current pulses occur roughly when the input is a maximum, so that the cosine may be approximated there by unity for the short duration of the pulse. Then,

$$I_1 = \frac{2}{T} \int_0^T i_D(t) \cos \omega t\, dt \approx \frac{2}{T} \int_0^T i_D(t)\, dt = 2 I_{\text{BIAS}}. \tag{6}$$

That is, the amplitude of the fundamental component is approximately twice the bias current, again in the limit of large V_1. The magnitude of the describing function is therefore

---

[2] The word "hypothetical" is here a euphemism for "wrong." However, even though the detailed waveforms shown are not strictly correct, the results and insights obtained are. In particular, this picture allows us to understand why the describing function transconductance for large drive amplitudes is essentially the same for bipolars and MOSFETs (both long- and short-channel), as well as for JFETs and vacuum tubes.

$$G_m = \frac{I_1}{V_1} \approx \frac{2I_{\text{BIAS}}}{V_1}. \tag{7}$$

It is important to note that the foregoing derivation does not depend on detailed transistor characteristics at any step along the way. Because no device-specific assumptions are used, Eqn. 7 is quite general and applies to MOSFETs (both long- and short-channel), as well as to bipolars, JFETs, GaAs MESFETs, and even vacuum tubes.

In deriving Eqn. 7, we have assumed that the drive amplitude, V_1, is "large." To quantify this notion, let us compute the G_m/g_m ratio for long- and short-channel MOSFETs and bipolar devices. For long-channel devices, the ratio of g_m to drain current I_{BIAS} may be written as

$$\frac{g_m}{I_{\text{BIAS}}} = \frac{2}{V_{gs} - V_t}, \tag{8}$$

so that

$$\frac{G_m}{g_m} = \frac{V_{gs} - V_t}{V_1}. \tag{9}$$

Evidently, "large" V_1 is defined relative to $(V_{gs} - V_t)$ for long-channel MOSFETs.

Repeating this exercise for short-channel devices yields[3]

$$\frac{g_m}{I_{\text{BIAS}}} = \frac{2}{V_{gs} - V_t} - \frac{1}{E_{\text{sat}}L + (V_{gs} - V_t)}, \tag{10}$$

which, in the limit of very short channels, converges to a value precisely half that of the long-channel case. Thus,

$$\frac{V_{gs} - V_t}{V_1} \leq \frac{G_m}{g_m} < \frac{2(V_{gs} - V_t)}{V_1}. \tag{11}$$

Finally, for bipolar devices we have

$$\frac{g_m}{I_{\text{BIAS}}} = \frac{1}{V_T}, \tag{12}$$

so that

$$\frac{G_m}{g_m} = \frac{2V_T}{V_1}. \tag{13}$$

In bipolar devices, large V_1 is therefore defined relative to the thermal voltage.

Although the equation for G_m is valid only for large V_1, practical oscillators usually satisfy this condition and so the restriction is much less constraining than one might think. We will also see in the next chapter that large V_1 is highly desirable for reducing phase noise, so one may argue that all well-designed oscillators automatically satisfy the conditions necessary to validate the approximations used. Nonetheless, it

---

[3] Here we have used the approximate, analytic model for short-channel MOSFETs introduced in Chapter 5.

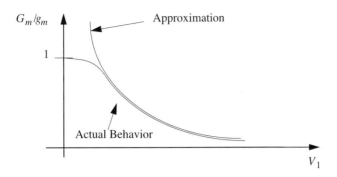

FIGURE 17.7. G_m/g_m versus V_1.

is important to recognize that G_m can never exceed g_m, so one must be careful not to misapply formulas such as Eqn. 13. To underscore this point, Figure 17.7 shows, in an approximate way, the actual behavior of G_m/g_m contrasted with the behavior as predicted by Eqn. 13. Although this equation applies strictly to the bipolar case, the overall behavior shown in Figure 17.7 holds generally.

Having presented numerous describing functions, we now consider two examples that illustrate how we can use them to analyze oscillators.

17.3.3 EXAMPLE 1: FUNCTION GENERATOR

To get good results with describing functions, it's important to satisfy the conditions used to derive them. That is, the circuit must be sufficiently low-pass or bandpass in nature to provide a near-sinusoidal drive to the nonlinearity. If this condition is not well satisfied, the results of describing function analysis cannot be expected to be accurate. Fortunately, many systems of practical interest do satisfy this requirement, and describing functions can be used successfully in other than purely academic settings. Oscillators for use in communications systems are particularly amenable to describing function analysis because high-Q resonators are often present, forcing nearly sinusoidal drives.

To illustrate the utility of describing functions, let's analyze an oscillator two different ways. As a specific example, consider the circuit of Figure 17.8, which is used as the basis for many laboratory function generators (i.e., the things that generate sine, square, and triangle waves, all in one instrument).

As seen in the figure, the oscillator consists of a Schmitt trigger and an inverting integrator. The output of the Schmitt is a square wave of amplitude B, while the output of the integrator is a triangle wave of amplitude E_M. Specifically, the output waveforms are as depicted in Figure 17.9. From this direct analysis in the time domain, we see that the period of oscillation is simply

$$T_{\text{osc}} = 4 \cdot \frac{E_M}{B}. \tag{14}$$

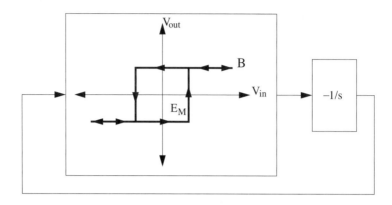

FIGURE 17.8. Function generator core.

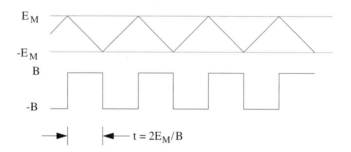

FIGURE 17.9. Function generator waveforms.

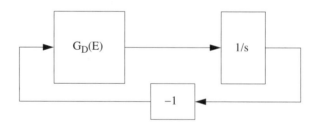

FIGURE 17.10. Describing function loop model for oscillator.

Now that we know the exact answer, let's see if we can obtain a reasonable prediction of the amplitude and frequency of oscillation from a describing function analysis. Before proceeding further, we note that some error is to be expected, since the waveform at the input to the nonlinearity (the Schmitt trigger) is not a particularly good approximation to a sinusoid. Nevertheless, a triangle wave's spectrum falls off as $1/\omega^2$, so perhaps the analysis won't be completely worthless.

To analyze this oscillator with describing functions, consider Figure 17.10's model for the loop. We have explicitly assigned the inversion its own separate block in order

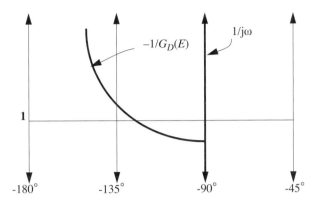

FIGURE 17.11. Gain phase plots for oscillator example.

to make the model consistent with the negative feedback–loop block diagrams we have used previously.

Recall that a necessary (but not sufficient) condition for oscillation is a unit loop transmission magnitude and a zero phase margin. For this system, that requirement translates to

$$G_D(E)\frac{1}{j\omega} = -1. \tag{15}$$

We can express this requirement a little differently:

$$\frac{1}{j\omega} = \frac{-1}{G_D(E)} = \frac{-1}{4B/\pi E \ (\angle[-\sin^{-1} R])} = -\frac{\pi E}{4B} \ (\angle[\sin^{-1}(E_M/E)]). \tag{16}$$

This last equation suggests a graphical technique for discovering possible oscillation frequencies and amplitudes: Plot the behavior of the purely linear stuff in the *gain phase plane,* and similarly plot the negative reciprocal of the describing function.[4] If any intersections occur, the corresponding frequencies and amplitudes are possible oscillatory states.

For the sake of simplicity, let $B = E_M = 1$. Then, applying this recipe to our specific example, we obtain plots in the gain phase plane (see Figure 17.11). There is only one intersection,[5] and it occurs at an amplitude $E = 1$ and at a frequency $\omega = 4/\pi$ rps, corresponding to a period of oscillation of

$$T_{\text{osc}} = \frac{\pi^2}{2} \approx 4.9 \text{ s.} \tag{17}$$

The exact analysis performed earlier shows that the actual amplitude is in fact unity for these values, while the oscillation period is 4 s. (In general, the intersection occurs when the amplitude is $E = E_m$, corresponding to an oscillation frequency ω

---

[4] Recall that the magnitude of the reciprocal is just the reciprocal of the magnitude. Similarly, the phase of the reciprocal is just the algebraic inverse of the original phase.

[5] Strictly speaking, we have a tangency rather than an intersection. Nevertheless, we see that the results of the describing function analysis are in reasonable agreement with the exact analysis.

of $4/\pi E_m$.) This level of agreement is fairly reasonable considering that a triangle wave, rather than a sinusoid, drives the nonlinearity.

We can glean some important insights from this type of analysis. If the goal is to achieve high spectral purity, then it is desirable for the nonlinearity to be relatively "soft." Furthermore, the linear elements in the loop should be low-pass (or perhaps bandpass) in nature to attenuate the distortion products that are generated by the non-linearity. Finally, one should use the *input* to the nonlinearity as the output signal, since it will have the lowest distortion. As an additional bonus, satisfying all of these requirements also guarantees conditions that favor highly accurate describing func-tion analyses.

As a closing note on this example, it should be mentioned that the sine-wave output in commercial function generators is obtained by *nonlinear* waveshaping of the triangle wave output, as discussed in an earlier chapter. The maximum operat-ing frequency for discrete implementations of such generators is typically around 30–50 MHz and is limited primarily by the quality of the square wave at higher fre-quencies. At higher frequencies, general-purpose function generators are replaced by special-purpose circuits, such as high-speed square-wave or pulse generators and tuned sine-wave oscillators, that use different architectures optimized for each par-ticular type of output waveform.

17.3.4 EXAMPLE 2: COLPITTS OSCILLATOR

Relaxation oscillators such as the function generator are rarely used in high-perfor-mance transceivers because they generate signals of inadequate spectral purity. Much more common are tuned oscillators, primarily for reasons that we may appreciate only after studying the subject of phase noise. For now, simply accept as an axiom the superiority of tuned oscillators. Our present focus, then, is the use of describing func-tions to predict the output amplitude of a typical tuned oscillator, such as the Colpitts circuit shown in Figure 17.12.[6]

We shall see in this chapter that a variety of oscillators differing in trivial details are named for their inventors. In keeping with standard practice, we will retain this naming convention, but the reader is advised to focus on operating principles rather than nomenclature.

The basic recipe for these oscillators is simple: Combine a resonator with an ac-tive device. The distinguishing feature of a Colpitts oscillator is the capacitively tapped resonator, with positive feedback provided by the active device to make os-cillations possible. In Figure 17.12, the resistance R represents the total loading due to finite tank Q, transistor output resistance, and whatever is driven by the oscilla-tor (presumably the oscillator's output is used somewhere). The current source is

---

[6] Edwin Henry Colpitts devised his oscillator in early 1915, while at Western Electric. His colleague Ralph Vinton Lyon Hartley had demonstrated *his* oscillator just a month earlier, on February 10th.

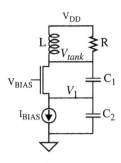

FIGURE 17.12. Colpitts oscillator
(biasing details not shown).

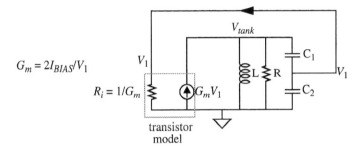

FIGURE 17.13. Describing function model of Colpitts oscillator.

frequently replaced by an ordinary resistor in practical implementations, but is used here to simplify (marginally) the analysis.

From our describing function derivation, we know that the transistor may be characterized by a large-signal transconductance G_m. For the sake of simplicity, we will ignore all dynamic elements of the transistor, as well as all parasitic resistances, although an accurate analysis ought to take these into account. The transistor also has a large-signal source–gate resistance, of course, which must be modeled as well. Taking a cue from describing functions, it seems reasonable to define this resistance as the ratio of the fundamental component of source current to the source–gate voltage. We've actually already found this ratio; it is simply $1/G_m$. Thus, we may model the oscillator with the circuit of Figure 17.13.

To simplify the analysis further, first reflect the input resistance R_i across the main tank terminals by treating the capacitive divider as an ideal transformer (see Chapter 3), so that we end up with a simple RLC tank embedded within a positive feedback loop. Note that the resulting circuit has zero phase margin at the resonant frequency of the tank; that will be the oscillation frequency in this particular case. Note also that the dependent current generator produces an output (sinusoid) whose amplitude is $G_m V_1 = 2I_{\text{BIAS}}$ at all times, so it may be replaced with an independent sine generator of this amplitude. Acting on these observations leads to the circuit of Figure 17.14, where C_{eq} is the series combination of the two capacitors,

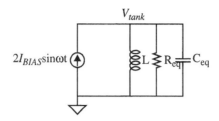

FIGURE 17.14. Simplified model
of Colpitts oscillator.

$$C_{eq} = \frac{C_1 C_2}{C_1 + C_2}, \tag{18}$$

and

$$\omega = \frac{1}{\sqrt{LC_{eq}}}. \tag{19}$$

Similarly, R_{eq} is the parallel combination of the original tank resistance R and the reflected large-signal input resistance of the transistor:[7]

$$R_{eq} \approx R \parallel \frac{1}{n^2 G_m}, \tag{20}$$

where n is the capacitive voltage divide factor,

$$n \equiv \frac{C_1}{C_1 + C_2}. \tag{21}$$

The amplitude V_1 is simply the amplitude V_{tank} of the tank voltage multiplied by the capacitive divide factor, so we may write

$$V_{tank} \approx \frac{V_1}{n}. \tag{22}$$

Now we have collected enough equations to get the job done. The amplitude of the tank voltage at resonance is simply the product of the current source amplitude and the net tank resistance:

$$V_{tank} \approx \frac{V_1}{n} \approx 2I_{BIAS} R_{eq}$$

$$\approx (2I_{BIAS}) \left[R \parallel \frac{1}{n^2 G_m} \right] = (2I_{BIAS}) \cdot \frac{R}{n^2 G_m R + 1}, \tag{23}$$

which ultimately simplifies to

$$V_{tank} \approx 2I_{BIAS} R(1-n). \tag{24}$$

---

[7] In this and related equations, the reason for the "approximately equals" symbol is that we are treating the capacitive divider as an ideal impedance transformer. As mentioned in Chapter 3, the approximation is good as long as the in-circuit Q is large.

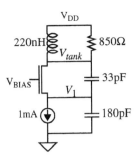

FIGURE 17.15. Colpitts
oscillator example.

Thus, the amplitude of oscillation is directly proportional to the bias current and the effective tank resistance. The loading of the tank by the transistor's input resistance is taken into account by the $(1 - n)$ factor and is therefore controllable by choice of the capacitive divide ratio. Since R also controls Q, it is usually made as large as possible, and adjustment of I_{BIAS} is consequently the main method of defining the amplitude.

As a specific numerical example, consider the ~60-MHz oscillator circuit that is sketched in Figure 17.15. For the particular element values shown, the capacitive divide factor n is about 0.155.[8] The expected oscillation amplitude (V_{tank}) is therefore about 1.4 V. Measurements made on a bipolar version of this circuit reveal an amplitude of 1.3 V, in good agreement with theoretical predictions. It is important to underscore again that this result is largely independent of the type of active device used to build the oscillator. The prediction was made for a MOSFET design and experimentally verified with a bipolar device.

Start-up, Second-Order Effects, and Pathologies

In the preceding analysis, nothing specific was mentioned about conditions for guaranteeing the start-up of oscillations. From the general root locus of Figure 17.1, however, it should be clear that a necessary condition is a greater-than-unity value of small-signal loop transmission. To evaluate whether start-up might be a problem, one should set the transconductance equal to its small-signal value (an appropriate choice, as the circuit is certainly in the small-signal regime before oscillations have started) and compute the loop transmission magnitude. If it does not exceed unity then the oscillator will not start up. To fix this problem, adjust some combination of bias current, device size, and tapping ratio.

In the case of the example just considered, let us identify the minimum acceptable transconductance for guaranteeing start-up. That minimum g_m, together with

---

[8] In practice, it is generally true that best phase noise performance tends to occur for a C_2/C_1 ratio of about 4, corresponding to $n = 0.2$. This rule of thumb can be put on a more rigorous theoretical basis by making use of the time-varying theory discussed in Chapter 18.

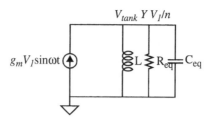

FIGURE 17.16. Start-up model
of Colpitts oscillator.

the given value of bias current, defines the width of the device. We use the model in
Figure 17.16. The amplitude of the voltage across the tank at resonance is

$$V_{\text{tank}} = \frac{V_1}{n} = g_m V_1 R_{\text{eq}} = g_m V_1 \left[R \parallel \frac{1}{n^2 g_m} \right],\tag{25}$$

which reduces to the following expression for the minimum transconductance:

$$g_m > \frac{1}{R(n - n^2)}.\tag{26}$$

With $n = 0.155$ and $R = 850\ \Omega$, the absolute minimum acceptable transconductance works out to approximately 9 mS. However, note that merely having enough
transconductance to achieve net unit loop gain with no oscillation is not sufficient to
make a good oscillator. Additionally, the describing function is accurate only in the
limit of large amplitudes, and therefore only if the small-signal transconductance is
substantially larger than the large-signal value. A reasonable choice for a first-cut
design is to select g_m to be five times the minimum acceptable value. Hence, we will
design for a 45-mS small-signal transconductance.

To estimate the necessary device width, initially assume that the gate overdrive
is small enough that the device conforms to square-law behavior. Then we may use
Eqn. 8 to estimate the overdrive as follows:

$$\frac{g_m}{I_{\text{BIAS}}} = \frac{2}{V_{gs} - V_t} \implies V_{gs} - V_t \approx 44\ \text{mV}.\tag{27}$$

This overdrive is indeed small compared with typical values of $E_{\text{sat}}L$ (e.g., 1–2 V), so
we will continue to assume operation in the long-channel regime. Solving for W/L in
this regime then yields a value of about 6000 for typical values of mobility and C_{ox}.
For a 0.5-μm channel length, then, the width should be roughly 3000 μm, which is
quite large. This large width is a consequence of the low bias current; a higher bias
current would permit the use of a substantially smaller device.

Aside from the neglect of start-up conditions in the foregoing development, several other simplifying assumptions were invoked to reduce clutter in the derivations.
Transistor parasitics were ignored, for example. We now consider how to modify the
analysis to take these into account.

The gate–drain and drain–bulk capacitances appear in parallel with the tank, and a first-order correction for their effect simply involves a reduction in the explicit capacitance added externally to keep the oscillation frequency constant. However, these capacitances are nonlinear, so distortion may be unsatisfactorily high if they constitute a significant fraction of the total tank capacitance. Temperature drift properties may also be affected.

The source–gate and source–bulk capacitances appear directly in parallel with C_2, and the same comments apply as for the other device capacitances.

One must also worry about the output resistance of the transistor, for it loads the tank as well. Many high-speed transistors have low Early voltages (e.g., 10–20 V or less), so this loading can be significant at times. In serious cases, cascoding (or some equivalent remedy) may be necessary to mitigate this problem. In other instances, this loading merely needs to be taken into account to predict the amplitude more accurately.

As a final comment on the issue of amplitude, it must be emphasized that there is always the possibility of amplitude instability, since feedback control of the amplitude is fundamentally involved. That is, instead of staying constant, the amplitude may vary in some manner (e.g., quasisinusoidally). This type of behavior is known as "squegging" and is the bane of oscillator designers.

To see how squegging might arise and to develop insights about how to prevent or cure it, we may employ the same analytical tools that we use to evaluate the stability of other feedback systems. Hence, we invoke once again the concepts of loop transmission, crossover frequency, and phase margin. The main subtlety is that we must evaluate these quantities in terms of the *envelope* of the RF signal. Another is that the nonlinearity of amplitude control renders our linearized analyses relevant only near the operating point assumed in the linearization. Proceeding with awareness of those considerations, we would cut the loop at some convenient point (while taking care to preserve all loadings, just as we must in evaluating any loop transmission) and then apply an RF signal to the input of the cut loop. The amplitude of this test signal should be chosen to be the same as the nominal amplitude that prevails in actual closed-loop operation, ensuring that we evaluate stability under conditions that correspond to normal operation. Given the nonlinear nature of amplitude control, it's also prudent to examine the loop transmission at several amplitudes in order to identify (or preclude) ranges of amplitudes that may result in squegging behavior.

Next, we have a choice of evaluating either the time- or frequency-domain response (or both). In the former, we would examine the loop transmission's response to a step change in *amplitude*. To evaluate the *envelope loop transmission* in the frequency domain, we apply a sinusoidally modulated RF carrier to the input of the cut loop and then sweep the frequency of the modulation, noting the gain and phase of the output modulation relative to the input modulation.

For tuned oscillators with drain-fed tanks, a natural choice is in fact to cut the loop at the drain. Inject an RF current into the tank at that point, and then let the RF current's amplitude undergo a step change. The tank by itself provides the equivalent of

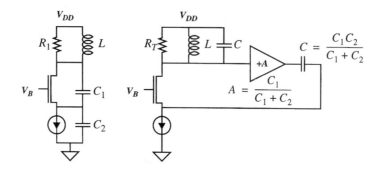

FIGURE 17.17. Colpitts oscillator and equivalent model
for evaluation of envelope loop transmission.

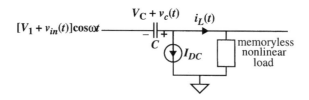

FIGURE 17.18. Capacitively coupled
circuit with nonlinear load.

single-pole filtering of the step, and the capacitive coupling into the source terminal of the transistor contributes additional dynamics.

To illustrate the procedure in detail, consider a Colpitts oscillator. To simplify analysis, we first make an equivalent circuit, shown on the right of Figure 17.17. That the two circuits are in fact equivalent is readily verified by comparing loop transmissions. The drain connection is a particularly convenient point to cut the loops for making this comparison. We see that the two circuits are in fact identical as long as the elements explicitly shown in the schematics include all device parasitics.

Having derived an equivalent circuit for computing the envelope loop transmission, we now decompose the loop transmission into individual pieces. First, we analyze the capacitively coupled circuit[9] shown in Figure 17.18. Here, the load current consists of DC and RF components:[10]

$$i_L(t) = I_{\text{DC}} + i_{\text{dc}}(t) + [I_{\text{OUT}} + i_{\text{out}}(t)] \cos \omega t. \qquad (28)$$

The quiescent values of DC current and RF current amplitude are I_{DC} and I_{OUT}, respectively, corresponding to an RF drive amplitude of V_1. Perturbing that drive amplitude by an amount $v_{\text{in}}(t)$ produces three effects in general. One is a change in the

---

[9] This analysis is an adaptation to MOS form of that presented by Kenneth K. Clarke and Donald T. Hess, *Communications Circuits: Analysis and Design,* Krieger, Malabar, FL, 1994.

[10] We use the term *DC* nonrigorously, to distinguish from RF components.

DC value of the voltage across the capacitor by an amount $v_c(t)$ owing to rectification by the nonlinear load. If the amplitude of the RF input voltage changes, that DC capacitor voltage generally changes as well.

Such rectification also changes the DC current through the nonlinear load by an amount $i_{dc}(t)$. Finally, there is a change in the amplitude $i_{out}(t)$ of the RF current flowing into the nonlinear load. That amplitude change results from direct action by $v_{in}(t)$ compounded by the change in the DC current through the load.

We wish to determine the small-signal admittance $i_{out}(s)/v_{in}(s)$, but doing so by inspection is nontrivial. Note that our very statement of the problem essentially presumes that small-signal analysis holds. If we validate this assumption by considering only cases where $|v_{in}(t)| \ll V_1$, then we may express each of the currents $i_{dc}(t)$ and $i_{out}(t)$ as a simple linear combination of the voltages $v_c(t)$ and $v_{in}(t)$. After Laplace transformation, we therefore have

$$i_{dc}(s) = G_{00}v_c(s) + G_{01}v_{in}(s) \tag{29}$$

and

$$i_{out}(s) = G_{10}v_c(s) + G_{11}v_{in}(s), \tag{30}$$

where the various constants G_{mn} are conductances to be determined later.

Because it remains true that

$$i_{dc}(s) = -sCv_c(s), \tag{31}$$

we may equate the two expressions for $i_{dc}(s)$ to obtain

$$-sCv_c(s) = G_{00}v_c(s) + G_{01}v_{in}(s) \implies v_c(s) = \frac{-G_{01}}{sC + G_{00}}v_{in}(s). \tag{32}$$

This equation states that the small-signal DC capacitor voltage is simply a low-pass–filtered version of the small-signal RF input amplitude.

Substitution of Eqn. 32 into Eqn. 30 then yields the desired small-signal relationship between input envelope voltage and output envelope current:

$$i_{out}(s) = G_{10}\frac{-G_{01}}{sC + G_{00}}v_{in}(s) + G_{11}v_{in}(s) \implies \frac{i_{out}(s)}{v_{in}(s)} = G_{11} - \frac{G_{01}G_{10}}{sC + G_{00}}. \tag{33}$$

After some rearrangement, this becomes

$$\frac{i_{out}(s)}{v_{in}(s)} = \frac{G_{11}(sC + G_{00}) - G_{01}G_{10}}{sC + G_{00}} = G_{11}\frac{\left(\dfrac{sC}{G_{00}} + 1\right) - \dfrac{G_{01}G_{10}}{G_{00}G_{11}}}{\dfrac{sC}{G_{00}} + 1}. \tag{34}$$

The derivation so far is completely general; Eqn. 34 is not limited to MOSFETs or bipolars. Without knowing anything about the various conductances, we can say that the admittance in question consists of a pole and a zero. This result makes physical sense, for we have one energy storage element (and therefore one pole). Furthermore, the capacitor provides a feedthrough path that emphasizes high-frequency content (both of the carrier and the envelope); that's the action of a zero. It's satisfying that the analysis presented so far passes this macroscopic reasonableness test.

Turning now to the task of figuring out what those conductances are, observe that the formal definitions for the various proportionality constants are readily obtained from the time-domain versions of Eqn. 29 and Eqn. 30:

$$G_{00} \equiv \left. \frac{di_{\text{dc}}}{dv_c} \right|_{v_{\text{in}}=0}, \tag{35}$$

$$G_{01} \equiv \left. \frac{di_{\text{dc}}}{dv_{\text{in}}} \right|_{v_c=0}, \tag{36}$$

$$G_{10} \equiv \left. \frac{di_{\text{out}}}{dv_c} \right|_{v_{\text{in}}=0}, \tag{37}$$

$$G_{11} \equiv \left. \frac{di_{\text{out}}}{dv_{\text{in}}} \right|_{v_c=0}. \tag{38}$$

Note that at least two of these conductances should be familiar. From its definition we see that G_{00} is simply the nonlinear load's small-signal ratio of DC current to DC voltage; hence G_{00} is the ordinary small-signal conductance evaluated at the bias point. Similarly, G_{11} is the change in the RF output current amplitude divided by the change in the RF input voltage amplitude, evaluated at constant capacitor drop. Thus, G_{11} is the describing-function conductance of the nonlinear load.

Two conductances we haven't encountered before involve the ratio of a DC term and an RF term. One, G_{01}, is the ratio of the change in rectified DC current divided by the change in the amplitude of the RF input voltage that produces that rectified current, evaluated at constant capacitor drop. The other, G_{10}, is the change in the amplitude of the RF output current divided by the change in DC capacitor voltage, for a constant-amplitude RF input voltage.

The last piece we need is a quantitative description of the envelope behavior of the drain tank. Specifically, consider a step change in the envelope of a sinusoidal drive current. The envelope response of the tank voltage will behave as a single-pole low-pass filter's response to a step voltage. Because the single-sided bandwidth of an RC low-pass filter is simply $1/RC$, we anticipate the corresponding time constant for an RLC bandpass filter to be $1/2RC$.[11] At resonance, the drain load thus contributes an *envelope impedance* given by

$$\frac{v_{\text{tnk}}(s)}{i_{\text{out}}(s)} = \frac{R_T}{2sR_TC + 1}. \tag{39}$$

---

[11] We emphasize *single-sided* to show more clearly the analogy between a low-pass and a bandpass filter. It is customary to measure a low-pass filter's bandwidth from DC to the positive-frequency -3-dB corner, rather than between the positive- and negative-frequency -3-dB corners. The pole time constant associated with that single-sided corner controls the risetime and is, of course, simply RC. For a bandpass filter, the single-sided bandwidth is $1/2RC$, meaning that the pole time constant that governs the envelope risetime is $2RC$.

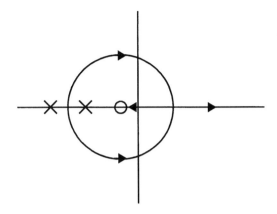

FIGURE 17.19. Possible root locus
for envelope feedback loop.

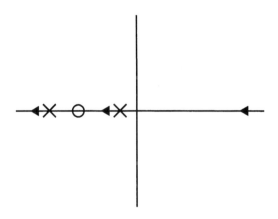

FIGURE 17.20. Another possible root locus
for envelope feedback loop.

The complete loop transmission is therefore

$$A \frac{i_{\text{out}}(s)}{v_{\text{in}}(s)} \frac{v_{\text{tnk}}(s)}{i_{\text{out}}(s)} = A \left(G_{11} \frac{\left(\dfrac{sC}{G_{00}} + 1 \right) - \dfrac{G_{01}G_{10}}{G_{00}G_{11}}}{\dfrac{sC}{G_{00}} + 1} \right) \frac{R_T}{2sR_TC + 1}. \tag{40}$$

Notice that we have two poles and a zero. Note also, somewhat ominously, that the envelope loop transmission is positive in sign. Positive feedback per se does not ensure instability, but we must avoid loop transmission magnitudes that are too large.

Figure 17.19 depicts one possible root locus corresponding to this loop transmission that reveals why squegging can occur. Since we don't know exactly where the zero might be, another possibility is as shown in Figure 17.20. In this case, the poles never become complex, but one of the poles can end up with a positive real part. If any pole enters the right half-plane then the envelope will be unstable. If there is a complex pole pair in that half-plane, the instability will be observed as a (quasi)sinusoidal modulation. If the poles are real, then the modulation will be relaxation-like in character.

Regrettably, we cannot make quantitative statements without considering a specific nonlinear load. Still more regrettably, a rigorous derivation is nigh impossible for real MOSFETs (it's hard enough for a bipolar transistor). Therefore, most practical evaluations of squegging require simulations at some point in the analysis. Discovering G_{01} and G_{10} through simulation is straightforward, though.[12] Thus, even though simple analytical expressions for these parameters may not be readily forthcoming, simulations will yield actual values for them without much trouble.

Even without performing such simulations, we can identify general strategies to stop squegging if it occurs. By analogy with the success of dominant pole compensation in ordinary amplifiers, we would consider increasing the tank Q. The attendant narrowing in bandwidth means that the pole it contributes to the amplitude control loop moves to a lower frequency (becomes more dominant). That forces crossover to occur at a lower frequency, where presumably there is greater phase margin.

If the resonator bandwidth cannot be practically narrowed, we may still reduce the amplitude loop's crossover frequency using other methods. For example, we could reduce the loop transmission by varying the capacitive tapping ratio to feed back less signal. And, of course, we always retain the option of combining strategies.

One possible difficulty is that many of these parameters are interdependent. Depending on how one achieves an increase in tank Q, for example, the envelope impedance of the tank could increase; this, in turn, would increase the loop transmission magnitude, frustrating efforts at stabilization. Thus, some deliberation is necessary to identify the best strategies for any given circuit.

In very stubborn cases, it may be necessary to impose external amplitude control (e.g., by explicitly measuring the amplitude, comparing it with a reference voltage, and then appropriately adjusting the bias current; see Figure 17.21). This decoupling of amplitude control from fundamental oscillator operation not only permits the exercise of additional degrees of freedom to solve the stability problem, it also allows one to design the oscillator without having to make compromises for such factors as start-up reliability (or speed) and amplitude stability.

Two additional terms that describe unwanted oscillator behaviors are *frequency pulling* and *supply pushing*. Both of these terms refer to shifts in oscillator frequency that are due to parasitic effects. Pulling may occur for numerous reasons that range from load changes to parasitic coupling from other periodic signals. Buffering and other isolation strategies help reduce pulling.

---

[12] The reader may reasonably ask why we should not simply simulate the entire oscillator. The answer is that squegging frequencies are usually quite a bit lower than the main oscillation frequency, so simulation times can quickly get out of hand. Using the results of a more rigorous analysis allows you to identify which simulations ought to be run.

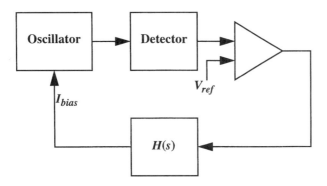

FIGURE 17.21. Oscillator with separate amplitude control.

Supply pushing reflects the unfortunate fact that oscillator frequency is not entirely independent of supply voltage. For example, device capacitances may change as a function of bias voltage, causing shifts in the oscillation frequency. Supply pushing is suppressed by selecting tank element values that swamp out such parasitic elements and by employing regulated supplies. It is important to filter the latter very well and also to watch out for $1/f$ fluctuations in the supply voltage, as these can cause (close-in) phase modulations of the oscillator.

17.4 RESONATORS

The previous describing function example analyzed a tuned oscillator. Since tuned circuits inherently perform a bandpass filtering function, distortion products and noise are attenuated relative to the fundamental component. Not surprising, then, is that the performance of these circuits is intimately linked to the quality of available resonators. Before proceeding onward to a detailed discussion of oscillator circuitry, then, we first survey a number of resonator technologies.

RESONATOR TECHNOLOGIES

Quarter-Wave Resonators

Aside from the familiar and venerable RLC tank circuit, there are many ways to make resonators. At high frequencies, it becomes increasingly difficult to obtain adequate Q from lumped resonators because required component values are often impractical to realize.

One alternative is to use a *distributed* resonator such as a quarter-wave piece of transmission line. To suggest why such a choice may be sensible, recall that Q is proportional to the ratio of energy stored to energy dissipated. Some distributed structures store energy in a volume, while dissipation is due mainly to surface effects (e.g., skin effect) within the volume. Hence, the volume/surface-area ratio is important in determining Q. This ratio may be made quite large for certain distributed structures, so high Q values are possible.

The required physical dimensions generally favor practical realization in discrete form in the UHF band and above. As an example, the free-space wavelength at 30 MHz is one meter, so that a quarter-wave resonator would be about ten inches. If the resonator is filled with a dielectric material other than air, the dimensions shrink as the square root of the relative dielectric constant.

Dimensions become compatible with ICs at mid-gigahertz frequencies. At 3 GHz, for example, the free-space quarter-wavelength is about one inch. With dielectric materials that are commonly available, quarter-wave IC resonators of about half an inch or so are possible.

If we terminate such a transmission line in a short, the input impedance is ideally an open circuit (limited by the Q of the line). At frequencies below the resonant frequency, the line appears as an inductor; above resonance, it appears as a capacitor. Hence, for small displacements about the resonant condition, the line appears very much like a parallel RLC network.

There is an important difference, however, between a shorted line and a lumped RLC resonator: the line appears as an infinite (or at least large) impedance at all odd multiples of the fundamental resonance. Sometimes, this periodic behavior is desired, but it can also result in oscillation simultaneously on multiple frequencies or a chaotic hopping from one mode to another. Additional tuned elements may be required to suppress oscillation on unwanted modes.

The oscillators in most cellular telephones use off-chip quarter-wave resonators in which a piezoelectric material, such as barium titanate, is used as the dielectric. The high dielectric constant of such materials allows the realization of physically small resonators that possess excellent Q-values (e.g., 20,000). It is virtually impossible to obtain such high Q-values using ordinary lumped elements.

Quartz Crystals

The most common non-RLC resonator is made of quartz. The remarkable properties and potential of quartz for use in the radio art were first seriously appreciated around 1920 by Walter G. Cady of Bell Laboratories.[13] Quartz is a piezoelectric material, and thus it exhibits a reciprocal transduction between mechanical strain and electric charge. When a voltage is applied across a slab of quartz, the crystal physically deforms. When a mechanical strain is applied, charges appear across the crystal.[14]

---

[13] W. G. Cady, "The Piezo-Electric Resonator," *Proc. IRE*, v. 10, April 1922, pp. 83–114. His first oscillator was somewhat complex, based as it was on a two-port piezoelectric filter.

[14] Piezoelectricity's mechanical-to-electrical transduction was discovered by Jacques and Pierre Curie (before the latter met and married Marie Sklodowska). See "Développement, par pression, de l'électricité polaire dans les cristaux hémièdres à faces inclinées" [Development, by pressure, of electrical polarization in hemihedral crystals with inclined faces], *Comptes Rendus des Séances de l'Académie des Sciences,* v. 91, 1880, pp. 294–5. Their friend, physicist Gabriel

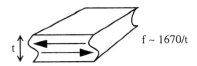

FIGURE 17.22. Illustration of
bulk shear mode.

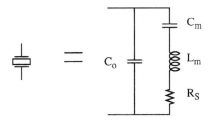

FIGURE 17.23. Symbol and
model for crystal.

Most practical quartz crystals used at radio frequencies[15] employ a bulk shear vibrational mode (see Figure 17.22). In this mode, the resonant frequency is inversely proportional to the thickness of the slab, according to the rough formula[16] in the figure (SI units assumed).

Even though quartz does not exhibit a particularly large piezoelectric effect, it has other properties that make it extremely valuable for use in RF circuitry. Chief among them is the exceptional stability (both electrical and mechanical) of the material. Furthermore, it is possible to obtain crystals with very low temperature coefficients by cutting the quartz at certain angles.[17] Additionally, the transduction is virtually lossless, and Q values are in the range of 10^4 to 10^6.[18]

An electrical model for a quartz resonator is shown in Figure 17.23. The capacitance C_0 represents the parallel plate capacitance associated with the contacts and the lead wires, while C_m and L_m represent the mechanical energy storage. Resistance R_S accounts for the nonzero lossiness that all real systems must exhibit.

To a very crude approximation, the resistance of well-made crystals is inversely proportional to the resonant frequency and generally follows a

Lippman, then predicted the existence of the inverse effect on thermodynamic grounds, with verification by the Curies shortly afterward.

[15] The crystals used in digital watches employ a torsional mode of vibration to allow resonance at a low frequency (32.768 kHz) in a small size.

[16] The formula neglects the influence of the other dimensions.

[17] It is also possible to obtain controlled, nonzero temperature coefficients. This property has been exploited to make temperature-to-frequency transducers that function at temperatures too extreme for ordinary electronic circuits.

[18] At lower frequencies, damping by air lowers Q significantly. The higher Q values correspond to crystals mounted inside a vacuum.

$$R_S \approx \frac{5 \times 10^8}{f_o} \tag{41}$$

relationship. This formula is a semiempirical one and should be used only if measurements aren't available.[19]

The values of C_m and L_m can be computed if R_S, Q, and the resonant frequency are given. In general, because of the extraordinarily high Q-values that quartz crystals possess, the effective inductance value will be surprisingly high, while the series capacitance value will be vanishingly small. For example, a 1-MHz crystal with a Q of 10^5 has an effective inductance of about eight henries (no typo here, that's really eight *henries*) and a C_m of about 3.2 fF (again, no typo). It is apparent that crystals offer significant advantages over lumped LC realizations, where such element values are unattainable for all practical purposes.

Above about 20–30 MHz, the required slab thickness becomes impractically small. For example, a 100-MHz fundamental mode crystal would be only about 17 μm thick. However, crystals of reasonable thickness can still be used if higher vibrational modes are used. The boundary conditions are such that only odd overtones are allowed. Because of a variety of effects, the overtones are not exactly integer multiples of the fundamental (but they're close, off by 0.1% or so in the high direction). Third and fifth overtone crystals are fairly common, and seventh or even ninth overtone oscillators are occasionally encountered. However, as the overtone order increases, so does the difficulty of guaranteeing oscillation on only the desired mode.

As another extremely crude rule of thumb, the effective series resistance grows as the square of the overtone mode. Hence,

$$R_S \approx \frac{5 \times 10^8}{f_o} N^2, \tag{42}$$

where f_o is here interpreted as the frequency of the Nth overtone.

Because the overtones are not at exact integer multiples of the fundamental mode, the crystal must be cut to the correct frequency at the desired overtone. Well-cut overtone crystals possess Q values similar to those of fundamental-mode crystals.

Quartz crystal fabrication technology is an extremely advanced art. Crystals with resonant frequencies guaranteed within 50 ppm are routinely available, and substantially better performance can be obtained, although at higher cost. The general chemical inertness of quartz guarantees excellent stability over time, and a judicious choice of cut in conjunction with passive or active temperature compensation and/or control can lead to temperature coefficients of well under 1 ppm/°C. For these reasons, quartz oscillators are nearly ubiquitous in communications equipment and instrumentation (not to mention the lowly wristwatch, where a one-minute drift in a month corresponds to an error of only about 20 ppm).

---

[19] This formula strictly applies only to "AT-cut" crystals operating in the fundamental mode.

Surface Acoustic Wave (SAW) Devices

Because quartz crystals operate in bulk vibrational modes, high-frequency operation requires exceedingly thin slabs. A 1-GHz fundamental-mode quartz crystal would have a thickness of only about 1.7 μm, for example. Aside from obvious fabrication difficulties, thin slabs break easily if the electrical excitation is too great. Because of their high Q, it is easy to develop large-amplitude vibrations with very modest electrical drive. Even before outright fracture occurs, the extreme bending results in a host of generally undesired nonlinear behavior.

One way to evade these limitations is to employ surface, rather than bulk, acoustic waves. If the material supports such surface modes then the effective thickness can be much smaller than the physical thickness, allowing high resonant frequencies to be obtained with crystals of practical dimensions.

Lithium niobate ($LiNbO_3$) is a piezoelectric material that supports surface acoustic waves with little loss, and has been used extensively to make resonators and filters at frequencies practically untouchable by quartz. Control of frequency to quartz-crystal accuracy is not yet obtainable at low cost, unfortunately, but performance is adequate to satisfy such high-volume, low-cost applications as automatic garage-door openers, which typically work around 250–300 MHz, as well as front-end filters for cellular telephones.

Sadly, neither quartz nor lithium niobate is compatible with ordinary IC fabrication processes. Also disappointing is the lack of any piezoelectric activity in silicon. Hence, no inherently high-Q resonator can be made with layers normally found in ICs.

17.5 A CATALOG OF TUNED OSCILLATORS

There seems to be no limit to the number of ways to combine a resonator with a transistor or two to make an oscillator, as will become evident shortly. In the examples that follow, only the most minimal explanations are usually provided, perhaps leaving the reader in doubt as to which topology is "best." It is generally true that, with sufficient diligence and care, just about any of these topologies can be made to perform well enough for a given application. When we consider the issue of phase noise in the next chapter, more rational selection criteria will become evident.

17.5.1 BASIC *LC* FEEDBACK OSCILLATORS

The basic ingredients in these oscillators are simple: one transistor plus a resonator. Many of the oscillators are named after the fellows who first came up with the topologies but, as we'll see, a more or less unified description of these designs is possible. Ignoring biasing details, the basic topologies are as follows.

As mentioned in the earlier example, a capacitive voltage divider off of the tank provides feedback to an amplifier in a Colpitts oscillator (Figure 17.24). Notice that

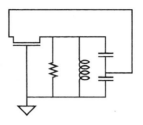

FIGURE 17.24. Colpitts oscillator
(biasing not shown).

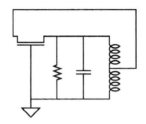

FIGURE 17.25. Hartley oscillator
(biasing still not shown).

the feedback is positive; you should sketch the root locus to convince yourself that this thing can oscillate. The locus will suggest why positive feedback topologies are generally favored with bandpass structures.

In alternative versions of the Colpitts, the feedback is from source back to the gate rather than from drain to source. That is, the transistor may be connected either as a source follower or as a common-source amplifier. Either way, there is net positive feedback.

The Hartley oscillator (Figure 17.25) is essentially identical to the Colpitts, but it uses a tapped inductor for feedback instead of a tapped capacitor. The Hartley oscillator has its origins in the very early days of radio, when tapped inductors were readily available. It is much less common today. One could also use a tapped resistor, in principle, but that particular configuration doesn't seem to have a name attached to it.

The Clapp oscillator (Figure 17.26) is a modified Colpitts oscillator, with a series LC replacing the lone inductor.[20] The Clapp oscillator is actually just a Colpitts oscillator with an additional tap on the capacitive divider chain, as is evident in the re-drawn schematic of Figure 17.27.

---

[20] See James K. Clapp, "An Inductive-Capacitive Oscillator of Unusual Frequency Stability," *Proc. IRE,* v. 36, 1948, pp. 356–8 and p. 1261. Clapp invented his modification of the Colpitts oscillator while working for the General Radio Corporation.

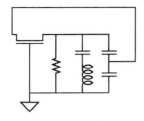

FIGURE 17.26. Clapp oscillator
(biasing *still* not shown).

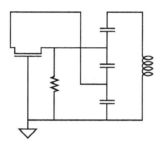

FIGURE 17.27. Re-drawn
Clapp oscillator.

The extra tap allows the voltage swing across the inductor (and divider chain) to exceed considerably that of either the drain or source – and therefore to exceed the supply and even device breakdown voltages. The larger signal energy helps overcome the effect of various noise processes (specifically, phase noise, as discussed in Chapter 18) and so improves spectral purity.

Of these topologies, the Colpitts is almost certainly the most commonly encountered. Its requirement of tapped capacitors is compatible with IC implementations, although the inductor is generally not (of course – you can't have everything). One other important reason for the popularity of the Colpitts configuration is that it is capable of excellent phase noise performance, as we'll see.

Another oscillator idiom actually owes its existence to the instability of some tuned amplifiers. Recall that it is possible for a common-source amplifier to have a negative input admittance if it operates with a tuned load below the resonant frequency of the load (so that it looks inductive).[21] This negative resistance can be used to overcome the loss in another resonant circuit to produce oscillations; see Figure 17.28.

The TITO oscillator uses a Miller-effect coupling capacitor. In many designs (particularly at very high frequencies), an explicit coupling capacitor is unnecessary; the device's inherent feedback capacitance is sufficient to provide the desired negative

---

[21] Satisfying this condition is not sufficient, however.

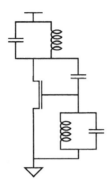

FIGURE 17.28. Tuned input–tuned output
(TITO) oscillator (bias details incomplete).

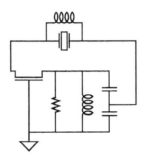

FIGURE 17.29. Colpitts
crystal oscillator.

resistance. This observation underscores the difficulty of using tuned circuits in both
the input and output circuits of nonunilateral amplifiers at high frequencies.

Because of its pair of tuned circuits, the TITO oscillator is theoretically capable
of producing signals with good spectral purity. However, its need of two inductors
makes this topology unattractive for IC implementation. An additional strike against
it is the need for careful tuning of two resonators if proper operation is to be obtained.

17.5.2 CRYSTAL OSCILLATOR POTPOURRI

Many crystal oscillators are recognizably derived from LC counterparts. In Figure 17.29, for example, the crystal is used in its series resonant mode (where it appears as a low resistance) to close the feedback loop only at the desired frequency.

The inductance across the crystal is frequently (but not always) needed in practical designs to prevent unwanted off-frequency oscillations due to feedback provided
by the crystal's parallel capacitance (C_0). The inductance resonates out this capacitor, so that only the series RLC arm of the crystal controls the feedback.

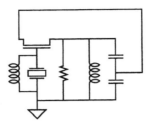

FIGURE 17.30. Modified
Colpitts crystal oscillator.

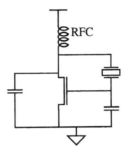

FIGURE 17.31. Pierce
crystal oscillator.

A variation on this theme is shown in Figure 17.30. In this particular configuration, the capacitive divider off the tank provides the feedback as in a classic *LC* Colpitts. However, the crystal grounds the gate only at the series resonant frequency of the crystal, permitting the loop to have sufficient gain to sustain oscillations at that frequency only. This topology is useful if one terminal of the crystal must be grounded.

Yet another topology is the Pierce oscillator, sketched in Figure 17.31.[22] In this oscillator, assume that the capacitors model transistor and stray parasitics, so that the transistor itself is ideal. Given this assumption, the only way to satisfy the zero–phase margin criterion is for the oscillation frequency to occur a bit above the series resonance of the crystal. That is, the crystal must look inductive at the oscillation frequency. This property confers the advantage that no external inductance is therefore

---

[22] Radio pioneer, entrepreneur, and Harvard professor George Washington Pierce made many contributions to the wireless art, of which the crystal oscillator bearing his name is but one example. See G. W. Pierce, "Piezoelectric Crystal Resonators and Crystal Oscillators Applied to the Precision Calibration of Wavemeters," *Proc. Amer. Acad. of Arts and Sci.,* v. 59, October 1923, pp. 81–106. Also see his U.S. Patent #2,133,642, filed 25 February 1924 and granted 18 October 1938. He made these developments soon after a personal demonstration by Cady of an early piezoelectric oscillator. Pierce had previously given the name "crystal rectifier" to point-contact diodes and had painstakingly disproven a thermally based explanation of their operation. His 1909 textbook, *Principles of Wireless Telegraphy,* was the first runaway best-seller of the technology's early days.

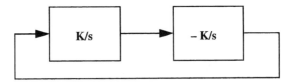

FIGURE 17.32. Quadrature oscillator block diagram.

required for this oscillator to function (the RF choke may be replaced by a large-valued resistor or current source). Hence, it is more amenable to integration than a Colpitts, for example, and particularly at low frequencies.

That the crystal must look inductive can be argued as follows. If we are to have a phase margin of zero and if the transistor's transconductance already provides a 180° phase shift, then the passive elements must supply the other 180°. There's no way for a two-pole RC network to provide 180° (close, but close doesn't count), so the crystal must look inductive.

Because the output frequency of a Pierce oscillator therefore does not coincide with the series resonance of the crystal, one must use a crystal that has been cut to oscillate at the desired frequency *with a specified load capacitance* (in this case, the value of the two capacitors in series).

As a final note on the Pierce oscillator, it happens to form the basis of many "digital" oscillators. An ordinary CMOS inverter, for example, can act as the gain element if biased to its linear region (e.g., with a large feedback resistor). Just add the appropriate amount of input and output capacitance, toss in the crystal from input to output, and chances are very good that you'll have an oscillator. Generally one or two stages of buffering (with more inverters, of course) are necessary to obtain full CMOS swings and also to isolate load changes from the oscillator core.

17.5.3 OTHER OSCILLATOR CONFIGURATIONS

In some applications, it is desirable to have two outputs in quadrature. One oscillator architecture that naturally provides quadrature outputs (at least in principle) uses a pair of integrators in a feedback loop (see Figure 17.32). From the magnitude condition, we can deduce that the frequency of oscillation is

$$\omega_{\text{osc}} = K. \tag{43}$$

Thus, tuning may be effected by varying the integrator gain. Furthermore, the desired quadrature relationship is obtained across any of the integrators.

In practice, unmodeled dynamics cause a departure from ideal behavior. Consider, for example, the effect of additional poles on the root locus. Rather than consisting of a purely imaginary pair, the locus with additional poles breaks away from the imaginary axis. Furthermore, these unmodeled parasitics tend to be rather unreliable, so that allowing the oscillation frequency to depend on them is undesirable. Despite

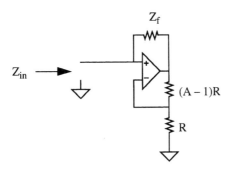

FIGURE 17.33. Generalized
impedance converter.

these obstacles, however, 1-GHz quadrature oscillators with reasonable quadrature phase (error under 0.5°) have been reported.[23]

17.6 NEGATIVE RESISTANCE OSCILLATORS

A perfectly lossless resonant circuit is very nearly an oscillator, but lossless elements are difficult to realize. Overcoming the energy loss implied by the finite Q of practical resonators with the energy-supplying action of active elements is one potentially attractive way to build practical oscillators, as in the TITO example.

The foregoing description is quite general, and covers both feedback and open-loop topologies. Among the former is a classic textbook circuit, the negative impedance converter (NIC). The NIC can be realized with a simple op-amp circuit that employs both positive and negative feedback. Specifically, consider the configuration of Figure 17.33.

If ideal op-amp behavior is assumed, it is easy to show that the input impedance is related to the feedback impedance as

$$Z_{\text{in}} = \frac{Z_f}{1 - A}. \tag{44}$$

If the closed-loop gain A is set equal to precisely 2, then the input impedance will be the algebraic inverse of the feedback impedance. If the feedback impedance is in turn chosen to be a pure positive resistance, then the input impedance will be a purely negative resistance. This negative resistance may be used to offset the positive resistance of all practical resonators to produce an oscillator.

As usual, the inherent nonlinearities of all real active devices provide amplitude limiting, and describing functions can be used to estimate the oscillation amplitude, if desired. Describing functions may also be used to verify that the oscillator will, in fact, oscillate.

---

[23] R. Duncan et al., "A 1 GHz Quadrature Sinusoidal Oscillator," *IEEE CICC Digest,* 1995, pp. 91–4.

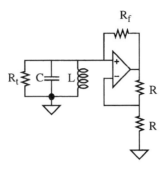

FIGURE 17.34. Negative
resistance oscillator.

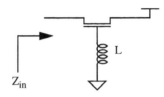

FIGURE 17.35. Canonical RF negative
resistance (biasing not shown).

As a specific example, consider the oscillator shown in Figure 17.34. In order to guarantee oscillation, we require the net resistance across the tank to be negative. Thus, we must satisfy the following inequality:

$$R_t > R_f. \tag{45}$$

The nonlinearity that most typically limits the amplitude at low frequencies is the finite output swing of all real amplifiers. Since there is a gain of 2 from the tank to the op-amp output, the signal across the tank will generally limit to a value somewhat greater than half the supply, corresponding to periodic saturation of the amplifier output.

At higher frequencies, it is possible for the finite slew rate of the amplifier to control the amplitude (partially, if not totally). In general, this situation is undesirable because the phase lag associated with slew limiting can cause a shift in oscillation frequency. In extreme cases, the amplitude control provided by slew limiting (or almost any other kind of amplitude limiting) can be unstable, and squegging can occur.

Finally, the various oscillator configurations presented earlier (for example, Colpitts, Pierce, etc.) may themselves be viewed as negative resistance oscillators.

A more practical negative resistance is easily obtained by exploiting yet another "parasitic" effect: Inductance in the gate circuit of a common-gate device can cause a negative resistance to appear at the source terminal; see Figure 17.35. A straightforward analysis reveals that, if C_{gd} is neglected, then Z_{in} has a negative real part

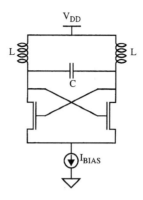

FIGURE 17.36. Simple differential
negative resistance oscillator.

for frequencies greater than the resonant frequency of the inductor and of C_{gs}. For frequencies much larger than that resonant frequency but much smaller than ω_T, the real part of Z_{in} is approximately

$$R_{in} \approx -\frac{\omega^2 L}{\omega_T} = -\frac{\omega}{\omega_T}|Z_L|. \tag{46}$$

The ease with which this circuit provides a negative resistance accounts for its popularity. However, it should be obvious that this ease also underscores the importance of minimizing parasitic gate inductance when a negative resistance is *not* desired.

A circuit that has become a frequently recurring idiom in recent years uses a cross-coupled differential pair to synthesize the negative resistance. "It is left as an exercise for the reader" to analyze this circuit, which is pictured in Figure 17.36.

As will be shown in the next chapter, spectral purity improves if the signal amplitudes are maximized (because this increases the signal-to-noise ratio). In many oscillators, such as the circuit of Figure 17.36, the allowable signal amplitudes are constrained by the available supply voltage or breakdown voltage considerations. Because it is the energy in the tank that constitutes the "signal," one could take a cue from the Clapp oscillator and employ tapped resonators to allow peak tank voltages that exceed the device breakdown limits or supply voltage, as in the negative resistance oscillator[24] of Figure 17.37.[25]

The differential connection might make it a bit difficult to see that this circuit indeed employs a tapped resonator, so consider a simplified half-circuit (Figure 17.38). In this simplified half-circuit, the transistors are replaced by a negative resistor, and

[24] J. Craninckx and M. Steyaert, "A CMOS 1.8GHz Low-Phase-Noise Voltage-Controlled Oscillator with Prescaler," *ISSCC Digest of Technical Papers,* February 1995, pp. 266–7. The inductors are bondwires stitched across the die.

[25] It is best to have a tail current source to constrain the swing, but we omit this detail in the interest of simplicity.

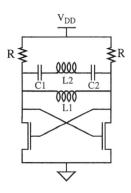

FIGURE 17.37. Negative resistance oscillator
with modified tank (simplified).

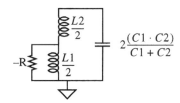

FIGURE 17.38. Simplified half-circuit
of negative resistance oscillator.

the positive resistors are not shown at all. Furthermore, the two capacitors are re-placed by their series equivalent, while the junction of the two inductors corresponds to the drain connection of the original circuit.

It should be clear that the swing across the equivalent capacitance (or across L_2) can exceed the supply voltage (and even the transistor breakdown voltage) because of the tapped configuration, so that this oscillator is the philosophical cousin of the Clapp configuration. Useful output may be obtained either through a buffer inter-posed between the oscillator core and load, or through a capacitive voltage divider to avoid spoiling resonator Q. As a consequence of the large energy stored in the tank with either a single- or double-tapped resonator, this topology is capable of excellent phase noise performance, as will be appreciated after the next chapter.

Tuning of this (and all other LC) oscillators may be accomplished by realizing all or part of C_1 or C_2 as a variable capacitor (such as the junction capacitor formed with a p+ diffusion in an n-well) and then tuning its effective capacitance with an appro-priate bias control voltage. Since CMOS junction capacitors have relatively poor Q, it is advisable to use only as much junction capacitance as necessary to achieve the desired tuning range. In practice, tuning ranges are often limited to below 5–10% if excessive degradation of phase noise is to be avoided. A simple (but illustrative) ex-ample of a voltage-controlled oscillator using this method is shown in Figure 17.39.

As a final comment on negative resistance oscillators, it should be clear that many (if not all) oscillators may be considered as negative resistance oscillators since, from

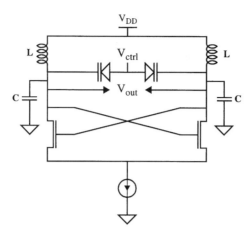

FIGURE 17.39. Voltage-controlled negative resistance oscillator (simplified).

the point of view of the tank, the active elements cancel the loss due to finite Q of the resonators. Hence, whether to call an oscillator a "negative resistance" type is actually more a philosophical decision than anything fundamental.

17.7 FREQUENCY SYNTHESIS

Oscillators built with high-Q crystals exhibit the best spectral purity but cannot be tuned over a range of more than several hundred parts per million or so. Since most transceivers must operate at a number of different frequencies that span a considerably larger range than that, one simple way to accommodate this lack of tuning capability is to use a separate resonator for each frequency. Clearly, this straightforward approach is practical only if the number of required frequencies is small.

Instead, virtually all modern gear uses some form of frequency synthesis, in which a single quartz-controlled oscillator is combined with a PLL and some digital elements to provide a multitude of output frequencies that are traceable to that highly stable reference. In the ideal case, then, one can obtain a wide operating frequency range and high stability from one oscillator.

Before undertaking a detailed investigation of various synthesizers, however, we need to digress briefly to examine an issue that strongly influences architectural choices. A frequency divider is used in all of the synthesizers we shall study, and it is important to model properly its effect on loop stability.

17.7.1 DIVIDER "DELAY"

Occasionally, one encounters the term "divider delay" in the literature on PLL synthesizers, in the context of evaluating loop stability. We'll see momentarily that the phenomenon is somewhat inaccurately named, but there is indeed a stability issue associated with the presence of dividers in the loop transmission.

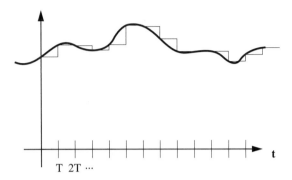

FIGURE 17.40. Sample-and-hold action.

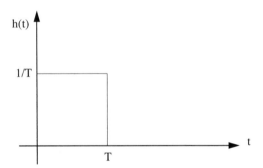

FIGURE 17.41. Impulse response of zero-order hold.

The use of a frequency divider generally implies that the phase detector is digital in nature.[26] As a consequence, knowledge about phase error is available to the loop only at discrete instants. That is, the loop is a sampled-data system. If a divider is present, the loop samples the phase error less frequently than might be implied by the VCO frequency. To model the PLL correctly, then, we need to account properly for this sampled nature.

In order to develop the necessary insight, consider a process in which a continuous-time function is sampled and held periodically; see Figure 17.40. The sample-and-hold (S/H) operation shown introduces a phase lag into the process. Mathematics need not be invoked to conclude that this is so; simply "eyeball" the sampled-and-held waveform and consider its time relationship with the original continuous-time waveform. You should be able to convince yourself that the best fit occurs when you slide the original waveform to the right by about half of a sample time.

More formally, the "hold" part of the S/H operation can be modeled by an element whose impulse response is a rectangle of unit area and T-second duration, as

---

[26] Although there are exceptions (e.g., subharmonic injection-locked oscillators), we will limit the present discussion to the more common implementations.

seen in Figure 17.41. This element, known formally as a zero-order hold (ZOH), has a transfer function given by [27]

$$H(s) = \frac{1 - e^{-sT}}{sT}. \tag{47}$$

The magnitude of the transfer function is

$$|H(j\omega)| = \frac{\sin \omega(T/2)}{\omega(T/2)}, \tag{48}$$

while the phase is simply

$$\angle[H(j\omega)] = -\omega(T/2). \tag{49}$$

The time delay is thus $T/2$ seconds. The same result is obtained by recognizing that a rectangular impulse response centered about zero seconds has zero phase by symmetry. Shifting that response to the right by $T/2$ seconds gives us the one shown in Figure 17.41. Independently of how we compute it, the fact that there is a delay is the reason for the term "divider delay." However, since the magnitude term is not constant with frequency, the term "delay" is not exactly correct.[28]

Now let's apply this information to the specific example of a PLL with a frequency divider in the loop transmission. From the expression for phase shift, we can see the deleterious effect on loop stability that dividers can introduce. As the divide modulus increases, the sampling period T increases (assuming a fixed VCO output frequency). The added negative phase shift thus becomes increasingly worse, degrading phase margin. As a consequence, loop crossover must be forced to a frequency that is low compared with $1/T$ in order to avoid these effects. Since the sampling rate is determined by the output of the *dividers* and hence of the frequency at which phase comparisons are made, rather than by the output of the VCO, a high division factor can result in a severe constraint on loop bandwidth, with all of the attendant negative implications for settling speed and noise performance. It is therefore common to choose loop crossover frequencies that are about a tenth of the phase comparison rate.

17.7.2 SYNTHESIZERS WITH STATIC MODULI

Having developed an understanding of the constraints imposed by the presence of a frequency divider in the loop transmission, we now turn to an examination of various synthesizer topologies.

The simplest PLL frequency synthesizer uses one reference oscillator and two frequency dividers, as shown in Figure 17.42. The loop forces the VCO to a frequency that makes the inputs to the PLL equal in frequency. Hence, we may write

$$\frac{f_{\text{ref}}}{N} = \frac{f_{\text{out}}}{M}, \tag{50}$$

---

[27] If you would like a quick derivation of the transfer function, note that the impulse response of the zero-order hold is the same as that of the difference of two integrations (one delayed in time). That should be enough of a hint.

[28] The magnitude is close to unity, however, for $\omega T < 1$.

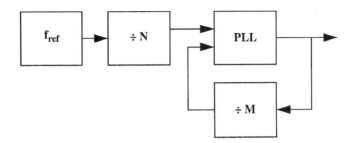

FIGURE 17.42. Classic PLL frequency synthesizer.

so that

$$f_{\text{out}} = \frac{M}{N} \cdot f_{\text{ref}}. \tag{51}$$

Thus, by varying the divide moduli M and N, any rational multiple of the input reference frequency can be generated. The long-term stability of the output (that is, the average frequency) is every bit as good as that of the reference, but stability in the shortest term (phase noise) depends on the net divide modulus as well as on the properties of the PLL's VCO and loop dynamics.[29]

Note that the output frequency can be incremented in steps of f_{ref}/N, and that this frequency represents the rate at which phase detection is performed in the PLL. Stability considerations as well as the need to suppress control-voltage ripple force the use of loop bandwidths that are small compared with f_{ref}/N. To obtain the maximum benefit of the (presumed) low noise reference, however, we would like the PLL to track that low noise reference over as wide a bandwidth as possible. Additionally, a high loop bandwidth will speed settling after a change in modulus. These conflicting requirements have led to the development of alternative architectures.

One simple modification that is occasionally used is depicted in Figure 17.43. For this synthesizer, we may write

$$f_{\text{out}} = \frac{M}{NP} \cdot f_{\text{ref}}. \tag{52}$$

The minimum output frequency increment is evidently f_{ref}/NP, but the loop compares phases at f_{ref}/N, or P times as fast as the previous architecture. This modification therefore improves the loop bandwidth constraint by a factor of P, at the cost of requiring the PLL to oscillate P times faster and the $\div M$ counter to run that much faster as well.

Yet another modification is the integer-N synthesizer of Figure 17.44. In this widely used synthesizer, the divider logic consists of two counters and a dual-modulus prescaler (divider). One counter, called the channel-spacing (or "swallow") counter,

---

[29] Within the PLL's loop bandwidth, the output phase noise will be M/N times that of the reference oscillator, since a phase division necessarily accompanies a frequency division. Outside of the PLL bandwidth, feedback is ineffective and the output phase noise will therefore be that of the PLL's own VCO.

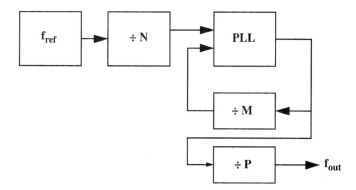

FIGURE 17.43. Modified PLL frequency synthesizer.

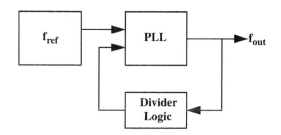

FIGURE 17.44. Integer-N frequency synthesizer.

is made programmable to enable channel selection. The other counter, which we'll call the frame counter (also known as the program counter), is usually fixed and determines the total number of prescaler cycles that comprise the following operation: The prescaler initially divides by $N + 1$ until the channel-spacing counter overflows, then divides by N until the frame counter overflows; the prescaler modulus is reset to $N + 1$, and the cycle repeats.

If S is the maximum value of the channel-spacing counter and F is the maximum value of the frame counter, then the prescaler divides the VCO output by $N + 1$ for S cycles, and by $F - S$ for N cycles, before repeating. The effective overall divide modulus M is therefore

$$M = (N + 1)S + (F - S)N = NF + S. \qquad (53)$$

The output frequency increment is thus equal to the reference frequency. This architecture is the most popular way of implementing the basic block diagram of Figure 17.42, and gets its name from the fact that the output frequency is an integer multiple of the reference frequency.

17.7.3 SYNTHESIZERS WITH DITHERING MODULI

In the synthesizers studied so far, the desired channel spacing directly constrains the loop bandwidth. An approach that eases this problem is to dither between two divide

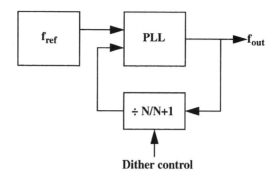

FIGURE 17.45. Block diagram for frequency
synthesizer with dithered modulus.

moduli to generate channel spacings that are smaller than the reference frequency;
this is shown in Figure 17.45.

As an illustration of the basic idea, consider that dividing alternately by, say, 4 and
then 5 with a 50% duty ratio is equivalent to dividing by 4.5 on average. Changing
the percentage of time spent on any one modulus thus changes the effective (average) modulus, so that the synthesized output can be incremented by frequency steps
smaller than the input reference frequency.

There are many strategies for switching between two moduli that yield the same
average modulus, of course, but not all of them are equally desirable because the
instantaneous frequency is also of importance. The most common strategy is used
by the fractional-N synthesizer, in which one divides the VCO output by one modulus (call it $N + 1$) every K VCO cycles, and by the other (N) for the rest of the time.
The average divide factor is thus

$$N_{\text{eff}} = (N + 1)\left(\frac{1}{K}\right) + N\left(1 - \frac{1}{K}\right) = N + \frac{1}{K}, \tag{54}$$

so that

$$f_{\text{out}} = N_{\text{eff}} f_{\text{ref}} = \left(N + \frac{1}{K}\right) f_{\text{ref}}. \tag{55}$$

We see that the resolution is determined by K, so that the minimum frequency
increment can be much *smaller* than the reference frequency. However, unlike the
other synthesizers studied so far, the phase detector operates with inputs whose frequency is *much higher* than the minimum increment (in fact, the phase detector is
driven with signals of frequency f_{ref}), thus providing a much-desired decoupling of
synthesizer frequency resolution from the PLL sampling frequency.

To illustrate the operation of this architecture in greater detail, consider the problem of generating a frequency of 27.135 MHz with a reference input of 100 kHz. The
integral modulus N therefore equals 271, while the fractional part ($1/K$) equals 0.35.
Thus, we wish to divide by 272 ($= N + 1$) for 35 out of every 100 VCO cycles (for
example), and by 271 ($= N$) for the other 65 cycles.

Of the many possible strategies for implementing this desired behavior, the most
common (but not necessarily optimum) one is to increment an accumulator by the

fractional part of the modulus (here, 0.35) every cycle. Each time the accumulator overflows (here defined as equalling or exceeding unity), the divide modulus is set to $N + 1$.

The residue after overflow is preserved, and the loop continues to operate as before. It should be apparent that the resolution is set by the size of the accumulator and is equal to the reference frequency divided by the total accumulator size. In our example with a 100-kHz reference, a 5-digit BCD (binary-coded decimal) accumulator would allow us to synthesize output steps as small as 1 Hz.

There is one other property of fractional-N synthesizers that needs to be mentioned. Because the loop operates by periodically switching between two divide moduli, there is necessarily a periodic modulation of the control voltage and hence of the VCO output frequency. Therefore, even though the output frequency is correct on *average,* it may not be on an *instantaneous* basis, and the output spectrum therefore contains sidebands. Furthermore, the size and location of the sidebands depend on the particular moduli as well as on loop parameters.

In practical loops of this kind, compensation for this modulation is usually necessary. This compensation is enabled by the fact that the modulation is deterministic – we know in advance what the control-line ripple will be. Hence, a compensating control voltage variation may be injected to offset the undesired modulation. In practice, this technique (sometimes called API, for *analog phase interpolation*) is capable of providing between 20-dB and 40-dB suppression of the sidebands. Achieving the higher levels of suppression (and beyond) requires intimate knowledge of the control characteristics of the VCO, including temperature and supply voltage effects, so details vary considerably from design to design.[30]

An alternative to this type of cancellation is to eliminate the periodic control-voltage ripple altogether by employing a more sophisticated strategy for switching between the two moduli. For example, one might randomize this switching to decrease the amplitudes of spurious spectral components at the expense of increasing the noise floor. A powerful improvement on that strategy is to use *delta-sigma* techniques to distribute the noise *nonuniformly.*[31] If the spectrum is shaped to push the noise out to frequencies far from the carrier, subsequent filtering can readily remove the noise. The loop itself takes care of noise near the carrier, so the overall output can possess exceptional spectral purity.

17.7.4 COMBINATION SYNTHESIZERS

Another approach is to combine the outputs of two or more synthesizers. The additional degree of freedom thus provided can ease some of the performance tradeoffs,

---

[30] See, for example, V. Mannassewitsch, *Frequency Synthesizers,* 3rd ed., Wiley, New York, 1987.

[31] The classic paper on this architecture is by T. Riley et al., "Sigma-Delta Modulation in Fractional-N Frequency Synthesis," *IEEE J. Solid-State Circuits,* v. 28, May 1993, pp. 553–9. The terms "delta-sigma" and "sigma-delta" are frequently used interchangeably, but the former nomenclature was used by the inventors of the concept.

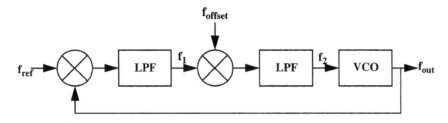

FIGURE 17.46. Offset synthesizer loop.

but at the expense of increased complexity and power consumption. The most common expression of this idea is to mix the output of a fixed frequency source with that of a variable one. The *offset synthesizer* (Figure 17.46) is one architecture that implements that particular choice.

With this architecture, the loop does not servo to an equality of output and reference frequencies, because the additional intermediate mixing offsets the equilibrium point. Without an intermediate mixing, note that the balance point would correspond to a zero frequency output from the low-pass filter that follows the (first and only) mixer. In the offset loop, then, the balance point corresponds to a zero frequency output from the final low-pass filter. Armed with this observation, it is a straightforward matter to determine the relationship between f_{out} and f_{ref}.

The low-pass filters selectively eliminate the sum frequency components arising from the mixing operations. Hence, we may write

$$f_1 = f_{out} - f_{ref}, \tag{56}$$

$$f_2 = f_1 - f_{offset}$$

$$= f_{out} - f_{ref} - f_{offset}. \tag{57}$$

Setting f_2 equal to zero and solving for the output frequency yields

$$f_{out} = f_{ref} + f_{offset}. \tag{58}$$

Thus, the output frequency is the sum of the two input frequencies.

An important advantage of this approach is that the output frequency is not a multiplied version of a reference. Hence, the phase noise similarly undergoes no multiplication, making it substantially easier to produce a low-phase noise output signal. A related result is that any phase or frequency modulation on either of the two input signals is directly transferred to the output without scaling by a multiplicative factor. As a consequence of these attributes, the offset synthesizer has found wide use in transmitters for FM/PM systems, particularly for GSM.

There are other techniques for combining two frequencies to produce a third. For example, one might use two complete PLLs and combine the outputs with a mixer. To select out the sum rather than the difference (or vice versa), one would conventionally use a filter. One may also ease the filter's burden by using a single-sideband

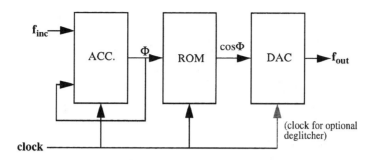

FIGURE 17.47. Direct digital frequency synthesizer.

mixer (also known as a complex mixer) to reduce the magnitude of the undesired component.[32] However, such loops are rarely used in IC implementations because of the difficulty of preventing two PLLs from interacting with each other. A common problem is for the loops to (attempt to) lock to each other parasitically through substrate coupling or through incomplete reverse isolation involving amplifiers and other circuitry. These problems are sufficiently difficult to solve that such dual-loop synthesizers are rarely used at present.

17.7.5 DIRECT DIGITAL SYNTHESIS

There are some applications that require the ability to change frequencies at a relatively high rate. Examples include frequency-hopped spread-spectrum systems, in which the carrier frequency changes in a pseudorandom pattern.[33] Conventional synthesizers may be hard-pressed to provide the fast settling required, so alternative means have been developed. The fastest-settling synthesizers are open-loop systems, which can evade the constraints imposed by the stability considerations of feedback systems (such as PLLs).

One extremely agile type of synthesizer employs direct digital synthesis (DDS). The basic block diagram of such a synthesizer is shown in Figure 17.47. This synthesizer consists of an accumulator, a read-only memory (ROM) lookup table (with integral output register), and a digital-to-analog converter (DAC). The accumulator accepts a frequency command signal f_{inc} as an input and then increments its output by this amount every clock cycle. The output therefore increases linearly until an overflow occurs and the cycle repeats. The output Φ thus follows a sawtooth pattern. A useful insight is that, since phase is the integral of frequency, the output of the accumulator is analogous to the integral of the frequency input command. The frequency of the resulting sawtooth pattern is then a function of the clock frequency, accumulator word length, and input command.

[32] We shall explore more fully the uses of SSB mixers in Chapter 19.
[33] This strategy is particularly useful in avoiding detection and jamming in military scenarios, for which it was first developed, because the resulting spectrum looks very much like white noise.

The phase output of the accumulator then drives the address lines of a ROM cosine lookup table that converts the digital phase values into digital amplitude values.[34] Finally, a DAC converts those values into analog outputs. Generally, a filter follows the DAC to improve spectral purity to acceptable levels.

The frequency can be changed rapidly (with a latency of only a couple clock cycles), and in a phase-continuous manner, simply by changing the value of f_{inc}. Furthermore, modulation of both frequency and phase are trivially obtained by adding the modulation directly in the digital domain to f_{inc} or Φ, respectively. Finally, even amplitude modulation can be added by using a multiplying DAC (MDAC), in which the analog output is the product of an analog input (here, the amplitude modulation) and the digital input from the ROM.[35]

The chief problem with this type of synthesizer is that the spectral purity is markedly inferior to that of the PLL-based approaches considered earlier. The number of bits in the DAC set one bound on the spectral purity (*very* loosely speaking, the carrier-to-spurious ratio is about 6 dB per bit), while the number of ROM points per cycle determine the location of the harmonic components (with a judicious choice of the n points, the first significant harmonic can be made to occur at $n-1$ times the fundamental). Since the clock will necessarily run much faster than the output frequency ultimately generated, these types of synthesizers produce signals whose frequencies are a considerably smaller fraction of a given technology's ultimate speed than VCO/PLL-based synthesizers. Frequently, the output of a DDS is upconverted through mixing with the output of a PLL-based synthesizer (or used as one of the inputs in an offset synthesizer) to effect a compromise between the two.

17.8 SUMMARY

We have examined how the amplitude of oscillation can be stabilized through non-linear means, and we extended feedback concepts to include a particular type of linearized nonlinearity. Armed with describing functions and knowledge of the rest of the elements in a loop transmission, both oscillation frequency and amplitude can be determined.

We looked at a variety of oscillators of both open-loop and feedback topologies. The Colpitts and Hartley oscillators use tapped tanks to provide positive feedback, whereas the TITO oscillator employs the negative resistance that a tuned amplifier with Miller feedback can provide. The Clapp oscillator uses an extra tap to allow resonator swings that exceed the supply voltage, and so permits signal energy to dominate noise.

Crystal oscillator versions of LC oscillators were also described. Since a quartz crystal behaves much like an LC resonator with extraordinarily high Q, it permits

---

[34] With a little additional logic, one can easily reduce the amount of ROM required by 75%, since one quadrant's worth of values is readily reused to reconstruct an entire period.

[35] One may also perform the amplitude modulation in the digital domain simply by multiplying the ROM output with a digital representation of the desired amplitude modulation before driving the DAC.

the realization of oscillators with excellent spectral purity and low power consumption. The Colpitts configuration oscillates at the series resonant frequency of the crystal and thus requires an LC tank. The Pierce oscillator operates at a frequency where the crystal looks inductive and therefore requires no external inductance. The off-resonant operation, however, forces the use of crystals that have been cut specifically for a particular load capacitance.

A random sampling of other oscillators was also provided, including a quadrature oscillator using two integrators in a feedback loop, as well as several negative resistance oscillators. Again, tapped resonators were seen to be beneficial for improving phase noise.

Finally, a number of frequency synthesizers were examined. Stability considerations force loop crossover frequencies well below the phase comparison frequency; phase noise considerations favor large loop bandwidths. Because the output frequency increment is tightly coupled to the phase comparison frequency in simple architectures, it is difficult to synthesize frequencies with fine increments while additionally conferring to the output the good phase noise of the reference. The fractional-N synthesizer decouples the frequency increment from the phase comparison rate, allowing the use of greater loop bandwidths. However, while phase noise is therefore improved, various spurious components can be generated owing to ripple on the control voltage. Suppression of these spurious tones is possible either by cancellation of the ripple (since it is deterministic in the case of the classical fractional-N architecture) or through the use of randomization or noise shaping of the spectrum.

PROBLEM SET FOR OSCILLATORS

PROBLEM 1   Consider the Colpitts oscillator of Figure 17.48.

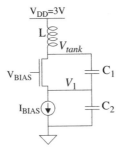

FIGURE 17.48. Colpitts oscillator
example for Problem 1.

(a) Assume that the inductor has a *finite Q*. Derive an expression for the minimum Q of the inductor such that the circuit just satisfies the conditions for oscillation. Express your answer in terms of the small-signal transconductance of the transistor.

(b) Now assume that the inductor has a Q of 10. If $n = C_1/(C_1 + C_2)$, provide an expression for the minimum n consistent with just satisfying the conditions for oscillation.

(c) Explain why these minima exist.

PROBLEM 2 This problem examines the start-up question from a somewhat different perspective than the previous problem. Consider the Colpitts oscillator shown in Figure 17.49.

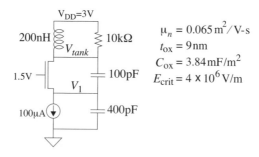

FIGURE 17.49. Colpitts oscillator example for Problem 2.

(a) Calculate the minimum W/L necessary for a start-up loop gain of 2. Assume long-channel operation and neglect transistor capacitances, body effect, and channel-length modulation.

(b) Repeat (a) but assume operation deep into the short-channel regime.

(c) Calculate the gate–source capacitance in both cases and calculate the impedance of these capacitances at 1 GHz.

PROBLEM 3 In the Clapp oscillator shown in Figure 17.50, calculate the oscillation frequency and the amplitude of oscillation at the gate, as well as across the inductor. Assume long-channel operation, and neglect all transistor parasitics.

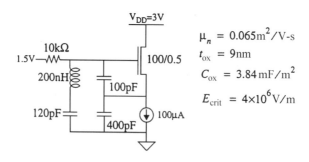

FIGURE 17.50. Clapp oscillator.

Also plot the shape of the drain current waveform. What is the peak drain current?

PROBLEM 4

(a) In the differential oscillator of Figure 17.51, calculate the oscillation frequency and amplitude across the LC tank. Assume an inductor Q of 10 and ignore all transistor parasitics except C_{gs}.

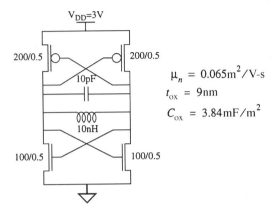

FIGURE 17.51. Negative resistance oscillator.

(b) What is the amplitude-limiting mechanism?
(c) Find the minimum supply voltage that guarantees start-up.
(d) Estimate the power dissipation of this oscillator. What limits it? Is this mechanism reliable?
(e) Is it possible to use a series tank instead of a parallel tank here? Explain.

PROBLEM 5 Suppose the capacitor in the previous problem is replaced with a series RC combination. Assume that the resistor can vary from small to large values.

(a) Derive an equivalent parallel RC network for this combination. What is the Q of the tank if the 10-nH inductor is perfect?
(b) Plot the frequency of oscillation and tank Q as a function of R.
(c) Using your answer to (b), what are the advantages and disadvantages of using this resistance variation as a tuning method?

PROBLEM 6 The text's derivation of oscillation frequency and amplitude for the Colpitts configuration glibly ignores any possible phase shift associated with G_m. Amend those derivations if the large-signal transconductance actually has the following form:

$$G_m = G_{m0}e^{-j\omega T_D}, \tag{P17.1}$$

where G_{m0} is the phase-free large-signal transconductance given in the text.

PROBLEM 7 Treating the mixers in an offset synthesizer as ideal multipliers and the oscillator inputs as perfect sinusoids, derive an expression for the loop transmission. Call the two filter transfer functions $H_1(s)$ and $H_2(s)$. Comment on how or if the input frequencies constrain the loop bandwidth.

PROBLEM 8   As mentioned in the text, an alternative to offset synthesizers is simply to take two oscillators and combine their outputs with a mixer. The mixer's output is then filtered to yield either the sum or difference frequency, whichever is desired. Qualitatively compare these two methods of synthesis.

PROBLEM 9   To ease the demands on filtering in the type of synthesizer described in the previous problem, a more complicated mixer is sometimes used to provide suppression of the undesired term prior to filtering. A synthesizer employing such a mixer appears as Figure 17.52.

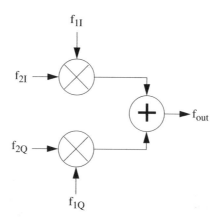

FIGURE 17.52. Combination synthesizer
with complex mixer.

The subscripts I and Q refer to "in-phase" and "quadrature," respectively, and indicate that there are two signals with a 90° phase difference between them at each frequency. What is the output frequency?

PROBLEM 10   The complex mixer of the previous problem can mix two frequencies and inherently produce only the sum or difference frequency with no additional filtering, at least in principle. The gains of the two paths must match, and the I and Q signals must be in perfect quadrature in order to cancel the undesired term. Derive an explicit expression for the ratio of undesired to desired output amplitude if there is only a gain mismatch in the two paths. Express your answer in terms of ε, where $1 + \varepsilon$ is the gain ratio.

PROBLEM 11   The offset synthesizer relies on filtering to select the difference component out of the various mixing operations. The filter bandwidths constrain the settling time of the loop, however. To ease this constraint, consider using the complex mixer of the previous two problems. Sketch a block diagram of an offset synthesizer that uses two sets of complex mixers. Comment on the filtering requirements relative to the classical offset synthesizer.

CHAPTER EIGHTEEN

PHASE NOISE

18.1 INTRODUCTION

We asserted in the previous chapter that tuned oscillators produce outputs with higher spectral purity than relaxation oscillators. One straightforward reason is simply that a high-Q resonator attenuates spectral components removed from the center frequency. As a consequence, distortion is suppressed, and the waveform of a well-designed tuned oscillator is typically sinusoidal to an excellent approximation.

In addition to suppressing distortion products, a resonator also attenuates spectral components contributed by sources such as the thermal noise associated with finite resonator Q, or by the active element(s) present in all oscillators. Because amplitude fluctuations are usually greatly attenuated as a result of the amplitude stabilization mechanisms present in every practical oscillator, phase noise generally dominates – at least at frequencies not far removed from the carrier. Thus, even though it is possible to design oscillators in which amplitude noise is significant, we focus primarily on phase noise here. We show later that a simple modification of the theory allows the accommodation of amplitude noise as well, permitting the accurate computation of output spectrum at frequencies well removed from the carrier.

Aside from aesthetics, the reason we care about phase noise is to minimize the problem of *reciprocal mixing*. If a superheterodyne receiver's local oscillator is completely noise-free, then two closely-spaced RF signals will simply translate downward in frequency together. However, the LO spectrum is not an impulse and so, to be realistic, we must evaluate the consequences of an impure LO spectrum.

In Figure 18.1, two RF signals heterodyne with the LO to produce a pair of IF signals. The desired RF signal is considerably weaker than the signal at an adjacent channel. Assuming (as is typical) that the front-end filter does not have sufficient resolution to perform channel filtering, downconversion preserves the relative amplitudes of the two RF signals in translation to the IF. Because the LO spectrum is of nonzero width, the downconverted RF signals also have width. The tails of the LO spectrum act as parasitic LOs over a continuum of frequencies. Reciprocal mixing is the heterodyning of RF signals with those unwanted components.

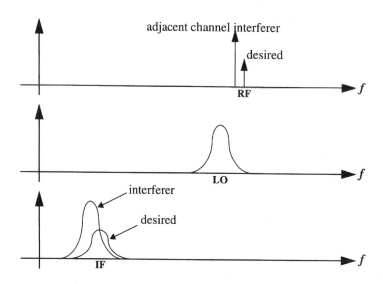

FIGURE 18.1. Illustration of reciprocal mixing due to LO phase noise.

As is evident from the figure, reciprocal mixing causes the undesired signal to overwhelm the desired signal in this particular case. Reduction of LO phase noise is essential to minimize the occurrence and severity of reciprocal mixing.

To place this subject in its proper context, we first identify some fundamental trade-offs among key parameters. These include power dissipation, oscillation frequency, resonator Q, and noise. After studying these tradeoffs qualitatively in a hypothetical ideal oscillator, we consider quantitatively how various noise processes corrupt the output spectrum of real oscillators.

The theoretical and practical importance of oscillators has motivated the development of numerous treatments of phase noise. The sheer number of publications on this topic underscores the importance attached to it. At the same time, many of these disagree on rather fundamental points, and it may be argued that the abundance of such conflicting research quietly testifies to the inadequacies of many of those treatments. Complicating the search for a suitable theory is that noise in a circuit may undergo frequency translations before ultimately becoming oscillator phase noise. These translations are often attributed to the presence of obvious nonlinearities in practical oscillators. The simplest theories nevertheless simply ignore the nonlinearities altogether, and frequently ignore the possibility of time variation as well. Such linear, time-invariant (LTI) theories manage to provide important qualitative design insights, but these theories are understandably limited in their predictive power. Chief among the deficiencies of an LTI theory is that frequency translations are necessarily disallowed, begging the question of how the (nearly) symmetrical sidebands observed in practical oscillators can arise.

Despite this complication, and despite the obvious presence of nonlinearities necessary for amplitude stabilization, the noise-to-phase transfer function of oscillators nonetheless may be treated as linear. However, a quantitative understanding of the frequency translation process requires abandonment of the principle of time invariance implicitly assumed in most theories of phase noise. In addition to providing a quantitative reconciliation between theory and measurement, the time-varying phase noise model presented in this chapter identifies an important symmetry principle, which may be exploited to suppress the upconversion of $1/f$ noise into close-in phase noise. At the same time, it provides an explicit accommodation of cyclostationary effects, which are significant in many practical oscillators, and of amplitude-to-phase (AM–PM) conversion as well. These insights allow a reinterpretation of why certain topologies, such as the venerable Colpitts oscillator, exhibit good performance. Perhaps more important, the theory informs design – suggesting novel optimizations of well-known oscillators and even the invention of new circuit topologies. We examine both tuned LC and ring oscillator circuit examples to reinforce the theoretical considerations developed, and then conclude by briefly considering practical simulation issues as well.

We first need to revisit how one evaluates whether a system is linear or time-invariant. This question rarely arises in the analysis of most systems, and perhaps more than a few engineers have forgotten how to tell the difference. Indeed, we find that we must even take care to define explicitly what is meant by the word *system*. We then identify some very general tradeoffs among such key parameters as power dissipation, oscillation frequency, resonator Q, and circuit noise power. Then, we study these tradeoffs qualitatively in a hypothetical ideal oscillator in which linearity of the noise-to-phase transfer function is assumed, allowing characterization by an impulse response. Although the assumption of linearity is defensible, we shall see that time invariance fails to hold even in this simple case. That is, oscillators are linear, time-varying (LTV) systems, where *system* is defined by the noise-to-phase transfer characteristic. Fortunately, complete characterization by an impulse response depends only on linearity, not time-invariance. By studying the impulse response, we discover that periodic time variation leads to frequency translation of device noise to produce the phase noise spectra exhibited by real oscillators. In particular, the upconversion of $1/f$ noise into close-in phase noise is seen to depend on symmetry properties that are potentially controllable by the designer. Additionally, the same treatment easily subsumes the cyclostationarity of noise generators. As we'll see, that accommodation explains why class-C operation of active elements within an oscillator can be beneficial. Illustrative circuit examples reinforce key insights of the LTV model.

18.2 GENERAL CONSIDERATIONS

Perhaps the simplest abstraction of an oscillator that still retains some connection to the real world is a combination of a lossy resonator and an energy restoration element. The latter precisely compensates for the tank loss to enable a constant-amplitude oscillation. To simplify matters, assume that the energy restorer is noiseless (see Figure 18.2). The tank resistance is therefore the only noisy element in this model.

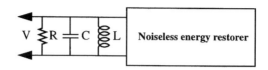

FIGURE 18.2. "Perfectly efficient" RLC oscillator.

To gain some useful design insight, first compute the signal energy stored in the tank:

$$E_{\text{sig}} = \tfrac{1}{2} C V_{\text{pk}}^2, \tag{1}$$

so that the mean-square signal (carrier) voltage is

$$\overline{V_{\text{sig}}^2} = \frac{E_{\text{sig}}}{C}, \tag{2}$$

where we have assumed a sinusoidal waveform.

The total mean-square noise voltage is found by integrating the resistor's thermal noise density over the noise bandwidth of the *RLC* resonator:

$$\overline{V_n^2} = 4kTR \int_0^\infty \left| \frac{Z(f)}{R} \right|^2 df = 4kTR \cdot \frac{1}{4RC} = \frac{kT}{C}. \tag{3}$$

Combining Eqn. 2 and Eqn. 3, we obtain a noise-to-carrier ratio (the reason for this "upside-down" ratio is simply one of convention):

$$\frac{N}{S} = \frac{\overline{V_n^2}}{\overline{V_{\text{sig}}^2}} = \frac{kT}{E_{\text{sig}}}. \tag{4}$$

Sensibly enough, one therefore needs to maximize the signal levels to minimize the noise-to-carrier ratio.

We may bring power consumption and resonator Q explicitly into consideration by noting that Q can be defined generally as being proportional to the energy stored divided by the energy dissipated:

$$Q = \frac{\omega_0 E_{\text{sig}}}{P_{\text{diss}}}. \tag{5}$$

Hence, we may write

$$\frac{N}{S} = \frac{\omega_0 kT}{Q P_{\text{diss}}}. \tag{6}$$

The power consumed by this model oscillator is simply equal to P_{diss}, the amount dissipated by the tank loss. The noise-to-carrier ratio is here inversely proportional to the product of resonator Q and the power consumed, and it is directly proportional to the oscillation frequency. This set of relationships still holds approximately

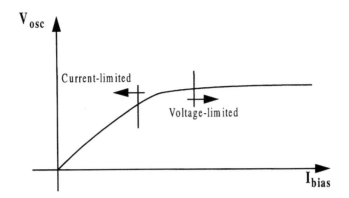

FIGURE 18.3. Oscillator operating regimes.

for many real oscillators, and it explains the traditional obsession of engineers with maximizing resonator Q, for example.

Other important design criteria become evident by coupling the foregoing with additional knowledge of practical oscillators. One is that oscillators generally operate in one of two regimes that may be distinguished by their differing dependence of output amplitude on bias current (see Figure 18.3), so that one may write

$$V_{\text{sig}} = I_{\text{BIAS}}R, \tag{7}$$

where R is a constant of proportionality with the dimensions of resistance. This constant, in turn, is proportional to the equivalent parallel tank resistance, so that

$$V_{\text{sig}} \propto I_{\text{BIAS}}R_{\text{tank}}, \tag{8}$$

implying that the carrier power may be expressed as

$$P_{\text{sig}} \propto I_{\text{BIAS}}^2 R_{\text{tank}}. \tag{9}$$

The mean-square noise voltage has already been computed in terms of the tank capacitance as

$$\overline{V_n^2} = \frac{kT}{C}, \tag{10}$$

but it may also be expressed in terms of the tank inductance:

$$\overline{V_n^2} = \frac{kT}{C} = \frac{kT}{1/\omega_0^2 L} = kT\omega_0^2 L. \tag{11}$$

An alternative expression for the noise-to-carrier ratio in the current-limited regime is therefore

$$\frac{N}{C} \propto \frac{kT\omega_0^2 L}{I_{\text{BIAS}}^2 R_{\text{tank}}}. \tag{12}$$

Assuming operation at a fixed supply voltage, a constraint on power consumption implies an upper bound on the bias current. Of the remaining free parameters, then, only the tank inductance and resistance may be practically varied to minimize the N/C ratio. That is, optimization of such an oscillator corresponds to minimizing L/R_{tank}. In many treatments, maximizing tank inductance is offered as a prescription for optimization. However, we see that a more valid objective is to minimize L/R_{tank}.[1] Since, in general, the resistance is itself a function of inductance, identifying (and then achieving) this minimum is not always trivial. An additional consideration is that, below a certain minimum inductance, oscillation may cease. Hence, the optimization prescription here presumes oscillation – and in a regime where the output amplitude is proportional to the bias current.

18.3 DETAILED CONSIDERATIONS: PHASE NOISE

To augment the qualitative insights of the foregoing analysis, let us now determine the actual output spectrum of the ideal oscillator.

Assume that the output in Figure 18.2 is the voltage across the tank, as shown. By postulate, the only source of noise is the white thermal noise of the tank conductance, which we represent as a current source across the tank with a mean-square spectral density of

$$\frac{\overline{i_n^2}}{\Delta f} = 4kTG. \tag{13}$$

This current noise becomes voltage noise when multiplied by the effective impedance facing the current source. In computing this impedance, however, it is important to recognize that the energy restoration element must contribute an average effective negative resistance that precisely cancels the positive resistance of the tank. Hence, the net result is that the effective impedance seen by the noise current source is simply that of a perfectly lossless LC network.

For a relatively small offset frequency $\Delta\omega$ from the center frequency ω_0, the impedance of an LC tank may be approximated by

$$Z(\omega_0 + \Delta\omega) \approx -j \cdot \frac{\omega_0 L}{2(\Delta\omega/\omega_0)}. \tag{14}$$

We may write the impedance in a more useful form by incorporating an expression for the unloaded tank Q:

$$Q = \frac{R}{\omega_0 L} = \frac{1}{\omega_0 GL}. \tag{15}$$

Solving Eqn. 15 for L and substituting into Eqn. 14 yields

$$|Z(\omega_0 + \Delta\omega)| \approx \frac{1}{G} \cdot \frac{\omega_0}{2Q\Delta\omega}. \tag{16}$$

[1] D. Ham and A. Hajimiri, "Concepts and Methods in Optimization of Integrated LC VCOs," *IEEE J. Solid-State Circuits*, June 2001.

Thus, we have traded an explicit dependence on inductance for a dependence on Q and G.

Next, multiply the spectral density of the mean-square noise current by the squared magnitude of the tank impedance to obtain the spectral density of the mean-square noise voltage:

$$\frac{\overline{v_n^2}}{\Delta f} = \frac{\overline{i_n^2}}{\Delta f} \cdot |Z|^2 = 4kTR\left(\frac{\omega_0}{2Q\Delta\omega}\right)^2. \tag{17}$$

The power spectral density of the output noise is frequency-dependent because of the filtering action of the tank, falling as the inverse-square of the offset frequency. This $1/f^2$ behavior simply reflects the facts that the voltage frequency response of an *RLC* tank rolls off as $1/f$ to either side of the center frequency and that power is proportional to the square of voltage. Note also that an increase in tank Q reduces the noise density when all other parameters are held constant, underscoring once again the value of increasing resonator Q.

In our idealized *LC* model, thermal noise causes fluctuations in both amplitude and phase, and Eqn. 17 accounts for both. The equipartition theorem of thermodynamics tells us that, in the absence of amplitude limiting, noise energy would split equally into amplitude and phase noise domains. The amplitude-limiting mechanisms present in all practical oscillators remove most of the amplitude noise, leaving us with about half the noise given by Eqn. 17.

Additionally, we are often more interested in how large this noise is relative to the carrier, rather than its absolute value. It is consequently traditional to normalize the mean-square noise voltage density to the mean-square carrier voltage and then report the ratio in decibels, thereby explaining the "upside down" ratios presented previously. Performing this normalization yields the following equation for phase noise:

$$L\{\Delta\omega\} = 10\log\left[\frac{2kT}{P_{\text{sig}}} \cdot \left(\frac{\omega_0}{2Q\Delta\omega}\right)^2\right]. \tag{18}$$

The units of phase noise are thus proportional to the log of a density. Specifically, they are commonly expressed as "decibels below the carrier per hertz," or dBc/Hz, specified at a particular offset frequency $\Delta\omega$ from the carrier frequency ω_0. For example, one might speak of a 2-GHz oscillator's phase noise as "−110 dBc/Hz at a 100-kHz offset." Purists may complain that the "per hertz" actually applies to the argument of the log, not to the log itself; doubling the measurement bandwidth does not double the decibel quantity. Nevertheless, as lacking in rigor as "dBc/Hz" is, it is common usage.

Equation 18 tells us that phase noise (at a given offset) improves as both the carrier power and Q increase, as predicted earlier. These dependencies make sense. Increasing the signal power improves the ratio simply because the thermal noise is fixed, and increasing Q improves the ratio quadratically because the tank's impedance falls off as $1/Q\Delta\omega$.

Because many simplifying assumptions have led us to this point, it should not be surprising that there are some significant differences between the spectrum predicted

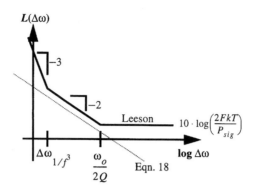

FIGURE 18.4. Phase noise:
Leeson versus Eqn. 18.

by Eqn. 18 and what one typically measures in practice. For example, although real spectra do possess a region where the observed density is proportional to $1/(\Delta\omega)^2$, the magnitudes are typically quite a bit larger than predicted by Eqn. 18 because there are additional important noise sources besides tank loss. For example, any physical implementation of an energy restorer will be noisy. Furthermore, measured spectra eventually flatten out for large frequency offsets rather than continuing to drop quadratically. Such a floor may be due to the noise associated with any active elements (such as buffers) placed between the tank and the outside world, or it can even reflect limitations in the measurement instrumentation itself. Even if the output were taken directly from the tank, any resistance in series with either the inductor or capacitor would impose a bound on the amount of filtering provided by the tank at large frequency offsets and thus ultimately produce a noise floor. Finally, there is almost always a $1/(\Delta\omega)^3$ region at small offsets.

A modification to Eqn. 18 provides a means to account for these discrepancies:

$$L\{\Delta\omega\} = 10\log\left[\frac{2FkT}{P_{\text{sig}}}\left\{1 + \left(\frac{\omega_0}{2Q\Delta\omega}\right)^2\right\}\left(1 + \frac{\Delta\omega_{1/f^3}}{|\Delta\omega|}\right)\right]. \qquad (19)$$

These modifications, due to Leeson, consist of a factor F to account for the increased noise in the $1/(\Delta\omega)^2$ region, an additive factor of unity (inside the braces) to account for the noise floor, and a multiplicative factor (the term within the second set of parentheses) to provide a $1/|\Delta\omega|^3$ behavior at sufficiently small offset frequencies.[2] With these modifications, the phase noise spectrum appears as in Figure 18.4.

It is important to note that the factor F is an empirical fitting parameter and therefore must be determined from measurements, diminishing the predictive power of the phase noise equation. Furthermore, the model asserts that $\Delta\omega_{1/f^3}$, the boundary between the $1/(\Delta\omega)^2$ and $1/|\Delta\omega|^3$ regions, is precisely equal to the $1/f$ corner

[2] D. B. Leeson, "A Simple Model of Feedback Oscillator Noise Spectrum," *Proc. IEEE*, v. 54, February 1966, pp. 329–30.

of device noise. However, measurements frequently show no such equality, and thus one must generally treat $\Delta\omega_{1/f^3}$ as an empirical fitting parameter as well. Also it is not clear what the corner frequency will be in the presence of more than one noise source, each with an individual $1/f$ noise contribution (and generally differing $1/f$ corner frequencies). Finally, the frequency at which the noise flattens out is not always equal to half the resonator bandwidth, $\omega_0/2Q$.

Both the ideal oscillator model and the Leeson model suggest that increasing resonator Q and signal power are ways to reduce phase noise. The Leeson model additionally introduces the factor F, but without knowing precisely what it depends on, it is difficult to identify specific ways to reduce it. The same problem exists with $\Delta\omega_{1/f^3}$ as well. Finally, blind application of these models has periodically led to earnest but misguided attempts by some designers to use active circuits to boost Q. Sadly, increases in Q through such means are necessarily accompanied by increases in F as well because active devices contribute noise of their own, so the anticipated improvements in phase noise fail to materialize. Again, the lack of analytical expressions for F can obscure this conclusion, and one continues to encounter various doomed oscillator designs based on the notion of active Q boosting.

That neither Eqn. 18 nor Eqn. 19 can make quantitative predictions about phase noise is an indication that at least some of the assumptions used in the derivations are invalid, despite their apparent reasonableness. To develop a theory that does not possess the enumerated deficiencies, we need to revisit, and perhaps revise, these assumptions.

18.4 THE ROLES OF LINEARITY AND TIME VARIATION IN PHASE NOISE

The preceding derivations have all assumed linearity and time invariance. Let's reconsider each of these assumptions in turn.

Nonlinearity is clearly a fundamental property of all real oscillators, as its presence is necessary for amplitude limiting. It seems entirely reasonable, then, to try to explain certain observations as a consequence of nonlinear behavior. One of these observations is that a single-frequency sinusoidal disturbance injected into an oscillator gives rise to two equal-amplitude sidebands that are symmetrically disposed about the carrier.[3] Since LTI systems cannot perform frequency translation and nonlinear systems can, nonlinear mixing has often been proposed to explain phase noise. As we shall see momentarily, amplitude-control nonlinearities certainly do affect phase noise – but only *indirectly,* by controlling the detailed shape of the output waveform.

An important insight is that disturbances are just that: perturbations superimposed on the main oscillation. They will always be much smaller in magnitude than the carrier in any oscillator worth using or analyzing. Thus, if a certain amount of injected

---

[3] B. Razavi, "A Study of Phase Noise in CMOS Oscillators," *IEEE J. Solid-State Circuits,* v. 31, no. 3, March 1966.

FIGURE 18.5. LC oscillator excited by current pulse.

noise produces a certain phase disturbance, we ought to expect that doubling the injected noise will double the disturbance. Linearity would therefore appear to be a reasonable (and experimentally testable) assumption *as far as the noise-to-phase transfer function is concerned.* It is therefore particularly important to keep in mind that, when assessing linearity, it is essential to identify explicitly the input–output variables. It is also important to recognize that this assumption of linearity is not equivalent to a neglect of the nonlinear behavior of the active devices. Because it is a linearization around the steady-state solution, it already takes the effect of device nonlinearity into account. This is precisely analogous to amplifier analysis, where small-signal gains are defined around a bias solution found using large-signal (nonlinear) equations. There is thus no contradiction here with the prior acknowledgment of nonlinear amplitude control. Any seeming contradiction is due to the fact that the word *system* is actually ill-defined. Most take it to refer to an assemblage of components and their interconnections, but a more useful definition is based on the particular input–output variables chosen. With this definition, a single circuit may possess nonlinear relationships among certain variables and linear ones among others. Time invariance is also not an inherent property of the entire circuit; it is similarly dependent on the variables chosen.

We are left only with the assumption of time invariance to re-examine. In the previous derivations we have extended time invariance to the noise sources themselves, meaning that the measures that characterize noise (e.g., spectral density) are time-invariant (stationary). In contrast with linearity, the assumption of time invariance is less obviously defensible. In fact, it is surprisingly simple to demonstrate that oscillators are fundamentally time-varying systems. Recognizing this truth is the main key to developing a more accurate theory of phase noise.[4]

To test whether time invariance holds, consider explicitly how an impulse of current affects the waveform of the simplest resonant system, a lossless LC tank (Figure 18.5). Assume that the system has been oscillating forever with some constant amplitude; then consider how the system responds to an impulse injected at two different times, as seen in Figure 18.6.

If the impulse happens to coincide with a voltage maximum (as in the left plot of Figure 18.6), the amplitude increases abruptly by an amount $\Delta V = \Delta Q/C$, but

---

[4] A. Hajimiri and T. Lee, "A General Theory of Phase Noise in Electrical Oscillators," *IEEE J. Solid-State Circuits,* v. 33, no. 2, February 1998, pp. 179–94.

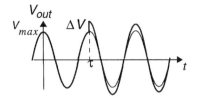 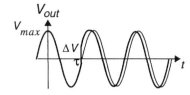

FIGURE 18.6. Impulse responses of LC tank.

because the response to the impulse superposes exactly in phase with the pre-existing oscillation, the timing of the zero crossings does not change. Thus, even though we have clearly changed the energy in the system, the amplitude change is not accompanied by a change in phase. In contrast, an impulse injected at some other time generally affects both the amplitude of oscillation and the timing of the zero crossings, as in the right-hand plot. Interpreting the zero-crossing timings as a measure of phase, we see that the amount of phase disturbance for a given injected impulse depends on when the injection occurs; time invariance thus fails to hold. An oscillator is therefore a linear yet (periodically) time-varying (LTV) system. It is especially important to note that it is theoretically possible to leave unchanged the energy of the system (as reflected in a constant tank amplitude of the right-hand response) if the impulse injects at a moment near the zero crossing when the net work performed by the impulse is zero. For example: a small positive impulse injected when the tank voltage is negative extracts energy from the oscillator, whereas the same impulse injected when the tank voltage is positive delivers energy to the oscillator; so just before the zero crossing, an instant may be found where such an impulse performs no net work at all. Hence the amplitude of oscillation cannot change, but the zero crossings will be displaced.

Because linearity (of noise-to-phase conversion) remains a good assumption, the impulse response still completely characterizes that system – even with time variation present. The only difference relative to an LTI impulse response is that the impulse response here is a function of *two* arguments, the observation time t and the excitation time τ. Noting that an impulsive input produces a step change in phase, the impulse response may be written as

$$h_\phi(t, \tau) = \frac{\Gamma(\omega_0\tau)}{q_{\max}} u(t - \tau), \tag{20}$$

where $u(t)$ is the unit step function. Dividing by $q_{\max}$, the maximum charge displacement across the capacitor, makes the function $\Gamma(x)$ independent of signal amplitude. This normalization is a convenience that allows us to compare different oscillators fairly. $\Gamma(x)$ is called the impulse sensitivity function (ISF), which is a dimensionless, frequency- and amplitude-independent function that is periodic in 2π. As its name suggests, the ISF encodes information about the sensitivity of the oscillator to an impulse injected at phase $\omega_0 t$. In the *LC* oscillator example, $\Gamma(x)$ has its maximum

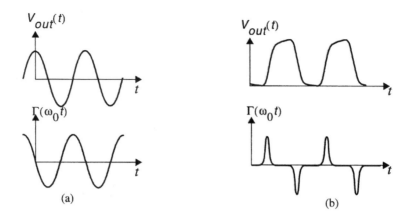

FIGURE 18.7. Example ISF for (a) an *LC* oscillator and (b) a ring oscillator.

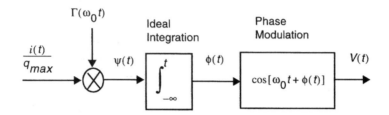

FIGURE 18.8. Equivalent block diagram of the
process described (in part) by Eqn. 21.

value near the zero crossings of the oscillation and a zero value at maxima of the oscillation waveform. In general, it is most practical (and most accurate) to determine $\Gamma(x)$ through simulation, but there are also analytical methods (some approximate) that apply in special cases.[5] In any event, to develop a feel for typical shapes of ISFs we shall consider two representative examples, first for an *LC* and a ring oscillator in Figure 18.7.

Once the impulse response has been determined (by whatever means), we may compute the excess phase due to an *arbitrary* noise signal through use of the superposition integral. This computation is valid here because superposition is linked to linearity, not to time invariance:

$$\phi(t) = \int_{-\infty}^{\infty} h_\phi(t, \tau) i(\tau)\, d\tau = \frac{1}{q_{\max}} \int_{-\infty}^{t} \Gamma(\omega_0 \tau) i(\tau)\, d\tau. \qquad (21)$$

The equivalent block diagram shown in Figure 18.8 helps us visualize this computation in ways that are familiar to telecommunications engineers, who will recognize a structure reminiscent of a superheterodyne system (more on this viewpoint shortly).

[5] F. X. Kaertner, "Determination of the Correlation Spectrum of Oscillators with Low Noise," *IEEE Trans. Microwave Theory and Tech.*, v. 37, no. 1, January 1989. Also see A. Hajimiri and T. Lee, *The Design of Low-Noise Oscillators,* Kluwer, Dordrecht, 1999.

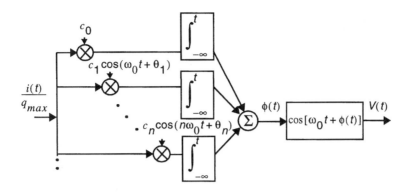

FIGURE 18.9. The equivalent system for ISF decomposition.

To cast this superposition integral into a more practically useful form, note that the ISF is periodic and therefore expressible as a Fourier series:

$$\Gamma(\omega_0\tau) = \frac{c_0}{2} + \sum_{n=1}^{\infty} c_n \cos(n\omega_0\tau + \theta_n), \qquad (22)$$

where the coefficients c_n are real and where θ_n is the phase of the nth harmonic of the ISF. (We will ignore θ_n in all that follows because we assume that noise components are uncorrelated and so their relative phase is irrelevant.) The value of this decomposition is that – like many functions associated with physical phenomena – the series typically converges rapidly, so that it is often well approximated by just the first few terms of the series.

Substituting the Fourier expansion into Eqn. 21 and then exchanging summation and integration, one obtains

$$\phi(t) = \frac{1}{q_{max}} \left[\frac{c_0}{2} \int_{-\infty}^{t} i(\tau)\, d\tau + \sum_{n=1}^{\infty} c_n \int_{-\infty}^{t} i(\tau) \cos(n\omega_0\tau)\, d\tau \right]. \qquad (23)$$

The corresponding sequence of mathematical operations is shown graphically in the left half of Figure 18.9. Note that the block diagram again contains elements that are analogous to those of a superheterodyne receiver. The normalized noise current is a broadband "RF" signal with Fourier components that undergo simultaneous down-conversions (multiplications) by a "local oscillator" signal that is the ISF, whose harmonics are multiples of the oscillation frequency. It is important to keep in mind that multiplication is a linear operation if one argument is held constant, as it is here. The relative contributions of these multiplications are determined by the Fourier co-efficients of the ISF. Equation 23 thus allows us to compute the excess phase caused by an arbitrary noise current injected into the system, once the Fourier coefficients of the ISF have been determined (typically through simulation).

We have already noted the common observation that signals (noise) injected into a nonlinear system at some frequency may produce spectral components at a different frequency. We now show that a linear but time-varying system can exhibit qualita-tively similar behavior, as implied by the superheterodyne imagery invoked earlier.

To demonstrate this property explicitly, consider injecting a sinusoidal current whose frequency is near an integer multiple m of the oscillation frequency, so that

$$i(t) = I_m \cos[(m\omega_0 + \Delta\omega)t], \tag{24}$$

where $\Delta\omega \ll \omega_0$. Substituting Eqn. 24 into Eqn. 23 (and noting that there is a negligible net contribution to the integral by terms other than when $n = m$) yields the approximation

$$\phi(t) \approx \frac{I_m c_m \sin(\Delta\omega t)}{2q_{\max}\Delta\omega}. \tag{25}$$

The spectrum of $\phi(t)$ therefore consists of two equal sidebands at $\pm\Delta\omega$, even though the injection occurs near some integer multiple of ω_0. This observation is fundamental to understanding the evolution of noise in an oscillator.

Unfortunately, we're not quite done. Equation 25 allows us to figure out the spectrum of $\phi(t)$, but we ultimately want to find the spectrum of the output voltage of the oscillator, which is not quite the same thing. However, the two quantities are linked through the actual output waveform. To illustrate what we mean by this linkage, consider a specific case where the output may be approximated as a sinusoid, so that $v_{\text{out}}(t) = \cos[\omega_0 t + \phi(t)]$. This equation may be considered a phase-to-voltage converter; it takes phase as an input and then produces from it the output voltage. This conversion is fundamentally nonlinear because it involves the phase modulation of a sinusoid.

Performing this phase-to-voltage conversion and assuming "small" amplitude disturbances, we find that the single-tone injection leading to Eqn. 25 results in two equal-power sidebands symmetrically disposed about the carrier:

$$P_{\text{SBC}}(\Delta\omega) \approx 10 \cdot \log\left[\frac{I_m c_m}{4q_{\max}\Delta\omega}\right]^2. \tag{26}$$

Note that the amplitude dependence is linear (the squaring operation simply reflects the fact that we are dealing with a power quantity here). This relationship has been verified experimentally for an exceptionally wide range of practical oscillators.

This result may be extended to the general case of a white noise source:

$$P_{\text{SBC}}(\Delta\omega) \approx 10 \cdot \log\left[\frac{\dfrac{\overline{i_n^2}}{\Delta f} \displaystyle\sum_{m=0}^{\infty} c_m^2}{4q_{\max}^2 \Delta\omega^2}\right]. \tag{27}$$

Together, Eqn. 26 and Eqn. 27 imply both upward and downward frequency translations of noise into the noise near the carrier, as illustrated in Figure 18.10. This figure summarizes what the preceding equations tell us: Components of noise near integer multiples of the carrier frequency all fold into noise near the carrier itself.

Noise near DC gets upconverted, with relative weight given by coefficient c_0, so $1/f$ device noise ultimately becomes $1/f^3$ noise near the carrier; noise near the carrier stays there, weighted by c_1; and white noise near higher integer multiples of the

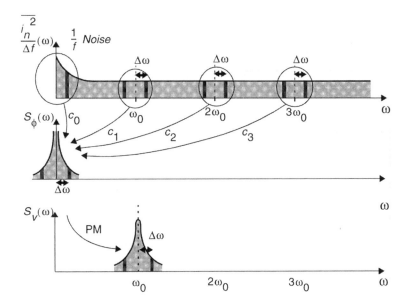

FIGURE 18.10. Evolution of circuit noise into phase noise.

carrier undergoes downconversion, turning into noise in the $1/f^2$ region. Note that the $1/f^2$ shape results from the integration implied by the step change in phase caused by an impulsive noise input. Since an integration (even a time-varying one) gives a white voltage or current spectrum a $1/f$ character, the power spectral density will have a $1/f^2$ shape.

It is clear from Figure 18.10 that minimizing the various coefficients c_n (by minimizing the ISF) will minimize the phase noise. To underscore this point quantitatively, we may use Parseval's theorem to write:

$$\sum_{n=0}^{\infty} c_m^2 = \frac{1}{\pi} \int_0^{2\pi} |\Gamma(x)|^2 \, dx = 2\Gamma_{rms}^2, \tag{28}$$

so that the spectrum in the $1/f^2$ region may be expressed as:

$$L(\Delta\omega) = 10 \cdot \log \left[\frac{\overline{\dfrac{i_n^2}{\Delta f}} \Gamma_{rms}^2}{2q_{max}^2 \Delta\omega^2} \right], \tag{29}$$

where Γ_{rms} is the rms value of the ISF. All other factors held equal, reducing Γ_{rms} will reduce the phase noise at all frequencies. Equation 29 is the rigorous equation for the $1/f^2$ region and is one key result of this phase noise model. Note that no empirical curve-fitting parameters are present in this equation.

Among other attributes, Eqn. 29 allows us to study quantitatively the upconversion of $1/f$ noise into close-in phase noise. Noise near the carrier is particularly important in communication systems with narrow channel spacings. In fact, the allowable channel spacings are frequently constrained by the achievable phase noise. Unfortunately, it is not possible to predict close-in phase noise correctly with LTI models.

This problem disappears if the new model is used. Specifically, assume that the current noise behaves as follows in the $1/f$ region:

$$\overline{i_{n,1/f}^2} = \overline{i_n^2} \cdot \frac{\omega_{1/f}}{\Delta\omega}, \tag{30}$$

where $\omega_{1/f}$ is the $1/f$ corner frequency. Using Eqn. 27, we obtain the following expression for noise in the $1/f^3$ region:

$$L(\Delta\omega) = 10 \cdot \log \left[\frac{\frac{\overline{i_n^2}}{\Delta f} c_0^2}{8 q_{max}^2 \Delta\omega^2} \cdot \frac{\omega_{1/f}}{\Delta\omega} \right]. \tag{31}$$

The $1/f^3$ corner frequency is then

$$\Delta\omega_{1/f^3} = \omega_{1/f} \cdot \frac{c_0^2}{4\Gamma_{rms}^2} = \omega_{1/f} \cdot \left(\frac{\Gamma_{dc}}{\Gamma_{rms}} \right)^2, \tag{32}$$

from which we see that the $1/f^3$ phase noise corner is not necessarily the same as the $1/f$ device–circuit noise corner; it will generally be lower. In fact, since Γ_{dc} is the DC value of the ISF, there is a possibility of reducing by large factors the $1/f^3$ phase noise corner. The ISF is a function of the waveform and hence is potentially under the control of the designer, usually through adjustment of the rise and fall time symmetry. This result is not anticipated by LTI approaches and is one of the most powerful insights conferred by this LTV model. This result has particular significance for technologies with notoriously poor $1/f$ noise performance, such as CMOS and GaAs MESFETs. Specific circuit examples of how one may exploit this observation are presented in the next section.

Another extremely powerful insight concerns the influence of cyclostationary noise sources. As alluded to earlier, the noise sources in many oscillators cannot be well modeled as stationary. Typical examples are the nominally white drain noise current in a FET and the shot noise in a bipolar transistor. Noise currents are a function of bias currents, and the latter vary periodically and significantly with the oscillating waveform. The LTV model is able to accommodate a cyclostationary white noise source with ease, since such a source may be treated as the product of a stationary white noise source and a periodic function:[6]

$$i_n(t) = i_{n0}(t) \cdot \alpha(\omega_0 t). \tag{33}$$

Here, i_{n0} is a stationary white noise source whose peak value is equal to that of the cyclostationary source, and the *noise modulation function* (NMF) $\alpha(x)$ is a periodic

[6] W. A. Gardner, *Introduction to Random Processes,* McGraw-Hill, New York, 1990.

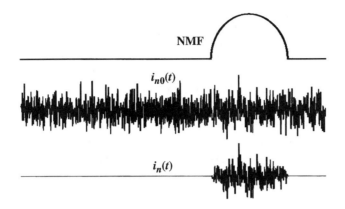

FIGURE 18.11. Cyclostationary noise as product of stationary noise and NMF.

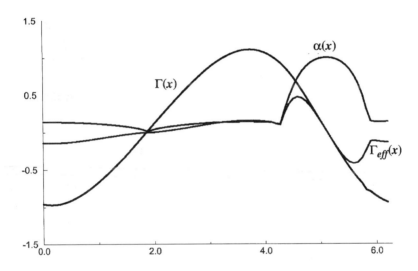

FIGURE 18.12. Accommodation of cyclostationarity.

dimensionless function with a peak value of unity. See Figure 18.11. Substituting the expression for noise current into Eqn. 21 allows us to treat cyclostationary noise as a stationary noise source, provided we define an effective ISF as follows:

$$\Gamma_{\text{eff}}(x) = \Gamma(x) \cdot \alpha(x). \tag{34}$$

Figure 18.12 shows $\Gamma(x)$, $\alpha(x)$, and $\Gamma_{\text{eff}}(x)$ for a Colpitts oscillator, all plotted over one cycle. The quasisinusoidal shape of $\Gamma(x)$ is perhaps to be anticipated on the basis of the ideal LC oscillator ISF examined earlier (where the output voltage and ISF were approximately the same in shape) but now in quadrature. The NMF is near zero most of the time, which is consistent with the Class-C operation of the transistor in a Colpitts circuit; the transistor replenishes the lost tank energy over a relatively narrow window of time, as suggested by the shape of $\alpha(x)$. The product of these two functions, $\Gamma_{eff}(x)$, has a much smaller rms value than $\Gamma(x)$, explicitly showing the exploitation of cyclostationarity by this oscillator.

This example underscores that cyclostationarity is thus easily accommodated within the framework already established. None of the foregoing conclusions changes as long as Γ_{eff} is used in all of the equations.[7]

Having identified the factors that influence oscillator noise, we're now in a position to articulate the requirements that must be satisfied in order to make a good oscillator. First, in common with the revelations of LTI models, both the signal power and resonator Q should be maximized, all other factors held constant. In addition, note that an active device is always necessary to compensate for tank loss and that active devices always contribute noise. Note also that the ISFs tell us there are sensitive and insensitive moments in an oscillation cycle. Of the infinitely many ways that an active element could return energy to the tank, the best strategy is to deliver all of the energy at once, where the ISF has its minimum value. Thus, in an ideal LC oscillator, the transistor would remain off almost all of the time, waking up periodically to deliver an impulse of current at the signal peak(s) of each cycle. The extent to which real oscillators approximate this behavior determines in large part the quality of their phase noise properties. Since an LTI theory treats all instants as equally significant, such theories are unable to anticipate this important result.

The prescription for impulsive energy restoration has actually been practiced for centuries, but in a different domain. In mechanical clocks, a structure known as an *escapement* regulates the transfer of energy from a spring to a pendulum. The escapement forces this transfer to occur impulsively, and only at precisely timed moments (coincident with the point of maximum pendulum velocity) that are chosen to minimize the disturbance of the oscillation period. Although this historically important analog is hundreds of years old – having been designed by intuition and trial and error – it was not analyzed mathematically until 1826 by Astronomer Royal George Airy.[8] Certainly its connection to the broader field of electronic oscillators has only recently been recognized.

Finally, the best oscillators will possess the symmetry properties that lead to small Γ_{dc} for minimum upconversion of $1/f$ noise. After examining some additional features of close-in phase noise, we consider in the following section several circuit examples of how to accomplish these ends in practice.

---

[7] This formulation might not apply if *external* cyclostationary noise sources are introduced into an oscillator, such as might be the case in injection-locked oscillators. For a detailed discussion, see P. Vanassche et al., "On the Difference between Two Widely Publicized Models for Analyzing Oscillator Phase Behavior," *Proc. IEEE/ACM ICCAD,* Session 4A, 2002.

[8] G. B. Airy, "On the Disturbances of Pendulums and Balances, and on the Theory of Escapements," *Trans. Cambridge Philos. Soc.,* v. 3, pt. I, 1830, pp. 105–28.

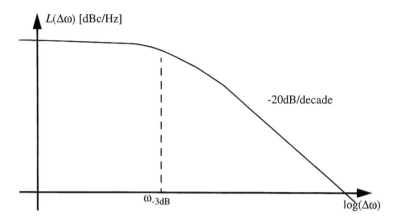

FIGURE 18.13. Lorentzian spectrum.

CLOSE-IN PHASE NOISE

From the development so far, one expects the spectrum $S_\phi(\omega)$ to have a close-in behavior that is proportional to the inverse cube of frequency. That is, the spectral density grows without bound as the carrier frequency is approached. However, most measurements fail to show this behavior, and this failure is often misinterpreted as being either the result of some new phenomenon or as evidence of a flaw in LTV theory. It is thus worthwhile to spend some time considering this issue in detail.

The LTV theory asserts only that $S_\phi(\omega)$ grows without bound. Most "phase" noise measurements, however, actually measure the spectrum of the oscillator's output voltage. That is, what is often measured is actually $S_V(\omega)$. In such a case, the output spectrum will not show a boundless growth as the offset frequency approaches zero, reflecting the simple fact that a cosine function is bounded even for unbounded arguments. This bound causes the measured spectrum to flatten as the carrier is approached; the resulting shape is *Lorentzian,*[9] as shown in Figure 18.13.

Depending on the details of how the measurement is performed, the -3-dB corner may or may not be observed. If a spectrum analyzer is used, the corner typically *will* be observed. If an ideal phase detector and a phase-locked loop were available to downconvert the spectrum of $\phi(t)$ and measure it directly, then no flattening would be observed at all. A -3-dB corner will generally be observed with real phase detectors (which necessarily possess finite phase detection range), but the precise value of the corner will now be a function of the instrumentation; the measurement will no longer reflect the inherent spectral properties of the oscillator. This lack

---

[9] W. A. Edson, "Noise in Oscillators," *Proc. IRE,* August 1960, pp. 1454–66. Also see J. A. Mullen, "Background Noise in Nonlinear Oscillators," *Proc. IRE,* August 1960, pp. 1467–73. A Lorentzian shape is the same as a single low-pass filter's power response; it just sounds more impressive if you say "Lorentzian."

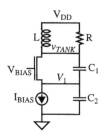

FIGURE 18.14. Colpitts
oscillator (simplified).

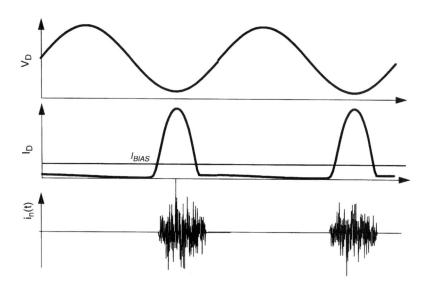

FIGURE 18.15. Approximate incremental tank voltage
and drain current for Colpitts oscillator.

of consistency in measurement techniques has been a source of great confusion in
the past.

18.5 CIRCUIT EXAMPLES

18.5.1 *LC* OSCILLATORS

Having derived expressions for phase noise at low and moderate offset frequencies,
it is instructive to apply the insights gained to practical oscillators. We first examine
the popular Colpitts oscillator and its relevant waveforms (see Figure 18.14 and Figure 18.15). An important feature is that the drain current flows only during a short
interval coincident with the most benign moments (the peaks of the tank voltage).
Its corresponding excellent phase noise properties account for the popularity of this

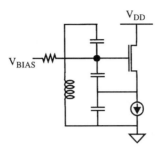

FIGURE 18.16. Clapp oscillator.

configuration. It has long been known that the least phase noise occurs within a certain narrow range of tapping ratios (e.g., a $3:1$ or $4:1$ C_2/C_1 capacitance ratio), but before LTV theory there was no theoretical basis for explaining a particular optimum.

The cyclostationary nature of the drain noise is evident in the figures. Because the noise is largest when the ISF is relatively small, the effective ISF (the product of the ISF and the noise modulating function) is much smaller than the ISF.

Both LTI and LTV models point out the value of maximizing signal amplitude. To evade supply voltage or breakdown constraints, one may employ a tapped resonator to decouple resonator swings from device voltage limitations. A common configuration that does so is Clapp's modification to the Colpitts oscillator (reprised in Figure 18.16). Differential implementations of oscillators with tapped resonators have recently made an appearance in the literature.[10] These types of oscillators are either of Clapp configurations or the dual (with tapped inductor). The Clapp configuration becomes increasingly attractive as supply voltages scale downward, where conventional resonator connections lead to V_{DD}-constrained signal swings. Use of tapping allows signal energy to remain high even with low supply voltages.

Phase noise predictions using the LTV model are frequently more accurate for bipolar oscillators owing to the availability of better device noise models. In Margarit et al. (see footnote 10), impulse response modeling (see Section 18.8) is used to determine the ISFs for various noise sources within the oscillator, and this knowledge is used to optimize the noise performance of a differential bipolar VCO. A simplified schematic of this oscillator is shown in Figure 18.17. A tapped resonator is used to increase the tank signal power, P_{sig}. The optimum capacitive tapping ratio is calculated to be around 4.5 (corresponding to a capacitance ratio of 3.5), based on simulations that take into account the cyclostationarity of the noise sources. Specifically, the simulation accounts for noise contributions by the base-spreading resistance

---

[10] J. Craninckx and M. Steyaert, "A 1.8GHz CMOS Low-Phase-Noise Voltage-Controlled Oscillator with Prescaler," *IEEE J. Solid-State Circuits,* v. 30, no. 12, December 1995, pp. 1474–82. See also M. A. Margarit, J. I. Tham, R. G. Meyer, and M. J. Deen, "A Low-Noise, Low-Power VCO with Automatic Amplitude Control for Wireless Applications," *IEEE J. Solid-State Circuits,* v. 34, no. 6, June 1999, pp. 761–71.

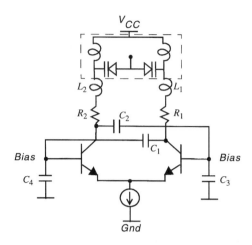

FIGURE 18.17. Simplified schematic
of the VCO in Margarit et al.

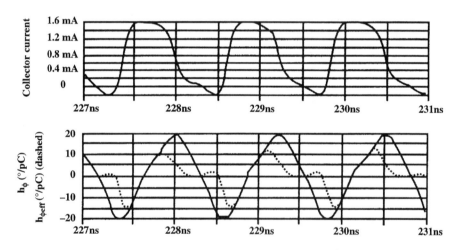

FIGURE 18.18. ISF for shot noise of each core transistor (after Margarit et al.).

and collector shot noise of each transistor as well as by the resistive losses of the tank elements. The ISFs (taken from Margarit et al., in which these are computed through direct evaluation in the time domain as described in Section 18.8) for the shot noise of the core oscillator transistors and for the bias source are shown in Figure 18.18 and Figure 18.19, respectively. As can be seen, the tail current noise has an ISF with double the periodicity of the oscillation frequency, owing to the differential topology of the circuit (the tail voltage waveform contains a component at twice the oscillator frequency). Noteworthy is the observation that tail noise thus contributes to phase noise only at even multiples of the oscillation frequency. If the tail current is filtered through a low-pass (or bandstop) filter before feeding the oscillator core,

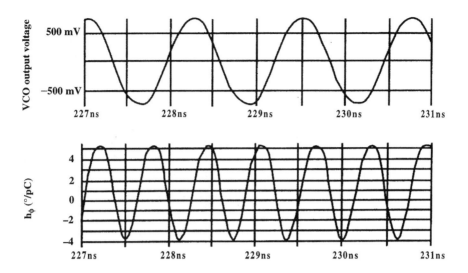

FIGURE 18.19. ISF for shot noise of tail current (after Margarit et al.).

then the noise contributed by the tail source can be reduced substantially; decreases of 10 dB or more have been noted.[11] Only the tail current's $1/f$ noise would remain as a noise contributor. The individual ISFs are used to compute the contribution of each corresponding noise sources, and the contributions are then summed.

The reduction of $1/f$ noise upconversion in this topology is clearly seen in Figure 18.20, which shows a predicted and measured $1/f^3$ corner of 3 kHz – in comparison with an individual device $1/f$ noise corner of 200 kHz. Note that the then-current version of one commercial simulation tool, *Spectre*, fails in this case to identify a $1/f^3$ corner within the offset frequency range shown, resulting in a 15-dB underestimate at a 100-Hz offset. The measured phase noise in the $1/f^2$ region is also in excellent agreement with the LTV model's predictions. For example, the predicted value of -106.2 dBc/Hz at 100-kHz offset is negligibly different from the measured value of -106 dBc/Hz. As a final comment, this particular VCO design is also noteworthy for its use of a separate automatic amplitude control loop; this allows for independent optimization of the steady-state and start-up conditions, with favorable implications for phase noise performance.

As mentioned, a key insight of the LTV theory concerns the importance of symmetry, the effects of which are partially evident in the preceding example. A configuration that exploits this knowledge more fully is the symmetrical negative resistance

---

[11] A. Hajimiri and T. Lee, in *The Design of Low-Noise Oscillators* (Kluwer, Dordrecht, 1999), describe a simple shunt capacitor across the tail node to ground. E. Hegazi et al., in "A Filtering Technique to Lower Oscillator Phase Noise" (*ISSCC Digest of Technical Papers,* February 2001), use a parallel tank between the tail source and the common source node to achieve a 10-dB phase noise reduction.

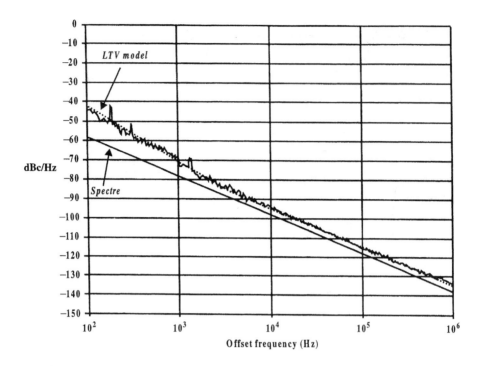

FIGURE 18.20. Measured and predicted phase noise of VCO in Margarit et al.

oscillator shown in Figure 18.21.[12] This configuration is not new by any means, but an appreciation of its symmetry properties is. Here it is the half-circuit symmetry that is important, because noise in the two half-circuits is only partially correlated at best. By selecting the relative widths of the PMOS and NMOS devices appropriately to minimize the DC value of the ISF (Γ_{dc}) for each half-circuit, one may minimize the upconversion of $1/f$ noise. Through exploitation of symmetry in this manner, the $1/f^3$ corner can be dropped to exceptionally low values, even when device $1/f$ noise corners are high (as is typically the case for CMOS). Furthermore, the bridgelike arrangement of the transistor quad allows for greater signal swings, compounding the improvements in phase noise. As a result of all of these factors, a phase noise of -121 dBc/Hz at an offset of 600 kHz at 1.8 GHz has been obtained with low-Q (estimated to be 3–4) on-chip spiral inductors consuming 6 mW of power in a 0.25-μm CMOS technology (see footnote 12). This result rivals what one may achieve with bipolar technologies, as seen by comparison with the bipolar example of Margarit et al. With a modest increase in power, the same oscillator's phase noise becomes compliant with specifications for GSM1800.

---

[12] A. Hajimiri and T. Lee, "Design Issues in CMOS Differential LC Oscillators," *IEEE J. Solid-State Circuits,* May 1999.

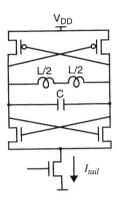

FIGURE 18.21. Simple symmetrical
negative resistance oscillator.

18.5.2. RING OSCILLATORS

As an example of a circuit that does not well approximate ideal behavior, consider a
ring oscillator. First, the "resonator" Q is poor because the energy stored in the node
capacitances is thrown away every cycle. Hence, if the resonator of a Colpitts oscillator may be likened to a fine crystal wine glass, the resonator of a ring oscillator is
mud. Next, energy is restored to the resonator during the edges (the worst possible
times) rather than at the voltage maxima. These factors account for the well-known
terrible phase noise performance of ring oscillators. As a consequence, ring oscillators are found only in the most noncritical applications, or inside wideband PLLs
that clean up the spectrum.

However, there are certain aspects of ring oscillators that can be exploited to
achieve better phase noise performance in a mixed-mode integrated circuit. Noise
sources on different nodes of an oscillator may be strongly correlated for various reasons. Two examples of sources with strong correlation are substrate and supply noise
arising from current switching in other parts of the chip. The fluctuations on the supply and substrate will induce a similar perturbation on different stages of the ring
oscillator.

To understand the effect of this correlation, consider the special case of having identical noise sources on all the nodes of the ring oscillator, as shown in Figure 18.22.
If all the inverters in the oscillator are the same then the ISF for different nodes will
differ only in phase by multiples of $2\pi/N$, as shown in Figure 18.23.

Therefore, the total phase resulting from all the sources is given by Eqn. 21 through
superposition:[13]

---

[13] A. Hajimiri, S. Limotyrakis, and T. H. Lee, "Jitter and Phase Noise in Ring Oscillators," *IEEE J.
Solid-State Circuits*, v. 34, no. 6, June 1999, pp. 790–804.

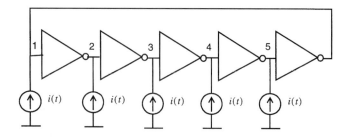

FIGURE 18.22. Five-stage ring oscillator with identical noise sources on all nodes.

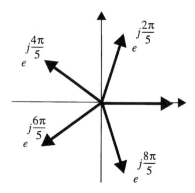

FIGURE 18.23. Phasors for noise contributions from each source.

$$\phi(t) = \frac{1}{q_{\max}} \int_{-\infty}^{t} i(\tau) \left[\sum_{n=0}^{N-1} \Gamma\left(\omega_0 \tau + \frac{2\pi n}{N} \right) \right] d\tau. \tag{35}$$

Expanding the term in brackets in a Fourier series, we observe that it is zero except at DC and multiples of $N\omega_0$. That is,

$$\phi(t) = \frac{N}{q_{\max}} \sum_{n=0}^{\infty} c_{(nN)} \int_{-\infty}^{t} i(\tau) \cos(nN\omega_0 \tau) \, d\tau, \tag{36}$$

which means that, for fully correlated sources, only noise in the vicinity of integer multiples of $N\omega_0$ affects the phase. Therefore, every effort should be made to maximize the correlations of noise arising from substrate and supply perturbations. This result can be achieved by making the inverter stages and the noise sources on each node as similar to each other as possible via proper layout and circuit design. For example, the layout should be kept symmetrical, and the inverter stages should be laid out close to each other so that substrate noise appears as a common-mode source.

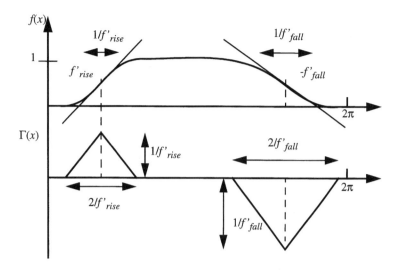

FIGURE 18.24. Derivation of approximate
analytical expression for ring oscillator ISF.

This latter consideration is particularly important in the case of a lightly doped substrate, because such a substrate may not act as a single node.[14] It is also important
that the orientation of all the stages be kept identical. The interconnecting wires between the stages must be identical in length and shape, and a common supply line
should feed all the inverter stages. Furthermore, loading on all stages should be kept
identical – perhaps by using dummy buffer stages as necessary, for example. Use of
the largest number of stages consistent with oscillation at the desired frequency will
also be helpful since (as a practical matter) fewer c_n coefficients will then affect the
phase noise. Finally, as the low-frequency portion of the substrate and supply noise
then dominates, one should exploit symmetry to minimize Γ_{dc}.

Another common conundrum concerns the preferred topology for MOS ring oscillators – that is, whether a single-ended or differential topology results in better jitter
and phase noise performance for a given center frequency f_0 and total power dissipation P. Facilitating an analysis of these choices is an approximate expression for
the ISF, as depicted in Figure 18.24. The ISF is approximated by two triangles, as
shown. The rms value of the ISF is given by

$$\Gamma_{rms}^2 = \frac{1}{3\pi}\left(\frac{1}{f'_{rise}}\right)^3(1 + A^3),\tag{37}$$

where f'_{rise} and f'_{fall} are the maximum slope during the rising and falling edges and
where A is the ratio of f'_{rise} to f'_{fall}.

[14] T. Blalack, J. Lau, F. J. R. Clement, and B. A. Wooley, "Experimental Results and Modeling of
Noise Coupling in a Lightly Doped Substrate," *IEDM Tech. Digest,* December 1996.

By coupling the ISF equation with the noise equations for transistors, one may derive expressions for the phase noise of MOS differential and single-ended oscillators. Based on these expressions, the phase noise of a single-ended (inverter chain) ring oscillator is found to be independent of the number of stages for a given power dissipation and frequency of operation. However, for a differential ring oscillator, the phase noise (jitter) grows with the number of stages. Hence even a properly designed differential CMOS ring oscillator underperforms its single-ended counterpart – with a disparity that increases with the number of stages. The difference in the behavior of these two types of oscillators with respect to the number of stages can be traced to the way they dissipate power. The DC current drawn from the supply is independent of the number and slope of the transitions in differential ring oscillators. In contrast, inverter-chain ring oscillators dissipate power mainly on a per-transition basis and therefore have better phase noise for a given power dissipation. However, in ICs with a large amount of digital circuitry, a differential topology may still be preferred because of its lower sensitivity to substrate and supply noise and lower noise injection into other circuits on the same chip. The decision of which architecture to use should be based on both of these considerations.

Yet another commonly debated question concerns the optimum number of inverter stages in a ring oscillator to achieve the best jitter and phase noise for a given f_0 and P. For single-ended CMOS ring oscillators, the phase noise and jitter in the $1/f^2$ region are not strong functions of the number of stages. However, if the symmetry criteria are not well satisfied or if the process has large $1/f$ noise (or both), then a larger N will reduce the jitter. This reduction results from the faster edge speeds that must accompany the use of a larger number of stages in order to achieve the same oscillation frequency. The faster edge speeds reduce the effect of asymmetries in rise and fall times and thereby reduce the upconversion of $1/f$ noise. The choice of the number of stages generally must be made on the basis of several design criteria, such as $1/f$ noise effect and the desired maximum frequency of oscillation, as well as on the basis of the influence of external noise sources (such as supply and substrate noise) that may not scale with N. A symmetry-based reduction in $1/f$ noise can significantly augment any reduction provided by the trap resetting that can attend operation of MOSFETs in the switching regime.[15]

The jitter and phase noise behavior is different for differential ring oscillators. Jitter and phase noise increase with an increasing number of stages. Hence, if the $1/f$ noise corner is not large and/or proper symmetry measures have been taken, then the minimum number of stages (3 or 4) should be used to give the best performance. This recommendation holds even if power dissipation is not a primary issue. It is not fair

[15] I. Bloom and Y. Nemirovsky, "1/f Noise Reduction of Metal-oxide Semiconductor Transistors by Cycling from Inversion to Accumulation," *Appl. Phys. Lett.,* v. 58, April 1991, pp. 1664–6. See also S. L. J. Gierkink et al., "Reduction of the 1/f Noise Induced Phase Noise in a CMOS Ring Oscillator by Increasing the Amplitude of Oscillation," *Proc. 1998 Internat. Sympos. Circuits and Systems,* v. 1, May 31–June 3, 1998, pp. 185–8.

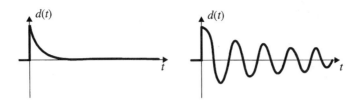

FIGURE 18.25. Overdamped and underdamped amplitude responses.

to argue that burning more power in a larger number of stages allows the achievement of better phase noise, because dissipating the same total power in a smaller number of stages with larger devices results in better jitter and phase noise – as long as it is possible to maximize the total charge swing.

18.6 AMPLITUDE RESPONSE

While the close-in sidebands are dominated by phase noise, the far-out sidebands are greatly affected by amplitude noise. Unlike the induced excess phase, the excess amplitude $A(t)$ due to a current impulse decays with time. This decay is the direct result of the amplitude-restoring mechanisms always present in practical oscillators. The excess amplitude may decay very slowly (e.g., in a harmonic oscillator with a high-quality resonant circuit) or very quickly (e.g., a ring oscillator). Some circuits may even demonstrate an underdamped second-order amplitude response. The detailed dynamics of the amplitude-control mechanism have a direct effect on the shape of the noise spectrum.

In the context of the ideal LC oscillator of Figure 18.5, a current impulse with an area Δq will induce an instantaneous change in the capacitor voltage, which in turn will result in a change in the oscillator amplitude that depends on the instant of injection (as shown in Figure 18.6). The amplitude change is proportional to the instantaneous normalized voltage change $\Delta V/V_{\max}$ for small injected charge $\Delta q \ll q_{\max}$:

$$\Delta A = \Lambda(\omega_0 t)\frac{\Delta V}{V_{\max}} = \Lambda(\omega_0 t)\frac{\Delta q}{q_{\max}}, \quad \Delta q \ll q_{\text{swing}}, \tag{38}$$

where the amplitude impulse sensitivity function $\Lambda(\omega_0 t)$ is a periodic function that determines the sensitivity of each point on the waveform to an impulse; it is the amplitude counterpart of the phase impulse sensitivity function $\Gamma(\omega_0 t)$. From a development similar to that for phase response, the amplitude impulse response can be written as

$$h_A(t, \tau) = \frac{\Lambda(\omega_0 t)}{q_{\max}}d(t - \tau), \tag{39}$$

where $d(t - \tau)$ is a function that defines how the excess amplitude decays. Figure 18.25 shows two hypothetical examples: $d(t)$ for a low-Q oscillator with overdamped response and for a high-Q oscillator with underdamped amplitude response.

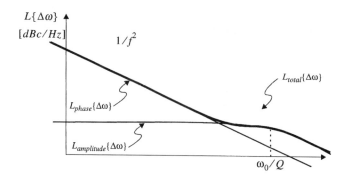

FIGURE 18.26. Phase, amplitude, and total sideband powers
for the overdamped amplitude response.

As with our evaluation of the phase response, we invoke a small-signal linear approximation here. Again, we are not neglecting the fundamentally nonlinear nature of amplitude control; we are simply taking advantage of the fact that amplitude noise will certainly be small enough to validate a small-signal linear approximation for any oscillator worth the analysis effort. We will assume without loss of generality that the amplitude-limiting system of most oscillators can be approximated as first or second order, again for small disturbances. The function $d(t - \tau)$ will thus typically be either a dying exponential or a damped sinusoid.

For a first-order system,

$$d(t - \tau) = e^{-\omega_0(t-\tau)/Q} \cdot u(t - \tau). \tag{40}$$

Therefore, the excess amplitude response to an arbitrary input current $i(t)$ is given by the superposition integral,

$$A(t) = \int_{-\infty}^{t} \frac{i(\tau)}{q_{max}} \Lambda(\omega_0 t) e^{-\omega_0(t-\tau)/Q} \, d\tau. \tag{41}$$

If $i(t)$ is a white noise source with power spectral density, then the output power spectrum of the amplitude noise $A(t)$ can be shown to be

$$L_{amplitude}\{\Delta\omega\} = \frac{\Lambda_{rms}^2}{q_{max}^2} \cdot \frac{\overline{i_n^2}/\Delta f}{2 \cdot (\omega_0^2/Q^2 + (\Delta\omega)^2)}, \tag{42}$$

where Λ_{rms} is the rms value of $\Lambda(\omega_0 t)$. If L_{total} is measured then the sum of both $L_{amplitude}$ and L_{phase} will be observed, and hence there will be a pedestal in the phase noise spectrum at ω_0/Q as shown in Figure 18.26. Also note that the significance of the amplitude response depends greatly on L_{rms}, which in turn depends on the topology.

As a final comment on the effect of amplitude-control dynamics, we note that an underdamped response would result in a spectrum with some peaking in the vicinity of ω_0/Q.

18.7 SUMMARY

The insights gained from LTI phase noise models are simple and intuitively satisfying: One should maximize signal amplitude and resonator Q. An additional, implicit insight is that the phase shifts around the loop generally must be arranged so that oscillation occurs at or very near the center frequency of the resonator. This way, there is a maximum attenuation by the resonator of off-center spectral components.

Deeper insights provided by the LTV model are that the resonator energy should be restored impulsively at the ISF minimum instead of evenly throughout a cycle, and that the DC value of the effective ISF should be made as close to zero as possible in order to suppress the upconversion of $1/f$ noise into close-in phase noise. This theory also shows that the inferior broadband noise performance of ring oscillators may be offset by their potentially superior ability to reject common-mode substrate and supply noise.

18.8 APPENDIX: NOTES ON SIMULATION

Exact analytical derivations of the ISF usually cannot be obtained for any but the simplest oscillators. Various approximate methods are outlined in the reference of footnote 4, but the only generally accurate method is a direct evaluation of the time-varying impulse response. In this direct method, an impulsive excitation perturbs the oscillator and the steady-state phase perturbation is measured. The timing of the impulse with respect to the unperturbed oscillator's zero crossing is then incremented, and the simulation is repeated until the impulse has been "walked" through an entire cycle.

The impulse must have a small enough value to ensure that the assumption of linearity holds. Just as an amplifier's step response cannot be evaluated properly with steps of arbitrary size, one must judiciously select the area of the impulse rather than blindly employing some fixed value (e.g. 1 C). If one is unsure whether the impulse chosen has been sized properly, linearity may always be tested explicitly by scaling the size of impulse by some amount and then verifying that the response scales by the same factor.

Finally, some confusion persists about whether the LTV theory properly accommodates the phenomenon of amplitude-to-phase conversion exhibited by some oscillators. As long as linearity holds, the LTV theory does accommodate AM-to-PM conversion – provided that an exact ISF has been obtained. This is due to the fact that changes in the phase of an oscillator arising from an amplitude change appear in the impulse response of that oscillator. A slight subtlety arises from the phase relationships among sidebands generated by these two mechanisms, however. Summed contributions from these two sources may result in sidebands with unequal amplitudes, in contrast with the purely symmetrical sidebands that are characteristic of AM and PM individually.

PROBLEM SET FOR PHASE NOISE

PROBLEM 1   Calculate the phase impulse response for a voltage-driven series RLC network.

PROBLEM 2   In every practical oscillator, the tank is not the only source of phase shift. Hence, the actual oscillation frequency may differ somewhat from the resonant frequency of the tank, for example. Using the time-varying model, explain why the oscillator's phase noise can degrade if such off-frequency oscillations occur.

PROBLEM 3   Assume that the steady-state output amplitude of the following oscillator is 1 V. Calculate the phase noise in dBc/Hz at an offset of 100 kHz from the carrier for the signal coming out of the ideal comparator (see Figure 18.27).

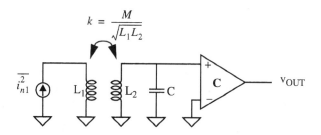

FIGURE 18.27. Oscillator with comparator.

Assume that $L_1 = 25$ nH, $L_2 = 100$ nH, $M = 10$ nH, and $C = 100$ pF. Further assume that the noise current is

$$\overline{i_{n1}^2} = 4kTG_{\text{eff}}\Delta f, \tag{P18.1}$$

where $1/G_{\text{eff}} = 50\ \Omega$. The temperature of the circuit is 300 K.

PROBLEM 4   Consider the CMOS Colpitts oscillator shown in Figure 18.28.

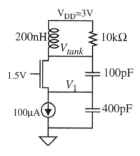

FIGURE 18.28. Colpitts oscillator example for Problem 4.

(a) Calculate V_{gs} and V_{ds} before oscillation starts. Calculate the steady-state tank oscillation amplitude and frequency. You may assume that the drain current consists of narrow pulses. Assume operation deep in the short-channel regime and ignore all transistor capacitances.

(b) Inductors used in practical oscillators typically have relatively low Q. Assume that an equivalent parallel resistance of 10 kΩ models energy losses in the inductor, as well as loading effects caused by a subsequent stage (not shown). Calculate the phase noise in the $1/f^2$ region due to this resistance, assuming that the transistor and current source are noiseless. Assume also that the capacitors are lossless. It may help to point out that the resistance in question introduces a stationary noise source.

PROBLEM 5   For the Colpitts oscillator of the previous problem, find the DC value of V_{gs} when the oscillator is in the steady state. Also find the conduction angle 2Φ, assuming for simplicity that the drain current waveform consists of triangular pulses (see Figure 18.29). *Hint:* Keep in mind that the average drain current must equal the bias current, I_{BIAS}.

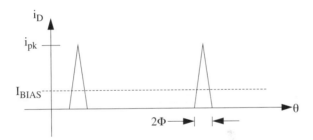

FIGURE 18.29. Assumed drain current
waveform for Problem 5.

Explain qualitatively how the phase noise in the $1/f^2$ region depends on the conduction angle.

PROBLEM 6   Assume a *noiseless* resistor in parallel with an otherwise lossless tank, and calculate the phase noise due to drain current noise for the Colpitts oscillator of Problem 4. Note that the drain current noise cannot be treated as stationary.

(a) Calculate $\alpha^2(\theta)$ and the effective value for Γ_{rms}, assuming a purely sinusoidal $\Gamma(\theta)$. Note that the drain current noise in any MOS device may be expressed as

$$\frac{\overline{i_{nd}^2}}{\Delta f} = 4kT\gamma\mu C_{ox}\frac{W}{L}(V_{gs} - V_t). \tag{P18.2}$$

Assume that the drain current waveform consists of the tips of a sinusoidal waveform, as shown in Figure 18.30.

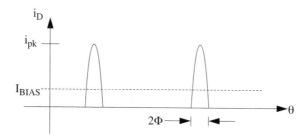

FIGURE 18.30. Assumed drain current
waveform for Problem 6.

(b) Use the calculated Γ_{rms} (effective) to calculate the phase noise due to NMOS drain current noise.
(c) Calculate F, the fitting parameter from the Leeson model.

PROBLEM 7 Find the voltage change, ΔV_1, across the tank caused by a current impulse of area Δq_2 injected at time $t = \tau$ into the middle node of a capacitively tapped tank, as shown in Figure 18.31.

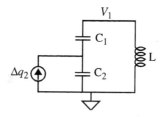

FIGURE 18.31. Tapped tank
and noise source.

(a) Find the equivalent charge Δq_{eq} injected in parallel with the inductor that would result in the same voltage change. Express your result in terms of the capacitive divide ratio $n = C_1/(C_1 + C_2)$.
(b) Find the power spectrum of that equivalent noise source, expressed in terms of the power spectrum of the original noise current source in parallel with C_2.

PROBLEM 8 Suppose that the tail current source of the oscillator of Figure 18.32 is actually implemented with a 1.3-kΩ resistor. Model the noise due to the resistor as an equivalent noise current source in parallel with the inductor. Assume that this noise is dominant over all other sources and calculate the phase noise resulting from this resistor's noise.

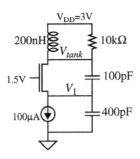

FIGURE 18.32. Colpitts oscillator
example for Problem 8.

PROBLEM 9 For the Colpitts oscillator of Problem 4, suppose the NMOS transistor possesses a $1/f$ noise corner of 200 kHz (this value is not atypical, unfortunately). Calculate the $1/f^3$ corner frequency of the phase noise. *Hint:* Calculate c_0 using

$$c_0 = \frac{1}{\pi} \int_0^{2\pi} \Gamma_{\text{eff}}(\theta)\, d\theta = \frac{1}{\pi} \int_0^{2\pi} \Gamma(\theta) \cdot \alpha(\theta)\, d\theta \qquad \text{(P18.3)}$$

and assume a triangular drain current waveform as in Problem 5.

PROBLEM 10 Re-consider the high-level model for an RLC oscillator (see Figure 18.33). Note that the active elements that keep the system in oscillation are no longer considered noiseless. In particular, model the "magic box" explicitly as possessing an equivalent input noise current and noise voltage. For simplicity, assume that these noise sources are white.

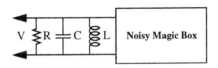

FIGURE 18.33. *RLC* oscillator.

(a) Derive an expression for the phase noise spectrum assuming LTI behavior.
(b) Re-derive this expression using the LTV model, assuming that the oscillation waveform is sinusoidal. Further assume that the noise sources are stationary.
(c) Compare your answers and provide an explicit expression for F, the Leeson fitting parameter.

CHAPTER NINETEEN

ARCHITECTURES

19.1 INTRODUCTION

Because of its high performance, the superheterodyne is the only basic architecture presently in use for both receivers and transmitters. One should not then infer, however, that all receivers and transmitters are therefore topologically identical, for there are many variations on a basic theme. For example, we will see that it may be desirable to use more than one intermediate frequency to aid the rejection of certain signals, leading to a question of how many IFs there should be, and what frequencies they should have. Answering those questions is known as *frequency planning,* and converging on an acceptable frequency plan generally involves a substantial amount of iteration.

An important constraint is that on-chip energy storage elements generally consume significant die area. Furthermore, they tend not to scale gracefully (if at all) as technology improves. Hence, the "ideal" integrated architecture should require the minimum number of energy storage elements, and there are continuing efforts even to eliminate the need for high-quality filters through architectural means. Complete success has been elusive, though, and one must accept that the desired performance frequently may be achieved only if external filters are used. It is not too much of an exaggeration to assert that architectures are essentially determined by available filter technology.

Once a basic architecture and its associated frequency plan have been chosen, other key considerations include how best to distribute the huge power gain (typically 120–140 dB for receivers) among the various stages. This is because important factors such as system noise, stability, and linearity are all strong functions of gain distribution.

The great diversity of existing architectures reflects the inability of any single one to satisfy all requirements of interest. So, after considering some universal system issues, what follows is a sampling of several representative receiver and transmitter architectures, along with a discussion of their attributes and limitations.

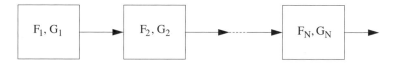

FIGURE 19.1. Cascaded systems for noise figure computation.

19.2 DYNAMIC RANGE

Dynamic range is one of the most basic system considerations, and we have already identified the two parameters, intercept and noise figure, that bound it. However, we need to extend our understanding to how the dynamic range of a cascade of systems depends on the intercepts and noise figures of the individual subsystems. We now develop the relevant combination rules in this section.

19.2.1 NOISE FIGURE OF CASCADED SYSTEMS

The overall noise figure of a cascade of systems depends on both the individual noise figures as well as their gains. The dependency on the gain results from the fact that, once the signal has been amplified, the noise of subsequent stages is less important. As a result, system noise figure tends to be dominated by the noise performance of the first couple of stages in a receiver.

How the individual noise figures combine to yield the overall noise figure is complicated by the variety of impedance levels typically found in the system. To develop an equation for the system noise figure, consider the block diagram of Figure 19.1, where each F_n is a noise factor and each G_n is a power gain (specifically, the *available gain*, the gain obtained with a matched load). Since noise factor depends on source resistance, one must compute the individual noise figures relative to the output impedance of the preceding stage to keep the calculation honest. This issue arises less frequently in discrete designs, where impedance levels are often standardized, but requires careful attention in IC implementations.

Noise factor may be expressed in several ways, but one form that is particularly useful for our task at hand is

$$F = \frac{R_s + R_e}{R_s} = 1 + N_e, \tag{1}$$

where R_e is a (possibly fictitious) resistance that accounts for the observed noise in excess of that due to R_s. The quantity N_e is thus an excess noise power ratio, equal to $F - 1$.

Reflecting this power ratio back to the input of the preceding stage simply involves a division by the available power gain of the previous stage. Reflecting the excess noise contribution of a given stage all the way back to the input thus requires division by the total available gain between that stage and the overall input.

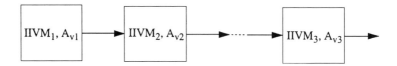

FIGURE 19.2. Cascaded systems for input intercept calculation.

The total noise factor is the sum of these individual contributions, and is therefore given by

$$F = 1 + F_1 - 1 + \frac{F_2 - 1}{G_1} + \frac{F_3 - 1}{G_1 G_2} + \cdots + \frac{F_N - 1}{\prod_{n=1}^{N-1} G_n}, \qquad (2)$$

which simplifies to

$$F = F_1 + \frac{F_2 - 1}{G_1} + \frac{F_3 - 1}{G_1 G_2} + \cdots + \frac{F_N - 1}{\prod_{n=1}^{N-1} G_n}. \qquad (3)$$

It is clear that the system noise figure is in fact dominated by the noise performance of the first few gain stages. Hence, in trying to achieve a good noise figure, most of the design effort will generally focus on the first few stages.

It should also be noted that the preceding equations must be used carefully if any of the stages is a mixer, because noise at a given IF can result from the translation of noise from two different frequencies. These equations will apply as long as the noise figure of the stages preceding a mixer is computed by examining the noise at both signal and image frequencies.

19.2.2 LINEARITY OF CASCADED SYSTEMS

The other figure of merit that bounds system dynamic range is the intercept point. Even though we have discussed only third-order intercepts, it should be mentioned that there are also instances in which the second-order intercept is a relevant linearity measure. A notable example is the degenerate case of a superheterodyne in which the IF is zero. We will study such *direct-conversion* receivers in greater detail shortly; we bring up the subject now merely to call attention to the fact that both second- and third-order (and possibly higher-order) intercepts may be useful measures of system linearity.

A difficulty in developing the desired equation is that the distortion products of one stage combine with those of a later stage in ways that depend on their relative phases. Hence, there is no simple, fixed relationship between the individual and over-all intercepts. However, it is possible to derive a conservative (worst-case) estimate by assuming that the amplitudes of the distortion products add directly. This choice in turn makes it most natural to express the gains as voltage ratios, in contrast with the use of power gains in the expression for system noise figure. This is sketched in Figure 19.2, where each A_{vn} is a voltage gain and each IIVM_n is an Mth-order input intercept voltage.

To facilitate the derivations, we use $V_{dM,n}$ to denote the Mth-order (intermodulation) distortion product at the output of the nth stage due to a voltage V applied to the input of that stage. Further note that, from the definition of an input intercept, the input-referred Mth-order IM distortion product may be written as

$$V_{dM} = \frac{V^M}{\text{IIV}M^{M-1}}. \tag{4}$$

Let us carry out the derivation for the specific case of the third-order intercept, and for a cascade of just two stages. The third-order IM at the output of the first stage is

$$V_{d3,1} = \frac{A_{v1}V^3}{\text{IIV}3_1^2}. \tag{5}$$

The third-order IM voltage at the output of the second stage is the sum of two components. One is simply a scaled version of the distortion produced by the first stage, and the other is the distortion produced by the second stage. Adding these directly together yields the following pessimistic estimate:

$$V_{d3,\text{tot}} = A_{v1}A_{v2}V_{d3,1} + A_{v2}V_{d3,2}. \tag{6}$$

The input-referred third-order distortion is found by dividing through by the total gain:

$$V_{d3\,\text{in,tot}} = \frac{A_{v1}A_{v2}V_{d3,1} + A_{v2}V_{d3,2}}{A_{v1}A_{v2}}. \tag{7}$$

Substituting Eqn. 4 and Eqn. 5 into Eqn. 7 yields

$$\frac{1}{\text{IIV}3_{\text{tot}}^2} = \frac{1}{\text{IIV}3_1^2} + \frac{A_{v1}^2}{\text{IIV}3_2^2}. \tag{8}$$

This last equation confirms that the later stages bear a greater burden because of the gain that precedes them. We can also see that the reciprocal IIV3 of a given stage, normalized by the total gain up to the output of that stage, contributes to the overall reciprocal input-referred intercept in root-sum-squared fashion.

Although Eqn. 8 applies strictly to a two-stage cascade, it is readily extended to an arbitrary number of stages as follows:

$$\frac{1}{\text{IIV}3_{\text{tot}}^2} = \frac{1}{\text{IIV}3_j^2} + \sum_{j=2}^{n}\left\{\frac{1}{\text{IIV}3_j^2}\prod_{i=1}^{j-1}A_{vi}^2\right\}. \tag{9}$$

One may follow a similar procedure to determine the overall input-referred intercept for distortion products of any order.

Having derived expressions for the noise figure and intercept of an arbitrary cascade of systems, we may now turn to an examination of both receiver and transmitter architectures.

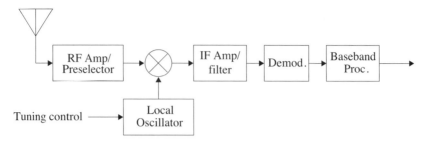

FIGURE 19.3. Superheterodyne receiver block diagram.

19.2.3 THE SINGLE-CONVERSION RECEIVER

Insight into important design issues, and an understanding of why alternative architectures were developed, may be obtained by studying the standard superheterodyne
block diagram introduced in previous chapters; see Figure 19.3. To distinguish this
basic architecture from more elaborate ones that employ more than one mixer and
intermediate frequency, this one is known as a *single-conversion* superheterodyne
receiver.

To appreciate some of the design tradeoffs involved, consider first the *image rejection* problem, caused by the fact that there are two input frequencies that can produce
an IF of a given frequency. Suppose, as is common, that the difference frequency
component out of the mixer is chosen as the IF. As a specific example, suppose the IF
is 70 MHz and we desire to tune in an RF signal at 800 MHz. With a corresponding
(assumed) local oscillator frequency of 870 MHz, an RF signal at 940 MHz would
also produce an IF signal at 70 MHz. This undesired signal is known as the *image* signal. Most typically, a filter at the front-end (known as a *preselector* or image-reject
filter) is used to attenuate greatly the image signal prior to mixing.

Note that the image signal is displaced from the desired frequency by twice the IF.
To make it easier to filter the image, it is therefore generally desirable to choose a relatively high IF. To allow the use of a fixed filter, the IF should be high enough so that
images never fall in band. In the case of ordinary AM radio, which spans 530 kHz to
1610 kHz, one would normally want to choose an IF of at least $(1610 - 530)/2$ kilohertz, or about 540 kHz. Unfortunately, 455 kHz evolved as the IF and has remained
(and will probably forever remain) the norm.[1] The preselector therefore can't be (and
isn't) a fixed filter in AM radios that use the historically conventional frequency plan.
In such receivers, the preselector filter tracks the LO to suppress the image.

Whereas a high IF is favored to relax the requirements on front-end filtering, a low
IF is preferred to reduce demands on the IF amplifier and filter. The resulting frequency plan is partly a consequence of balancing these factors against one another.

---

[1] FM broadcast radio is in better shape. It spans the 88–108-MHz band (in the United States), and
the typical IF is 10.7 MHz, so that no image of a legitimate FM radio signal coincides with another
legitimate FM radio signal.

Another part of frequency planning involves the choice of LO frequencies since, once again, there are two possible values that produce a given IF from a given RF. Normally, it is desirable to choose an LO range that corresponds to the RF range plus the desired IF, rather than the RF range minus the desired IF. Although both choices are valid, the former choice, known as *high-side injection,* reduces the ratio of maximum to minimum LO frequency required, easing oscillator design. Continuing with our AM radio example, our LO choices would be one that spans 75 kHz ($= 530 - 455$ kHz) to 1155 kHz ($= 1610 - 455$ kHz), or one that spans 985 kHz to 2065 kHz. The former choice requires a tuning range in excess of 15 : 1, while the latter choice requires only about a 2.1 : 1 tuning range. It is much easier to design an oscillator to span a 2.1 : 1 frequency range, particularly if the frequency variation is accomplished by varying just a single capacitance in a tank (which is the most common method). Since the resonant frequency of a tank is proportional to the inverse square root of the *LC* product, the capacitance must vary over a range that is the square of the frequency range. Obtaining a $(2.1)^2$ range in capacitance is much easier to achieve than a $(15)^2$ ratio, either mechanically or electronically. For this reason, high-side injection is the nearly universal choice.

19.2.4 UPCONVERSION

The primary motivation for choosing the difference frequency component out of the mixer is to perform a downconversion from RF to a lower-frequency IF. The implicit assumption is that such a frequency lowering makes it easier to realize high-quality IF filters and to obtain the requisite gain. As a consequence, the vast majority of superheterodyne receivers are of this type. In many cases, however, other design considerations may influence the frequency plan.

One option is to choose an IF that is actually *higher* than the RF. Such a choice greatly reduces the image rejection problem and therefore relaxes front-end filtering requirements considerably. Another significant benefit is a reduction in the fractional tuning range required of the LO. Thus, if either of these considerations is important, an upconversion architecture may be preferred.

Let's carry out a design exercise to see how upconversion might be applied to the design of an AM receiver (perhaps, even, to permit a fully integrated solution). Rather than choosing the traditional IF of 455 kHz, suppose we select, somewhat arbitrarily, an IF of 5 MHz. To cover the entire AM band, we might want the LO to tune from 5.530 MHz to 6.610 MHz, for an LO tuning ratio of just 1.2 : 1. Constructing an oscillator that tunes over a 20% range is relatively easy.

With this choice of IF, the image frequency is 10 MHz away from the desired RF signal. It is easy to build a fixed filter that rolls off above 1610 kHz to provide excellent attenuation by 10.530 MHz, so the image rejection problem largely disappears.

We see that upconversion eases demands on the performance of both the local oscillator and the preselector. The price paid, of course, is that channel selection has to occur at a higher frequency (the IF), and this places greater demands on elements

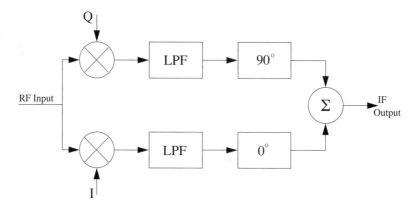

FIGURE 19.4. Image-reject mixer.

in the IF chain. Furthermore, the receiver is now susceptible to interference from sources within the relatively large bandwidth of the front end, so linearity requirements become more severe (perhaps much more so). Choosing to use upconversion therefore depends on whether the increased susceptibility to interferers is acceptable in exchange for improved image rejection and simplified LO design.

19.2.5 DUAL CONVERSION

We've noted that a low intermediate frequency is demanding of preselector and LO performance, but places no great demands on IF filtering. We've also noted that a high IF increases the difficulty of realizing the channel-selective IF filters, but mitigates the image rejection problem and therefore relaxes requirements on the preselector. A high IF also happens to simplify LO design owing to the reduction in required oscillator tuning range.

A *dual-conversion* receiver uses two IFs in an effort to combine the benefits of both downconversion and upconversion architectures by using both techniques together. In a dual-conversion superheterodyne receiver, a first mixer produces a high IF to take care of the image rejection issue, while a second mixer and low IF ease the channel selection problem. Some receivers employ a third IF to provide even greater flexibility in the tradeoff. Most modern high-performance receivers implemented in discrete technologies use a dual-conversion architecture, with the frequency plan determined in large part by what high-quality filters are available at reasonable cost.

19.2.6 THE IMAGE-REJECT RECEIVER

The image-reject receiver employs a *complex mixer* that exploits the relationship between the desired signal and the undesired image to cancel the image during the mixing process. Thus, no preselector filter is needed (at least in principle), and design of the overall architecture can proceed with diminished concern for the image problem. In particular, a relatively low IF may be chosen to ease the requirements on IF filtering, analog-to-digital conversion, and subsequent baseband processing. See Figure 19.4.

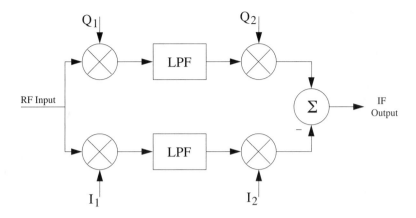

FIGURE 19.5. Weaver architecture.

To perform this miracle, the RF signal feeds two mixers. Two LO signals, in quadrature, drive the other mixer ports. The desired difference frequency component is the same in both mixer outputs, but the image terms are in quadrature. In the classic implementation shown, a constant-gain, frequency-independent 90° phase shifter allows simple addition of the two signals to cancel the undesired image terms.

Because it is somewhat difficult to build broadband quadrature phase shifters, the alternative architecture (due to Weaver[2]) shown in Figure 19.5 is often attractive. In the Weaver architecture, a pair of quadrature mixing operations eliminates the need for the phase shifters. As long as the two pairs of LO signals are truly in quadrature and the gains in the two paths are perfectly matched, there will be perfect cancellation of the unwanted image.

Effect of Gain and Phase Errors

Complete image rejection in the preceding architectures depends on perfect phase quadrature and perfect gain matching. Of course, it is more realistic to assume that neither condition is satisfied in practice. The ultimate rejection of the image then depends on the levels of matching achieved. To quantify the matching requirement, first consider explicitly the effect of gain errors in the block diagram of Figure 19.4. Specifically, but without loss of generality, let the I and Q signals model all the error as follows:

$$I = B \cos(\omega_{LO} t), \tag{10}$$

$$Q = A \sin(\omega_{LO} t). \tag{11}$$

If the RF input signal is $\cos(\omega_{RF} t)$, then the output of the I mixer is

$$\frac{B}{2}[\cos(\omega_{RF} t + \omega_{LO} t) + \cos(\omega_{RF} t - \omega_{LO} t)] \tag{12}$$

---

[2] D. K. Weaver, "A Third Method of Generation and Detection of Single-Sideband Signals," *Proc. IRE,* December 1956, pp. 1703–5.

and the output of the Q mixer is

$$\frac{A}{2}[\sin(\omega_{RF}t + \omega_{LO}t) - \sin(\omega_{RF}t - \omega_{LO}t)]. \tag{13}$$

The low-pass filters remove the sum frequency component, while the 90° phase shifter converts (minus) sine to cosine.

The sensitivity of the difference frequency component in Eqn. 13 to the sign of the argument is ultimately the source of this mixer's image rejection property, since an RF signal above the LO frequency by an amount equal to the IF produces an IF signal with a sign opposite to that produced by mixing down the image. Phase shifting, and then summing with a cosine-mixed version of the LO, results in a reinforcement of the desired component and a simultaneous suppression of the image.

For an RF signal above the LO, the overall output may be expressed as

$$\tfrac{1}{2}[A\cos(\omega_{IF}t) + B\cos(\omega_{IF}t)]; \tag{14}$$

the image RF signal produces an output given by

$$\tfrac{1}{2}[-A\cos(\omega_{IF}t) + B\cos(\omega_{IF}t)]. \tag{15}$$

Let us define the *image rejection ratio* (IRR) as the power ratio of the desired output to the undesired output. Then

$$\text{IRR}_{\text{gain}} = \left[\frac{A+B}{A-B}\right]^2 = \left[\frac{1+B/A}{1-B/A}\right]^2 = \left[\frac{1+(1+\varepsilon)}{1-(1+\varepsilon)}\right]^2 \approx \frac{4}{\varepsilon^2}; \tag{16}$$

in the last approximation, it is assumed that the gain error ε is much smaller than unity. As we shall see shortly, the gain error of any practical image-reject mixer *must* satisfy this inequality in order to meet realistic performance specifications. Hence, the approximation is guaranteed to be valid in all cases of practical interest.

One may carry out an analogous derivation assuming perfect gain matching but imperfect 90° phase shift. In that case, the IRR due to a departure $\Delta\phi$ (in radians) from quadrature may be expressed as

$$\text{IRR}_{\text{phase}} = 1 + 4(\cot\Delta\phi)^2 \approx \frac{4}{(\Delta\phi)^2}. \tag{17}$$

Again assuming small gain and phase errors, the net IRR may be computed using the sum of the errors given by Eqn. 16 and Eqn. 17:

$$\text{IRR}_{\text{tot}} \approx \frac{4}{(\Delta\phi)^2 + \varepsilon^2}. \tag{18}$$

It is quite difficult to achieve much better than about a 0.1% gain error and a 1° phase error, particularly at high frequencies, without some form of calibration. These numbers correspond to an IRR of about 41 dB. More typically, image rejection is rarely significantly better than about 35 dB. Because most receivers require much larger (e.g. 80-dB) image rejection, the image-reject architecture alone rarely provides sufficient rejection. Autocalibration can help close the gap, but it is unlikely to

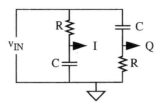

FIGURE 19.6. Passive analog
quadrature generator.

provide an additional 40 dB of image rejection. Hence, additional filtering is usually required.

Aside from that limitation, an additional drawback of the image-reject mixer is that mixers are noisy and can be power-hungry, so adding another one can be unattractive from the perspective of power consumption and dynamic range.

Quadrature Generators

The image-reject mixer requires the ability to generate quadrature phase shifts. Although it isn't difficult to provide a 90° shift over a broad frequency range, it is much harder to maintain a constant amplitude response at the same time. In fact, no finite network can provide both a constant phase shift and a constant gain magnitude over an infinite frequency range, so one must settle for approximations that work well only over some limited range.

The *RC–CR* network shown in Figure 19.6 is a popular method for generating quadrature signals over a narrow frequency band. The phase shift is 90° at all frequencies, but the magnitude response is not constant.

The phase shift of the *I* branch is zero at DC, and asymptotically heads down toward −90° as the frequency approaches infinity. The phase shift of the *Q* branch starts off at +90° at DC, and heads downward toward zero with the same functional shape as the *I* branch. Hence, even though the phase shift of *each RC* branch is not constant, the 90° *difference* between them is.

Unfortunately, the amplitude of each output changes dramatically with frequency, since the *I* branch is low-pass in nature whereas the *Q* branch is a high-pass filter. The amplitudes of the two outputs are equal only at the pole frequency:

$$\omega = \frac{1}{RC}. \tag{19}$$

There is, of course, a 3-dB attenuation at this frequency, so one must accept this significant loss when using this network. Furthermore, this network is thermally noisy, and its contribution to overall noise figure must be kept under control.

The design of this quadrature generator is straightforward in principle: the resistor and capacitor values are chosen to set the pole frequency equal to the operating frequency. In so doing, the effect of parasitic loading must be taken into account. Even so, an amplitude mismatch is all but inevitable because of component tolerances. If necessary, one may restore amplitude equality with a variable gain amplifier in an

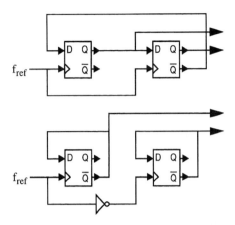

FIGURE 19.7. Digital quadrature generators.

AGC loop or with a limiter. Clearly, this type of compensation is effective only in generating one frequency at time, such as in producing a quadrature LO.

In such single-frequency applications, it may be appropriate to consider alternative quadrature generation methods that automatically provide equal-amplitude outputs. Quadrature oscillators, consisting of two integrators in a feedback loop, can inherently produce equal-amplitude sine waves in quadrature (see Chapter 17).

In those instances where square-wave signals are acceptable, one may use digital circuits to eliminate the amplitude matching problem. Ring oscillators, in which differential gain stages are used (in a two- or four-stage ring, for example), can also produce excellent quadrature outputs. Such an oscillator must be embedded within a PLL in order to set the frequency.

Other digital options that do not require a PLL involve clocked circuits (see Figure 19.7). The first circuit produces quadrature outputs whose frequency is one fourth that of the clock, so the clock frequency may have to be exceedingly high. One attribute of the circuit, though, is that it is insensitive to the duty cycle of the incoming clock, since state changes occur only on the rising edge.

The output of the second circuit is at one half the clock frequency, so this circuit demands less of the circuitry. However, since state changes occur on both rising and falling edges of the clock, the quality of the quadrature outputs is now sensitive to the duty cycle of the clock. If the clock and its inverse are generated with the crude method shown, further degradation results from the propagation delay of the inverter.

Various other phase-shifting methods have evolved to handle those cases where one must provide a quadrature relationship to a relatively broadband signal. A common example is the passive *RC* network shown in Figure 19.8.

The geometric mean of the *RC* frequencies is chosen equal to the desired center frequency. The detailed selection of the individual stage *RC* time constants to satisfy an arbitrary gain constancy constraint is quite involved, but an extremely crude rule of thumb often suffices for a first-cut design. Each stage provides reasonably constant

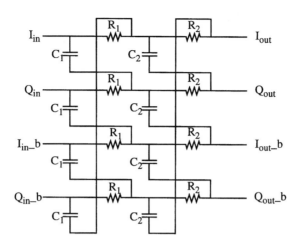

FIGURE 19.8. Two-stage broadband quadrature generator.

gain (within about 0.2 dB) over roughly a 10% bandwidth, so the required number of stages is determined by the bandwidth over which a constant gain is required. The two-stage design shown can provide relatively constant gain over approximately a $\pm 20\%$ bandwidth about the center frequency by staggering the two time constants. For example, to design for a 1-GHz center frequency, one might choose R_1C_1 to correspond to about 900 MHz, while R_2C_2 might be selected to correspond to about 1.1 GHz.[3] A significant disadvantage of this network, however, is its attenuation and high noise.

This type of network provides multiple output phases and thus is known as a *polyphase filter.* It is exceptionally good at providing quadrature outputs yet has a versatility that extends well beyond that function, as we'll see shortly.

To understand first why the polyphase network can generate quadrature so well, consider redrawing the network as shown in Figure 19.9 (only one stage is shown for clarity). Suppose that the input signals are not quite in quadrature. As a specific example, suppose that Q_{in} is closer to I_{in} than it should be. If we imagine this imperfection as a slight clockwise rotation of Q_{in} and $Q_{in\_}b$ in the octagonal layout on the right, then we see that Q_{out} moves a little clockwise from where it should be, but so does I_{out}. This circuit thus improves quadrature as signals pass through it.[4] Cascading such stages provides additional improvement. The ultimate limitations

---

[3] This network is evidently due to M. J. Gingell, "Single Sideband Modulation Using Sequence Asymmetric Polyphase Networks," *Electrical Communication,* v. 48, 1973, pp. 21–5. Practical examples that function over a decade frequency range for SSB generation using Hartley's phasing method may also be found in various editions of the *ARRL Handbook for Radio Amateurs,* from about 1981 to 1992. See also British Patent #1,174,710 (filed 7 June 1968, granted 17 December 1969) as well as its U.S. counterpart, #3,559,042 (filed 19 May 1969, granted 26 January 1971).

[4] As a consequence, the requirements for component matching are looser in the early stages than in the later ones.

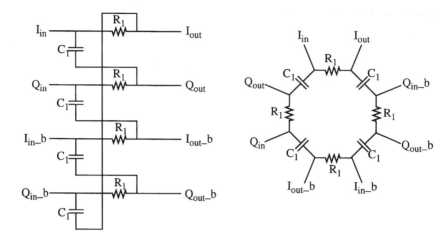

FIGURE 19.9. Two equivalent ways to draw a single-stage polyphase filter.

are imposed by mismatch in the components and in the loads on the various output nodes. These loads include, of course, ever-present parasitics. As long as one accommodates all of these effects in the design of the networks, excellent quadrature is assured.[5] For this reason, the polyphase network is extremely popular among hams for SSB generation using the phasing method of Hartley.[6] There, a multistage cascade of filters provides good quadrature over the baseband bandwidth, and a separate polyphase filter is often used to generate good quadrature for the mixers.

The same properties that make it attractive for quadrature and SSB generation make the polyphase filter useful for image rejection. Just as SSB generation requires (in effect) the ability to distinguish between negative and positive frequency baseband components, image rejection requires the ability to distinguish between signals that are negatively and positively offset from a carrier. However, an ordinary filter by itself cannot make such a distinction; the magnitude response is a function of frequency alone. An ordinary image-reject filter may thus have to exhibit extraordinary steepness in order to function (as in Carson's SSB generation method).

Another way of using ordinary filters to provide image rejection is to decompose a signal into its separate orthogonal components through a quadrature downconversion, leaving us with two real signals. A pair of real filters may then process those in-phase and quadrature signals separately from that point on. This method is the

---

[5] Additionally, one must always bear in mind the lossiness (and attendant noise) of these networks.

[6] In discrete implementations, the Weaver method has not found wide use for SSB generation. Recall that, unlike the methods of Carson and Hartley, mixer offsets in a Weaver modulator produce an in-band tone. This artifact is generally much more objectionable than carrier leakage, and the cancelling of multiple offsets that is needed to eliminate it is more bother than most wish to bear. Consequently, the phasing method has shown enduring popularity among amateur radio enthusiasts. The polyphase filter's invention has largely taken care of what was once the most difficult part: designing the phase shifter.

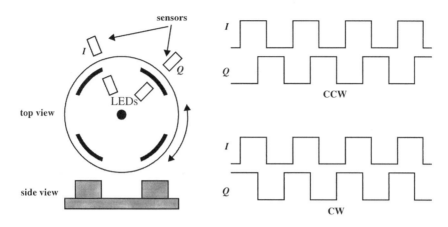

FIGURE 19.10. Idealized computer mouse (only one of two wheel assemblies shown).

essence of the Weaver architecture, and fundamentally it seeks to implement a *complex filter* with a pair of real filters.

Another option is simply to implement a complex filter directly. Such a filter operates on two (or more) inputs simultaneously and generally produces two (or more) outputs. The most natural and common choice is to drive a complex filter with in-phase and quadrature components that represent the real and imaginary parts of a complex signal. Such a filter's magnitude response is a function not only of frequency but also of the phase relationships between the two inputs. This dual sensitivity means that such a filter potentially responds differently to positive and negative frequency components, allowing us to discriminate against (say) an image signal.

Conventional demonstrations of complex filters usually begin with the theory of complex signals – invoking Hilbert transforms, if not a host of trigonometric identities. The reader is encouraged to investigate that rigorous approach independently; it's good for you, like broccoli. Here, however, we pursue a somewhat different approach and study how a conventional computer mouse works.

A *mouse*?

Yes, and here's why. A mouse needs to respond not only to speed (frequency), but also to direction (the sign of that frequency); a mouse that cannot distinguish between upward and downward motions is not a very useful device. These are the same requirements that we must satisfy to generate SSB as well as to reject images. So, if we understand how a mouse works, we'll have developed the key insights we need to understand complex filters.

Construction details vary among manufacturers, but Figure 19.10 captures the essential features of a basic mechanical mouse (real mice have more teeth). In this imagined implementation of a mouse, optically opaque teeth are disposed around a wheel that are set up to interrupt intermittently the light paths of a pair of photoemitters and photodetectors as the wheel rotates. Thus, when one emitter–detector path

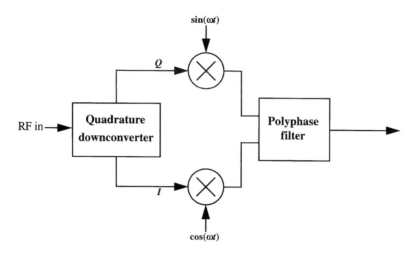

FIGURE 19.11. Polyphase image-reject filter.

is fully obscured (corresponding to a logic zero, say), the other is on the verge of un-blocking. Depending on the direction of rotation, the I signal will lead or lag the Q signal. For example, if the wheel is rotating counterclockwise, the rising edge of I precedes that of Q by 90°. If the rotation is clockwise, then the rising edge of I lags that of Q by 90°. This arrangement thus produces a pair of output signals that tells us both the frequency and direction of rotation. These attributes are precisely those used by a polyphase filter.[7]

Observe that if we have access to only one of the outputs then the only infor-mation conveyed is the magnitude of the frequency (speed of mouse motion). Any subsequent processing of the mouse signal is therefore limited to operations that take frequency – and only frequency – as an input. However, with access to *both* I and Q we could build a filter that would, for instance, make the mouse respond only to mo-tion in one direction. Disregarding the undesirability of such a mouse, we recognize the deeper principle: Knowledge of both I and Q enables us to perform the desired discrimination against, say, a negative-frequency "image."

As a specific electronic example, consider the application of a polyphase filter dis-played as Figure 19.11. This system first uses a pair of quadrature mixers to resolve an input signal into its in-phase and quadrature components. The polyphase filter then accepts *both* components as filter inputs and processes them as an ensemble.

---

[7] You may see reference to terms such as "sequence asymmetry" and the like. The term *sequence* may be taken as synonymous with *frequency* for the most part. Strictly speaking, what is meant is the sequence of vectors encountered as the phasor rotates. Hence, a positive sequence might be defined as I, Q, I_b, and Q_b (corresponding to a counterclockwise rotation in our mouse diagram). *Asym-metry* refers to the possibility that the frequency response magnitude may be different for positive and negative frequencies (sequences). Finally, the term *polyphase* derives from the multiplicity of input or output phases that such a filter processes.

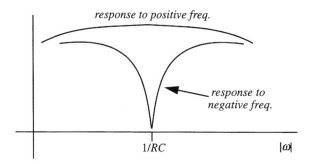

FIGURE 19.12. Idealized frequency response
of polyphase image-reject filter.

As with the mouse example, the filter has the ability to treat negative and positive frequency components differently. In this particular case, the frequency response is indeed different for positive and negative frequencies, as seen in Figure 19.12.

We see that the polyphase filter used this way is actually a bandstop filter. By choosing the zero location(s) properly, we may suppress the negative-frequency components and thereby reject the undesired image. This property has been exploited to enable good performance from a low-IF (also known as a near-zero IF, or NZIF) architecture.[8] As noted earlier, one may cascade several stages to obtain broader band rejection.

One impediment to doing so is that deriving the transfer functions by hand for polyphase filters of varying order is an unpleasant exercise. It's bad enough for the second order, but it's completely hopeless for the third order and above. It's so unpleasant, in fact, that we won't even leave it to the reader as a homework problem, tempting as that may be. Many published examples seem to have been designed by trial and error (at least, these publications are conspicuously silent about the specific methods used). Instead of making you (or us) derive them, we will simply provide explicit formulas that have been kindly derived elsewhere.[9] In all that follows, the transfer functions are defined in terms of the complex representations of signals:

$$V_{in}(t) = I_{in}(t) + jQ_{in}(t), \tag{20}$$

$$V_{out}(t) = I_{out}(t) + jQ_{out}(t), \tag{21}$$

For the first-order filter, the transfer function is

$$G_1(j\omega) \equiv \frac{V_{out}(j\omega)}{V_{in}(j\omega)} = \frac{1 + \omega RC}{1 + j\omega RC}. \tag{22}$$

---

[8] J. Crols and M. Steyaert, *CMOS Wireless Transceiver Design,* Kluwer, Dordrecht, 1998.

[9] H. Kobayashi et al. cite the use of *Mathematica* to perform the actual derivation in their paper, "Explicit Transfer Function of RC Polyphase Filter for Wireless Transceiver Analog Front-End," *Proceedings of APASIC,* 2002.

The second-order filter's transfer function is a little more involved, but still tractable:

$$G_2(j\omega) = \frac{(1 + \omega R_1 C_1)(1 + \omega R_2 C_2)}{1 - \omega^2 R_1 C_1 R_2 C_2 + j\omega(R_1 C_1 + R_2 C_2 + 2R_1 C_2)}. \tag{23}$$

The transfer function for the third-order filter, on the other hand, is big enough that we can't even write it down in one go. We need to break it into pieces:

$$G_3(j\omega) = \frac{N(j\omega)}{D(j\omega)}, \tag{24}$$

where

$$N(j\omega) = (1 + \omega R_1 C_1)(1 + \omega R_2 C_2)(1 + \omega R_3 C_3). \tag{25}$$

The denominator is

$$D(j\omega) = D_R(j\omega) + jD_I(j\omega), \tag{26}$$

with

$$D_R(j\omega) = 1 - \omega^2[R_1 C_1 R_2 C_2 + R_2 C_2 R_3 C_3 \\ + R_1 C_1 R_3 C_3 + 2R_1 C_3(R_2 C_1 + R_2 C_2 + R_3 C_2)] \tag{27}$$

and

$$D_I(j\omega) = \omega[R_1 C_1 + R_2 C_2 + R_3 C_3 + 2(R_1 C_2 + R_2 C_3 + R_1 C_3)] \\ - \omega^3 R_1 C_1 R_2 C_2 R_3 C_3. \tag{28}$$

Note that all of these filters exhibit notches at negative frequencies, as in the first example. Observe further that the notch locations are simple RC products, where each RC is that of an individual section. These observations help to converge on a final design much more quickly by guiding the choice of an initial set of component values. It is common practice (but by no means essential or necessarily optimum) to choose all the capacitors equal in value and then to vary only the resistance values. As a further guide to designing broadband polyphase filters, choose the geometric mean of the zero frequencies equal to the center frequency of the filter. Then, vary the ratio of the zero frequencies and investigate the worst-case stopband rejection. If insufficient, reduce the ratio and add more cascaded sections. In general, you will need to supplement this work with an accommodation of component tolerances and mismatch. Monte Carlo simulation techniques work well for these relatively simple structures.

19.2.7 DIRECT CONVERSION

The upconversion and dual-conversion receiver architectures seek to solve the image rejection problem by using a high IF to displace the image enough to allow a simple filter to provide the necessary rejection. An alternative choice is to use a downconversion architecture in which the intermediate frequency is zero. Since a signal and its image are separated by twice the IF, a zero IF implies that the desired signal is its own image. Therefore, goes the argument, there is no image to reject and hence no

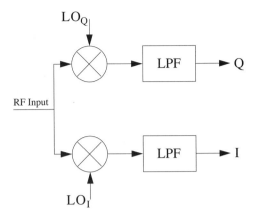

FIGURE 19.13. Direct-conversion receiver.

need for a front-end image-reject filter; front-end filtering requirements thus become especially easy to satisfy.[10] Furthermore, with a zero IF, all subsequent baseband processing can take place at the lowest possible frequency. For example, relatively low-frequency analog-to-digital conveters (ADCs) and digital-signal processing engines can implement the channel filter, as well as perform demodulation and ancillary housekeeping functions. Hence, no external filters are required (in principle, anyway). Furthermore, the flexibility inherent in digital approaches opens the possibility for a "universal" receiver, one that can accommodate many different standards with one piece of hardware; see Figure 19.13.

The most general direct-conversion receiver requires *two* mixers and LOs. The reason is that the phase of the LO with respect to the incoming RF signal is important. If the phases are coincident (or anticoincident), the demodulated signal is of maximum strength. If the phase relationship happens to be a quadrature one, the demodulated signal is zero.

In order to accommodate arbitrary phase relationships between the RF and LO signals, then, a mixer must be augmented with another one driven by an LO in quadrature with the first one. By combining the outputs of these two mixers, correct demodulation is possible with arbitrary input phase. Mismatch in the I and Q paths is not nearly as serious here as in an image-reject architecture because – rather than attempting to reject the image – this architecture actually *exploits* the "image." The only effect of mismatches is some distortion.

With all of these attributes, it seems as if the direct-conversion receiver has no peer, especially for amenability to IC implementation. Indeed, the ubiquitous pager is often a direct-conversion receiver. The simple architecture (and simple signaling scheme) thus appears to permit a highly integrated, low-cost realization.

---

[10] It is nonetheless generally good practice to use some type of filter in any case – to avoid, for example, out-of-band interferers overloading the front end.

However, there are several serious drawbacks to direct conversion, and these impediments have thus far stymied efforts to use this architecture for more sophisticated applications, despite a considerable number of earnest attempts. Among these problems is an unfortunate, extreme sensitivity to DC offsets (both internal as well externally induced) and $1/f$ noise: with a zero IF, offsets and $1/f$ noise represent error components within the same band as the desired signal. Consider, for example, the problem of detecting a 10-μV input signal (this value is not atypical). Offsets (and $1/f$ noise) that are small compared with this value are not easily achieved. Hence, noise figures tend to be rather poor, and it is easy for offsets to dominate the output and overload subsequent stages. The only way that pagers have managed to evade these problems to some degree is through the use of relatively unsophisticated two-tone signaling. The resulting spectrum has little DC energy, so simple AC coupling largely solves the problem there.

Unfortunately, straightforward AC coupling is not a panacea. A low cutoff frequency is necessary if the modulation happens to have significant spectral components at low frequencies. A low cutoff frequency in turn forces the use of relatively large capacitors, possibly requiring the use of off-chip elements. A more serious difficulty is that such a network recovers slowly from overload. For example, a 100-Hz cutoff frequency implies settling times on the order of 10 ms. Such slow recovery could cause the receiver to drop bits. While some of these problems may be addressed at the system and protocol level by mandating the use of modulation methods that minimize low-frequency spectral energy, that approach is clearly practical only in new systems, where modulation schemes have not already been codified in a standard.

Other methods for alleviating the problems with capacitive coupling include the use of active offset cancellation. Time-division multiple-access (TDMA) systems, for example, inherently provide users with intervals of time in which they are doing nothing. Offsets may be measured and removed during these periodic idle times.[11] An alternative is to use two sets of mixers so that, at any given moment, one is being used and the other is having its offsets cancelled. The two sets of mixers exchange places periodically.

An important consideration in all of these proposed methods is that the offset cancellation capacitors must be large enough for kT/C noise to be negligible. Frequently, the required capacitances are a good fraction of a nanofarad, so the die area consumed can often be unattractive.

Another difficulty is intolerance of front-end nonlinearity. Any even-order distortion produces a DC offset that is signal-dependent, and thus represents another "noise" term. Front-end LNAs, for example, must therefore be designed to have exceptionally high IIP2 (more relevant for this architecture than IIP3, although it's useful to have both numbers). This requirement usually forces a significant increase in front-end power dissipation since, all other factors held constant, elevating bias

---

[11] J. Sevenhans et al., "An Integrated Si Bipolar RF Transceiver for a Zero IF 900MHz GSM Digital Mobile Radio Front-End of a Hand Portable Radio," *Proc. IEEE CICC,* May 1991, pp. 7.7.1–4.

levels improves linearity. Furthermore, differential structures are almost certainly a necessity in the front end because their symmetry reduces even-order distortion. However, use of differential circuits involves a doubling of power consumption.

Yet another problem is that of LO radiation. Since the LO is the same frequency as that of the RF input signal, LO energy can find its way to the antenna and radiate, causing interference to other receivers. Worse yet, the LO can cause interference to its *own* receiver. Depending on the phase relationship between the LO component appearing at the RF port of the mixer and that at the normal LO port, yet another DC "noise" component will appear in the baseband signal as a result of the mixing action. Since the LO power is generally stronger than the RF signal (perhaps by many orders of magnitude), this self-rectification of LO energy is a significant problem indeed. Extraordinary isolation must be achieved to prevent the DC offset from dominating the output of the mixer. Furthermore, the amount of LO radiation that does leak back into the front end often finds its way back through a path that may be quite sensitive to factors such as the proximity of the antenna to nearby objects. Finally, leakage during transmit of the PA output back to the LO can cause pulling of the LO due to parasitic coupling. This feedback can cause the generation of a whole host of highly objectionable spurs.

In summary, the direct-conversion receiver thus needs an exceptionally linear LNA, two exceptionally linear mixers, two LOs (operating at or near the RF, which may be a relatively high frequency), a method for obtaining a quadrature relationship between the two LO signals, extraordinary isolation of the energy from these two LOs, and a method for achieving submicrovolt offsets and $1/f$ noise. These requirements are difficult to satisfy simultaneously.

One other point deserves mention. The direct-conversion receiver gets much of its gain at one frequency. On a power basis, this might be as high as 10^{12} or even higher. Thus, it becomes *critically* important to avoid any parasitic feedback loops in order to avoid oscillation. Because of the large gain, extraordinary input–output isolation is required to guarantee a parasitic loop transmission well below unity. Again, it can be challenging to provide this level of isolation. Sustained, dedicated engineering effort is finally bearing fruit, with many successful commercial examples proving that these problems – though difficult – are not intractable. We will study one particular implementation in Section 19.6.4.

19.3 SUBSAMPLING

There has been a fair amount of activity recently on the development of subsampling architectures. This class of receivers exploits the large ratio of carrier frequency to bandwidth that characterizes almost all RF links. Satisfying the Nyquist sampling criterion requires only that we sample at twice the signal bandwidth, not at twice the carrier. So, in principle, we can sample the RF directly, but with a much lower sampling frequency than the RF (see Chapter 13).

The fly in the ointment here is that all of the noise within the bandwidth of the front end folds into the subsampled baseband. To avoid this problem, one would need RF

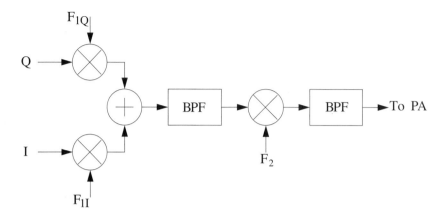

FIGURE 19.14. Superheterodyne transmitter.

filters whose bandwidths were impractically narrow. If we had such filters in the first place, we would structure receivers very differently indeed.

Because of this unfortunate noise-folding property, subsampling architectures typically exhibit *terrible* noise figures (e.g., 30 dB!). As a consequence, designers of such receivers are forced instead to direct the reader's attention to linearity. The careful reader will note, however, that comparable "linearity" may be obtained simply by preceding an ordinary architecture with an attenuator, as mentioned in the context of certain mixers (e.g., subsampling and potentiometric mixers). This equivalency should leave one with some skepticism about the general utility of subsampling. Hence, although research continues, success again remains elusive.

19.4 TRANSMITTER ARCHITECTURES

As one might suspect, transmitter architectures are generally the inverses of corresponding receiver architectures. For example, a traditional ordinary superheterodyne (now often simply shortened to "heterodyne") transmitter is sketched in Figure 19.14. The heterodyne transmitter allows modulation at a low frequency to be translated up to RF in steps, but it does require filters that may be difficult to implement in integrated form.

To reduce the filtering requirements, one may employ direct (up)conversion, as shown in Figure 19.15. As mentioned previously, feedback from the PA can perturb the LO (since they are at the same frequency) if this architecture is implemented literally. Huge reductions in LO pulling by the PA can be achieved if an offset synthesizer is used instead of an LO operating directly at the desired RF carrier frequency; see Figure 19.16.

In this architecture, the output carrier frequency is not the same as the modulation carrier frequency F_1, reducing greatly the pulling problems discussed earlier. Furthermore, any modulation on the input is transferred directly to the carrier without

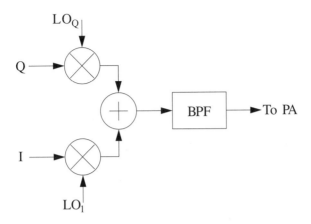

FIGURE 19.15. Direct-conversion transmitter.

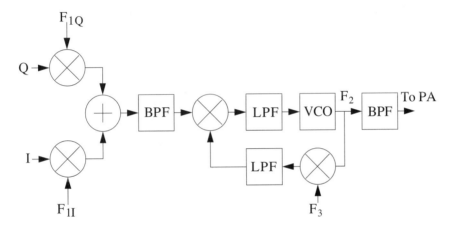

FIGURE 19.16. Transmitter with offset frequency synthesizer.

any scaling. Synthesizer loops that employ frequency dividers in the feedback path multiply any phase noise at the reference input by the divide factor N. Such noise enhancement can be highly objectionable, so the offset synthesizer's lack of any such multiplication is a potentially significant advantage.

19.5 OSCILLATOR STABILITY

An issue that requires some discussion is that of LO frequency stability. In the AMPS analog cellular telephone system, for example, the channel spacing is 30 kHz while the carrier frequency is approximately 900 MHz. Since the channel spacing is therefore roughly 30 ppm of the carrier, the LO frequency must be controlled to within, say, 3 ppm if the error is to be an acceptably small fraction of the channel spacing. Needless to say, it is rather challenging to achieve – let alone sustain over variations in temperature and supply voltage – this level of accuracy.

Stable voltage regulators are routinely used to solve the voltage sensitivity problem, but eliminating the effect of temperature variation remains troublesome. One straightforward solution is to enclose the oscillator in a thermostatically controlled environment. Such "ovenized" oscillators may exhibit excellent stability, but they generally consume too much power for most portable applications.

An alternative technique exploits the repeatability of a crystal's drift with temperature. The actual temperature is continuously measured, digitized with an ADC, and then fed to a calibration ROM. The output of the ROM drives a DAC, which in turn drives a varactor to compensate for the drift.[12] This open-loop correction method is widely used in cellular telephones, and it typically provides about an order-of-magnitude improvement in net drift over a temperature range of 0°C to 40°C.

A recent approach is to employ closed-loop methods in which a digital signal processor (DSP) continuously estimates the frequency error by examining features of the downconverted signal (typically at baseband) and retunes the oscillator to minimize the error.[13] This technique eliminates the thermometer, ADC, and ROM. However, blind estimation methods are not always robust, so best performance results when provisions are explicitly made in the protocol to accommodate this type of closed-loop frequency control.

19.6 CHIP DESIGN EXAMPLES

With that background material in place, we now direct more sustained attention to three detailed examples that will put the knowledge of the foregoing chapters together in context.

19.6.1 GPS RECEIVER

An example[14] that highlights several relevant design issues is a receiver for the global positioning system (GPS). Before describing the design itself, we need to know a little bit about the GPS signal structure. The GPS signal is centered around a carrier of 1.57542 GHz, and it consists of two spread-spectrum signals.[15] The stronger C/A (coarse acquisition) code is also the narrower in bandwidth. The weaker, broader-band P (precision) code is intended for military use and won't be discussed here any further.

---

[12] In less sophisticated versions, the "ROM" and "DAC" are simply comparators that switch capacitors into and out of the circuit.

[13] In some systems, the "features" examined can include a set pattern specifically transmitted to facilitate this type of correction.

[14] This section is largely adapted from D. K. Shaeffer and T. H. Lee, *The Design and Implementation of Low-Power CMOS Radio Receivers,* Kluwer, Dordrecht, 1999.

[15] There are actually two GPS frequency bands. Aside from the L1 band mentioned, there is an L2 band at 1.2276 GHz. However, the C/A mode is broadcast only in the L1 band. The precision P code is transmitted in both bands, whose carrier frequencies are integer multiples of the P code's 10.23-MHz chip rate.

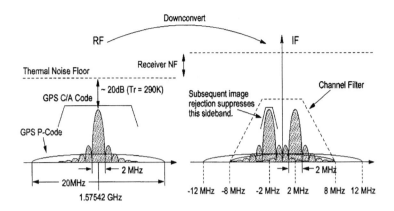

FIGURE 19.17. GPS signal structure and receiver frequency plan.

The satellites sending these signals are constrained in power and so a typical received signal is of the order of −130 dBm, assuming a relatively healthy satellite and an unobstructed path between transmitter and receiver (aging reduces the transmit power over time, further reducing link margins). Over the ∼2-MHz bandwidth of the C/A code signal's main lobe, the integrated noise power density ends up equal to about −111 dBm.[16] Thus, though stronger, the C/A code signal is not strong; it's about 20 dB *below* the thermal noise floor at the antenna of most receivers (and worse indoors and under shadowed conditions). See Figure 19.17. What makes detection possible nonetheless is the large ratio between the 1.023-MHz chip rate and the 50-Hz symbol rate. The ratio is called the *spreading* (or *processing*) *gain,* and just as wideband FM enjoys low noise upon demodulation, so does a spread spectrum signal. The boost in postcorrelation SNR is equal to the process gain. In this case, the process gain of $10 \log_{10}(1.023 \text{ MHz}/50 \text{ Hz})$ is about 43 dB, permitting a postcorrelation SNR in excess of +20 dB (before accounting for receiver noise figure). That large a value is readily demodulated. However, it is all too easy to lose 10–20 dB (or more) from shadowing effects and other propagation vagaries.[17] Consequently, we must minimize the noise figure of the receiver to make full use of every last decibel of SNR.

A basic design decision in every superheterodyne receiver is the choice of IF or, equivalently, what image signal to tolerate. Preferably, one can identify a low enough IF (to ease A/D converter requirements and the design of other frequency- or speed-sensitive blocks) that nonetheless corresponds to a benign image. In the case of GPS, the choice is made easier by the presence of the P code. That frequency band is occupied *only* by the P code, and thus no strong image signal can reside there. We are

---

[16] Recall that thermal noise has a density of −174 dBm/Hz (at the standard reference temperature of 290 K). The 2-MHz bandwidth corresponds to 63 dBHz, yielding the −111-dBm value cited.

[17] Indoor operation is essentially precluded, as well as operation in canyons (natural or urban). One essentially requires unobstructed line of sight to three or four satellites.

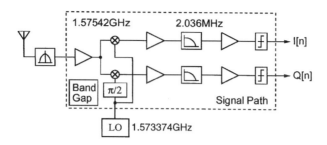

FIGURE 19.18. Receiver block diagram.

therefore free to select a low IF without incurring a difficult image rejection problem. Furthermore, the relatively large bandwidth of the P code provides us with a generously wide transition band for any putative IF filter. Here, we select an IF of 2.036 MHz, leaving 3–8 MHz as a filter transition band.

With this choice the architecture becomes quite simple, as is evident from the receiver block diagram (Figure 19.18). After quadrature downconversion, the IF signal is amplified, then filtered with low-pass filters possessing a 2-MHz corner frequency, further amplified, and then fed to 1-bit A/D converters (comparators). A 2-bit A/D is more common in commercial implementations, but a 1-bit A/D keeps things simple.[18]

The LNA is the same as one of the examples studied in Chapter 12. In this particular design (built in a now trailing-edge 0.5-μm process technology), the 2.5-V supply voltage is large enough to permit stacking of stages to save power by sharing bias current. The differential LNA drives a second differential common-source amplifier, with which it is in cascade signalwise but stacked vertically biaswise. See Figure 19.19. As is evident from the accompanying noise figure plot of Figure 19.20, this design achieves about 2.4 dB NF at the GPS L1 carrier frequency. The power consumed is about 12 mW (and the mixer and LO drivers consume another 3 mW per side). Better LNA performance at about one half to one third this power is available from process technologies currently in use. A further halving of power is possible by shifting to a single-ended architecture, but then package parasitics become much more important. This problem is not impossible to solve, but it adds another level of risk that must be fully comprehended before selecting an input stage architecture.

The output of the two-stage LNA is capacitively coupled to a passive MOS mixer, whose LO port is driven by a 1.573374-GHz signal generated by an on-chip PLL (and an external crystal). The output of the mixer in turn drives an IF amplifier that is designed for extremely linear operation in order to maximize jamming margin.

The easiest way to understand the linearization technique used in this amplifier is to note that input transistor M_1 (and its differential twin) is operated at a constant current, I_{BIAS}. To first order, the gate–source voltage of M_1 is therefore constant,

---

[18] Use of a 1-bit quantizer costs 3 dB, relative to an infinitely fine quantizer. The more commonly used 2-bit quantizer, combined with appropriate AGC to maximize use of the available quantization levels, reduces the loss to about 0.7 dB. Thus, we are throwing away about 2.3 dB of margin (relative to that achieved in common implementations) by using so crude a quantizer.

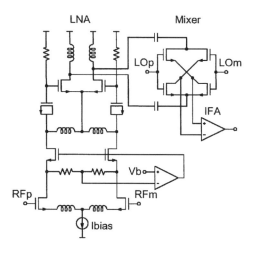

FIGURE 19.19. LNA and mixer.

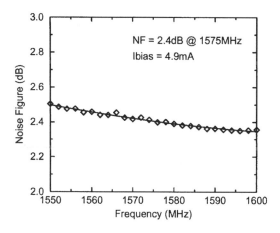

FIGURE 19.20. LNA noise figure plot.

implying that a differential voltage applied to the input terminals (I_{np}, I_{nm}) appears entirely across the source degeneration resistor. If the resistor is linear then the circuit performs a theoretically perfectly linear transconduction. Consequently, the drain current of M_3 contains a signal current term that is linearly proportional to the applied input voltage. To the extent that M_3 and M_4 match, this signal current appears at the drain of M_4 as well and ultimately appears as an output voltage at the drain of M_4. Transistor M_2 closes a feedback loop necessary to drive M_3 with the correct gate voltage for M_1 to satisfy Kirchhoff's current law at its drain node. The success of this linearization technique is evident in the accompanying transfer plot of Figure 19.21, which shows essentially constant gain for input amplitudes up to approximately 150 mV while consuming about 4 mW.

The output of this amplifier then feeds the channel filter. Considerations here once again include power consumed, noise generated, and linear range, although this list

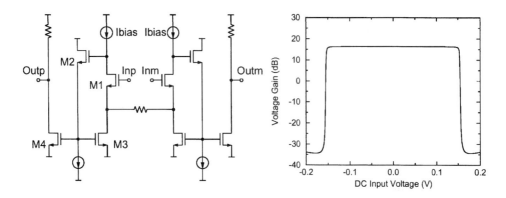

FIGURE 19.21. IF amplifier and transfer characteristic.

is hardly comprehensive. So many filter topologies exist that it is impractical to consider them all here. However, we can make a few relevant observations. One is that – by the cascaded noise figure formula – we require a certain *power* gain ahead of the filters to moderate the noise figure. For a given voltage gain ahead of the filter, this requirement translates into the need to operate into a sufficiently low impedance in order to maximize dynamic range.

Another observation is that filters based on doubly terminated passive *LC* ladder filters tend to have relatively low sensitivity to parameter variation (unless operating near the limits of the technology).[19] Combined with the criterion elucidated in the previous paragraph, we wish to select an input termination that is suitably small in order to maintain consistency with the dynamic range goals. Then, to reduce the power consumed while obtaining this dynamic range, one should choose filter topologies that allow the achievement of the desired performance with the minimum filter order. Finally, the most power-efficient circuits should be used to realize the filter sections.

The fifth-order elliptic filter shown in Figure 19.22 is the passive *LC* prototype on which the active IF filter is based. It is designed to provide a 3.5-MHz passband edge and a minimum stopband attenuation of 77 dB at about 10 MHz. The elliptic filter (named for the appearance of elliptic functions in their design) was shown by Cauer to provide the most dramatic transition from passband to stopband – given a specification on allowable passband ripple and minimum stopband attenuation – for a given filter order.[20] Alternatively, it allows us to satisfy those constraints with the minimum order.

[19] For a more precise statement of what is meant by *sensitivity* and of which sensitivities are and are not minimized, see Harry J. Orchard, "Inductorless Filter," *Electronics Letters,* v. 2, no. 6, June 1966, pp 224–5, as well as the much-misquoted "Loss Sensitivities in Singly and Doubly Terminated Filters," *IEEE Trans. Circuits and Systems,* v. 26, no. 5, May 1979, pp. 293–7.

[20] Wilhelm Cauer, "Ein Interpolationsproblem mit Funktionen mit positivem Realteil" [An Interpolation Problem with Functions with Positive Real Part], *Mathematische Zeitschrift,* v. 38, 1933, pp. 1–44. Probably more accessible (in several senses of the term) is his posthumously published *Synthesis of Linear Communication Networks* (McGraw-Hill, New York, 1958). Cauer invented this class of filters in the 1930s to reduce the number of elements required to build filters for the

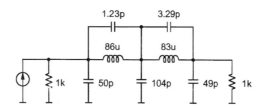

FIGURE 19.22. Elliptic IF filter prototype.

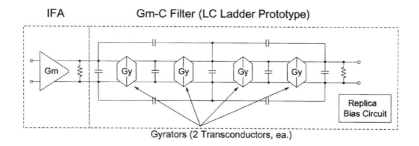

FIGURE 19.23. IF filter architecture.

In the actual filter, the inductors are replaced by gyrators, which are two-port elements capable of converting available capacitors into otherwise unavailable inductors.[21] Gyrators are usually implemented as two cross-connected transconductors, each converting one port's voltage into another's current. See Figure 19.23. By thus exchanging port voltages and currents, the gyrator presents at one port an impedance that is proportional to the reciprocal of an impedance connected to its other port. In the present example, differential amplifier structures are used throughout. Some economies might be afforded by using single-ended implementations, but we will not consider those here.

The gyrator-based inductor, along with its corresponding noise model, thus appears as shown in Figure 19.24. The dynamic range of a receiver can be constrained very easily by the linearity of the filters. If preceded by sufficient gain, a filter will be overdriven and produce nonlinear behavior. Consequently, power-efficient and

German telephone system. Legend has it that Bell Labs engineers, upon reading of Cauer's invention in a patent, rushed over to the New York Public Library to bone up on the then- and (still-) obscure mathematics of elliptic functions so they could understand what he had done. The underlying intuition is straightforward enough (use the response notches of imaginary zeros to produce a rapid transition to stopband), but figuring out exactly where all the poles and zeros should go is a decidedly nontrivial problem.

[21] The gyrator was first named and studied as an element by the famous Dutch theoretician, Bernard D. H. Tellegen, in "The Gyrator, a New Electric Element," *Philips Research Reports,* v. 3, 1948, pp. 81–101. However, gyration of capacitors into inductors is actually the basis of many early FM generators that predate the explicit naming of these elements by Tellegen. Countless "reactance tube" modulators were used before the word *gyrator* entered the engineering lexicon. Tellegen formalized and extended the concept.

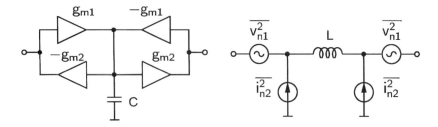

FIGURE 19.24. Gyrator architecture and noise model.

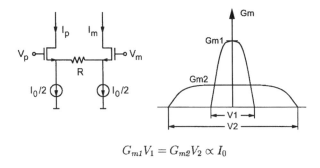

$$G_{m1}V_1 = G_{m2}V_2 \propto I_0$$

FIGURE 19.25. Degenerated differential pair.

linear transconductors are essential to realizing good filters of this type. In the present context of a GPS receiver, the filter we are designing is vast overkill, and significant power and die area reductions may be achieved by seeking a more modest design. Nevertheless, artificially inflating the requirements gives us an excuse to identify some design concepts that are broadly applicable.

Assuming that we do seek linear transconductors, a quick survey of the literature reveals several popular candidates. One is the venerable degenerated differential pair (Figure 19.25). To first order, with or without degeneration, the edge of the linear region is defined by a differential twist sufficient to cause a significant fraction of the bias current to be steered to one side. Degeneration does nothing to change this fundamental behavior; all it does is increase the input differential voltage required to reach this limit. Consequently, the product of transconductance and linear input voltage range stays roughly constant – and proportional to the bias current. Stated another way, the mere presence of a limiting bias current assures a bounded linear range.

A variation on that theme replaces the fixed linear degeneration resistor with triode-operated MOSFETs whose gates are tied to the differential pair's inputs. This way, as the differential twist increases, the value of the effective degeneration resistor decreases, partially offsetting the effects of nearly complete current steering. This simple trick can provide large improvements over fixed degeneration. However, the optimum conditions are relatively narrow and so the improvements available in practice are not as large as theory predicts.

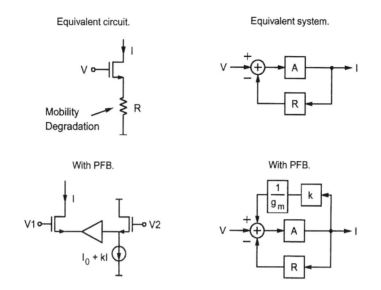

Equivalent circuit.

Equivalent system.

Mobility Degradation

With PFB.

With PFB.

FIGURE 19.26. Positive feedback for mobility degradation compensation.

A third option exploits the allegedly square-law behavior of a MOSFET. A differential amplifier constructed out of two truly quadratic devices possesses a linear differential gain (the second-order output terms due to the quadratic have opposite signs, because of the symmetry, and thus cancel). The practical problem with this approach is that real MOSFETs (even ones that have been elongated to channel lengths considerably in excess of the process minimum) do not behave as square-law devices at large gate overdrives owing to vertical-field mobility degradation. Because this mechanism causes transconductance to drop as the gate overdrive increases, mobility degradation acts very much as if there were built-in source degeneration. Thinking about the effects of mobility degradation in this fashion allows us to devise a potential solution. Degeneration is a form of negative feedback, so positive feedback should be able to counteract it; see Figure 19.26. A proper choice of the positive feedback constant k can theoretically remove the effects of mobility degradation.

To facilitate a fair comparison among alternatives, one may define a figure of merit Γ, which combines transconductance G_m per power consumed P_D in addition to linearity (as measured by IP3) and noise:

$$\Gamma = \frac{G_m}{P_D} \frac{V_{IP3}^2}{\varepsilon},$$ (29)

where

$$\varepsilon = \frac{\overline{i_{n,\,out}^2}}{4kTBG_m}.$$ (30)

The plots of Figure 19.27 provide a comparison of the three methods. For the resistively degenerated case, Γ is plotted as a function of device width, with degeneration resistance as a parameter. For the other two cases, Γ is plotted as a function of the

724   **CHAPTER 19** ARCHITECTURES

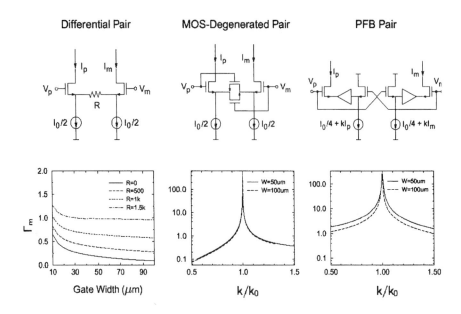

FIGURE 19.27. Summary of linear transconductors.

deviation from the condition that maximizes linearity. The relative width of those plots thus yields insight about the sensitivity of the circuit to small departures from the assumptions made during the design.

We see that the MOS-degenerated pair can improve the figure of merit by more than two orders of magnitude, but obtaining this large an improvement requires unrealistically tight control over parameters. A 5% deviation from optimum leads to figures of merit not very different from those obtained from ordinary degenerated transconductors. The positive-feedback–corrected transconductor, on the other hand, exhibits a considerably smaller sensitivity to parameter variation. An order-of-magnitude improvement is obtained even with deviations from the optimum in excess of 5%. This robustness explains the reason for choosing this architecture over the other two for this design.

The circuit implementation of this idea is shown in Figure 19.28. Transistors M_2, M_5, and M_6 form a loop similar that used in the linearized IF amplifier. As such, M_2's gate–source voltage would be constant in the absence of M_{10}. Transistor M_1 is the main transconductor (along with its mirror image on the right-hand side). Its output current is mirrored to M_{10}, whose drain current adds to that biasing M_2. This closes the positive feedback loop, as required of this method. The mirror gain is adjusted to maximize linearity.

The accompanying plot of transconductance as a function of input voltage shows the large improvements possible. The lower curve shows the transconductance variation obtained without positive feedback. As seen, the transconductance drops by 5% somewhere below a 250-mV input voltage. With compensation, the transconductance increases by less than 1% at that same input voltage. Even with 10% variation in the width of M_{10}, the greatly improved flatness of G_m is evident.

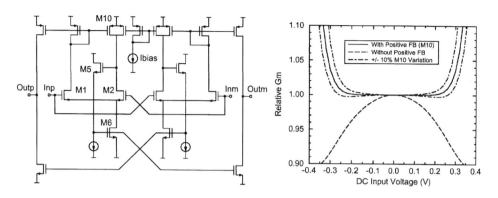

FIGURE 19.28. Linearized transconductor with positive feedback (simplified).

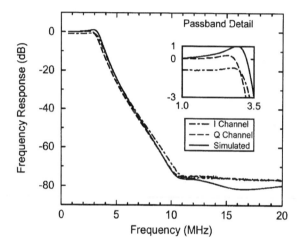

FIGURE 19.29. 3.5-MHz IF filter response.

Using that transconductance in the filter gyrators leads to the overall IF filter response graphed in Figure 19.29. Each of the two filters consumes 10 mW in the core, with an additional 1 mW consumed in the replica bias used to set the termination resistance values. Thanks to the linearization, the filters exhibit a 60-dB spurious-free dynamic range. Nevertheless, this filter remains the chief constraint on the overall linearity of this GPS receiver.

The plots of Figure 19.30 show measured two-tone IP3 and 1-dB blocking performance results. The IP3 plot looks pathological because it is. The linearization techniques are effective at lower amplitudes, but they make higher-order nonlinearities more prominent by comparison. Thus, at low amplitudes, the third-order IM term does grow with the expected slope of 3 but then exhibits a steeper slope at higher powers. Thus, one may compute rather different IP3 values depending on the source power used in the measurement. Because of the possibility of selective reporting (either to show your design in the best light, or that of your competitor in the worst), it is best to provide a plot.

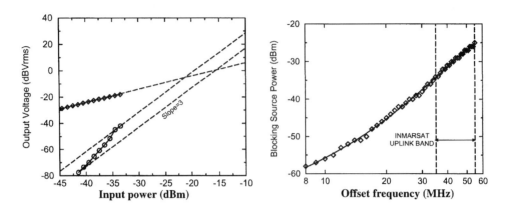

FIGURE 19.30. Two-tone linearity test and 1-dB blocking performance.

The blocking performance is measured with no front-end filtering of any kind. Even without helpful filtering, a 1-dB desensitization occurs for −35-dBm source powers in the potentially troublesome INMARSAT uplink band. This value is the result of reciprocal mixing due to LO phase noise. If the blocking performance were to require improvement, one could employ simple front-end filters (because the potential blockers are well removed from the center frequency in this case) or improve the phase noise (or some combination of the two). Given the difficulty of the latter, a prudent design choice would simply be to add some elementary filtering between the antenna and the front end. In fact, the antenna's limited bandwidth could be turned to advantage here. It should be straightforward to obtain 20-dB improvements in blocking performance with little effort – to the point where no credible threat of such strong blockers exists.

The overall receiver noise figure (graphed in Figure 19.31) is little affected by the circuitry downstream of the LNA and mixer combination. If coherent detection is combined with infinitely fine quantization then the noise figure would be a tiny bit above 4 dB. With the 1-bit slicer actually used, we incur a penalty yet manage just over a still excellent 6-dB noise figure. If noncoherent detection is used in combination with that slicer, the noise figure increases to nearly 9 dB.

The five-stage limiting amplifiers are more or less conventional and consume a total of about 1.5 mW per side. The latch, drivers, and bandgap reference together consume another 5 mW per side for a total consumption of about 79 mW. The synthesizer consumes the balance of the 115-mW chip power.

The die photo shown as Figure 19.32 depicts the floorplan of the GPS receiver. The I and Q channels bound the upper and lower parts of the die, and the input is supplied to the chip in the middle of the left edge of the die. The synthesizer occupies the right-hand portion of the die between the I and Q channel circuitry. Outputs are available from the upper right and lower left corners of the chip. The sixteen on-chip spiral inductors peacefully coexist with very little mutual coupling, as anticipated from the spacing.

Chip performance is summarized in Table 19.1.

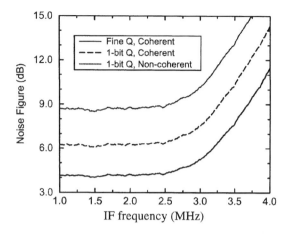

FIGURE 19.31. Overall receiver spot noise figure as a function of frequency.

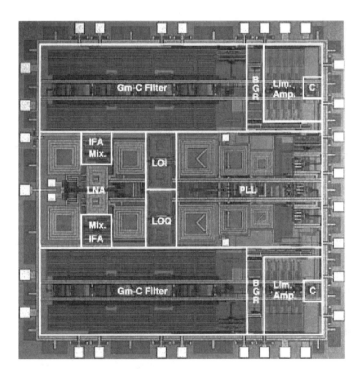

FIGURE 19.32. Die photomicrograph.

Balanced-to-Unbalanced Conversion

In the GPS chip just described, the LNA is fed by a differential signal – begging the question of how one might produce it. At lower frequencies, a common solution would simply be to drive the differential amplifier with a single-ended signal and to ground the other input. However, it's preferable not to break symmetry, for

Table 19.1. *Summary of specifications*

Signal path performance	
LNA noise figure	2.4 dB
LNA S11	≤ -15 db
Coherent receiver NF	4.1 dB
IIP3 (filter-limited)	-16 dBm @ -43 dBm P_s
Peak SFDR	60 dB
Filter cutoff frequency	3.5 MHz
Filter PB peaking	≤ 1 dB
Filter SB attenuation	≥ 52 dB @ 8 MHz
	≥ 68 dB @ 10 MHz
Pre-filter G_p	19 dB
Pre-filter A_v	32 dB
Total G_p	$\sim$94 dB
Total A_v	$\sim$122 dB
Noncoherent output SNR	15 dB
PLL performance	
Loop bandwidth	5 MHz
Spurious tones	≤ -42 dBc
VCO tuning range	240 MHz ($\pm$7.6%)
VCO gain constant	240 MHz / V
LO leakage @ LNA	< -53 dBm
Power/technology	
Signal path	79 mW
PLL / VCO	36 mW
Supply voltage	2.5 V
Die area	11.2 mm^2
Technology	0.5-μm CMOS

then we lose the advantages that motivate the choice of a differential topology in the first place.

To solve this problem, we consider an extremely versatile element known as a *hybrid*. A so-called 180° hybrid is useful for performing single-ended to differential conversion (and vice versa); hence, it can serve as a splitter or a combiner. The first widespread use of a hybrid was in telephony, allowing duplex communication to take place over a single wire pair. The classic telephone hybrid is a broadband multi-tap transformer wound on a soft iron core. In distributed form, a narrowband hybrid consists of a closed path (e.g., a circular loop) of transmission line whose electrical circumference is $3\lambda/2$. Three taps (labeled A, B, C, and D in Figure 19.33) are separated from each other by $\lambda/4$. A signal supplied to A splits between the clockwise and counterclockwise paths. In traveling to point B, the clockwise-going signal has been shifted by $\lambda/4$ and the counterclockwise signal shifted by $5\lambda/4$, so they add in phase. In traveling to C the signals are shifted by $\lambda/2$ and λ, respectively, leading to cancellation; no signal emerges from that tap. Finally at tap D, the signals have shifted $3\lambda/4$ in each direction, leading once again to addition in phase. Note that the

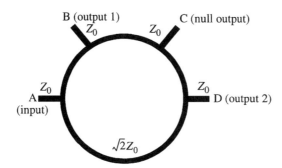

FIGURE 19.33. Ring (or rat-race) hybrid.

signals emerging from B and D are shifted in phase from each other by $\lambda/2$. Thus, a single-ended input at A becomes a differential signal between B and D. And, by reciprocity, differential inputs supplied at B and D combine to produce a single-ended output at A.

It is important to note that the characteristic impedance of the ring proper must differ from that of each of the taps if an impedance match is to be preserved. The precise relationship can be derived by recognizing that the power splits evenly between the two output taps. Therefore, the source driving point A sees an equivalent load impedance of $Z_0/2$ that is effectively driven in parallel through two paths, each of which has an impedance we'll call Z_{ring}. Recall that a quarter-wave line can be used as an impedance match if its characteristic impedance is chosen as the geometric mean of the source and load impedances. Here, the source impedance is Z_0 and the effective load impedance is $Z_0/2$, so we may write

$$\frac{Z_{\mathrm{ring}}}{2} = \sqrt{Z_0 \cdot \frac{Z_0}{2}} = \frac{Z_0}{\sqrt{2}} \implies Z_{\mathrm{ring}} = \sqrt{2}Z_0. \tag{31}$$

A hybrid of this type is known as a ring (or rat-race) hybrid.[22] It should be clear that a circular shape is not strictly necessary. The only requirements are that the total perimeter and tap locations satisfy the various wavelength criteria and that the impedance of the ring proper be $\sqrt{2}$ times that of the taps. These criteria imply that such a hybrid is, of necessity, a narrowband element. Typical useful bandwidths are on the order of 15–30%.

When a hybrid is used to produce differential outputs from a single, ground-referenced one (or vice versa), it is also sometimes known as a *balun* (for "balanced-to-unbalanced" converter; rhymes with *gallon*).[23] It is often mispronounced "bail-un."

---

[22] It is sometimes called a hybrid ring, but in this context *hybrid* is the noun and *ring* is the adjective, so "ring hybrid" is more grammatically correct.

[23] The term "unbalanced" refers to a port whose two terminals are not at the same potential with respect to ground (usually, one of these terminals actually *is* ground). A balanced port's two signals are the same with respect to ground (ignoring phase). Occasionally you may find reference to an *unbal,* which is a balun used backwards, but most engineers just use the term *balun* for both.

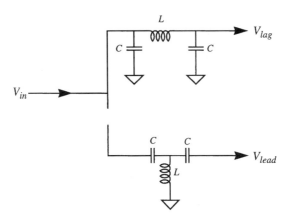

FIGURE 19.34. Lumped 180° hybrid.

The ring hybrid's diameter is on the order of $\lambda/2$ and so we may prefer to consider a lumped alternative for use at lower frequencies, where the transmission line hybrid might occupy an unacceptably large area. As an example, the free-space wavelength of a GPS signal is approximately 19 cm. Typically, the transmission line would normally be realized in microstrip form. With the dielectric constants of common PC board materials, we can expect a rough halving of wavelengths. Consequently, a microstrip ring hybrid would have a diameter of about 5 cm. That size is certainly within reason but is also a bit on the large side, so it is worthwhile considering a lumped alternative (much as we might consider lumped alternatives to distributed transmission line delay elements). The lumped balun shown in Figure 19.34 is actually a diplexer (i.e., a network that splits signals into two frequency bands),[24] but it does not make explicit use of its frequency-selective properties. The low-pass and high-pass filters produce frequency-dependent lagging and leading phases, respectively. At each filter's corner frequency, the magnitude of the phase shift is 90°. Even though the phase shift of each filter path certainly varies with frequency, the *difference* between the output phases is a constant 180° over a broad frequency range (although the *amplitudes* are strictly equal at only one frequency). This lumped hybrid thus shares with its distributed cousin the limitation of relatively narrowband operation.

If R_S is the source resistance and if a load of value R_L connects the two outputs to each other, then we wish to choose the characteristic impedance of each filter equal to the geometric mean of those source and termination resistances:

$$\left(\sqrt{\frac{L}{C}} = \sqrt{R_S R_L} \right) \implies \frac{L}{C} = R_S R_L. \tag{32}$$

The other equation needed to complete the design derives from choosing the corner frequencies of the filters equal to the center frequency of operation for the hybrid:

---

[24] Not to be confused with a *duplexer,* which is a device for enabling simultaneous two-way communications. A diplexer may be used as a duplexer, but they are not the same thing.

$$\omega_0 = \frac{1}{\sqrt{LC}}. \tag{33}$$

Solving for each element yields

$$C = \frac{1}{\omega_0 \sqrt{R_S R_L}}, \tag{34}$$

$$L = \frac{\sqrt{R_S R_L}}{\omega_0}. \tag{35}$$

Succinctly stated, choose the inductors and capacitors to have an impedance equal to the geometric mean of the source and load resistances at the center frequency of operation. Thus, for a 1.575-GHz hybrid driven by 50 Ω and terminated in 100 Ω (50 Ω from each output to ground), the element values are about 7.2 nH and 1.4 pF. Either discrete components, microstrip equivalents, package parasitics (e.g., bond-wires or lead inductances) or on-chip elements (or some combination of these) may be used to implement this hybrid. To preserve good noise figure, one simply requires that these elements have low loss.

If there is any parasitic capacitance in parallel with the load resistance, it may be "resonated away" with a suitable inductance (in principle). In a similar fashion, par-asitics in series with the load may also be removed (again, over a narrow frequency band).

19.6.2 WLAN EXAMPLE

Before describing illustrative 5-GHz WLAN receiver and transceiver implementa-tions, it's worthwhile saying at least a bit about standards, propagation, and signal structure – for all of these factors influence architectural decisions at least indirectly.

As we noted early in this textbook, there is a "sweet spot" of spectrum for mo-bile wireless, roughly spanning 500 MHz to 5 GHz, that combines the attributes of reasonable propagation with reasonable antenna length. Note that Friis's famous propagation formula shows that free-space attenuation is a constant per *wavelength*. On that basis alone, one would expect 5-GHz signals to propagate more poorly than those at (say) 2.4 GHz, mitigating interference to a certain extent. Published measure-ments are not entirely in agreement with each other, with some showing significantly worse propagation at 5 GHz than do other studies. The author's own experience is that 5-GHz signals indeed suffer significantly worse attenuation inside his own home. Your mileage may vary.

To help compensate for unfavorable and variable propagation and to support higher data rates at the same time, IEEE 802.11a (and the latest modification to the similar European WLAN standard, HiperLAN2, which is converging with 802.11a) uses or-thogonal frequency division multiplexing (OFDM). In OFDM a carrier is subdivided into several individually modulated orthogonal subcarriers (each carrier's spectrum has a null at the center of all the other carriers), all of which are subsequently trans-mitted in parallel. The OFDM signal is generated by computing the inverse FFT of modulated subsymbols and is recovered by an FFT, followed by QAM demapping.

52 carriers per channel

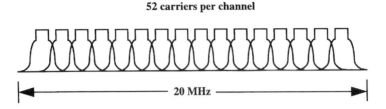

FIGURE 19.35. Subchannel structure for 802.11a.

In 802.11a, the carriers are separated by 20 MHz and subdivided into 52 subchannels, each about 300 kHz wide (not factoring in guardbands; see Figure 19.35). Forty-eight of these subchannels are used for data; the remaining four are for (forward) error correction.

This subdivision provides a convenient means for accommodating a variety of data rates, permitting different levels of service and providing adaptive behavior in the face of changing propagation conditions. At the lowest data rate, binary phase shift keying (BPSK) encodes 125 kbps of data per channel, resulting in a 6-Mbps data rate. Using quadrature phase shift keying (QPSK), the data rate doubles to 250 kbps per channel, yielding a 12-Mbps data rate. With 16-level quadrature amplitude modulation (16-QAM) the rate increases further, to 24 Mbps. The standard also supports rates beyond 24 Mbps using higher-order QAM, propagation conditions permitting. For example, the standard defines the use of 64-QAM to permit a theoretical increase to 54 Mbps. Multiple channels may also be combined to provide still higher aggregate data rates, of the same order as Fast Ethernet.

Although HiperLAN2 and 802.11a do share some superficial similarities, there are differences. For example, HiperLAN2 specifies minimum sensitivities of −85 dBm at 6 Mb/s and −68 dBm at 54 Mb/s, corresponding to 3-dB more stringent requirements than for 802.11a.

Networking in general requires some strategy for deciding who gets access to the shared medium. IEEE 802.11's medium access control (MAC) protocol uses a contention resolution mechanism that traces its heritage to that of 802.3 Ethernet. In the latter, each transmitter first listens to the channel to establish whether the medium is free by sensing whether a carrier is already present. If the medium appears free, the transmitter then sends data while simultaneously monitoring its own transmissions. Hearing anything other than the intended transmission is assumed to arise from a collision with data from another transmitter. Upon detecting a collision the transmitter ceases operation and then tries sending the data again after waiting a random backoff interval. This protocol, known as CSMA/CD for "carrier sense multiple access with collision detection," is simple and works remarkably well for wired LANs. However, it requires the ability both to sense the presence of a carrier before transmitting and to detect corrupted data during transmission. In the wired case, the relatively low attenuation of the medium assures that transmitted and received signals are similar in amplitude, facilitating this detection. But wireless propagation involves considerably more – and more variable – attenuation than through a cable. As a result, both carrier sensing and collision detection will fail in numerous ways. As one simple example, consider three linearly arrayed WLAN nodes labeled A, B, and C. Suppose that B can communicate with both A and C but, because of shadowing, A and C are unable to communicate directly with each other. In this case, it is possible for both A and C to attempt communication with B simultaneously, each unaware of the presence of the other. Both carrier sense and collision detection fail in such an instance.

Note also that, aside from being largely ineffective, listening while transmitting would impose a severe implementation penalty for wireless nodes because receive and transmit circuit blocks could no longer be shared. Because of such problems, the 802.11 WLAN MAC protocol differs in several important respects from 802.3. It includes a slot reservation mechanism and does not require that a transmitter also listen to its own transmissions. The resulting scheme is called carrier sense multiple access with collision *avoidance* (CSMA/CA). Here, a node first listens before transmitting, just as in CSMA/CD. If it detects no carrier signal, it can safely conclude only that the medium *might* be free. However, there are two additional possibilities: Either an out-of-range station may be in the process of requesting a slot, or such a station may already be using a slot reserved for it.

To reserve a slot, a WLAN node sends to the intended receiver a *request to send* (RTS) message specifying the duration of the requested slot. At the same time, other receivers within range of the sender also note the request. The intended receiver replies with a *clear to send* (CTS) message confirming the duration of the slot, with other stations noting this information as well. All of the stations within range of both sender and receiver use the information contained in the RTS and CTS packets to refrain from transmitting during the requested transfer slot. At the end of the transfer the receiver transmits an acknowledgment (ACK) packet.

The RTS and CTS frames are designed to be short in order to minimize the probability of collision during their transmission. If one occurs nonetheless – or if an RTS does not result in a CTS for some reason – then a random backoff interval prior to retransmit is used, just as in the 802.3 Ethernet wired LAN standard.

Although CSMA/CA functions well, it burdens the transceivers with considerable overhead, causing 802.11 WLANs to have slower performance than that of an otherwise equivalent Ethernet LAN. Under favorable conditions the 802.11 MAC is about 70% efficient, so data throughput at "54" Mbps is, at best, not quite 40 Mbps in actual practice. Additional inefficiencies in drivers, combined with propagation vagaries, may reduce typical throughput to about 25–30 Mbps. A similar reduction is experienced with 802.11b systems, where "11" Mbps links generally supply about 6 Mbps in practice. As a rule of thumb, then, it seems that sustained transfer rates of about half the specified values are to be expected in practice.

Performance Requirements for a 5-GHz HiperLAN WLAN Receiver

To determine the precise target values, we first compute the specifications for both HiperLAN (the original, not HiperLAN2) and 802.11a separately and then select the more stringent of the two in every case. Here we reduce the specification set to frequency range, noise figure, maximum input signal level (or input-referred 1-dB compression point), and limits on spurious emissions.

For frequency range, we choose to cover only the lower 200-MHz band, largely out of laziness. The upper 100-MHz domain is not contiguous with that allocation, so its coverage would complicate somewhat the design of the synthesizer. Furthermore, that upper 100-MHz spectrum is not globally available. Hence the choice here is to span the range from 5.15 GHz to 5.35 GHz.

Required noise figures in general are a function of modulation and therefore of data rate. The original HiperLAN system specifies a minimum receive sensitivity of −70 dBm. Given its 23.5-MHz channel bandwidth and assuming that we require a 12-dB predetection SNR, we may compute a worst-case allowable receiver noise figure as

$$\text{NF} < -70 \text{ dBm} - 10\log_{10}(23.5 \text{ MHz}) - 12 \text{ dB} + 174 \text{ dBm/Hz} = 18.3 \text{ dB}. \quad (36)$$

We conservatively select a 10-dB maximum noise figure as the design goal for the present example. The 8-dB margin assures the ability to absorb manufacturing tolerances and to continue functioning with worse propagation than assumed in setting the minimum standards.

As stated previously, 802.11a specifies a value of −30 dBm (for a 10% *packet* error rate).[25] Because HiperLAN's −25-dBm specification is the more severe of the two, we select it as the target maximum input level. Converting these specifications into a precise IIP3 target or 1-dB compression requirement is nontrivial. However, as a conservative rule of thumb, the 1-dB compression point of the receiver should be about 3–4 dB above the maximum input signal power level that must be tolerated successfully. Based on this approximation, we target a worst-case input-referred 1-dB compression point of −21 dBm.

Finally, the spurious emissions generated by the receiver must not exceed −57 dBm for frequencies below 1 GHz, and −47 dBm for higher frequencies, in order to comply with FCC regulations.

Receiver Implementation – Architectural Considerations

A low-IF architecture[26] possesses many of the putative attributes of a homodyne receiver (namely, relaxed speed demands on IF circuit blocks), but it has lower sensitivity to DC offsets and $1/f$ noise. The tradeoff, however, is that the image rejection

[25] There are (too) many ways of specifying performance. Error rates for bits, packets, blocks, and frames are all in use. Needless to say, it's important to keep track of what particular error rate is being reported before comparing numbers. For an excellent overview of system-level WLAN considerations, see H. Ahmadi, A. Krishna, and R. LaMaire, "Design Issues in Wireless LANs," *J. High-Speed Networks,* v. 5, no. 1, 1996, pp. 87–104.

[26] The example described here owes a great deal to the Ph.D. work of Hirad Samavati and Hamid Rategh, as summarized in "A Fully Integrated 5GHz CMOS Wireless LAN Receiver, *ISSCC Digest of Technical Papers,* February 2001, pp. 208–9.

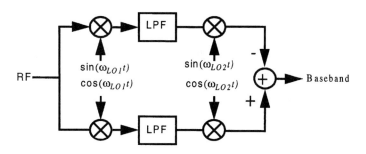

FIGURE 19.36. Weaver architecture.

problem reappears. If the goal is to avoid the use of expensive filters, then the burden of image rejection must be borne architecturally. Recall the Weaver architecture (Figure 19.36), which we first encountered as a method for single-sideband generation and demodulation. The Weaver architecture's ability to resolve the difference between negative and positive frequencies also endows it with the ability to resolve the difference between a signal and its image.

Since a signal and its image may thus be distinguished by their differing phase, cancellation of the image signal while simultaneously passing the RF signal is possible. But as with any system that is reliant on miraculous cancellations, a high degree of image rejection depends on exquisite matching of gains and phase throughout the receiver chain. If the radian phase matching error ε and fractional gain mismatch θ are both small, we've seen that the image rejection ratio (defined as the power ratio of signal to image) may be expressed approximately as

$$\text{IRR} \approx \frac{4}{\varepsilon^2 + \theta^2}. \tag{37}$$

To underscore the tight requirements on matching, consider that errors of 0.1% in gain and 1° in phase bound the IRR to below 41 dB. At a carrier frequency of 5 GHz, note that a 1° phase error corresponds to a time mismatch of less than 0.6 ps, or the time it takes light to travel about 200 μm in free space. Perhaps it is thus not surprising that image rejection ratios for 5-GHz receivers typically fall short of 41 dB. In fact, values generally lie within the range of 25–35 dB. Unfortunately, higher values may be needed for practical systems. Automatic calibration techniques can improve the practically achievable image rejection ratios, as will be shown in a later example, but for now we shall consider some alternatives.

With a simple modification the Weaver architecture readily provides quadrature outputs, as is needed for many modulation types, and it is this architecture that is used in this receiver. The modification involves replacing the second set of modulators by two pairs of quadrature mixers and then properly combining their contributions (see Figure 19.37). The result is in fact another exploitation of the polyphase concept for complex signal processing.

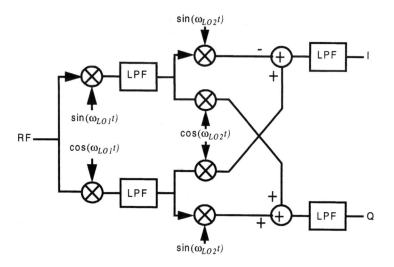

FIGURE 19.37. Quadrature Weaver architecture.

Choice of the Weaver architecture's two local oscillator frequencies represents another degree of freedom. Here our ultimate goal is to heterodyne down to a zero-frequency baseband. Of the infinite combinations of first and second LO frequencies that would accomplish this goal, a convenient choice uses frequencies that are 16/17 and 1/17 that of the RF input, respectively. This arrangement has several attributes. One is that the second LO is readily derived from the first LO through a simple binary divider. Another is that the image signal then happens to lie within the downlink spectrum of an existing satellite system and thus is relatively weak, thanks to statutory emission limits. Imperfect image rejection therefore has less serious consequences than would otherwise be the case.

The overall receiver architecture is shown in Figure 19.38. As seen, it consists of an LNA combined with an integral tracking notch filter controlled by a PLL, a quadrature Weaver image-reject core, and AC-coupled baseband buffers. In addition, the receiver contains a frequency synthesizer that provides coverage for the 200-MHz frequency span. The synthesizer provides quadrature outputs at both 16/17 and 1/17 the RF input frequency, as required by the Weaver demodulator. It also replaces a standard flip-flop–based divider with an injection-locked frequency divider to reduce synthesizer power. Implementation details for these and other blocks are discussed in the following sections.

LNA with Tracking Notch Filter

Because mismatches in a Weaver circuit inevitably degrade IRR, many practical receivers need to supplement the image rejection with other means. One way to augment IRR is through the use of an external image-reject bandpass filter, of course,

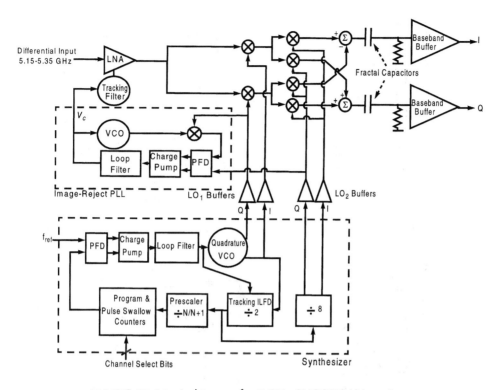

FIGURE 19.38. Architecture for 5-GHz CMOS WLAN receiver.

but the lossiness of all real filters degrades noise figure, decibel for decibel. Moreover, such a solution frequently runs counter to the cost-conscious philosophy that motivates consideration of the architecture in the first place. Another method is to implement some form of autocalibration, as mentioned earlier. Yet another alternative is to use a notch filter, which is more easily integrated than a conventional bandpass filter. This ease of integration stems from the fact that a deep notch may be provided by simple low-order networks. The major drawback of this approach is the need for tuning, owing to the narrowness of the notch. Hence, automatic tuning is mandatory if a notch filter is to be used for image cancellation.[27]

In order to save area and power, the notch filter is merged here with an otherwise standard source-degenerated low-noise amplifier. Because the sensitivity requirements are not extreme, the low-noise amplifier need not exhibit extraordinarily low

---

[27] Automatic tuning is a time-honored technique. For good examples of modern implementations, see V. Aparin and P. Katzin, "Active GaAs MMIC Bandpass Filters with Automatic Frequency Tuning and Insertion Loss Control," *IEEE J. Solid-State Circuits,* v. 30, October 1995, pp. 1068–73; J. Macedo and M. Copeland, "A 1.9-GHz Silicon Receiver with Monolithic Image Filtering," *IEEE J. Solid-State Circuits,* v. 33, no. 3, March 1998, pp. 378–86; and M. Copeland et al., "5-GHz SiGe HBT Monolithic Radio Transceiver with Tunable Filtering," *IEEE Trans. Microwave Theory and Tech.,* v. 48, no. 2, February 2000, pp. 170–81.

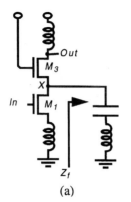
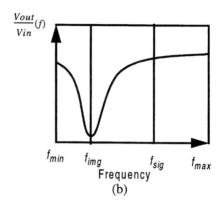

FIGURE 19.39. (a) Image-reject LNA; (b) input–output transfer function.

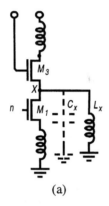
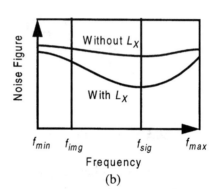

FIGURE 19.40. (a) Improving the noise figure of a
standard LNA; (b) noise figure versus frequency.

noise figures. Consequently, the focus is on low power consumption and providing sufficient linearity. To understand how the notch may be implemented with minimal overhead, first consider modifying the transfer function of this LNA by a series LC resonator, as shown in Figure 19.39.

The frequency of the LC circuit's series resonance is chosen equal to that of the image. At the image frequency, the LC circuit's low impedance shunts signal current away from M_3, thus reducing the gain at that frequency. Regrettably, the impedance of the LC circuit at the *signal* frequency is still finite, so noise figure and gain suffer.

As shown in Figure 19.40, parasitic capacitance at node X further degrades the noise performance of the cascode structure. This parasitic capacitance may be almost as large as C_{gs} (and extrapolation of scaling trends leads to pessimism about the likelihood of future improvement). This parasitic capacitance C_X lowers the impedance at node X and reduces the gain of the cascode structure. The presence of this capacitance increases the noise contribution of M_3 while simultaneously reducing the signal contribution of M_1. To reduce the resulting noise figure penalty, this

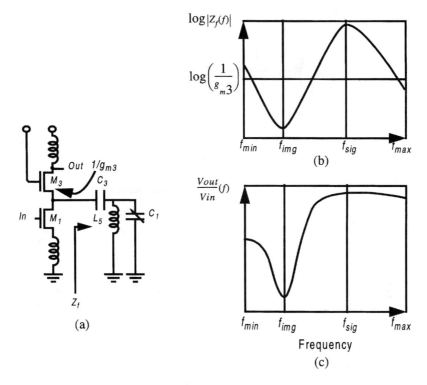

FIGURE 19.41. (a) Circuit diagram of the LNA with filter; (b) input impedance of the filter versus frequency; (c) transfer function of the LNA–filter combination.

capacitance must be nullified. Ignoring for the moment the issue of biasing, the problem is largely solved (at least for narrowband circuits) by placing an inductor in parallel with this parasitic capacitance. In Figure 19.40, noise figure is plotted versus frequency, showing the improvement obtained with the help of the inductor.

With a little reflection, we recognize that it should be possible to synthesize a passive network that combines the generation of the notch (for image suppression) with the neutralization of the parasitic capacitance. Doing so results in the circuit shown in Figure 19.41. The filter comprises an inductor, a capacitor, and a varactor. As desired, the filter has a low impedance at the frequency of the image and a high impedance at the frequency of the signal. Optimistically assuming lossless elements, the input impedance Z_f of the filter can be written as

$$Z_f(s) = \frac{S^2 L_5 (C_3 + C_1) + 1}{s^3 L_5 C_1 C_3 + s C_3}.$$ (38)

The filter thus has purely imaginary zeros at

$$\omega_z = \pm \frac{1}{\sqrt{L_5 (C_3 + C_1)}}$$ (39)

and purely imaginary poles at

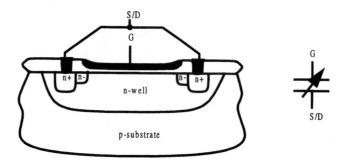

FIGURE 19.42. Accumulation-mode varactor.

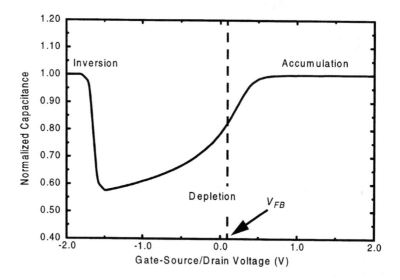

FIGURE 19.43. Small-signal varactor capacitance as a function of bias.

$$\omega_p = \pm \frac{1}{\sqrt{L_5 C_1}}. \tag{40}$$

The location of the pole–zero pair on the imaginary axis is controlled by an accumulation-mode varactor (Figure 19.42) whose small-signal tuning characteristics are shown in Figure 19.43. This structure is inherently available in all CMOS processes, and it exhibits Q-frequency products in the range of 100–200 GHz for the 0.25-μm process used here.

Also shown in Figure 19.41 is the input impedance of the filter, $|Z_f|$, as a function of frequency. The resistance looking into the source of the cascode device, $1/g_{m3}$, has also been marked on the same graph for comparison. For frequencies close to the location of the zero, the filter has an impedance lower than $1/g_{m3}$ and steals the AC current away from M_3, thus reducing the LNA gain. Near the pole frequency, $|Z_f|$ is larger than $1/g_{m3}$ and the LNA gain is consequently high. As seen from the figure,

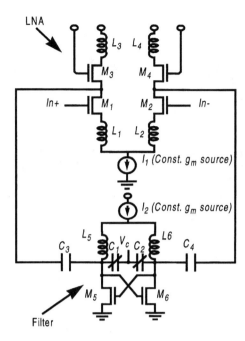

FIGURE 19.44. Simplified circuit diagram
of the LNA–filter combination.

the resulting overall transfer function has a narrow valley, so for correct image cancellation the zero must occur at the correct frequency. On the other hand, the peak is fairly broad and so the exact location of the pole is less important.

The third-order filter thus not only boosts image rejection but also diminishes the effect of the parasitic capacitance at node X. As a result, the filter simultaneously provides good image rejection and good noise performance. To be rigorous we would slightly modify Eqns. 38–40 to include the effect of this parasitic capacitance (as well as nonzero loss throughout the network). Even so, the foregoing argument is still valid in its essential features.

Figure 19.44 shows in greater detail the combined LNA–filter as actually implemented. A differential architecture is chosen for its better rejection of on-chip interference and for its insensitivity to parasitic inductance between the common-source connection and ground. This last consideration is particularly important for 5-GHz circuits, where the ~30-Ω reactance of a 1-pF stray capacitance or 1-mm bondwire is hardly negligible. To achieve the desired linearity, the LNA consists of only one stage, which is formed by transistors M_1 through M_4. Inductive degeneration is employed in the sources of M_1 and M_2 to produce a real term in the LNA's input impedance, as we've seen.

Capacitors C_1 through C_4 together with inductors L_5 and L_6 form a differential version of the third-order filter. Accumulation-mode MOS varactors C_1 and C_2, varied by control voltage V_c, accommodate the requirement for precise tuning of the notch.

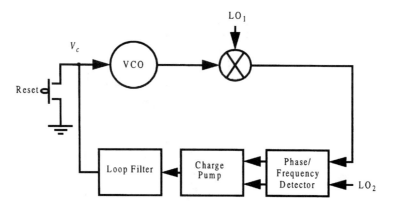

FIGURE 19.45. Image-reject PLL.

The negative resistance generated by the cross-connected differential pair, M_5–M_6, deepens the notch by cancelling filter losses arising mainly from the finite Q of the inductors.[28] A constant-g_m biasing source decreases the sensitivity of this negative impedance (and consequently of the LNA gain) to temperature and process variation. Here, bias current I_2 is chosen to boost the Q by a factor of 5. This value is high enough to provide a significant deepening of the notch but is not nearly high enough to endanger loop stability with variations in process, temperature, and supply voltage. For example, the circuit tolerates more than a tripling of the nominal bias current without instability.

A low-power image-reject phase-locked loop (IR PLL) generates the control voltage for the notch filter; see Figure 19.45. This PLL is a simple offset synthesizer that achieves lock when the internal IR PLL's VCO frequency equals the *difference* between the two LO frequencies. To prevent parasitic locking at the sum frequency instead, the lock range is restricted and acquisition always starts from the low-frequency side (using the reset switch shown), assuring that the loop first encounters the desired difference frequency condition.

The IR PLL's VCO (Figure 19.46) and the LNA's notch filter are topologically identical, differing only in bias current. Hence tuning the VCO to the image frequency also tunes the notch frequency, assuring process independence of the notch location. The LNA proper consumes 6.7 mW and exhibits a noise figure of 4.3 dB, using a standard 0.25-μm digital CMOS process. Considering that this design operates at triple the frequency of the GPS example and consumes half the power, the NF achieved seems quite reasonable (even taking into account the more advanced process technology). The IR PLL adds 3.1 mW for a total consumption here of just under 10 mW. The image rejection enhancement provided by the notch filter is 16 dB.

---

[28] The noise inevitably contributed by any active element must be considered as well. Fortunately, in this case the noise contributed by the negative resistance cell only moderately offsets the noise improvement due to neutralization of the parasitic capacitance, and there is thus a net improvement in NF overall.

To Loop Mixer

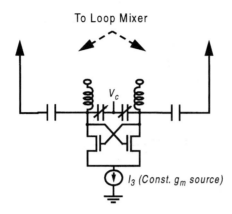

FIGURE 19.46. Schematic of IR PLL VCO.

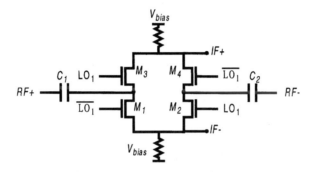

FIGURE 19.47. Passive ring mixer.

Mixers

The six mixers of the quadrature Weaver architecture are implemented two ways. Since CMOS transistors are good voltage-mode switches, the first pair of mixers are simple passive ring mixers for good linearity and low power (Figure 19.47). The outputs of this first pair drive a quad of Gilbert-type mixers (Figure 19.48). Although these mixers exhibit worse linearity than do passive rings, the attenuation provided by the passive mixers relaxes the requirements. More relevant is that their differential current mode nature makes it easy to implement addition and subtraction of the output signals. Finally, the gain provided by these active mixers is desirable by itself and also for reducing the noise figure contribution of subsequent stages. The common-source connection of the input transistors is grounded to reduce supply voltage requirements and also to mitigate any second-order distortion that might contribute to the generation of beat components.[29]

---

[29] B. Razavi, "Design Considerations for Direct-Conversion Receivers," *IEEE Trans. Circuits and Systems,* v. 44, June 1997, pp. 428–35.

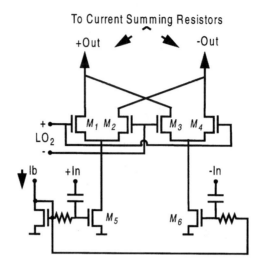

FIGURE 19.48. Gilbert-type double-balanced mixer.

AC Coupling for Offset Mitigation

The outputs of the second set of mixers is AC coupled to the baseband circuitry. Although AC coupling confers relative freedom from offsets compared to an otherwise equivalent single-conversion homodyne receiver, important issues remain nonetheless. For example, the coupling capacitors must be linear. In addition, the pole frequency of the coupling network must be high enough to assure sufficiently fast settling (including recovery from overload transients) yet low enough to avoid causing excessive intersymbol interference. This latter consideration usually demands the use of coupling capacitors that are relatively high in value. Here we use 15-pF lateral-flux capacitors to produce a 5-kHz corner frequency without consuming excessive die area. The capacitance density is boosted by a factor of 3.5 (relative to a standard parallel plate sandwich with the same number of metal layers), to 700 af/μm^2. This boost factor increases as lithography continues to scale and is also a function of the particular fractal geometries chosen. Here, the layout is based on a Minkowski sausage, chosen for its reasonable boost factor as well as its ability to fill a rectangular space.

We've noted that a subsidiary benefit of exploiting lateral flux is a reduction in bottom plate capacitance. In the present case, the bottom plate capacitance per terminal is only 8% of the total value. This value is lower than that found in many processes with special structures dedicated to enhance analog and RF performance.

The crude AC coupling used here is adequate for the GMSK modulation specified in the original HiperLAN system for the high–data rate format (recall that GMSK is also used in the GSM system). However, we should note that settling time requirements are more stringent for most OFDM-modulated systems, so direct coupling would almost certainly be a requirement there.

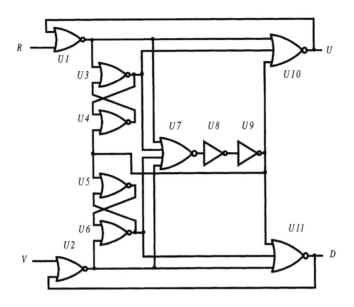

FIGURE 19.49. Conventional phase-frequency detector.

Frequency Synthesizer

The local oscillator signals are generated by an integer-N frequency synthesizer. The loop employs a conventional phase-frequency detector with the standard additional delay (via U_8 and U_9) in the reset path to mitigate dead-zone effects arising from runt pulses. See Figure 19.49.

Not shown is additional, "plain vanilla" digital circuitry for generating low-skew complementary versions of the U and D outputs used to drive the charge pump. To reduce power, the feedback divider is implemented as a cascade of an injection-locked frequency divider and a more conventional prescaler. The injection-locked divider (Figure 19.50) has a free-running frequency of approximately 2.5 GHz, nominally half the synthesized output frequency. As described in Chapter 16, such a circuit consumes considerably less power than an analogous flip-flop–based divider owing to the former's use of resonant circuits. The tradeoffs are an increase in die area consumption and a reduction in operational frequency range. Since most commercial systems are narrowband in nature, this latter limitation does not preclude the use of such circuits. Here, cross-coupled differential pair M_1–M_2 synthesizes a negative resistance to overcome the loss in the LC drain network and thereby sustain oscillation; the output of the divider is taken from the drains of these transistors. Obtaining the desired "divide by 2" action is facilitated by enhancing the second-order nonlinearity in the loop. Such a nonlinearity produces an intermodulation component at a frequency equal to the difference between the frequency of the oscillator and the injection signal. If these frequencies are in a precise 2:1 ratio, then a self-consistent solution to the loop equations can exist and synchronization results.

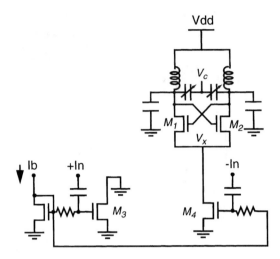

FIGURE 19.50. Injection-locked frequency divider.

In this differential circuit, the common-source node V_x contains a strong spectral component at twice the oscillation frequency. Interpreting this double-frequency component as the signature of a second-order nonlinearity in action, we expect that the injection of a 5-GHz signal into this node will result in synchronization at half the injection frequency. This conjecture has been confirmed in a quantitative analysis of the circuit.[30]

The same analysis reveals that maximizing the locking range requires maximizing the tank inductance L. However, power consumption is inversely related to the *tank impedance* at resonance and hence to the QL product. Unfortunately, there is no guarantee that maximizing L automatically maximizes QL at the same time. Hence, the divider implemented here maximizes the inductance subject to a (somewhat arbitrary) power consumption limit of 1 mW. This power is one fifth that consumed by a conventional flip-flop divider built in this technology.

Because the optimization objective chosen does not directly accommodate a specification on the tuning range, there is always a danger of insufficiency. To solve this problem, the center frequency of the divider is made to track automatically the frequency of the VCO. Now, rather than having to accommodate the entire tuning range of the receiver, the divider's tuning range is required to accommodate only component mismatches – a considerably simpler task. The tuning capacitance is implemented as an accumulation-mode varactor, as is the tuning element in the LNA notch filter.

To minimize spurs, the phase detector drives a differential charge pump that is designed for low leakage and low feedthrough of up and down command pulses as well as for the removal of all sources of systematic offset (Figure 19.51). Although the

---

[30] H. Rategh and T. Lee, "Superharmonic Injection-Locked Frequency Dividers," *IEEE J. Solid-State Circuits,* v. 34, June 1998, pp. 813–21.

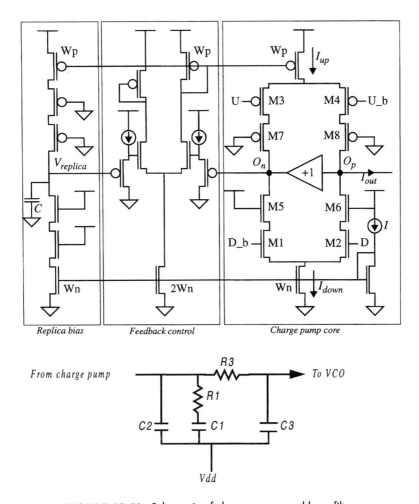

FIGURE 19.51. Schematic of charge pump and loop filter.

charge pump output is taken as a single-ended signal, a bootstrapping buffer forces the unused output in the charge pump core to the same voltage as the main output. Furthermore, a replica bias circuit assures minimum sensitivity of the pump current mismatch to the common-mode output voltage. If the charge pump output voltage differs from V_r, an op-amp adjusts the pull-up current until equality is restored. The loop order is also increased to four to enhance filtering of the control voltage, as seen in the figure. As a result of these combined strategies, all synthesizer spurs are below the -70-dBc noise floor of the instrumentation and well below any values needed to meet performance objectives.

Note that the loop filter is referenced to the supply voltage, not to ground. This choice is intentionally made because the control voltage for the varactors is itself actually referenced to V_{dd} as well. Returning the loop filter to ground would thus result in the injection of any supply noise (and we must assume that there is always supply noise) into one of the most sensitive nodes in the circuit.

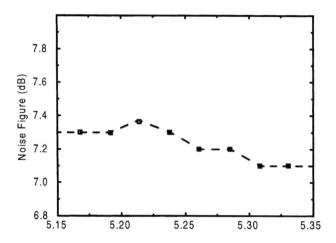

FIGURE 19.52. Measured receiver noise figure.

Thanks in part to the use of the resonant frequency divider, the complete synthesizer consumes 25 mW, including the VCO and all conventional dividers (whose power consumption now dominates). Compare this value with the ~40 mW consumed in the synthesizer for the GPS receiver example, which operates at one third the frequency. Even accounting for the differences in process technology, the 25-mW dissipation seems favorable.

The measured phase noise of the synthesizer is −134 dBc/Hz at the center of the adjacent channel (22-MHz offset). Integrating this level of VCO noise over two adjacent channels yields −58 dBc, which implies that an adjacent channel interferer can be 48 dB stronger than the desired signal while maintaining a 10-dB signal-to-interference ratio.

Finally, the settling time after changing channels is under 35 μs, comfortably faster than the 1-ms requirement.

Performance Measurements

The overall receiver noise figure is plotted as a function of frequency in Figure 19.52, which shows that the stages after the LNA increase the cascaded noise figure by about 3 dB to a value of about 7.2 dB. Despite the relatively large second-stage contribution, this noise figure remains well below the 18-dB target value and still comfortably below the 10-dB figure for 802.11a.

The image rejection is seen from Figure 19.53 to lie between 50 dB and 53 dB over the entire band. About 16 dB of this rejection is due to the notch filter in the LNA, and another 35 dB comes from the Weaver architecture itself. These values are robustly achieved without implementing calibration of any kind. Some combination of autocalibration and additional prefiltering could be used to augment performance if the specifications on image rejection were significantly tighter.

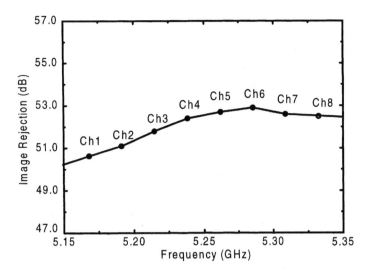

FIGURE 19.53. Measured image rejection.

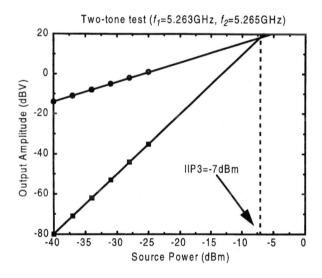

FIGURE 19.54. Two-tone intermodulation test results.

Linearity is evaluated with a two-tone test in Figure 19.54. The input-referred IP3 is −7 dBm, with a −1-dB compression point of −18 dBm. The latter value is comfortably better than the −21-dBm target. This performance is obtained at relatively low bias currents, thanks to the high linearity of short-channel MOSFETs.

A revealing and practically relevant test is a 1-dB blocking desensitization evaluation. Because desensitization is the result of gain compression that accompanies large signals (interferers or not), blocking performance is closely related to the 1-dB

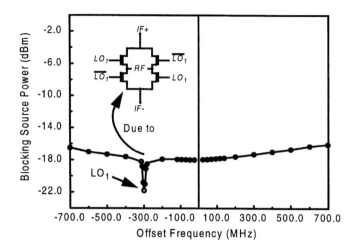

FIGURE 19.55. Measured 1-dB blocking performance.

Table 19.2. *Summary of 5-GHz CMOS WLAN receiver characteristics*

	Achieved	Required
Signal path performance		
Noise figure	7.2 dB	18.3dB
Voltage gain	26 dB	
S_{11}	< −14 dB	
Image rejection (filter only)	16 dB	
Image rejection (total)	53 dB	
Input-referred IP3	−7 dBm	
1-dB compression point	−18 dBm	−21 dBm (est.)
LO_1 leakage to RF	−87 dBm	−47 dBm
LO_2 leakage to RF	−88 dBm	−57 dBm
Power dissipation		
Synthesizer	25.3 mW	
Divide-by-8 (for LO_2)	6.0 mW	
Signal path	18.5 mW	
Image-reject PLL	3.1 mW	
LO buffers	5.0 mW	
Biasing	0.9 mW	
Total power	58.8 mW	
Supply voltage	1.8 V	

compression point. As a zeroth-order approximation, in fact, in-band 1-dB desensitization performance may be taken as equal to the 1-dB compression point. As seen in Figure 19.55, the receiver generally tolerates blockers larger than −18 dBm (a value that happens to equal the 1-dB compression point, validating the rule of thumb) over the entire frequency range. Since HiperLAN specifies that receivers tolerate in-band

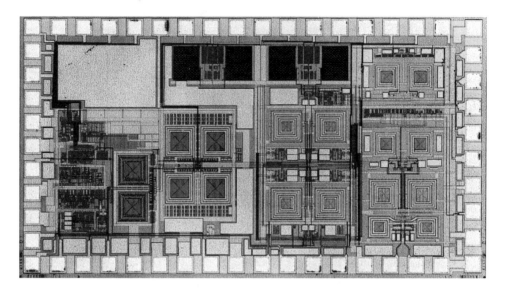

FIGURE 19.56. Die micrograph (0.25-μm technology).

blockers as large as -25 dBm, there is evidently substantial margin. Note that, at a frequency equal to that of the first LO, there is a noticeable dip in the blocking performance (but to a still satisfactory -22 dBm). The passive ring mixer used to implement the RF mixers is the reason for this diminished performance. When a strong blocker at a frequency LO_1 appears at the RF port, it generates a DC voltage at the output of the RF mixers. The resulting bias shift reduces the gain.[31]

The overall characteristics of the receiver are summarized in Table 19.2, and the photomicrograph of the 4-mm^2 die is shown in Figure 19.56. As with the GPS receiver, the mutual proximity of the eighteen spiral inductors poses no significant cross-coupling problem. Coupling of inductor signals into the substrate is minimized by the global use of differential circuits. The suppression is further augmented by the use of a patterned ground shield under each inductor.

19.6.4 IEEE 802.11A DIRECT-CONVERSION WLAN TRANSCEIVER

Our final example[32] is a direct conversion transceiver for 5-GHz 802.11a WLAN, a standard whose main specifications are summarized in Table 19.3. From the channel bandwidth and specified minimum receive sensitivity we can estimate a maximum NF

---

[31] Samavati et al., op. cit. (see footnote 26).

[32] The author gratefully acknowledges the kindness of Iason Vassiliou and his colleagues at Athena Semiconductor for providing the figures used in this section. The accompanying text naturally owes much to the corresponding paper by I. Bouras et al., "A Digitally Calibrated 5.15GHz–5.825GHz Transceiver for 802.11a Wireless LANs in 0.18μm CMOS," *ISSCC Digest of Technical Papers*, February 2003.

Table 19.3. *Summary of IEEE 802.11a specifications*

Operating frequencies	5.15–5.35 GHz
	5.725–5.825 GHz
TX/RX	TDD
Number of RF channels / spacing	12 / 20 MHz
RF channel bandwidth	16.6 MHz
Modulation	OFDM-QAM
Subcarriers	52
Data rates	6 Mbits/s (BPSK),
	54 Mbits/s (64 QAM)
Minimum RX sensitivity (54 Mbits/s)	−65 dBm
Minimum TX EVM (54 Mbits/s)	−25 dB
P_{out} (maximum)	5.15–5.25 GHz: 40 mW
	5.25–5.35 GHz: 200 mW
	5.725–5.825 GHz: 800 mW

of around 18.8 dB, assuming an 18-dB predetection SNR requirement. For 64-QAM, this SNR value theoretically allows better than a 10^{-3} bit-error rate (BER), assuming no other impairments. As with the previous HiperLAN example, it is prudent to target a considerably lower NF value because there are always "other impairments" – particularly with direct conversion.

Despite its many theoretical attributes, direct conversion's implementation is not without serious difficulties. These problems are primarily a consequence of a homodyne's inherently high sensitivity to DC and low-frequency signals. For example, typical input-referred DC offsets of well-matched CMOS differential pairs are of the order of 1 mV. This value corresponds to a power of about −47 dBm into 50 Ω, which is *much* more than the typical power of RF signals processed by circuits near the antenna. Additionally, as seen in Figure 19.57, radiation from the local oscillator may couple back into the RF input port with a random phase, producing additional and variable DC offset after mixing. Local oscillator energy can re-enter through the antenna or past the LNA (or some other point in the front-end electronics). This mechanism is of concern because LO power is generally within 10 dB of 0 dBm. A 40-dB level of isolation might still leave us with a quite substantial −30-dBm self-interference term, and even-order nonlinearities can also create signal-dependent offsets.[33] Regardless of origin, these offsets may change dramatically when the LO

---

[33] We thus also worry about *second*-order intercept in a direct-conversion receiver, because any intermodulation between two RF signals of nearly equal frequency will result in an output component close to DC. This example underscores the need to revisit figures of merit (and augment or discard them) whenever the context changes. Third-order intercept is not the only applicable linearity measure.

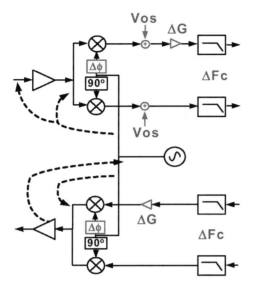

FIGURE 19.57. Illustration of some
problems with direct conversion.

frequency changes value during frequency hopping[34] or channel selection, making
offset removal additionally challenging. Finally, $1/f$ noise is unfortunately of a nature to produce the maximum negative effect in homodyne receivers. Regrettably,
CMOS exhibits large $1/f$ noise. Even if offsets and noise aren't large enough to
overload subsequent stages, they represent a form of self-interference that can readily reduce sensitivity to poor values.

Similar impairments afflict direct-upconversion transmitters as well. Coupling
from the LO directly to the output stage can perturb the modulated signal ahead of
the power amplifier. The output of the power amplifier may also couple back to the
LO, perturbing the latter and causing systemic problems. This feedback problem is
especially serious in designs that integrate a high-power transmit amplifier on the
same die. If an LO with its power of 0 dBm causes difficulties, it is doubtful that
problems would diminish with a 20–30-dBm-output PA.

Implementing successful and cost-effective solutions to this collection of problems
has been sufficiently difficult that widespread commercialization has taken place only
recently. To combat these problems often requires some form of autocalibration, since
open-loop techniques are generally insufficient. The block diagram of Figure 19.58
reveals numerous switches and digital-to-analog converters (DACs) whose function
is to enable the requisite calibration capability. As the diagram shows, the receiver
has provisions for nulling offsets by introducing compensating offsets at the outputs
of the receive mixers. Other DACs allow for adjustment of the corner frequency of
the channel filters.

---

[34] Frequency hopping is not part of 802.11a, but it is widely used in other systems such as Bluetooth.

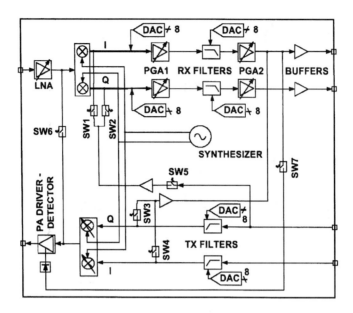

FIGURE 19.58. Transceiver block diagram.

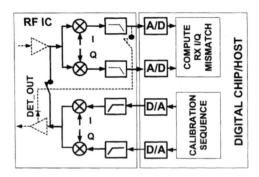

FIGURE 19.59. Illustration of calibration method.

In implementing autocalibration, there's always the question of how and when. An answer to the latter almost universally includes power-up. In systems that do not demand 100% duty cycle of transceiver activity, there is also an opportunity to perform calibrations between frames or other dwell times, for example. In this manner, compensation can track variations in temperature, supply voltage, and ambient conditions.

The answer to *how* calibration might be performed is given in Figure 19.59. Several classes of corrections are applied, but we will focus on the most difficult. To fix impairments such as I/Q mismatch and LO leakage, this transceiver uses a configurable *loop-back* architecture.[35] In the first step of the calibration, the dotted path

---

[35] The calibration method outlined here is described by J. K. Cavers and M. W. Liao, "Adaptive Compensation for Imbalance and Offset Losses in Direct Conversion Transceivers," *IEEE Trans. Vehicular Tech.*, v. 42, November 1993, pp. 581–8.

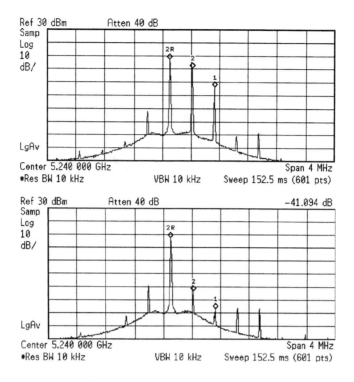

FIGURE 19.60. Transmit sideband rejection:
performance before and after calibration.

shown in the figure is active. An on-chip detector samples the power amplifier's output and then feeds it to the interface drivers. After A/D conversion, the transmitter *I/Q* mismatch and LO leakage are measured. That information enables the computation of an appropriate digital predistortion algorithm, which "undoes" those transmit impairments.

Having fixed the transmit path, the switches are now configured as shown in bold to permit the calibration of the receive path. A digital calibration sequence again drives the chip, and the output of the receive path is characterized. Once the receiver's *I/Q* mismatch is determined and nulled, the transceiver is ready for communication.

Underscoring the success of this method, the plots in Figure 19.60 show the transmit sideband performance before and after calibration. In the uncalibrated state, sideband rejection is only 21 dB and LO leakage is a frighteningly large −7 dBc. Calibration improves sideband rejection to 54 dB and reduces LO leakage to a more manageable −41 dBc.

Another way of assessing the effects of calibration is to look directly at the quality of quadrature (easily tested by feeding the system with symbols representing polar modulation). As is obvious from Figure 19.61, the uncalibrated transceiver has offset and gain errors. Rather than the polar modulation locus being a circle with its center at (0, 0), it's an ellipse whose axes are not coincident with the basis vectors and whose center is offset. After calibration, these defects are largely absent.

Before calibration

After calibration

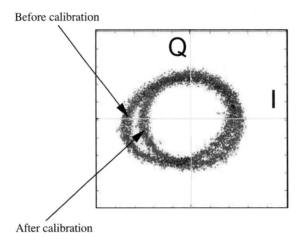

FIGURE 19.61. Quadrature quality (offset and gain): performance before and after calibration.

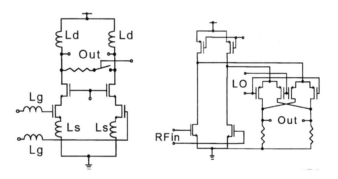

FIGURE 19.62. LNA and mixer schematic.

Having briefly reviewed the calibration method, we now examine some of the key circuit blocks used in the transceiver.

The LNA shown in Figure 19.62 is a conventional inductively degenerated differential cascode amplifier, as we've seen throughout this textbook. Planar spirals are used for all but the gate inductors (for which bondwires are used). To provide coarse gain-ranging, a series switch–resistor combination permits the controllable shunting of the drain load. When the switch is closed, the gain of the LNA drops to 10 dB from its 18-dB maximum. The noise figure is approximately 3 dB in the lower band and rises to 4.5 dB in the upper band. These values are well below the maximum tolerable noise figures computed earlier, leaving plenty of margin to absorb the noise contributions of subsequent stages as well as a host of other nonidealities.

The mixer shown next to the LNA is a PMOS folded cascode version of a Gilbert-type mixer. The use of PMOS devices in the current-mode switching quad reduces flicker noise. The potentially large flicker noise of the NMOS pair that receives the

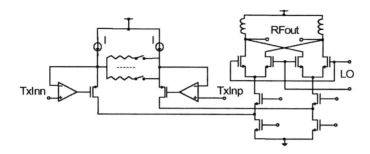

FIGURE 19.63. Transmit mixer.

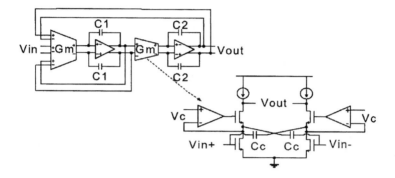

FIGURE 19.64. Chebyshev baseband filter schematic.

RF input is translated away from baseband and is thus of minor consequence. Direct grounding of the common-source connection of the NMOS pair improves IP2. Measured IIP3 exceeds 4 dBm, and the overall SSB NF of the mixer is better than 12 dB.

Like the receive mixer, the transmit mixer (Figure 19.63) is a folded cascode Gilbert-type. Because large-signal performance is of primary importance for a transmitter while noise figure is mostly irrelevant, design evolves along somewhat different paths than for the receive mixer. The feedback connection with the op-amps assures that the differential input signal voltage appears across the source degeneration resistors. The input PMOS stage thus acts as a near-ideal transconductance, as evidenced by a +20-dBm IIP3. An array of switches and resistors enables programmable attenuation over a 27-dB range.

The baseband filters are implemented as fourth-order 9-MHz g_m–C Chebyshev types; see Figure 19.64. The op-amps within the g_m stages produce a better approximation to a virtual ground load for the bottom NMOS common-source differential pair. By reducing the effective load impedance for that pair by a factor equal to the op-amp gain A, the overall stage gain is boosted by nearly that same factor, much as would be obtained in a cascade. However, the dynamics of the op-amp do not appear directly in cascade, reducing stability degradation in feedback connections (e.g., in

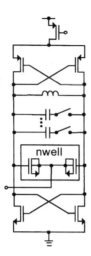

FIGURE 19.65. Doubly symmetric
voltage-controlled oscillator.

this filter). The filter corner frequency is controllable through variation in transconductance, and its adjustment is included as part of the overall calibration suite.

The synthesizer uses a doubly symmetric VCO, as described in Chapter 18. A complementary pair of negative-g_m cells makes oscillation possible by overcoming tank loss. Proper selection of the PMOS–NMOS width ratio suppresses flicker-noise upconversion by enforcing the symmetry criterion elucidated by LTV phase noise theory. An array of switches and tank capacitors provides tuning in discrete steps, with continuous variation within those intervals provided by n-well accumulation-mode varactors. See Figure 19.65. The varactor gates are tied to the outputs of the oscillator, and the common drain–source–body connection is driven by the control voltage so as to avoid loading the critical oscillator nodes with transistor parasitics.

The VCO manages to produce a phase noise figure of better than −120 dBc/Hz at 1-MHz offset by using on-chip inductors with Q-values somewhat south of 10. To cover the noncontiguous 5-GHz WLAN band with a single VCO is possible, but providing a large tuning range generally entails a tradeoff with such other desirable parameters as supply voltage sensitivity and phase noise. The tradeoffs may be severe enough to make that choice unappealing. Consequently, separate VCOs (operating at 2.6 and 2.9 GHz) provide the necessary coverage in this integer-N PLL design. Separate frequency doublers provide the 5.2- and 5.8-GHz outputs, and separate (passive) second-order polyphase filters produce quadrature outputs. Use of the frequency doubler reduces problems from unwanted coupling. For example, leakage from the PA will be less likely to perturb a VCO operating at half the frequency. This technique is as useful today as when a half-frequency oscillator was used in superhets in the 1920s for similar reasons.[36]

---

[36] See Harry W. Houck, U.S. Patent #1,686,005, filed 3 March 1923, granted 2 October 1928. Houck, Armstrong's assistant and close friend, invented the second harmonic superheterodyne in order to

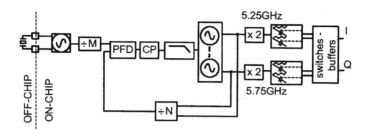

FIGURE 19.66. Synthesizer for LO generation.

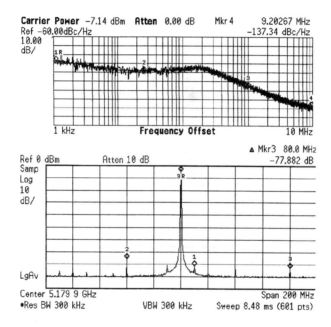

FIGURE 19.67. PLL spectra (close-in and farther out).

The PLL's third-order loop filter is integrated with the rest of the synthesizer (see Figure 19.66) and is designed to provide selectable loop bandwidths of 150 kHz and 500 kHz. The PLL output spectrum plots of Figure 19.67 show that the phase noise, integrated from 1 kHz to 10 MHz, is below 0.8° rms. The reciprocal mixing thus causes very little degradation in BER or, equivalently, produces very little increase in the SNR required to achieve a given BER. The far-out spectrum is clean, with all spurs below −65 dBc.

Measured RF-to-baseband performance is shown in Figure 19.68. This plot is largely a characterization of the baseband filters. The system provides better than 40 dB of rejection at 20 MHz and over 70 dB at 40 MHz.

solve the problem of LO radiation. We have Houck to thank for preserving the vast bulk of Armstrong's papers and equipment. See http://users.erols.com/oldradio/ for an online museum.

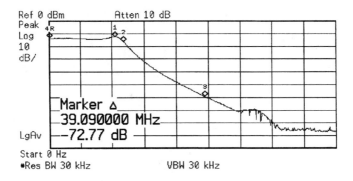

FIGURE 19.68. RF-to-baseband performance.

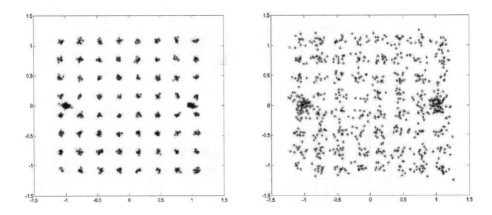

FIGURE 19.69. Receive constellations at −55 dBm (left) and −75 dBm (right).

Receive sensitivity is better than −75 dBm – for a BER below 10^{-5} – when measured directly at the input terminals of the chip itself. At −55 dBm, the receive EVM is better than −33 dB (2.2%).[37] See Figure 19.69.

The OFDM signal presents serious challenges for the power amplifier because of its large peak-to-average ratio. The peaks can be so large, in fact, that it isn't even necessarily meaningful to cite a particular peak-to-average value. Instead, it is preferable to present a complementary cumulative distribution function (CCDF), which plots probability on the vertical axis and power above the average value on the horizontal axis. From such a plot for 802.11a we can determine, for example, that the power will exceed the average value about 40% of the time (because the mean and median powers are different) and by 7 dB about 0.1% of the time.[38] Thus, if we build a power amplifier with 7 dB of compression headroom, it will go into compression

---

[37] When converting to and from dB for EVM, use 20 log, not 10 log (EVM is not a power-related quantity).

[38] See e.g. Bob Cutler, "RF Testing of Wireless LAN Modems," Agilent Technologies, April 2001, http://www.tmintl.agilent.com/images_agilent/us/WLANbbtest.pdf.

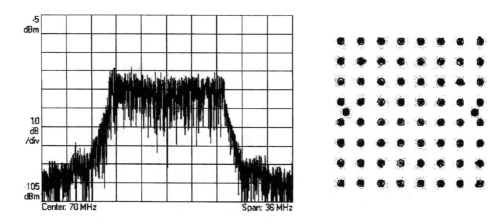

FIGURE 19.70. Transmit spectrum and EVM plot.

Table 19.4. *Summary of specifications*

Vdd	1.8 V / 3.3 V / I/O
TX/RX mode power	302 mW / 248 mW
RX overall NF	5.2 dB
RX max/min gain	79 dB / 20 dB
RX IIP3 max/min gain	−18 dBm / −8 dBm
SSB phase noise (1 MHz)	−115 dBc/Hz
Integrated phase noise	0.8° (1 kHz–10 MHz)
Supported bands	5.15–5.35 GHz, 5.7–5.8 GHz
TX P −1 dB	6 dBm
Technology	1P6 CMOS 0.18-μm
Die size	4.5 mm × 4.1 mm
Package	MLF-64

about 0.1% of the time. The same sorts of plots reveal that a 4-dB headroom will result in compression about 5% of the time. The linearity constraints are so tight that many vendors simply reduce transmit power until they are able to meet the tight EVM specification (5.6%) for 802.11a. (The corresponding number is a generous 35% for 802.11b, which uses DSSS.)

The transmitter EVM plot is shown in Figure 19.70. At an output power of −3 dBm, the measured EVM is better than −33 dB (again, 2.2%), 8 dB better than the specified value of −25 dB (5.6%).

The performance of this transceiver is summarized in Table 19.4. The total die area is about 18.5 mm², and the chip is built in a 0.18-μm CMOS technology.

We close this chapter with a die photo (Figure 19.71) showing the floorplan of this chip. The reader will note from this micrograph (or from the higher-resolution one that adorns the cover of this book) that there is a fair amount of unused chip area. The reason is simply that the die photo is of a test chip, the penultimate step prior to a final design.

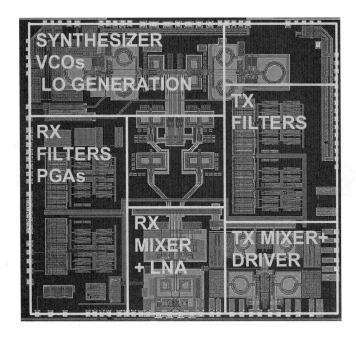

FIGURE 19.71. Die micrograph (courtesy Athena Semiconductor).

19.7 SUMMARY

This chapter has built on the foundation laid by the previous chapters to provide the basic information needed to construct receivers and transmitters. We've seen that the search for a fully integrated receiver reduces to the quest for an architecture that does not require an external filter. Both the zero-IF (i.e., direct-conversion) and low-IF (e.g., image-reject) receivers are potential candidates, but the lack of convincing existence proofs underscores the need for continued research. Linearity requirements of the former are almost ludicrously stringent, while the achievable image rejection of the latter falls far short of values typically needed by most systems. Perhaps more so than at any other level, the tradeoffs at the architectural level are particularly serious and existing solutions unsatisfying.

PROBLEM SET FOR ARCHITECTURES

PROBLEM 1   Derive Eqn. 17, the image rejection ratio due to imperfect quadrature.

PROBLEM 2

(a) Design a simple RC–CR quadrature generator for a 1-GHz center frequency. First select the capacitance so that the kT/C noise is $1.6 \times 10^{-11}\,\mathrm{V}^2$, and then determine the necessary resistance from the center frequency specification. Is this resistance value reasonable? Explain.

(b) Suppose we may ignore any possible practical difficulties associated with the resistor value. However, we discover that the capacitors in this technology have a bottom plate capacitance to substrate that is 30% of the main capacitance. Compute the gain and phase error at 1 GHz resulting from the bottom plate problem. You may assume that the substrate is at ground potential and is a superconductor.

(c) Estimate the image rejection ratio if all the errors in an image-reject architecture come from this phase shift network.

PROBLEM 3   Derive an expression for the IRR of the image-reject mixer without making any simplifying assumptions. Verify that your expression reduces to the ones given in the text in the limit of small errors.

PROBLEM 4   The low-pass filters in the image-reject mixer would appear to be superfluous because even the sum frequency components are theoretically rejected by the architecture. Explain why the filter is nonetheless important in practical mixers of this type.

PROBLEM 5   The text alluded to the usefulness of the offset frequency synthesizer for use in the direct conversion receiver. Consider the transmitter architecture shown in Figure 19.72.

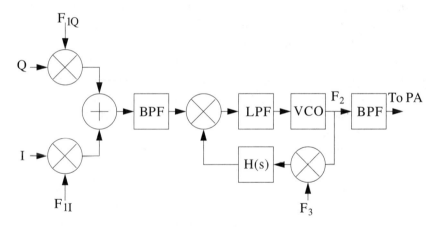

FIGURE 19.72. Transmitter with offset frequency synthesizer.

(a) For this transmitter architecture, how is the output carrier frequency, F_2, related to F_1 and F_3 if $H(s)$ is a low-pass filter?
(b) Repeat part (a) if $H(s)$ is a high-pass filter.
(c) Discuss the advantages and disadvantages of the two choices.

PROBLEM 6   Assume a cubic nonlinearity and derive a relationship between IIP2 and IIP3 in terms of the power series coefficients.

PROBLEM 7   Following a development analogous to the one in the text, derive an expression for the input *second*-order (voltage) intercept for a cascade of systems in which individual gains and intercepts are known.

CHAPTER TWENTY

RF CIRCUITS THROUGH
THE AGES

20.1 INTRODUCTION

As mentioned in Chapter 1, there just isn't enough room in a standard engineering curriculum to include much material of a historical nature. In any case, goes one common argument, looking backwards is a waste of time, particularly since constraints on circuit design in the IC era are considerably different from what they once were. Therefore, continues this reasoning, solutions that worked well in some past age are irrelevant now. That argument might be valid, but perhaps authors shouldn't prejudge the issue. This final chapter therefore closes the book by presenting a tiny (and nonuniform) sampling of circuits that represent important milestones in RF (actually, radio) design. This selection is somewhat skewed, of course, because it reflects the author's biases; conspicuously absent, for example, is the remarkable story of radar, mainly because it has been so well documented elsewhere. Circuits developed specifically for television are not included either. Rather, the focus is more on early consumer radio circuits, since the literature on this subject is scattered and scarce. There is thus a heavy representation by circuits of Armstrong, because he laid the foundations of modern communications circuitry.

20.2 ARMSTRONG

Armstrong was the first to publish a coherent explanation of vacuum tube operation. In "Some Recent Developments in the Audion Receiver,"[1] published in 1915, he offers a plot of the V–I characteristics of the triode, something that the quantitatively impaired de Forest had never bothered to present.[2] In this landmark paper, Armstrong follows up on the circuit implications of those characteristics to explain the

---

[1] *Proc. IRE,* v. 3, 1915, pp. 215–47. It should be noted that he had first published triode characteristics somewhat earlier, in an article in *Electrical World* (12 December 1914, pp. 1149–52).

[2] De Forest's contemporaneous paper, "The Audion – Detector and Amplifier" (*Proc. IRE,* 1914, pp. 15–36), stands in remarkable contrast to Armstrong's. Nowhere in the 20+ page paper can one find any quantitative information about how one uses a triode Audion, although there are a couple

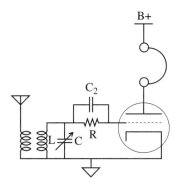

FIGURE 20.1. Grid-leak detector.

operation of an AM demodulator and the first positive-feedback amplifier, and also to show how the latter may also function as an oscillator and heterodyning element (mixer). We now study these early achievements of the 24-year-old inventor.

20.2.1 THE GRID-LEAK AM DEMODULATOR

In this demodulator, re-drawn with modern symbols as Figure 20.1, a process called "grid-leak detection" is used.[3] For those unfamiliar with vacuum tube operation, it is perhaps helpful to regard the triode as a depletion-mode n-channel junction FET, with the cathode, grid, and plate analogous to source, gate, and drain, respectively.[4] In "normal" operation, the grid potential is negative, so that negligible grid current flows. However, just as with such a FET, a small positive voltage on the grid forward-biases a diode, causing grid current to flow. The grid-leak detector uses this diode to perform the demodulation. Although Armstrong did not invent the circuit, its principles of operation were shrouded in mystery until publication of his paper.

In this circuit, the signal from the antenna is transformer-coupled to a resonant LC filter, which performs channel selection. A coupling capacitor, C_2, connects this tuned RF signal to the grid. A positive-going input signal forward-biases the grid cathode diode, causing grid current to flow and charging C_2 toward the peak input level; C_2 then discharges slowly (relative to the carrier frequency), as a consequence of grid current and of leakage in the capacitor (represented by R), until the next positive-going input pulse occurs. The voltage across C_2 (as well as the average grid

---

of handsome photographs of equipment. Typical of the intellectual quality of the writing are statements such as "Positive or negative charges impressed on the grid cause a diminution in [the plate] … current," along with the seemingly incompatible assertion that the triode Audion exhibits "freedom from … distortion." Armstrong explained how the former could be true, but not at the same time as the latter. De Forest saw no issue that needed further discussion.

[3] In radio parlance, "detection" and "demodulation" are usually synonymous.

[4] Indeed, vacuum tube circuits translate quite gracefully into JFET form with very few modifications other than supply voltages.

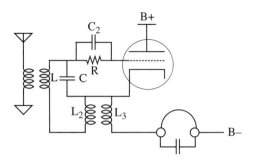

FIGURE 20.2. First regenerative amplifier
(adapted from Armstrong).

voltage) therefore follows the modulation envelope. The triode then gains up this demodulated signal just as in any other common-cathode amplifier, leaving a healthy audio signal in the headphones.

As another trivia note, "B+" is the standard notation for the plate supply. Why "B"? – because "A" is reserved for the filament supply. There is also a "C" supply, for grid bias in those circuits that use it.

20.2.2 THE REGENERATIVE AMPLIFIER AND DETECTOR

In the same paper, Armstrong presents the first positive feedback amplifier, with which one could obtain much more gain from a single stage than had previously been possible. His amplifier is shown in Figure 20.2, again re-drawn with modern symbols.

Here, coupled inductors L_2 and L_3 provide the positive feedback. The plate current is sampled through L_3, while L_2 provides the feedback to the grid circuit. The amount of feedback may be adjusted through various means (such as varying the relative spacing and orientation of L_2 and L_3) to control the gain (and, simultaneously, the Q). When the system is on the verge of instability, the overall gain can be considerably larger than without regeneration. Furthermore, this circuit can simultaneously demodulate through grid-leak detection.

In the same paper, Armstrong describes alternative methods for achieving regeneration. Among these is the use of an inductive plate load impedance, acting in concert with feedback through the plate–grid capacitance. He was thus the first to recognize, and then exploit, the destabilizing properties of this combination.

20.2.3 OSCILLATOR AND MIXER

The circuit of Figure 20.2 can also be used as an oscillator simply by overcoupling L_2 and L_3. Such a compact generator of continuous waves was considered somewhat of

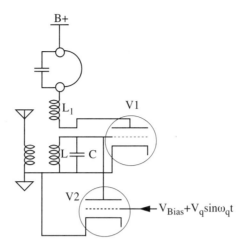

FIGURE 20.3. Early superregenerative receiver (with separate quench oscillator).

a miracle at the time because spark apparatus, arc oscillators, and large rotating machines were then the main sources of RF energy. Furthermore, mixing action takes place between the oscillation and any RF input signal because of the nonlinearity of the triode. Thus, demodulation can occur through homodyne detection, with a conversion gain roughly proportional to the amplitude of the oscillation (within limits). If the oscillation frequency differs from the RF, then the mixing action leads to heterodyne conversion and is particularly useful for detecting CW code signals. Working with heterodyne converters undoubtedly set the stage for Armstrong's invention of the superheterodyne two or three years later.

20.2.4 THE SUPERREGENERATIVE AMPLIFIER

While fooling around with the regenerative amplifier, Armstrong discovered superregenerative amplification.[5] In his paper of 1922, he describes the basic principles and then proceeds to show a number of circuits for realizing superregenerative amplifiers. In all of these examples, the quench oscillator is separate from the main amplifier, as shown in Figure 20.3.

In this circuit, V_1 is a regenerative oscillator/detector with enough feedback from L_1 to L to ensure oscillations when V_2 is removed from the circuit.[6] Device V_2 is driven by a low-frequency oscillator and acts as a voltage-controlled resistance across the tank, periodically killing oscillations in V_1. When V_2 is in its high-resistance state, oscillations grow exponentially from initial conditions established by the input signal.

---

[5] "Some Recent Developments on the Regenerative Receiver," *Proc. IRE,* 1922.

[6] The slightly different feedback connection shown here was by far the most popular way of achieving regeneration.

Because this circuit trades the log of gain for bandwidth, as discussed in Chapters 1 and 9, exceptionally large gains are possible from a single stage. Armstrong describes the attainment of approximately 50 dB of power gain from one such stage at 500 kHz, about 30 dB greater than what one normally obtains from a conventional regenerative amplifier, allowing the elimination of one or two stages.

Later, engineers discovered that a separate quench oscillator is not strictly necessary, although better performance is possible if one is available. The intermittent oscillations characteristic of a superregenerative amplifier can be produced in a standard Armstrong oscillator with certain choices of grid-leak element values.[7] It is possible for so much negative bias to build up from the oscillations that the tube actually cuts off for a time until the grid-leak resistor discharges the grid-leak capacitor. Hence, the same circuit as shown in Figure 20.2 can quench itself, leading to its use as a superregenerative amplifier and detector (through nonlinear action, as usual). As should be evident by now, this is an exceptionally versatile circuit. Modern replications using n-channel JFETs function similarly well and are excellent hobby circuits for weekend experimentation.

20.3 THE "ALL-AMERICAN" 5-TUBE SUPERHET

Armstrong's superheterodyne underwent considerable refinement during the 1920s and 1930s. Unprecedented ease of operation conferred by the single required tuning control, coupled with circuit improvements and cost reductions made possible by better vacuum tubes, made the superhet the dominant architecture by 1930.

For over two decades, the standard low-end consumer AM table radio used a complement of five vacuum tubes (with only four in the actual signal path), as shown in the schematic of Figure 20.4.[8] The filament voltages sum to about 120 V, allowing a direct connection to the AC line and so eliminating the need for a costly power transformer.[9] The chassis also is actually connected to one side of the AC line, so that miswiring can inadvertently lead to the chassis being connected to the hot side of AC power. Hence, an isolation transformer should always be used when servicing these units, else touching the chassis could cause your eyeballs to counterrotate at 60 Hz.

This circuit evidently originated in an applications note published in the 1940s by RCA, which had developed (and naturally wanted to sell) the tubes used in the circuit. I have not been able to locate a primary source, however, so this conjecture remains to be verified.

[7] Actually, Armstrong himself describes this effect in his 1915 paper, but in the context of conditions to avoid if one wants to construct good oscillators.
[8] Early versions of this circuit used the larger octal 12SA7, 12SK7, 12SQ7, 50L6, and 35Z5 tubes in place of the miniature 12BE6, 12BA6, 12AV6, 50C5, and 35W4, respectively.
[9] The power rectifier tube's filament has a tap so that a 6.3-V pilot light ("power on" indicator) may be hooked up to it. The tap also feeds the anode of the rectifier, providing a bit of resistance between the AC power line and the B+ supply in order to limit peak current.

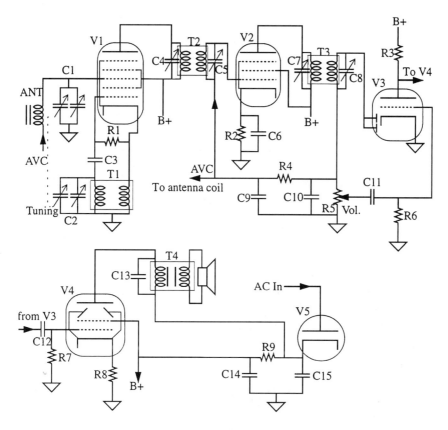

FIGURE 20.4. "All-American" 5-tube superhet (typical schematic).

The first tube (V_1), a 12BE6 pentagrid converter, acts as both local oscillator and mixer. This "autodyne" circuit is an improvement over the Armstrong oscillator/mixer in that internal cascoding within the tube (provided by incrementally grounded grids 2 and 4) reduces coupling between the oscillator and RF ports, making independent tuning easier to achieve.[10]

In the autodyne circuit, the local oscillator is an echo of Armstrong's original regenerative circuit. The cathode current couples back to the first grid through transformer T_1, whose tuned secondary controls the oscillation frequency and thus the channel selected, as in all superhets. Simultaneous tuning of a simple bandpass filter at the RF input port aids image rejection. The tuning capacitors for both circuits are mechanically linked ("ganged"), so that the consumer only has to turn one knob to change frequency.

Grid 2 is incrementally grounded and acts as a Faraday shield to isolate the oscillator and RF circuits, as discussed earlier. The RF signal feeds grid 3, and nonlinear

---

[10] Some purists reserve "autodyne" for circuits in which the tube has signals applied to only one grid. I am not one of them.

interaction within the tube performs the mixing action. Grids 4 and 5 are incrementally grounded; they remove the Miller effect and suppress secondary electron emission, respectively.

Note the absence of an RF amplifying stage. Strictly speaking, one is not needed for ordinary broadcast AM radio because *atmospheric* noise in the 1-MHz band is much larger than the electronic noise generated by any practical front-end circuit. Hence, an RF amplifier would not improve the overall noise performance of a receiver in this band. Indeed, its gain would be somewhat of a liability, as large signals could then overload the front end.

The output of the first stage is coupled through a double-resonant IF bandpass filter to a single IF amplifier, a 12BA6 (V_2), operating at 455 kHz. The 12BA6 is a pentode and thus behaves much like a cascode, allowing one to use filters on both the input and output ports without worrying about detuning or instability from feedback.

Demodulation and audio amplification take place in V_3, a 12AV6, which contains two diodes and a triode within one glass envelope. The diodes[11] are used in an ordinary envelope detector (rearranged slightly to accommodate a grounded cathode), and the triode amplifies the demodulated audio. Because there are no tuned circuits connected to the input and output ports, the Miller feedback of an ordinary triode is acceptable.

The demodulated output is sent to two destinations. One is the output power amplifier, V_4. The other is an additional low-pass filter, the output of which is the average of the demodulated output. This signal is used to control automatically the gains of the autodyne and IF amplifier as a function of received signal strength. The greater the demodulated output, the more negative the bias fed back to those stages, reducing their gain. This automatic gain control (AGC) or automatic volume control (AVC)[12] thus reduces potentially jarring variations in output amplitude as one tunes across the dial.

Device V_4 is a 50C5 beam-power tube used in a Class A audio configuration. Transformer coupling provides the necessary impedance transformation to deliver roughly a watt of audio into the speaker. The power rectifier used to generate the B+ plate supply for the other tubes is a 35W4 (V_5). In some later versions, this tube was replaced by a semiconductor rectifier.

With minor variations, the All-American was widely copied, and clones could be found all over the world. Once it caught on, high manufacturing volumes drove down the cost of these five particular tube types, so anyone designing a new radio intended for a cost-sensitive application tended to use the same tubes, and thus used similar circuits. Variations among different versions were really quite slight (e.g., small resistor or capacitor value differences, absence or presence of cathode resistor bypass capacitors, etc.).

---

[11] A common variant grounds one of the diodes, while another uses both in separate audio and AGC functions.

[12] AVC was yet another invention of the inestimable Harold Wheeler.

20.4 THE REGENCY TR-1 TRANSISTOR RADIO

The first portable transistor radio became available in time for Christmas in 1954 and was the result of an effort by a young Texas Instruments (TI) to create a mass market for transistors. Up to this time, the only commercial use for transistors had been in hearing aids. As the father of the project, Patrick Haggerty, later noted, the idea was that "a dramatic accomplishment by [us would] awaken potential users to the fact that ... we were ready, willing, and able to supply [transistors]."[13] TI arranged a deal with a small company called IDEA (Industrial Development Engineering Associates) to perform some cost-reducing modifications of TI's first-pass circuit (principally designed by Paul D. Davis and Roger Webster) and then to manufacture the radios through IDEA's Regency Division. The task was challenging, as no one had much expertise with transistors yet. To make a tough job even more difficult, the germanium transistors then available were quite poor by today's standards (f_Ts of only a few megahertz at best, and βs of 10–20), and their cost was high. Compounding those difficulties was the lack of miniature components. IDEA eventually had to subcontract out the manufacture of some of the capacitors to a university professor who set up a little side business. The poor quality of the small speakers also gave the engineers infinite grief, as did numerous other design and production difficulties. It was quite a struggle to cram all of the circuitry into a case small enough to fit in a (large) shirt pocket.

Cost was a big problem as well, and the expensive transistors dominated it. It was determined early on that no more than four transistors could be used or IDEA would not be able to make a profit at the targeted sale price of $49.95 (at a time when an All-American 5-tuber could be purchased for about $15). The four transistors accounted for about half the materials costs.

As seen in Figure 20.5, four transistors were enough. In this circuit, the first transistor, Q_1, functions as an autodyne converter. A common-base oscillator configuration is used, with transformer coupling between collector and emitter circuits providing the feedback necessary for oscillation.

The incoming RF signal is tuned using a mechanism called "absorption," evidently developed by the German company Telefunken around World War I.[14] In this technique, an LC tank is coupled to the input circuit so that it shorts out (absorbs) signals at all frequencies other than the resonant frequency of the tank. The RF signal can pass to the base of Q_1 only when this shorting disappears, at the tank's resonant frequency. The nonlinearity of the base-emitter diode provides the mixing action. Hence, in addition to the local oscillator signal, the collector current also has a component at the sum and difference heterodyne terms. The difference signal is then fed to the first IF amplifier, Q_2, through an LC bandpass filter tuned to the IF of 262 kHz.

---

[13] The fascinating story of how the first transistorized portable radio came to be is told in detail by Michael Wolff in *IEEE Spectrum,* December 1985, pp. 64–70.

[14] "Funken" means "spark," giving you an idea of when the company came into being.

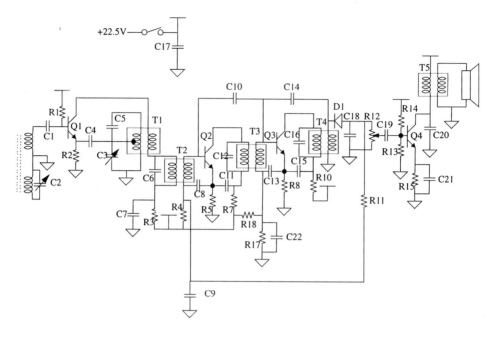

FIGURE 20.5. The Regency TR-1 portable transistor radio.

The unusually low IF allows the low-f_T transistors to provide useful amounts of gain, but it exacerbates an already bad image rejection problem.[15] The variable capacitor in the absorptive LC front-end tank is ganged with the LO variable capacitor. The degree of image rejection achieved here is best described as adequate.

The second IF amplifier, Q_3, is connected in a manner essentially identical to Q_2. The large C_μ values (probably about 30–50 pF) are partially cancelled by positive feedback through C_{10} and C_{14}, in an homage to the neutrodyne circuit of the 1920s.[16]

Demodulation is performed with a standard envelope detector, followed by a single stage of audio amplification. Transformers couple signals into the detector and out of the audio amplifier. Automatic gain control action is provided in a familiar manner: the demodulated audio is further RC-filtered (here by R_{11} and C_9), and the resulting negative-polarity feedback signal controls the gain of only the first IF stage by varying its bias.

The huge success of the TR-1 had important consequences beyond establishing TI as a leader in the semiconductor business.[17] Of particular significance is that IBM

---

[15] This IF is not without precedent, however. Early vacuum tube superhets used a wide variety of IF values, with 175, 262, 455, and 456 kHz among the more common. The absurd 1-kHz difference between the last two smacks of an effort to evade patent litigation problems. The 455-kHz IF eventually won out as the (near) universal standard.

[16] As mentioned in Chapter 1, Harold Wheeler invented neutralization while working for Hazeltine.

[17] Texas Instruments had another major triumph in 1954, with the production of the first silicon transistors. Gordon Teal, who had joined TI from Bell Laboratories, tells the story of being the last

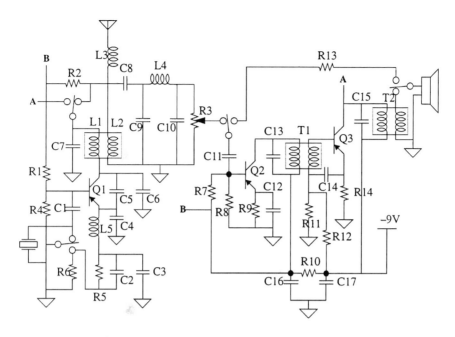

FIGURE 20.6. Three-transistor CB walkie-talkie (T/R switch shown in receive mode).

quickly abandoned development of new vacuum tube computers, with Thomas Watson, Jr. reasoning that if transistors were mature enough to show up in high-volume consumer gear, they were ready for prime time. As he later told the story, every time one of his subordinates expressed doubt about transistors, he'd give him a TR-1, and that usually settled the argument.[18]

20.5   THREE-TRANSISTOR TOY CB WALKIE-TALKIE

Another engineer at Texas Instruments, Jerry Norris, was responsible for developing the first toy walkie-talkie, in 1962.[19] This widely copied and ingenious circuit uses a single-transistor superregenerative detector, followed by two stages of audio amplification in receive mode (see Figure 20.6). When transmitting, the superregenerative stage becomes a stable crystal-controlled 27-MHz oscillator, amplitude-modulated by an audio amplifier built out of the other two transistors. The speaker doubles as a microphone in this mode.

speaker at a conference where everyone who went before him predicted that it would take a few more years to produce silicon transistors. He wowed the audience by pulling a few silicon devices out of his pocket and announcing their availability. The other speakers were partially right, though: TI had a monopoly on silicon transistors for four more years. For more on this fascinating story, see Teal's reminiscences in *IEEE Trans. Electron Devices,* July 1976. (The entire issue is devoted to histories told by the principals themselves.)

[18] Wolff, op. cit. (see footnote 13).

[19] See J. Norris, "Three-Transistor CB Transceiver," *Electronics World,* November 1962, pp. 38–9.

20.5.1   RECEIVE MODE

Transistor Q_1, a 100-MHz f_T 2N2189, does all the RF work in this circuit. The antenna is physically shorter than ideal, so loading coil L_3 helps "lengthen" it, electrically speaking.[20] The incoming RF signal is coupled through a transformer to the collector circuit of Q_1. Capacitor C_5 provides positive feedback from collector to emitter in a Colpitts oscillator configuration. As oscillations build up exponentially from whatever initial conditions are established by the input signal, the emitter-base diode causes the emitter voltage swing to be asymmetrical, with negative-going signals experiencing a higher impedance than positive-going ones. One consequence of this asymmetry is that the *average* emitter voltage heads to more negative values. Capacitor C_3 charges up to this average voltage through RF choke L_5. Resistor R_6 discharges C_3, but at a relatively slow rate (the C_2–R_5 network has such a large time constant that the voltage across it may be considered constant). The DC value of the emitter voltage therefore heads so far downward that the oscillation actually terminates until C_3 discharges sufficiently to restart oscillations, resulting in an intermittent oscillation whose quench frequency is too high to be audible. While this self-quenching action hardly results in optimum superregeneration (in particular, selectivity suffers mightily), it permits a one-transistor circuit to perform remarkably well.

Another consequence of the asymmetrical emitter voltage is that the transistor amplifies the modulated RF signal asymmetrically. Hence, the collector current contains a component roughly proportional to the modulation itself. A low-pass filter consisting of C_9, L_4, and C_{10} removes the RF component, passing only the modulation to the two-transistor audio amplifier.

20.5.2   TRANSMIT MODE

Although Q_1 acts as a self-quenched LC oscillator in the receive mode, FCC regulations require much more frequency stability and accuracy during transmit than can be provided by a simple LC network.[21] Hence, a quartz crystal is used to control the frequency of oscillation in the transmit mode. Capacitor C_1, which bypasses the base to ground in the receive mode, acts as a coupling capacitor to grounded crystal XTAL during transmit. The crystal shorts the base to ground only at its series resonance, and therefore permits oscillations only at that frequency. Resistor R_6 is shorted out during transmit to prevent intermittent oscillations.

---

[20] An electrically short dipole has a net capacitive reactance, so a series inductance permits better power transfer. To improve radiation (and, by reciprocity, sensitivity), a decent ground plane is essential. One can increase greatly the range of many a cheap walkie-talkie simply by providing a ground plane (just touching the battery can make a noticeable improvement). Even wrapping aluminum foil around part of the case and connecting it to the ground connection of the circuit helps a great deal, since contact with the body increases its effective dimensions.

[21] Channels are spaced 10 kHz apart in the 27-MHz band, so frequencies must be accurate to better than about 50 ppm, requiring the use of crystal-controlled oscillators.

An audio signal derived from the two-transistor audio amplifier varies the effective collector supply voltage of the oscillator, thereby effecting amplitude modulation. The speaker functions as a microphone in this mode, and the amplified audio signal is available at the collector of Q_3. Because of transformer T_2's connection to the battery, the collector voltage of Q_3 consists of a DC component on which the audio modulation is superimposed. This DC + AC signal supplies the collector voltage to Q_1 through bypass capacitor C_1 and inductor L_1, which here behaves simply as a short.

The oscillator amplitude is roughly proportional to the collector supply voltage, so the varying supply voltage amplitude-modulates the carrier. Although the distortion from this process hardly meets the standards of high-fidelity audio, it is certainly adequate for voice communications and most definitely adequate for a toy.

Because this simple circuit provides such large gain with so few transistors, it dominates the low-end walkie-talkie market, having been copied and modified countless times by toy manufacturers. Every superregenerative walkie-talkie[22] whose circuit I've traced has just one transistor doing all of the RF work, just as in Norris's original version. As with the All-American Five, variations among different manufacturers are relatively minor. The relatively expensive and bulky audio coupling transformer is not used in modern implementations since transistors now provide enough gain without it. The choke L_4 is also generally omitted, ordinary RC filters being adequate. Other than these simple modifications, other superregenerative walkie-talkie circuits are quite similar.

---

[22] It is trivial to determine whether a superregenerative circuit is used, even without tracing the schematic. Superregenerative receivers have a raucous hiss unless a reasonably strong signal is being received. That is, noise is always present, and diminishes in amplitude only as the signal strength increases. This behavior is quite different from that of a superheterodyne, for example, where the background noise is roughly constant in amplitude and independent of the received signal strength.

INDEX

AAC algorithm, 57
absorption, RF signal tuning by, 771
AC coupling, for offset mitigation, 744
accumulation-mode varactor, 129f
acknowledgment (ACK) packet, from WLAN
  receiver, 733
acoustic resonators, 87n1
A/D converters, in GPS, 718
adaptive bias, in power amplifiers, 528
addition rule, for risetime, 263
adjacent channel power ratio (ACPR), of power
  amplifiers, 536–8
admittance
  correlation, 367
  optimum source admittance, 369
  of parallel RLC tank, 87
admittance relation, reflection coefficient relation to,
  362
Advanced Mobile Phone Service (AMPS), 47, 49
  characteristics of, 48t
AF signals, 16n21, 17
Agilent, 410n11
Airy, George, 676
Aitken, Hugh, 8n16, 11n24, 22
Alexanderson, Ernst F. W., 7, 8
Alexanderson alternator, 8
"All-American Five" (radio)
  development of, 31, 768–70
  schematic of, 769f
alternator technology, history of, 7, 22
aluminum foil, as walkie-talkie cover, 774n20
aluminum interconnect, 121
AM, bandwith of, 21n39
AM broadcast, first by Fessenden, 8, 42
AM broadcast radios
  circuit for, 62
  demodulation in, 63
  phasor representation for, 63
  standard IF for, 408
  vestigial sideband signal for, 63
AM demodulator
  Armstrong's explanation for, 765
  conventional, 63
  envelope detector as, 62f

American National Standards Institute (ANSI), 228
American Telephone & Telegraph Co., 12, 48n22, 445,
  513
American Wireless Telephone and Telegraph, 6
America's Cup yacht race, 6
amplifiers
  bipolar common-emitter type, 402
  cascaded, 297–306
  cascoded, *see* cascode amplifiers
  common-gate type, 268f, 311f, 374f, 402f
  common-source type, 109, 291–2, 373f, 377f
  crossover distortion in, 558–9, 612, 613f
  in degenerate differential cascade, 756
  design example for, 275–6
  distributed type, 282, 304–6
  feedback-biased, 268f
  feedforward amplifiers, 443, 444f, 515–18
  gain of, 435n42
  high-frequency type, 270–313
  IF of, 698, 720f
  inverting amplifiers, 201, 460f, 461f, 462f, 479f
  load-pull experiments on, 553–5, 556
  narrowband type, 400
  negative feedback amplifiers, 444–5
  neutralization of, 294–7
  noise performance of, 354, 360, 361, 737
  noninverting amplifiers, 460f
  open-circuit time constants in design of, 253
  optimum gain per stage in, 299–301
  oscilloscope deflection type, 274
  parametric, 406n5, 435n42
  positive-feedback amplifiers, 442f, 446
  RF power type, *see* power amplifiers
  saturating, 612, 613f
  shunt-peaked, 279f
  shunt-series type, 282–8, 374f, 401
  with single-tuned load, 293f, 309
  source-coded, 295
  superregenerative type, 218, 301–3, 309, 767–8
  with T-coil bandwidth enhancement, 281
  tuned, 290–4
  two-port bandwidth enhancement of, 279–82
  unilateralization of, 294–7
  with voltage-sensitive capacitance, 199f

Ampliphase, 521, 522, 523f, 524
amplitude modulation (AM)
    conversion to PM, 306–7, 589, 661
    mathematical expressions for, 61–4
    symbols for, 59f
    see also AM
amplitude noise, 687–8
amplitude-shift keying (ASK), 64, 74
AM–PM conversion, 306–7, 661, 689
    in power amplifiers, 526, 532
Ampère's law, 117
    bandwidth and, 90–2
AM radio
    IF of, 698
    LO choices for, 699
    static in, 21
    vacuum tubes in, 768
AM receiver, upconversion in design of, 699
AM signals, demodulator for, 435
AM table radio, early model ("All-American Five"),
    768
analog circuits, 190, 221
Analog Devices, 356
analog multiplier, as phase detector, 574–6
analog phase interpolation (API) technique, 651
analog-to-digital converters (ADCs), 700, 711, 716
angle modulations, FM and PM, 70–3
antennas, diversity for, 82
antiresonant frequency, of parallel *RLC* tank, 88
Antognetti, P., 183, 198
AppCAD program, 410n11
architectures, 694–763
    for chip design, 716–61
    for direct conversion, 710–13
    for dual-conversion receivers, 700
    dynamic range in, 695–6
    for image-reject receiver, 700–10
    performance measurements in, 748–51
    problem set for, 762–3
    for single-conversion receiver, 698–9
    for subsampling, 713–14
    for transmitters, 714–15
    for upconversion, 699–700
    for WLANs, 731–40
arc tangents, role in phase shift, 275
arc technology
    history of, 22
    in industrial illumination, 7n14
Armstrong, Edwin Howard, 1, 17, 20, 21, 22, 44, 301,
    303, 404, 758n36, 764–6
    grid-leak AM demodulator of, 765–6
    history of inventions of, 13–15, 18–20, 764–70
    in legal battle, 21
    regenerative oscillator and, 21, 441–2, 443
Armstrong, Marian, 21
asymptotic angle rule, 456f
Atalla, M., 168
Athena Semiconductor, 751n32, 762f
atmosphere
    additive white Gaussian noise (AWGN) in, 56
    attenuation vs. frequency of, 79f
    as conductor, 37
atmospheric noise, 770
attenuation, sources of, 430n33

attenuation constant, 210, 231
    of transmission lines, 230
atto (prefix a), 228
audio coupling transformer, in walkie-talkies, 775
audions, as de Forest vacuum tubes, 11, 12, 13, 764
audio signal, in walkie talkies, 775
audio systems, noise in, 334
auditory system, of humans, 5n10
autodyne circuit, of "All-American 5-tube superhet,"
    769f
automatic gain control (AGC), in All-American Five,
    770
automatic volume control (AVC), in All-American
    Five, 770
autotransformer, 296
available noise power, 335
Avantek, 390

back-gate bias effect, in MOSFETs, 188–9, 240, 331,
    414n28
Ballantine, Stuart, 40
ballasting, for thermal runaway, 550
balun converter, 729, 730
bandgap reference principle, 319f
bandgap references
    classic, 320–4
    design example for, 321
    narrowing of, 321n9
bandgap voltage references, in CMOS technology,
    318–25
bandwidth
    in cascaded amplifiers, 298–9
    estimation of, *see* bandwidth estimation
    formulas for calculation of, 239–46
    risetime, delay, and, 259–65, 267
    of shunt–series amplifiers, 286–8
    T-coil enhancement of, 281f, 281–2, 282f
    two-port enhancement of, 279–82
    zeros as enhancers of, 270, 276–8
bandwidth estimation, 233–70
    machine computation of, 233
    open-circuit time constants (OCτs) method,
        234–54
    problem set for, 266–9
    by risetime measurement, 255
    short-circuit time constants (SCτs) method,
        254–8
Bardeen, John, 168
barium oxide, in filaments, 24
Barke, E., 131n11, 132
Barkhausen oscillation criteria, 564
barretter, 8
base-emitter (BE), 314
Battjes, Carl, 289–90
Battjes f_T doubler, 290f
beam forming, from multiple antennas, 82
beam-forming electrodes, 31
beam-power structure, 31f
Bell, Alexander Graham, 34, 39, 45n18, 228n4, 443
Bell Laboratories, 21, 43, 168, 349n28, 443, 632,
    720n20, 722n17
Bell System, cellular service of, 47
Belrose, J. S., 33n47
Bessel functions, 70

bias circuits
    constant-g_m, 325–8, 328, 332f
    supply-independent, 317–18
biasing, 314–33, 740
    problem set for, 331–3
Big Bang, echoes of, 338n10, 407n7
bilinear transformation, 222
binary-coded decimal (BCD), 651
binary frequency-shift keying (BFSK), 74–5, 76
binary modulation, 60
binary phase-shift keying (BPSK), 74–5, 732
biological effects, from directional antennas, 82
bipolar circuits, biasing based on, 314
bipolar devices
    breakdown phenomena in, 549–50
    high-speed, 321
    thermal runaway in, 550
bipolar technologies, 186
bipolar transistors, 316
    describing functions for, 614–17
    invention of, 167
    in mixers, 416
    noise model for, 358f, 359
bistable noise, 347; see also popcorn noise
bit, origin of term, 44
bit-error rate (BER)
    in detection, 77
    SNR for, 759, 760
Black, Harold S.
    feedforward amplifier of, 443, 517
    negative feedback amplifier of, 444–5
blackbody radiation, noise with, 338
black boxes
    responses of, 489
    two-port, 221
Bluetooth system, 54–5, 753n34
Bode approximation, 571
Bode plots, 453, 491
body effect, see back-gate bias effect
Boltzmann's constant, 335, 336
bondwire inductors, 144–6, 165
Bose, J. C., 4
Bose detector, 4n7
bounded-input bounded-output (BIBO) definition, of
    stability, 450, 472
Branly, Edouard, 2
Branly's coherer, 3f
Brattain, Walter, 168
Braun, Ferdinand, 4n7
brickwall bandwidth, 337
British Edison Company, 9
British Marconi, 10
broadcasting, origin of term, 42
broadcast radio
    Ampliphase role in, 521
    economic development of, 44
    frequency span for, 85t
    modulation in, 72
Brokaw cell, 320f, 321, 322, 323, 324
BSIM4, MOSFET modeling by, 190, 191
Bucher, Elmer E., 40
buckshot, noise from, 344
Burns, Ken, 22
burst noise, 347

cable, coaxial system of conductors in, 116
cable TV equipment, power levels in, 231
C/A code, for GPS signal, 716, 717
"CAD Model for Threshold and Subthreshold
    Comduction in MOSFETs" (Antognetti et al.),
    183
Cady, Walter G., 632, 639n22
calibration method, 754f, 755f
canyons, effect on GPS, 717n17
capacitance
    equations for, 123, 130, 143, 162–3
    gate-to-channel, 175, 176
    of interconnect, 130–6
    lateral, 124–5
    of MOSFETs, 172–3
    overlap type, 174
    propagation delay from, 210
capacitive degeneration, in mixers, 418n34
capacitively loaded source followers, 377n12
capacitive voltage divider, 99
capacitors, 122–36
    fractal-based, 126
    interdigitated, 127
    metal–metal type, 123
    neutralizing type, 16
    parallel plate capacitors, 122, 123f
    woven, 127–8
carbon film resistors, flicker noise in, 345
carborundum detectors, 5, 6n11, 9, 11, 20
carrier sense multiple access with collision detection
    (CSMA/CD), 732, 733
cars, radio-controlled, 19
Carson, John R., 21n39, 42, 43, 64, 72
Carson's filter method, 65, 68
Carson's rule, 72
Carson's SSB generation method, 706
Cartesian feedback, in power amplifiers, 526–8
Cartesian modulation, 60
cascaded amplifiers, 297–306, 531
    bandwidth shrinkage in, 297–9
cascade systems
    delay in, 259–61
    for intercept calculation, 696
    linearity of, 696–7
    noise figure of, 359f, 695–6
    risetime of, 261–3
cascode amplifiers, 240f, 248f, 251f, 257f, 295f, 313
cascoding technique, 248–9
cascoding transistor, 403
"cascomp" circuit, 425
cathodes
    basic principles of, 23–5
    oxide-coated, 24
catwhisker, as detector part, 5, 6, 9
Cauer, Wilhelm, 720n20
CB walkie-talkie, schematic of, 773f
cell phones
    crystal radio compared to, 40
    internal parts of, 41f
    lithium niobate use in, 635
    predictive coders for, 57
    U.S. market for, 48n22
cellular systems
    first-generation, 47–8

cellular systems (cont.)
  second-generation, 48–52
  third-generation, 52
center frequency
  of double-tapped resonator, 103, 104
  of phase-locked loops, 599
channels, of mobile wireless, 45
channel capacity, definition of, 56–7
channel capacity theorem, 43
channel coding, 57
channel-length modulation (CLM), 200, 240
charged-device model (CDM), for ESD protection,
    162
charge pump
  as alternative for op-amp loop filter, 747f
  for phase-locked loops, 590
Chebyshev baseband filter, 757f
Chebyshev polynomials, 104
chip
  design examples for, 716–61
  die photo of, 761, 762f
  performance summary of, 726, 728t
chip rate, of IS-95 CDMA, 50
Chireix, Henri, 521
Chireix outphasing, of power amplifiers, 521–4
choke, load of amplifier by, 513, 543
Christmas Eve demonstration, by Fessenden, 8
circadian rhythms, of humans, as injection locking,
    561n3
circuit design, economics of, 1
circuit noise, phase noise from, 673
circuit noise power, 661
circuit switching, packet switching replacement of,
    52n28
circular spiral inductors, 136, 138
Clapp, James K., 636n20
Clapp oscillator, 636, 637f, 643, 644, 654, 656, 679f
Class AB power amplifiers, 494, 502–3, 557, 558, 559
  design example for, 545
  modulation of, 512–39
Class A power amplifiers, 494–7, 528, 556, 557, 559
  design example for, 545, 555
  drain voltage and current for, 496f
  load-pull contour example of, 553–5
  modulation of, 512–39
Class B power amplifiers, 494, 497–9, 557
  drain voltage and current for, 498f
  modulation of, 512–39
Class C power amplifiers, 499–502, 520, 541, 551,
    557, 558
  design example for, 545, 551n56
  drain voltage and current for, 500f
  modulation of, 512–39
Class D power amplifiers, 503–5, 541
  drain voltage and current for, 504f
Class E power amplifiers, 505–7, 541
  design example for, 546–7
  modulation of, 512
  waveforms for, 506f
Class F power amplifiers, 507–10, 541
  alternative topology of, 511
  drain voltage and current for, 509
  inverse type, 510–11
  modulation of, 512–39

clear to send (CTS), from WLAN mode, 733
clocks, escapement in, 676
closed-loop systems
  stability of, 451
  transfer function of, 461, 485
clutches, loop transmissions and, 461–2
CMOS circuit blocks, dimensions of, 204
CMOS technology
  bandgap voltage references in, 324–5, 329, 332
  bipolar circuit in, 289–90
  bipolar transistors in, 316
  in deep-submicron phase, 507
  differential-pair noise in, 363
  differential pairs in, 752
  diodes in, 316
  distributed amplifiers in, 306
  frequency divider in, 565
  g_m cell in, 427f
  junction capacitors in, 644
  metal layers in, 158
  mixers, 404–40
  noise performance in, 371, 674, 753
  resistor options in, 120–2
  RF integrated circuit design in, 221–32
  ring oscillators in, 197–8, 685, 686
  scaling laws in, 197–8
  switches in, 417, 427–33
  vertical parallel plate structure in, 128
coaxial cable, 118, 119f
  power-handling capacity of, 229–30
Cochrun, B. L., 259
code-division multiple access (CDMA), 50, 56, 76
coherer
  Branly's, 2, 3
  in wireless technology, 35
coins, as cystal detectors, 5n9
color television
  hue encoding in, 306–7
  PLL-like circuit in, 562
Colpitts, Edwin Henry, 620n6
Colpitts crystal oscillator, 638f, 655
Colpitts oscillator, 635–6, 642, 654, 655, 656, 661,
    679f, 690f, 691, 693f, 774
  crystal in, 639f
  describing function analysis of, 620–31
  example of, 623f
  noise in, 675–6
  simplified model of, 622f
  startup model of, 624f
Columbia University, 211
combination synthesizers, 651–3, 658
common-gate amplifiers, 268f, 311f
common-source (CS) amplifier, 109, 291–2, 373f,
    377f
  with single tuned load, 291–2
communications systems, oscillators for, 617
commutating multiplier, as phase detector, 576–7
comparator(s), 73
  describing function for, 612
  oscillator with, 612
compensating capacitance, in neutrodyne amplifier, 16
compensation, root-locus examples and, 470–7
complementary cumulative distribution function
    (CCDF), 760

complementary metal-oxide silicon processes, *see* CMOS technology
complementary to absolute temperature (CTAT), 316, 318, 321
complex digital modulation systems (CDMA), 536–8
complex filters, 707
complex pole, angle near, 457f
compression algorithms, 57n33
computer mouse, 707–8
condensers, 16
conductors
  coaxial system of, 118
  losses from, 220
constant-*k* lines, 206n4
The Continuous Wave (Aitken), 22
continuous-wave (CW) signal, 14
continuous-wave (CW) transmitters, 42
control theory, 441
conversion gain, of mixers, 414, 416, 440
Coolidge, W. D., 23
coplanar transmission line, 158, 159f
coplanar waveguide, 158
copper
  in interconnect, 121
  in rectifiers, 168n2
  skin depth of, 117
copper oxide rectifiers, 168n2
copper pyrites, in detectors, 4n7
cordless phones, bands for, 80
Coriolis force, 74
corporate power combiner, 533–4
correlation admittance, 367
correlation coefficient, 368
Costas loop, for SC-AM demodulation, 65f, 69
coupled inductors, 146
coupling coefficient, 150, 154, 158
cross-coupled NOR gates, in phase detectors, 581
cross-modulation, of LNAs, 390
crossover distortion, in amplifiers, 558–9, 612
crossover frequency, of phase-locked loops, 593–4, 599, 655
cross-quad MOSFET, 424f
CRTs, 32
crystadyne receiver, of Losev, 20, 21f
crystal
  definition in modern electronics, 5n8
  temperature-induced drift of, 718
"The Crystal Detector" (Douglas), 22
crystal detector(s), 5f, 10, 435
  invention and history of, 5–6
crystal diode, 15
crystal oscillators, 638–40, 654–5
crystal radio, 40, 113, 435
  early model of, 6f
"crystal rectifier," 639n22
Curie, Jacques, 632n14
Curie, Marie Sklodowska, 632n14
Curie, Pierre, 632n14
current mirrors, problem based on, 198, 200
cutoff frequency, of lumped lines, 215–16
cyclostationary noise sources, 420, 661, 674, 675–6
  external, 676n7
cyclotron, 7n14

damping ratio, for feedback systems, 470
Darlington pair, as f_T doubler, 289f
Davis, Paul D., 771
dBc, dbm, and dBw units, definitions of, 227, 228
DC current, for oscillators, 626, 627
DC transmission, of power, 9
"dead zone" problem, 583
de Bellescize, H., 560, 561, 579
decibel, unit version of, 228n4
de Forest, Lee, 6, 9, 10–11, 17, 21, 22, 764
de Forest triode audion, 11, 12f, 23, 27
delay
  in cascade systems, 259–61
  in distributed systems, 204
delay line, lumped, 215f
delay spread, multipath propagation, 81–2
delta-sigma technique, for noise distribution, 651n31
demodulation, 73
  in "All-American" 5-tube superhet, 770
  in early transistor radio, 772
demodulators, as detectors, 72, 435n40
depletion-mode devices, 765
describing functions, 611–31
  catalog for, 612–13
  for MOS and bipolar transistors, 614–17
  in oscillator analysis, 620–31
desensitization, of LNAs, 390
destabilization, suppression of, 307
detectors, as demodulators, 435n40
detuning, suppression of, 307
deviation ratio, 732
DIBL, *see* drain-induced barrier lowering (DIBL)
Dicke, R. H., 338n10
die photo, of GPS receiver, 726, 727f
differential, "addition" law, 362
differential pair, as f_T doubler, 289f
digital modulations, 73–7
digital process technology, ring oscillators in, 596
digital quadrature generators, 704f
digital signal processor (DSP), 716
digital systems, phase-locked loops for, 560
digital-to-analog converter (DAC), 653, 654, 716, 753
digital transmitter, block diagram for, 57f
digital watches, quartz crystals in, 633n15
diode noise model, 357f
diode-ring mixers, 434–7
diodes
  behavior of, 314–16
  in CMOS technology, 316
  gain from, 435n42
  structure of, 23f
diplexer, 730
dipole antenna, effective area of, 78n54
direct-conversion receiver, 560, 696
  architecture for, 710–13
  drawbacks of, 712, 753f
direct digital frequency synthesizer, 653f
direct digital synthesis, 653
direct-sequence spread spectrum (DSSS), 50
discriminators, 73
distortion, from amplifiers, 443
distributed active transformer (DAT), 534–6
distributed amplifiers, 282, 304–6
distributed resonators, 631

distributed systems
   lumped systems link to, 204–5
   problem set for, 218–20
dithering moduli, synthesizers with, 649–51
divider "delay," in PLL synthesizers, 645
Doherty, W. H., 529, 530
Doherty and Terman–Woodyard composite amplifiers,
   528–30
Dolbear, Amos, 34, 35f, 37, 49
Donald Duck effect, 69
Doppler shift, 77
double-balanced mixers, 417, 419–27, 436–7, 439
   with low headroom, 421f
   passive, 427–33
double-sideband noise figure (DSB-NF), 407
double-sideband suppressed-carrier (DSB-SC) signal,
   invention of, 42–3, 64, 563
double-tapped resonant match, 103f, 107
double-tapped resonator, 103–4
double-tuned discriminator, 73
Douglas, A., 22
downconversion, from RF to IF, 699
drain current
   envelope impedance and, 628
   expression for, 171–3, 191
   noise of, 339–40, 361, 365, 691
   for transistors, 615
   waveform of, 501f, 691F, 692f
drain diode zener breadown, in MOS devices, 548–9
drain–gate capacitance, 372
   resistance for, 245, 267
drain-induced barrier lowering (DIBL), 193, 240
   cumulative effects of, 186, 201
drain modulation, in power amplifiers, 514f, 515
drain noise current, 367
drain–source punchthrough, in MOS devices, 548
driving-point impedance, of iterated structures, 205–6
drywall, attenuation from, 80
dual-conversion receivers, architecture for, 700
"duck test" version, of Occam's razor, 203n1
Dunwoody, Henry Harrison Chase, 5, 11
duplexer, 730n24
   in WLANs, 53n30
dynamic range, in architectures, 695–6

Early Radio Wave Detectors (Phillips), 22
Early voltage, 173, 201, 320n7
eddy currents
   induction of, 141, 142
   loss, 141, 142, 151
Edison, Thomas A., 9–10, 23
Edison effect, 10
effective isotropically radiated power (EIRP), 84
802.3 Ethernet, 732, 733
802.11 WLAN
   medium access control of, 732
   specification summary for, 752t
   transceiver for, 751–62
Eimac, 24n41
Einstein, Albert, 20, 40
electrodynamic equations (Maxwell), 2
electromagnetic theory, texts on, 116n1
electromagnetic waves, Maxwell's prediction of, 2
electromigration effects, 166

Electronic Principles (Gray and Searle), 259
electrons, 10
   cathode emission of, 26n42
   thermionic emission of, 13
electrostatic discharge (ESD), 161–2
   protection circuit, 161f
Elmore delay, 260
Elmore risetime, 263, 264, 265n6
Elwell, Cyril, 7, 12n25
The Empire of the Air (Lewis), 22
energy storage elements, 694
envelope detector, 111–13, 435
   for AM demodulation, 62f
envelope elimination and restoration (EER), in power
   amplifiers, 520–1
envelope feedback
   of oscillators, 629f
   of power amplifiers, 515–17
envelope impedance, 628
envelope loop transmission, 625, 626
envelope probability density function (PDF), 529–30
epi noise, 340
error coefficient, 464
error-vector magnitude (EVM), 538, 540f
   of transmitter, 761
Ethernet LAN, 732, 733
Europe
   cellular systems in, 48–9
   chroma in color TV of, 562
   WLAN standard in, 731
exa (prefix E), 228
exclusive-OR gate, as phase detector, 577–9
eyeball technique, for noise measurement, 355

Faraday shield, 163, 769
Faraday's law, 203
The Father of Radio (de Forest), 22
FCC Report and Order (February 2003), 538
Federal Communications Commission (FCC), 72,
   447, 561, 734, 774
Federal Telephone and Telegraph Co., 7n14, 12n25
Feedback Amplifier Principles (Rosenstark), 259
feedback-biased amplifier, for resistance calculation,
   268f
feedback factor, 445
feedback systems, 441–92
   with additive noise sources, 449f
   clutches and loop transmissions in, 461–2
   compensation in, 477–81, 487–8
   error series in, 464–6
   errors in, 462–6
   examples of, 470–7
   first-order low-pass sytems, 466–7
   frequency- and time-domain characteristics of,
      466–9
   gain and phase margins of, 451–3
   gain reduction in, 478–81
   history of, 441
   lag compensation in, 481–3
   lead compensation in, 484–6, 487, 489
   loop transmission in, 462–6
   low-pass in, 490
   modeling of, 459–61
   negative, desensitivity of, 446–50, 488

op-amp follower in, 463f, 488
problem set for, 488–92
root-locus rules for, 458, 470–7
root-locus techniques for, 453–8
second-order low-pass sytems, 468–9
slow rolloff compensation in, 486–7
stability-criteria summary for, 459
stability of, 450–1, 473–4, 476–7
feedforward amplifiers, 443, 444f, 517–18
feedforward correction, 443
femto (prefix f), 228
Fermi level, 175
Fessenden, Reginald, 7, 8, 14, 41, 404n1
Fessenden barretter, 7, 8, 10
FETs, *see* field-effect transistors
FFT algorithms, 423
field-effect transistors (FETs)
charge control in, 376
depletion-mode devices, 765
early work on, 29, 167–8
enhancement-mode transistor
noise from, 674
shot noise from, 343
thermal noise in, 339, 428
fifth-order elliptic filter, 720, 721f
filament, of power rectifier tube, 768n9
fill factor, 139
filters
Chebyshev, 757
complex, 707
effects on architectures, 694
IF of, 698
LNA combined with, 741f
noise bandwidth of, 361
polyphase, 705
first-order phase-locked loops, 567–8
Fleming, John Ambrose, 9–10, 23
Fleming valve, 10f
flicker noise, 344–7, 756
in junctions, 347
in MOSFETs, 346
flip-flop–based divider, 565, 745, 746f
flux stealing, 125f
FM (frequency modulation)
angle modulations in, 70–3
bandwidth of, 21n39
legal battle over, 21
phase linearity importance in
FM devices, gain linearity in, 558
FM/PM systems, offset synthesizer use in, 652
FM radio, 292n5
frequency spans for, 85t
IF of, 698n1
signal of, 698n1
Fokkema, J. T., 132
Fokkema equation, 132
"Formulas for the Skin Effect" (Wheeler), 120
forward error correction (FEC), for signals, 83
Foster–Seeley discriminators, 73
four-diode double-balanced mixer, 434
4046 CMOS phase-locked loop, characteristics of, 596–9
Fourier components, of signals, 591, 671
Fourier series, 577, 612, 684

Fourier transforms, 260, 261, 430
four-resonator system, in invention of radio, 39
"four sevens" patent (#7777), of Marcoini, 7n12
fractal, based on Koch islands, 126f
fractal capacitor, based on Minkowski sausage, 127
France, color TV system in, 562n6
Franklin, Benjamin, 32–3
"free-energy" radio, 5
free-space transmission formula, 78
frequency detector, 745f
frequency diversity, from multiple antennas, 82
frequency divider
injection-locked, 565f, 745
resonant, 748
frequency-division duplexing (FDD), 45, 47
time-division duplexing compared to, 54f
frequency-division multiple access (FDMA), of
mobile wireless, 45, 47
frequency-domain linearization, of oscillators, 611
frequency-domain simulators, for IP3 computation, 393
frequency-hopped spread-spectrum (FHSS), for
WLANs, 53
frequency hopping, 50, 753
frequency modulation, *see* FM
frequency planning, 694, 699
frequency pulling, by oscillators, 630, 714
frequency-selective fading, 82
frequency-shift keying (FSK), 42, 49, 74, 75
detection for, 77
frequency synthesizer, *see* synthesizers
Friis, Harold T., 78, 348n28
Friis free-space transmission formula, 78, 83
Friis path loss, 79n57
fringe correction, for circular capacitance
fringing, in circular capacitors, 163t
f_T doubler
bandwidth enhancement with, 288–90
Darlington and Battjes type, 308
Fuller, Leonard, 7n14
function generator, oscillator as, 617–20

GaAs MESFETs, 616
noise performance in, 674
GaAs substrate, 142
GaAs technology, 186
amplifiers in, 552
gain, in LNA design, 364
gain and phase errors, 701–3
gain margin, computation of, 452
gain-phase margin, in feedback stability, 451–3
gain phase plane, of oscillator
galena detectors, 4, 6
Galvin Manufacturing Corp. (Motorola), 44n13
gamma (γ), *see* propagation constant
GAMMA, in SPICE level-1 models, 189
Γ-plane, 222
garage-door openers
lithium niobate use in, 635
radio-controlled, 19
gate capacitance, voltage dependency of, 199
gate capacitors, of MOS devices, 128
gate current
from hot electrons, 187
noise, 343n19, 365, 372

gate noise, 341–2, 371
Gaussian distribution, 81, 335, 355
Gaussian minimum-shift keying (GMSK), 49, 75
  for HiperLAN system, 744
Gauss's law, 203
General Admiral Apraskin (battleship),
  wireless-assisted rescue of, 37
General Electric Co., 7, 13
general packet radio service (GPRS), 52
generator, curve for, 125
germanium transistors, early types of, 771
"getters," use in cathode production, 24–5
giga (prefix G), 228
Gilbert, Barrie, 417n22, 574
Gilbert bipolar "multi-tanh" arrangement, 426
Gilbert gain cell, 519n24
Gilbert multipliers, 577
Gilbert-type mixers, 75, 417n22, 427, 743, 744f, 756
  linearity of, 422
Gingell, M. J., 706n3
Ginzton, E. L., 304
Global Positioning System (GPS)
  free-space wavelength of, 730
  frequency used in, 379n13, 538, 716n15
  receiver for, 717–31, 748, 751
  signal structure for, 717f
Global System for Mobile Communications, 49
Goddard, Robert, 12, 441
gold doping, of bipolar transistors, 347
golden ratio (golden section), 206
Grabel, A., 259
Gray, P. E., 259
Greenhouse's method, 137
grid, of early triode, 27
grid-leak AM demodulator, Armstrong's explanation
  of, 765–6
grid-leak detector, 765f
Groupe Spéciale Mobile (GSM)
  base-station power amplifiers of, 518, 538, 652
  GMSK use in, 744
  parameters for, 48–9, 50t, 52, 75
gyrators, architecture of, 722f, 725

Haggerty, Patrick, 771
half sections, *m*-derived, 217–18
hams
  polyphase network use by, 706
  as term for radio amateurs, 42
handheld mobile telephones, 47n21
Handie-Talkie AM transceiver, 44
handoff concept, in mobile telephones, 45
harmonic distortion, estimation of, 394n29
harmonic telegraph, of Bell, 45n18
Hartley, Ralph Vinton Lyon, 43, 66, 620n6
Hartley modulator, 67, 68
Hartley oscillator, 620n6, 636, 654
Hartley phasing method, 706
Hawks, Ellison, 34n49
Hazeltine, Louis, 16, 296n9, 772n16
Hazetine's circuit, 16
Heaviside, Oliver, 2n2, 211
Heising, Raymond A., 43, 66n43, 513
Heising modulator, 513, 514f, 515n18
heptode, 31

Herrold, Charles "Doc," as radio pioneer, 41–2
Hertz, Heinrich, 1–2, 22, 34, 35, 36, 37, 39
Hertzian waves, 32, 35, 38n57, 39
heterodyne principle, 14
Hewlett, W. R., 304
Hewlett-Packard, 374n8, 410n11
  waveguide bands of, 86
hexagon spiral inductors, 139t
hexode, 31
high-device-count circuits, 1
high-frequency amplifier design, 270–313
high-frequency performance, problem set for,
  308–13
high-side injection, LO frequency and, 699
Hilbert transforms, 67n46, 707
HiperLAN, 732
  performance needs for receiver of, 733–4, 750–1
HiperLAN2, as European WLAN, 731, 732
Hogge, C. R., 584
Hogge's implementation, 587
Hogge's phase detector, 584, 585, 586
hole–electron pairs, from impact ionization, 187
hollow spiral inductors, 138
homodyne receivers, 390n23, 560, 561, 579
hot-carrier theory, 339
Houck, Henry W., 758n36
HSPICE, 423
Hughes, David Edward, 34, 37
human-body model (HBM), for ESD testing, 162
hybrid parameters, 225
hybrid ring, 729n22
hyperabrupt junctions, 130n8

IC components
  passive, 114–66
  problem set for, 163–6
IC technology, transistors in, 370
IC varactors, 130
ideal diode law, 315
IEEE 802.11 (WiFi), 52–4
  subchannel structure for, 732f
IEEE 802.15.4 (ZigBee), parameters of, 55t
IF, *see* intermediate frequency (IF) signal
IIP2
  estimation of, 402
  IIP3 relation to, 763
IIP3, 408
  estimation of, 395–6, 397–8, 399, 401, 439, 757
  of narrowband LNA, 394–5
IIV3 intercept, 697
image-reject filter, 736–7
  polyphase, 708f, 709f
  of single-conversion receiver, 698
image-reject mixer, 700f, 703
image-reject phase-locked loop (IR PLL), 742,
  743f
image-reject receiver
  architecture for, 700–10
  gain and phase errors in, 701–3
  quadrature generators of, 703–10
image rejection, measurement of, 749f
image-rejection problem
  of single-conversion receiver, 698
  solution of, 748

image-rejection ratio (IRR)
    augmentation of, 736
    definition of, 702, 735
    mismatch degradation of, 736
    problem for, 763
image resistance, 96
image signal
    of mixers, 406
    of single-conversion receiver, 698
IM distortion, third-order, 516
impedances
    characteristic (Z_0), 206
    driving point, 215n8
    of lossy transmission line, 207–8
    of RLC network, 87, 93–102
    Smith chart for data on, 222–3
    standardized (50 or 75 ohms), 229
    transformation of, 309, 375f
    transient (pulse) type
impedance transformation ratio, of double-tapped
        resonator, 103
Improved Mobile Telephone Service (IMTS), 47
impulse response, moments of, 259, 260f
impulse sensitivity function (ISF), 669, 671, 675, 689
    for LC oscillator, 670, 676
    for ring oscillator, 685
    for shot noise, 680, 681
impulsive energy restortion, 676
IM3 power, 398, 422, 512, 518
incandescent light bulb, 9, 23
Independent Radio Manufacturers Association
        (IRMA), 16
indoor operation, of GPS, 717n17
inductance
    formulas for, 146–8
    lumped, 162, 211
    propagation delay from, 210
inductive degeneration, in mixers, 418
inductive source degeneration method, 376, 385
inductors
    active, 361f
    formulas for, 140f
    gyrator-based, 721
    lossiness of, 92
    in passive IC networks, 136–48, 165
    in passive RLC networks, 87
Industrial Development Engineering Associates
        (IDEA), 771
industrial-scientific-medical (ISM) bands, 80n58, 84
information theory, 44
injection-locked divider, 745, 746f
injection-locked oscillators (ILOs), 561, 564
    nomenclature for, 564n8
injection locking
    in phase-locked loops, 563–6
    of power amplifiers, 532
    unwanted, 563n7
INMARSAT link band, 726
input impedance
    in LNA design, 364
    in vacuum tubes and MOSFETs, 375
input intercept, calculation of, 696
input–output impedances, of shunt-series amplifiers,
        284–6

input–output transfer function, 442
input resistance, calulation for, 241–2
input-to-error transfer function, 464
insertion power gain, 229
Institute of Electrical and Electronics Engineers
        (IEEE), 228, 751
integer-N frequency synthesizer, 649f
integrated circuits, *see* IC
intercept point, third-order (IP3), *see* IP3 intercept
interconnect
    capacitance of, 130–6
    problem on, 164
    properties of, 114, 231
    at radio frequencies, 114–20
    as transmission line, 204
    use at high frequencies, 158–62
interleaved transformer, 153
intermediate frequency (IF) signal, 14–15, 16n21, 405,
        406, 694
    filter architecture, 721f, 725
    mixing down to, 404, 431
intermediate resistance, 96
intermodulation (IM) product, 391, 412
International Business Machines (IBM), 772
inversion layer, induction in MOSFETs, 170
ion implantation, source–drain diffusions defined by,
        121
IP3 intercept, 408, 518, 536
    of Gilbert-type mixers, 408, 422
    in linearity measurement, 723, 725
    simulation of, 398–9, 422
iron pyrites, in crystal detectors, 5–6
ISF, *see* impulse sensitivity function
IS-54 TDMA cellular system, 76
IS-54/IS-136, parameters for, 50t
IS-95 CDMA, 50
    parameters of, 51t
isolation, as mixer parameter, 407–9
iterated structures, driving-point impedance of, 205

Jasberg, J. H., 304
Jell-O™ transistors, 292
JFETs
    describing function transconductance for, 615n2,
        616
    n-channel use in, 769
    from vacuum tube circuits, 765n4
jitter noise, of phase-locked loops, 573, 586
jitter peaking, in second-order phase-locked loops,
        570–1
Johnson, J. B., 334
Johnson noise, 335
JPEG algorithm, 57
junctions, flicker noise in, 347
junction capacitance, in electronically tuned circuits,
        129
junction capacitors, 130, 164
junction FET (JFET), 168–9
junction transistor, 168

Kahn, Leonard, 520
Kahn EER system, 520f
Kahng, D., 168
KAPPA parameter, in MOSFET modeling, 191

Kennedy, John F., 39, 404n2
keying, as radio lexicon, 42
Kirchhoff's current law (KCL), 203, 204, 710
Kirchhoff's laws, 202
Kirchhoff's voltage law (KVL), 115, 116, 203, 204
Koch curve, 125
Koch islands, 125
    fractal based on, 126
Ku and Ka designations, for radar bands, 84

Lackawanna Ferry, 445
ladder networks, 206f, 266f
    as ideal transmission lines, 206
lag compensation, in feedback systems, 481–3
Lamarr, Hedy, 53n31
Langmuir, Irving, 13
Laplace transforms, 211n5, 483, 572
large-signal excitation, of coherers, 3
large-signal impedance matching, in power amplifiers, 551
lateral flux capacitor, 124f, 127
Latour, Marius, 17
Lawrence, Ernest O., 7n14
LC circuits, series resonance of, 736
LC oscillator
    current-pulse excitation of, 668
    phase noise from, 678–82
    waveforms for, 670
LC tank, impulse responses of, 669
lead compensation, in feedback systems, 484–6
leakage inductance, 151
leaky peak detector, for AM demodulation, 62f
LEDs
    discovery of, 20
    GaN blue type, 1, 9, 14
Leeson model, for phase noise, 666, 667
Lévy, Lucien, 404n2
Lewis, Tom, 22
light bulb, invention of, 9, 23
lightly doped drain (LDD)
    in drain engineering, 195
    in MOS devices, 187
Lilienfeld, Julius, 167, 168
linear amplification and nonlinear components
    (LINC), in power amplifiers, 521
linear-envelope operation, of power amplifiers, 520
linearity
    of cascaded systems, 696
    in circuit analysis, 404
    in low-noise amplifiers, 390
    of mixers, 407–9
    parameters of, 409
    role in phase noise, 667–78
linearization, of power amplifiers, 515–28
linear mixers, 411–16
linear oscillators, 610–11
linear time-invariant (LTI) noise analysis, 72n49
linear time-invariant (LTI) theories, 660, 661, 667,
    669, 674, 676, 679, 689
linear time-varying (LTV) noise model, 674, 677, 679,
    681, 758
Lippman, Gabriel, 632n14
liquid barretter, 8
lithium niobate, use in resonators and filters, 635

L-match
    attempt, 104, 105
    in cascade with parallel tank, 432
    series inductance of, 288
L-match circuit(s), 94–6, 108, 109, 544
    π-match circuit as cascade of, 96–8, 104, 105–6,
        108
LM309 5-V regulator IC, 318
LNA, see low-noise amplifier
L-network, shunt capacitance of, 288
load-pull characterization, of power amplifiers, 551–3
local oscillator (LO)
    in direct-conversion receiver, 711
    frequency change in, 404, 699, 714
    harmonics of, 410, 427n32
    impure spectrum of, 659
    leakage measurement for, 755
    phase noise from, 659–60, 726
    power transfer from, 435n42
    radiation from, 713
    RF signal mixing with, 417n23, 660
    signals in mixers, 428
    stability of, 715
Lodge, Oliver, 2, 7, 34–5, 36, 37, 39
Loebner, E., 22
loktals, 31
long-channel MOSFETs, 169, 179–80, 184, 326, 414
Loomis, Mahlon, 32–3, 37
loop architecture, phase-locked, 566f
loop-back architecture, 754
loop filters
    in phase-locked loops, 588–96
    schematic of, 747f
loop transmission
    calculation of, 450n9, 481
    clutches and, 461–2
    feedback system for, 488f
Lorentzian spectrum, 677f
Losev, Oleg I., 1, 22
Losev crystadyne receiver, 20–1
lossless algorithms, 57n33
lossy algorithms, 57n33
lossy transmission lines, lumped model for, 207
low-frequency gain, of shunt-series amplifiers,
    284–6
low-headroom cascade, 330f, 331f
low-noise amplifier (LNA)
    design of, 364–403
    differential type, 387–90
    for direct-conversion receiver, 713
    energy re-entrance through, 752
    in filter combination, 741f
    function of, 364
    in GPS architecture, 718, 727
    IIP2 of, 712
    image-reject type, 738f
    with inductive source degeneration, 378f
    linearity and large-signal performance in, 390–7
    linearity with short-channel MOSFET, 395–7
    mixer noise and, 421
    mixer schematic of, 756f
    noise of, 356, 357, 362, 400
    noise figure plot for, 719f
    notch filter in, 745

power match vs. noise match in, 373–80
problem set for, 400–3
shunt-series broadband for, 401
single-ended type, 384–7, 400f, 401
spurious-free dynamic range in, 397–9, 401
with tracking notch filter, 736–42
transfer function of, 738, 739f
low-pass filter (LPF), noise bandwidth of, 337
LSB component, of AM, 67
LTV, *see* linear time-varying (LTV) noise model
lumped delay line, 215f
lumped lines
    cutoff frequency of, 215–16
    termination of, 216–17
lumped model, for lossy transmission line, 207
lumped *RLC* model, of transmission-line segment
lumped systems, 202–20

"magic box," in phase noise, 693
magnesium getter, in vacuum tube, 24–5
magnetizing inductance, 151
Marconi, Guglielmo, 2, 3–4, 7, 10, 22, 41
    "four sevens" patent of, 7n12
    inventions of, 37, 38–9
Marconi's coherer, 4f
Marconi Wireless Telegraph Corporation of America, 38
Margarit, M. A., 679, 680f, 682
Massachusetts Institute of Technology (MIT), 44n11, 234
Massobrio, G., 183, 198
maximum power transfer theorem, 94
Maxwell, James Clerk, 1–2, 34, 441
Maxwell's equations, 34, 135, 136, 202–4, 211n6
McCandless, H. W., 11
McCullough, Frederick, 24n41
Medhurst, R. G., 147
medium access control (MAC), of IEEE 802.11, 732
Meijs, N. v.d., 132
Meijs equation, 132
Meijs–Fokkema (MF) equation and method, 132, 133t, 134
MESFETs, describing functions for, 615–16
Meta-Software Corporation, 423n26
"A Method for the Determination of the Transfer
    Function of Electronic Circuits" (Cochrun &
    Grabel), 259
MF equation, 132, 133t, 134
micro-arcs, from carbon resistors, 345
microhenries, inductance given in, 147
microstrip lines, 220
microwave, band designations for, 86t
microwave circuits and systems, design of, 221
microwave oven
    bands for, 80
    and signal attenuation, 80
microwave radiation, in universe, 338n10
microwave receivers, noise from, 338n19
Miller, John M., 244
Miller effect, 29, 248, 249, 283, 286, 287, 294, 386, 387, 447, 537, 654, 770
minimum-shift keying (MSK), 75
Minkowski sausage, fractal capacitor based on, 127
mirror galvanometer, in Fleming valve, 10

mirrors, in PMOS, 317–18
mixers, 364, 404–40
    architecture of, 743
    complex, 700
    conversion gain of, 406, 416, 440
    degeneration in, 418
    diode-ring mixers, 434–7
    in direct-conversion receiver, 711
    double-balanced, 417, 419–27, 436–7, 439, 744f
    fundamentals of, 405–11
    Gilbert, 417n22, 420–7
    history of, 766–8
    IP3 of, 408, 422
    isolation in, 407–9
    linear, nonlinear systems as, 411–16
    linearity of, 407–9
    linearization techniques for, 423–7
    with linearized transconductance, 419f
    multiplier-based, 416–33
    noise in, 756
    potentiometric mixers, 427–8, 714
    problem set for, 437–40
    RF transconductors for, 418f
    single-balanced, 417–19, 440
    square-law mixer, 413–16
    for subsampling, 433–4, 714
    three-port, 412n13
    transmit type, 757f
    two-port, 413–16
mixing function, definition for, 430
mobile telephones, capacity of, 45
Mobile Telephone Service (MTS), 45–7, 46f
mobile wireless, bands for, 83
mobility degradation
    compensation for, 723f
    in MOS devices, 189–90
modems, early, 74
modulation, mathematical study of, 42
modulation index, 70, 72
monomial formula, for inductors, 139, 140f
Monte Carlo simulation techniques, 710
Moore's law, 83
MOS capacitor, in CMOS processes, 128
MOS devices
    breakdown phenomena in, 548
    describing functions for, 614–17
    history of, 167–8
    mobility degradation in, 189–90
    noise model for, 358f
    physics of, 167–201
    problem set for, 198–201
    substrate current in, 187
    temperature variation in, 189
    thermal runaway in, 550
    threshold reduction in, 187–8, 190
MOS differential oscillators, ISF equation for, 686
MOSFETs
    accumulation MOSFETs, 128–9
    back-gate bias in, 188–9
    capacitances of, 172–6
    capacitive input impedances in, 375
    capacitors, 176
    cascomp, 426f
    channel-length modulation in, 173

MOSFETs (cont.)
  comprehensive model for, 238
  cross-quad, 424f
  describing functions for, 614–17
  diode-connected, 318
  drain current in, 171–3, 361
  dynamic elements of, 172–6, 179
  flicker noise in, 346
  gate structure of, 169, 170, 375
  high-frequency performance of, 176–9
  input impedance in, 373
  inversion layer in, 170
  long-channel MOSFETs, 169, 179–80, 184, 198,
      326, 414, 615n2, 616
  n-channel MOSFETs, 170, 171f
  noise model of, 365, 366, 370
  open-circuit time constants for, 238–46
  operation in weak inversion, 180–8
  oscillators in, 614–17
  physics of, 169–80
  power gain of, 169
  quad operation in, 318
  in short-channel regime, 183–7, 378, 395, 415, 512,
      615n2, 615, 748
  shot noise in, 343
  SPICE modeling of, 183, 190–4
  square-law behavior of, 723
  subthreshold model equations for, 182–3
  as switches, 159–61, 325
  terminal capacitances of, 177t
  thermal noise in, 179, 339–42
  transconductance of, 395, 615n2
  transit-time effects in, 178
  triode region in, 427, 500n9, 722
  two-port noise parameters of, 368t
MOS11 model (Philips), 372, 373
MOS transistor
  describing functions for, 614–17
  use as resistor, 121
"motorboating," low-frequency oscillation as, 490
Motorola, 44, 47
  waveguide bands of, 86
mouse, see computer mouse
MPEG algorithm, 57
Mth-order input intercept voltage, 696
multipath propagation, signal impairment from, 80
multipliers
  commutating, 576–7
  mixers based on, 416–33
  as phase detectors, 575f, 576f, 578
multiplying digital-to-analog converter (MDAC),
      654
multi-tanh transconductor, 440
Murphy's law, 544
mutual conductance, calculation of, 158
mutual inductance, 150

Nakamura, Shuji, 9n19
Narda, waveguide bands of, 86
narrowband amplifiers, 400
  design of, 270
narrowband passive mixer, 431f
narrowband PM, 71f
NASA, waveguide bands of, 86

National Bureau of Standards, 244
National Semiconductor Corp., 318
National Television Standards Committee (NYSC),
      562
n-channel junction FET, 169
n-channel MOSFET, 169, 170
near-far problem, in CDMA systems, 51
near-zero IF (NZIF) architecture, 709
negative feedback
  for modulation linearity, 516f
  noise and, 448–50
negative feedback amplifier, 444–5
negative feedback systems
  desensitivity of, 446–50
  disconnected, 451f
negative impedance converter (NIC), 641
negative resistance behavior, in triodes, 29n43, 30
negative resistance oscillators, 641–2, 657, 683
neutralization, 294–7
  as bandwidth improvement, 253
neutrodyne amplifier, 15–17
neutrodyne kits, 16
NIC, see negative impedance converter
Nichia Chemical, 9n19
nichrome resistors, 121
Nielsen's equation, 359
NMOS devices
  channel charge density in, 170
  current mirrors of, 198, 318
  hole–electron pairs in, 187
  mirrors in, 595n20
  noise in, 756–7
  upward threshold shifts in, 187
NMOS transistors, 246, 326
  drain current of, 198, 333
Nobel Prize winnners, 4n7, 168, 338n10
Noe, J. D., 304
noise, 334–63
  atmospheric, 770
  bandwidth for, 336
  calculations of, 352–4
  cyclostationary sources of, 420, 661, 674, 675–6
  definition of, 334
  drain current noise, 339–42
  flicker noise, 344–7
  gate noise, 341–2
  jitter noise, 573, 576
  models of, 357–8
  negative feedback and, 448–50
  optimization limitations of, 351
  optimization method for, 383
  parameters for, 364
  phase noise, see phase noise
  pink noise, 344
  popcorn noise, 347–8
  power-constrained optimization of, 380–4
  problem set for, 358–63
  role in wireless systems, 43–4
  shot noise, 342–4
  thermal noise, 339–42
  two-port theory of, 348–52
  typical performance of, 356–7
  unit for, 228
noise bandwidth, 335

noise factor, 348–50, 352t, 402
  derivation of, 695–6
  of Gilbert-type mixers, 420–7
noise figure (NF), 351, 352f, 397
  of cascaded systems, 695–6, 748
  degradation source for, 387
  derivation of, 364
  of GPS receiver, 726
  of mixers, 406–7
  of receivers, 717, 748f
  in subsampling architecture, 714
noise modulation function, 674–5
noise power, spectral density of, 335
noise temperature, 351–2, 352f
noise-to-carrier ratio, 662
noise-to-phase transfer function, 668
nonquasistatic (NQS) effects, in MOSFETs, 178–9,
  372
Nordic Telephone System (NMT-450), 48
Norris, Jerry, 773, 775
North American Digital Cellular (NADC), 49
Norton equivalent model, noise in, 336
notch filters, automatic tuning of, 737
NTSC color television, hue encoding in, 306–7
Nuvistor, as apex of vacuum tube evolution, 32
Nyquist, Harry, 43, 334
Nyquist noise, 335
Nyquist sampling criterion, 302, 433, 713
Nyquist stability test, 451, 452

Occam's razor, "duck test" version of, 203n1
octagonal spirals, 136, 139t
offset oscillator, 658
offset quadrature phase-shift keying (OPQSK), 51,
  76
offset synthesizer, 652, 763
Ohm's law, 115, 335
OIP3 intercept, 408
on-chip detector samples, 755
on-chip energy storage elements, 694
on-chip inductors, 136–44, 682
on–off keying (OOK), in wireless, 42, 64
op-amp circuit, in feedback systems, 463f, 492,
  641
OP-27 device (Analog Devices), noise performance
  of, 356
open-circuit time constants (OCτs) method
  accuracy of, 236–7
  advantages of, 259
  all-poly transfer function in, 252
  application of, 237
  for bandwidth estimation, 234–54
  design example for, 246–53
  for high-bandwidth amplifiers, 279, 290
  incremental model for, 239f
  risetime addition and, 263
  summary of, 253
  superbuffer for, 269f
open-loop system
  with additive noise sources, 450f
  with distortion, 465f, 466f
  in synthesizers, 653
optical fiber system, gain–delay tradeoff in, 303
optical phonons, scattering by, 183

optimum source admittance, 369
OP-215 JFET, noise performance of, 356
orthogonal frequency division multiplexing (OFDM),
  54, 76–7
  in European HiperLAN2, 731, 744
  peak-to-average ratio of, 760
oscillations
  frequency of, 661
  low-frequency, as "motorboating," 490
  start-up of, 623–31
oscillator(s), 13, 408n9, 441, 610–58
  amplitude of, 610, 631f, 669, 775
  Clapp oscillator, 636, 643, 644, 654, 655, 679
  Colpitts crystal oscillator, 635, 638f
  Colpitts oscillator, 620–31, 635–6, 642, 654, 655,
    656, 661, 675–6, 678, 690f, 691, 774
  comparator with, 690
  crystal oscillators, 632–4
  describing functions for, 610, 611–31
  differential implementation of, 679
  frequency synthesis in, 645–51
  function generator, 617–20
  function loop model for, 618f
  gain phase plots for, 619f
  Hartley oscillator, 620, 636, 654
  history of, 766–8
  injection-locked, 563; see also injection-locked
    oscillator (ILO)
  LC feedback type, 635–8
  linear, 610–11
  local, see local oscillator
  negative resistance oscillators, 641–5, 657
  noise from, 660, 678–87
  offset type, 658
  operating regimes for, 658, 663f
  output spectrum of, 664
  phase-locked loops in, 595–6, 645, 683, 704
  phase noise in, 565–6, 637
  Pierce crystal oscillator, 639–40, 642
  problem set for, 655–8
  quadrature oscillator, 640–1, 704
  quartz-controlled, 562n5, 632–4, 645
  ring oscillators, 595–6, 687, 704
  RLC oscillator, 638, 693f
  root locus for, 610, 611f
  stability of, 715–16
  superregenerative amplifier as, 305
  TITO oscillator, 637–8, 654
  tuned oscillators, 635–40, 659
  vacuum tubes in, 441
  voltage-controlled, 65, 758f
oscilloscope
  phase distortion in, 274
  probe with, 277f, 278f
  vacuum tube distributed amplifiers in, 305
oscilloscope deflection amplifier, 274
outphasing modulator, for power amplifiers, 522f
output compression point, 408n8
output resistance, calculation for, 240, 242
output third-order intercept point (OIP3), 408
output tuning inductance, computation of, 385
overlap capacitances, 174
oxide cathodes, 24–5
oxygen, effect on wireless signals, 79

packet switching, in wireline systems, 52
parallel plate capacitors, 122
parallel resonant tank circuit, problem on, 164
parallel *RLC* tank
    admittance of, 87
    bandwidth of, 90–2
    branch currents of, 90
    characteristic impedance of, 89–90
    Q parameter of, 88–90
    resonant frequency of, 88
    ringing in, 91–2
parametric amplifier, 435n42
parametric converters (amplifiers), 406n5
parasitic bipolar devices, 314, 328
parasitic capacitances and inductances
    effect on amplifier performance, 307
    neutralization of, 742n28
parasitics, reduction of, 204
parasitic transistors, bias circuits from, 314
Parseval's theorem, 673
passive double-balanced mixers, 427–33
passive IC components, 114–66
    capacitors, 122–36
    inductors, 136–48
    interconnect, 114–20, 158–62
    problem set for, 163–6
    resistors, 120–2
    transformers, 148–58
passive ring mixer, 743f
passive *RLC* networks, 87–113
    problem set for, 107–13
path transfer function, derivation of, 436
patterned ground shield (PGS), in spiral inductors,
    143–4
P code, for GPS signal, 716, 717
peaking problem, in phase-locked loops, 571n11
pentode, characteristics of, 30f
Penzias, Arno, 338n10
Percival, W. S., 304
periodic steady-state (PSS) analysis, 399
perveance, 26
peta (prefix P), 228
phase-alternating line (PAL) system, 306–7, 562
phase detector, 574–9
    analog multiplier as, 574–6
    commutating multiplier as, 576–7
    exclusive-OR gate as, 577–9
    with extended range, 582f
    frequency detector vs., 563
    Hogge's phase detector, 584, 585, 586
    in phase-locked loop circuits, 566–7, 574–604
    sequential, 579–88
    set–reset (SR) flip-flop type, 580–1
    for spur minimization, 745
    triwave type, 586f
    XOR type, 579
phase detector constant, 578
phase detector gain, 579
phase detector I, 596
phase detector II, 597, 599
phase distortion, errors caused by, 274
phase errors, 580
    in AM–PM conversion, 306–7
phase-frequency detectors, 583, 745f

phase-locked loops (PLLs), 464, 490, 560–609
    with active loop filter, 603f
    for AM, 65, 560
    architecture of, 566f
    closed-loop phase transfer for, 571f
    control-line ripple in, 592–6
    design examples for, 596–604
    first-order type, 567–8
    history of, 560–6
    injection locking in, 563–6
    linearized, 566–7, 572f
    loop bandwidth of, 648n29
    loop filters in, 588–96
    noise problems of, 571–4
    noise rejection on input, 573–4
    noisy, 606, 607
    with offset, 605
    peaking problem in, 571n11
    problem set for, 604–9
    quartz-controlled oscillator combined with, 632–4,
        638, 654–5
    ring oscillators in, 595–6, 683, 704
    as sampled-data systems, 650
    second-order type, 569–71, 599, 604, 606
    spectra from, 759f
    as synthesizers, 647, 648, 649f
    uses of, 560
    voltage-controlled oscillators in, 566–8, 572, 574,
        582, 589, 608, 648, 742
phase locking, 565
phase margin, computation of, 452, 593, 599
phase modulation (PM)
    AM conversion to, 306–7
    angle modulations of, 70–3
phase noise, 565n10, 573, 655, 659–93
    amplitude response as, 687–8
    from circuit noise, 673f
    close-in, 677–8
    equation for, 665
    of ideal oscillator, 637
    Leeson model for, 666, 667
    linear time-invariant analysis of, 667–78
    prediction of, 679
    problem set for, 690–3
    in ring oscillators, 683
    of synthesizer, 748
    theory of, 594n18
    VCO production of, 758
phase shifting, 274, 702
phase-shift keying (PSK), 74, 75
    detection for, 77
phase transfer function, derivation of, 569
phasor representation, for PM, 71f
phasors
    for AM representation, 63–4
    for noise contributions, 684
Philips Electronics N.W., 372, 373n6
Phillips, V., 22
phone-line data modem, 58
phosphor, designation for, 32
phosphorus getter, in vacuum tube, 24–5
photodetectors, in computer mouse, 707
photoelectric effect, Einstein's, 20
photoemitters, in computer mouse, 707

photophone, Bell's invention of, 34, 39
"A Physical Model for Planar Spiral Inductors on
    Silicon" (Yue et al.), 157
Pickard, Greenleaf Whittier, 5–6
pico (prefix p), 228
Pierce, George Washington, 639n22
Pierce crystal oscillator, 639–40, 642, 655
piezoelectricity
    Curies' studies on, 632n14
    of quartz crystals, 632–3
pilot signal, in stereo FM radio, 563
π-match circuit, 96–8, 104, 105–6, 108
    with transformed right-hand L-section, 97
pink noise, 344; see also flicker noise
$\pi/4$-QPSK, 76
planar spiral inductors, 135–44, 162n32
planar transformers, analytical models for, 155–8
planar triode, 27f
Planck's constant, 338
PLLs, see phase-locked loops
PMOS devices
    folded cascade version, 756
    noise in, 346
    threshold reductions in, 187
PMOS mirror, 317–18, 325, 329, 595
PMOS–NMOS width ratio, selection of, 758
PMOS switches, 160
p–n–p parasitic substrate, in CMOS, 316f
Poisson's equation, 25
polar feedback, in power amplifiers, 524–6
polarization diversity, in satellite communications, 82–3
polar modulation, 66
    symbol constellation for, 59–60
pole-zero doublet, 277, 278f
police radio receivers, 44
poly emitters, 120
polyphase
    for complex signal processing, 735
    derivation of term, 708n7
polyphase filter, 705, 706f, 708
    design of, 710
polysilicon ("poly") resistors, 120, 165
popcorn noise, 347–8
Popov, Alexander, 35–6, 37, 39
port, variable definitions for, 225f, 226f
portable radio, history of, 21, 771n13
portable wireless, bands for, 83
positive feedback amplifiers, 446
positive feedback systems, root-locus rules for, 454–8
postdistortion, in power amplifiers, 518–20
potentiometric mixers, 427–8
Poulsen, Valdemar, 7, 14
Poulsen Tikker, 14
power
    consumption in LNA design, 364
    DC transmission of, 9
    handled by coaxial cable, 229–30
    published gain figures for, 414n18
    units for, 227–9
power-added efficiency (PAE), of power amplifiers,
    547
power amplifiers (PAs), 76, 493–559
    adoptive bias in, 528
    breakdown phenomena in, 548–50

    Cartesian feedback in, 526–8
    cascading of, 297–306, 531–2
    Class A, 494–7, 512–39, 528, 541–5, 553–5, 555,
        556, 559
    Class AB, 494, 502–3, 512–39, 545, 558, 559
    Class B, 494, 497–9, 512–39, 530, 543, 545
    Class C, 494, 499–502, 520, 528, 541, 545, 558
    Class D, 503–5, 541
    Class E, 505–7, 512–39, 541, 546–7
    Class F, 507–10, 512–39, 541
    design examples for, 541–7
    design summary for, 555
    Doherty and Terman–Woodyard composite, 528–30
    efficiency boosting of, 528–30
    envelope elimination and restoration in, 520–1
    envelope feedback for, 515–17
    feedforward, 517–18
    injection locking of, 532
    instability, 548
    inverse Class F, 510–11
    large-signal impedance matching in, 551
    linearization of, 515–28, 557
    load-pull characteristics of, 556
    load-pull contour example of, 551–3
    model for, 494f
    modulation of, 512–39
    outphasing and LINC of, 521–4
    output model for, 557f
    performance metrics for, 536–9
    polar feedback in, 524–6
    power-added efficiency of, 547
    power boost by combining of, 532–6
    pre- and postdistortion in, 518–20
    problem set for, 555–9
    pulsewidth modulation of, 531
    push–pull amplifiers, 558
    summary of characteristics, 540–1
    thermal runaway in, 550
power control, essentiality for CDMA, 51n27
power gain
    available, 229
    definitions of, 228–9
    from insertion, 229
    published figures for, 414n18
    of transducers, 229
Practical Wireless Telegraphy (Bucher), 40
predistortion, in power amplifiers, 518–20
Preece, William, 39n59
preselector, of single-conversion receiver, 698
problem sets
    for architectures, 762–3
    for bandwidth estimation, 266–9
    for biasing and voltage references, 331–3
    for distributed systems, 218–20
    for feedback systems, 488–92
    for high-frequency amplifier design, 308–13
    for IC components, 163–6
    for low-noise amplifiers, 400–3
    for mixers, 437–40
    for MOS devices, 198–201
    for noise, 358–63
    for oscillators, 655–8
    for passive RLC networks, 107–13
    for phase-locked loops, 604–9

problem sets *(cont.)*
for phase noise, 690–3
for power amplifiers, 555–9
for Smith chart and S-parameters, 231–2
processing gain, in GPS signal, 717
production beam power tube, 30n44
propagation constant (γ), 207, 208–9, 214
relationship to line parameters, 209–11
proportional to absolute temperature (PTAT)
of resistance, 121
voltage as, 318, 319, 321, 322, 325
proximity effect, 119, 141
Public Broadcasting System (PBS), 22
pulsewidth modulation (PWM), of power amplifiers, 531
Pupin, Michael, 211
Pupin coils, 211
push–pull amplifiers, 558

quadrature, qualilty of, 755, 756f
quadrature amplitude modulation (QAM), 54, 60, 66, 538, 732
quadrature generators, for image-reject receivers, 703–10
quadrature mixers, 735
quadrature oscillators, 640–1
quadrature phase-shift keying (QPSK), 50, 75, 732
offset modification of, 76
symbols for, 76
quadrature Weaver architecture, 736
Qualcomm, 50
quality factor (Q)
alternate formula for, 111
bandwidth and, 90–2
of double-tapped resonator, 103
of gate capacitors, 129
of inductors, 136
of oscillators, 622n7, 659
of parallel RLC tank, 88–90
of quartz, 633
of resonators, 631, 661
in RF applications, 126–7
ringing and, 91–2
quantizers, 718n18
quantum effect, in carborundum LEDs, 20
quarter-wave resonators, 631–5
quartz crystals
fabrication of, 634
in oscillators, 632–4, 638, 654–5
piezoelectricity of, 632n14
as resonators, 562n5, 632–4
symbol and model for, 633
quasiperiodic steady-state (QPSS) analysis, 399
quaternary modulation, 60–1
quench oscillator, in early superregenerative receiver, 767
Quinn, Pat, 425

radar
classification systems for, 85
superheterodyne use in, 21
radio
history of, 1–39
invention of, 32–9

radioactivity, discovery of, 10
radio amateurs ("hams"), 42
radio astronomy, sideband information role in, 407n7
Radio Corporation of America (RCA), 15, 19, 21, 30, 32, 44, 303, 521, 768
Radio Day (Russia), 36
radio electronics, developments in, 15
radio frequency, *see* RF
radiotelegraphy, 3, 22
transition to radiotelephony, 8, 40
radiotelephony, 8
Radiotelephony for Amateurs (Ballantine), 40
rainfall, effect on wireless signals, 79
random telegraph signals (RTS), 347
Rategh, Hamid, 565
ratio detectors, 73
Rayleigh distribution, 80, 81f
Raytheon
Radarange, 80n58
waveguide bands of, 86
RC diffusion line, 219
reactance tubes, *see* gyrators
read-only-memory (ROM) lookup table, 653, 654, 716
receive mode, in walkie-talkies, 774
receiver
architectures for, 697, 733–4
with coherer, 3f
direct-conversion type, 700
image-reject type, 700–10
implementation of, 734–6
single-conversion type, 698–9
for subsampling, 713–14
superregenerative, 767f
receive sensitivity, 760
reciprocal mixing, minimization of, 659
rectifiers, cuprous oxide in, 168n2
rectifying detectors, 8
reflection coefficient, 213, 221–2, 226
admittance relation to, 362
reflex circuits, 17–18
reflex receiver, 17f
Regency TR-1 transistor radio
history of, 771–3
schematic of, 772f
regeneration, Armstrong's patent on, 21
regenerative amplifier/detector/oscillator, invention of, 13–15, 441–2, 766
regenerative radios, de Forest's sale of, 17
regenerative receiver, 12, 15
request to send (RTS) message, from WLAN mode, 733
resistances, for open-circuit time constants calculations, 244–53
resistive attenuator, noise figure of, 358–9
resistive network, noisy, 353f
resistors
of fixed linear degeneration, 722
flicker noise in, 345–6
negative, 301, 302
thermal noise in, 336f, 337f, 357f
resonant circuits, amplifier function in, 108
resonant frequency, of parallel RLC tank, 88
resonant RLC networks, 92–3

resonators, 631–5, 659
  distributed type, 631
  in oscillators, 620
  quarter-wave resonators, 631–5
  quartz-crystal resonators, 632–4
  SAW devices, 635
RETMA, 32
return difference, 447n8
RF, band designations for, 85t
RF carriers, sine-wave type, 7
RF circuits and systems, 190, 221, 403
  history of, 15–17, 22, 764–75
  inductors in, 136–48
  noise in, 372
  passive components of, 15–17
  quartz crystals in, 632–3
  RLC networks in, 87
RF ID tags, bands for, 80
RF negative resistance, 642f
RF power amplifiers, 493–559
RF rectifiers, 10
RF signals, 17
  demodulation of, 8
  mixer translation of, 404
  in reciprocal mixing, 659
RF switch, high-linearity, 160f
RF-to-baseband performance, 759–60
RF transconductors, for mixers, 420f
ring hybrid, 729n22
ring mixer, passive, 743, 751
ring oscillators, 595–6, 687, 704
  phase noise in, 683
  waveforms for, 670
ripple components, in phase-locked loops, 592–6, 655
risetime
  addition rule for, 263
  bandwidth relations with, 259–65, 267
  of cascade systems, 261–3
risetime addition rule, 263
RLC networks, 111f, 271
  electrostatic discharge in, 161
  impedance of, 87, 93–102, 232, 272
  passive components in, see passive RLC networks
  phase noise in, 690
  in series, 92
RLC oscillator, 638, 693f
  "perfectly efficient," 663f
RLC tank circuits, 87–113, 621
  admittance of, 294
  parallel, 87–92
  in series, 92
Rogers, Edward S., 24n41
Röntgen, Wilhelm Konrad, 10
root-locus techniques, for feedback systems, 453–8
Rosenstark, S., 259
Round, Henry J., 8–9, 14, 20
Russia, 35
Russian Physical and Chemical Society, 35

Sakurai equation and method, 132, 133t, 134, 135, 136, 164
sample-and-hold action, in synthesizers, 646f
sample-and-hold circuit, 362f
Sarnoff, David, 15, 19, 21, 22

satellite communications
  phase-locked loops in, 563
  polarization diversity in, 82–3
satellite links, 78
saturating amplifier, describing function for, 612, 613f
saturation, in power amplifiers, 493
Saturn, rings of, 441
SAW, see surface acoustic wave (SAW) devices
sawtooth generators, 561
scaling laws, in CMOS technology, 197–8, 370
scattering parameters, see S-parameters
Schmitt trigger, 613f, 617
Scholten, A. J., 365n1, 369
Schottky, Walter, 334, 344, 404n2
Schottky diode, 5
Schottky noise, 342
scientific instrumentation, phase-locked loops in, 563
screen grid, 29, 31
SCR-536, as AM transceiver, 44
Searle, C. L., 259
second-order intercept, for linearity measurement, 390n23, 696, 752n33
second-order phase-locked loops, 569–71
  jitter peaking in, 570–1
seconds, number in year, 345n21
selenium, photosensitivity of, 34
self-inductance, calculation of, 158
self-restoring device, barretter as, 8
semiconductive amplifier, 168
semiconductor, early types of, 5, 7, 8, 167, 435n41
semiconductor detector, first patent for, 4n7
Semiconductor Device Modeling with SPICE
  (Massobrio & Antognetti), 183, 198
semi-infinite conductive block, 116f
sensitivity, definition of, 720n19
sequence asymmetry, 708n7
Séquential Couleur avec Mémoire (SECAM), 562n6
sequential phase detectors, 581–3, 590
series-peaked RLC network, 308f
series resistance
  calculation of, 141
  of on-chip spiral inductors, 141
series RLC networks, 92
set-reset (SR) flip-flop, as phase detector, 580–1
Shannon, Claude, 43, 44n11, 56–61
Shannon channel capacity theorem, 77
Shannon logarithmic factor, 58
Shockley, William, 167–8
short-channel MOSFETs, 183–7, 187, 320, 415, 512
short-channel NMOS devices, 339, 342
short-circuit time constants method
  for bandwidth estimation, 254–8
  interpretations of, 256
shot noise, 342–4, 680
shunt and double-series peaking, 280f
shunt capacitance, 142
shunt-peaked amplifiers, design of, 271–5, 290, 308
shunt peaking
  design example for, 275–6
  summary of, 275t
shunt–series amplifiers
  design of, 284–8
  input–output impedances of, 284–6
  low-frequency gain of, 284–6

sichrome resistors, 121
sidebands, of mixers, 406
sideband suppressed-carrier AM (SC-AM), 64, 69
sigma-delta technique, for noise distribution, 651n31
Signal Corps Radio (SCR), 44n14
signals
    attenuation of, 79
    propagation in wireless systems, 77–83
    spectra of, 411n12
signal-to-noise ratio (SNR)
    bit rate increase by, 58, 59
    in CDMA systems, 51
    increased data rate and, 77
signum function (sgn), 576
silicided polysilicon resistors, 120
silicon
    lack of piezoelectric activity in, 635
    Losev's experiments with, 21
    lossiness due to, 158
silicon detectors, early, 5, 9
silicon dioxide, as low-loss material, 159
silicon transistors, development of, 772n17
sine-wave RF carrier, 7
single-balanced mixers, 417–19, 436f, 440f
single-conversion superheterodyne receiver,
    architecture for, 698–9
single-diode mixers, 435
single-ended low-noise amplifier, 384–7, 400f, 401
single-ended oscillators, ISF equation for, 686
single-sideband (SSB) forms
    of AM, 43, 65, 66, 69, 70
    polyphase network for, 706
single-sideband large carrier (SSB-LC), 70
single-sideband noise figure (SSB NF), 407
skin depth, 117, 142, 165
    application to resistance calculation, 118
skin effect, 114–15, 140, 152
Sklodowska [Curie], Marie, 632n14
slope demodulation, of FM, 19
slope detectors, 73
slow rolloff compensation, in feedback systems,
    486–7
small-signal varactor capacitance, 740
Smith, P. H., 224n1
Smith chart, 214, 221–32, 351n29
    problem set for, 231–2
    with series RC example, 224
"snow," as TV noise, 334
solenoid, single-layer type, 147f
solid-state amplifiers, 20–1
"Some Recent Developments in the Audion Receiver"
    (Armstrong), 764
source admittance, optimum, 350–1
source coding, 57
source–drain diffusions, resistors from, 120–1
source followers
    capacitively loaded, 309f
    inductively loaded, 310f
    resistance of, 243
Soviet Union, 20
S-parameters, 221, 225
    problem set for, 231–2
spark-gap detector, of Hertz, 2
spark plug, Hertz's apparatus as, 2

spark telegraphy, transition to carrier radiotelephony,
    40, 42
spectral masks, 538, 539f
spectral regrowth, 538
Spectre, 681
spectrum analyzer, noise density determination by,
    355
Sperry, waveguide bands of, 86
SPICE
    in amplifier studies, 558
    IP3 estimation using, 393–4, 398–9, 422
    use for bandwidth estimation, 236, 248, 249, 250,
        251, 252
    use for frequency response, 220
    use for MOSFET noise, 371
SPICE level-1 models
    for MOSFETs, 189, 191, 195–7, 238
    parameters for, 197t
SPICE level-2 models, for MOSFETs, 182–3, 191
SPICE level-3 models
    deficiencies of, 195
    for MOSFETs, 183, 190–1, 191–5, 199, 308
    parameters in, 194t
spiral inductors, 136–44, 163, 388n21
spot noise
    density of, 336
    of GPS receiver, 727f
spreading gain, in GPS signal, 717
spread spectrum modulation, 44
spurious-free dynamic range (SPDR), of LNAs,
    397–9
spurs
    of mixers, 409–11
    in phase-locked loops, 592
square-law mixer, 413–16, 438
square-law regime, of MOSFET current, 181
squares, counting of, 122
square spiral inductors, 139t
square waves, 577, 578
stacked transformers, 154f, 155f, 156–7
standing wave ratio (SWR), 219
static moduli, synthesizers with, 647–9
static phase error, 580
stereo FM demodulation, phase-locked loops in, 563
stereo FM radio, phase-locked loop circuits in, 563
Stone, John, 39
strontium oxide, in filaments, 24
subcarrier, in color television, 562
subharmonic injection-locked oscillators, 646n26
"Subhistories of the Light-Emitting Diode" (Loebner),
    22
subsampling, architecture for, 713–14
subsampling mixers, 433–4
substrate, losses from, 220
substrate current
    from hot electrons, 187
    in MOS devices, 187
substrate p–n–p, in CMOS technology, 316
"success has many fathers," as Kennedy statement, 39
superbuffer, for OCτ calculation, 269f
superheterodyne AM radio, with heptode, 31
superheterodyne receiver, 17n35, 404, 659, 694, 698f
    block diagram for, 15f, 405f
    invention and use of, 768

oscillator signal for, 560
second harmonic, 758n35
superheterodyne transmitters, 694, 714f
superregenerator, 18–20
superregenerator amplifiers, 218, 301–3, 309
    Armstrong's invention of, 19, 767
superregenerator detector, use in toys, 775n22
supply pushing, by oscillators, 630
suppressor grid, 29, 30
surface acoustic wave (SAW) devices, as resonators, 635
swallow counter, in integer-N synthesizer
sweep oscillators, TV use of, 561
switches
    MOSFETs as, 159–61, 325
    resistance of, 439
    transistors as, 505
synchronous motor, Tesla's invention of, 159–61
synthesizers, 610–58
    combination synthesizers, 651–3
    with dithering moduli, 649–51
    divider "delay" in, 645–7
    integer-N synthesizer, 649
    for LO generation, 759f
    offset frequency synthesizer, 715f
    output spectra of, 591f, 591
    phase-locked loop with, 607, 608f, 648f
    with static moduli, 647–9
    VCO in, 758
syntony, definition of, 35
Syntony and Spark (Aitken), 22, 35n53

Talwalker, N., 160f
tank circuit, 87–92, 102
tank resistance, as noise source, 661
tantalum, in light bulbs, 23
tapped capacitor resonator, as impedance matching network, 99–102, 104, 106, 108
tapped inductor match, 102–3, 104, 106–7
tapped transformer, 153f, 156f, 158
T-coil, bandwidth enhancement by, 281, 282f, 304, 305n13
Teal, Gordon, 772n17
Tektronix, 290, 305n13
Telefunken, 771
telegraphone, 7
telephone
    repeater amplifier in, 12
    transmission lines in, 728, 729f
television, 764
    amplifier development for, 271, 303
    color television, 32
    early amplifiers for, 62
    filter omission in, 63n39
    frequency spans for, 85t
    phase-locked loops in, 561–2
    phosphors for, 32
    "snow" as noise in, 334
    sweep oscillators in, 561
    UHF, 47, 83
Tellegen, Bernard D. H., 721n21
temperature
    effect on junction voltage, 315
    on-chip measurement of, 550

temperature-to-frequency transducer, 633n17
tera (prefix T), 228
Terman, F. E., 530
Terman–Woodyard amplifier, 528–30
termination parasitics, T-match circuits and, 99
Tesla, Nikola, 7, 37, 39
Tesla coil, 37–8, 92n6
tetrode tubes, properties of, 29
Texas Instruments (TI), 771, 772n17, 773
Thackeray, D., 22
thermal noise, 334–42, 354
    with conductance, 179
    density of, 717n16
    epi noise as, 340
    in MOSFETs, 339–42
    with substrate resistance, 340f
thermal runaway, in power amplifiers, 550
Thévenin representation, 336, 430
third-order intercept (IP3)
    estimation methods for, 393–5
    for linearity measurement, 390–1, 392, 696, 697
Thomson, J. J., 9
thorium, in filaments, 24
three-point method, for IP3 estimation, 394
three-port mixers, 412n13
threshold reduction, in MOS devices, 187–8, 190
threshold voltage, increases in, 200
tickler coil, in regenerative amplifier/detector, 13
time constants, 274
time-dependent dielectric breakdown (TDDB), in MOS devices, 548
time-division duplexing (TDD)
    frequency-division duplexing compared to, 54f
    in WLANs, 53
time-division multiple access (TDMA) systems, 34, 49, 712
time-domain simulators, simulation of mixer IP3 with, 422–3
time–frequency duality concept, 43
time invariance, in circuit analysis, 404
time-varying theory, 633n8
TITO oscillator, 637–8, 654
T-match circuits, 98–9, 108
topological routes, for f_T enhancement, 288
toys
    radio-controlled, coherer use in, 4–5, 7n14
    superregenerator use in, 775n22
    see also walkie-talkies
track-and-hold subsampling mixer, 433
transatlantic SSB wireless, 43
transceivers
    block diagram for, 754f
    switches for, 159–60
transconductance, 414, 415
    equations for, 28, 326, 381, 391, 394
    large-signal, 614
    linearization of, 437
    of mixers, 419, 427
    plot for, 724–5
    of short-channel MOSFETs, 395
transconductor
    common-gate type, 438f
    cross-connected, 721
    cross-quad type, 425f, 438

transconductor *(cont.)*
  large-signal type, 614f
  linear, 724f
  multi-tanh, 440
transcontinental telephone service, inauguration of,
    443
transducer, temperature-to-frequency type, 633n17
transducer power gain, 229
transfer function, 273, 297, 431, 478, 490, 709
  derivation of, 647n27
transformers
  ideal, 149f, 152
  lossless, 150f, 151f
  monolithic, 152–5
  in passive *RLC* networks, 87, 148–58
transistors
  cascoding, 403
  describing functions for, 614–17
  invention of, 1
  output resistance of, 625
  parasitic cancellation in, 87
  as power amplifier, 556
  in radios, 771
  shot noise of, 680
  silicon in, 772n17
  as switches, 505, 743
  velocity saturation effect on dynamics, 185–6
transistor amplifiers, open-circuit time constants for,
    238
transistor radio, history of, 771–3
transmission lines, 207–11
  with arbitrary terminations, 212–14
  artificial, 214–18, 219
  attentuation of, 443
  characteristic impedance of, 89n4, 90n5
  cutoff frequency of, 215–16
  equations for, 214
  of finite length, 212–14
  frequency-independent delay of, 211
  ideal, as infinite ladder network, 206
  lossless type, 218, 219
  lossy type, 207–8
  lumped lines, 215–16
  with matched terminations, 212–14
transmit antenna gain factor, 78
transmit mixer, 757f
transmit mode, in walkie-talkies, 774–5
transmit sideband rejection, 755f
transmitter(s)
  architectures for, 697, 714–15
  direct-conversion type, 715f
  EVM of, 761
  of Hertz, 2
  with offset frequency synthesizer, 715, 763f
transponders, bands for, 80
transresistance, 414
Travis discriminator, 73
triangle waves, 577n14
triode, structure and properties of, 13–14, 23, 28f, 764
triode vacuum tube, 9, 27
  invention of, 11, 29
triwave phase detector, 586, 587, 588
tuned amplifiers, 290–4
tuned input–tuned output (TITO) oscillator, 637–8

tuned radio-frequency (TRF) receiver, 15–17
tungsten, in light bulbs, 23
two-diode mixers, 435–6
two-port bandwidth enhancement, of amplifiers,
    279–82
two-port mixers, 412f, 413–16
two-port noise theory, 348–52
  parameters for, 366
two-tone intermodulation, test results for, 749f
two-wire line, 119f
Tyne, Gerald, 11

UHF television, 47, 83
"ultraviolet catastrophe," 338
ultrawideband (UWB), 56
ultrawideband systems, spectrum allocation for, 538
unilaterilization, of amplifiers, 294–7
U.S. Army Signal Corps, 14, 44
units, definitions of, 227–9
"universal" equations, for impedance transformation,
    309
University of California (Berkeley), 7n14
unlicensed national information infrastructure (UNII),
    band for, 84, 85t
unsilicided "poly," resistivity of, 120, 165
upconversion, architecture for, 699–700, 753
USB component, of AM, 67

vacuum diode, 22
vacuum tubes, 410n10
  in AM broadcast radios, 768
  in amplifiers, 305
  Armstrong's work on, 764
  basic information on, 22–32
  capacitive input impedances in, 375
  charge control in, 376
  de Forest's work on, 11
  describing function transconductance for, 615n2
  development of, 9–12
  FETs and, 28
  as oscillators, 441
  in superhets, 772n15
  V–I characteristics of, 25–32
vacuum tube computers, 773
vacuum tube superregenerator, 18n36
van der Ziel, Aldert, 179, 341n16
varactors, 129, 406n5, 758
  in accumulation mode, 740
Vassiliou, Iason, 751n32
velocity saturation, effect on transistor dynamics,
    185–6
vertical parallel plate (VPP), in CMOS technology,
    128
very large-scale integration, *see* VLSI technologies
vestigial sideband (VSB) signal, for AM modulation, 63
V–I characteristics
  of MOSFETs, 100
  of vacuum tubes, 25–32
video amplifiers, shunt peaking in, 275
VLSI circuits, passive components in, 87
VLSI technologies, source–drain diffusion in, 121
volt, 228
voltage
  of diodes, 27

proportional to absolute temperature (PTAT), 318, 319, 321, 322
references for, *see* voltage references
voltage-controlled oscillators (VCOs), 65, 594–5
  characteristics of, 597–9
  in 4046 CMOS device, 597
  in phase-locked loop circuits, 566–8, 572, 574, 582, 589, 608, 648, 742
  phase noise in, 608f, 682f, 758
voltage gain, calculation for, 240–1, 242–4, 557
voltage references, 314–33
  problem set for, 328–33
voltages, bias, *see* biasing
Volterra series, 391n24

walkie-talkies
  receive mode of, 774
  as toys, 19
  transmit mode of, 774–5
  wartime use of, 44
war, radar secrecy concerns in, 85n63
Ward, William Henry, 33, 37, 39
Wardenclyffe tower, 37n56
Watson, Thomas, Jr., 443, 773
watt, 228
wave equation (Maxwell), 203
Weaver, Donald K., 68, 701n2
Weaver architecture, 701, 735f, 748
  quadrature of, 736
Weaver's SSB architecture, 68, 69, 706
Webster, Roger, 771
Western Electric, 620n6
Westinghouse, 16, 442
Wheeler, Harold A., 15, 120, 296n8, 770n12, 772n16
Wheeler formula, 138, 139, 140f, 148, 155
"When Tubes Beat Crystals: Early Radio Detectors" (Thackeray), 22
White, Stanford, 37n56
white drain noise current, in FET, 674
white noise, 335, 653n33, 664, 672, 674
Whittier, John Greenleaf, 5
wideband FM, noise-reducing qualities of, 44
Widlar, Bob, 318
Wilkinson power combiner, 533
Wilson, Robert, 338n10
wing electrode, 25
wireless communication, 32, 34
  inventors of, 32–9
wireless fidelity, 53n29
wireless local-area networks (WLANs), 52–3, 77, 81n62
  architectures for, 731–49

bands for, 52–3, 77, 81n62
EVM of, 539
parameter summary for, 750
power amplifiers for, 533
receiver for, 537f, 750f
wireless personal network (WPAN), goal of, 55
wireless systems
  history of, 40–52
  noncellular, 52–4
  signal propagation in, 77–83
wire loop, inductance of
"Wollaston wire," in liquid barretter, 8, 10
Wood's metal, in detectors, 5
Woodyard, J. R., 530
Worcester Polytechnic Institute, 443
"The Work of Hertz" (Lodge), 35
World War I, 7, 14, 15, 21, 38, 40, 42, 771
World War II, 44, 53n31, 85, 435n41
woven capacitors, 127–8

XOR phase-detector, 596, 597, 604
  use in design examples, 596
X-rays, discovery of, 10
X-ray tube, 23, 24

year, seconds in, 345n21
yocto (prefix y), 228
yotta (prefix Y), 228
Yuan formula and method, 131–2, 133t, 134
Yue, C. P., 157

zener diode, 318
zepto (prefix z), 228
zero damping, of feedback systems, 468
zero-making capacitor, 594
zero-order hold (ZOH), impulse response of, 646f, 647
zero-peaked amplifier, 310, 310f
zero-peaked common-source amplifier, 278f
zero phase error, in phase-locked loops, 568
zero-power receivers, 113
zeros
  as bandwidth enhancers, 270, 271–8
  in root-locus techniques, 458
zero value, *see* open-circuit time constants (OCτs) method
zetta (prefix Z), 228
ZigBee, parameters for, 55t
zigzag wire electrode, in triode, 11
zincite-crystal diodes, 20
Z parameters, 225
Z-plane maps, 222